Fundamentals of Thermodynamics
(with Technical Notes for Engineers)

Fundamentals of Thermodynamics
(with Technical Notes for Engineers)

Dr. Nikhilesh Mukherjee
Former Head of R&D (Home Care and Water)
Hindustan Unilever Limited, India

Allied Publishers Pvt. Ltd.
Noida • Mumbai • Kolkata • Chennai • Bangalore

ALLIED PUBLISHERS PRIVATE LIMITED

D-5, Sector-2, **Noida**–201 301
Ph. Nos.: 0120-4320295/2542557/4352866 • E-mail: delhi.books@alliedpublishers.com

17 Chittaranjan Avenue, **Kolkata**–700072
Ph.: 033-22129618 • E-mail: cal.books@alliedpublishers.com

15 J.N. Heredia Marg, Ballard Estate, **Mumbai**–400001
Ph.: 022-42126969 • E-mail: mumbai.books@alliedpublishers.com

No. 25/10, Commander-in-Chief Road, Ethiraj Lane (Next to Post Office)
Egmore, **Chennai**–600008
Ph.: 044-28223938 • E-mail: chennai.books@alliedpublishers.com

P.B. No. 9932, No. 15, 3rd Floor (Next to Vijaya Bank), 5th Cross, Gandhinagar,
Karnataka, **Bangalore**–560009, Ph.: 080-41530285 / 22386239
E-mail: bngl.books@alliedpublishers.com

Website: www.alliedpublishers.com

© 2025, Author

ISBN: 978-93-89934-55-7

No part of the material protected by this copyright notice may be reproduced or utilized in any form or by any means, electronic or mechanical including photocopying, recording or by any information storage and retrieval system, without prior written permission from the copyright owners. The views expressed in this volume are of the individual contributors, editor or author and do not represent the view point of the centre.

Published by Sunil Sachdev and printed by Ravi Sachdev at Allied Publishers Pvt. Ltd.,
D-5, Sector-2, Noida–201 301

Acknowledgement

The author extends his gratitude to all those who contributed to making this information accessible. This includes researchers, archivists, and institutions that have dedicated their time and resources to preserving and sharing valuable data and imagery. Their contributions have not only enhanced the quality of this work but have also played a significant role in advancing the field as a whole.

Furthermore, the author would like to express sincere thanks to colleagues, mentors, and peers who provided guidance, feedback, and support throughout the creation of this book. Their insights and encouragement have been instrumental in shaping the final outcome. The collaborative spirit of the academic and creative communities continues to inspire the author, and this work stands as a testament to the power of collective effort in the pursuit of knowledge.

I would like to acknowledge the invaluable support and guidance provided to me by Allied Publishers, one of the leading and most esteemed publishers in India, throughout the process of writing and publishing this book. Their dedication and expertise have been instrumental in bringing this project to fruition. I am extremely grateful for their assistance and for believing in the potential of this book.

Acknowledgement

[illegible]

[illegible]

[illegible]

Preface

This book aims to provide a comprehensive yet accessible understanding of thermodynamics, tailored to the specific needs of professional engineers across various fields. It is carefully curated to encompass the fundamental knowledge necessary for engineers to excel in their respective roles, excluding superfluous or easily obtainable information. An attempt has been made to incorporate pertinent technical notes with the fundamentals provided in the earlier part of the book. This might help readers to see a topic from a more comprehensive angle. This can lead to a deeper and more useful understanding by enabling them to see how these concepts are applied in the real world in addition to the theoretical aspects.

Contents

PART–II
Fundamentals of Thermodynamics with Technical Note for Engineers

PART–I

Thermodynamics

CHAPTER 1

Introduction

HOW DID THERMODYNAMICS EVOLVE?

General

Classical thermodynamics and statistical thermodynamics are the two broad branches of thermodynamics. Classical thermodynamics is the traditional approach that deals with macroscopic systems and quantities such as temperature, pressure, and volume, while statistical thermodynamics provides a more microscopic understanding of thermodynamic properties based on the statistical behaviour of particles. While classical thermodynamics remains an essential and foundational subject, the development of statistical thermodynamics has significantly enriched our understanding of the underlying microscopic processes that govern thermodynamic behaviour.

Thermodynamics is the study of the relationship between heat, work, temperature, and energy. The laws of thermodynamics describe how energy changes in a system and whether it can perform useful work on its environment.

ENERGY'S ROLE IN THERMODYNAMICS

Energy changes in a system tells whether a thermodynamic system can perform useful work on its environment.

What is Energy?

According to Wikipedia, energy is a quantitative property of a physical system or body that can be transferred to perform work, or in the form of heat or light. It can be understood as the ability of a physical system to exert force against an object moving along a path. The conservation of energy principle states that energy is a conserved quantity, which means it can be converted in form but cannot be created or destroyed. The SI unit of energy is joule. A joule is the amount of energy exerted when one Newton of force is applied over a one-meter displacement.

Kinetic energy is the energy possessed by a moving object, whereas potential energy is the energy conserved by a stationary object.

Mass and energy are inextricably linked. Because of mass-energy equivalence, any item with mass (called rest mass) has an equivalent amount of energy (called rest energy), and any additional energy (of any type) gained by the object above that rest energy increases the object's total mass in the same way that it increases its total energy. With a sensitive enough scale, the increase in energy from heating an object could theoretically be quantified as a slight increase in mass.

KEY PILLARS OF THERMODYNAMICS

The key pillars and concepts on which thermodynamics stands can be traced back to the evolution of thermodynamics and the development of key concepts and laws. Here is a simplified overview:

- **Laws of Thermodynamics**: The foundation of thermodynamics lies in the four laws of thermodynamics. These laws describe the fundamental principles governing energy and heat transfer within a system and its surroundings.
- **Kinetic Theory of Gases**: The kinetic theory of gases, developed in the 19th century, provided a foundation for understanding the behaviour of gases in terms of the motion of their particles. This theory laid the groundwork for statistical mechanics, a precursor to statistical thermodynamics.
- **Gibbs' Contributions**: The contributions of Josiah Willard Gibbs were highly significant. Gibbs introduced crucial concepts that enriched classical thermodynamics and laid the groundwork for statistical thermodynamics.

Gibbs made key contributions to the understanding of phase transitions and equilibrium in systems, particularly through his development of the Gibbs phase rule. This rule provided a means to predict the number of degrees of freedom in a system at equilibrium, based on the number of components and phases present. His phase rule was instrumental in understanding the behaviour of complex systems and remains a foundational concept in classical thermodynamics.

- **Entropy**: The concept of entropy has enriched thermodynamics in several significant ways. Initially introduced in the context of classical thermodynamics, entropy has revolutionized our understanding of the behaviour of systems.
- **Quantification of Disorder**: Entropy, often described as a measure of disorder or randomness in a system, provided a quantitative metric for characterizing the nature of energy dispersal within a system. This concept allowed thermodynamic principles to be applied to a broader range of phenomena, beyond the traditional study of heat and work.
- **Second Law of Thermodynamics**: The introduction of entropy led to the formulation of the second law of thermodynamics, which states that the total entropy of an isolated system can never decrease over time. This law provided a fundamental understanding of the direction of natural processes, emphasizing that systems tend to evolve towards states of higher entropy or greater disorder.

Thermodynamics is a branch of physics that deals with a system's energy and work. It was created in the nineteenth century, when scientists were learning how to build and operate steam engines. Thermodynamics only deals with a system's large-scale response, which can be observed and measured in experiments. The kinetic theory of gases describes gas interactions on a small scale. Some principles are better understood in terms of thermodynamics, while others are better explained by kinetic theory.

Thermodynamics is governed by three fundamental laws. The first, second, and third laws of thermodynamics have traditionally been identified simply by an ordinal identification. After the first three laws were established, a more fundamental statement was later referred to as the zeroth law.

Each law leads to the definition of thermodynamic properties, which aid in understanding and predicting the behaviour of a physical system. The laws of thermodynamics refer to a set of scientific laws that define physical quantities that describe thermodynamic systems. Relationships between different parameters for thermodynamic processes, such as thermodynamic work and heat, are established by the laws. Empirical facts present by them serve as a foundation for excluding certain phenomena, like perpetual motion. In general, they are the fundamental laws of physics and can be applied to natural sciences, not just thermodynamics.

CHAPTER 2

The International System of Units [SI units]

The International System of Units, or SI, is a decimal and metric unit system that was established in 1960 and has been updated on a regular basis since then. The SI has official status in most countries, including the United States, Canada, and the United Kingdom, though these three countries are among the few that continue to use their traditional systems to varying degrees. Nonetheless, the SI "has been used around the world as the preferred system of units, the basic language for science, technology, industry, and trade," owing to its nearly universal acceptance.

The SI is made up of seven basic units of measure. These are the second unit (symbol s), the metre unit (m, measure), the kilogram unit (kg, measure), the ampere unit (A, measure of electric current), the kelvin unit (K, measure of thermodynamic temperature), the mole unit (mol, measure of mass), and the candela unit (cd, measure of luminous intensity).

The system can hold an infinite number of these units. These units are called "conjecture derived units". They are always represented as product of powers of base units. We have assigned names and symbols to 22 of these units.

The imperial and US customary measurement systems are the only other types of measurement systems that are still widely used around the world. Other, less common systems of measurement are occasionally used in specific parts of the world. In the United Kingdom, the metric system is used in business and technology, but imperial units are still commonly used by the public. For example, all UK roads use the imperial system for distance, but newer signs also include metric.

The metric or SI system does have legal standing in the US, and all customary units are defined in metric terms and have been for over 100 years.

Table 1: SI Base Units

Symbol	*Name*	*Quantity*
S	Second	Time
M	Metre	Length
Kg	Kilogram	Mass
A	Ampere	Electric current
K	Kelvin	Thermodynamic temperature
Mol	Mole	Amount of substance
Cd	Candela	Luminous intensity

Credit: Wikipedia.

CONTROLLING POWER

The General Conference on Weights and Measures (CGPM) established and maintains the SI. In practice, the CGPM follows the recommendations of the Consultative Committee for Units (CCU), the actual body conducting technical deliberations on new scientific and technological developments related to unit definition and the SI. The International Committee for Weights and Measures reports to the CCU, which reports to the CGPM.

CHAPTER 3

Linear, Log and Exponential Scale

WHAT DOES LINEAR, LOG AND EXPONENTIAL SCALES MEAN?

What is Linear Scale?

A linear scale is a straight line with equal divisions that shows the relationship between a map or drawing's scale and the actual length on the ground.

What is Log Scale?

Log scale, or logarithmic scale, is a way to display numerical data on a graph that uses a logarithm function to transform value.

A log scale is a nonlinear scale that uses a base value raised to a power to represent each unit of length. For example, on a log scale, 10, 100, 1000, 10000, and 100000 are equally spaced values.

What it's used for?

Log scales are useful for displaying data that has a wide range of values or numbers that are growing exponentially. They are often used to show rates of change, such as how much something increases or decreases.

Why *pH* is a Log Scale?

The pH scale is logarithmic because it's a convenient way to represent the wide range of hydrogen and hydroxide ion concentrations in a solution:

Range of Values

The concentration of hydrogen ions in a solution can range from $10°$ to 10^{-14} moles per liter. The logarithmic scale compresses this range into a more manageable scale of 0 to 14.

Intuitive Understanding

A decrease of one *pH* unit corresponds to a tenfold increase in acidity. For example, a solution with a pH of 5 is ten times more acidic than a solution with a *pH* of 6.

What is Exponential Scale?

An exponential scale is a non-linear scale that increases or decreases at an exponential rate. It's often used to display data that grows or declines rapidly, such as the spread of a disease or the growth of a population.

Here are some things to know about exponential scales:

Exponential Growth

Exponential growth occurs when a quantity grows at a rate directly proportional to its present size. For example, if something is three times as big as it is now, it will be growing three times as fast as it is now.

Logarithmic Scales

Data that is thought to display exponential growth is often plotted on semilog graphs, which have one logarithmic scale and one linear scale.

Straight Lines on a Log Scale

If you see a straight line in a log-scaled chart, something grows exponentially.

Different Bases

You can use any base to represent exponential growth, but it's just a matter of scaling. For example, you can use a base-2 log scale to make the doubling time easy to see on the graph.

Exponential vs Log Scale

An exponential scale increases quickly by large numbers, while a logarithmic scale increases in a factor of the base of the logarithm. The main difference between the two is that exponential and logarithmic functions are inverse functions of each other:

Example

Logarithmic Scale Example

On a log scale, the numbers 1, 10, 100, 1000 would be spaced equally apart, such as 1, 2, 3, 4.

Exponential Scale Example

On an exponential scale, the numbers increase at a geometric rate, such as 2, 4, 8, 16.

HOW TO CHOOSE BETWEEN LINEAR, LOG AND EXPONENTIAL SCALES?

Applying different scales like linear, log, and exponential scale in data visualization can help to better represent the data and highlight trends or patterns that may not be apparent on a linear scale.

To Apply Different Scales

1. **Linear scale**: A linear scale is the most common type of scale used in data visualization. It represents data in a direct, proportional way. To use a linear scale, simply plot your data points on a linear axis with equal intervals.
2. **Log scale**: A log scale is useful when you have a wide range of data points or when the data is highly skewed. To use a log scale, simply change the axis to a logarithmic scale. This will compress the data points at the lower end of the scale and expand the data points at the higher end of the scale.
3. **Exponential scale**: An exponential scale is often used when the data points are growing or decaying at an exponential rate. To use an exponential scale, you can simply plot the data points on an axis with an exponential scale.

To Choose a Scale, Consider the Following Factors

- ***The distribution of your data:*** If your data is evenly distributed, a linear scale may be appropriate. If your data is highly skewed or has a wide range, a log or exponential scale may be more appropriate.
- ***The type of trend or pattern you want to highlight:*** Different scales can emphasize different trends or patterns in the data. A log scale can help to highlight small changes in a large dataset, while an exponential scale can emphasize rapid growth or decay.
- ***The viewers:*** Consider who will be looking at your data visualization and what type of scale would be most easily understood by them. Linear scales are usually the most intuitive, but log and exponential scales may be more appropriate for certain types of data.

In general, it may be helpful to experiment with different scales and see which one best represents your data and helps to communicate your message effectively.

HOW TO CHOOSE BETWEEN LOG BASE E AND LOG BASE 10 SCALE?

In logarithmic scale, "base *e*" refers to the natural logarithm, which is denoted as ln. The base e logarithm is the logarithm with base *e*, where *e* is the Euler's number approximately equal to 2.718. Base e logarithm is commonly used in mathematical and scientific calculations.

Similarly, "base 10" refers to the common logarithm, which is denoted as log. The base 10 logarithm is the logarithm with base 10, which is frequently used in various applications. The common logarithm is widely used in engineering, physics, and computer science.

When choosing between using a natural logarithm (ln, base *e*) or a common logarithm (log, base 10) in a logarithmic scale, it depends on the context of the problem and what you are trying to achieve. Here are some factors to consider when deciding which scale to use:

1. *Natural Logarithm (ln, base e)*

- Use ln when dealing with exponential growth or decay, as it is often the natural choice for continuous and exponential functions.
- ln is commonly used in calculus, differential equations, and other mathematical calculations where e is a natural fit.
- ln is often used in scientific research and modeling.

2. *Common Logarithm (log, base 10)*

- Use log when working with orders of magnitude or powers of 10, as it simplifies calculations involving factors of 10.
- log is commonly used in engineering, physics, finance, and other fields where base 10 is more practical.
- log scale can make data easier to interpret and visualize, especially when dealing with large ranges of values.

Ultimately, the choice between ln and log scale depends on the specific application and convenience for the problem at hand. It's important to consider the properties of each logarithmic scale and how they relate to the data or equations you are working with.

LOG AND EXPONENTIAL SCALES BOTH REPRESENT NON-LINEAR DATA. HOW TO INTERPRET BETWEEN LOG FIT AND EXPONENTIAL FIT DATA?

When interpreting whether the data fits a log or exponential scale, there are a few key differences to consider:

1. **Logarithmic Scale**

- In a log scale, the relationship between the variables is such that one variable increases or decreases logarithmically with respect to the other variable.
- In a log fit, the data points will form a straight line when plotted on a log scale.
- The relationship between the variables in a log fit is typically of the form $y = a \times \log(bx + c)$ or $y = a \times \log(x) + b$.

2. **Exponential Scale**

- In an exponential scale, the relationship between the variables is such that one variable increases or decreases exponentially with respect to the other variable.
- In an exponential fit, the data points will form a curve when plotted on a linear scale.
- The relationship between the variables in an exponential fit is typically of the form $y = a \times e^{bx}$.

To determine whether the data fits a log or exponential scale, you can consider the shape of the data when plotted on a linear scale and a log scale. If the data points form a straight line on a log scale, then it is likely a log fit.

If the data points form a curve on a linear scale, then it is likely an exponential fit.

Additionally, you can also perform a regression analysis to determine the best-fitting model for the data.

CHAPTER 4

Heat and Work

WHAT IS HEAT?

Heat is the transfer of thermal energy between objects due to a temperature difference. It is described by the first law of thermodynamics ($Q = U + W$), where it is the sum of the internal energy (a state function) and work (a path function). Since heat depends on its path for travel between two points, it is a property in transit that cannot be contained or stored. It can only be transferred from one object to another or converted to other forms of energy.

Heat and Work Both are Path Functions

What is Path Function?

Path functions are properties or quantities whose values depend on the transition of a system from the initial state to the final state. The two most common path functions are heat and work. On the left image heat, Q has more than one option to flow from A to B. Similarly on the right image on a PV diagram work has a choice between two points A and C to follow either path ADC or ABC. The work done by path $ADC > ABC$ is shown by diagonal lines. Both heat and work have numerous ways to travel between two points.

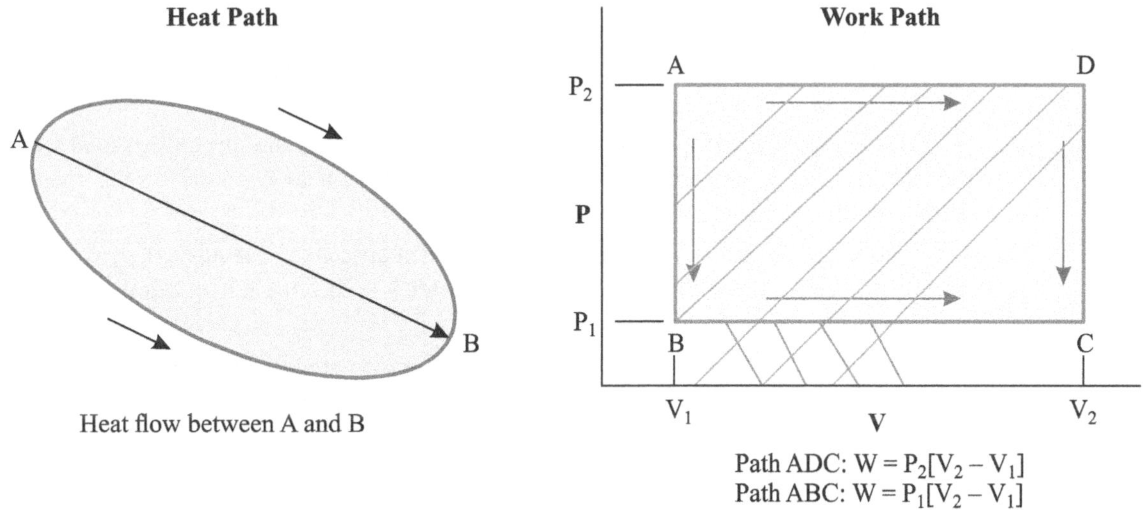

Figure 1

Since heat and work are path functions with multiple choices to follow, the quantity of heat and work in transit cannot be measured. Both heat and work are inexact differentials. Since the quantity of heat and work cannot be measured, they are not thermodynamic properties. Rather heat and work are dependent on how the thermodynamic system changes.

Heat is Neither Intensive nor Extensive Property

Physical properties of materials and systems in thermodynamics can often be categorized as being either intensive or extensive, according to how the property changes when the size (or extent) of the system changes. An intensive property is one whose magnitude is independent of the size of the system, whereas an extensive quantity is one whose magnitude is additive for subsystems. Since heat is always in transit it is not a thermodynamic property it does not fall in either of these groups.

Heat is Always in Transit

Temperature is a consequence of specific heat. Unequal specific heat everywhere makes heat spread and increase entropy. The specific heat plays a crucial role in determining heat content of a substance. Heat flows in the direction of heat content of substances. The flow of heat is determined by the difference in heat content or energy level within a system. Heat always flows from regions of higher heat content to regions of lower heat content [high to low energy] until a level of equilibrium is reached.

The temperature gradient refers to the variation in temperature across a given space or between two points. When there is a temperature difference, heat tends to move from regions of higher temperature to regions of lower temperature. This movement of heat occurs in order to achieve thermal equilibrium, where the temperatures become balanced.

The second law of thermodynamics is a fundamental principle that describes the direction of heat flow. It states that heat spontaneously flows from a higher-temperature region to a lower-temperature region until thermal equilibrium is reached. This principle is based on the fact that heat naturally tends to disperse and distribute itself evenly.

Temperature Gradient

The temperature gradient refers to the variation in temperature across a given space or between two points. When there is a temperature difference, heat tends to move from regions of higher temperature to regions of lower temperature. This movement of heat occurs in order to achieve thermal equilibrium, where the temperatures become balanced.

Obvious Question Arises What Generates *dt*?

The temperature gradient in a substance is a consequence of its specific heat the specific heat is atomic arrangement and heat content of a substance. The atomic arrangement of molecules in a substance determines how they interact and transfer heat energy. When there is a temperature gradient in a substance, it means that there is a variation in temperature across different regions of the substance.

Heat naturally flows from regions of lower specific heat [hotter or less resistant to heat transfer] to cooler regions of higher specific heat. Heat content of a substance, $Q = m\ Cp\ dt$.

The equation $Q = m\ Cp\ dt$ describes the transfer of heat, where Q represents the amount of heat transferred, m represents the mass of the material being heated, Cp represents the specific heat capacity of the material, and dt represents the change in temperature.

This equation, shows that heat (Q) is directly proportional to the specific heat capacity (Cp) of the material at constant mass and dt, the temperature gradient. It means that the greater a substance's heat storage capacity Cp, the greater the amount of heat it can store.

It can be looked into another way, $dt = Q/m\ Cp$, for same Q and mass, dt is inversely proportional to Cp. It is very common observation that in a fridge at same Q different items like tomatoes and potatoes have different temperatures. This is because they have different specific heats.

In summary, a substance with a low specific heat feels hotter than a substance with a high specific heat. This is because a substance with low specific heat requires less energy to raise its temperature, so it heats up more quickly and feels hotter to the touch. In contrast, a substance with high specific heat requires more energy to raise its temperature, so it heats up more slowly and feels cooler to the touch.

Therefore, it can be concluded that the specific heat plays a crucial role in determining heat content of a substance. Temperature is a consequence of specific heat. Unequal specific heat everywhere makes heat to spread and increase entropy.

Modes of Heat Transfer

There are three ways that heat can move: conduction, convection, and radiation.

Radiation

Heat radiates when it travels from its source to another object directly as infrared energy waves. In this way, heat from the Sun reaches Earth. Heat is actually radiated from all hot objects towards all colder objects. The cooler object's molecules accelerate upon impact due to heat waves. Faster-moving molecules inside an object result in a higher temperature. The electromagnetic spectrum, which is made up of various energy waves, includes infrared waves. The electromagnetic spectrum is made up of all forms of energy that can travel in waves, such as microwaves, radio waves, ultraviolet waves, x-rays, light, and heat.

Example: One instance of energy transfer by radiation is the Earth's heating caused by the Sun. Another example would be a room heated by an open-hearth fireplace.

Conduction

Conduction is another method of heat transfer. When any form of energy, such as heat, comes into contact with matter like anything that can be touched physically, atoms and molecules start to move. When one atom or molecule starts moving, it will collide with another to start moving as well. These then cause other molecules to collide and start moving. This is how heat is transferred through matter by conduction.

For Instance: Conduction is what heats a pot's handle when its bottom is the only part in contact with the stove. The heat from the stove causes the molecules in contact with it to start moving. These molecules collide with other molecules in the pot, which collide with still others, and so on, until every molecule in the pot, including the handle, is in motion. One can feel the heat when they touch the pot handle. Through conduction, the heat has transferred from the stove to the cook's hand.

Convection

Heat transfer by convection is characterized by the bulk movement of molecules in fluids like gases and liquids. The bulk heat transfer is caused by the fluid's motion, but the initial heat transfer between the object and the fluid occurs by conduction.

Example: Our body's method of transferring heat is an illustration of convection. Convection is the process by which heat is removed from our bodies by moving air. The air will take up our heat and rise. Cooler air enters to replace the warmed air surrounding us and absorbs more of our body heat. We feel colder the faster this air moves.

Conduction *vs* Convection: How they Compete?

Conduction is the transfer of thermal energy through direct contact. Convection is the transfer of thermal energy through the movement of a liquid or gas.

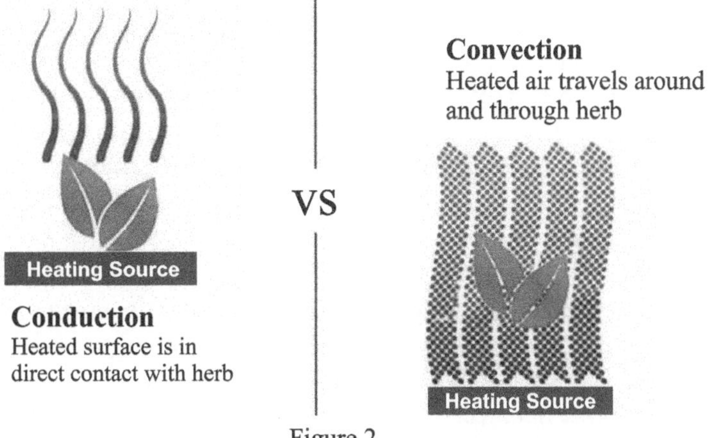

Figure 2

Convection and conduction are thought to compete for domination in a body of fluid that is heated from below its container. The liquid in contact with the wall surface will turn into vapour if heat conduction is too high, or [wall temperature minus surface temperature] is too high. Film boiling would take the place of convection boiling. The buoyancy that drives will decrease. Convectional heat transfer will give way to radiative heat transfer. Thus, a key factor in convective heat transfer is the intensity of conductive heat transfer.

Heat *vs* Temperature

Heat is the internal energy of a substance, which includes both kinetic and potential energy, while temperature is a measure of the average kinetic energy of the particles in a substance. It's important to note that temperature alone does not provide a complete picture of the energy content of a substance, as it does not account for potential energy stored in chemical bonds. This distinction is crucial in understanding the behaviour of materials and the transfer of energy in various physical processes. Another name of potential energy is latent heat. Heat or total thermal energy has both sensible and latent heat. Temperature has only sensible heat or kinetic energy.

Heat *vs* Thermal Energy

Heat is the transfer of heat energy from a hot body to a cold body. Heat is always in transit. It is not a property of either hot or cold bodies. Thermal energy is the internal energy of a substance. It is stored energy in a substance. It is the property of the substance. It gets generated at the transition of energy within atom or molecules.

Radiation Heat Transfer

Heat transfer from a body with a high temperature to a body with a lower temperature when bodies are not in direct physical contact with each other or when they are separated in space, is called heat radiation. Heat radiation (as opposed to particle radiation) is the transfer of internal energy in the form of electromagnetic waves.

For most bodies on the Earth, this radiation lies in the infrared region of the electromagnetic spectrum.

Electromagnetic Radiation

Electromagnetic radiation (EMR) refers to the waves of the electromagnetic field, propagating (radiating) through space, carrying electromagnetic radiant energy. It includes radio waves, microwaves, infrared, (visible) light, ultraviolet, X-rays, and gamma rays.

Electromagnetic radiation consists of electromagnetic waves, which are synchronized oscillations of electric and magnetic fields. In a vacuum, electromagnetic waves travel at the speed of light. In homogeneous, isotropic media,[an isotropic medium is one whose electromagnetic properties are the same in all directions] the oscillations of the two fields are perpendicular to each other and perpendicular to the direction of energy and wave propagation, forming a transverse wave.

The position of an electromagnetic wave within the electromagnetic spectrum can be characterized by either its frequency of oscillation or its wavelength.

Electromagnetic waves of different frequencies are called by different names since they have different sources and effects on matter. In order of increasing frequency and decreasing wavelength, these are radio waves, microwaves, infrared radiation, visible light, ultraviolet radiation, X-rays, and gamma rays.

What Happens at the Transition of Energy Within Atom or Molecules?

Electrons in atoms and molecules can change (make transitions) energy levels by emitting or absorbing a photon (of electromagnetic radiation), whose energy must be exactly equal to the energy difference between the two levels. If an atom, ion, or molecule is at the lowest possible energy level, it and its electrons are said to be in the ground state. If it is at a higher energy level, it is said to be excited, or any electrons that have higher energy than the ground state are excited. Such a species can be excited to a higher energy level by absorbing a photon whose energy is equal to the energy difference between the levels. Conversely, an excited species can go to a lower energy level by spontaneously emitting a photon equal to the energy difference.

WHAT IS WORK?

Work, regardless of the type of energy, is a means of transporting energy. Work transports energy between two points, either internally or externally, whether it is thermal energy, mechanical energy, chemical energy, or electrical energy. Work as a means of transporting energy is a fundamental principle in thermodynamics, and it's fascinating to think about how different types of energy can be transferred through work in various physical processes.

One of the primary processes by which a thermodynamic system can interact with its surroundings and exchange energy is thermodynamic work. Heat transfer by radiation is an exception. In radiation the mode of heat transfer is electromagnetic radiation. This exchange produces externally measurable macroscopic forces on the system's surroundings, which can result in mechanical work, such as lifting a weight, or changes in electromagnetic or gravitational variables. The environment can also do work on a thermodynamic system, which is measured using an opposite sign convention.

Heat transfer is frequently estimated calorimetrically in experimental practice, by changing the temperature of a known quantity of calorimetric material substance.

In addition to the transfer of energy as work, thermodynamics allows for the transfer of energy as heat. The first law of thermodynamics relates changes in the internal energy of a closed (no matter transfer) thermodynamic system to those two modes of energy transfer, as work and as heat.

Adiabatic work is performed without the transfer of matter or heat. In thermodynamics. In an adiabatic system there is no heat flow into the system from the surroundings nor there is any heat rejection by the system to the surroundings. Adiabatic systems use its internal energy to perform thermodynamic work.

Work done on and work done by a thermodynamic system must be distinguished by examining their precise mechanisms. When the surroundings do work on the system by a compression is called 'Work on the system' Work performed on a thermodynamic system is designated by + W. Work done by the system on the surroundings such as expansion is designated by - *W*.

Non-Thermodynamic Work

Shaft work, stirring, and rubbing, are not thermodynamic work because they do not change the volume of the system in relation to its resisting pressure.

Types of Work

- **Mechanical work**: Involves the displacement of an object against an external force. This type of work is typically done in mechanical devices such as engines or pumps.
- **Electrical work**: Involves the movement of charged particles through a circuit, resulting in the generation of electrical energy. Examples include the work done by batteries, generators, and electrical motors.
- **Chemical work**: Involves the breaking or formation of chemical bonds in a reaction. This type of work is associated with chemical processes such as electrolysis, combustion, or synthesis reactions.
- **P-V work**: Involves the expansion or compression of a gas in a container. This type of work is commonly seen in thermodynamic processes such as piston-cylinder systems or turbine and compressor operations.
- **Surface work**: Involves changes in the surface area of a system. This type of work is relevant in systems with interfaces, such as in adsorption processes or phase transitions.
- **Magnetic work**: Involves the movement of magnetic fields or magnetic dipoles within a system. This type of work is seen in processes involving magnetic materials or devices such as magnetic motors or generators.
- **Gravitational work**: Involves lifting or lowering a mass against the force of gravity. This type of work is commonly encountered in lifting objects, potential energy calculations, or gravitational potential energy conversions.

PV Work can Further be Divided as Isothermal, Isobaric, Adiabatic and Isochoric Work

- **Isothermal PV work** occurs when the temperature remains constant during the process, such as in an ideal gas undergoing a reversible expansion or compression at a constant temperature.
- **Isobaric PV work** occurs when the pressure remains constant during the process, such as in a gas undergoing a reversible expansion or compression at a constant pressure.
- **Adiabatic PV work** occurs when there is no heat exchange with the surroundings during the process, such as in a gas undergoing a reversible expansion or compression without any heat transfer.
- **Isochoric PV work** occurs when the volume remains constant during the process, such as in a rigid container where the gas is compressed or expanded without changing its volume.

Modes of Work Transfer

Work Transfer in Thermodynamics

An energy outlet is required for every thermodynamic system. When a system is enclosed in a closed box that does not allow it to share or change its temperature, pressure, or mass, the system enters a dynamic equilibrium. The free energy decreases to zero, while the entropy reaches the maximum and consumes the enthalpy, $dG = dH - TdS$. Isolated systems are such closed systems.

The issue arises from the adiabatic system, which states that it cannot transfer heat or mass but can only transfer work.

Different Types of Work Transfer

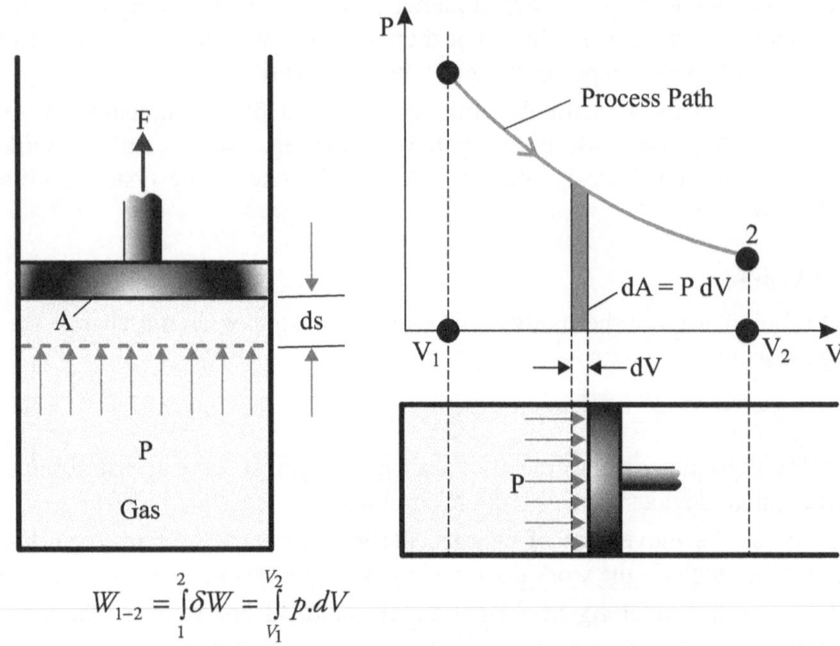

$$W_{1-2} = \int_1^2 \delta W = \int_{V_1}^{V_2} p.dV$$

Figure 3: PdV Work or Displacement Work

- **Electrical work of a capacitor:** (1/2) CV^2 where C is the capacitance and V is the voltage.
- **Electrical work for a resistor:** $i^2 R t$ where i is the current through the resistor, R is the resistance and t is the time of operation.
- **Shaft work:** (2 *pi N T t*)/(60) where N is the shaft speed (in revolutions per minute) and T is the torque. Also, t is the time of operation.
- **Gravitational work:** *mgh* where m is the mass, g is the acceleration due to gravity and h is the height by which the mass is raised or lowered.
- **Surface tension work:** (surface tension) × (increase in area of the film).

VdP Work

Detail

VdP Signifies Work in a Turbine

Isentropic Shaft Work of Turbine

The system boundary of closed *vs* open system: A very important concept.

The difference between *PdV* and *VdP* arises from the boundary work. A piston-cylinder closed system is just permeable to energy and not mass. It does not have to push any mass coming out. Therefore, *PdV* provides internal energy for work.

On the other hand, in an open system, which is also permeable to mass in addition to heat the steam has to supply both *PdV* for internal energy and *VdP* to push back the fluid which is already inside the turbine. Here, *PdV* and *VdP* both are required.

Why *VdP* Applies to Signify Work in the Turbine?

There are two types of thermodynamic work (1) *PdV* when volume changes at constant pressure and (2) *VdP* work when pressure changes at constant volume.

We are generally familiar with (1).

The first point to note since *VdP* work occurs at constant volume it is isentropic work. There is no entropy generation.

A typical example of *VdP* work is a power plant turbine. The job of *PdV* work when the steam expands gets over at the boundary of the turbine.

VdP work begins at the boundary of the turbine when the steam has to push the volume of steam already present inside the turbine. This is a compression work.

There is a constant volume of steam already inside the turbine that gets pushed by *VdP* work. This is called isentropic work on the turbine.

Since the volume of steam inside the turbine remains constant there is no generation of entropy.

VdP work in a turbine is very small compared to *PdV* work.

SI Unit for Heat and Work

The SI unit for heat and work is Joule.

What is Joule?

Joule is a unique unit. It represents many forms of energy. Joule is the unit of energy in the International System of Units (SI). It is named after the English physicist James Prescott Joule. One Joule is equal to the work done when the force of one Newton acts over a distance of one meter.

Why Joule is a Unique Unit?

The joule is a unique unit in that it can represent various forms of energy. It is a versatile unit that can be applied to measure heat, work, kinetic energy, potential energy, chemical energy, and mechanical energy. This versatility makes it a fundamental and widely used unit in physics and engineering for quantifying different types of energy. The Joule has unique dimensional units (ML^2T^{-2}). In each case, the dimensional units of mass (M), length (L), and time (T) fit with the dimensional unit of Joule, demonstrating how Joule can represent various forms of energy. Joule represents all forms of energy because it is a measure of the capacity to do work or the transfer of heat.

How Joule Represents Various Forms of Energy

Heat

In the context of heat, one joule is the energy transferred when the heat produced by one watt of power is dissipated for one second.

Work

In the context of work, one joule is the work done when a force of one Newton moves its point of application one meter in the direction of the force.

Other Forms of Energy

When considering kinetic energy, potential energy, chemical energy, and mechanical energy, the dimensional unit of Joule (ML^2T^{-2}) can be applied as follows:

- **Kinetic energy**: 1 Joule = 1/2 mass velocity^2 (ML^2T^{-2})

- **Potential energy**: 1 Joule = mass gravitational acceleration height (ML^2T^{-2})⁻
- **Chemical energy**: 1 Joule = amount of energy released by a chemical reaction (ML^2T^{-2})
- **Mechanical energy:** 1 Joule = work done by a force of one Newton acting over a distance of one meter (ML^2T^{-2})

In each case, the dimensional units of mass (*M*), length (*L*), and time (*T*) fit with the dimensional unit of Joule, demonstrating how Joule can represent various forms of energy.

CHAPTER 5

Temperature

FUNDAMENTALS

Temperature is a measure of the average kinetic energy of the particles in a substance, while heat is the sum of internal energy and work energy as per the 1st law of thermodynamics. Internal energy includes both kinetic and potential energy.

It's important to note that temperature alone does not provide a complete picture of the energy content of a substance, as it does not account for potential energy stored in chemical bonds. This distinction is important in understanding the behaviour of materials and the transfer of energy in various physical processes. Another name of potential energy is latent heat. Heat or total thermal energy has both sensible and latent heat. Temperature has only sensible heat or kinetic energy.

A classic example is phase change. In a phase change, the potential energy of the molecules remains constant. The temperature remains constant. The increase in kinetic energy causes the molecules to move more freely, therefore changing its phase. In solid-liquid or liquid-gas phase transitions, energy is used to break the bonds between molecules (potential energy), rather than increasing their kinetic energy. During these transitions, the temperature remains constant as the energy added is used in overcoming the forces holding the molecules together, rather than increasing their kinetic energy.

Temperature is our sensory feeling of how hot or cold a substance is. In thermodynamics, temperature is a measure of the average energy of the translational motions of matter's particle constituents (atoms and molecules).

Temperature is defined by the equation, $E = 1/2$ KbT [for one degree of freedom] where: E is kinetic energy in joules (J), $kb = 1.3806504 \times 10^{-23}$ J/K is the Boltzmann constant, T is the kinetic temperature in kelvins (K) Boltzmann constant is a physical constant that relates the average relative kinetic energy of particles in a gas with the temperature of the gas. E is directly proportional to T [temperature in kelvin]. But there is a catch here, only ideal gases have predominantly KE. Since there are no intermolecular attractions in ideal gases, they have practically no potential energy.

Therefore, the internal energy of ideal gases signifies temperature. It does not apply to real gases. Precisely the translational kinetic energy symbolises temperature. The linear collisions of moles increase temperature.

IS TEMPERATURE A FORM OF ENERGY?

Temperature $T = 2E/Kb$, T signifies only external motion of atoms and more particularly the translational motions of molecules.

WHAT HAPPENS WHEN AN OBJECT IS HEATED?

Average energy of the translational, vibrational and rotational motions of matter's particle constituents (atoms and molecules) increases as KE is proportional to T. T increases until the object reaches a thermal equilibrium with surrounding. At equilibrium, each degree of freedom will have on average the same energy, $\frac{1}{2}\, Kb\, T$, Kb is Boltzmann constant, and T is the kinetic temperature in kelvins (K) Temperature is a state function and heat is a path function. Temperature is an intensive property. For an ideal gas its internal energy is its temperature. Since there is no potential energy in ideal gases internal energy of ideal gas = Kinetic energy and it is proportional to temperature. Precisely the translational kinetic energy symbolises temperature. The linear collisions of moles increase temperature.

A dehumidifier is a classic example of temperature acting as a kinetic energy supplier. It is frequently debated whether a humidifier can cool a hot/humid environment. No, it does not.

Explanation

The job of a dehumidifier is to remove excess humidity. This gives a sense of comfort and affects indirectly the perception of temperature. The reason is a lower humidity gives more sweat evaporation and it makes you feel cooler. A dehumidifier does not change the temperature.

The dehumidifier removes moisture. Moisture carries latent heat out. Latent heat is primarily potential energy. It is primarily energy stored in intermolecular bonds.

In summary, a dehumidifier primarily removes a part of potential energy leaving behind kinetic energy unchanged so there is practically no cooling by dehumidifier.

It may be noted that a very important point is that latent heat transfer always happens at a constant temperature when the liquid and the vapor phase reach thermal equilibrium. The latent heat primarily removes or adds potential energy during phase change and leaves behind kinetic energy, which is temperature $KE = 1/2\ KT$, T is temperature and the temperature remains constant.

ZEROTH LAW OF THERMODYNAMICS

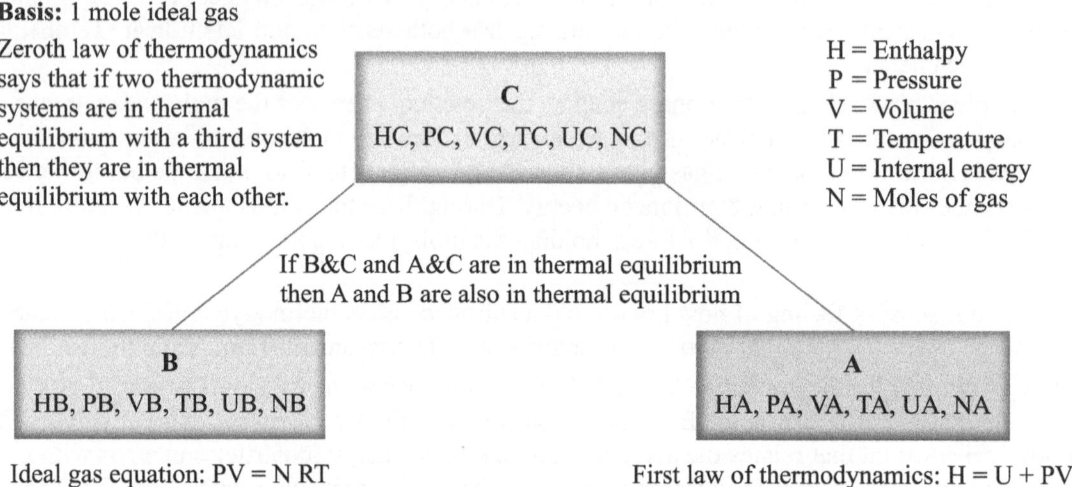

Figure 1: Zeroth Law of Thermodynamics

According to the law, if two substances A and B are in thermal equilibrium with another substance C, then A and B are also in thermal equilibrium.

We are more concerned with the first and second laws of thermodynamics. The Zeroth Law is not widely discussed among us, even though it is a very important thermodynamic law that explains thermal equilibrium and is fundamental to understanding temperature.

Proof of Zeroth Law of Thermodynamics

What is Thermal Equilibrium?

A thermal equilibrium between two physical systems is which does not allow net flow of thermal energy between them when they are connected by a path permeable to heat. Objects in thermal equilibrium have same temperature.

Let us imagine, three thermodynamic systems, A, B and C, all with equal mass separated by heat permeable wall. Assume, both A and B are independently in thermal equilibrium with the third system C. We have to prove A and B are in thermal equilibrium.

If A and C are in thermal equilibrium

$$dT\ AC = 0$$

$$Cp\ A = Cp\ C \text{ [for 1 mole mass]} \qquad \dots (1)$$

Similarly, if B and C are in thermal equilibrium

$$Cp\,B = Cp\,C \qquad \ldots (2)$$

From equation [1] and [2]

$$Cp\,A = Cp\,B$$

For 1 mole gas, therefore

$$dT\,AB = 0$$

Therefore, A and B must be in thermal equilibrium.

A LIST OF SOME IMPORTANT TEMPERATURES

Triple Point Temperature

The triple point of a substance is the temperature and pressure at which its three phases (gas, liquid, and solid) coexist in thermodynamic equilibrium. The triple point of water was used to define the kelvin, the International System of Units' base unit of thermodynamic temperature. The single pressure and temperature combination that allows liquid water, solid ice, and water vapor to coexist in stable equilibrium occurs at exactly 273.15 K (0.0100°C; 32.0180°F) and a partial vapor pressure of 611.657 pascals.

Saturation Temperature

The boiling point is simply referred to as the saturation temperature. The term "saturation" refers to the temperature at which a liquid must be in order to boil and enter the vapor phase, based on its saturation pressure. At a given temperature and pressure, saturation describes a condition in which a mixture of vapor and liquid can coexist. A substance is called saturated when it does not have any more heat capacity to absorb heat. Further supply of heat goes to break the intermolecular bonds and perform PV work of expansion as the liquid becomes vapor. The saturation temperature or boiling point is the temperature at which liquid transits to vapor at constant temperature in a state of thermal equilibrium at a given pressure.

Superheat Temperature

Superheat temperature is the temperature of a vapor above its saturation temperature at a particular pressure. It is the additional heat energy added to a vapor after it has reached its boiling point or saturation temperature.

Critical Temperature

The temperature above which a gas, under any pressure, cannot condense into a liquid is known as the critical temperature for a pure substance. At the critical point, the liquid expands and loses density until the liquid and vapor's densities are equal, eliminating the phase boundary. Each material has a specific critical point that is determined by the forces that act between its molecules and the structure of its molecules.

Supercritical Temperature

The temperature [and pressure] at which a fluid converts to a supercritical fluid is referred to as supercritical temperature. A supercritical fluid (SCF) is any substance that exists at temperatures and pressures above its critical point but lacks distinct liquid and gas phases.

Reduced Temperature

The reduced temperature is simply the temperature of the gas divided by its critical temperature. Similarly, the reduced pressure is simply the pressure of the gas divided by its critical pressure. As for the compressibility of gases, the principle of corresponding states indicates that any pure gas at the same reduced temperature, Tr, and reduced pressure, Pr, should have the same compressibility factor. Tc and Pc are known as the critical temperature and critical pressure of a gas.

Inversion Temperature

In thermodynamics and cryogenics, the inversion temperature is the critical temperature below which a non-ideal gas (all gases in reality) expanding at constant enthalpy experiences a temperature decrease and above which it experiences a temperature increase. The temperature change is known as the Joule-Thomson effect, and it is used in gas liquefaction. At inversion temperature there is no cooling or increase in temperature. The Joule Thomson coefficient = dt/dp changes sign at inversion temperature [t is temperature and p are pressure]. Real gases tend to behave as ideal gases at inversion temperature.

Boyle Temperature

It is the temperature at which a real gas behaves like an ideal gas over a range of pressure. At this temperature, the attractive and repulsive forces acting on the gas particles balance out. Gases with a higher Boyle temperature are easier to liquefy, while gases with a lower Boyle temperature are more difficult to liquefy.

Equilibrium Temperature

If there is no net transfer of thermal energy between two physical systems connected by a heat-permeable path, then they are in thermal equilibrium. The rule governing thermal equilibrium is the zeroth law of thermodynamics. When a system's internal temperature remains steady and uniform over time, it is said to be in thermal equilibrium with itself.

Absolute Zero Temperature

Absolute zero is the temperature at which matter particles (molecules and atoms) have the lowest energy. Some may believe that at absolute zero, particles lose all energy and cease to move. This is not the case. There is something known as zero-point energy, which means that particles retain some energy even after all of their energy has been removed. The vibrational energy that molecules retain even at absolute zero temperature is referred to as zero-point energy. Absolute zero is called 0 kelvin. 0 kelvin is –273.15 degc. This is the start point of Kelvin scale.

SI UNIT OF TEMPERATURE

Kelvin

Kelvin (K) is the base unit of thermodynamic temperature measurement in the International System of Units (SI). The 2018 General Conference on Weights and Measures decided that effective from May 20, 2019, the unit would be defined such that the Boltzmann constant would be equal to 1.380649×10^{-23} joule per kelvin. This unit was originally defined as 100/27, 316 of the triple point (equilibrium among the solid, liquid, and gaseous phases) of pure water. The kelvin is also the fundamental unit of the Kelvin scale. Kelvin scale is also called 'absolute' scale of temperature.

An absolute temperature scale has as its zero point absolute zero (−273.15° on the Celsius temperature scale and −459.67° on the Fahrenheit temperature scale), the theoretical temperature at which the molecules of a substance have the lowest energy; hence, all values on such a scale are nonnegative. The difference between the freezing and boiling points of water is 100 degrees in both the Kelvin and Celsius scales; thus, the Kelvin degree has the same magnitude as the Celsius degree.

How Kelvin scale [K] Developed? How Kevin Scale Addressed Various Issues?

How Absolute Zero Temperature Became the Lower end of Kelvin Scale?

Kelvin scale was built up with no reference to a cold or hot object. It centers only on the energy in a molecule and thus it became the thermodynamic scale of temperature. The idea for the Kelvin scale came of a relationship between the volume and the temperature of a gas. Scientists theorized that the volume of gas should become zero at a temperature of minus –273.15 C.

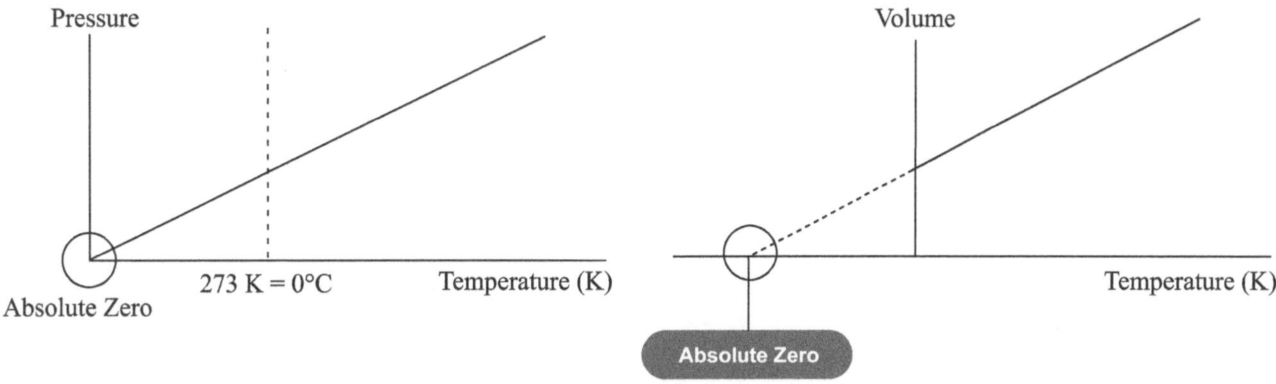

At triple point of water volume and pressure become zero. Molecules have zero energy
The triple point of water is - 273.16 K at 611.73 Pa.

Figure 2

Detail

Charles's law plots of temperature and volume and temperature and pressure are shown above. A remarkable feature of these plots is that if the straight line is extrapolated down to a value of zero volume, the corresponding temperature (the intercept on the temperature axis) is –273.15 degrees Celsius, which corresponds to 0 kelvin, the absolute zero temperature. This relationship shows that absolute zero is the lowest possible temperature because it corresponds to a gas volume of zero, and it is physically impossible to have a volume less than zero. On the right-side plot when pressure vs. temperature data is extrapolated to a pressure of zero, it intercepts the temperature axis at exactly –273.15 degrees Celsius (absolute zero temperature). This observation makes practical sense because, at absolute zero, the thermal molecular motion of the gas molecules nearly ceases.

CHAPTER 6

Pressure

Pressure is a type of potential energy per unit volume. For example, pressure can be used in a fluid to measure energy density.

Pressure is a state variable. Pressure is a measure of the force exerted on a given area, often resulting from the collisions of gas molecules in a closed system. In thermodynamics, pressure plays an important role in determining the work done by a system as it changes volume and temperature. The relationship between pressure, volume, and temperature is described by the ideal gas law, which helps in understanding the behavior of gases under different conditions. Pressure can have a significant impact on the physical properties of substances, affecting their boiling points, melting points, and even their states of matter.

According to kinetic theory, gases are made up of tiny particles that collide with each other and the container's wall at random, resulting in elastic collisions. At constant temperature, there is no kinetic energy loss. Pressure is the impact of these collisions perpendicular to the container's wall per unit area. Moving molecules have the same kinetic energy at a constant temperature, though their speed may differ if the molecules' masses differ. In other words, pressure is the average kinetic energy of molecules. Pressure = $Nm\ v^2/3V$ [N is the number of molecules, m is the mass, v^2 is the average square of the speed, and V is the volume].

THE PRESSURE IS A STATE FUNCTION

The kinetic energy per unit volume (KE/V) varies with the state of the substance and can be a factor in determining pressure. V is volume. KE is kinetic energy. The change in KE/V as a substance transitions from one state to another, such as from a gas to a solid, can affect the pressure exhibited by the substance regardless of the path of transition from one state to another. Ultimately, pressure is a state function because it is determined by the properties of the system in a given state and is independent of the path followed to reach that state of matter. In gases, the pressure is determined by the number of gas molecules in a given volume and their average kinetic energy. In liquids, the pressure is influenced by the density and arrangement of molecules, as well as external forces acting on the liquid. In solids, the pressure is determined by the arrangement and bonding of atoms or molecules within the material.

Static Pressure

Static pressure is exerted by a fluid that is not moving. It's caused by the weight of the fluid above it and is measured perpendicular to the fluid surface. It generally applies to liquid.

Dynamic Pressure

Dynamic pressure is the kinetic energy per unit volume. It is one of the terms of Bernoulli's equation, which can be derived from the conservation of energy for a fluid in motion. Remember the pressure as such is not an energy.

PRESSURE AND KINETIC ENERGY RELATIONSHIP

The kinetic energy of a gas is described by the equation $PV = 2/3\ KE$, where:

P: is the pressure, V: is the volume, KE: is the average kinetic energy per unit volume

Pressure is a measure of energy per unit volume because its units are the same as the units of energy per unit volume:

How?

- ***Pressure:*** Pressure is defined as force per unit area, or $P = F/A$.

- ***Energy per unit volume:*** The expression for energy per unit volume is $E/V = \text{kg m/m}^2\ \text{s}^2$.
- **Units:** The units of pressure and energy per unit volume are both $\text{kg m/m}^2\ \text{s}^2$.

Pressure is a type of potential energy per unit volume as said above. For example, pressure can be used in a fluid as a measure of energy density, kinetic energy density, and potential energy density.

HERE ARE SOME COMMON THERMODYNAMIC EQUATIONS RELATED TO PRESSURE

Equation of State for Ideal Gases

A thermodynamic equation that relates state variables, such as pressure, volume, temperature, or internal energy, to describe the state of matter.

$$PV = nRT$$

This is a formula for thermodynamics pressure, where P is pressure, V is volume, T is temperature, n is the number of molecules, and R is the universal gas constant.

Equation of State for Real Gases

The equation of state for a real gas is the **van der Waal's equation**, which is written as $(P + an^2/v^2)(V - nb) = nRT$. In this equation:

- ***P***: is pressure
- ***V***: is volume
- ***T***: is temperature
- ***R***: is the universal gas constant
- ***a* and *b***: are constants of the gas

The **Peng–Robinson equation** of state is another equation that can be used to model real gases and some liquids.

CHAPTER 7

Thermal and Thermodynamic Equilibrium

GENERAL

The term "equilibrium" refers to a state of balance. The concept of balance in thermodynamics includes not only a balance of forces, but also a balance of other influences.

Each type of influence is related to a different aspect of thermodynamic (complete) equilibrium.

TYPES OF EQUILIBRIUM

Thermal equilibrium = temperature equality.

Mechanical equilibrium is defined as an equality of pressure phases.

Chemical equilibrium is defined as an equality of chemical potentials.

Individual types of equilibrium must exist for thermodynamic equilibrium to exist. For a thermodynamic system to be in equilibrium, all intensive (temperature, pressure) and extensive thermodynamic properties (*U, G, A, H, S*, etc.) must be constant. Hence, the total change in any of those properties must be zero at equilibrium. *U* is internal energy, *G* is Gibbs free energy, *A* is Helmholtz free energy, *H* is enthalpy and *S* is entropy. These are all state variables.

Thermal Equilibrium

Heat is the transfer of energy from a high to a low temperature. The system (or set of systems) is said to be in thermal equilibrium when these temperatures balance out and heat stops flowing. Thermal equilibrium implies that no matter is flowing into or out of the system. Thermal equilibrium is used by the zeroth law of thermodynamics to define how two different systems can be said to be at the same temperature. When molten rock, for example, rises from a volcano, it emits heat into the atmosphere until the rock and the atmosphere reach the same temperature. All systems eventually reach thermal equilibrium, though some will take much longer than others. Knowing that interacting systems tend to have the same temperature enables important applications in all fields of science. For example, a substance's specific heat capacity can be determined by immersing it in water and measuring the temperature after a certain period of time.

Classic Example: Earth's Thermal Equilibrium

It is critical for the Earth to maintain thermal equilibrium in order to maintain a constant temperature. The incoming solar energy must be balanced, which means that the Earth must radiate just as much heat as it receives. The greenhouse effect slows heat transfer from Earth to space, increasing the planet's temperature. The natural greenhouse effect is required for the Earth's temperature to be habitable by humans and other animals, and without it, the Earth would be far too cold. However, as levels of carbon dioxide, methane, and nitrogen dioxide in the atmosphere rise, the greenhouse effect increases, causing slightly less heat to be released from the Earth than is absorbed in the Earth's energy flows. This slight deviation from equilibrium is causing a warming climate, which is causing dangerous changes to the planet.

Thermodynamic Equilibrium

Thermodynamic equilibrium is a thermodynamic concept. It is a single thermodynamic system's internal state or a relationship between several thermodynamic systems connected by permeable or impermeable walls. There are no net macroscopic travels of matter or energy within or between systems in thermodynamic equilibrium.
The existence of gradients of various types within or across a thermodynamic system is zero at thermodynamic equilibrium. Thermodynamic equilibrium effectively halts a thermodynamic process.

To be in equilibrium, all intensive (temperature, pressure) and extensive thermodynamic properties (*U, G, A, H, S*, and so on) must be constant. As a result, at equilibrium, the total change in any of those properties must be zero.

It can also be demonstrated that, at equilibrium, the total free energy of the system G(total) must take a minimum value, reinforcing the fact that dG (total) = 0. The minimum Gibbs energy criterion for equilibrium is a restatement of the second law of thermodynamics, which states that the entropy of a system in equilibrium must be maximum when all possible states for equilibrium are considered.

Not to mention, thermodynamic equilibrium is a broader idea than thermal equilibrium. The system is said to be in thermodynamic equilibrium if the conditions for following three equilibrium is satisfied: Mechanical equilibrium, Chemical equilibrium, and Thermal equilibrium.

Examples

Thermal Equilibrium

A Pot on the Stove

When the stove is turned off, the pot will eventually reach thermal equilibrium as the heat energy becomes evenly distributed. It will then cool to the same temperature as the room around.

A Cup of Hot Tea

When placed on a table, the tea will eventually reach thermal equilibrium with the surrounding temperature.

Thermodynamic Equilibrium

Phase Equilibrium

A system is in phase equilibrium when there are no phase transformations, such as melting or vaporization. Phase equilibrium is a part of thermodynamic equilibrium, which is concerned with the internal state of a thermodynamic system.

Phase Transition

A phase transition is the physical process of a medium changing from one state to another. For example, a liquid may become a gas when heated to its boiling point.

At phase transition the free energy changes reach minimum and entropy change reaches maximum.

In summary, thermal equilibrium and thermodynamic equilibrium are related concepts in thermodynamics, but they refer to different states of a system.

Thermal Equilibrium

Thermal equilibrium occurs when two or more systems (or parts of the same system) are at the same temperature, and there is no net heat transfer between them.

Thermodynamic Equilibrium

Thermodynamic equilibrium is a broader concept that encompasses thermal equilibrium as well as mechanical and chemical equilibrium. A system is in thermodynamic equilibrium when it is in a state where macroscopic properties (like pressure, temperature, and chemical potential) do not change over time.

A system in thermodynamic equilibrium satisfies three conditions:

- **Thermal Equilibrium**: No temperature gradients exist.
- **Mechanical Equilibrium**: No pressure gradients exist; forces are balanced.
- **Chemical Equilibrium**: No chemical reactions are occurring, and concentrations of species are constant.

Gibbs free energy reaches minimum and entropy reaches maximum.

CHAPTER 8

Le Chatelier's Dynamic Equilibrium

LE CHATELIER'S PRINCIPLE AND THERMODYNAMICS

Le Chatelier's principle plays a role in thermodynamics. It states that a system at equilibrium will respond to any applied stress in a way that tends to counteract the stress and restore equilibrium. In the context of thermodynamics, this principle can be used to predict how changes in temperature, pressure, or concentration will affect the equilibrium position of a chemical reaction and how the system will respond to the changes. It is a fundamental concept in understanding the behavior of chemical systems in thermodynamics.

In simple words, when any system at equilibrium for a long period of time is subjected to a change in concentration, volume, or pressure and temperature, (1) the system changes to a new equilibrium, and (2) this change partly counteracts the applied change. In other words, when a settled system is disturbed, it will adjust to diminish the change that has been made to it.

Explanation

Let us take a simple reaction in equilibrium and see what happens according to Le Chatelier's principle when concentration, pressure, and temperature are changed.

$$A\,[g] + 2\,B\,[g] \leftrightarrow C\,[g] + D\,[g]$$

Case 1

What would happen if you changed the conditions by increasing the concentration of A?

Let us write the equation

$$A\,[g] + 2\,B\,[g] \leftrightarrow C\,[g] + D\,[g]$$

According to Le Chatelier, the position of equilibrium will move in such a way as to counteract the change. That means that the position of equilibrium will move so that the concentration of A decreases again – by reacting it with B and turning it into $C + D$. That is by producing more C and D. The position of equilibrium moves to the right.

$$A\,[g] + 2\,B\,[g] \rightarrow C\,[g] + D\,[g]$$

The rate of forward reaction goes up; you get more C and D regardless if the reaction is exothermic or endothermic.

Law of mass action: the law of mass action says that the rate of the chemical reaction is directly proportional to the product of the activities or concentrations of the reactants. It explains and predicts the behaviors of solutions in dynamic equilibrium. Specifically, it implies that for a chemical reaction mixture that is in equilibrium, the ratio between the concentration of reactants and products is constant

Case 2

What would happen if you changed the conditions by decreasing the concentration of A?

According to Le Chatelier, the position of equilibrium will move so that the concentration of A increases again. That means that more C and D will react to replace the A that has been removed. The position of equilibrium moves to the left.

Position of equilibrium moves to left

$$A\,[g] + 2\,B\,[g] \leftarrow C\,[g] + D\,[g]$$

The rate of reverse reaction goes up. You get more *A* and *B*. But the ratio between the concentration of reactants and products remains constant as per the law of mass action.

This is essentially what happens in many batch process reactions like esterification. As the reaction proceeds and the concentration of reactants diminish the reaction tends to reverse. The standard practice is to remove one of the products of the reaction as soon as it is formed. If for example, you removed C as soon as it was formed, the position of equilibrium would move to the right to replace it. If you keep on removing it, the equilibrium position would keep on moving rightwards—turning this into a one-way forward reaction.

Case 3

What would happen if you changed the conditions by increasing the pressure?

According to Le Chatelier, the position of equilibrium will move in such a way as to counteract the change.

$$A\,[g] + 2\,B\,[g] \leftrightarrow C\,[g] + D\,[g]$$

That means that the position of equilibrium will move so that the pressure is reduced again. Pressure is caused by gas molecules hitting the sides of their container. The more molecules you have in the container, the higher the pressure will be. The system can reduce the pressure by reacting in such a way as to produce fewer molecules. *C* and *D* will increase.

In this case, there are 3 molecules on the left-hand side of the equation, but only 2 on the right. By forming more *C* and *D*, the system causes the pressure to reduce. Increasing the pressure on a gas reaction shifts the position of equilibrium towards the side with fewer molecules.

The reaction shifts the equilibrium to the right to generate more *C* and *D* and reduce the total number of molecules to counteract the pressure increase.

$$A\,[g] + 2\,B\,[g] \rightarrow C\,[g] + D\,[g]$$

Case 4

What would happen if you changed the conditions by decreasing the pressure?

The equilibrium will move in such a way that the pressure increases again. It can do that by producing more *A* and *B* molecules.

$$A\,[g] + 2\,B\,[g] \leftrightarrow C\,[g] + D\,[g]$$

In this case, the position of equilibrium will move towards the left-hand side of the reaction.

$$A\,[g] + 2\,B\,[g] \leftarrow C\,[g] + D\,[g]$$

Case 5

What happens if there are the same number of molecules on both sides of the equilibrium reaction and pressure increased?

In this case, increasing the pressure has no effect whatsoever on the position of the equilibrium. Because you have the same numbers of molecules on both sides, the equilibrium can't move in any way that will reduce the pressure again.

Case 6

Le Chatelier's Principle with a Change of Temperature

For this, you need to know whether heat is given out or absorbed during the reaction. Assume that our forward reaction is exothermic (heat is evolved):

$$A\,[g] + 2\,B\,[g] \leftrightarrow C\,[g] + D\,[g],\ \Delta H = -250\ \text{kj/mol}$$

This shows that 250 kJ is evolved (hence the negative sign) when 1 mole of *A* reacts completely with 2 moles of *B*. For reversible reactions, the value is always given as if the reaction was one-way in the forward direction. The back reaction (the conversion of *C* and *D* into *A* and *B*) would be endothermic by exactly the same amount.

What would happen if you changed the conditions by increasing the temperature?

According to Le Chatelier, the position of equilibrium will move in such a way as to counteract the change.

$$A\,[g] + 2\,B\,[g] \leftrightarrow C\,[g] + D\,[g],\ \Delta H = -250 \text{ kj/mol}$$

That means that the position of equilibrium will move so that the temperature is reduced again. Suppose the system is in equilibrium at 300°C, and you increase the temperature to 500°C. How can the reaction counteract the change you have made? How can it cool itself down again? To cool down, it needs to absorb the extra heat that you have just put in. In the case we are looking at, the back reaction absorbs heat. The reverse endothermic reaction becomes dominant.

$$A\,[g] + 2\,B\,[g] \leftarrow C\,[g] + D\,[g],\ \Delta H = -250 \text{ kj/mol}$$

The position of equilibrium therefore moves to the left. The new equilibrium mixture contains more *A* and *B*, and less *C* and *D*.

To note, in a reversible reaction where the forward reaction is exothermic, an increase in temperature reverses the reaction.

Case 7

What would happen if you changed the conditions by decreasing the temperature?

The equilibrium will move in such a way that the temperature increases again.

$$A\,[g] + 2\,B\,[g] \leftrightarrow C\,[g] + D\,[g],\ \Delta H = -250 \text{ kj/mol}$$

Suppose the system is in equilibrium at 500°C and you reduce the temperature to 400°C. The reaction will tend to heat itself up again to return to the original temperature. It can do that by favouring the exothermic forward reaction.

$$A\,[g] + 2\,B\,[g] \rightarrow C\,[g] + D\,[g],\ \Delta H = -250 \text{ kj/mol}$$

The position of equilibrium will move to the right. More *A* and *B* are converted into *C* and *D* at the lower temperature.

In summary, Le Chatelier's principle states that when an external stressor is applied to a system in equilibrium, the system will shift in a way that counteracts the stressor:

Concentration

Adding more reactants shifts the equilibrium to the right, while adding more products shifts it to the left.

Temperature

Increasing the temperature of an endothermic reaction shifts the equilibrium to the right, while increasing the temperature of an exothermic reaction shifts it to the left.

Pressure

Increasing the pressure of a reaction with more gas molecules on one side shifts the equilibrium to the side with fewer gas molecules. Decreasing the pressure shifts the equilibrium to the side with more gas molecules.

CHAPTER 9

Role of Arrhenius and van der Waals in Thermodynamics

THE STORY OF TWO LEGENDS: ARRHENIUS AND VAN DER WAALS ABOUT "INTRAMOLECULAR BONDS" AND "INTERMOLECULAR BONDS"

Arrhenius is known for his work in chemical kinetics, while van der Waals is known for his work in molecular interactions and the development of the van der Waals equation.

Briefly

Intramolecular Bonds

Intramolecular atomic bonds exist within a molecule. These are chemical bonds.

Intermolecular Bonds

Intermolecular bonds are the bonds between molecules held by intermolecular bonds. These are physical bonds.

Behind every thermodynamic property like boiling point, melting point, surface tension, viscosity, heat capacity, latent heat etc the intermolecular forces determine the values.

Types of Intramolecular Bonds

- Metallic bonds
- Covalent bonds
- Ionic bonds

Types of Intermolecular Bonds

- Dipole-dipole interaction
- Hydrogen bonds
- London dispersion forces

Arrhenius and van der Waals

In the field of molecular structure, two names stand out: (1) Arrhenius and (2) van der Waals. Arrhenius's contributions focused on "intramolecular atomic bonds," which are fundamental to kinetics, and van der Waals's contributions focused on "intermolecular bonds," which are fundamental to thermodynamics. Both of these legends made significant contributions to our understanding of molecular structure and bonding, and their work continues to have a profound impact on the field of chemistry. Their theories and principles have laid the groundwork for much of our current understanding of molecular behaviour and paved the way for further advances in the field.

Arrhenius contributed to thermodynamics **by developing the Arrhenius equation**, which relates the rate constant (K) of a chemical reaction to the free energy change (ΔG) of the reaction. The equation is expressed as:

$$\ln K = -\Delta G/RT$$

Where:

- K is the rate constant
- ΔG is the free energy change
- R is the gas constant
- T is the temperature

By rearranging the equation, we can express the relationship between ΔG and K as:

$$\Delta G = -RT \ln K$$

Therefore, Arrhenius's work in developing the Arrhenius equation made a significant contribution to our understanding of the relationship between the rate constant and free energy in chemical reactions.

van der Waals is one of the architects of thermodynamics who understood the role of molecular structure and their intermolecular interactions that define many thermophysical properties and phenomena in real life, such as boiling point, freezing point, critical point, surface tension, viscosity, etc

van der Waals' equation of state for real gases

$[P + an^2/v^2]\,[V - nb] = RT$ extends the ideal gas law to include non-zero gas molecule sizes and interactions (both of which vary depending on the substance). It fits pressure-volume-temperature data for a real gas better than the ideal gas equation does. The improved fit is obtained by introducing two parameters (designated "a" and "b") that must be determined experimentally for each gas.

As a result, the equation can model the transition from liquid to gas and vice versa. It also generates simple analytic expressions for real-world properties that shed light on their behavior.

Other notable contributions of van der Waals have been

Critical Point and Phase Transitions: van der Waals introduced the concept of a critical point in a substance, which represents the conditions at which a substance undergoes a phase transition. This work laid the foundation for the study of phase transitions and critical phenomena in thermodynamics.

- **Theory of corresponding state**: van der Waals discovered through experiments that most gases have comparable properties close to the critical point. Following that, he put forth the "Theory of Corresponding State." The simplest definition of corresponding state in one sentence is "For most gases, when the ratio of actual temperature to critical temperature or the ratio of actual pressure to critical pressure is the same, they exhibit similar thermodynamic properties."
- **Thermodynamic Potentials**: He made significant contributions to the understanding of the thermodynamic potentials, such as the Helmholtz and Gibbs free energies, and their application in describing the equilibrium and stability of thermodynamic systems.
- **Nobel Prize in Physics**: In 1910, van der Waals was awarded the Nobel Prize in Physics for his work on the equation of state for gases and liquids.

These contributions have had a profound impact on the development of thermodynamics and have greatly enhanced our understanding of the behaviour of gases and liquids.

CHAPTER 10

Arrhenius's Equation

The Arrhenius equation is an expression that provides a relationship between the rate constant (of a chemical reaction), the absolute temperature, and the A factor (also known as the pre-exponential factor; can be visualized as the frequency of correctly oriented collisions between reactant particles). It provides insight into the dependence of reaction rates on the absolute temperature.

The most commonly used equation that relates kinetics to thermodynamics is the Arrhenius equation:

$$k = Ae^{\frac{-E_a}{RT}}$$

where:

k is the rate constant of the reaction

A is the pre-exponential factor or frequency factor

Ea is the activation energy of the reaction

R is the gas constant

T is the temperature in Kelvin

RATE CONSTANT AND TEMPERATURE

The Arrhenius equation shows how the rate constant of a reaction is affected by the temperature and the activation energy. It helps establish the relationship between kinetics (rate constant) and thermodynamics (temperature and activation energy).

In the Arrhenius equation, the units of the pre-exponential factor *A* are identical to those of the rate constant and will vary depending on the order of the reaction. If the reaction is first order it has the units: s^{-1}.

K is the number of collisions that result in a reaction per second, A is the number of collisions (leading to a reaction or not) per second occurring with the proper orientation to react, and $e^{-Ea/(RT)}$ is the probability that any given collision will result in a reaction.

Additionally, the concept of equilibrium constant (*K*) from thermodynamics can be related to the rate constant (*k*) through the relationship known as the Gibbs free energy (ΔG) change.

The rate of a reaction roughly doubles for every 10°C rise in temperature. This is a widely used rule of thumb, but it's not always true, especially when a strong covalent bond needs to be broken.

The Arrhenius equation can be used to roughly estimate the rate of reaction increase for a given temperature range and activation energy. The equation assumes that the reaction rate depends exponentially on temperature.

ACTIVATION ENERGY AND TEMPERATURE

Activation energy is the energy required for a chemical reaction to occur. It is the minimum amount of energy that molecules must have in order to react. The relationship between activation energy and temperature is that as temperature increases, the activation energy required for a reaction decreases. Hen molecules have more kinetic energy at higher temperatures, they are able to overcome the activation energy barrier more easily and react faster. This is because the molecules are moving faster and colliding more frequently, increasing the likelihood of successful collisions that lead to a reaction.

In contrast, at lower temperatures, molecules have less kinetic energy and may not have enough energy to overcome the activation energy barrier. This results in slower reaction rates and a longer time for the reaction to occur.

Therefore, temperature and activation energy have an inverse relationship - as temperature increases, the activation energy required for a reaction decreases, leading to faster reaction rates. This relationship is represented by the Arrhenius equation, which describes the exponential relationship between temperature and reaction rate constant.

Arrhenius Equation

$$k = Ae^{\frac{-E_a}{RT}}$$

Where:

k = rate constant

A = pre-exponential factor

Ea = activation energy

R = gas constant

T = temperature in Kelvin

From the Arrhenius equation, we can see that as temperature increases, the exponential factor ($-Ea/RT$) decreases. This implies that at higher temperatures, the exponential factor becomes smaller, making it easier for the reaction to occur. In other words, higher temperatures lower the effective activation energy, leading to an increase in the rate constant and faster reaction rates. Conversely, at lower temperatures, the exponential factor becomes larger, requiring a higher activation energy for the reaction to proceed, resulting in slower reaction rates.

Example

Problem: Calculate change in Ea of a reaction if temperature increases from 25 degc to 50 degc

The Arrhenius equation can be used to calculate the change in activation energy (Ea) if the reaction temperature is increased [for example] from 25°C to 50°C. The Arrhenius equation is given by:

$$k = A \times \exp(-Ea/RT)$$

where:

k = rate constant

A = pre-exponential factor

Ea = activation energy

R = gas constant (8.314 J/(mol × K))

T = temperature in Kelvin

By taking the natural logarithm of both sides of the equation, we can isolate the activation energy:

$$\ln(k) = \ln(A) - Ea/RT$$

We can calculate the activation energy at 25°C (298K) first:

$$\ln(k_1) = \ln(A) - Ea/RT1$$

$$\ln(k_1) = \ln(A) - Ea/(8.314 \times 298)$$

Similarly, we can calculate the activation energy at 50°C (323K):

$$\ln(k_2) = \ln(A) - Ea/(8.314 \times 323)$$

By rearranging these equations, we can isolate the activation energy terms:

$$Ea = -R \times T_1 \times \ln(k_1/A)$$

$$Ea = -R \times T_2 \times \ln(k_2/A)$$

Subtracting the second equation from the first, we can find the change in activation energy:

$$\Delta Ea = Ea_2 - Ea_1$$

CHAPTER 11

Laws of Thermodynamics

First **Law of Thermodynamics**: This law connects heat transfer to thermodynamics by stating that the total energy of an isolated system is constant. This means that the amount of heat transferred into or out of a system is directly related to changes in the internal energy of the system and the work energy. The sum of internal energy and the work energy is the total energy called enthalpy.

Second Law of Thermodynamics: This law connects heat transfer to thermodynamics by introducing the concept of entropy, which is a measure of the disorder or randomness of a system. The second law states that in any heat transfer process, the total entropy of a closed system will always increase, providing a direction for the flow of heat and establishing the principle of irreversibility.

Third Law of Thermodynamics: This law connects heat transfer to thermodynamics by stating that as a system approaches absolute zero temperature, its entropy approaches a minimum possible value. This law helps to define the behaviour of systems at extreme low temperature conditions, providing insight into the relationship between heat transfer and thermodynamic properties.

Zeroth Law of Thermodynamics: This law connects heat transfer to thermodynamics by establishing the concept of thermal equilibrium and the transitive property of temperature. It states that if two systems are each in thermal equilibrium with a third system, they are also in thermal equilibrium with each other. This law allows for the definition and measurement of temperature, which is essential for understanding heat transfer processes.

Renumbering of Thermodynamic Laws

The numbering of the laws of thermodynamics is indeed puzzling. The term 'zeroth law' was coined in the 1930s, long after the first, second, and third laws had become widely accepted. The zeroth law was introduced much later.

General Thoughts

The Zeroth Law should have been placed at the beginning as it establishes the concept of temperature and thermal equilibrium, which is foundational to the other laws.

Remember the first law defines temperature as average kinetic energy $KE = 1/2\ K\ T$. K is constant. The second law defines entropy $S = Q/T$ in terms of temperature.

The Zeroth Law states that if two systems are in thermal equilibrium with a third system, then they are in thermal equilibrium with each other. This law therefore, establishes the concept of temperature and the basis for the measurement of temperature.

Without this fundamental principle, the other laws would have no basis for comparison or measurement.

Therefore, it would make more sense to reorder the laws as follows:

1. Zeroth Law – Law of equilibrium
2. First Law – Energy conservation
3. Second Law – Entropy always increases in irreversible systems
4. Third Law – No entropy at absolute zero

This reordering would align the laws in a more logical and foundational progression, with the Zeroth Law serving as the essential starting point for understanding the behaviour of thermodynamic systems.

Relationship of Thermodynamic Laws with Key Concepts of Thermodynamics

- **Energy conservation**: The first law of thermodynamics states that energy cannot be created or destroyed, only transferred or transformed. This is consistent with the principle of energy conservation in physics.
- **Entropy:** The second law of thermodynamics states that the entropy of a closed system can never decrease over time. Entropy is a measure of the disorder or randomness in a system, and the second law of thermodynamics is related to the increase in entropy that occurs in natural processes.
- **Absolute zero**: The third law of thermodynamics states that the entropy of a perfect crystal at absolute zero is zero. This is related to the concept of absolute zero, the lowest possible temperature at which the entropy of a system reaches a minimum.
- **Equilibrium**: The zeroth law of thermodynamics states that if two systems are in thermal equilibrium with a third system, they are in thermal equilibrium with each other. This concept of thermal equilibrium is related to the idea of reaching a state of balance or stability in a system, where no further changes occur.

CHAPTER 12

Kinetic Theory of Gases and Equations

Gas molecules are studied by using a theoretical model called kinetic theory of gases. The model assumes that gas molecules are very small in comparison to the distance between them. The molecules are constantly moving at random and collide with each other and the walls of any container.

Gas molecules have mass, momentum, and energy. The density of a gas is the sum of the mass of the molecules divided by the volume that the gas occupies. The pressure of a gas is a measurement of the linear momentum of the molecules. When gas molecules collide with the walls of a container, the molecules impart momentum to the walls, resulting in a force that can be measured. The pressure is defined as force divided by area. The temperature of a gas is a measure of the gas's mean kinetic energy.

The molecules are in constant random motion, and there is an energy (mass x square of velocity) associated with it. Higher temperature means greater motion.

The location of the molecules relative to each other in a solid is nearly constant. However, in a gas, molecules can move around and interact with one another and with their surroundings in a variety of ways. As earlier stated, molecular motion always includes a random component. The entire fluid can be made to move in a controlled manner (flow). The ordered motion is superimposed or added to the molecules' normal random motion. There is no distinction between the random and ordered components at the molecular level. As the static pressure, we measure the pressure produced by the random component. Dynamic pressure is the pressure produced by ordered motion. Bernoulli's equation tells us that the sum of the static and dynamic pressure is the total pressure.

When a gas is at rest, all of the molecules' motion is random, and the pressure we detect is the total pressure of the gas. When the gas is moved or flows, some of the random components of velocity are changed in favour of the directed motion. The directed motion is referred to as "ordered," as opposed to the disordered random motion.

In summary, the following assumptions are made when applying kinetic theory to ideal gases, the gas is made up of extremely fine particles.

Because of their small size, the total volume of the individual gas molecules is negligible in comparison to the volume of the gas container.

This is equivalent to saying that the average distance between gas particles is large in comparison to their size, and that the time elapsed during a collision between particles and the container's wall is negligible in comparison to the time between successive collisions.

Because there are so many particles, a statistical approach to the problem is well justified. This assumption is known as the thermodynamic limit.

KEY EQUATIONS OF KINETIC THEORY OF GASES

- **Ideal Gas Law**: $PV = nRT$, where P is the pressure, V is the volume, n is the number of moles of gas, R is the ideal gas constant, and T is the temperature.
- **Root–mean-square speed**: The average velocity of gas molecules in a gas sample is given by the equation $vrms = sqrt(3RT/m)$, where R is the ideal gas constant, T is the temperature, and m is the molar mass of the gas.
- **Maxwell-Boltzmann distribution**: Describes the distribution of speeds of gas molecules in a gas sample. The distribution function is given by $f(v) = 4\pi\ (m/(2\pi RT))^{3/2} \times v^2 \times e^{(-mv^2/2RT)}$, where m is the molar mass, v is the speed, and T is the temperature.
- **Kinetic energy**: The average kinetic energy of a gas molecule is given by the equation $KE = 3/2kT$, where k is Boltzmann's constant and T is the temperature.

- **Pressure of a gas**: The pressure exerted by a gas in a container is given by the equation $P = 1/3n(mv^2)$, where n is the number of gas molecules, m is the mass of a gas molecule, and v is the average speed of the gas molecules.

HOW KINETIC THEORY OF GASES FITS INTO THERMODYNAMICS?

Thermodynamics and kinetic theory of gases to be complementary and interconnected. Thermodynamics provides a macroscopic view of systems and their behaviour in terms of heat and work interactions, whereas the kinetic theory of gases provides a microscopic explanation of individual gas particle behaviour. The first and second laws of thermodynamics, for example, provide a framework for understanding and predicting the behaviour of macroscopic systems, including gases. The kinetic theory of gases, on the other hand, aids in explaining the underlying mechanisms at the molecular level that give rise to the macroscopic behaviour described by thermodynamics. The two are related because the macroscopic behaviour of gases, as described by thermodynamics, emerges from the microscopic interactions and movements of individual gas particles, as described by kinetic theory.

In this sense, the two approaches are linked and provide a comprehensive understanding of gas behaviour as well as heat and work interactions. Overall, the relationship between thermodynamics and the kinetic theory of gases is mutually beneficial and enriching, with each providing valuable insights into different aspects of gas behaviour and interactions with heat and work.

CHAPTER 13

Ideal Gas Theory

An ideal gas is one in which both the volume of molecules and the forces between the molecules are so small that they have no effect on the gas's behaviour.

The ideal gas theory is very important for process analysis. The volume of an ideal gas, the pressure it exerts, and its temperature are the three parameters that define it.

The definition of ideal gas is a special state of matter that can be delimited by a system boundary.

KEY POINTS

- **All particles are assumed** to have rest mass ($m > 0$; the particles are not photons).
- **The particle count is low** in comparison to the system volume.
- **Particle volume is negligible** in comparison to system volume.
- **Collisions between particles are much less likely** than collisions with the system boundary.

The practical benefit of treating real gases as ideal (under certain conditions) is that it results in a simple equation of state with only one constant. The ideal gas equation of state can be written as follows:

$$PV = m\,RT \text{ or } P\,v = RT \qquad \ldots (1)$$

where P is the pressure in Pa, V is the gas volume in m^3, m is mass of gas in kg, T is gas temperature in K, R is known as the gas constant and is given in $J/kg\ K$, v is mass specific volume in m^3/kg, v is molar specific volume in m^3/k mol, and $\mathcal{R}$ is the universal gas constant of 8.134 J/mol K.

Take note that the gas constant is unique to each gas and is determined by the universal gas constant and the gas's molecular mass (M).

$$R = \mathcal{R}/M \qquad \ldots (2)$$

The thermal equation of state" of the ideal gas expresses the relationship between pressure, specific volume, and temperature.

The ideal gas equation can be expressed in terms of internal energy, specific volume, and temperature. The equation of state in this case is known as the **Caloric equation of state**. The internal energy of an ideal gas, in particular, is temperature dependent. The following is the caloric equation of state for a monoatomic ideal gas:

$$u = 1.5\,RT \qquad \ldots (3)$$

[this equation can also be written as $du/dT = 1.5\,R$ or $Cv = 1.5\,R$]

u denotes molar specific internal energy.

Because $h = u + P\,v$, the enthalpy of a monoatomic ideal gas is given by

$$h = 2.5\,RT \qquad \ldots (4)$$

[this equation can also be written as $dh/dT = 2.5\,R$] or $Cp = 2.5\,R$]

Combining the two caloric equations of state from equations (3) and (4), the known **Robert Meyer equation** for ideal gas can be derived as follows:

$$Cp - Cv = R \text{ or}$$

$$Cp = Cv + R \qquad \ldots (5)$$

It should be noted that the internal energy of an ideal gas is only a function of temperature. As a result, the ideal gas's specific heat is $Cv = 1.5\ R$ and $Cp = 2.5\ R$. The adiabatic exponent is the ratio of specific heat at constant pressure and constant volume.

$$Y = Cp/Cv \quad \ldots (6)$$

The adiabatic exponent for ideal gas is 1.4 for monoatomic gas and 5/3 = 1.67 for diatomic gas. If one of *P, v*, or *T* is constant, there are some special cases. At a constant temperature, the volume of a given quantity of ideal gas varies inversely with pressure (this is known as Boyle's law), describing a volume change as follows:

$$P_1 V_1 = P_2 V_2 \quad \ldots (7)$$

where the subscripts denote the initial and final states.

Analysts use eq. (7) in a variety of situations, including selecting an air compressor, calculating the consumption of compressed air in reciprocating air cylinders, and determining the time required to store air. Charles' law applies if the process is under constant pressure or volume:

$$v_1/T_1 = v_2/T_2 \text{ and } p_1/T_1 = p_2/T_2 \quad \ldots (8)$$

If the number of moles of an ideal gas in an enclosed volume does not change, the combined ideal equation of state is as follows:

$$P_1\, v_1/T_1 = p_2\, v_2/T_2 \quad \ldots (9)$$

Isothermal, isochoric, [constant volume] isobaric, [constant pressure] and polytropic are the basic processes with ideal gas table below contains a description of each process as well as the relevant equation.

Table 1

Process	*Relationship between ρ, V, T*	*Work (W)*	*Internal-energy change ($U_2 – U_1$)*	*Heat (Q)*
Constant pressure	ρ = constant $\frac{V_2}{T_1} = \frac{V_2}{T_2}$	$\rho(V_2 - V_1)$	$mc_v(T_2 - T_1)$	$mc_p(T_2 - T_1)$
Constant volume	V = constant $\frac{\rho_1}{T_1} = \frac{\rho_2}{T_2}$	0	$mc_v(T_2 - T_1)$	$Q = U_2 - U_1$ $Q = mc_v(T_2 - T_1)$
Isothermal	$T = c$ $\rho_1 V_1 = \rho_2 V_2$	$\rho_1 V_1 \ln\left(\frac{V_2}{V_1}\right)$	0	$Q = W$ $Q = \rho_1 V_1 \ln\left(\frac{V_2}{V_1}\right)$
Polytropic	ρV^n = constant $\frac{T_2}{T_1} = \left(\frac{V_1}{V_2}\right)^{n-1}$ $\frac{\rho_2}{\rho_1} = \left(\frac{T_2}{T_1}\right)^{\frac{n}{n-1}}$	$\frac{\rho_1 V_1 - \rho_2 V_2}{n-1}$	$mc_v(T_2 - T_1)$	$Q = W + U_2 - U_1$
Adiabatic	As for polytropic with $n = \gamma$	As for polytropic with $n = \gamma$	$mc_v(T_2 - T_1)$	0

Table 2: The Polytropic Index Varies According to the Type of Process

Polytropic Index	*Relation*	*Effects*
$n < 0$	–	Negative exponents reflect a process where work and heat flow simultaneously in or out of the system. In the absence of forces except pressure, such a spontaneous process is not allowed by the second law of thermodynamics; however, negative exponents can be meaningful in some special cases not dominated by thermal interactions, such as in the processes of certain plasmas or if there are other forms of energy (e.g. chemical energy) involved during the process (e.g. explosion).
$n = 0$		Equivalent to an isobaric process (constant pressure)
$n = 1$		Equivalent to an isothermal process (constant temperature), under the assumption of ideal gas law.
$1 < n < \gamma$	–	Under the assumption of ideal gas law, heat and work flows go in opposite directions (K > 0), such as in vapor compression refrigeration during compression, where the elevated vapour temperature resulting from the work done by the compressor on the vapour leads to some heat loss from the vapour to the cooler surroundings.
$n = \gamma$	–	Equivalent to an isentropic process (adiabatic and reversible, no heat transfer), under the assumption of ideal gas law.
$\gamma < n < \infty$	–	Under the assumption of ideal gas law heat and work flows go in the same direction (K < 0), such as in an internal combustion engine during the power stroke, where heat is lost from the hot combustion products, through the cylinder walls, to the cooler surroundings, at the same time as those hot combustion products push on the piston.
$n = +\infty$		Equivalent to an isochoric process (constant volume)

Source: Wikipedia.

IDEAL GAS LAW BASED ON EQUATIONS

There are several other equations that are derived from the ideal gas law equation $PV = nRT$:

1. Density form of the ideal gas law:

   $$\rho = (n/M) \times (P/RT)$$

   where
   ρ = density of the gas,
   M = molar mass of the gas.

2. Molar volume form of the ideal gas law:

   $$Vm = V/n = RT/P$$

   where
   Vm = molar volume of the gas.

3. The specific gas constant form:

   $$R \text{ specific} = R/M$$

   where
   R specific = specific gas constant,
   M = molar mass of the gas.

Do ideal gases experience critical point?

The answer is no.

Explanation

A critical point is a specific set of conditions (temperature and pressure) at which a substance transitions between being a gas and a liquid. This transition occurs at the critical point because the distinction between the gas phase and the liquid phase disappears, and the substance behaves as a supercritical fluid.

In real gases, the critical point occurs because of the intermolecular forces between gas molecules. The intermolecular forces progressively diminishes as the system approaches critical point.

As the temperature and pressure of a gas are increased, the kinetic energy of the gas molecules also increases. At the same time, the pressure compresses the gas molecules closer together, increasing the density of the gas. When the temperature and pressure reach the critical point, the kinetic energy of the gas molecules is so high that they can no longer be held together by intermolecular forces, causing the substance to transition to a supercritical fluid state.

IDEAL GAS

Ideal gases do not experience a critical point because they are assumed to have no intermolecular forces and occupy zero volume. In an ideal gas, the gas molecules are considered to have no volume and do not interact with each other. As a result, ideal gases do not exhibit the same condensation and liquefaction behaviors as real gases, and therefore do not reach a critical point. Ideal gases also do not have the same temperature and pressure conditions at which the distinction between the gas phase and the liquid phase disappears, as they do not have intermolecular forces to overcome.

CHAPTER 14

Real Gases

Real gases are nonideal gases with molecules that occupy space and interact, and thus do not obey the ideal gas law. The following are the three key points to comprehend the behaviour of real gases:

KEY POINT

Compressibility Effects

For most applications, such a detailed analysis is unnecessary, and the ideal gas approximation can be used with reasonable accuracy. Real-gas models, on the other hand, must be used near the condensation point of gases, near critical points, at extremely high pressures, to explain the Joule-Thomson effect, and in other unusual cases. The compressibility factor Z describes the deviation from ideality.

Explanation

Compressibility Effects of Ideal Gases vs Real Gases Key Points

Compressibility is the measure of how much a substance can be reduced in volume when pressure is applied to it. In the case of ideal gases, there are several key factors that influence their compressibility:

A. Ideal gases do not have any intermolecular forces between particles, meaning they do not experience any significant attraction or repulsion. This lack of interaction allows ideal gases to easily compress when pressure is applied.
B. Ideal gases follow the ideal gas law, which states that the pressure and volume of a gas are inversely proportional, while the temperature and volume are directly proportional. This relationship allows ideal gases to compress uniformly and predictably when pressure is changed.
C. The compressibility factor, also known as the Z factor, is a measure of how closely a real gas behaves like an ideal gas. In the case of ideal gases, the compressibility factor is equal to 1, indicating that they follow ideal gas behavior perfectly.
D. Ideal gases have perfectly elastic collisions, meaning that no energy is lost during collisions between gas particles. This allows ideal gases to easily compress and expand without losing any energy in the process.
E. At low pressures and high temperatures, real gases tend to behave more like ideal gases and have higher compressibility. This is because the intermolecular forces between particles become less significant as the distance between particles increases.
F. As pressure increases, real gases deviate more from ideal gas behavior and have lower compressibility. This is due to the fact that the volume of the gas becomes more restricted, leading to interactions between particles that disrupt the uniform compression of the gas.
G. The van der Waals equation of state is a modified form of the ideal gas law that accounts for the finite size of gas particles and intermolecular forces. This equation provides a more accurate representation of real gas behavior, including their compressibility properties.
H. The compressibility of real gases is affected by factors such as temperature, pressure, and molecular size. These factors can lead to deviations from ideal gas behavior and influence the compressibility of the gas under different conditions.

In summary, the compressibility of ideal gases is defined by their lack of intermolecular forces, perfect elastic collisions, and adherence to the ideal gas law. Real gases, on the other hand tend to behave as ideal gas at low pressure and high temperature.

Details

Compressibility of Real Gases

Compressibility Factor

The compressibility factor (Z), also known as the compression factor or the gas deviation factor in thermodynamics, describes the deviation of a real gas from ideal gas behaviour. It is simply defined as the ratio of a gas's molar volume to that of an ideal gas at the same temperature and pressure. It is a useful thermodynamic property for accounting for real-world gas behaviour by modifying the ideal gas law. **In general, the closer a gas is to a phase change, the lower the temperature, or the higher the pressure, the greater the deviation from ideal behaviour**. Compressibility factor values are typically calculated using equations of state (EOS), such as the virial equation, which take compound-specific empirical constants into account.

Before compressibility can be calculated for a gas that is a mixture of two or more pure gases (for example, air or natural gas), the gas composition must be known.

Alternatively, the compressibility factor for specific gases can be calculated using generalised compressibility charts that plot Z as a function of pressure at constant temperature.

Explanation

At high pressure and low temperature, closer a gas to phase change, gaseous atoms or molecules get closer to each other and van der Waal forces operates. So, molecules or atoms start attracting each other. Hence a gas deviates the most from its ideal behaviour.

Significance of Compressibility Factor

In mathematical framework Z is

$$Z = pV/n\,RT$$

where p is the pressure, n is the number of moles of gas, T is the absolute temperature, and

R is the gas constant, V is unit volume

For an ideal gas, according to the definition, $Z = 1$. In general, the value of Z increases with pressure and decreases with temperature. At high pressures, molecules collide more frequently. This allows repulsive forces between molecules to have an effect, causing the molar volume of the real gas (Vm) to be greater than the corresponding ideal gas (Vm) ideal gas $= RT/p$, causing Z to exceed one. When pressures are low, molecules can move freely. In this case, attractive forces prevail, resulting in $Z < 1$. The greater the deviation from Z, the closer the gas is to its critical point or boiling point. Hydrogen and Helium gases are two exceptions with Z exceeding one [$Z > 1$] at all temperature and pressure due to dominance of repulsive forces between molecules.

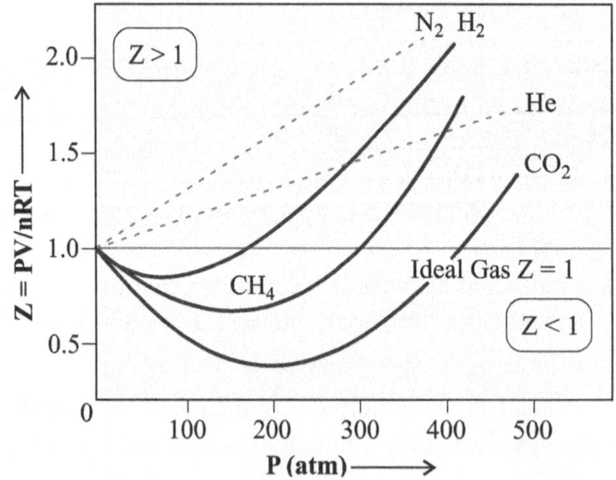

Figure 1: Compressibility Factor Z *vs.* Pressure

In summary, molecules of ideal gases are considered to have zero volume and zero intermolecular attraction this is not true for real gases and this is the fundamental difference between an ideal and a real gas. The van der Waals equation is an equation of state that corrects for two properties of real gases, $P = RT/(v{-}b) - a/v^2$, P = pressure, R = universal gas constant, T = absolute temperature, V = molar volume 'a' is a measure of the average attraction between particles, and '*b* 'is the volume excluded from *V* by one particle. Both *a* and *b* in the van der Waals's equation are the size and molecular structure-dependent.

The above is a *Z vs P* graph. On the central straight red line parallel to x-axis $Z = 1$ that is the ideal gas line. Above the red line, $Z > 1$, and below it is $Z < 1$. It may be observed in this graph, deviations of the compressibility factor, *Z*, from unity are due to attractive and repulsive intermolecular forces. The attractive force between molecules initially makes the gas more compressible than an ideal gas at low pressure. Thereafter, as the pressure increases at a given temperature and pressure, repulsive forces tend to make the volume larger than for an ideal gas; when these forces dominate *Z* is greater than unity. When attractive forces dominate, *Z* is less than unity. When repulsive forces dominate, *Z* is greater than unity. The relative importance of attractive forces decreases as temperature increases.

Compressibility of Gases: Concept of Corresponding States

In simple words, the theory of corresponding states say that at reduced temperature and pressure, all gases have the same compressibility factor *Z*. This means that at constant fractions of kinetic energy (temperature) and potential energy (pressure) relative to their critical values, all gases exhibit similar behavior in terms of compressibility. This concept allows for the comparison of different gases under similar conditions, making it easier to study their properties and behavior in a standardized way.

The theory of corresponding states allows for the prediction of the compressibility factor *Z* of a gas at any temperature by considering the reduced temperature and reduced pressure. By comparing the temperature and pressure of a gas to its critical temperature and pressure, one can determine the relative ratios known as reduced temperature and reduced pressure. These ratios can then be used to calculate the compressibility factor *Z*, providing valuable information about the behavior of the gas under different conditions. This makes it easier to compare and analyze the properties of different gases in a standardized way.

The interesting part is that theory of corresponding states provides insight into the complex interplay between kinetic and potential energy in determining the behavior of real gases. It underscores the importance of considering both factors in understanding the deviations from ideal gas behavior at different thermodynamic conditions.

Details

Background

The simplest definition of corresponding state in one sentence is "For most gases, when the ratio of actual temperature to critical temperature or the ratio of actual pressure to critical pressure is the same, they exhibit similar thermodynamic properties."

What is Corresponding State Theory?

van der Waals discovered through experiments that most gases have comparable properties close to the critical point. Following that, he put forth the "Theory of Corresponding State."

According to the theory of corresponding states, the thermodynamic properties of most gases **become similar or "correspond"** to one another when the temperature is expressed as a ratio of the critical temperature (reduced temperature, *Tr*). According to the theory of corresponding states, **the various thermodynamic properties, including enthalpy, specific heat, compressibility factor, and phase behaviour, approach similar values** when gases are compared at the same reduced temperature (*Tr*). The theory of corresponding states makes use of the idea of reduced variables, such as reduced temperature and reduced pressure. It is possible to compare various substances and analyse their behaviour close to the critical point by expressing the temperature and pressure in terms of these condensed variables.

The theory of corresponding states is primarily applicable to simple, non-polar gases where intermolecular forces are primarily due to London dispersion forces. For complex molecules with additional intermolecular forces like H bonds, the theory may not accurately predict their behaviour.

Generalized Compressibility Chart

This can be done if the coordinates are modified such that $P_R = P/P_{cr}$ and $T_R = T/T_{cr}$ where P_R and T_R are the reduced pressure and reduced temperature.

The Z factor is the same for all gases at the same reduced pressure and temperature.

This is called the principle of corresponding states.

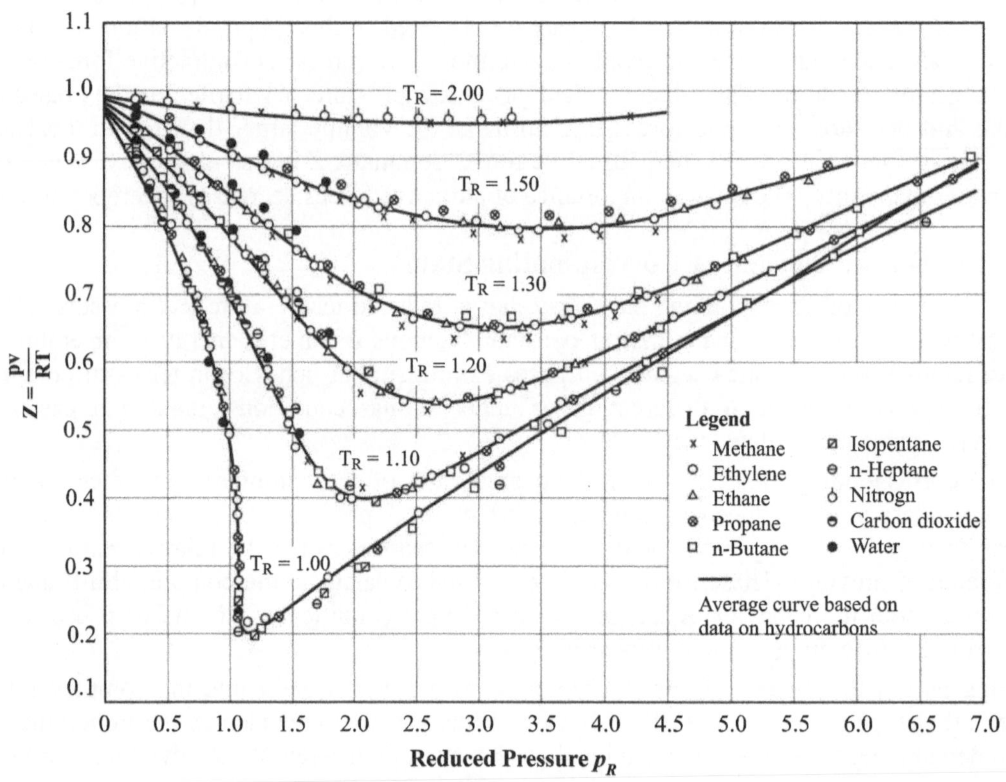

Figure 2

Explanation

The theory of corresponding states is a concept in thermodynamics and statistical mechanics that aims to describe the behavior of fluids across different temperatures and pressures by normalizing the properties of the fluid with respect to critical point properties.

The theory was first proposed by Johannes van der Waals in the late 19th century and builds upon his equation of state for real gases. The idea behind the theory is that the properties of different fluids at their critical points are similar, and therefore can be normalized to a universal form by dividing by critical point properties such as critical temperature, critical pressure, and critical volume.

By using this normalized form, the theory of corresponding states allows for the comparison of the behavior of different fluids at various conditions such as temperature and pressure. This is particularly useful for predicting the phase behavior of fluids, such as the vapor-liquid equilibrium, at conditions beyond the critical point.

The theory of corresponding states has been applied to a wide range of fluids, including gases, liquids, and supercritical fluids, and has been found to be particularly successful in predicting the behavior of simple fluids. However, it may not always provide accurate predictions for complex fluids or mixtures.

Reduced Properties

In the theory of corresponding states, the concept of reduced properties is used to normalize the properties of fluids with respect to their critical point properties. Reduced properties are dimensionless ratios of the actual properties of a fluid to the corresponding critical point properties.

The most common reduced properties used in the theory of corresponding states are:

Reduced temperature (T_r):

$$T_r = T/T_c$$

where T is the temperature of the fluid and T_c is the critical temperature of the fluid. By using reduced temperature, the behavior of fluids at different temperatures can be compared and correlated.

Reduced pressure (P_r):

$$P_r = P/P_c$$

where P is the pressure of the fluid and P_c is the critical pressure of the fluid. Reduced pressure allows for the comparison of the behavior of fluids at different pressures.

Reduced volume (V_r):

$$V_r = V/V_c$$

where V is the volume of the fluid and V_c is the critical volume of the fluid. Reduced volume is used to compare the behavior of fluids at different volumes.

By using these reduced properties, the properties of different fluids can be normalized and compared at various conditions. This allows for the prediction of phase behavior, such as vapor-liquid equilibrium, at conditions beyond the critical point. The theory of corresponding states has been successful in predicting the behavior of simple fluids using reduced properties, although it may not always provide accurate predictions for complex fluids or mixtures.

Fundamentals

The fundamental reason why all gases have the same compressibility factor (Z) at their respective reduced temperature and reduced pressure is because of the principle of corresponding states. According to the principle of corresponding states, at the same reduced conditions (i.e., reduced temperature and reduced pressure), all gases will exhibit similar behavior in terms of their equation of state.

When a gas is at reduced conditions (where the reduced temperature and pressure are equal to unity), the compressibility factor of the gas (Z) is found to be approximately 1. This means that the gas behaves ideally under these conditions, following the ideal gas law $PV = RT$.

The reason why all gases exhibit the same compressibility factor at their respective reduced conditions is that the reduced properties normalize the behavior of gases with respect to their critical point properties. At reduced conditions, the effects of the specific gas and its intermolecular interactions are less pronounced, and the behavior of gases becomes more similar across different gas species.

CHAPTER 15

Intra and Intermolecular Bonds: Thermodynamics

The specific heat of a substance is a measure of energy stored in intramolecular bonds. The latent heat is a measure of energy stored in intermolecular bonds.

BRIEF INTRODUCTION TO INTRA AND INTERMOLECULAR BONDS

The equipartition theorem is a fundamental principle in thermodynamics that states that each degree of freedom of a system will have an average energy of $kT/2$, where k is the Boltzmann constant and T is the temperature of the system.

In the context of interatomic bonds, the equipartition theorem can be used to describe the energy distribution among vibrational, rotational, and translational modes of the atoms in a molecule. This energy distribution affects the strength and stability of the bonds within the molecule.

Intermolecular bonds, on the other hand, refer to the forces of attraction or repulsion that exist between molecules. These bonds are weaker than intramolecular bonds and are responsible for the physical properties of substances such as melting and boiling points, viscosity, and solubility.

Interatomic bonds, such as covalent bonds, metallic bonds, and ionic bonds, hold atoms together within a molecule through the sharing or transfer of electrons. The energies associated with these bonds contribute to the overall stability and structure of the molecule. The equipartition theorem can be used to study the distribution of energy among various degrees of freedom within the molecule's atoms.

For example, in a diatomic molecule like hydrogen gas (H_2), the atoms are connected by a covalent bond. The energy associated with this bond contributes to the vibrational, rotational, and translational motions of the atoms. According to the equipartition theorem, each of these degrees of freedom will have an average energy of $kT/2$ at a given temperature.

Intermolecular bonds, including hydrogen bonds, van der Waals forces, and dipole-dipole interactions, occur between molecules rather than within them. These bonds are typically weaker than intramolecular bonds but are still important in determining the physical properties of substances.

The interactions between molecules influence properties such as boiling points, melting points, and solubility. For example, hydrogen bonding between water molecules gives rise to its relatively high boiling point compared to other similar-sized molecules.

Thermodynamics provides a framework for understanding the behavior of substances based on the interactions of their constituent molecules. In the context of intra and intermolecular bonds, thermodynamics offers valuable insights into how these bonds influence the physical and chemical properties of materials.

In conclusion, thermodynamics offers a powerful framework for studying the roles of intra and intermolecular bonds in determining the behavior of substances. By considering the energy changes associated with these bonds, thermodynamics provides valuable insights into the physical and chemical properties of materials, shedding light on processes like phase transitions, reactions, and equilibrium phenomena in the context of molecular interactions.

CHAPTER 16

System-Surroundings and Control Volume in Thermodynamics

In practice, the system and its surroundings are "everything." This is because the total energy in the system and its surroundings is constant. Even though energy can flow between these two, the total energy remains constant. As a result, the system and its surroundings are referred to as the "universe." We can discuss changes in the system and its surroundings as changes in the universe because energy is constant in the universe (from a chemistry thermodynamics standpoint).

System + Surroundings = Universe

The concept of defining a system and its surroundings is critical in thermodynamics because it serves as the foundation for many different types of descriptions and calculations.

A thermodynamic system is a body of matter and/or radiation that is surrounded in space by walls with defined permeabilities that keep it isolated from its surroundings. Other thermodynamic systems or physical systems that are not thermodynamic systems could exist in the environment. When a thermodynamic system's wall is described as 'permeable' to all matter, all radiation, and all forces, it may be purely fictitious. Various sets of thermodynamic state variables can be used to fully describe the state of a thermodynamic system.

A thermodynamic system is a quantity of matter or a region in a space of interest in a nutshell. The surrounding mass or region is referred to as the surroundings, and the boundary is the surface that separates the system from the surroundings. Depending on the type of exchange that occurs between the system and its surroundings, a thermodynamic system can be (1) isolated, (2) adiabatic, (3) closed, or (4) open.

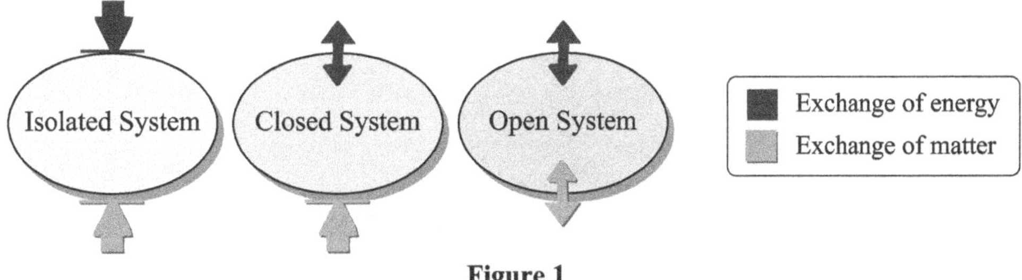

Figure 1

CLOSED SYSTEM

A closed thermodynamic system is surrounded by walls that are impermeable to substances, while at the same time allowing heat exchange with the environment. A closed system does not allow mass exchange, so mass within the system is always conserved.

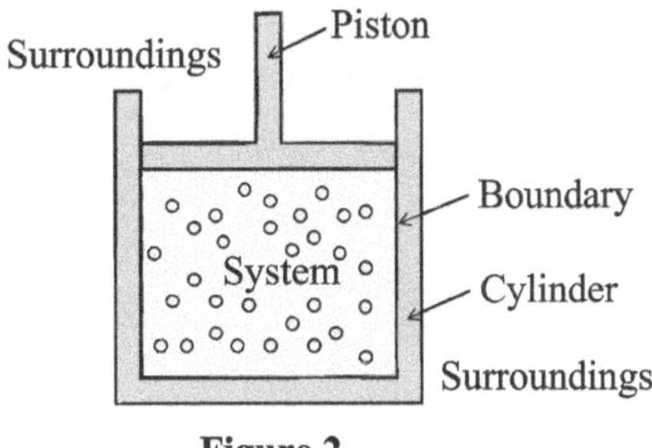

Figure 2

If no energy exchange is allowed, it is called an isolated system.

In the piston-cylinder device shown below, the boundary is movable, but there is no mass across the boundary, assuming negligible gas leakage.

In this case, the above device is classified as a closed system.

ISOLATED SYSTEM

Isolated thermodynamic systems have heat-sensitive walls that are impermeable to all forms of matter and forces. In other words, an isolated system is impermeable to both energy and matter.

OPEN SYSTEM

An open system is capable of exchanging energy and matter with its surroundings. The open system is illustrated below as an example.

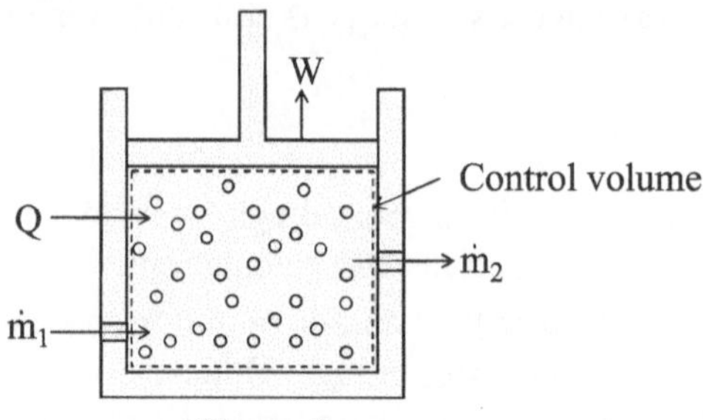

Figure 3

This device is similar to the one shown above, but it is classified as an open system due to the presence of mass flow crossing the boundary. A steady flow system exists when the boundary is stationary and the inlet and outlet mass flow rates are equal and do not change over time. The main components of a thermal power plant, for example, such as the boiler, turbine, condenser, and pump, are frequently treated as steady flow systems.

CONTROL VOLUME

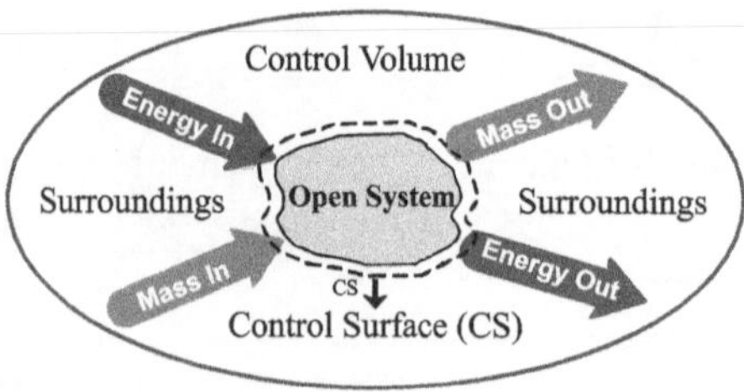

Figure 4

A control volume is a fixed region in space chosen for the thermodynamic study of mass and energy balances for flowing systems. The boundary of the control volume may be a real or imaginary envelope. The control surface is the boundary of the control volume.

At a steady state, a control volume can be thought of as an arbitrary volume in which the mass of the fluid system remains constant. As a fluid moves through a control volume Mass in = Mass out. At steady-state and in the absence of work and heat transfer, the energy within the control volume remains constant.

ASSUMPTIONS

What are the assumptions?

1. **Conservation of mass:** The total mass within the control volume remains constant, with no mass entering or leaving the control volume.

2. **Conservation of energy:** Energy cannot be created or destroyed within the control volume, only transferred in or out as work or heat.
3. **Steady-state conditions:** The properties of the fluid within the control volume do not change with time.
4. **Uniform properties:** Properties such as density, temperature, and velocity remain constant throughout the control volume.
5. **No internal heat generation:** Heat is not generated within the control volume by internal sources.
6. **Negligible kinetic and potential energy changes:** Changes in kinetic and potential energy of the fluid within the control volume are small and can be neglected.
7. **No leaks or losses:** There are no leaks in the control volume and no losses of mass or energy through its boundaries.
8. **No chemical reactions:** Chemical reactions do not occur within the control volume.

These assumptions help simplify the analysis of fluid flow and thermodynamic processes within the control volume and determine important parameters such as mass flow rate, energy transfer, and work done by the fluid.

CHAPTER 17

Intensive and Extensive Properties

Physical or chemical properties are often referred to as "intensive" or "extensive." Intensive properties according to IUPAC are those that have a magnitude that does not depend on the size of the system. Examples of intensive properties include temperature, refractive index, density, and hardness.

Extensive property or quantity refers to properties that add up over multiple subsystems. Examples of extensive properties include entropy, mass, volume, etc.

COMPOSITE OR DERIVED PROPERTY

The ratio of two extensive properties on a single object/system becomes an intensive property. This is also called derived property. For instance, density is the ratio between an object's mass and volume which are extensive properties.

ROLE OF INTENSIVE AND EXTENSIVE PROPERTIES IN THERMODYNAMICS

Intensive and extensive properties are important in thermodynamics because they help to characterize the state of a system and provide valuable information about its behavior.

As said earlier, intensive properties are independent of the size or mass of the system, such as temperature, pressure, and density. These properties describe the system's intrinsic characteristics and are useful for comparing different systems regardless of their size.

On the other hand, extensive properties depend on the size or mass of the system, such as volume, mass, and energy. These properties scale with the size of the system and provide information about the overall amount of a particular substance in the system.

Intensive and extensive properties are connected with state functions in thermodynamics. State functions are properties of a system that depend only on the current state of the system and not on the path it took to reach that state. Intensive and extensive properties are both examples of state functions, as they describe the state of the system at a specific point in time and space.

By understanding and utilizing intensive and extensive properties in thermodynamics, one can better analyze and predict the behavior of systems, as well as make informed decisions about their design and operation.

Typical Examples and Explanation

Intensive Properties

Temperature: Temperature does not depend on the amount of a substance. In a system at thermal equilibrium, the temperature is an intensive property, meaning it does not depend on the amount of substance present. This is because at thermal equilibrium, the average kinetic energy of the particles in the system is constant throughout the entire substance, and it is this average kinetic energy that we measure as temperature. Therefore, at thermal equilibrium, the temperature cannot be partitioned or divided into different parts within the substance. This fundamental principle arises from the conservation of energy, as dictated by the laws of physics. When an object is at a constant temperature, its kinetic energy (KE) remains constant. This means that the speed (v) of the object is inversely proportional to the mass (m) of the object, as given by the equation: $KE = 1/2\ mv^2$.

If the speed of the object were to increase, the kinetic energy would also increase, violating the conservation of energy principle. To maintain the constant kinetic energy, the mass of the object must decrease as the speed increases (and vice versa) to keep kinetic energy constant. So, the temperature does not depend on mass. This explains how temperature is an intensive property.

Pressure: Pressure is an intensive property because it is not affected by the size or mass of the sample. It is also constant at a given temperature within a system. This means that if you were to divide a given volume of gas into numerous smaller samples, the pressure would be the same in each of those smaller samples.

One way to understand this is through the ideal gas law, which states that pressure (P) is equal to the number of moles of gas (n) times the gas constant (R) times the temperature (T) divided by the volume (V) of the gas, expressed as $P = n\,RT/V$. This equation can be rearranged as $P = n/V\,RT$ or, $P = \rho RT$, where ρ is the molar density of the gas. The molar density of an ideal gas is constant at STP. At STP, 1 mole of an ideal gas = 22.4 liters. Therefore *P*, pressure is constant for ideal gases at a constant temperature and it does not depend on the mass. That explains why pressure is an intensive property.

Density: Density is an intensive property. Density is a ratio of two extensive properties, mass and volume. Such intensive properties which arrive from the ratio of two extensive properties are called derived properties. The explanation is simple. For an ideal gas, the ratio of mass to volume is constant under specific conditions. This is what the density is known as. However, it's important to note that for real gases, this relationship may not hold true under all conditions.

Extensive property: An extensive property is a physical quantity with a value proportional to the size of the system it describes, or the amount of matter in the system. For example, a sample's mass is an extensive property that is determined by the amount of substance present. The related intensive quantity is density, which is independent of the amount. Water has a density of about 1 g/mL whether it's in a drop or a swimming pool, but the mass is different in each case.

Dividing one extensive property by another extensive property usually results in an intensive value—for example, mass (extensive) divided by volume (extensive) yields density (intensive).

Examples of Some of the Extensive Properties

Enthalpy, *H*, Entropy, *S*, Gibbs energy, *G*, Heat capacity, *Cp*, internal energy, *U*.

CHAPTER 18

Heat Capacity and Specific Heat

Heat capacity and specific heat have been discussed in various sections of the book because they are important topics in thermodynamics.

Specific heat is the energy storage in a molecule and that it is the **intra-molecular** energy demand of a molecule for atomic motions. This is opposed to latent heat which signifies the **intermolecular** energy storage and is the intermolecular energy demand of a substance.

Specific heat provides sensible heat, $Q = m\ C_p\ dt$ [m is mass, C_p is specific heat and *dt* is temperature difference].

Specific heat depends on the number of atoms in a molecule. Larger the number of atoms in a molecule the larger is the specific heat. Heat of vaporization or latent heat depends on how strong is the intermolecular attraction. The stronger the intermolecular attraction the larger the latent heat.

GENERAL

Heat affects different substances in different ways. If left in direct sunlight on a hot day, a metal chair can become quite hot to the touch. When exposed to the same amount of sunlight, an equal mass of water will not become nearly as hot.

This means that water has a high heat capacity (the amount of heat required to raise the temperature of an object by one degree Celsius). Metals are not as resistant to temperature changes as water is.

The specific heat of a substance is defined as the amount of energy required to raise the temperature of one gramme of the substance by one degree Celsius.

When a substance, especially a gas, is heated in a closed vessel that inhibits expansion, its specific heat capacity (specific heat capacity at constant volume) may be much lower than when the substance is allowed to expand during heating (specific heat capacity at constant pressure). Typically, *Cp* and *CV* are used to represent these two values.

The heat capacity ratio is their quotient, $y = Cp/Cv$.

WHY SPECIFIC HEAT IS AN IMPORTANT CONCEPT?

A good understanding of specific heat is important for engineers for several reasons:

- **Energy Transfer**: Specific heat helps engineers understand how energy is transferred in various engineering processes.
- **Thermal Design**: Engineers need to design systems and devices that can handle heat effectively. Understanding specific heat allows them to select appropriate materials that can withstand high temperatures or efficiently transfer heat without degradation.
- **Temperature Control**: Specific heat is essential for designing heating and cooling systems, as engineers need to know how much heat energy is required to achieve a certain change in temperature.
- **Phase Transitions**: Specific heat also helps engineers understand phase transitions, such as the transition from solid to liquid or liquid to gas. The sensible heat is a measure of the specific heat of the substance at the given pressure. When a substance is heated its temperature increases until it becomes saturated with energy and reaches its heat capacity. At saturation any further heating causes a phase transition from a single to a two phase system at thermodynamic equilibrium that holds two phases at the same temperature.
- **Thermal Storage**: Specific heat is important in thermal energy storage systems, where heat is stored and released later for various applications.

Specific Heat and Thermodynamics

The specific heat is essential for understanding thermal equilibrium and the behavior of substances at the molecular and atomic level.

Thermodynamics is the study of energy and heat transfer, and specific heat is a fundamental property that helps quantify the amount of heat required to change the temperature of a substance. It is directly related to the molecular and atomic behavior of a substance because it describes the energy content and behavior of individual particles within the substance.

Every thermodynamic law can be indirectly related to specific heat because specific heat is a property that describes the amount of heat energy required to raise the temperature of a substance by a certain amount. The specific heat is the heat storage capacity of a substance.

Specific Heat is Primarily Associated with the First Law of Thermodynamics because it relates to the transfer of heat and the change in the internal energy of a system. It helps quantify the heat transfer in terms of the temperature change and the amount of substance present.

The Second Law of Thermodynamics, on the other hand, deals with the concept of entropy and equilibrium at phase transition.

A FEW EQUATIONS THAT RELATE SPECIFIC HEAT WITH THERMODYNAMIC LAWS

The equation for specific heat capacity (*c*)**,** $q = mc\Delta T$, where q is the heat energy transferred, m is the mass of the substance, c is the specific heat capacity, and ΔT is the change in temperature.

- **The equation for heat capacity** (*C*)**:** $Q = CT$, where Q is the heat energy transferred and T is the temperature change. Heat capacity is the amount of heat energy required to raise the temperature of an object by a given amount.
- **The relationship between specific heat capacities at constant pressure (*Cp*) and constant volume (*Cv*): $Cp - Cv = R$, where *R* is the specific gas constant.** This equation relates the specific heat capacities of a substance at constant pressure and constant volume.
- **The equation for enthalpy change (ΔH):** $\Delta H = nCp\Delta T$, where ΔH is the enthalpy change, n is the number of moles of the substance, Cp is the molar specific heat capacity at constant pressure, and ΔT is the change in temperature.
- **The equation for internal energy change (ΔU):** $\Delta U = nCv\Delta T$, where ΔU is the internal energy change, n is the number of moles of the substance, Cv is the molar specific heat capacity at constant volume, and ΔT is the change in temperature.
- **The equation for entropy change (ΔS):** $\Delta S = \int(dq \div T)$, where ΔS is the entropy change, dq is the infinitesimal amount of heat transferred, and T is the temperature at which the heat transfer occurs. This equation relates the heat transfer and temperature change to the change in entropy of a system.

CHAPTER 19

Thermodynamic Processes

FUNDAMENTALS

In a nutshell, thermodynamic processes involve different types of heat and work transfer. Typically, there are four thermodynamic processes. [1] isobaric or constant pressure process [2] isothermal or constant temperature process [3] adiabatic process; [4] isochoric or constant volume process. The constant pressure process uses heat from the outside to keep the pressure constant. Isothermal processes maintain a constant temperature by absorbing heat from the surroundings and returning it to the surroundings, whereas adiabatic processes use their internal energy in the process. Internal energy increases during a constant pressure process. Internal energy remains constant during the constant temperature process. Increase or decrease of internal energy in adiabatic process depends on compression and expansion. There is no work in a process with constant volume.

The change in the internal energy of a system is the $U = Q - W$,

U is internal energy

Q is heat and

W is work.

Therefore, constant pressure process produces maximum work followed by constant temperature process. Then comes the adiabatic process. Finally, it is the turn for isochoric process which does not perform any work.

Thermodynamic processes follow the equation $PV^n = C$. In this case, P is pressure, V is volume, n is the polytropic index, and C is a constant.

WHAT DOES POLYTROPIC EQUATION SIGNIFY?

Polytropic index 'n' signifies the PV work done by a thermodynamic system.

The polytropic process equation can be used to describe a variety of expansion and compression processes, including heat transfer. The term "polytropic" was coined to describe any reversible process on any open or closed system of gas or vapour that involves both heat and work transfer and maintains a specified set of properties constant throughout the process. The expression relating the system's properties throughout the process is referred to as the polytropic path. There are an infinite number of reversible polytropic paths between two given states; the most common polytropic path is $TdS/dT = C$, which is a constant that equals zero in an adiabatic process. T is temperature, and S is entropy.

In different thermodynamic processes, n takes on different values. Some examples are provided below. The polytropic index measures the work done by the system.

A word of caution: Many gas compression and expansion processes can be approximated by polytropic methods. In each case, the polytropic coefficient must be determined experimentally by measuring heat and work transfer, as well as the initial and final conditions.

Polytropic Work Transfer

$n = 0$, constant pressure or isobaric process, $W = P\,[V_2 - V_1]$

$n = 1$, constant temperature or isothermal process, $W = Q = RT \ln V_2/V_1$

$n = y$, reversible adiabatic process, $W = [P_1V_1 - P_2V_2]/[Y - 1]$, $[Y = Cp/Cv]$

$n =$ infinity, constant volume or isochoric process, $W = 0$

Thermodynamic Work on *Pv* Diagram

It may be clearly noticed that there are two ends of thermodynamic processes. The top end that tells the maximum work and the bottom end that signifies zero work.

The area under each line in the above graph represents the *PV* work. It may be noted that for isochoric process it is a vertical line that shows zero work.

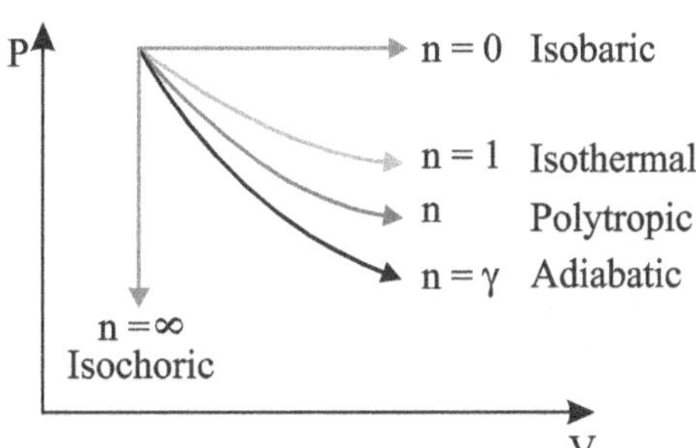

Figure 1: Thermodynamic Process

What is *y* [gamma]? What does it Signify?

Y [gamma] is a ratio of *Cp* and *Cv*.

Cp and Cv

The *Cp* and *Cv* are the specific heats of an ideal gas at constant pressure and at constant volume. These indicate the quantity of heat that can increase the temperature of 1 gram mass by 1°C.

$$Cp > Cv$$

Cp at constant pressure involves heat to raise temperature of 1 gram mass by 1 degc. *Cp* involves heat required for both work of expansion and internal energy for 1 gram mass to raise temperature by 1 degc.

Cv at constant volume signifies heat required to raise temperature of 1 gram mass by 1 degc. *Cv* supplies heat for increase in internal energy when 1 gram mass is heated by 1 degc.

This explains why *Cp* > *Cv*.

According to the 1st law of thermodynamics, delta *H* = delta *U* + delta *W* [*H* is enthalpy, *U* is internal energy and *W* is work.

This means following. At constant pressure, the heat absorbed is not only increasing internal energy but also goes for doing work. At constant volume since delta-*V* = 0 heat goes to increase only internal energy delta *W* = 0. *Cp* signifies total energy called enthalpy *H* that is sum of internal energy *U* and work energy *W*. *Cv* signifies internal energy. Hence *Cp* > *Cv*

Therefore, $Cp/Cv = dH/dU$

Cp/Cv Ratio: Thermodynamic Significance of Cp/Cv

Cp/Cv is a measure of how much internal energy gets converted to work in any adiabatic compression/expansion process. Every change in internal energy impacts entropy changes and brings irreversibility in the process except in isentropic adiabatic process.

Typical Values of CP/CV

In summary, from the 1st law of thermodynamics $U = Q - W$, that if there is a change in internal energy (ΔU), it means that either heat (Q) or work (W) or both are involved. If ΔU is positive, [constant pressure process] it implies an increase in internal energy, which can be due to the system absorbing heat or work being done on the system. If ΔU is constant it means no change in internal energy [constant temperature process]. If ΔU is negative, it implies a decrease in internal energy in the case of diabatic process it is system's consumption of its own internal energy. Therefore, a change in internal energy can indeed result in a change in work done.

Table 1

Gas	*Ratio of Specific Heats*
Acetylene	1.3
Air, Standard	1.4
Ammonia	1.32
Carbon Dioxide	1.28
Carbon Monoxide	1.4
Chlorine	1.33
Ethane	1.18
Helium	1.66
Helium	1.66
Hydrogen	1.41
Methane	1.32
Natural Gas (Methane)	1.32
Nitrogen	**1.4**
Oxygen	1.4
Propane	1.12
Steam	1.28
Sulphur dioxide	1.26

Isobaric *vs* Isothermal Process

Between constant pressure or isobaric process and isothermal process, in isobaric process pressure is maintained constant by supplying additional heat that increases the internal energy of the gas. Therefore, the work done by an isobaric process is purely a function of pressure and not limited to atmospheric pressure as in the case of isothermal process. In isothermal process the internal energy stays constant.

Isothermal *vs* Adiabatic Process

An infinite amount of heat is available in isothermal processes from the surroundings to maintain a constant temperature. This means that the system can exchange as much heat as needed to keep the temperature constant, allowing for unlimited work to be done.

On the other hand, in an adiabatic process, no heat is exchanged with the surroundings. The work done in an adiabatic process is limited by the available internal energy or the specific heats (*Cp* and *Cv*) of the gas. This limitation on the available energy means that the adiabatic process can do only as much work as the internal energy allows, making it less than the theoretical maximum for the isothermal process.

Isochoric Process

As explained. There is no work in isochoric process.

Finally, the work done by Isobaric process > Isothermal process > Adiabatic process.

Isothermal and Adiabatic Process: Pros and Cons

Isothermal Process

In thermodynamics, an isothermal process is a type of process in which the temperature T of a system remains constant: $\Delta T = 0$. This typically occurs when a system is in contact with an outside thermal reservoir, and a change in the system occurs slowly enough to allow the system to be continuously adjusted to the temperature of the reservoir through heat exchange (see quasi-equilibrium). In contrast, an adiabatic process is where a system exchanges no heat with its surroundings ($Q = 0$). In an isothermal process) $dT = 0$. In an isothermal process the change in the internal energy $dU = 0$. This applies to ideal gases only which do not have potential energy in the absence of intermolecular attractions. To summarize, Typically there are two types of occurrences that can cause this process to occur. If a system comes into contact with a thermal reservoir from the outside, the system gradually adjusts its temperature to that of the reservoir through heat exchange to preserve thermal equilibrium. In another phenomenon, there is no heat transfer between a system and its surroundings. The temperature of the system is altered during this procedure to keep the heat consistent [adiabatic process].

Where Does the Isothermal Process Apply

Essential conditions for isothermal process are: [1] The container should be perfectly conducting to the surroundings. [2] The process must be carried out very slowly so that there is sufficient time for exchange of heat with the surroundings so that temperature remains constant.

Constraints of Isothermal Process

For an ideal gas, in an isothermal process, $\Delta U = 0 = Q - W$, so $Q = W$

The reason is for an ideal gas by definition there is no interactions between particles, no intermolecular forces, so pressure change at constant temperature does not change internal energy.

However, this does not apply to real gases.

Real gases have intermolecular interactions, attractions between molecules at low pressure, and repulsion at high pressure. Their internal energy changes with a change in pressure, even if the temperature is constant.

In an isothermal process, the temperature is kept constant (unchanged) as the pressure increases during compression. Although it is impossible to build a machine that will compress isothermally. This needs a machine that must be slow yet it can compress and generate pressure which is impossible.

Furthermore, although isothermal compression cannot actually be attained in practice, it is often used as the basis for compression with other compression processes like as interstage coolers in compressors.

What does it mean?

The isothermal process does not apply to real gases since there is a change in internal energy. On the contrary, your choice would be an adiabatic process where the source of energy is internal energy change.

Do Real Gases Achieve Isothermal Thermodynamic Conditions?

The answer is "no."

What are the Implications?

Phase change cannot occur at constant temperature.

In isothermal conditions, for real gases, dT is not dU, implying that the change in internal energy does not equal the temperature change.

This implies that in isothermal conditions when $dT = 0$, real gases still have residual internal energy.

What does this mean?

The first law of thermodynamics states that delta U = delta Q – delta W, where U represents internal energy, Q represents heat, and W represents work. When delta U = 0, delta Q equals delta W. In other words, in isothermal conditions, all heat is converted into work. This is the necessary condition for a phase change at constant temperature and pressure, where all heat converts to work.

What is the significance of this?

Real gases do not undergo phase transitions at constant temperatures.

Explanation

Unlike ideal gases, which lack intermolecular interactions and potential energy, real gases have both.

To summarise, ideal gases contain only kinetic energy in their internal energy in the absence of intermolecular attractions. The Internal energy of real gases includes both potential and kinetic energy. The kinetic energy is proportional to the $KE = 1/2\ KT$, where k is constant, and T is temperature. As a result, for ideal gases, dU equals zero when $dT = 0$. Since they have only kinetic energy. For real gases when $dT = 0$, dU is not zero. Because real gases have both potential and kinetic energy in the internal energy which together are not proportional to temperature.

The requirements for an isothermal process are $dT = 0$ and $dU = 0$.

Source of Energy and Work for an Isothermal Process

The heat [energy] comes from the surrounding and cause changes in the system. The energy then goes back to the surrounding and do work on the surrounding.

Adiabatic Process

An adiabatic process on the contrary is a fast process. A process without transfer of heat to or from a system, so that $Q = 0$, is called adiabatic, and such a system is said to be adiabatically isolated. The assumption that a process is adiabatic is a frequently made simplifying assumption. For example the compression of a gas within a cylinder of an engine is assumed to occur so rapidly that on the time scale of the compression process, little of the system's energy can be transferred out as heat to the surroundings. Even though the cylinders are not insulated and are quite conductive, that process is idealized to be adiabatic. The same can be said to be true for the expansion process of such a system.

Work Transfer by Adiabatic Process

In thermodynamics, an adiabatic process is one that occurs without the transfer of heat or mass between the thermodynamic system and its environment. In contrast to an isothermal process, an adiabatic process only transfers energy to its surroundings as work.

Without a doubt, it is difficult to grasp this concept when most of the books do not explain it. Like said earlier, every thermodynamic system requires an energy outlet. While adiabatic systems cannot transfer heat, they can transfer work.

A system enters a dynamic equilibrium when it is enclosed in a closed box that does not allow it to share or change its temperature, pressure, or mass. For such processes, the free energy falls to zero, while the entropy increases and consumes the enthalpy, $dG = dH - TdS$. The process stops as the system goes to equilibrium. Isolated systems are examples of closed such systems.

Let us take two real world examples of adiabatic processes and examine how they transfer work.

Turbine in Power Plant: Adiabatic Work Transfer

The superheated steam at its pressure and enthalpy expands adiabatically between the turbine inlet and the condenser by consuming its own internal energy. All superheat is extracted out the superheated steam between these two points and converted into work energy that spins the turbine. From the first law of thermodynamics is given as $\Delta U = Q - W$ where ΔU is the change in internal energy of a system, Q is the net heat transfer (the sum of all heat transfer into and out of the system), and W is the net work done.

For adiabatic process, $Q = 0$, therefore $\Delta U = -W$

At the condenser at constant pressure and temperature the saturated steam cools to water by rejecting the heat in a cooling tower. While steam condenses to water at constant temperature and pressure in the condenser it does a positive work $+ W$ and raises the water temperature of the cooling tower.

Note: Condenser works at constant temperature and pressure and it is not an adiabatic process.

Compressor: Adiabatic Work Transfer

When compressor compresses a gas, it becomes hot because air compression reduces the gap between gas molecules and increases collisions which generate heat.

As the gas becomes hot its density reduces and it expands. The problem is unless this heat is removed it becomes increasingly difficult to compress the expanding gas. So, there are interstage isothermal coolers. Compressor works adiabatically, $dQ = 0$. Compressor transfers work to isothermal interstage coolers at ambient temperature. Compressed gas cools down to near ambient temperature by doing a positive work $+ W$. The isothermal coolers while absorbs this heat of compression cannot increase temperature since the coolers run in equilibrium with surrounding under isothermal conditions. The heat gets rejected eventually to surrounding.

CHAPTER 20

First Law of Thermodynamics

FIRST LAW OF THERMODYNAMICS: INTRODUCTION TO KINETIC AND POTENTIAL ENERGY, ENTHALPY AND INTERNAL ENERGY

The first law of thermodynamics is the application of the conservation of energy principle to heat and thermodynamic processes. The first law of thermodynamics is typically regarded as the easiest to understand.

The first law of thermodynamics states that the change in internal energy of a system, dU is equal to heat Q minus work done by the system, W

$$dU = Q - W$$

There are four pillars of 1st law of thermodynamics. [1] Heat [2] Enthalpy [3] Internal energy and [4] Work.

Enthalpy is the total energy. It is the sum of internal energy and work.

Internal energy is the sum of kinetic energy and potential energy.

Heat and work are the modes of transport of energy in thermodynamics.

POTENTIAL AND KINETIC ENERGY

Potential Energy

One of the two types of energy in the universe is potential energy. It's simple, albeit a little harder to grasp intuitively: it's a type of energy that has the ability to do work but is not actively doing so or exerting any force on other objects. An object's potential energy is found in its position, not in its motion. It is the positional energy.

When objects are moved out of their equilibrium locations, they gain energy.

When things are moved from their equilibrium positions, they gain energy that was previously stored in them before they were pushed out of equilibrium by elastic rebound, gravity, or chemical reactions. An object like an archer's bow, which stores the energy released by pulling back the bowstring, is a good example of this. The kinetic energy generated by the potential energy held in the pullback is accountable for the energy released upon release.

Because it is more evident that moving objects have energy, understanding kinetic energy is intuitively easier.

When potential energy is released and propelled into motion by gravity or elastic forces, among other catalysts, kinetic energy is formed.

Kinetic Energy

The energy of motion is known as kinetic energy. The kinetic energy of an object is increased when work is done on it and it accelerates. The motion (measured as velocity) and mass of the object are the most fundamental components that determine kinetic energy.

While mass is a universal metric, an object's motion can take several forms, including rotation around an axis, vibration, translation, or any combination of these and other motions.

Kinetic energy is divided into three types: vibrational, rotational, and translational.

Unsurprisingly, items vibrating create vibrational kinetic energy. Moving objects generate rotational kinetic energy, whereas objects colliding with one another generate translational kinetic energy.

Basic Distinction between *KE* and *PE*

The basic distinction between potential and kinetic energy is that one represents what can be and the other represents what is. In other terms, potential energy is energy that is stored and ready to be released, whereas kinetic energy is energy that is being used for movement.

Another significant distinction is velocity. This is the foundation of kinetic energy, yet it has nothing to do with potential energy. In fact, for determining the quantity of kinetic energy for any given object, velocity is the most significant portion of the equation, and it is absent from the potential energy equation.

These fundamental types of energy are complementary to one another, notwithstanding their differences.

When potential energy is released, it always converts to kinetic energy. Kinetic energy is required for an item to store energy as potential in some form. A rock on the edge of a cliff, for example, does not require kinetic energy to store the potential energy that will propel it down the cliff face. However, kinetic energy is required to get the rock to the edge. As a result, the rock needs it to maintain its potential energy.

Potential Chemical Energy

Chemical potential energy is held in molecular bonds (sometimes called chemical bonds). When these ties are broken, the potential energy held inside them is released, resulting in varied degrees of kinetic energy, depending on the strength of the bonds.

Potential Gravitational Energy

Due to the potential for the force of gravity to move and draw the object towards the Earth, gravitational potential energy is stored in the object. The quantity of potential gravitational energy contained in an object is precisely proportional to its mass and, more critically, its height above the earth.

HOW DO KINETIC AND POTENTIAL ENERGY SWITCH PLACES?

The first energy lesson is about understanding internal energy. The fundamental one-line definition of internal energy of a substance is it is the energy of its electrons. Every atom has an atomic orbital. In simple words an atomic orbital of a substance tells you the location of an electron and its behavior. Thermodynamics monitors the energy associated with the movement and the location of electrons which is the internal energy of the substance. As the name implies, kinetic energy is associated with electron motion, whereas potential energy is associated with electron location in an atom. When a real gas is made to cool adiabatically without doing any work at constant internal energy as in Joule Thomson effect, it does so by converting its kinetic energy to potential energy.

Example: How do KE and PE Swap Roles?

As shown in the attached image, a projectile is launched into the air with only its internal energy, E1. When it is launched upward, kinetic energy is added, propelling it up against the resistance of earth's gravitational forces and air resistance.

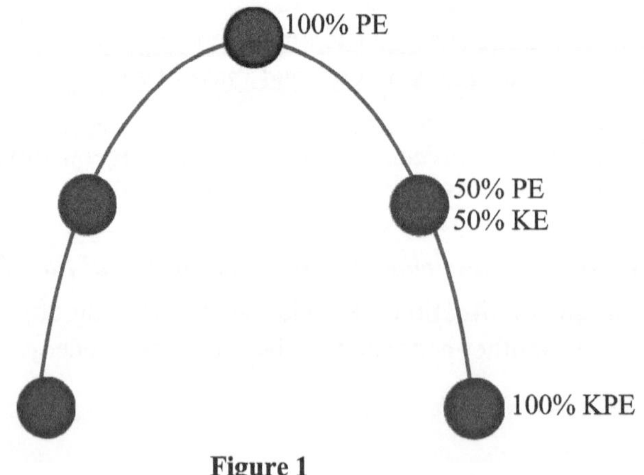

Figure 1

A point reaches when earth's gravitational forces neutralize the kinetic energy. The projectile starts descending again it is a free fall of the projectile against the air resistance. As it loses height, its potential energy diminishes and converts to kinetic energy.

In addition, when objects move through a fluid, such as air, the fluid exerts a frictional force on the moving object. The frictional force from a fluid is called a drag force. Friction drag force causes objects to slow down as they move through a fluid, such as air. Therefore, frictional forces have their own effects in slowing down velocity of an object in the air.

Finally, when the projectile touches the earth, it has only its internal energy E_2 at the temperature of the projectile. Its temperature is controlled by its specific heat. Balance, all its kinetic energy has been lost as frictional heat. Remember, internal energy is energy of electrons and it does not become zero even at –273.15 degc. So, energy can never be zero.

$E_1 > E_2$. $[E_1 - E_2]$ = Frictional loss of energy. E_2 is less than E_1 since when the free fall of the object begins from the highest point [*PE*] it is already at a lower energy level having lost energy in friction than when it was launched up in the air with kinetic energy E_1.

Therefore, the projectile at zero time and when it returns and touches the earth has only its internal energy at its temperature.

Another example of *KE* and *PE* swap: Joule Thomson effect of real gases.

The cooling of the gas takes place when the gas expands at the throttling valve adiabatically at constant enthalpy by converting its kinetic energy to potential energy.

The Carnot cycle is an excellent example of constant internal energy.

The true example of a process where in the internal energy which consist of kinetic and potential energy is approximated to remain constant is the cyclic process of Carnot cycle.

Internal energy only depends on the temperature of the object and the temperature of the object depends on its specific heat.

Basics of the 1st Law of Thermodynamics

The change in internal energy of a system, dU is equal to heat Q minus work done by the system, W

$$dU = Q - W \quad \dots (1)$$

The first law makes use of the key concepts of internal energy, heat, and system work. It is used extensively in the discussion of heat engines. The standard unit for all these quantities would be the joule, although they are sometimes expressed in calories or BTUs.

The first law is usually written as $\Delta U = Q + W$ in chemistry textbooks. The conservation of energy principle finds its thermodynamic expression in the same law, naturally. Simply put, W is defined as the work performed on the system rather than the work produced by the system. In physics, adding heat to a volume of gas and utilizing the gas's expansion to accomplish work is a common scenario, as in the pushing down of a piston in an internal combustion engine. It purely arises from the sign convention of work. The sign convention of work followed as $-W$ when the system does work on the surroundings and $+W$ when the surrounding does work on the system.

There is another way to express the 1st law of thermodynamics in terms of enthalpy H [total energy]

$$H = U + PV \quad \dots (2)$$

where U is internal energy and P and V stand for pressure and volume. Since enthalpy can be defined in terms of three other precisely definable state variables, it can therefore be considered a precisely measurable state variable.

From above, equation (1) dU = delta $Q - d(PV)$

delta $Q = dU + d(PV)$

delta $Q = dU + PdV$ [at constant pressure]

Since, $H = U + PV$ from equation (2)

Therefore, delta $Q = dH$ at constant pressure

Therefore, both equations (1) and (2) are parallel equations for the expression of the 1st law of thermodynamics.

Internal Energy and Enthalpy: Two Close Concepts

Heat energy in a thermodynamic system is known as internal energy when studied under constant volume, and enthalpy of a system when studied under constant pressure.

Enthalpy is the sum of internal energy and work energy that exists in all substances. Internal energy is the kinetic and potential energy of a substance. When a substance performs thermodynamic work, it borrows energy from the environment or releases energy to the environment, which either adds to or subtracts from the internal energy. The total energy, or enthalpy, is expressed as $H = U + PdV$, at constant pressure where H represents enthalpy, U represents internal energy, P represents pressure, and dV represents volume changes if the substance is a gas. While both internal energy and work are forms of energy, they are treated differently in the above expression because internal energy is state function and work is path function. The internal energy U cannot be measured because the energy of electrons can't be measured. Internal energy is always expressed as internal energy change, ΔU. The change in enthalpy is measured by the difference between final enthalpy (enthalpy of products) and initial enthalpy (enthalpy of reactants).

Enthalpy change = Final enthalpy – Initial enthalpy, $\Delta H = Hf - Hi$

The internal energy and enthalpy both increase with temperature. The kinetic energy of particles in a system increases as temperature rises. This generates internal energy and eventually increases enthalpy (H).

INTERNAL ENERGY: SUMMARY

There are two parts to this. [1] it is the contributions of energy of electron's movement overcoming attractions and repulsions they undergo while they are in motion within the nucleus. This is the potential energy component arising from the location of an electron in the nucleus and [2] the kinetic energy of electrons or the vibrations of electrons as they undergo at a temperature within the nucleus.

$K.E.$ total = $K.E.$ translation + $K.E.$ rotation + $K.E.$ vibration

$P.E.$ total = $P.E.n\text{–}n + P.E.e\text{–}e + P.E.e\text{–}n + P.E.IMF$

n = nucleus, e = electron,

n–n is positively charged nucleus to nucleus repulsion

e–e repulsion between two negative charges of electrons

e–n represents the attraction of the nucleus atom (+) for the electron cloud (–). IMF is the intermolecular forces.

Significance of $H = U + W$, H is enthalpy, U is internal energy and W is work.

$H = U + W$ is an expression of the 1st law of thermodynamics. The equation says that the total energy of system H is the sum of U and W is always constant. Often, we ask ourselves when both U and W are forms of energy, why they are separately accounted for.

The answer is while U and W both express energy, they are not the same. While U is a state function, W is a path function. Though it is true that both represent energy but they work differently.

U and W are accounted for sedately to understand the contribution of each. One key difference is U is sum of kinetic and potential energy.

U tells us how stored energy is available in a system to do adiabatic work. While W is the process of energy transfer. W is associated with entropy generation while U is a stored energy.

Let's take some examples to understand how U and W contribute differently in thermodynamic processes.

In a constant volume process, $H = U$ sine there is no work.

In a constant pressure process, $H = U + PdV$

In an isothermal process, $H = W$, since temperature is constant and $dU = 0$

In adiabatic process, $U = Q - W$. since. $Q = 0$

$$U = -W$$

All these above equations tell us how within a given total energy H how U and W contribute differently. This helps us determine how a process works on the ground and the energy distribution in a process.

CHAPTER 21

Second Law of Thermodynamics

REVERSIBLE AND IRREVERSIBLE PROCESSES AND CARNOT CYCLE

There are two concepts in thermodynamics: reversible and irreversible processes. Reversible processes are important because they affect process efficiency. A reversible process in thermodynamics is a slow thermodynamic process that remains in equilibrium with its surroundings, resulting in no energy losses. Reversible processes are ideal processes that never appear in nature. On the contrary, irreversible processes are real-world processes in nature in which the system constantly loses energy to its surroundings that cannot be recovered.

DETAILS

Reversible Process – Meaning

A reversible process is one in which no energy is lost to the environment; in other words, the system is in equilibrium with its surroundings. You cannot bring back the energy that has left your system and gone to the surroundings. In reality, no process is completely reversible, so no machine is 100% efficient. Because your cold reservoir temperature can never be zero, even a Carnot heat engine cannot achieve 100 percent efficiency = $[Th–Tc/Th] \times 100$. You can never go below the ambient temperature; *Tc* can never be lower than the ambient temperature. The question is, why do we need to know what is a reversible process if we can't achieve it? The answer is that it makes you aware of an important concept and makes you wonder, "Can I do something about it?" The answer is that it raises your awareness of an important concept and prompts you to consider, "Can I do anything to reduce the irreversibility of my machine?" It sticks with you and poses the question, "How can I improve the efficiency of my machine?"

As said earlier, a thermodynamic process is reversible if it can be reversed in such a way that both the system and its surroundings return to their original states, with no other changes occurring elsewhere in the universe. It means that at the end of the reverse process, both the system and its surroundings are returned to their initial states.

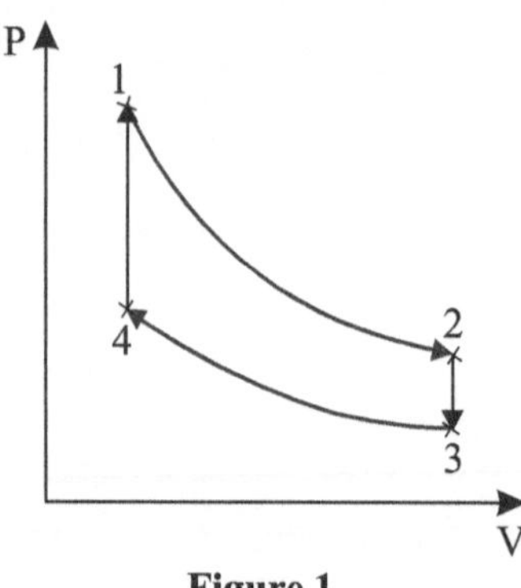

Figure 1

The system has transitioned from state 1 to state 2 in the figure. The reversible process can be completely reversed, leaving no trace that the system had undergone a thermodynamic change. All changes in state that occur in the system during the reversible process are in thermodynamic equilibrium with each other.

Internally Reversible Process

If there is no irreversibility within the system's boundaries, the process is internally reversible. A system goes through a series of equilibrium states in these processes, and when the process reverses, the system goes through exactly the same equilibrium states while returning to its initial state.

Externally Reversible Process

No irreversibility occurs outside the system boundaries during an externally reversible process. Heat transfer between a reservoir and a system is an externally reversible process if the system and reservoir have the same temperature at the point of contact.

Only when two conditions are met, a process become reversible.

- Dissipative force must be absent.
- The process must occur in an infinitely small time.

In other words, a reversible process is one that can be completely reversed. This means that the system's final properties can be perfectly reversed back to the original properties. Only if the changes in the process are infinitesimally small can the process be perfectly reversible. Because it is impossible to trace these extremely small changes in such a short period of time in practice, the reversible process is also an ideal process. The reversible process produces changes that are in equilibrium with one another.

Irreversible Process

Irreversible processes occur as a result of deviating from the line 1–2 and returning to the initial state of the process by the line 3–4 in the curve. It may be noted that line 3–4 in the above curve reduces the amount of overall work done. An irreversible process is one that deviates from equilibrium in thermodynamic terms. In terms of pressure and volume, it occurs when the pressure (or volume) of a system change so dramatically and instantly that the volume (or pressure) does not have time to stabilize. This is why a more rapid process is irreversible.

Typical Examples of Reversible Processes

Carnot Cycle

Reversibility in the different processes in the Carnot cycle.

Boiler

Heat is transferred from the high-temperature reservoir to the boiler's water. Because the temperature of the water is only infinitesimally lower than the temperature of the high-temperature reservoir, the heat transfer process is reversible. Because the temperature of the reservoir remains constant '*TH*,' the temperature of the water must also remain constant. A pure substance's phase transition from liquid to vapor at constant pressure in a boiler is, of course, an isothermal process. As a result, the first process in the Carnot cycle is a reversible isothermal process.

Turbine

The process in a turbine is so fast that it occurs without heat transfer and is thus adiabatic. The temperature of the working fluid decreases from a high-temperature reservoir to a low-temperature reservoir during this process. The turbine is isentropic due to internal reversibility.

Condenser

After rejecting its heat to the low-temperature reservoir, the working fluid flowing through the condenser is completely condensed in the third process. This is a reversible isothermal process because the temperature '*TL*' of the working fluid in the condenser is infinitesimally higher than that of the low-temperature reservoir.

Pump

The fourth and final process in the Carnot cycle occurs in the pump, where work is done on the pump-to-pump water from condenser pressure to boiler pressure. The turbine powers the pumping system. This is a reversible adiabatic process in which the temperature of the working fluid rises from low to high temperatures.

Typical Example of Irreversible Process

Irreversible processes are actual processes that take place in a finite amount of time with real substances and cause energy dissipation due to changes in internal temperature, pressure, or velocity. The Joule Thomson effect is a typical example of an irreversible process. Throttling is generally regarded as an irreversible process due to excessive frictional heat generation and heat radiating out through the throttling valve.

What is Entropy?

Entropy is a way of thermal energy dissipation by disordered molecules in a system without performing assigned work.

The second law states that there exists a useful state variable called entropy. The change in entropy (dS} is equal to the heat transfer (delta Q) divided by the temperature (T), dS = delta Q/T. For a given physical process, the entropy of the system and the environment will remain constant if the process can be reversed. If we denote the initial and final states of the system by "i" and "f", then: $S_f = Si$ [reversible process]

The ideal scenario would involve forcing a flow through a narrow pipe as an illustration of a reversible process. (A boundary layer loss-free state is ideal). The pressure, temperature, and velocity of the flow would alter as it passes through the constriction, but they would revert to their initial values after the constriction. The gas would revert to its initial state, and the system's entropy would have changed by zero. According to the second law, the system's and the environment's entropy must rise if the physical process is irreversible, and the final entropy must surpass the initial entropy.

The reason heat moves from a hot object to a cold object is explained by the application of the second law. Assume for the moment that heat is transferred from object 1 (which is hot) at temperature T_1 to object 2 (which is cold) at temperature T_2. We will refer to the ultimate equilibrium temperature for both objects as T_f, and the quantity of heat transferred as Q. As the heat is dispersed from the hot object, its temperature fluctuates. We will refer to the hot object's average temperature during the process as Th, which is equal to the average of T_1 and T_f.

$$Th = [T_1 + T_f]/2$$

Similarly, for the cold object, the final temperature is T_f and the average temperature during the process is T_c which is the average of T_f and T_2.

$$T_c = [T_2 + T_f]/2$$

Th will always be greater than T_c, because T_1 is greater than T_2.

$$Th > TC$$

The entropy change for the hot object will be ($-Q/Th$), with the minus sign applied because the heat is transferred away from the object.

$$dS\,[h] = -Q/Th$$

For the cold object, the entropy change is (Q/Tc), positive because the heat is transferred into the object.

$$dS\,[c] = Q/Tc$$

So, the total entropy change for the whole system would be given by the equation,

$$Sf - Si = -Q/Th + Q/Tc$$

with Si and Sf being the final and initial values of the entropy. The term (Q/Tc) will always be greater than ($-Q/Th$) because Th is greater than Tc. Therefore, Sf will be greater than Si, as the second law predicts.

Heat and Entropy

In general, thermal energy or heat is often associated with the generation of entropy. However, there can also be cases where mechanical or chemical energy transformations contribute to entropy generation or changes. Specifically, friction and other forms of mechanical energy conversion can lead to heat generation and, consequently, entropy increase. Similarly, chemical reactions can involve changes in the arrangement and distribution of molecules, which can also result in entropy changes. So, while thermal energy is commonly associated with entropy generation, there are instances where other forms of energy play a role as well. The entropy plays a crucial role in phase change' When the entropy of a phase exceeds its capacity to hold the disorder the phase changes to another phase that can accommodate the excess disorder. A typical case is phase transformation of water to vapor by expansion of water.

Why Does Heat Move? Heat-Entropy

Let us question the role of Reynolds number [$Re = Du\rho/\mu$] in heat transfer. Is it just a dimensionless number or more than that? When we boil milk, we see the Heat-Entropy-Reynolds number triangle every day. In a heat exchanger, the same thing happens. Your milk will not boil if any of them is missing. You see the 2nd law of thermodynamics working in your kitchen every day.

Reynolds Number is a Consequence of Entropy

Reynolds number is a ratio of = Inertial forces/Viscous forces

The more the kinetic energy there is, the greater the entropy and the lower the viscous force. Viscous forces are nothing more than a measurement of how molecules diffuse energy. As a result of having more kinetic energy, [more disorder] there is more entropy, which means less viscous forces and thus a higher Reynolds number.

Let us imagine a shell and tube heat exchanger. It has two closed systems one is shell and the other is tube compartment separated by a heat permeable metal wall which does not allow substance transfer.

Let us look at the 2nd law of thermodynamics

There are two fundamental laws which any energy transport system has to satisfy [1] Energy flow from high energy level to low energy level and [2] entropy of universe [system + surrounding] always increases.

Heat Exchanger and 2nd Law of Thermodynamics

Let us see how heat transfer meets these fundamental laws for energy transport in heat exchanger.

Imagine, hot fluid in container red flows to cold fluid in container blue [image below] through a heat permeable connection which does not allow any matter to flow between containers. Both containers are closed systems. A shell and tube heat exchanger consists of similar two closed systems separated by heat permeable metal wall.

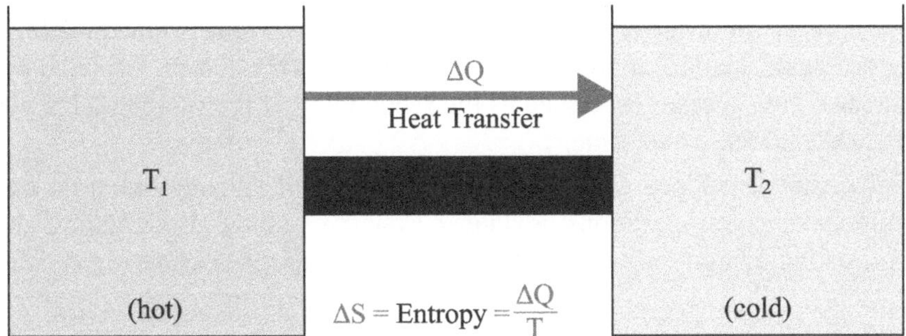

There exists a useful thermodynamic variable called entropy (S). A natural process that starts in one equilibrium state and ends in another will go in the direction that causes the entropy of the system plus the environment to increase for an irreversible process and to remain constant for a reversible process.

$S_f = S_i$ (reversible) $S_f > S_i$ (irreversible)

Figure 2

By definition, heat is an energy which is randomly moving in the direction of a higher disorder or higher entropy. Heat has no fixed destination. Heat is always in transit. Heat does not belong to any particular substance or system.

With this background, let us see what is happening in a heat exchanger.

We can write, $Q = \Delta U + T\Delta S + W$ [Q is heat [total energy], U is internal energy, T is temperature in kelvin and S is entropy, W is work [W can have +ve or –ve sign].

Q does not carry W, so equation can be re-written as, $Q = \Delta U + T\Delta S$, Q carries useful internal energy and useless entropy.

Hot fluid loses heat and becomes cold as it flows to lower temperature. Heat Q goes from high temperature T_2 to a lower temperature T_1 which means entropy goes from low to high value. The overall entropy. $S_1 = Q/T_1 < S_2 = Q/T_2$.

Therefore, we can see the heat transfer meets this requirement of the second law of thermodynamics also.

Entropy and Disorder

Entropy is often associated with disorder because it measures the number of microscopic arrangements or states that a system can occupy. In a highly ordered state, there are fewer possible arrangements, and therefore, the entropy is lower. On the other hand, in a disordered state, there are more possible arrangements, resulting in a higher entropy. Heat is a randomly dispersed disordered thermal energy rushing in all directions wherever it finds a lower temperature. Therefore, all heat is not available to perform thermodynamic work. The heat which is not available to perform work is entropy.

Clausius and Boltzmann's Concepts of Entropy

Clausius did not explicitly discuss entropy in terms of disorder or randomness. Instead, he formulated entropy as a measure of the unavailability of a system's energy to perform work. Clausius' concept of entropy focused on the second law of thermodynamics, which states that the total entropy of an isolated system always increases over time.

It was Boltzmann who introduced the idea of entropy as a measure of disorder or randomness at the microscopic level, based on the statistical behavior of particles. Boltzmann's conceptualization helped to bridge the gap between the macroscopic definition of entropy proposed by Clausius and the underlying molecular-level processes.

So, while Clausius did not directly address entropy in terms of disorder or randomness, Boltzmann's contributions to statistical mechanics provided a more nuanced interpretation that has become essential to our modern understanding of entropy.

However, the essence is both concepts converge into a single point "the more the disorder or randomness the more is the distraction of useful thermal energy on its path to convert into work".

Introduction

Clausius defined entropy as a measure of the amount of energy in a system that is unavailable to do work. He stated that in any energy transfer or transformation, if no energy enters or leaves the system [isolated system], the entropy of the system will always increase. This is often expressed as the second law of thermodynamics, stating that the entropy of an isolated system will always increase over time.

On the other hand, Boltzmann developed a statistical interpretation of entropy based on the idea that entropy is related to the number of microscopic configurations or states a system can have. Boltzmann's definition of entropy is more probabilistic in nature, stating that entropy is a measure of the number of possible ways energy can be distributed among the particles in a system.

In essence, Clausius's definition focuses on the macroscopic aspect of entropy as a measure of the unavailability of energy for work, while Boltzmann's definition delves into the microscopic level and links entropy to the probabilistic distribution of energy among particles. Both definitions are complementary and provide different perspectives on the concept of entropy in thermodynamics.

Details

Clausius' concept of entropy is related to the energy flow in a system and its surroundings. According to Clausius, the entropy of a system and its surroundings always increases in any irreversible process. This aligns with the understanding that entropy is a measure of the system's dispersal of energy or its tendency to move towards equilibrium. In an isolated system, where there is no energy exchange with the surroundings, the entropy will naturally increase as the system moves towards equilibrium. This aligns with the concept that entropy always increases in isolated systems.

On the other hand, Boltzmann's concept of entropy is related to the microscopic level and the number of possible microstates of a system. According to Boltzmann, entropy (S) is proportional to the natural logarithm (ln) of the number of microstates (W) available to a system in a particular macrostate. This aligns with the idea that entropy is a measure of the system's disorder or randomness. As the number of microstates increases, the system becomes more disordered, leading to an increase in entropy.

Boltzmann's Entropy Formula

$S = kb \ln W$, where S represents entropy, kb is the Boltzmann constant, and W is the number of microstates or possible arrangements of a system.

As the number of microstates (W) increases, the entropy (S) also increases. This means that as disorder or randomness in a system increases, the number of possible ways the system can be arranged also increases. Consequently, there is a greater opportunity for energy or heat to be dispersed and transferred within the system, leading to a higher likelihood of energy loss and reduced efficiency in performing assigned work.

This means that the increased disorder or entropy can result in a higher chance for energy dissipation, reducing the capacity for heat to be effectively utilized for doing assigned work.

CHAPTER 22

Introduction to State and Path Function

STATE FUNCTION

In thermodynamic equilibrium, a **state function, function of the state**, or **point function** for a thermodynamic system is a mathematical function relating several state variables or state quantities (that describe the equilibrium states of a system) that depend only on the current equilibrium thermodynamic state of the system. A state function describes the equilibrium states of a system, thus also describing the type of system. Whenever we define a substance one of the first things mentioned is the state of the compound. "State" refers to temperature, pressure, and the amount and type of substance present. Once the state has been established, state functions can be defined. Thermodynamic variables such as the internal energy (U), volume, pressure, temperature, entropy (S), and free energy (G), all these quantities can be used to specify the state of a system. They are properties of the current state of the system, and they do not depend on the way the system got to that state. State properties refer to the characteristics of a substance at specific temperature, pressure and the amount of matter and they can include both atomic and molecular level properties. There properties remain constant under these specific conditions.

Explanation

The energy of a system, such as water, depends on its current state, which is determined by factors like temperature, pressure, and composition. In the case of water, its energy, specifically its enthalpy H, can vary depending on whether it is in a liquid or gaseous state. When water is in the liquid state, at ambient pressure, its enthalpy is $H_1 = 419$ KJ/kg [about]. However, when it undergoes a phase change between water to vapor [state 1 to state 2] and becomes saturated steam, its enthalpy H_2 becomes much higher than H_1 KJ/kg, $H_2 = 2257$ KJ/kg about]. This difference in enthalpy illustrates how the energy of water depends on its state. This concept applies to other systems as well. Whether it is a gas, solid, or liquid, the energy of a system is influenced by its state, and changes in energy can occur when the system undergoes transformations or phase changes. Therefore, energy is indeed a state function, as it is determined by the current characteristics of the system and not by the specific pathway it took to reach that state. Enthalpy between water and saturated steam always goes from 419 KJ/kg to 2257 KJ/kg regardless of the path between two points 1 and 2.

Path Function Examples

Heat and Work

Path functions depend on the path taken to reach one state from another. Examples of path functions include work and heat. In contrast to path functions, state functions are independent of the path taken. Heat and work are path functions.

Heat

Heat is a dispersed energy rushing in all directions of a lower temperature. Between two points at every point heat delta Q is different because the specific heat is different. So, heat between two points have multiple paths to reach from one point to another.

Work

Between two points there can be multiple pressure P and volume V options for work PV to travel between two points.

DIFFERENTIAL AND INEXACT DIFFERENTIAL

Quantities whose values are independent of path are called state functions, and their differentials are exact (*dP, dV, dG, dT* ...). Quantities that depend on the path followed between states are called path functions, and their differentials are inexact (delta *W*, delta *Q*). When we integrate an exact differential, the result depends only on the final and initial points, they are the state functions. They don't depend on the path chosen.

STATE FUNCTIONS EXAMPLES

Enthalpy and Entropy

Enthalpy

Enthalpy is expressed by equation $dH = Cp\ dT$

dH is a change in enthalpy, Cp is the specific heat at constant pressure dT is temperature change. Therefore, enthalpy change is a product of specific heat and temperature both of which are state functions.

All heat capacities are state functions since it does not matter HOW the heat was added or HOW the temperature was changed. Rather, the net change in heat and temperature determines a substance's heat capacity.

Temperature is an intensive property. The temperature is a state function the net change in the temperature only depends on the initial and final states of the system. This explains how enthalpy is a state function.

Enthalpy is a function of two state properties specific heat and temperature.

Entropy

Entropy is a function of temperature and pressure. Both temperature and pressure are state functions. Both temperature and pressure depend on initial and final values.

Equations

$$\text{delta } S = Cp \ln T_2/T_1 - R \ln P_2/P_1$$

Both T and P are state functions. 1 and 2 are two points. Hence entropy is a state function.

CHAPTER 23

Phase Change Thermodynamics

WHY AND HOW DOES PHASE CHANGE OCCUR?

Fundamentally, energy levels differ between the two phases. Both phases of a phase molecule have the same energy level at maximum symmetry. A phase change can occur in two ways [1] when two existing phases in a system find out a point of pressure and temperature where they have minimum molecular symmetry and [2] when two existing phases in a system find out a point of temperature and pressure where both phases have the minimum free energy.

The first one is called the second-order phase change. The second one is called the first-order phase change.

Matter exists in four primary states: solid, liquid, gas, and plasma. When matter changes from one state to another, phase changes take place. This process occurs when a sizable amount of energy is gained or lost. Phase change is also impacted by pressure and temperature. When matter undergoes a phase change, heat is either absorbed or released. This process, referred to as heat transfer, can occur in three different ways: conduction, convection, and radiation.

Phase changes occur when the intermolecular energy between molecules is changed by an external force (heating, cooling, etc.), which breaks the bonds between the molecules. After de-clustering, the molecules move into a stable phase and rearrange to form new molecules. When water cools, it becomes ice, and when it heats up, it becomes vapor.

Issues with Two Phase Heat Transfer

Unsteady heat transfer post a phase change can affect heat transfer coefficient by causing fluctuations in temperature and pressure within the system. This can lead to reduced efficiency in heat transfer and potentially cause damage to the equipment involved. The presence of two phases can also lead to increased resistance to heat transfer, as the different physical properties of the two phases can lead to uneven distribution of heat. While the increase in heat transfer rate due to latent heat in a two-phase heat transfer may initially seem advantageous, the potential negatives of two-phase heat transfer, such as unsteady heat transfer, uneven temperature distribution, and potential equipment damage, can outweigh the benefits. It's important to carefully weigh these factors and consider the potential drawbacks when designing heat exchangers or systems involving two-phase heat transfer.

In addition, the unsteadiness in temperature distribution can also affect the heat transfer coefficient by causing unpredictable changes in the flow of heat through the system. This can make it difficult to accurately predict and control the heat transfer process, leading to potential inefficiencies and safety concerns.

Thermodynamics of Phase Change

Enthalpy, Entropy and Gibbs free energy play crucial role in phase change.

Enthalpy (H) is a measure of the total heat energy in a system, including both the internal energy (U) and the work done by or on the system. During a phase change, the change in enthalpy (ΔH) is primarily due to the change in the internal energy of the substance as the liquid turns into vapor.

Entropy (S) is a measure of the degree of disorder or randomness within a system.

At constant temperature breaking intermolecular bonds during phase change increases entropy. The opposite happens when vapor condenses into liquid.

Gibbs Free Energy (G)

Free energy (G) is measure of available energy post phase change after accounting for entropy losses (S) at temperature T.

$$dG = dU - TdS$$

At $dG = 0$, the two phases are in equilibrium.

The value of ΔG can be positive or negative depending on the specifics of the phase transition.

When dG is negative ($\Delta G < 0$), it indicates that the phase transition is spontaneous.

When ΔG is positive ($\Delta G > 0$), the phase transition is not spontaneous, and external energy input is required for the transition to occur. This usually occurs when the system is at a lower temperature than the transition temperature, and the enthalpy change dominates over the entropy change.

TWO-PHASE FLOW AND SLUG FLOW

Difference between Phase Change and Multiphase

Phase Change

It is physical process. Phase change is an energy-driven process. Change can take place in either direction. Water has so little energy at 0 degrees Celsius that it converts to ice, a lower energy phase. Water has so much energy at 100 degrees Celsius that it turns to vapour, a higher energy phase. Water's kinetic energy is so low at 0 degrees Celsius that molecules coalesce into solid ice, which acts as a centre for potential energy via H-bonds. Similarly, the kinetic energy is so high at 100 degrees Celsius that molecules break H-bonds and move to a stable high energy centre for kinetic energy, which is vapour.

In summary, a phase change occurs in order for a phase to maintain maximum stability in a given situation. At the phase boundary, both phases stay at equilibrium. For, a single component system as water with two phases, $C = 1, P = 2$, the degree of freedom $F = C - P + 2 = 1 - 2 + 2 = 1$. Only one independent variable, temperature or pressure, decides the location of the phase boundary in the phase diagram.

Multiphase

A multiphase flow is defined as one in which more than one phase (i.e., gas, solid and liquid) occurs.

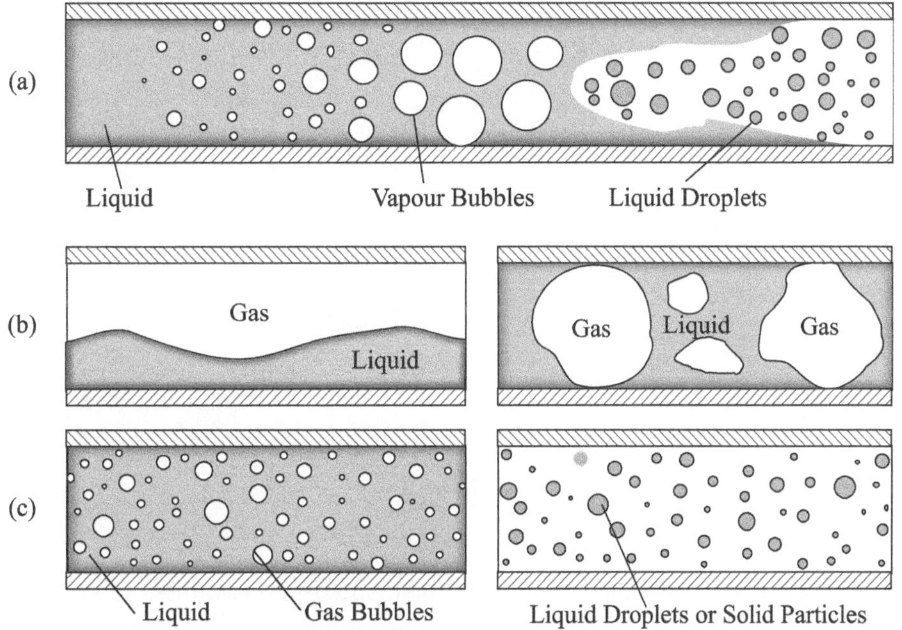

Figure 1: Different Regimes of Two-Phase Flows

Image Credit: Research Direct.

(a) transient two-phase flow with the transition from liquid to vapour flow, (b) separated or segregated two-phase flow with large regions of gas, (c) dispersed two-phase flow with gas bubbles (or liquid droplets) dispersed in a liquid or solid particles or droplets dispersed in gas.

Almost all oil and gas wells, as well as the associated pipeline systems that collect produced fluids and transport them to treatment facilities, exhibit multiphase flow. Because these fluids have significantly different physical properties, multiphase flow is much more complicated than single-phase flow. The major issues are associated with the main difficulties encountered in describing multiphase flow: concurrently, the flowing liquid and gas can have different geometrical arrangements, called flow properties [1] since a two-phase system is compressible, its density changes with temperature and pressure [2] additionally, a new type of energy loss occurs called slippage loss. The slippage loss occurs as a result of the gas phase slipping through the liquid due to the large specific gravity difference between phases.

Most multiphase correlations assume a homogeneous flowing mixture and employ the general energy equation written for this hypothetical fluid.

Significance of prediction of flow patterns: Flow in any channel requires design, development, and optimization. It is vital to predict the flow phases, like flow patterns primarily based on the interfaces formed between the phases. This know-how enables the prediction of the pressure drop and heat transfer characteristics based totally on the flow rate, channel size and operating conditions. From the pressure drop data, the acceptable float characteristics can be decided to decrease the incidence of corrosion, erosion, or scale formation, all of which can lead to excessive friction.

Slug Flow

A Slug flow is a two-phase flow. It is more particularly a liquid-vapor flow. Slug flow occurs due to two different densities of the phases.

In a multiphase flow in a pipe, the lighter fluid stays as bubbles dispersed within, and pushing along, the heavier fluid. The word slug normally refers to the heavier, slower-moving fluid, but sometimes to the bubbles of lighter fluid. There are also small bubbles within the liquid, but many of these have coalesced to form large bubbles until they span much of the pipe. In gas-liquid mixtures, slug flow is similar to plug flow, but the bubbles are generally larger and move faster. As flow rates increase, slug flow becomes churn flow.

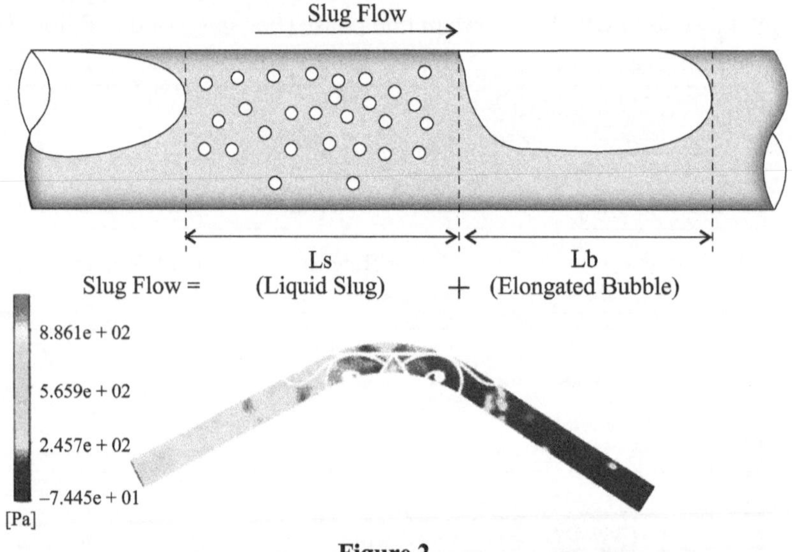

Figure 2

Image Credit: Research Direct.

To summarize, the intermittent gas-liquid flow could be divided into two distinct systems: plug and slug flow. Each regime can provide its particular shape and hydrodynamics properties. The main characteristic of the slug flow is the intermittent passage of air-laden liquid eruptions followed by elongated air bubbles called cells. Interactions with neighbouring cells form or break liquid slugs.

DIFFERENT DIMENSIONS OF THE SLUG FLOW

Every fluid wants to reach the shape of a sphere because a sphere has the minimum energy. A sphere is the most stable shape for any fluid.

The problem is created by the surface tension at the interphase of two fluids. Surface tension is nothing but a stretch of the surface by the intermolecular forces at the interphase of two fluids. In the bulk, a liquid molecule experiences 360-degree pull from all sides.

Slug Flow

Interestingly you can see bubbles of gases floating in the liquid causing a slug flow.

The obvious question arises how the light bubbles survive the hydrostatic pressure of the liquid inside the liquid. The answer is a gas has no option for any different shape. It will break and again will reform the sphere that is the minimum energy shape.

The answer is the surface tension.

Surface Tension

Gases have no surface tension because molecules are far apart to cause any wall tension.

This is making air or gas fold itself into its minimum energy shape which is a sphere.

Water cannot fold itself because of its large intermolecular forces pulling in different directions at 360 degrees.

Viscosity

There is another dimension. That is the effect of viscosity on gas bubbles. The shearing of the surface of the bubble by non-Newtonian liquids elongates the bubbles like a bullet that pushes a slug.

Consequences of Slug Flow

Oil and gas transportation is an important activity for the oil industry worldwide. Pipelines are primarily used to transport oil and gas from wells or production areas to refinery facilities. These pipelines are subjected to severe environmental and operational stresses that can result in severe dynamic reactions that lead to pipe failure and fatigue.

Slug flow is one of the greatest threats to the pipeline industry.

This flow system can cause rapid fluctuations in mass distribution and pressure disturbances along the pipe, which can lead to violent vibrations. These vibrations create transient stresses that lead to fatigue failure, localized buckling, and even excessive bending and failure. These fluctuating stresses in pipelines can significantly disrupt normal operation, accelerate pipeline corrosion, and cause significant damage to pipe walls. In fact, it has been reported that many accidents have occurred due to vibration caused by internal two-phase flow.

Effect of Temperature on Slug Flow

The temperature would primarily influence the viscosity and solubility of the gas.

The following events are anticipated:

1. **Viscosity:** The viscosity of the liquid phase decreases as temperature rises. As a result, the liquid flows more easily and forms smaller slugs as the temperature rises.
2. **Gas fraction:** As the temperature rises, so does the gas fraction in slug flow. This is because the solubility of gas in liquids decreases with increasing temperature. The gas bubbles congregate and form larger pockets as the temperature rises, resulting in larger slugs.
3. **Slug length and frequency:** As temperature rises, the length of liquid slugs shortens, while slug formation frequency increases. This is due to the lower viscosity of the liquid phase, which allows it to flow more easily and form smaller slugs.
4. **Flow instability:** At extremely high temperatures, the liquid evaporates rapidly, causing the liquid phase to decrease and the gas phase to increase. This destabilises the flow, which can lead to flow reversal or other system instabilities.

CHAPTER 24

Triple Point and Critical Point Thermodynamics

TRIPLE POINT AND CRITICAL POINT

A substance has two phase properties: triple point and critical point. At the triple point, the phases of the substance, solid, liquid and gases coexist. They achieve a thermodynamic equilibrium when all of their state variables reach a common point on the phase diagram and are identical. A critical point in a phase diagram is the end point where the phase boundary between saturated liquid and vapour breaks and the two phases combines into a single supercritical phase with two phases reaching equal density.

In summary, the triple point represents the unique conditions at which all three phases of a substance coexist in equilibrium, while the critical point marks the end of the distinction between the liquid and gas phases, leading to the formation of a supercritical phase.

SIGNIFICANCE OF TRIPLE POINT AND CRITICAL POINT

The triple and critical points are significant in thermodynamics because they provide important information about the behaviour and properties of substances, particularly in relation to phase transitions.

Triple Point Significance in Thermodynamics

- **Unique conditions**: The triple point represents the unique set of temperature and pressure conditions at which all three phases of a substance (solid, liquid, gas) can coexist in equilibrium. This provides critical information about the properties of a substance and helps in the accurate determination of its phase diagram.
- **Calibration and standardization**: The triple point of water is used in the calibration of thermometers, providing a fixed reference point for temperature measurements. This makes it a crucial standard in the field of thermodynamics and metrology.
- **Understanding phase transitions**: The triple point helps in understanding phase transitions and the behaviour of substances under different conditions of temperature and pressure. It provides insight into the transitions between various phases of a substance and the stability of these phases.

Critical Point Significance in Thermodynamics

- **Understanding phase behaviour**: The critical point holds significance as it represents the end point of the liquid-gas phase boundary. At this point, the distinction between the liquid and gas phases disappears, and the substance behaves as a supercritical fluid. This helps in understanding the behaviour of substances under extreme conditions.
- **Engineering applications**: The characteristics of supercritical fluids are utilized in various industrial and engineering applications, such as in supercritical fluid extraction, where the unique properties of supercritical fluids are harnessed for extraction and separation processes.

Example

The Water is a Classic Case to Study Triple Point and Critical Point

Triple Point of Water

As explained, in thermodynamics, the triple point of a substance is the unique combination of temperature and pressure at which the solid phase, liquid phase, and gaseous phase can all coexist in thermodynamic equilibrium. The single combination of pressure and temperature at which liquid water, solid ice, and water vapor can coexist in a stable equilibrium occurs at exactly 273.1600 K (0.0100°C; 32.0180°F) and partial vapor pressure of 611.657 pascals (6.11657 mbar; 0.00603659 atm). By international agreement, the triple point of water has been used to define the kelvin, the base unit of thermodynamic temperature.

Phase Equilibrium Lines

The lines between different phases signify a set of points where the phases are in equilibrium. The sublimation curve shows when both solid and vapor can exist. The fusion curve represents when both solid and liquid are in equilibrium. If the fusion curve points toward the left (like in a water phase diagram), the liquid phase is denser than the solid phase. The vaporization curve stands for vapor and liquid equilibrium line. The gas-liquid–solid triple point of water corresponds to the minimum pressure at which liquid water can exist. At pressures below the triple point (as in outer space), solid ice when heated at constant pressure is converted directly into water vapor in a process known as sublimation. Above the triple point, solid ice when heated at constant pressure first melts to form liquid water, and then evaporates or boils to form vapor at a higher temperature. The vaporization curve ends at the critical point.

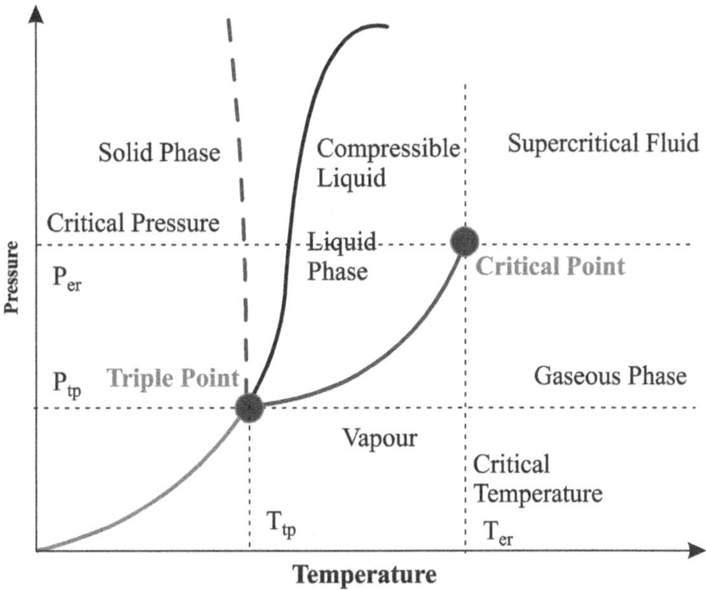

Figure 1: Phase Diagram of Water

Critical Point

The critical point is the end of a phase envelope. At the critical point, defined by a critical temperature *Tc* and a critical pressure pc, phase boundaries disappear. The critical point is the highest temperature and pressure at which a pure material can exist in vapor/liquid equilibrium. At temperatures higher than the critical temperature, the substance cannot exist as a liquid, no matter what the pressure. At temperatures and pressures higher than the critical point, the substance is considered a fluid—something neither gas nor liquid. At pressures lower than the critical pressure (but at higher temps), the substance is considered a gas.

The image is temperature-steam entropy [*TS*] diagram with x-axis entropy and y-axis temperature. On the left side of the bell curve, there is single-phase water. The fractional numbers at the bottom of the bell curve represent the dryness fraction for steam as steam proceeds to the right side in the bell. Inside the bell two phases, water and vapor coexist and on the right side, it is the one-phase superheated region for steam. The topmost point of the bell curve, is the critical point. Above the critical point, there is the supercritical region.

As water is heated at saturation temperature it begins conversion into vapor. This is the region within the bell curve.

The temperature gets [isothermal] fixed in the saturated region as heat goes to supply latent heat to water to convert to vapor. A dryness fraction = 1 on the right contour of bell curve [dryness fraction = 100] Steam is 100% dry. The two-phase region ends. During the journey of two phases inside the bell curve entropy increases as more water converts to randomly moving vapor.

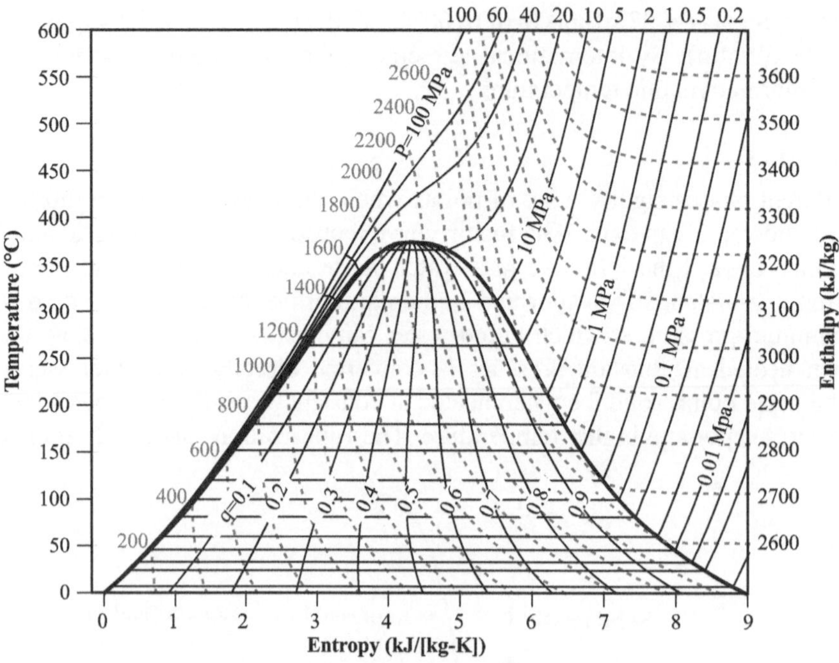

Figure 2: T-S Diagram of Steam

When steam is pressured outside the bell curve on right to generate enthalpy, the volume of steam reduces for the same mass of water and that limits the available space for steam particles to move around, the entropy reduces. The same happens till the critical point.

Thermal Instability Near the Critical Point

In the vicinity of the critical point, the physical properties of the liquid and the vapor change dramatically. As the temperature is raised, the vapor pressure increases, and the gas phase becomes denser. With two phases being in equilibrium, the liquid expands as van der Walls forces become smaller leaving only repulsive forces to dominate making the water less dense until, at the critical point, the densities of liquid and vapor become equal, eliminating the boundary between the two phases. In water, the critical point occurs at 647.096 K (373.946°C; 705.103°F) and 22.064 megapascals (3,200.1 psi; 217.75 atm). As a consequence, the water's properties change dramatically.

For instance, liquid water under normal conditions is nearly incompressible, has a low thermal expansion coefficient, has a high dielectric constant, and is an excellent solvent for electrolytes. Near the critical point, all these properties change into the exact opposite: water becomes compressible, expandable, a poor dielectric, a bad solvent for electrolytes, and prefers to mix with nonpolar gases and organic molecules.

Beyond Critical Point

Above the critical temperature, the molecules have too much kinetic energy for the intermolecular attractive forces to hold them together in a separate liquid phase.

Critical Point Phase Change – 2nd Order Phase Change

A critical point is a property of a pure substance. Every pure substance has a critical composition at a fixed temperature of pressure for example water has a critical point at 374 degc and 217.7-atmosphere pressure. The critical point is a point where the two phases have minimum molecular structural symmetry.

Let us elaborate on the second-order phase change.

Let's start with the basics. When the energy difference between two phases is the smallest, a phase change occurs. They achieve equilibrium. Entropy or disorder is important whether the phase change is first or second order.

There are two points to be made here. The first is that a phase is more disordered at high temperatures than at low temperatures due to molecular motions.

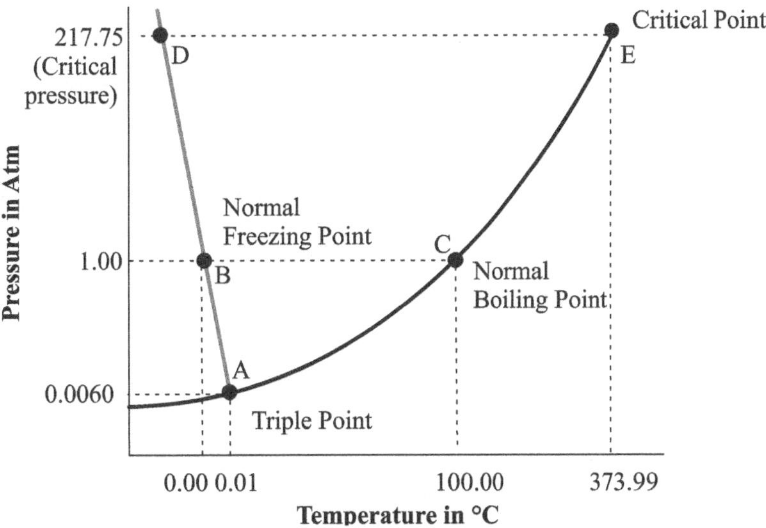

Figure 3: Phase Diagram of Water

Due to spontaneous symmetry breaking, the high-temperature phase typically contains fewer symmetries or more entropy than the low-temperature phase. This minimizes the free energy for phase change.

This happens abruptly when two phases interact and reach critical temperature and pressure. Critical point phase change is an example of 2nd order phase change. Phase change at a critical point does not involve the role of latent heat. At a critical point, all phases exist at minimum free energy with degrees of freedom $F = 0$.[$F = C - P + 2$, at the critical point there are 3 phases [1] Sub-critical vapor, [2] Sub-critical liquid, and [3] critical fluid, $F = C - P + 2 = 1 - 3 + 2 = 0$].

Therefore, the critical point does not have an option or freedom to reverse its phase. A superfluid is a unique collection of particles that has no kinetic energy and no viscosity. The superfluid component has zero viscosity and zero entropy. Since all the atoms in the superfluid are in the same energy level the total entropy of the system is zero.

WATER-VAPOR PHASE CHANGE

First-Order Phase Change

The water-vapor phase change is a first-order phase change.

Any transition involving latent heat in which the system absorbs the heat but the temperature remains constant is referred to be first order.

- The system does not fully shift to the new phase in a seamless manner (regions of both phases exist simultaneously).
- The state variables of the system have a discontinuity (such as temperature or pressure).

Melting ice and boiling water are two common instances of first-order phase transitions (not all ice melts instantaneously nor does all water turn to steam immediately).

What triggers 1st order phase change?

The fundamental unit of matter we meet on a daily basis is an atom, which is made up of protons, neutrons, and electrons in particular ratios. Atoms have a strong interaction with one another and can produce many phases of matter. In a nutshell, a phase of matter is a group of interacting constituents with macroscopic properties that are not accessible through constituent analysis. A phase is the existence of molecules at the lowest free energy.

So, when a huge number of ingredients are present, how do they decide which phase to form?

The answer is they all want to achieve minimum free energy. That is how different phases arise. $\Delta G = \Delta U - T\Delta S$ or, $\Delta G = 0$.

When two phases in equilibrium are struggling to maintain equilibrium as in the case when heat is applied to water, or you are adding enthalpy to water, the entropy will force more intermolecular bonds of water to break to reduce its free energy and let water-vapor stay in equilibrium by pulling down the energy of water.

The first-order phase change is not a property of a pure substance as we saw at the critical point. The degree of freedom for 1st order phase change, $F = C - P + 2 = 1 - 2 + 2$ [C = Components, P = Phases and F = Degrees of freedom] 1st order phase change can occur at any temperature or pressure. In other words, 1st order phase change has the freedom to change its location by independently varying either-or pressure, $F = 1$. By varying either temperature or pressure independently while holding the other constant, there are infinite possibilities for a phase change.

CHAPTER 25

Thermodynamics and Kinetics

Thermodynamics and kinetics are two important aspects in understanding and predicting chemical reactions. Thermodynamics predicts the 'feasibility' ['free energy'] and activation energy predicts the 'rate of a reaction'.

Thermodynamics deals with the energy changes and equilibrium of a chemical reaction. It helps determine whether a reaction is thermodynamically favourable or unfavourable by providing information about the stability of the products and reactants.

Thermodynamics gives insights into the direction and extent of a reaction, as well as the equilibrium constant (*K*). It provides information about the feasibility and spontaneity of a reaction, but it does not provide information about the rate at which the reaction will occur.

Kinetics focuses on the rate at which a reaction proceeds. It provides information about the mechanism of the reaction, the factors that affect the reaction rate (such as temperature, concentration, and catalysts), and the rate constant (*k*). Kinetics explains the "how" of a reaction, including the specific steps involved, the activation energy required, and the reaction rate expression.

The rate equation or rate law derived from kinetic studies can be influenced by the thermodynamic parameters, such as temperature, concentration, and activation energy.

The most commonly used equation that relates kinetics to thermodynamics is the Arrhenius equation:

$$k = A \times e\frac{-E_a}{RT}$$

where:

k is the rate constant of the reaction

A is the pre-exponential factor or frequency factor

E_a is the activation energy of the reaction

R is the gas constant

T is the temperature in Kelvin

The Arrhenius equation shows how the rate constant of a reaction is affected by the temperature and the activation energy. It helps establish the relationship between kinetics (rate constant) and thermodynamics (temperature and activation energy).

In the Arrhenius equation, the units of the pre-exponential factor A are identical to those of the rate constant and will vary depending on the order of the reaction. If the reaction is first order it has the units: s^{-1}.

K is the number of collisions that result in a reaction per second, A is the number of collisions (leading to a reaction or not) per second occurring with the proper orientation to react, and $e^{-Ea/(RT)}$ is the probability that any given collision will result in a reaction.

Additionally, the concept of equilibrium constant (*K*) from thermodynamics can be related to the rate constant (*k*) through the relationship known as the Gibbs free energy (ΔG) change:

$$dG = -RT \times \ln(K)$$

By considering the rate constants at equilibrium, thermodynamics can provide insight into the relationship between the equilibrium constant and the rate constant.

While there isn't a single equation that combines thermodynamics and kinetics, these equations illustrate how thermodynamic parameters can influence the rate of a chemical reaction.

FREE ENERGY AND EQUILIBRIUM CONSTANT

The basic criteria for determining whether a given reaction will occur are as follows:

> dS univ > 0, dG sys < 0 [univ stands for universe and sys stands for system], applicable under constant temperature and pressure conditions.

The magnitude of the reaction quotient Q versus the equilibrium constant K should follow Q as follows: when $Q < K$, the reaction will proceed to the right, resulting in the net conversion of reactants to products. If $Q > K$, the reaction will proceed to the left, resulting in a net conversion of products to reactants. If $Q = K$, the system is in equilibrium and no net reaction occurs.

The interpretation of the equation, $dG = -RT \times \ln(K)$ is if the products and reactants are in their standard states and dG[std] < 0, then $K > 1$, and products are favoured over reactants when the reaction is at equilibrium. Conversely, if dG[std] > 0, then $K < 1$, and reactants are favoured over products when the reaction is at equilibrium. If dG [std] = 0, then $K = 1$, and neither reactants nor products are favoured when the reaction is at equilibrium.

CHAPTER 26

Thermodynamics and Chemical Reactions

Any chemical reaction has two aspects [1] 'How fast' and [2] 'How long'. The first one is dealt by the kinetics and the second by thermodynamics. Thermodynamics can tell you only that a reaction should go because the products are more stable (have a lower free energy) than the reactants. Thermodynamics has nothing to do with time. Kinetics, on the other hand, can tell you how fast the reaction will go but doesn't tell you anything about the final state of things once it gets there. Thermodynamically favourable can be kinetically unfavourable.

GENERAL

A chemical reaction is a process that leads to the chemical transformation of one set of chemical substances to another, $A + B = C + D$. A chemical reaction involves breaking of bonds and forming of bonds by changes in the positions of electrons. The substance initially involved in a chemical reaction are called reactants. Chemical reactions are usually characterized by a chemical change, and they yield one or more products. Chemical reactions are described with chemical equations, which symbolically present the starting materials, end products, and sometimes intermediate products and reaction conditions. Chemical reactions happen at a characteristic reaction rate at a given temperature and chemical concentration. Typically, reaction rates increase with increasing temperature because there is more thermal energy available to reach the activation energy necessary for breaking bonds between atoms. Reactions may proceed in the forward or reverse direction until they go to completion or reach equilibrium. Reactions that proceed in the forward direction to approach equilibrium are often described as spontaneous, requiring no input of free energy to go forward. Non-spontaneous reactions require input of free energy to go forward.

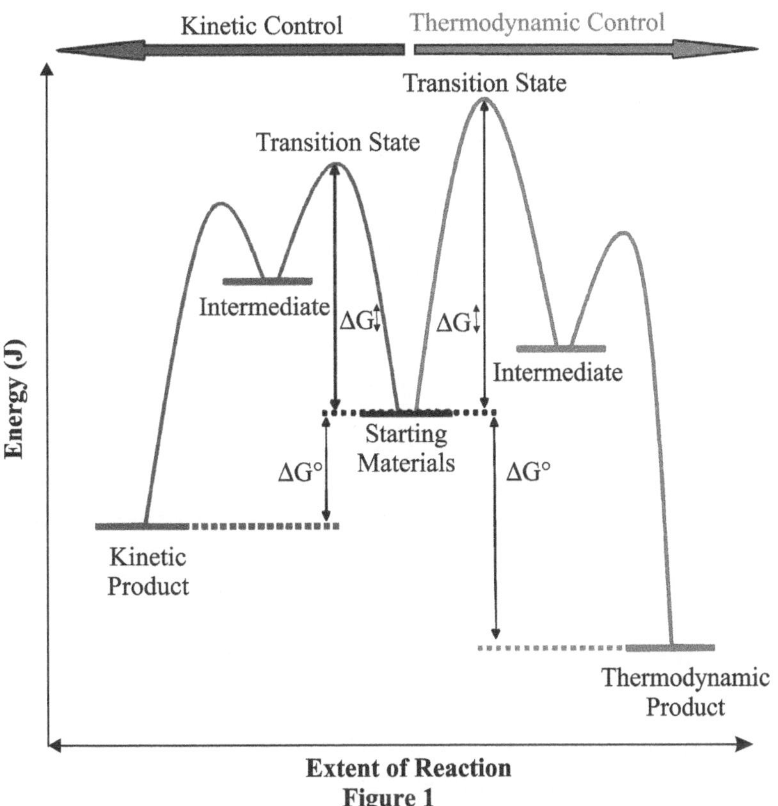

Figure 1

A reaction is thermodynamics controlled or the reaction is controlled by the kinetics can decide the composition in a reaction product mixture when competing pathways lead to different products. The distinction is relevant when product A forms faster than product B because the activation energy for product A is lower than that for product B, yet product B is more stable. In such a case A is the kinetic product and is favoured under kinetic control and B is the thermodynamic product and is favored under thermodynamic control. The conditions of the reaction, such as temperature, pressure, or solvent, affect which reaction pathway may be favored: either the kinetically controlled or the thermodynamically controlled one. Note this is only true if the activation energy of the two pathways differs, with one pathway having a lower Ea (energy of activation) than the other.

X is the kinetic reaction because it is faster with low activation energy and Y is the thermodynamic reaction.

A reaction can be thermodynamically favorable but still kinetically unfavourable.

Because thermodynamics deals with state functions, it can be used to describe the overall properties, behavior, and equilibrium composition of a system. It is not concerned with the particular pathway by which physical or chemical changes occur, however, so it cannot address the rate at which a particular process will occur. Although thermodynamics provides a significant constraint on what can occur during a reaction process, it does not describe the detailed steps of what actually occurs on an atomic or a molecular level. In many cases, we will encounter reactions that are strongly favored by thermodynamics but do not occur at a measurable rate. In contrast, we may encounter reactions that are not thermodynamically favored under standard conditions but nonetheless do occur under certain nonstandard conditions.

Example: A simple typical example is graphite and diamond are both forms of carbon, but graphite has a lower free energy. Therefore, thermodynamics favor conversion of diamond into graphite. Even though the reaction is thermodynamically favorable, it is slow. It's just plain too difficult to get the diamond to break all of its bonds and re-form them in the different, more stable graphite configuration.

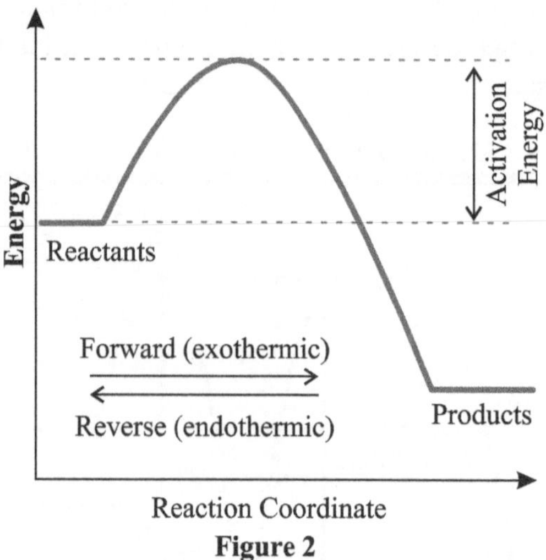

Figure 2

To answer question what dominates, kinetics or thermodynamics in a chemical reaction, we must go back to the first reaction diagram featuring the hump between reactants and products and look at the height of the hump. **That hump is a measure of how difficult it is to get the reaction to go in any reasonable amount of time**. For diamond, the hump for the conversion into graphite is high. Even though the reaction should go thermodynamically, it does not because it is kinetically unfavorable.

Quick review of kinetics and thermodynamics of chemical reactions:

Kinetics

Kinetics is the speed at which reactions take place. The rate of a typical first order reaction is (–) $d[A]/dt = Kx\ A$. Here K is first-order rate constant having dimension 1/time, $[A](t)$ is concentration at a time t and $[A]0$ is the initial concentration. [Kept it simple]

Some Key Factors

Reactant Concentrations

Reactant concentrations usually make the reaction happen at a faster rate by increasing collisions per unit time. Some reactions, however, have rates that are independent of reactant concentrations. These are called zero order reactions.

Surface area available for contact between the reactants: [in particular solid ones in heterogeneous systems]

Larger surface areas lead to higher reaction rates.

Pressure

Increasing the pressure decreases the volume between molecules and therefore increases the frequency of collisions between the molecules.

Activation Energy

Activation energy, which is defined as the amount of energy required to make the reaction start and carry on spontaneously. Higher activation energy implies that the reactants need more energy to start than a reaction with a lower activation energy. Temperature, which hastens reactions if raised, since higher temperature increases the energy of the molecules, creating more collisions per unit time. Catalysts change the pathway (mechanism) of a reaction which in turn increases the speed of a reaction by lowering the activation energy needed for the reaction to take place.

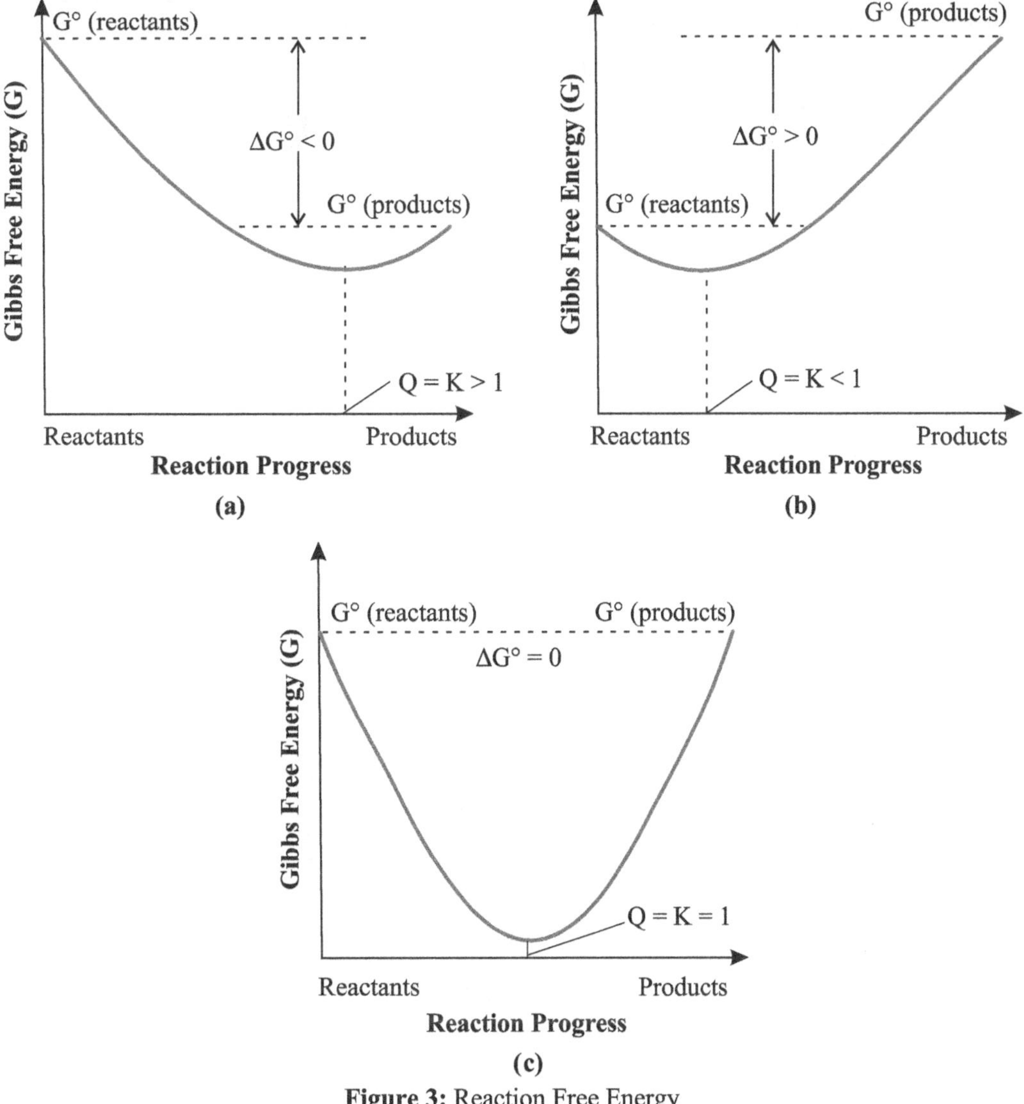

Figure 3: Reaction Free Energy

Thermodynamics simply say, substances combine chemically when their total energy contents change from a higher level to a lower level. In other words, there is consumption of energy. The change in total energy content is known as free energy. Thus, if the free energy of reactants is greater than products, the reaction will proceed. When $\Delta G = 0$ the reaction is at equilibrium.

In general, to conclude, short reaction times favour kinetic control, whereas longer reaction times favour thermodynamic reaction control. Low temperatures will enhance the selectivity under both sets of conditions.

In summary, at fundamental level thermodynamics and kinetics both play roles in a chemical reaction. At the primary level, three factors influence a reaction from going from reactants to products: [1] temperature, [2] pressure, and [3] concentration. At the secondary level, these three factors generate three requirements: [1] activation energy, [2] enthalpy, and [3] entropy. The activation energy determines how fast or slow a reaction can proceed. This is the kinetics part of the reaction. Thermodynamics is the study of equilibrium and essentially free energy. The change in free energy in a chemical reaction is the result of three factors: [1] enthalpy, [2] entropy, and [3] temperature.

WHO CONTROLS A REACTION?

Thermodynamics would control a reaction or kinetics is decided by the composition of a reaction product mixture when competing pathways lead to different products. If in a reaction there are two products X and Y. If X has lower activation energy the kinetics would favour X since X would form faster than Y.

Yet X may not be a stable product unless the Gibbs free energy change [dG] is negative. So not necessarily even if kinetics meets the activation energy level the thermodynamics can still put a stop on the reaction if dG is not negative,

So, in a chemical reaction, there could be a kinetic product and the other could be a thermodynamic product. Thermodynamics can favour a reaction yet it could be such a slow reaction that it would not proceed.

Thus, thermodynamics and kinetics are inextricably linked.

CHAPTER 27

Exergy: Quality of Energy

The four components of exergy are potential, kinetic, chemical and physical exergy. If changes in elevation and speed are minimal, the chemical and physical changes are considered significant. The maximum theoretical useful work that a system can achieve while interacting with a reference environment at equilibrium is known as physical exergy. The difference between a system's chemical composition and that of the reference environment is known as chemical exergy. For fuels, chemical exergy is important. By using the first and second laws of thermodynamics, one can obtain an equation for exergy balance.

$$ExQ + \sum i \; \dot{m}i \; exi = \sum e \dot{m}e \; exe + ExW + ExD$$

where the subscripts i and e denote the inlet and outlet of a control volume, ExD is the exergy destruction, Q is heat, ex is exergy. Exergy quantities are evaluated with respect to a reference environment. The intensive properties of the reference environment in part determine the exergy of a stream or system. The exergy of the reference environment is zero. The reference environment is in stable equilibrium, with all parts at rest relative to one another. No chemical reactions can occur between the environmental components. The reference environment acts as an infinite system and is a sink and source for heat and materials. It experiences only internally reversible processes in which its intensive state remains unaltered.

Energy and exergy balances can be written for a general process or system. An energy balance for a system may be written as

Energy input − Energy output] = [Energy accumulation] .... [1]

Energy input and output refer respectively to energy entering and exiting through system boundaries. Energy accumulation refers to buildup (either positive or negative) of the quantity within the system. Clearly, energy is subject to a conservation law.

By combining the conservation law for energy and non-conservation law for entropy (which states that entropy is created during a process due to irreversibilities), an exergy balance can be obtained:

[Exergy Input − Exergy output − Exergy consumption] = Exergy accumulation ... [2]

Exergy is consumed due to irreversibilities. Exergy consumption is proportional to entropy creation. Equations (1) and (2) demonstrate an important difference between energy and exergy: Energy is conserved, whereas exergy, a measure of energy quality or work potential, can be consumed.

DETAILS

Exergy is defined as the amount of work (= entropy-free energy) a system can perform when it is brought into thermodynamic equilibrium with its environment. **Exergy is the quality of energy.** Exergy is the available energy to do maximum work. Once energy is lost that equivalent of work is lost. So, there is a destruction of exergy. The exergy destruction is maximum when a system reaches equilibrium since entropy is at peak at this point.

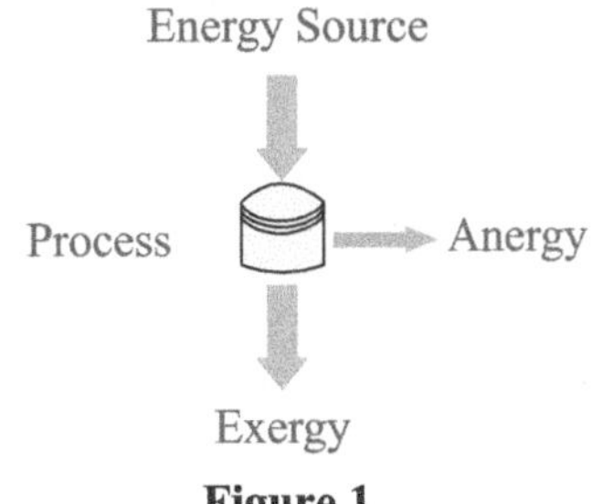

Figure 1

In simple words, the exergy of a system is the maximum useful work possible during a process before the system reaches equilibrium with a heat reservoir, reaching maximum entropy. In real-life in processes which are irreversible, there is always some entropy generation and correspondingly loss of work potential of energy.

When thermal energy does work it loses some energy while doing the work in the disorder that thermal energy generates. This disorder is known as entropy. The lost energy is called 'Anergy'. After providing for anergy the amount of energy left in the system is available to perform work. This energy is called 'Exergy'.

Therefore, the exergy is work performed in a system by energy free from entropy before the system reaches thermodynamic equilibrium with its environment.

Exergy is the theoretical limit for the work potential that can be obtained from a source or a system at a given state when interacting with a reference (environment) at a constant condition.

Energy is conserved but not exergy: 1st law of thermodynamics *vs* 2nd law of thermodynamics

It is seen from this description that exergy is dependent on the state of the total system (= system + reservoir) and not depend entirely on the state of the system. Exergy is therefore not a state variable.

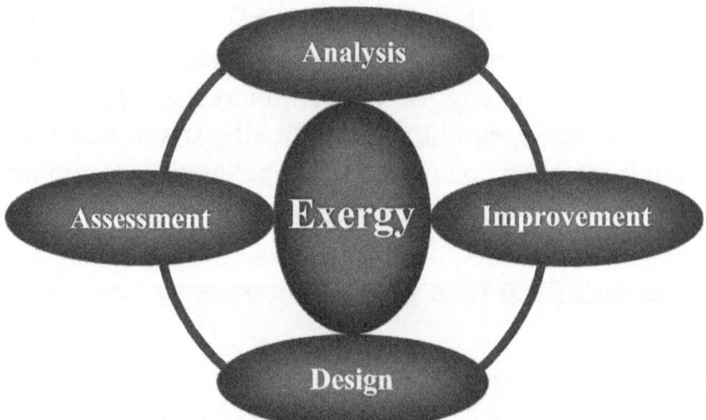

Figure 2: Significance of Exergy

Exergy analyses are very convenient methods for assessing the performance of energy conversion systems. Exergy analysis helps in finding the type, location, and magnitude of energy losses in a system.

EXERGY AND THERMODYNAMICS

The First Law of Thermodynamics states that in any process energy is conserved for a steady-state flow process,

$$H_0 - H_1 = Q + W \qquad \dots (1)$$

The Second Law of Thermodynamics says that energy transformations in which entropy is reduced are not possible. From the definition of entropy,

$$S_0 – S_1 > \text{or} = Q/T_0, \text{ or } T_0 [S_0 – S_1] > \text{or} = Q \qquad \dots (2)$$

If you combine the above two equations,

$$T_0 [S_0 – S_1] > \text{or} = [H_0 – H_1] – W$$

The maximum work available from a process is, therefore,

$$W = [H_1 – H_0] – T_0[S_1 – S_0]$$

W is exergy or the work available post entropy losses.
W is maximum when the entropy change is minimum or near zero and the opposite.

Let us call it the exergy equation.

EXERGY AND FREE ENERGY

There is a razor thin difference between exergy and free energy.

Look at the exergy equation, $W = [H_1 - H_0] - T_0[S_1 - S_0]$, W is exergy. If you see this equation, it is same as Gibbs free energy equation,

$$dG = dH - TdS$$

W, the exergy has replaced free energy change G. But free energy and exergy are not same. Gibbs equation says free energy is energy available in a system at a constant temperature [isothermal] and pressure [isobaric] conditions to do work after providing energy for entropy losses.

The amount of exergy in a system is not dependent on whether or not it's an isothermal or isobaric process. It could be any type of process and it will still have the same amount of exergy regardless. The same cannot be said for the Gibbs free energy.

Exergy and free energy are related though – conceptually.

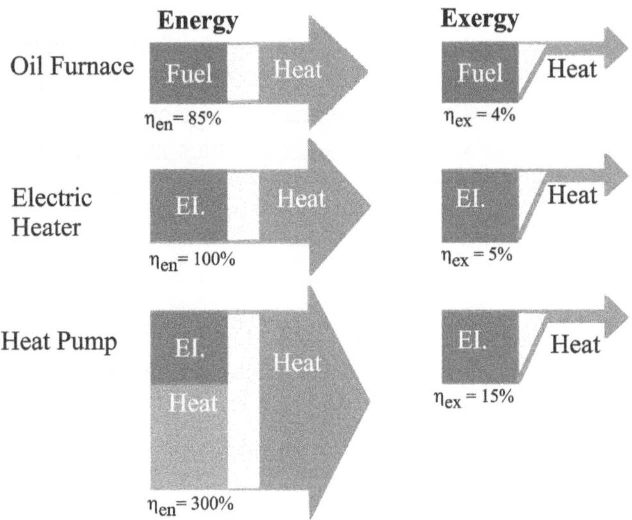

Figure 3: Economics of Exergy

The primary contributors to exergy destruction are irreversibilities associated with chemical reactions, heat transfer, mass transfer, mixing, and friction in a process.

Exergy is Not a State Function

Exergy does not depend strictly on two points. Exergy depends on the state of both system and surroundings.

Irreversibilities

Any process [closed system] operating above ambient temperature loses energy as entropy. Suppose a process generates heat Q, at temperature T_1. Entropy generation in process $S = Q/T_1$. The process then transfers the entropy to surrounding the entropy at lower temperature T_2, the entropy generation in the surrounding, $S_2 = Q/T_2$. Since $T_2 < T_1$, $S_2 > S_1$, the system losses energy to the environment.

When the system transfers energy to surroundings that energy is permanently lost. While a system can be reversed the surrounding can't. You lost work potential of energy to the surroundings that cannot be reversed. This is irreversibility.

Equilibrium

A system is said to be in equilibrium when both system and surroundings have the same entropy. At equilibrium, a system is fully reversible and you get maximum work.

Exergy destruction [*EXd*] is expressed as

$EXd = Tx\ dS$ [gen]

Exergy destruction *EXd* is a product of temperature *T* and entropy generation *dS* [gen]

There are three cases:

$EXd > 0$: irreversible process

$EXd = 0$: reversible process

$EXd >$ impossible process

At equilibrium exergy = 0

Equilibrium and Reversibility are the most confusing concepts of thermodynamics

Are equilibrium and a reversible process the same – the answer is no.

More about equilibrium: The total entropy of the system + surrounding is always increasing. When the system and surrounding both reach maximum entropy, there is equilibrium. At equilibrium, both the system and surrounding have run out of energy. Therefore, because there is maximum entropy, there is maximum disorder at equilibrium. The energy is more spread out at maximum entropy. A system has the least energy at the highest entropy. A system has maximum stability at the highest entropy. The principle of minimum energy which is essentially a restatement of the second law of thermodynamics states that for a closed system, with constant external parameters and entropy, the internal energy will decrease and approach a minimum value at equilibrium.

REVERSIBLE PROCESS

Entropy is zero in a reversible process. It is just the opposite of equilibrium.

If a system is undergoing a change from a state *A* to state *B*, the entropy change is given by

$dS = dQ/T$

And for a reversible process, the surrounding undergoes an equal and opposite change as the system. Therefore, for a reversible process, + *dS* of system = –*dS* of surrounding = 0

Which means that entropy change for a reversible system [*dS* (sys]–*dS* (surr)] = 0.

Example: A process has 100% exergy destruction. How would you define the process?

Is it closer to equilibrium or away from equilibrium?

Explanation

The answer is the more you go near-equilibrium there is more entropy generation. At equilibrium, both the system and surrounding would reach their maximum entropy. When the total entropy has reached the maximum value there is no energy available to do any more work. Both the systems and the surroundings have run out of energy. Since as per the 2nd law of thermodynamics total entropy can only increase and it cannot decrease when both system and surrounding have the highest amount of entropy the process stays there and we call it equilibrium.

100% destruction of exergy means the system has only entropy. Exergy = 0. The system has no useful energy to do work. The system is dead.

Let us go back to the fundamental equation of exergy that we have discussed above.

$W = [H_1 – H_0] – T_0[S_1 – S_0]$

W is exergy or the work available post entropy losses. At zero exergy, $W = 0$

Therefore, $[H_1 – H_0] = T_0[S_1 – S_0]$

What does it mean?

It means the system's all energy creation between state 1 and 0 has got locked up in entropy creation between state 1 and 0. It means the system has no energy to do any work. Therefore, there is 100% destruction of exergy or the exergy of the system is zero.

Entropy will always increase in an irreversible process. The problem lies with reactions that generate heat. The extra entropy due to exothermic reactions will push the reaction towards equilibrium and slow down the reaction if you do not remove the heat. As you approach equilibrium and a higher state of entropy there is more disorder and hence more disorder of energy as well and spread out of energy and the system is moving towards a low energy stable state. You do not want that. You want the reaction to move forward and do not reach equilibrium. You cannot go back from low to high energy because of the 2nd law of thermodynamics does not allow it. Low energy can only be further lower. The only thing you have under your control delay the process of the reaction reaching equilibrium until you have achieved your conversion.

CHAPTER 28

Typical Specific Questions

WHAT DO *CP* AND *CV* SIGNIFY?

An important point right in the beginning is solids and liquids because of their tight molecular network, can't perform thermodynamic work *PV*. Their total energy is their internal energy. But in the case of gases, they can expand when heated and therefore they consume a part of the total energy. *Cp* or the specific heat of a gas at constant pressure. The total energy of a gas is the sum of the internal energy, called *Cv* and work energy. *Cv*, the specific heat of a gas at constant volume. *Cv* symbolises the internal energy. Therefore, gases have two specific heats, *Cp* and *Cv*. While solids and liquids have just a specific heat.

Specific Heat Constant Pressure: *Cp*

Cp is the amount of heat energy released or absorbed by the unit mass of the substance with the change in temperature at constant pressure. In other words, it is the heat energy transfer at a constant pressure between a system and its surroundings.

Specific Heat Constant Volume: *Cv*

During a small change in the temperature of a substance, *Cv* is the amount of heat energy released or absorbed by the unit mass of the substance with the change in temperature at constant volume. In other words, it is the heat energy transfer between system and surroundings when there is no change in volume.

What is the Meaning of This?

Cp: When heat is transferred at a constant pressure with volume increasing, the gas expands. To create space for expansion the gas has to do *PV* work to push the surrounding. This is additional energy required by the gas compared to a situation where there is no expansion of gas.

Cv: At constant volume, since there is no expansion, there is no work, heat just goes to increase the internal energy of molecules. For gases, $Cp > Cv$. Because *Cp* signifies total energy and *Cv* represents a part of total energy which is internal energy.

Cp represents the change in total energy (H) concerning temperature (T) at constant pressure and is given by $Cp = dH/dT$. On the other hand, *Cv* represents the change in internal energy (U) concerning temperature at constant volume and is given by $Cv = dU/dT$.

The difference between *Cp* and *Cv* denoted as [*Cp* – *Cv*,] represents the work done (*PdV*) by an ideal gas at constant pressure. This relation comes from the first law of thermodynamics equation, $H = U + PV$, at constant pressure.

$$Cp/Cv = y = \text{Adiabatic index}$$

The ratio *Cp/Cv* signifies the internal energy available in an adiabatic process as a proportion to the total energy *CpT* at constant pressure. This ratio is also known as the adiabatic index for an ideal gas, and it is a measure of the amount by which the pressure and density of a gas vary in an adiabatic process. For diatomic gases, the adiabatic index is typically 1.4, meaning the internal energy available in an adiabatic process is $Cv = Cp/1.4$. It should be noted that *Cp* is always greater than *Cv*.

Thus, it can also be said that the heat capacity ratio is the ratio between the enthalpy to the internal energy: $y = Cp/Cv = H/U$ [Enthalpy divided by internal energy]

Cp/Cv's Role in the Control Valves and Compressors

Control valves play an important role in regulating the flow of gases in various industrial processes. When gases pass through a control valve, they often experience changes in pressure, temperature, and volume. Understanding the thermodynamic properties of the gas, such as *Cp* and *Cv*, is essential for optimizing the performance of control valves.

In the case of compressors, these devices are used to increase the pressure of gases, typically for transportation. During the compression process, gases experience temperature and pressure changes, which are influenced by their thermodynamic properties. *Cp* and *Cv* play a key role in determining the energy requirements and efficiency of compressors.

In adiabatic processes, where no heat exchange occurs with the surroundings, *Cp* and *Cv* are particularly important. The specific heat capacities of gases at constant pressure and volume provide insights into how gases respond to changes in pressure and temperature without heat transfer. The ratio of specific heats ($\gamma = Cp/Cv$) is a crucial parameter for describing the behavior of gases during adiabatic processes.

For an ideal gas undergoing adiabatic compression or expansion, the relationship between pressure, temperature, and volume is governed by the adiabatic equation: $P_1V_1^{\gamma} = P_2V_2^{\gamma}$. This equation illustrates how changes in pressure and volume affect the temperature of the gas during adiabatic processes.

The *Cv* value is utilized in calculating the work done during compression or expansion processes in compressors. The change in specific internal energy (ΔU) of the gas is determined using $\Delta U = Cv\ (T_2 - T_1)$. This energy change is directly related to the compression or expansion work performed by the compressor, highlighting the importance of understanding the thermodynamic properties of gases.

In conclusion, *Cp* and *Cv* are fundamental to the analysis of control valves and compressors, particularly in adiabatic processes. These properties help predict temperature, pressure, and energy changes during compression and expansion, enabling the optimization of system performance and efficiency in various industrial applications.

Specific Heat *vs* Latent Heat

Specific heat is held by the motion of atoms in a molecule. Latent heat is held by the intermolecular bonds of a cluster of molecules. Atomic bonds store the specific heat of a substance. Intermolecular bonds store the latent heat. Specific heat signifies total energy or enthalpy. Latent heat signifies the internal energy of a molecule. When a liquid is heated its entropy begins to increase. The liquid accommodates the increase in entropy by converting into vapor and the process proceeds until the saturation temperature reaches and all intermolecular bonds have broken.

Heat and Kinetic Energy

Heat and kinetic energy are two related but distinct concepts in the field of thermodynamics and physics. While both are forms of energy, they have different characteristics and play different roles in the behavior of matter.

Definition

Heat is the transfer of energy from one object to another due to a temperature difference. It is typically transferred through conduction, convection, or radiation.

Kinetic energy, on the other hand, is the energy of motion possessed by an object. It is a form of mechanical energy and depends on the object's mass and velocity.

Source and Form

- Heat originates from the internal energy of a system, which is the sum of its kinetic and potential energy. It is transferred between objects as a result of temperature gradients.
- Kinetic energy, on the other hand, is a type of energy associated with the motion of an object. It can exist in various forms such as translational, rotational, or vibrational kinetic energy.

Units

Heat is typically measured in units of joules (*J*) or calories (cal), with 1 calorie being the amount of heat required to raise the temperature of 1 gram of water by 1 degree Celsius.

Kinetic energy is also measured in joules (J) and depends on the mass and velocity of the object according to the formula $KE = 0.5 \times m \times v^2$, where m is the mass of the object and v is its velocity.

Effects on Matter

- Heat can cause changes in the temperature, phase, or chemical composition of a substance. For example, heating a solid can melt it into a liquid, or heating a gas can increase its pressure.
- Kinetic energy determines the speed and behavior of particles in a system. It influences the overall motion of objects and can be converted into other forms of energy, such as potential energy or thermal energy.

Relationship

- Heat and kinetic energy are related in the sense that heat transfer can result in an increase in the kinetic energy of particles within a system. This leads to an increase in the overall temperature of the system.
- However, it is important to recognize that heat and kinetic energy are distinct forms of energy that have different origins and effects on matter.

In summary, heat is a form of energy transfer related to temperature differences, while kinetic energy is the energy of motion possessed by an object. Understanding the distinction between these two forms of energy is essential for comprehending the behavior of matter in various physical systems.

Heat and Work

Energy can be transferred from one system to another in two different ways: through work and through heat. In the study of thermodynamics, it's critical to understand the difference between work and heat. Whereas work is the movement of mechanical energy between two systems, heat is the transfer of thermal energy between systems. The way thermodynamic processes operate depends critically on the distinction made between microscopic motion (heat) and macroscopic motion (work). Although they are not the same, heat can be converted into work and vice versa (see the mechanical equivalent of heat). The second law of thermodynamics sets a limit on the quantity of heat that can be converted into work by introducing the concept of entropy, whereas the first law states that both heat and work contribute to a system's total internal energy.

Heat and Enthalpy

An energy transfer caused by a temperature differential is known as heat. The change in heat content in a system at constant pressure is known as enthalpy. Heat and enthalpy are only equivalent at constant pressure.

Heat and Internal Energy

Internal energy results from the motion of the particles that make up a substance and is defined as the total of a substance's kinetic and potential energy.

Heat is the energy that moves from one object to another and is typically produced when two objects have different temperatures.

Internal energy refers to kinetic and potential energy, whereas heat refers to the thermal energy in transit of an object. This is important to keep in mind when making this distinction.

Which law of thermodynamics connects heat and temperature?

The answer is the first law $U = Q - W$, U is internal energy, Q is heat and W is work.

Internal energy is kinetic energy + potential energy. KE is related to temperature by the expression $KE = 1/2$ kT. K is a constant.

Therefore, the average KE is proportional to temperature T.

But there is a catch here, only ideal gases have predominantly KE. Since there are no intermolecular attractions in ideal gases, they have practically no potential energy.

Therefore, the internal energy of ideal gases signifies temperature. It does not apply to real gases. Precisely the translational kinetic energy symbolises temperature. The linear collisions of moles increase temperature. This is how the first law links heat and temperature.

Which thermodynamic state function combines 1st and 2nd law of thermodynamics?

Free energy combines 1st and 2nd law of thermodynamics, $dG = dH + TdS$, H is enthalpy [1st law], S is entropy [2nd law] and T is temperature.

Heat and Free Energy

Two thermodynamic concepts—free energy and enthalpy—are used to describe how heat energy and chemical reactions in a thermodynamic system relate to one another. The quantity of work that a thermodynamic system is capable of performing is known as free energy, or thermodynamic free energy. Stated differently, the quantity of energy that is accessible within a thermodynamic system to carry out thermodynamic tasks is known as free energy. The total energy content of a thermodynamic system is represented by the thermodynamic quantity enthalpy, on the other hand. The primary distinction between enthalpy and free energy is that the former provides the total energy that can be used to carry out thermodynamic work, while the latter provides the total energy of a thermodynamic system after accounting for entropy losses which can be transformed into work.

Heat and Temperature

The first law defines heat. $dU = [Q - W]$, $Q = [dU + W]$. The heat in a system gets distributed between the internal energy and work. The translational kinetic or linear collisions of molecules produce hotness or cold ness in a body called temperature.

Temperature and Internal Energy

The fundamental equation is $KE = 1/2\ Kb\ T$ [Kb is Boltzmann constant]

$T = 2\ KE/Kb$, T is proportional to KE

This equation applies to ideal gases.

Internal energy consists of both kinetic and potential energy. In ideal gases there is very little potential energy. Ideal gases therefore, have predominantly the kinetic energy. Precisely, translational kinetic energy symbolizes temperature. Internal energy is not proportional to temperature for real gases.

Why do intermolecular bonds break at the liquid's saturation temperature?

This explains why a liquid boil at saturation temperature. The explanation is that at saturation temperature, all intermolecular bonds break, leading to a maximum increase in entropy. At saturation temperature, Gibbs free energy is at its lowest at constant pressure and temperature.

Detail

When a liquid is heated, intermolecular bond breaking occurs first, leading to the generation of entropy. The liquid's entropy rises with temperature. In its search for more space, the liquid begins to turn into vapour to accommodate the increase in entropy.

This initiates the liquid's phase change from liquid to vapour. This process continues at constant pressure and temperature until all intermolecular bonds have broken and the entropy reaches its maximum value, or Q evaporation/T.

Vapour and liquid maintain thermodynamic equilibrium with minimum Gibbs free energy throughout the phase transition, which is characterised by the dryness fraction of vapour increasing from zero to one at constant temperature and pressure at all times with Gibbs free energy near zero.

It may be noted that the Gibbs function G is pressure [enthalpy] and temperature [entropy] dependent. During phase change at constant temperature and pressure G remains constant which means $dG = 0$.

The pressure, temperature and composition, all of these state variables of both phases remain constant at phase change. This satisfies the requirements for thermodynamic equilibrium.

BASIC THERMODYNAMIC EQUATIONS

1. **Internal energy (*U*):** $U = Q - W$; the energy of a system that results from the motion and interactions of molecules within it.
2. **Enthalpy (*H*):** $H = U + PV$; the total amount of energy in a system plus the energy required to create the system's volume against external pressure.
3. **Heat (*Q*):** $Q = mc\Delta T$; the transfer of thermal energy between two bodies at different temperatures.
4. **Work (*W*):** $W = -P\Delta V$; the transfer of energy through the performance of mechanical work.
5. **Free energy (*G*):** $G = H - TS$; the energy in a system that can be used to do work at constant temperature and pressure.
6. **Entropy (*S*):** $\Delta S = Q/T$; a measure of the disorder or randomness of a system.

CONFLICTS IN THERMODYNAMICS

One of the greatest conflicts in thermodynamics was the controversy over the second law of thermodynamics. This law states that the entropy of a closed system will always increase over time, leading to the idea of the "arrow of time" and the irreversibility of certain processes. However, there have been debates and conflicts over the interpretation and implications of this law, with some arguing that it is not absolute and can be violated under certain conditions. This conflict has led to ongoing discussions and research in the field of thermodynamics.

The Conflict Still Exists

The conflict over the second law of thermodynamics still exists in the scientific community. While the second law is widely accepted as a fundamental principle of thermodynamics, there are still debates and discussions about its interpretation and implications in certain scenarios, such as in quantum mechanics and the behavior of complex systems. Researchers continue to explore and investigate the limits and extensions of the second law, leading to ongoing debates and conflicts in the field of thermodynamics.

Another Conflict

Another conflict in thermodynamics is the controversy surrounding the concept of entropy and its relationship to information theory. Some researchers argue that entropy is not just a measure of disorder or randomness in a system, as traditionally defined in thermodynamics, but also a measure of the information content or uncertainty in a system. This has led to debates about the connections between thermodynamics and information theory, and how the two fields can be integrated to provide a more comprehensive understanding of physical systems.

The ongoing discussions and disagreements regarding the interpretation of entropy and its implications for thermodynamics and information theory illustrate another significant conflict in the field.

ZERO-POINT ENERGY: UNDERSTANDING ELECTRON BEHAVIOUR IN ATOMS

Energy Never Dies

1. An atom is the smallest unit of matter that contains the properties of an element.
2. An atom consists of a nucleus, which contains protons and neutrons, surrounded by electrons in various energy levels.
3. Electrons are negatively charged particles that orbit the nucleus. Protons are positively charged particles located in the nucleus. Neutrons are neutral particles also located in the nucleus.
4. Electrons are responsible for bonding interactions and chemical reactions. Protons determine the element of the atom. Neutrons stabilize the nucleus.
5. Electrons are spread out around the nucleus in specific energy levels or shells. Protons and neutrons are tightly packed in the nucleus.
6. Electrons are held in specific energy levels around the nucleus by electrostatic attraction. They move in orbits based on their energy level.

7. Different energy levels in an atom are represented by letters such as *K, L, M,* etc. These levels represent the distance of the electron from the nucleus. These are called quantom levels of energy. .

   The energy levels in an atom are quantized, meaning they can only have specific discrete values or quantum levels.

   The energy levels in an atom are typically labeled with the letters *K, L, M,* etc., with K being the closest to the nucleus and M being the furthest away.
8. Energy levels determine the behavior of electrons and their interactions with other atoms.
9. Electrons can move up to higher energy levels when they absorb energy from an external source.
10. Excited electrons can return to their ground state or lower energy levels, releasing excess energy in the form of photons.
11. Even at absolute zero temperature, the ground state energy of an atom, which is mostly potential energy, still exists.
12. Energy never dies, even at absolute zero. It simply exists in different forms.
13. This concept is referred to as zero-point energy and is a fundamental principle of quantum mechanics.

APPLICATION OF ZERO POINT ENERGY IN THERMODYNAMICS AS REFERENCE POINT

In thermodynamics, the concept of zero point energy is used as a reference point for calculating the energy of a system. Zero point energy refers to the minimum possible energy that a system can have, even at absolute zero temperature

Explanation

Where ZPE is a reference temperature and why?

Following are not zero at absolute zero

1. Internal energy is not zero due to non-zero potential energy contributions.
2. Vibrational motion of molecules is minimized, but not necessarily zero due to zero-point energy and quantum mechanical effects.
3. Potential energy arising from intermolecular forces or interactions between particles may also not be zero.

Examples

Three typical cases.

Enthalpy Change

At absolute zero temperature, the kinetic energy of molecules in a system is minimized, but the potential energy arising from interactions between molecules (such as electrostatic forces, chemical bonding, etc.) can still contribute to the overall internal energy of the system.

This potential energy component can prevent the internal energy from reaching zero at absolute zero temperature.

The presence of non-zero potential energy at absolute zero allows for the retention of internal energy in the system, which is why enthalpy may still have a non-zero value at this temperature.

Enthalpy includes both the internal energy of the system and the energy associated with maintaining constant pressure, so it can account for the potential energy contributions even at extremely low temperatures.

Specific Heat

In the calculation of specific heat capacity, zero point energy is accounted for in the vibrational motions of the atoms or molecules in a substance.

At absolute zero temperature, the atoms or molecules in a substance do not come to a complete stop but continue to vibrate due to zero point energy.

These vibrational motions contribute to the overall internal energy of the substance and play a role in determining its specific heat capacity. When calculating specific heat capacity, the zero point energy of these vibrational motions is considered as it represents the baseline energy that the substance possesses at the lowest possible temperature.

By taking into account the zero point energy of vibrational motions, thermodynamic calculations can more accurately predict the behavior of substances at different temperatures and provide insights into their thermal properties.

Binding Energy

Zero point energy is considered as a reference point is in the calculation of the binding energy of a molecule or atom.

The binding energy is the energy required to break apart a molecule or atom into its individual components. To calculate the binding energy, one must take into account the zero point energy of the individual components when they are separated, as this represents the minimum energy that the system can have.

MASS TO ENERGY [JOULE] CONVERSION

In thermodynamics, mass is a fundamental property that plays an important role in understanding and analyzing the behavior of systems.

Mass is used to characterize the amount of material present within a system, which in turn affects its energy balance, heat transfer, and overall thermodynamic properties.

Mass is often considered in conjunction with other properties such as temperature, pressure, and volume to describe the state of a system.

The conservation of mass principle, which states that mass cannot be created or destroyed in an isolated system, is a key concept in thermodynamics that helps in analyzing and predicting the behavior of systems undergoing different processes.

In thermodynamic equations and calculations, mass is typically included as a parameter to account for the amount of substance involved in a particular thermodynamic process. For example, the mass of a gas in a system can influence its pressure, volume, and temperature, which are essential parameters in thermodynamic analyses.

Some Key Applications of Mass

Mass influences various thermodynamic properties and phenomena.

For instance

1. **Specific heat capacity**: Mass affects the specific heat capacity of a substance, which is the amount of heat required to raise the temperature of a unit mass of the substance by one degree Celsius. Different substances have different specific heat capacities, and mass is a crucial factor in determining the total amount of heat required to change the temperature of a system.
2. **Phase changes**: Mass can also influence phase changes in thermodynamic systems. For example, the amount of mass present in a substance can determine the temperature at which a phase transition, such as melting or boiling, occurs. The mass of a substance can also affect the amount of heat required for a phase change to take place.
3. **Energy transfer**: Mass plays a key role in energy transfer processes within thermodynamic systems. The amount of mass present in a system can affect the heat transfer mechanisms, such as conduction, convection, and radiation. Mass also influences the rate at which energy is transferred within a system, which is essential for understanding the thermal dynamics of a system.

Einstein's concept of "Mass Energy Equivalence"

Einstein introduced the concept of mass-energy equivalence in his theory of special relativity, which revolutionized our understanding of the relationship between mass and energy. In Einstein's famous equation, $E = mc^2$, he proposed that mass and energy are interchangeable, and that mass can be converted into energy, and vice versa.

Einstein's definition of mass was based on the idea that mass is a form of energy, and that a particle's rest mass contributes to its total energy content. By equating mass to energy in his equation, Einstein showed that mass and energy are fundamentally connected and can be converted into each other. This concept has had profound implications in the field of physics, particularly in understanding nuclear reactions and the behavior of particles at high speeds.

Einstein's recognition of the equivalence of mass and energy was motivated by his desire to explain the nature of the universe in a more unified and comprehensive way. By establishing this relationship, he provided a deeper understanding of the fundamental connections between matter and energy, and paved the way for new insights into the structure of the cosmos.

Conversion of Mass (Kg) to Energy (Joule)

In the context of Einstein's mass-energy equivalence, mass can be expressed in terms of energy using the units of energy. In Einstein's famous equation, $E = mc^2$, mass is typically measured in kilograms (kg) and energy is measured in joules (J). This means that mass can be converted into energy by multiplying it by the speed of light squared (c^2) to determine the equivalent energy content.

So, while mass traditionally has units of kilograms, in the context of the mass-energy equivalence, mass can be expressed in units of energy by using the appropriate conversion factor (c^2) to account for the relationship between mass and energy.

Kg to Joule conversion

The conversion of kilograms (kg) to joules (J) involves using Einstein's famous equation, $E = mc^2$, where E represents energy, m represents mass in kilograms, and c represents the speed of light in a vacuum.

To convert mass in kilograms to energy in joules, you can use the formula:

Energy (E) = mass (m) × (speed of light)^2

Here's the mathematical representation of the conversion:

1 kilogram (kg) of mass × (speed of light)^2 = energy in joules

Given that the speed of light in a vacuum is approximately 3×10^8 meters per second, the speed of light squared (c^2) is equal to $(3 \times 10^8)^2 = 9 \times 10^{16}$ (m^2/s^2).

INVERSION TEMPERATURE AND BOYLE'S TEMPERATURE

In one line Boyle's temperature is temperature at which van der Waals constants 'a' and 'b' compensate each other. Inversion temperature is the temperature at which Joule Thomson coefficient changes the sign.

Background

Ideal Gas

An ideal gas is a theoretical gas made up of point particles that don't interact with each other. The ideal gas law is a simplified equation of state that describes the behavior of ideal gases.

Real Gases

Real gases deviate from ideal behavior at low temperatures and high pressures because the intermolecular forces are more significant.

A real gas behaves most like an ideal gas at low pressure and high temperature because under these conditions, the intermolecular forces between gas molecules become negligible, and the volume occupied by the molecules themselves is insignificant compared to the total volume, closely resembling the assumptions of an ideal gas where no intermolecular forces exist and molecules have no volume.

Explanation

Low Pressure

When pressure is low, the gas molecules are spread far apart, minimizing interactions between them, which means the attractive forces between molecules are less significant.

High Temperature

At high temperatures, the kinetic energy of the molecules is much greater than the potential energy due to intermolecular forces, effectively negating their influence.

Key points about ideal gas behavior

No intermolecular forces

An ideal gas is defined as having no attractive or repulsive forces between its molecules.

Negligible molecular volume

In an ideal gas, the volume of the molecules is considered negligible compared to the total volume occupied by the gas.

Boyles Temperature

At Boyle's temperature a real gas behaves as ideal gas for wide range of temperature and pressure.

van der Waals equation

$$(P + an^2/V^2) \times (V\text{-}nb) = RT$$

Reason: At Boyle's temperature van der Waals constant 'a' and b compensate each other.

Boyle's temperature vs. Inversion temperature

Explanation

Boyle's temperature

The temperature at which normal gases behave like ideal gases. This is because there are no attractive or repulsive forces at this temperature.

Inversion temperature

The temperature at which non-ideal gases change from decreasing to increasing in temperature. This is because the sign of the temperature change flips at zero.

Relation between Boyle's temperature and Inversion temperature

There is no direct mathematical relationship between Boyle's temperature and inversion temperature.

However, both Boyle's temperature and inversion temperature are related to the behavior of gases and how they respond to changes in pressure and temperature.

While there may not be a specific mathematical formula that relates these two concepts directly, they are both important in understanding the properties of gases under specific conditions.

Boyle's temperature is determined based on the behavior of a gas according to Boyle's law ($P_1V_1 = P_2V_2$), For real gases at Boyle's temperature van der Waals constants "a" and "b" compensate each other.

While inversion temperature is determined based on the sign change in the Joule-Thomson coefficient of a gas.

THERMODYNAMICS OF METAL ALLOYING

Alloying of metals: Alloying metals is the process of mixing two or more chemical elements to create a new material, called an alloy. Alloys can be made from metals or nonmetals, but most often include at least one metallic element.

Alloys are formed in order to enhance the properties of the individual elements, such as increasing strength, hardness, or resistance to corrosion. They can also be used to create materials with specific properties, such as being lightweight or having a specific color.

In general, alloying is done for many reasons, typically to increase strength, increase resistance to corrosion or reduce cost.

Thermodynamics: Thermodynamic of metal alloying broadly involves (1) identification of stable phases and phase transformations with respect to temperature during the process of alloying (2) alloying reactions: Alloying is a process of combining two or more elements to form a solid solution with specific properties. Alloying reactions can involve the dissolution of one element in the crystal lattice of another, leading to the formation of a homogeneous alloy (3) diffusion: Diffusion plays a crucial role in the alloying of metals by enabling the homogenization of alloying elements within the matrix of the base metal. When two or more metals are combined to form an alloy, diffusion processes facilitate the movement of atoms or ions from the alloying elements into the base metal and vice versa and (4) energy transfer: During the manufacturing of alloys, energy transfer occurs in various forms, such as heat, mechanical work, and electromagnetic radiation, to facilitate the mixing of different elements and the formation of a homogeneous alloy.

The focus on this note is on the studies of Phase diagram of metal alloys: In metal alloys, phase diagrams provide valuable information about the different phases that can exist in the alloy, the temperature ranges at which these phases are stable, and the composition of the alloy at different temperature conditions. This information is crucial for understanding the microstructure and properties of the alloy, such as its strength, hardness, ductility, and corrosion resistance.

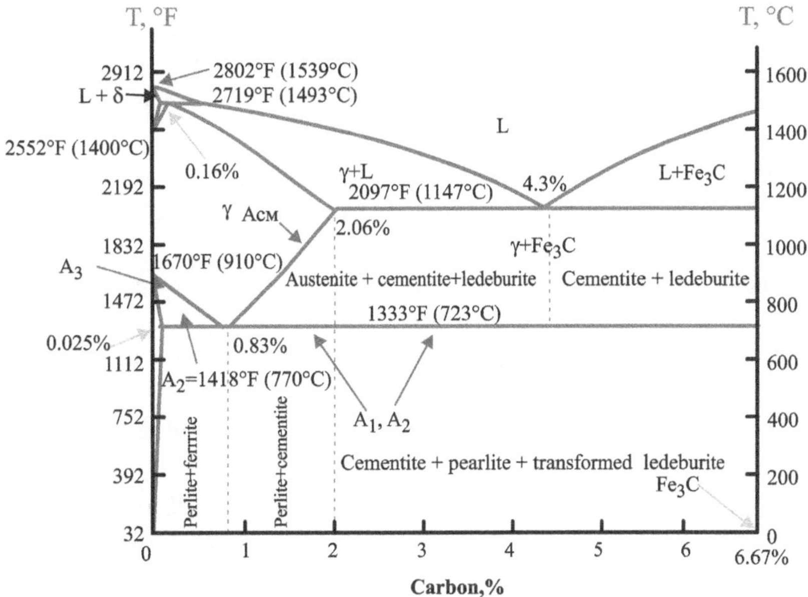

Figure 1: Fe - C (Iron - Carbon) Phase Diagram

Iron is a commonly studied material in the context of phase diagrams, particularly in the field of metallurgy. One of the most well-known phase diagrams for iron is the iron-carbon phase diagram, which shows the different phases that can exist in iron-carbon alloys as a function of temperature and carbon composition.

In the iron-carbon phase diagram, the two main elements of interest are iron (Fe) and carbon (C). The diagram typically consists of regions representing different phases such as austenite, ferrite, cementite, and various types of steels and cast irons.

Please refer to the phase diagram
Following phases are involved in the transformation, occurring with iron-carbon alloys:

Explanation

- **L** – Liquid solution of carbon in iron;
- **δ-ferrite** – Solid solution of carbon in iron.

Maximum concentration of carbon in δ-ferrite is 0.09% at 2719ºF (1493ºC) – temperature of the peritectic transformation. The crystal structure of δ-ferrite is BCC (cubic body centered).

- **Austenite** – Interstitial solid solution of carbon in γ-iron.

Austenite has FCC (cubic face centered) crystal structure, permitting high solubility of carbon – up to 2.06% at 2097ºF (1147ºC).

Austenite does not exist below 1333°F (723°C) and maximum carbon concentration at this temperature is 0.83%.

- **α-ferrite** – solid solution of carbon in α-iron.

  α-ferrite has BCC crystal structure and low solubility of carbon – up to 0.025% at 1333°F (723°C).
  α-ferrite exists at room temperature.

- **Cementite** – iron carbide, intermetallic compound, having fixed composition Fe_3C.

  Cementite is a hard and brittle substance, influencing on the properties of steels and cast irons.

The following phase transformations occur with iron-carbon alloys:

Alloys, containing up to 0.51% of carbon, start solidification with formation of crystals of δ-ferrite. Carbon content in δ-ferrite increases up to 0.09% in course solidification, and at 2719°F (1493°C) remaining liquid phase and δ-ferrite perform peritectic transformation, resulting in formation of austenite.

Alloys, containing carbon more than 0.51%, but less than 2.06%, form primary austenite crystals in the beginning of solidification and when the temperature reaches the curve ACM primary cementite stars to form.

Iron-carbon alloys, containing up to 2.06% of carbon, are called steels.

Alloys, containing from 2.06 to 6.67% of carbon, experience eutectic transformation at 2097°F (1147°C). The eutectic concentration of carbon is 4.3%.

In practice only hypoeutectic alloys are used. These alloys (carbon content from 2.06% to 4.3%) are called cast irons. When temperature of an alloy from this range reaches 2097°F (1147°C), it contains primary austenite crystals and some amount of the liquid phase. The latter decomposes by eutectic mechanism to a fine mixture of austenite and cementite, called **ledeburite**.

All iron-carbon alloys (steels and cast irons) experience eutectoid transformation at 1333°F (723°C). The eutectoid concentration of carbon is 0.83%.

When the temperature of an alloy reaches 1333°F (733°C), austenite transforms to pearlite (fine ferrite-cementite structure, forming as a result of decomposition of austenite at slow cooling conditions).

Critical temperatures

- **Upper critical temperature (point)** A_3 is the temperature, below which ferrite starts to form as a result of ejection from austenite in the hypoeutectoid alloys.
- **Upper critical temperature (point)** A_{CM} is the temperature, below which cementite starts to form as a result of ejection from austenite in the hypereutectoid alloys.
- **Lower critical temperature (point)** A_1 is the temperature of the austenite-to-pearlite eutectoid transformation. Below this temperature austenite does not exist.
- **Magnetic transformation temperature** A_2 is the temperature below which α-ferrite is ferromagnetic.

Phase compositions of the iron-carbon alloys at room temperature

- **Hypoeutectoid steels** (carbon content from 0 to 0.83%) consist of primary (proeutectoid) ferrite (according to the curve A_3) and pearlite.
- **Eutectoid steel** (carbon content 0.83%) entirely consists of pearlite.
- **Hypereutectoid steels** (carbon content from 0.83 to 2.06%) consist of primary (proeutectoid)cementite (according to the curve A_{CM}) and pearlite.
- **Cast irons** (carbon content from 2.06% to 4.3%) consist of proeutectoid cementite C_2 ejected from austenite according to the curve A_{CM}, pearlite and transformed ledeburite (ledeburite in which austenite transformed to pearlite).

Summary

Phases

Explanation

- **L** – Liquid solution of carbon in iron.
- **δ-ferrite** – Solid solution of carbon in iron
- Austenite – interstitial solid solution of carbon in γ-iron. Y gamma iron is a stable phase at high temperatures and is one of the key phases in iron-carbon alloy systems, along with alpha iron (α-Fe) and delta iron (δ-Fe).
- **α-ferrite** – solid solution of carbon in α-iron.
- **Cementite** – iron carbide, intermetallic compound, having fixed composition Fe_3C.
- **Hypoeutectoid, eutectoid and hypereutectoid**

The term "eutectoid" comes from the word "eutectic," which refers to a specific composition in a phase diagram where a liquid phase separates into two different solid phases upon cooling. In the context of steel, the eutectoid composition represents a specific carbon content (around 0.76%) in the iron-carbon phase diagram where austenite transforms into pearlite upon cooling.

The eutectoid transformation in steel is analogous to a eutectic transformation in other material systems, where a single phase (austenite in the case of steel) transforms into two distinct phases (ferrite and cementite in the case of pearlite) at a specific composition and temperature. The naming convention simply reflects this similarity in the phase transformation behavior between eutectic and eutectoid systems.

Thermodynamics

Challenges

The process of metal alloying involves mixing two or more metals to form a new alloy. This process can be described in terms of enthalpy (H), entropy (S), and free energy (G) changes.

- **Enthalpy (ΔH):** The enthalpy change during alloying refers to the heat absorbed or released during the mixing process. It depends on the specific metals being alloyed and their respective enthalpies of mixing. The enthalpy change can be positive (endothermic) or negative (exothermic), depending on whether heat is absorbed or released during alloying.
- **Entropy (ΔS):** The entropy change reflects the disorder or randomness in the alloying process. Upon mixing metals, the randomness generally increases, leading to a positive entropy change. This increase in entropy contributes to the thermodynamic driving force for alloy formation.
- **Free energy (ΔG):** The free energy change is a measure of the overall spontaneity of the alloying process. The Gibbs free energy change (ΔG) takes into account both the enthalpy and entropy changes. A negative ΔG indicates that the alloying process is favorable and spontaneous; energy is released and the disorder increases.

Thermodynamic Challenges

Obtaining a sufficient free energy change for a spontaneous process is a challenge in commercial alloying. This limitation affects the choice of alloying elements by considering that the enthalpy change should not exceed the product of temperature (T) and entropy change (ΔS), which is denoted as $T\Delta S$.

The challenge in alloying lies in finding alloying elements that contribute to the desired properties without causing a large enthalpy change that cannot be compensated by the entropy change, limiting the feasibility of the process.

Role of entropy: High entropy increases the more random distribution of the energy. In essence . . . "a system becomes more stable when its energy is spread out in a more disordered state" free from any internal stress.

A system that is more disordered in space will tend to have more disorder in the way the energy is arranged as well. In essence "a system becomes more stable when its energy is spread out in a more disordered state".

High entropy can mitigate the issue of obtaining enough free energy for alloy formation to some extent by incorporating a large entropy contribution.

The high entropy contribution increases the overall entropy change (ΔS) in the free energy equation ($\Delta G = \Delta H - T\Delta S$), allowing for a larger enthalpy change (ΔH) without making the free energy change positive.

AZEOTROPE THERMODYNAMICS

Summary

One can look at an Azeotrope from two perspectives.

The first perspective involves the formation of an azeotrope when two non-ideal liquids form a mixture with a boiling point that is different from that predicted by Raoult's Law due to intermolecular attractions between the molecules. This can result in the formation of a mixture with a constant boiling point, known as an azeotrope.

The second perspective involves looking at the azeotrope formation in terms of maximizing entropy and minimizing free energy to achieve thermodynamic equilibrium. When two liquids are mixed to form an azeotrope, the entropy of the system is increased as the degree of disorder in the system increases. At the same time, the free energy of the system is minimized to reach a stable state of equilibrium. The formation of an azeotrope represents a balance between maximizing entropy and minimizing free energy to produce an inseparable mixture of the two liquids.

Explanation

Azeotropes are a special type of mixture of two or more liquids that have constant boiling points and compositions. In other words, an azeotrope is a mixture of liquids that behaves like a single substance when it comes to boiling, such that the vapor produced has the same composition as the liquid mixture.

Thermodynamics plays an important role in the creation of azeotropes as it governs the behavior of the components in the mixture under changing conditions.

Azeotropes formation depends on a combination of factors including enthalpy, entropy, and molecular interactions between the components.

The formation of an azeotrope occurs when the composition of the vapor phase is the same as that of the liquid phase, leading to a constant boiling point mixture. This means that the vapor and liquid phases are in equilibrium at a specific composition, known as the azeotropic composition. At this composition, the components cannot be separated by simple distillation.

The formation of an azeotrope is a result of the specific interactions between the components in the mixture. These interactions can lead to deviations from ideal behavior and can impact the thermodynamic properties of the system.

The enthalpy of mixing, which reflects the heat released or absorbed during mixing, can also play a significant role in the formation and stability of azeotropes.

At a fundamental level, the formation of an azeotrope is driven by the intermolecular forces between the molecules of the components in the mixture. These forces determine how the components interact with each other and with the surrounding environment, which ultimately influences the behavior of the mixture.

One of the key thermodynamic properties that come into play in the creation of azeotropes is the vapor-liquid equilibrium (VLE) behavior of the components. At a certain temperature and pressure, the components in the mixture reach a point where the composition of the vapor phase is the same as that of the liquid phase. This leads to the formation of an azeotrope, which is characterized by a constant boiling point and composition.

Additionally, thermodynamics also governs the phase behavior of the components in the mixture, such as their solubility in each other and the enthalpy of mixing. These properties determine how the components interact and coexist in the mixture, which can influence the formation of an azeotrope.

Role of Entropy and Free Energy

The entropy and free energy play important roles in the creation of azeotropes.

Entropy, which is a measure of the disorder or randomness of a system, influences the mixing behavior of the components in the mixture. When two liquids are mixed together, their entropy increases due to the increased number of ways the molecules can arrange themselves in the mixture. This increase in entropy can either favor or disfavor the formation of an azeotrope, depending on the specific interactions between the components.

Free energy, on the other hand, determines the spontaneity of the mixing process. Azeotropes are formed when the Gibbs free energy of mixing is minimized, indicating that the components are favorably interacting with each other in the mixture. The free energy change associated with mixing can be influenced by the enthalpy of mixing (which reflects the heat released or absorbed during the mixing process) and the entropy of mixing.

In the context of azeotropes, entropy and free energy can affect the phase behavior of the components in the mixture, their solubility, and the stability of the azeotropic composition. Overall, entropy and free energy are crucial thermodynamic parameters that govern the formation and properties of azeotropes.

At the azeotropic composition, the entropy of mixing is maximized, leading to a more stable equilibrium between the liquid and vapor phases. This means that the system is in a state of maximum disorder.

Similarly, the Gibbs free energy reach a minimum point, indicating a stable equilibrium between the phases.

The tendency for the two phases to not separate is due to the fact that the system reaches a point of thermodynamic equilibrium where the composition in both phases is identical, preventing further separation of the components through distillation.

ROLE OF ENTROPY IN PSYCHROMETRY: 'ENTROPY RULES PSYCHROMETRY'

"Give a salute to Clausius and Boltzmann for introducing the concept of Entropy, which has greatly enhanced our understanding and explanations in various fields, particularly in Psychrometry."

It is not exaggeration to say Entropy rules Psychrometry.

Entropy influences the direction and efficiency of processes in psychrometry, making it an important factor to consider when studying air and water vapor mixtures.

Entropy reaches maximum in Psychrometry at saturation. The specific volume of wet air is maximum at saturation because the density of wet air is less than dry air. This is one of the fundamental understandings of Psychrometry.

This makes Gibbs free energy near zero and makes the two phases in equilibrium.

This is the fundamental answer to the thermal equilibrium at saturation point that makes DBT WBT and Dew point temperature same at saturation. All thermodynamic potentials reaches minimum at saturation point.

At saturation, the air-water vapor system is at its maximum entropy because it represents a state of maximum disorder and randomness. The specific volume of wet air is maximum at saturation because the density of wet air is less than dry air. Correspondingly, the system is at minimum free energy in a state of thermal and thermodynamic equilibrium.

The Helmholtz free energy, Gibbs free energy, enthalpy, and internal energy are thermodynamic potentials that are at a minimum at thermodynamic equilibrum.

Furthermore, at the dew point temperature, the rates of evaporation and condensation are equal, maintaining a state of equilibrium between the vapor and liquid phases. This equilibrium ensures that temperature remains constant during the phase change.

During condensation, kinetic energy of the water vapor molecules decreases as they come into closer proximity, leading to the formation of potential energy stored in the intermolecular bonds as they reunite to form liquid water.

Why Wet Air Can't be Cooled Below Dew Point

When air is cooled to its dew point temperature, saturatin will occur as the air reaches its saturation point.

Any further cooling will result in a reduction in the vapor pressure, which will reduce the rate of condensation

This equilibrium between evaporation and condensation processes maintains the dew point temperature as the lowest temperature at which air can be cooled while still maintaining a state of thermal equilibrium.

Explanation

Below the dew point temperature, if the air is cooled further, the reduction in vapor pressure with temperature will lead to a decrease in the rate of condensation.

This reduction in condensation will eventually restore equilibrium by ensuring that the rate of evaporation is once again equal to the rate of condensation.

This dynamic balance between evaporation and condensation processes helps to maintain the thermal equilibrium at the dew point temperature.

A NOTE ON CRYOGENICS: H_2 GAS AS AN EXAMPLE

What is Cryogenics

Cryogenics is the branch of physics that deals with the production and effects of very low temperatures. The purpose of cryogenics is to create and maintain extremely low temperatures to study the behavior of materials at these temperatures, as well as to store and transport materials that require low temperatures for preservation.

Cryogenic temperatures are typically below –150 degrees Celsius (–238 degrees Fahrenheit) and can sometimes approach close to –273 degrees Celsius (–460 degrees Fahrenheit), which is absolute zero.

Achieving and maintaining such low temperatures involves the use of special materials and equipment that can withstand the extreme cold.

Process

Cryogenic temperatures are obtained through a process called cryogenic cooling, which involves the liquefaction and subsequent cooling of gases such as nitrogen, helium, and hydrogen. These gases can be liquefied at normal atmospheric pressure through compression and cooling. Once in liquid form, they are further cooled to achieve cryogenic temperatures.

Thermodynamics

The thermodynamic process involved in cryogenics is the Joule-Thomson effect, which is the process in which a gas is allowed to expand through a valve and cools as a result of the decrease in pressure. This cooling effect is used to lower the temperature of gases to cryogenic levels.

Some of the thermodynamic properties that change in cryogenic temperatures include entropy, enthalpy, and specific heat capacity. At low temperatures, the molecular motion of gases slows down, leading to a decrease in their entropy and enthalpy. This makes the gases more stable and less likely to interact with their surroundings. The specific heat capacity of gases also decreases at cryogenic temperatures, meaning that it takes less energy to raise their temperature.

In other words, in cryogenics, the low enthalpy and entropy take the liquid phase far away from equilibrium and provide security against reverting to the gas phase. The low specific heat allows for easy conversion to the gas phase without requiring much heat.

Liquefaction of H_2 Gas

Cryogenic cooling of hydrogen involves lowering the temperature of hydrogen gas to very low temperatures, below its inversion temperature where JT coefficient changes sign which is about 205k or – 68 degc. Since H_2's inversion temperature is below ambient temperature and at ambient temperature H_2's JT coefficient is negative it is cooled below inversion temperature.

Explanation and Thermodynamics

If hydrogen gas is not cooled below its inversion temperature, the compressibility factor of H_2 remains greater than 1, indicating the dominance of repulsive forces. This causes hydrogen to exhibit a negative Joule-Thomson coefficient. Cooling the gas below the inversion temperature allows hydrogen to enter a region where the Joule-Thomson coefficient changes sign from negative to positive. Once the Joule-Thomson coefficient becomes negative, it provides cooling mainly due to the reduction of kinetic energy, bringing molecules closer together and allowing intermolecular attractive forces to dominate. This change in the sign of the Joule-Thomson coefficient is essential for effective cooling during the expansion process, ultimately leading to the achievement of lower cryogenic temperatures.

Process

Obtainment of cryogenic temperatures: Cryogenic temperatures are obtained through various methods such as using cryocoolers, cryogenic liquids (such as liquid nitrogen or helium), or cryogenic refrigeration systems. These systems work by extracting heat from the hydrogen gas to lower its temperature.

Thermodynamic Process: The cryogenic cooling process of hydrogen involves several thermodynamic principles:

- **Compression**: The hydrogen gas is compressed to increase its temperature and pressure. This process increases the internal energy of the gas.
- **Cooling**: The compressed hydrogen gas is then cooled using cryogenic methods to lower its temperature. As the gas is cooled, its internal energy decreases, causing the molecules to slow down and eventually condense into a liquid form.
- **Expansion:** The cooled and liquefied hydrogen is then expanded to reduce its pressure and temperature further. This expansion process allows the hydrogen to reach even lower temperatures, nearing cryogenic levels.
- **Joule-Thomson Effect:** During expansion, the Joule-Thomson effect occurs, which states that a gas will either heat up or cool down when expanded without exchange of heat with the surroundings, depending on the gas's initial temperature and pressure. In the case of cryogenic cooling of hydrogen, the expansion results in a significant temperature decrease.

Changes in Thermodynamic Properties

During the cryogenic cooling process of hydrogen, several thermodynamic properties change:

- **Temperature**: The temperature of the hydrogen gas decreases significantly as it is cooled to cryogenic levels, reaching temperatures below –150 degrees Celsius.
- **Pressure:** The pressure of the hydrogen gas may vary during the compression and expansion stages of the process, but it is typically reduced as the gas is cooled and liquefied.
- **Phase Change:** The hydrogen gas undergoes a phase change from a gas to a liquid state as it reaches cryogenic temperatures. This phase change is accompanied by a release of latent heat.

In conclusion, the cryogenic cooling of hydrogen involves lowering the temperature of hydrogen gas to very low levels below its inversion temperature using various thermodynamic processes and then expansion.

CHAPTER 29

Thermodynamics in Chemical Processes

STRIKING A BALANCE: THE ART OF PROCESS SELECTION IN ENGINEERING

As a process engineer, the responsibility of choosing a process is an important aspect of our job. It is not simply a matter of selecting a process that looks good on paper or has been successful in the past. There are numerous factors that must be taken into consideration before deciding on a particular process for a given project.

One of the primary considerations when choosing a process is thermodynamics. This involves determining if the reaction or separation process is feasible based on the laws of thermodynamics. This includes considerations such as heat transfer, energy requirements, and equilibrium conditions. It is essential to ensure that the process chosen is thermodynamically sound and will result in the desired products.

Kinetics is another important factor that must be considered when selecting a process. This involves studying the rates of reactions and how they are affected by various factors such as temperature, pressure, and catalysts. Understanding the kinetics of a reaction is crucial in determining the optimal conditions for running the process efficiently and effectively.

In addition to thermodynamics and kinetics, process engineers must also evaluate the thermodynamic processes involved in the chosen process. This includes considerations such as heat and mass transfer, fluid flow, and reaction mechanisms. It is essential to understand how these processes work together to ensure the successful operation of the chosen process.

Safety is another key consideration when choosing a process. Process engineers must evaluate potential hazards and risks associated with the operation of the chosen process. This includes considerations such as handling of hazardous materials, reaction runaway, and equipment failures. It is essential to design the process in a way that minimizes risks and ensures the safety of personnel and the environment.

Environmental considerations are also important when selecting a process. Process engineers must evaluate the potential impact of the chosen process on the environment, including factors such as emissions, waste generation, and resource consumption. It is essential to choose a process that is environmentally sustainable and minimizes its impact on the surrounding ecosystem.

Finally, cost considerations are also a crucial factor in the process selection process. Process engineers must evaluate the cost of implementing and operating the chosen process, as well as the potential return on investment. It is essential to choose a process that is cost-effective and provides the desired results within budget constraints.

In conclusion, the process engineer plays a critical role in selecting the appropriate process for a given project. By carefully considering factors such as thermodynamics, kinetics, thermodynamic processes, safety, environment, and cost, process engineers can ensure the success of the chosen process and the overall project. It is essential to approach process selection with a thorough understanding of these factors and a commitment to finding the best solution for each unique situation.

HEAT ENGINE

It's a device that transforms heat into useful work. A portion of the heat input from the hot reservoir is converted into work in a heat engine, and the remaining heat is released as waste heat into the cold reservoir.

CARNOT CYCLE

Carnot's focus in developing his heat engine was to maximize efficiency by minimizing irreversibilities and achieving a reversible process. By selecting the four arms of his machine to represent four distinct processes (isothermal expansion, adiabatic expansion, isothermal compression, and adiabatic compression), Carnot aimed to create an idealized cycle that could serve as a benchmark for the maximum possible efficiency of a heat engine.

In a reversible process, the system can be returned to its original state without loss of energy, thereby maximizing efficiency. By carefully designing the Carnot cycle to incorporate only reversible processes and minimize irreversibilities such as friction and heat loss, Carnot demonstrated the theoretical maximum efficiency that could be achieved in converting heat into work.

One of the key aspects of the Carnot cycle is that it is a closed cycle where the system returns to its initial state at the end of the cycle. This means that there is zero net change in internal energy from the beginning to the end of the cycle. This property of the Carnot cycle highlights the reversibility of the processes involved and distinguishes it as an idealized and efficient heat engine cycle.

By constructing a cycle in which the system undergoes a series of processes and returns to its original state with zero net change in internal energy, Carnot demonstrated the potential for maximizing efficiency and minimizing energy losses in a heat engine. This concept of a closed cycle with zero internal energy change between the start and end points is a defining characteristic of the Carnot cycle and underlines its significance as a benchmark for idealized thermodynamic processes.

What is Carnot Cycle

Carnot cycle is based on the principle of reversible processes as it goes through a cycle. When a fluid passes through various processes and then eventually returns to the same state it began with, the system is said to have undergone a cyclic process. For a reversible process, no change occurs in the total entropy produced, as said above i.e., the entropy of the system plus the entropy of the surroundings, ΔS [total] = 0.

Four Stages in the Carnot Cycle

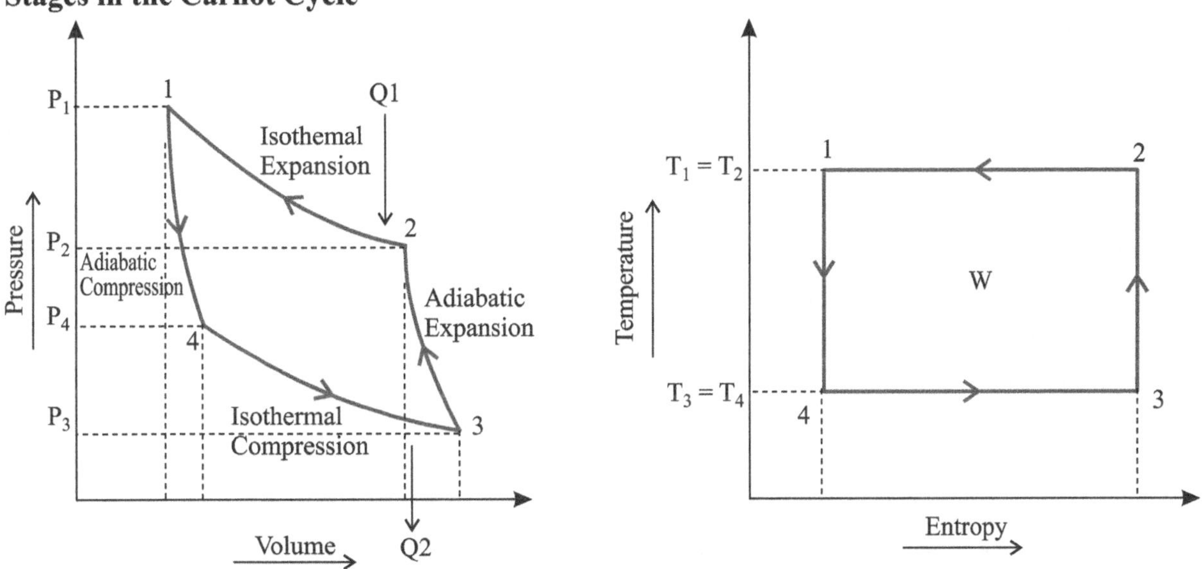

Figure 1: *P-v* and *T-s* Diagrams of a Carnot

Point 1–2: Isothermal Expansion. Heat is transferred reversibly from the high-temperature reservoir at constant temperature T_1 (isothermal heat addition or absorption). During this step the gas is allowed to expand, doing work on the surroundings. Although the pressure drops from points 1 to 2 the temperature of the gas does not change during the process because it is in thermal contact with the hot reservoir at T_1, and thus the expansion is isothermal.

Point 2–3: Isentropic (reversible adiabatic) expansion of the gas (isentropic work output). For this step the gas in the engine is thermally insulated from both the hot and cold reservoirs. Thus, they neither gain nor lose heat, an 'adiabatic' process. The gas continues to expand by reduction of pressure, doing work on the surroundings and losing an amount of internal energy equal to the work done. The gas expansion without heat input causes it to cool to the "cold" temperature, T_3.

Point 3–4: Isothermal Compression. Heat transferred reversibly to low temperature reservoir at constant temperature T_3. (Isothermal heat rejection) (3 to 4 on now the gas in the engine is in thermal contact with the cold reservoir at temperature T_3. The surroundings do work on the gas causing an amount of heat energy Q_2 to leave the system to the low temperature reservoir.

Point 4–1: Adiabatic reversible compression. Once again, the gas in the engine is thermally insulated from the hot and cold reservoirs, and the engine is assumed to be frictionless, hence reversible. During this step, the surroundings do work on the gas. its internal energy, compressing it, and causing its temperature to rise back to T_1 due solely to the work added to the system, but the entropy remains unchanged. At this point the gas is in the same state as at the start of step 1.

Efficiency Calculation

There is a misconception that hot and cold reservoirs do not produce entropy. Both reservoirs produce entropy.

The hot reservoir generates positive entropy and the cold reservoir produces exactly the same amount of negative entropy, therefore there is no net loss of energy from either of the cold or hot reservoirs.

The two adiabatic reversible stages do not produce entropy and these isentropic stages can be easily understood.

Calculation

Qh is hot reservoir enthalpy. *Th* is hot reservoir temperature, *Qc* is cold reservoir enthalpy, *Tc* is cold reservoir temperature.

From the first law of thermodynamics

$$Qh = Qc + W \qquad \dots (1)$$

From the second law of thermodynamics

$$Qh/Th + Qc/Tc = 0 \qquad \dots (2)$$

[since the process is reversible]

Assumptions

Let $Qh = 2749$ kj/kg and $Th = 300$ degc [from steam table]

Let cold reservoir temperature = 30 degc

From equation [2]

$$2749/573 + Qc/303 = 0$$

$$4.79 + Qc/303 = 0$$

$$Qc = -4.79 \times 303 = -1453 \text{ kj/kg}$$

From the equation [1] $W = Qh - Qc = 2749 - [-1453] = 4202$ kj

Efficiency % = $[Th-Tc]/Th = [573-303]/573 = 47\%$

Entropy in hot reservoir = + 2749/573 = + 4.79 kj.kg-degc

Entropy in cold reservoir = –1453/303 = –4.79 kj/kg-degc

So, both hot and cold reservoirs generate entropy. The hot reservoir generates positive entropy and the cold reservoir produces exactly the same negative entropy, therefore there is no net loss of energy from either of the cold or hot reservoirs.

The first fundamental point in a Carnot cycle is that it only works for ideal gases without potential energy. Ideal gases have no intermolecular attraction to generate potential energy. Ideal gases have only kinetic energy, kinetic energy = 1/2 *Kb T*. kinetic energy is proportional to temperature *T*. *kb* is a constant. The internal energy can be equated to temperature in isothermal processes such as in the Carnot cycle only when the energy is kinetic. In summary, temperature is proportional to energy when the energy is kinetic energy that is provided by only ideal gasses.

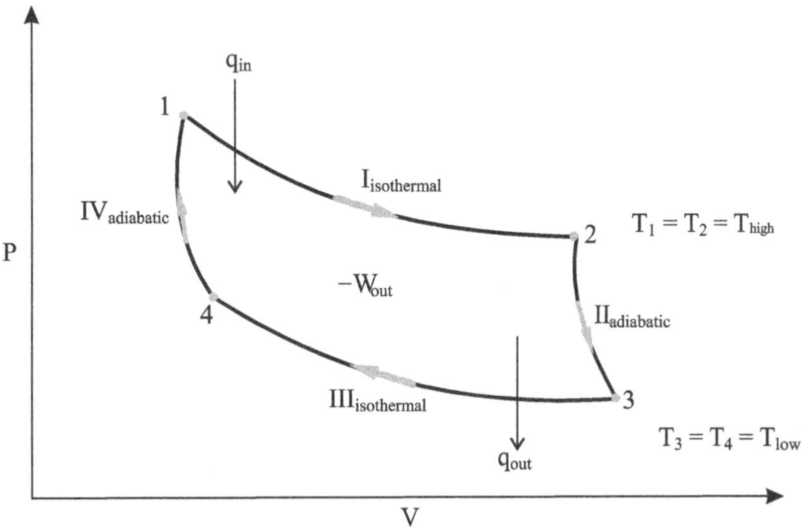

Figure 2: Thermodynamic of Carnot Cycle

$Cv \times dT = dU$ (internal energy change)

$Cp\ dT - Cv\ dT = R\ dT$ (R is gas constant). $R\ dT$ is the work. $dH - dU$ = delta Work (1st law of thermodynamics).

The hot reservoir supplies the total energy H in stage 1. The internal energy U is consumed in stage 2, in the adiabatic expansion of the gas to perform the work. That leaves behind the work energy. Work energy transfers to stage 3 in the isothermal compression of the steam. The steam condenses to liquid by releasing the latent heat. This is the cold reservoir.

Note, that an adiabatic process can only transfer work. Adiabatic expansion (stage 2) to isothermal compression (stage 3) is work transfer. The larger the $(dH - dU)$ the larger is the work performed in the cycle. $(dH - dU)$ produces 'work' in the Carnot cycle. This is the role of Cp and Cv in a Carnot cycle to generate "work" between hot and cold reservoirs.

Free Energy in Carnot Heat Engine

There are no irreversibilities in a Carnot cycle. In thermodynamics, energy losses are associated with irreversibilities In summary, the change in internal energy in a Carnot cycle, $dU = 0$.

From the first law of thermodynamics, $U = Q - W$ with dU being zero, $Q = W$, Q is heat, U is internal energy and W is work. This means all heat that is provided in an ideal reversible Carnot engine is spent in performing useful work.

From the second law of thermodynamics, the free energy change $dG = dH - TdS$ Since $TdS = 0$, T is temperature, H is enthalpy and S is entropy. $dS = 0$. The change in free energy $dG = dH$ or total energy. This means a Carnot heat engine can deliver all the input energy in performing useful work. In other words a Carnot heat engine has the maximum free energy to perform useful work. Often it is debated if Carnot machine can have 100% efficiency, $(1 - Tc/Th) \times 100$. The answer is no. Because both maximum Th that is infinity and minimum Tc, that is absolute zero temperature are not achievable.

Well, one should not confuse with Carnot's mission for developing a benchmark heat engine. Carnot simply provided means for an ideal reversible cyclic machine with no exergy losses.

Note Further

Infinity is not a number. Zero/Infinity or Infinity/Zero is not defined. Carnot cycle only says if you have a cyclic reversible machine the machine converts 100% of the input heat into work with zero exergy losses.

RANKINE CYCLE

The Rankine cycle, named after Scottish engineer William John Macquorn Rankine, is a thermodynamic cycle that is commonly used in power plants to generate electricity. It is a modification of the Carnot cycle, which is the most efficient cycle possible for a heat engine operating between two temperature extremes.

One notable difference between the Rankine cycle and the Carnot cycle is the addition of a phase change process in the Rankine cycle. This phase change process typically involves the use of a liquid-vapor mixture (such as water and steam) in the cycle, allowing for more efficient energy transfer between the heat source and the working fluid.

In the Rankine cycle, the working fluid is typically pumped to a higher pressure before entering the boiler, where heat is added to convert the liquid into a vapor. The vapor then expands through a turbine to produce work, and the remaining vapor is condensed back into a liquid in the condenser before being pumped back to the boiler to repeat the cycle.

This addition of the phase change process in the Rankine cycle allows for greater energy transfer efficiency compared to the Carnot cycle, as the latent heat of vaporization can be utilized to increase the efficiency of the cycle. Additionally, the Rankine cycle is more practical for real-world applications, as it allows for the use of water/steam as the working fluid, which is readily available and environmentally friendly.

Overall, the Rankine cycle represents a significant advancement in the field of thermodynamics and heat engine design, allowing for the efficient conversion of heat into mechanical work in power generation applications. Its versatility and practicality have made it a cornerstone of modern power generation technology.

Construction of Rankine Cycle

The Rankine cycle is a representation used to predict the performance of steam turbine systems.

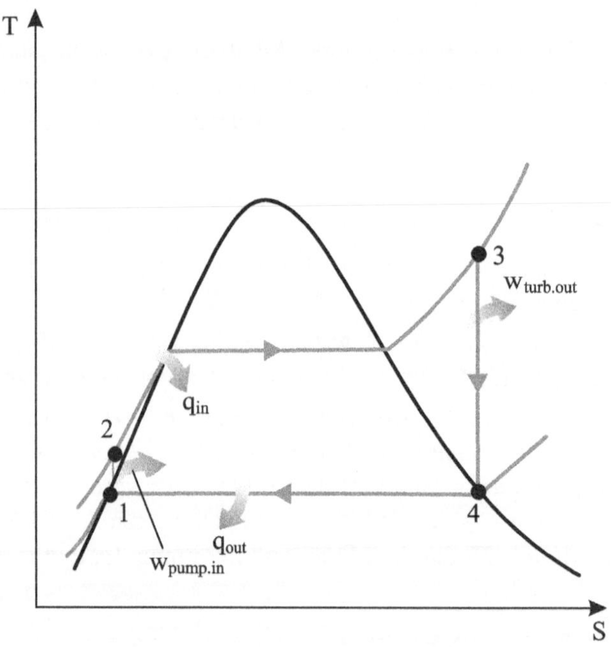

Figure 3: Rankine Cycle on Temperature-Entropy Diagram

Let us follow the above *T–S* diagram. Let us assume that the boiler feed pump is located at point (1) on the above diagram.

The boiler feed pump (1) does work on the water increasing its pressure and temperature adiabatically. The high-pressure water flows into the steam boiler (2).

The entropy increase through the pump is neglected (1–2).

The liquid water is warmed up at constant pressure until it reaches the superheated steam temperature (2–3).

- **Isentropic Expansion:** The superheated steam expands through flowing the turbine (3-4) at constant entropy.
- **Isobaric Heat Rejection**. Along the isobar (6–1) the heat content of the wet steam is rejected in the condenser.
- **Isentropic Compression**: Condensed liquid flows to the suction side of the boiler feed pump and the cycle begins again.
- **Entropy Creation:** Maximum entropy is created along (2–3) when water converts to steam.

Numerous studies have indicated that neither the rate of heat input nor the power output in a thermal power plant is treated as a fixed parameter. Minimizing the entropy generation does not lead to an improved thermal efficiency. It is found that the thermal efficiency inversely correlates with specific entropy generation, and minimization of specific entropy generation is identical to maximization of thermal efficiency. Specific entropy generation is defined as the total entropy generation rate per unit flowrate of the fuel.

CARNOT AND RANKINE: THE MISSIONS OF TWO LEGENDS

In one line, Carnot cycles conceived $dE = 0$ while Rankine cycle works on dP = constant.

Carnot and Rankine had different missions when they conceived their heat engines. Carnot wanted to conceive an ideal machine that could be the benchmark for efficiency. Rankine conceived a practical machine that could run in real life. Carnot conceived internal energy change $dE = 0$, $Q = W$. Rankine removed the constraints of Carnot and conceived dP = Constant.

Carnot Engine

The purpose of Carnot in conceiving a hypothetical heat engine was to get all heat available as work or maximise work in the machine. He realized it was only possible in a reversible machine and accordingly, he conceived what is known as the Carnot engine. Therefore, he conceived a machine that could run while it was in equilibrium with its surroundings. From the first law of thermodynamics, $E_2 - E_1 = Q - W$, E is internal energy, Q is heat and W is work. Carnot conceived a hypothetical slow thermodynamic process that cannot transfer internal energy outside the system and the system remains always in equilibrium with outside surroundings. Thus, $dE = 0$ and $Q = W$. It means the entire amount of Q is available for work W. This is called an isothermal system. Since Carnot conceived a reversible machine, there are no entropy losses of energy in the Carnot cycle, E_1 and E_2 in a Carnot cycle are equal and as a consequence dE, the change in the internal energy of Carnot cycle is zero.

Rankine Engine

Rankine converted Carnot's idea into a practical working heat engine. He realized a real-life machine cannot be reversible. He also realized that one major constraint of the Carnot engine is its isothermal heat addition and heat rejection which limits the hot reservoir temperature and thus limits the capability of the engine to produce the maximum amount of work. Rankine conceived an isobaric heat addition/rejection instead of an isothermal process of Carnot. An isobaric process does maximum thermodynamic work since it receives heat from outside in proportion to the work it does to keep the pressure constant. Outside heat continuously compensates for dP in the heat engine and keeps dP = Constant. Therefore, while Carnot focussed on $dE = 0$, Rankine conceived dP = Constant. Thus, while Carnot got maximum efficiency Rankine achieved the maximum volume of work.

This opened the opportunity for Rankine to choose any temperature (superheat) and maximize the hot reservoir temperature that the system can tolerate. This is how Rankine converted Carnot's ideal machine into a real practical machine.

What did Rankine Lose?

While Rankine made a notable departure by conceiving a two-phase cold reservoir [condenser] to squeeze out the latent heat but still lost 60–65% efficiency, in the heat recovery in the condenser at constant pressure due to cooling water temperature restricted to ambient temperature.

THERMODYNAMICS OF RANKINE CYCLE

The two main types of thermodynamic cycles are power cycles, which generate net power output, and refrigeration and heat pump cycles, which consume net power input. There are two types of thermodynamic power cycles: gas cycles and vapour cycles. The working fluid in a gas cycle never changes phases during the cycle. In vapour cycles, the working fluid exits the system in two phases: one as a liquid phase at one point in the cycle and another as a vapour phase at another. Water is the working fluid in steam power plants' vapour power cycles.

Rankine Cycle

In summary, the Rankine cycle is an ideal cycle. The fluid passes through the four components without irreversibilities and pressure drops. The ideal Rankine cycle consists of the following four processes, as shown on the *T–S* diagram above.

1–2: Isentropic compression in a pump

2–3: Constant pressure heat addition in a boiler

3–4: Isentropic expansion in a turbine

4–1: Constant pressure heat rejection in a condenser

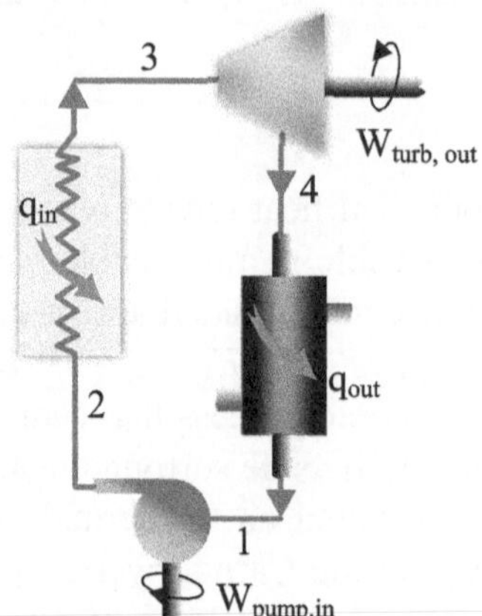

Figure 4: Schematic View of Rankine Cycle
Image Credit: Thermodynamic e-book.

Since each of the four elements of the perfect Rankine cycle is a steady-flow device, each of the four processes that comprise the cycle can be examined as a steady-flow process. Water's kinetic and potential energy changes are negligible in comparison to the heat and work terms, and are therefore disregarded. The four components' energy analyses are listed below.

Energy Balance

Pump (process 1–2): Pump pressurized the liquid water from the condenser prior to going back to the boiler. Assuming no heat transfer with the surroundings, the energy balance in the pump is

$$W\text{ pump, in} = h_2 - h_1$$

Boiler (process 2–3): Liquid water enters the boiler and is heated to superheated state in the boiler. The energy balance in the boiler is,

$$Q\text{ in} = h_3 - h_2$$

Turbine (process 3–4): Steam from the boiler, which has an elevated temperature and pressure, expands through the turbine to produce work and then is discharged to the condenser with relatively low pressure. Neglecting heat transfer with the surroundings, the energy balance in the turbine is

$$W\text{ turbine, out} = h_3 - h_4$$

Condenser (process 4–1): Steam from the turbine is condensed to liquid water in the condenser. The energy balance in the condenser is,

Q out $= h_4 - h_1$

For the whole cycle, the energy balance can be obtained by summarizing the four energy equations above. It gives,

(Q in-Q out) – (w turbine, out – w pump, in) = 0

The thermal efficiency of the Rankine cycle is determined from

η^{th} = w net, out/Q in = 1 – Q out/Q in

Where the net work output from the cycle is

W net, out = w turbine, out – w pump, in

Rankine Cycle Efficiency Calculation

A Case Study:

Given Data

Condenser Pressure: 10 kPa

Boiler Pressure: 16 MPa

Steam temperature at the turbine inlet: 600°C

The actual data are:

$P_1 = 9$ kPa $T_1 = 38$ degc

$P_2 = 16$ MPa

$P_3 = 15.9$ MPa $T_3 = 45$ degc

$P_4 = 15.2$ MPa $T_4 = 625$ degc

$P_5 = 15$ MPa $T_5 = 600$ degc

$P_6 = 10$ kPa

The pump has an efficiency of 85% and the turbine has an efficiency of 80%.

The mass flow rate is 20 kg/s.

Working

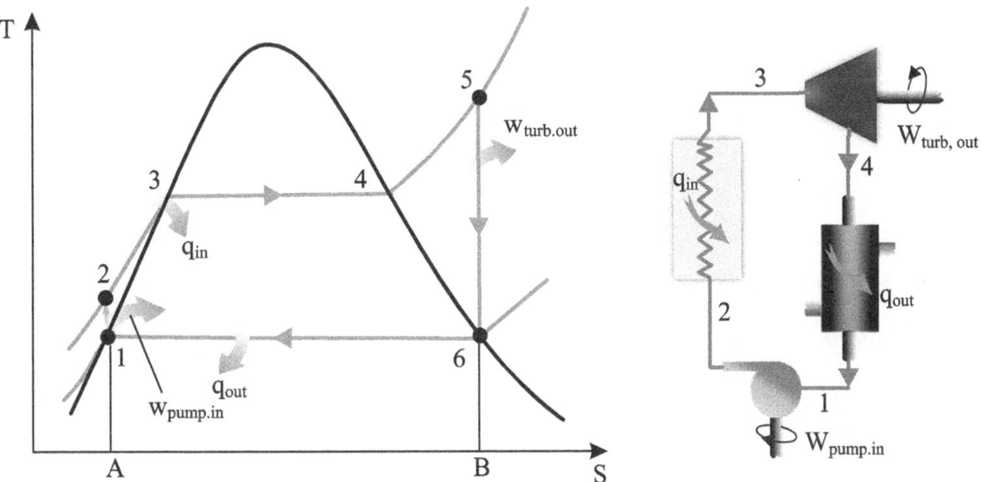

Figure 5: *T–S* and Schematic Diagrams

Image Credit: Thermodynamic e-book.

The schematic and the *T–S* diagram of this cycle is shown above.

Assumptions

All the components in the cycle operate at steady state.

Kinetic and potential energy changes are negligible.

Work Output

The net work output of the cycle equals the difference between the turbine work output and the pump work input.

$$w_{net,out} = w_{turb,out} - w_{pump,in}$$

The enthalpy at each state needs to be determined first, which can be obtained from the steam table.

State 1: Compressed liquid water

$P_1 = 9$ kPa $T1 = 38°C$ (given)

$v_1 = 0.001008$ m^3/kg

State 3: Compressed liquid water

$P_3 = 15.9$ MPa $T3 = 45°C$ (given)

$h_3 = 187.8$ kJ/kg

State 4: Superheated vapor

$P_4 = 15.2$ MPa $T_4 = 625°C$ (given)

$h_4 = 3641.8$ kJ/kg

$s_4 = 6.7961$ kJ/(kg-K)

State 5: Superheated vapor

$P_5 = 15$ MPa $T_5 = 600°C$ (given)

$h_5 = 3582.3$ kJ/kg

$s_5 = 6.6776$ kJ/(kg-K)

State 6: Saturated mixture

$P_6 = 10$ kPa (given)

$s_{6s} = s_5 = 6.6776$ kJ/(kg-K)

$h_{6s} = 2114.9$ kJ/kg

where s_{6s} and h_{6s} are the entropy and enthalpy at state 6 if process 5–6 is isentropic.

Since the isentropic efficiencies of the pump and turbine are given, the turbine work and the pump work can be determined as following:

$$\eta_{turb} = \frac{w_a}{w_s} - \frac{w_a}{h_5 - h_{6s}}$$

$$\rightarrow w_{turb,out} = w_a = \eta_{turb}(h_5 - h_{6s})$$

$$= 0.80(3,582.3 - 2,114.9) = 1,173.9 \text{ kJ/kg}$$

$$\eta_{pump} = \frac{w_s}{w_a} = \frac{v(P_2 - P_1)}{w_a}$$

$$\rightarrow w_{pump,in} = w_a = \frac{v(P_2 - P_1)}{\eta_{pump}}$$

$$= \frac{0.001008(16,000 - 9)}{0.85} = 19.0 \text{ kJ/kg}$$

Therefore, the net work output per mass is

$$w_{net,out} = w_{turb,out} - w_{pump,in}$$
$$= 1,173.9 - 19.0 = 1,154.9 \text{ kJ/kg}$$

The power produced by the new system is determined from

$$W_{net,out} = \dot{m}w_{net,out}$$
$$= 20(1,154.9)$$
$$= 23,098 \text{ kW}$$

Thermal Efficiency

Heat input to the cycle from the boiler is determined from

$$q_{in} = h_4 - h_3 = 3,641.8 - 187.8 = 3,454 \text{ kJ/kg}$$

The thermal efficiency of the cycle is

$$\eta_{th} = w_{net,out}/q_{in} = 1,154.9/3,454 = 33\%$$

Reference

Thermodynamic E-Book.

JOULE THOMSON EFFECT OVERVIEW

Joule: Thompson effect is an interplay of heat, work, and internal energy at constant enthalpy.

The Joule: Thomson effect, also known as the Joule-Kelvin effect, is a thermodynamic phenomenon that occurs when a real gas undergoes a throttling process, or expansion, **without the addition or removal of heat and work**. This effect has significant practical applications in various engineering fields, including refrigeration, natural gas processing, and cryogenics.

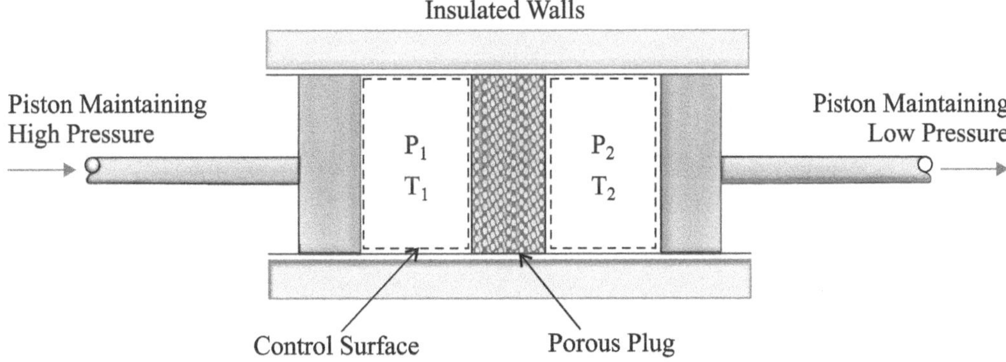

Figure 6: Porous Plug

The key point is Joule-Thomson effect is an isenthalpic process. The joule-Thomson effect exemplifies the principle of energy conservation, particularly when considering the constancy of internal energy (*U*) and work (*W*) while keeping enthalpy constant. This articulates the fundamental concept of conserving energy during adiabatic expansion, and the implication that the internal energy and work within the system remain unaltered. By acknowledging the Joule-Thomson effect as a representation of the preservation of energy and the constancy of enthalpy, it underscores the crucial role of this principle in thermodynamics and its implications for real-world engineering applications.

JT effect exemplifies the behavior of real gases during expansion. Unlike ideal gases, real gases do not always behave predictably when subjected to changes in pressure and temperature. The Joule-Thomson effect helps engineers understand how real gases will behave under different conditions, allowing them to design more efficient and effective systems. As said above, ***JT* is an isenthalpic process**. The *JT* effect is an outstanding thermodynamic principle that provides a deep understanding of the isenthalpic process. It is a fascinating phenomenon where a gas undergoes adiabatic expansion or throttling, meaning there is no heat transfer to the surroundings and also there is no work transfer. Instead, the change in internal energy occurs at the expense of the thermodynamic *PV* work.

The most important point is the constancy of internal energy (U) and work (W) while keeping enthalpy constant. In *JT* expansion the sum of $U + W$ remains constant.

Furthermore, although the *JT* effect is an adiabatic process, it is not isentropic due to internal irreversibilities such as friction as the gas is forced through the pores of the porous plug or throttled through a tiny opening of the throttling valve. This distinction is crucial in understanding the intricacies of the process and its practical applications in various engineering fields.

Additionally, the *JT* coefficient plays a crucial role in understanding the effect. This coefficient describes how the temperature of a gas changes when it undergoes expansion. **The inversion temperature** is another important concept in the *JT* effect. This temperature represents the point at which a gas transitions from exhibiting a cooling effect to a heating effect during expansion.

Finally, it is important to **consider the exceptions to the JT effect**, particularly in the case of certain gases such as **hydrogen, helium, and neon**. These gases do not exhibit the same behavior as other gases when subjected to the *JT* effect.

What is the Joule-Thomson Effect?

In thermodynamics, the *JT* effect describes the temperature change of a real gas or liquid when it is forced through a valve or porous plug while keeping it insulated so that no heat is exchanged with the environment. This procedure is called a throttling process or *JT* process. At room temperature, all gases except hydrogen, helium, and neon cool upon expansion by the *JT* process. Under similar conditions, there is an increase in temperature for H_2, He, and Ne.

It is important to study the cooling produced in the JT expansion in refrigeration and also in the petrochemical industry, where the cooling effect is used to liquefy gases. It is also used in many cryogenic applications.

Fundamentals

Real Gas vs Ideal Gas

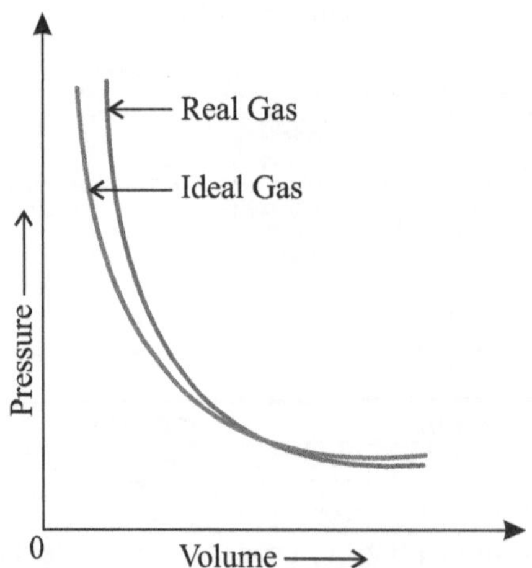

Figure 7: Fundamental to the JT Effect

The deviation of real gas from ideal gas behavior occurs due to the assumption that if pressure increases the volume decreases. The volume will approach a smaller number but will not be zero because the molecules will occupy some space that cannot be compressed further. It has been observed, at high pressure, the measured volume of the gas is more than the calculated volume. But at low pressure, the calculated and the measured volumes approach each other. So, it can be stated that the real gases do not follow Boyle's law perfectly under all conditions of temperature and pressure.

Hence, the volume of the gas molecules for real gases is a significant fraction of the volume of the container at higher pressure.

When we use the ideal gas law, we make a couple of assumptions:

- We ignore the volume taken up by the imaginary ideal gas molecules.
- We ignore that the gas molecules do attract or repel each other.

However, we know that in real life, gases are made up of atoms and molecules that take up some finite volume, and we also know that atoms and molecules interact with each other through intermolecular forces. One way we can look at how accurately the ideal gas law describes our system is by comparing the molar volume of real gas with the molar volume of an ideal gas.

Let n be the number of moles, V is the volume of real gas, the molar volume of real gas = V/n

Refer to the ideal gas equation, $PV = n\,RT$. The molar volume = RT/P

The ratio of the molar volume of real gas/molar volume of ideal gas = $V/n/RT/P = PV/n\,RT$

This ratio is called compressibility of gas and is expressed by the symbol Z

$$Z = PV/n\,RT$$

$$Z = 1 \text{ for ideal gases}$$

Compressibility factor Z for gases: Very important concept for JT expansion.

Thus, the compressibility factor of gas tells us how far the gas is from ideal behavior and this has a huge meaning in cryogenics.

There are two factors [1] high pressure and [2] low temperature that make a real gas deviate from ideal behavior $PV = n\,RT$.

Effect of pressure on the gas compressibility: Contributes to excluded volume

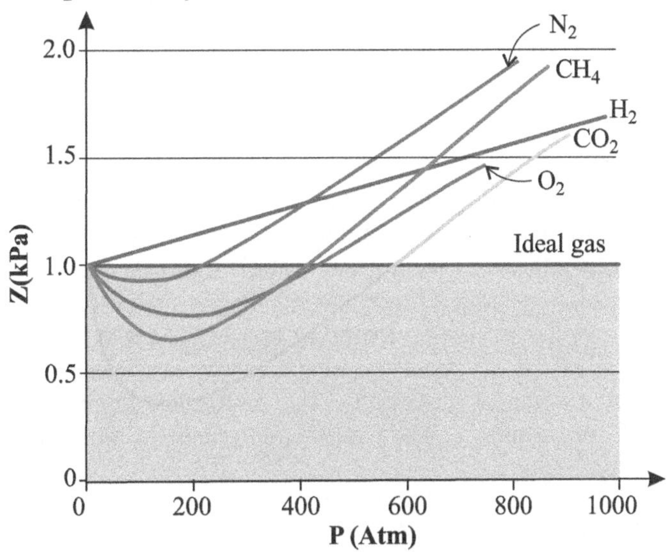

Figure 8: Pressure Brings Molecular Repulsive Forces into Dominance

This graph shows the compression factor Z over a range of pressures for many gases. For all of the real gases [every gas is a real gas] in this graph, you might notice that the shapes of the curves look a little different for each gas, and most of the curves only approximately resemble the ideal gas line at Z equals 1 over a limited pressure range. Also, for all the real gases Z is sometimes less than 1 at very low pressures, which tells us that the molar volume of real gases is less than that of an ideal gas. As you increase the pressure past a certain point that depends on the gas, Z gets increasingly larger than 1. That means, at high pressures, the molar volume of real gases is larger than the ideal gas.

The next question that automatically arises is, why?

At high pressures, the gas molecules get more crowded and the amount of space between the molecules is reduced. This makes the repulsive forces coming to dominance tend to make the volume larger than for an ideal gas; when these forces dominate Z is greater than unity.

The error in molar volume gets worse the more compressed the gas becomes, which is why the difference between Z for the real and ideal gas increases with pressure.

Effect of temperature on the compressibility of gases: Contributes to intermolecular forces.

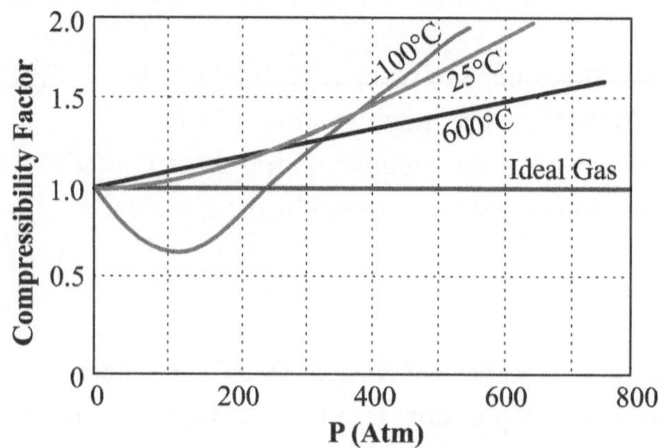

Figure 9: Nitrogen at Several T

At low temperatures the attractive forces dominate

You may notice that the compressibility of N_2 [the molar volume of real gas/ideal gas volume ratio] of N_2 is less than 1 stands for an ideal gas at –100 degc. This ratio goes near 1 at 25 degc and then shoots up with $Z > 1$ at 600 degc. You would also notice that at high pressure the Z value of N_2 is >1 at all temperatures.

What does it mean?

Explanation

The pressure we measure comes from the force of the gas molecules arising from the kinetic energy of the molecules at that temperature hitting the walls of the container. At a relatively larger temperature, a gas molecule has more kinetic energy to overcome the intermolecular attractions and hit the wall giving more P and consequently also more V because of the larger spread of molecules. This results in a much larger positive deviation of real gases with a bigger positive $Z > 1$ at high temperatures. This reverses at low temperature and low pressure the condition that applies to the *JT* effect. The intermolecular forces become much more prominent at low temperatures because the molecules have less kinetic energy to overcome the intermolecular attractions. The molecules come closer. The compressibility reduces mostly $Z < 1$. Here, the attractive forces dominate. The relative importance of attractive forces decreases as temperature increases.

The van der Waals [VDW] equation

The van der Waals [VDW] equation is the simplest equation to model the behavior of real gases. VDW equation corrects the ideal gas equation of state $PV = n\,RT$ by including the effect of gas molecule volume and intermolecular forces.

The fundamental equation for ideal gases is $PV = n\,RT$

Let us assume P is pressure and V is the volume of an ideal gas.

The van der Waals equation corrects for two properties of real gases: the excluded volume of gas particles and attractive forces between gas molecules.

The van der Waals equation is presented as:

$$(P + an_2/V_2)\,(V - nb) = n\,RT$$

P, equals measured pressure

V, equals the volume of the container

n, equals moles of gas

R, equals gas constant

T, equals temperature (in Kelvin)

Compared to the ideal gas law, the *VdW* equation includes "correction" to the pressure term, $[an^2/V^2]$ which accounts for the measured pressure being lower due to attraction between gas molecules. The "correction" to the volume, $[nb]$ subtracts out the volume of the gas molecules from the total volume of the container to get a more accurate measure of the empty space available for the gas molecules. a and b are measured constants for a specific gas (and they might have some slight temperature and pressure dependence).

At low temperatures and low pressure, the correction for volume is not as important as the one for pressure, so Z is less than 1. At high pressures, the correction for the volume of the molecules becomes more important so Z is greater than 1. At some range of intermediate pressure, the two corrections cancel out and the gas appears to follow the relationship given by the ideal gas.

JT Effect: How does it Work?

Mechanism of JT Effect

JT effect is approximated as an Isenthalpic process. It's an interplay of heat, work, and internal energy at constant enthalpy.

How is *JT* isenthalpic?

It is an experimental fact that *JT* is an isenthalpic process backed by thermodynamics.

What is 'Isenthalpic'

A process becomes isenthalpic when there is no change in the enthalpy of the process. The constancy of enthalpy means the sum of internal energy, U and work, W remains constant.

How *JT* is unique

There are many adiabatic processes while generally, in most of them, $dQ = 0$ but there is adiabatic work transfer to surroundings. Let's look at two common adiabatic processes [1] compressors and [2] turbines in power plants. While both are adiabatic processes, they do transfer 'work' to their surroundings. Compressors transfer adiabatic work to isothermal interstage coolers and the turbine transfers adiabatic work to condenser.

JT is unique. It retains the 'work' within the system and uses the internal energy to perform the work that causes cooling in JT effect.

From the first law of thermodynamics,

$$H = U + PV$$

where U is internal energy, P is pressure, and V is volume the change in PV represents the work done.

The net work is done when a mass m of the gas expands.

$$W = mV_1P_1 - m\, V_2P_2$$

as the gas expands from P_1V_1 to P_2V_2, it does a W amount of work.

From the first law of thermodynamics dU = delta Q – delta W

In the *JT* process, delta $Q = 0$ as the process is adiabatic

$$dU = -\text{delta } W$$

$$mU_2 - mU_1 = -mP_2V_2 + mP_1V_1$$

$$mU_2 + mP_2V_2 = mU_1 + mP_1V_1$$

This gives $mH_1 = mH_2$,

$H_1 = H_2$ meaning Isenthalpic process.

In summary, the *JT* effect contains heat and work within the system. The 'work 'consumes internal energy at constant enthalpy that does the cooling. While individually both *U* and *W* change, the sum of $U + W$ remains constant.

Why and when a process becomes isenthalpic?

An isenthalpic process is one where there is no change in total energy or enthalpy.

Joule-Thomson cooling formula

You know initial pressure of gas P_i in bar. Find out the final pressure of the gas (P_f) in bar.

Next, determine the Joule Thomson coefficient (μ) in Kelvin per bar.

Next, use the formula from above

$= T_f = T_i + \mu \times (P_i - P_f)$.

Finally, calculate the final temperature (T_f) in Kelvin.

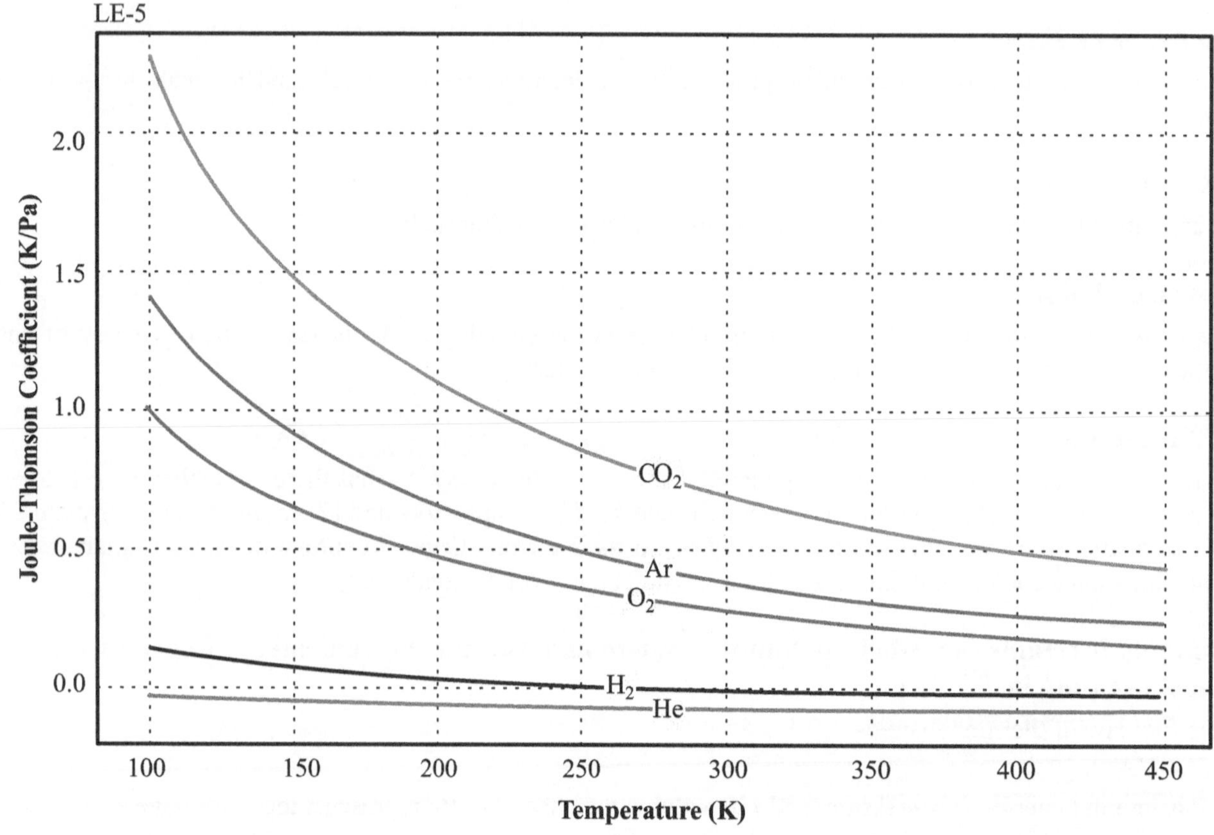

Figure 10: Joule-Thomson Coefficient *vs* Temperature

How JT Does Isenthalpic Cooling?

Joule-Thomson cooling porous plug or throttling valve can have pores in the range from 5 to 125 micrometers. The upstream pressure and temperature remain constant. As the gas leaks through the these micro pores there is significant drop in kinetic energy downstream that does cooling called JT cooling. The gas leaks into a large volume effortlessly without doing any work. Therefore, we always say in *JT* cooling there is no "work" involved. But never understand, why?

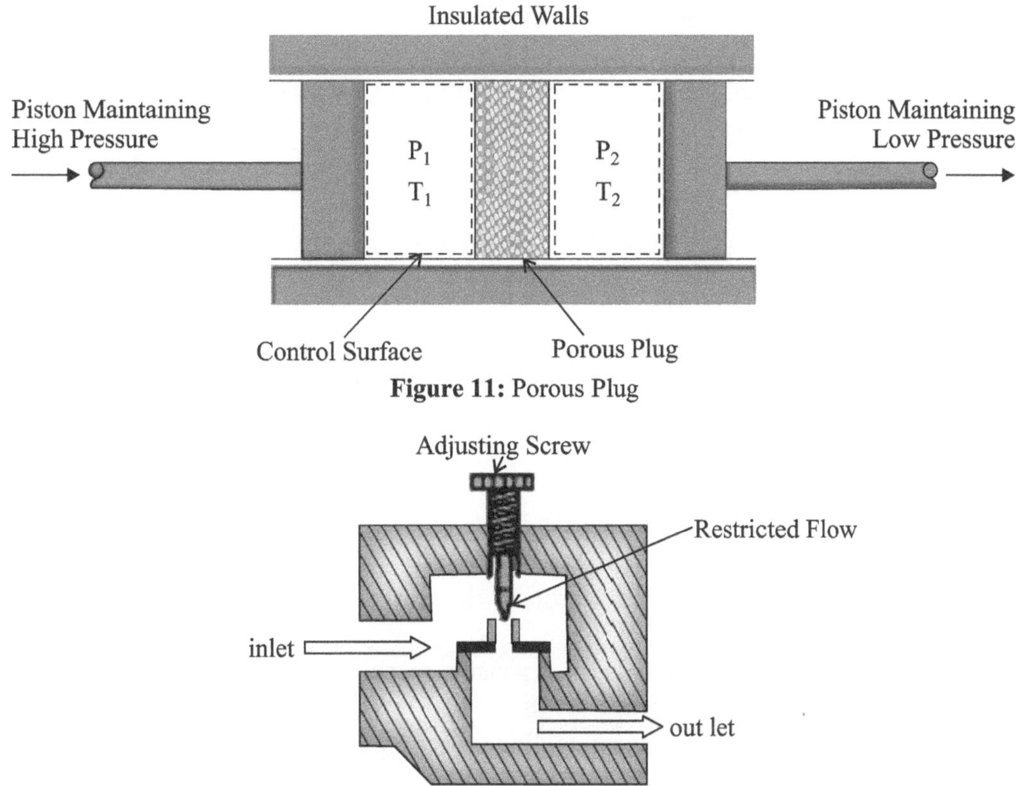

Figure 11: Porous Plug

Figure 12: Throttling Valve

To sum up (1) large reduction in kinetic energy across the throttling valve gives the cooling at no cost of energy because of effort less expansion of small volume of gas into a much larger space suddenly involves "no work". This explains why *JT* cooling is "isenthalpic".

Explanation

An isenthalpic process occurs when suddenly a real gas is made to expand into a large space which precisely defines what follows a throttling process.

Since the gas expands into a large space it does so without much pushing the surroundings and practically no work.

Thus the "Work" component of enthalpy $H = U + W$ remains constant. H is enthalpy U is internal energy, W is work.

The second thing that follows the expansion of gas in large space is dramatic reduction of molecular collisions that reduces the temperature or internal energy.

Overall in equation, $H = U + W$ the sum remains constant except the value of each changes.

Summary

To summarize, in an isenthalpic process such as the Joule-Thomson expansion, the gas undergoes a sudden expansion into a large space through a throttling valve.

This expansion occurs without the gas doing much work on the surroundings, leading to a constant enthalpy.

The decrease in pressure and temperature during the expansion results in a reduction of molecular collisions, leading to a decrease in internal energy.

Despite these changes in internal energy and work, the total enthalpy of the system remains constant.

Throttling *vs* Expansion

Are expansion and throttling the same?

The answer is no.

A porous plug and throttling valve consists of pores/tiny openings in the range of 5 to 100 micrometers through which high pressure gas at a given temperature is forced suddenly into a large space.

The key difference is throttling is isenthapic and expansion can be isentropic.

Expansion and throttling are related processes, but they are not the same.

Expansion (turbine)

In a turbine, expansion of a fluid can produce useful work output and be close to a reversible process. In expansion it is more gradual than a sudden expansion by a throttling valve with considerable frictional losses.

Throttling (refrigeration)

Throttling is an expansion process that is highly irreversible. In throttling, a fluid's pressure and temperature are reduced suddenly when it passes through a throttling valve. This is due to the fluid expanding into a larger space and frictional effects.

Explanation

Throttling valves are used in many systems, including refrigeration and air conditioning.

For example, in a refrigerator, a throttling valve reduces the pressure and temperature of the refrigerant so it can cool the evaporator.

In gas-expansion processes, the gas expands in an expansion machine after being compressed and rejecting heat. This method has a higher cooling intensity than throttling.

Gas expansion is isentropic that is more efficient cooling process than throttling which is isenthalpic.

Joule Thomson coefficient μ and inversion temperature [An important concept]

The rate of change of temperature T concerning pressure P in a Joule–Thomson process (that is, at constant enthalpy H) is the Joule–Thomson (Kelvin) coefficient

This coefficient can be expressed in terms of the gas's volume V, its heat capacity at constant pressure Cp, and its coefficient of thermal expansion alpha as,

$$\mu = dT/dP = V/Cp\ [\text{alpha} \times T - 1]$$

The Joule–Thomson coefficient of an ideal gas is zero. In real gases, the Joule–Thomson coefficient is different from zero and depends on pressure and temperature. For $\mu\ JT > 0$, temperature decreases, and for $\mu\ JT < 0$, temperature increases during an expansion.

The temperature at which the sign of the Joule–Thomson coefficient changes is the inversion temperature, which itself depends on pressure. The inversion temperature of most gases is above ambient temperature, but for hydrogen, the inversion temperature is about $T = -80$°C. The values for the Joule–Thomson coefficient help to realize the changes occurring during processes.

Inversion temperature and compressibility of a gas is central to *JT* effect

The inversion temperature and JT coefficient are related subjects. This has been discussed together above. Few more lines about inversion temperature. Some of them may be the repetition of what is already said above.

The inversion temperature in thermodynamics and cryogenics is the temperature below which a non-ideal gas (all gases in reality) that is expanding at constant enthalpy will experience a temperature decrease, and above which will experience a temperature increase. This temperature change is known as the Joule–Thomson effect, Helium, hydrogen, and Neon are three gases whose Joule–Thomson inversion temperatures at a pressure of one atmosphere are very low (e.g., about 45 K, −228°C for helium). Thus, these gases warm when expanded at constant enthalpy at typical room temperatures. On the other hand, nitrogen, and oxygen, the two most abundant gases in the air, have inversion temperatures of 621 K (348°C) and 764 K (491°C) respectively: these gases can be cooled from room temperature by the Joule–Thomson effect.

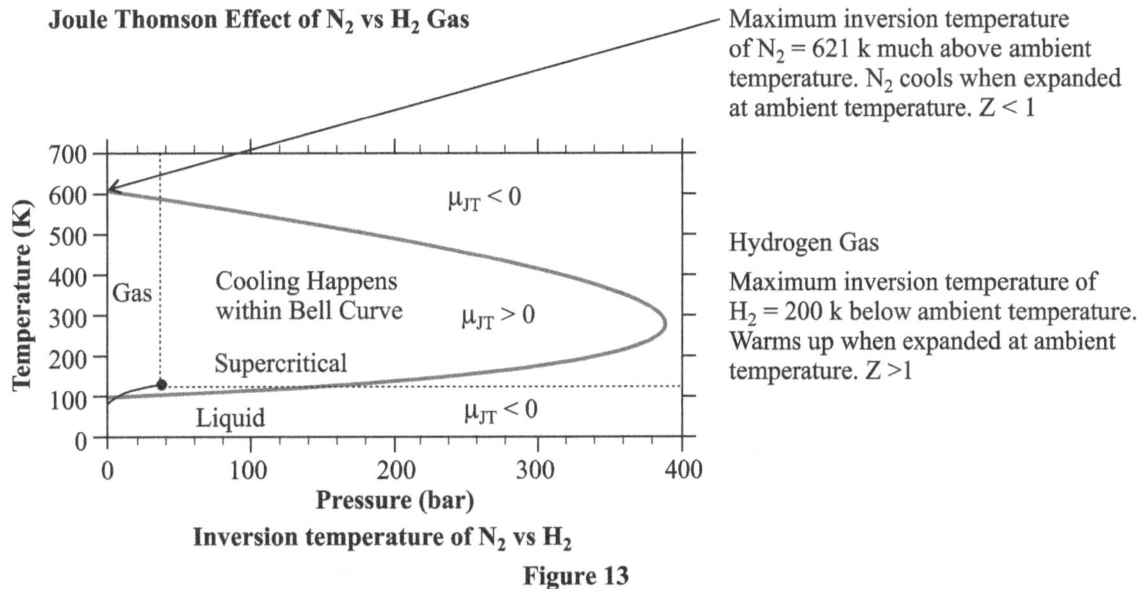

Figure 13

H_2, *He* and *Ne* Don't Cool by JT Expansion

All real gases except H_2, *He*, and *Ne* cool, when expanded at constant enthalpy below, their inversion temperature. For example, the N_2 gas has an inversion temperature of 621 k, much above the normal ambient temperature. Those gases that cool are represented by the positive *JT* coefficient and those that don't by the negative *JT* coefficient.

All real gases those which have $Z < 1$, have molar volume < ideal gas. They have intermolecular attraction. When they expand their KE slows down because they have to expand against attractive forces. Reduction in *KE* reduces temperature. PE gains energy. Total internal energy = *KE* + *PE* remains constant.

H_2 is an exception [HE and Ne are the other two exceptions]

The inversion temperature of H_2 gas is 200 k much below the ambient temperature [image below]. When H_2 is expanded at room temperature the gas becomes hot instead of cooling. But H_2 does show a *JT* effect if it is expanded below 200 k.

Explanation

What triggers this temperature change behind the scenes is intermolecular attractions and repulsions.

All molecules have some attractions and repulsions working between them. This arises because the electrons are moving around the nucleus continuously and this movement is unevenly distributed and changing continuously. When at some point there is more electron density at a particular location, that particular point acquires a negative charge, and another end becomes positively charged to balance it. These are called temporary dipoles.

The opposite charge ends produce attraction and similar charge ends produce repulsion when the molecules come closer. Unless a molecule is ionic in nature there is a balance of the positive and negative charges.

What difference is made that makes H_2 cool if expanded < 200 k and heat up if expanded at 298 k, the ambient temperature?

At 200 k H_2 has relatively less kinetic energy, molecules are closer and molecules have relatively more intermolecular attractions compared to 298 K [ambient temperature] that H_2 exploits for its cooling below its inversion temperature. On the contrary, at 298 K [ambient temperature] H_2 molecules have more kinetic energy, more collisions, more pressure, and more repulsion, $Z > 1$, see the image.

The compressibility *Z* of H_2 further increases when H_2 is expanded because of lack of intermolecular attractions its volume becomes even larger than ideal gas. This gas expansion needs more translational kinetic energy which is supplied by the potential energy. This kinetic energy sends molecules farther apart adding *KE* to the gas that heats H_2.

Table 1: van der Waals Constants for Gas Molecules

Substance	*a(L²-atm/mol²)*	*b(L/mol)*
He	0.0341	0.02370
Ne	0.211	0.0171
Ar	1.34	0.0322
Kr	2.32	0.0398
Xe	4.19	0.0510
H_2	0.244	0.0266
N_2	1.39	0.0391
O_2	1.36	0.0318
Cl_2	6.49	0.0562
H_2O	5.46	0.0305
CH	2.25	0.0428
CO_2	3.59	0.0427
CCl_4	20.4	0.1383

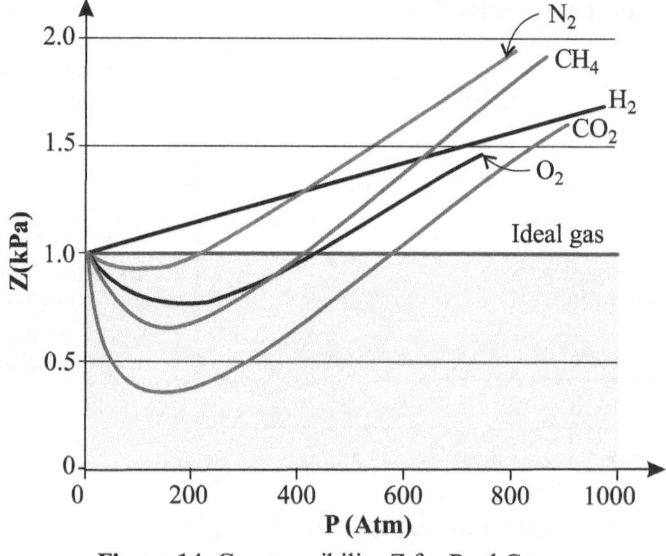

Figure 14: Compressibility Z for Real Gases

Table 2: Inversion Temperature

Gas	*Maximum Inversion Temperature [K]*
Helium-4	45
Hydrogen	205
Neon	250
Nitrogen	621
Air	603
Carbon monoxide	652
Argon	794
Oxygen	761
Methane	939
Carbon dioxide	1500
Ammonia	1994

HABER PROCESS OF AMMONIA MAKING

Haber Process

Haber process is a chemical process used to produce ammonia (NH_3) from nitrogen (N_2) and hydrogen (H_2) gases. This process was developed by German chemist Fritz Haber in the early 20th century and is still widely used in the production of ammonia for various industrial applications, such as fertilizers and explosives.

The Haber process takes place in a reactor at high temperature (400–450°C) and high pressure (150–300 atm) in the presence of a catalyst, typically iron, promoted with potassium oxide. The catalyst helps speed up the reaction without being consumed in the process.

The reaction for the production of ammonia in the Haber process can be represented by the following equation:

$$N_2 + 3H_2 \rightarrow 2NH_3$$

The nitrogen gas is obtained from the air through a process called air separation, while the hydrogen gas is typically produced by steam reforming of natural gas or water gas shift reaction.

The gases are mixed in the reactor and then compressed to the required pressure. The mixture then passes through the catalyst bed where the reaction takes place. The ammonia produced is then cooled and condensed into liquid form for further processing and storage.

The unreacted gases are recycled back into the reactor to increase the overall efficiency of the process. The production of ammonia in the Haber process is exothermic, meaning it releases heat. This heat is typically used to produce steam, which can be used for various purposes within the plant.

Thermodynamics

Energy vs Entropy Competition

By the Gibbs free energy equation, $\Delta G = \Delta H - T\Delta S$, G is Gibbs free energy, H is enthalpy, T is temperature and S is entropy. When $\Delta G = 0$, the system is in equilibrium when $\Delta G < 0$, the system is spontaneously moving forward, and When $\Delta G > 0$, the system is moving in the reverse direction. When a system approaches equilibrium, energy tends to a minimum and entropy tends to a maximum. Usually, the two tendencies favor different configurations of the body. The energy that wants to go to a low stable level is countered by entropy by creating disorder.

There are several examples when energy and entropy act as competitors: The Haber synthesis of ammonia is a classic case.

Thermodynamic Issues in the Haber Process

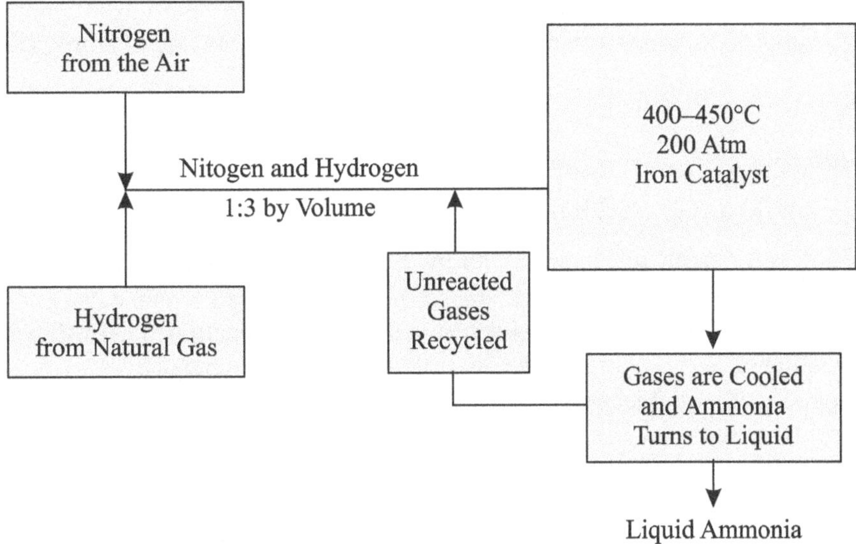

Figure 15

The Haber Process combines nitrogen from the air with hydrogen derived mainly from natural gas (methane) into ammonia. The reaction is reversible and the production of ammonia is exothermic.

$N_2(g) + 3H_2(g) \leftrightarrow 2NH_3(g)$, $\Delta H = -92$ KJ/mole is not a favorable reaction from the point of thermodynamics because there is a decrease in entropy in the forward direction of the reaction involving 4 moles of reactants converting into 2 moles of products. There is a decrease in entropy, $\Delta G > 0$. The reaction in the forward direction is not viable. Coming back to the fundamental Gibbs equation, $\Delta G = \Delta H - T\Delta S$, at standard state this reaction is not viable because of entropy gain or negative sign of entropy even though the reaction is exothermic with negative enthalpy makes $\Delta G > 0$. Enthalpy or energy gets a beating and entropy wins to produce $\Delta G > 0$. This is a wonderful case of how energy and entropy compete.

How the Battle between Energy and Entropy is Won to Produce Ammonia?

Fight on Various Fronts

The Temperature

Equilibrium Considerations

$$N_2(g) + 3H_2(g) \leftrightarrow 2NH_3(g),\ \Delta H = -92 \text{ KJ/mole}$$

You need to shift the position of the equilibrium as far as possible to the right to produce the maximum possible amount of ammonia in the equilibrium mixture.

The forward reaction (the production of ammonia) is exothermic.

According to Le Chatelier's Principle, [when a settled system is disturbed, it will adjust to diminish the change that has been made to it] this will be favored if you lower the temperature. The system will oppose it and respond by moving the position of equilibrium to counteract this - in other words by producing more heat. This in turn will produce more products. To get as much ammonia as possible in the equilibrium mixture, you need as low a temperature as possible. However, in reality, this is not possible.

Issues with a Reduction in Reaction Temperature

The lower the temperature you use, the slower the reaction becomes.

Compromise

400–450°C is a compromise temperature for producing a reasonably high proportion of ammonia in the equilibrium mixture (even though it is only 15%) but in a very short time with a reasonable rate of conversion.

The Pressure

Equilibrium Considerations

$$N_2(g) + 3H_2(g) \leftrightarrow 2NH_3(g), \Delta H = -92 \text{ KJ/mole}$$

Notice that there are 4 molecules on the left-hand side of the equation, but only 2 on the right.

According to Le Chatelier's Principle, if you increase the pressure the system will oppose it and counter it by favoring the reaction which produces fewer molecules that is the forward reaction. That will cause the pressure to fall. In other words, the generation of NH_3 would increase that is what we want to happen.

To get as much ammonia as possible in the equilibrium mixture, you need as high a pressure as possible. 200 atmospheres is a high pressure used.

Rate Considerations

Increasing the pressure brings the molecules closer together. In this particular instance, it will increase their chances of hitting and sticking to the surface of the catalyst where they can react. The higher the pressure the better in terms of the rate of a gas reaction.

The Catalyst

Equilibrium Considerations

The catalyst does not affect whatsoever on the position of the equilibrium. Adding a catalyst doesn't produce any greater percentage of ammonia in the equilibrium mixture. Its only function is to speed up the reaction.

FUEL CELL: HYDROGEN PRODUCTION

A fuel cell is an electrochemical device that converts chemical energy from a fuel into electricity through a chemical reaction. It operates much like a battery, but unlike a battery, it does not need to be recharged, as long as fuel is supplied. Fuel cells have the potential to revolutionize the way we generate power and are considered a clean and efficient alternative to traditional combustion-based power generation methods.

Fuel cells produce electricity by combining fuel, such as hydrogen, with oxygen from the air. This reaction generates electric current, heat, and water as by products, making fuel cells a clean and environmentally friendly power source. They are also highly efficient, with some types of fuel cells achieving efficiencies of up to 60%.

Fuel cells have a wide range of applications, from powering small electronic devices to providing electricity for residential buildings and even powering vehicles like cars, buses, and trains. They offer a quiet operation, low emissions, and high reliability, making them an attractive option for a sustainable energy future.

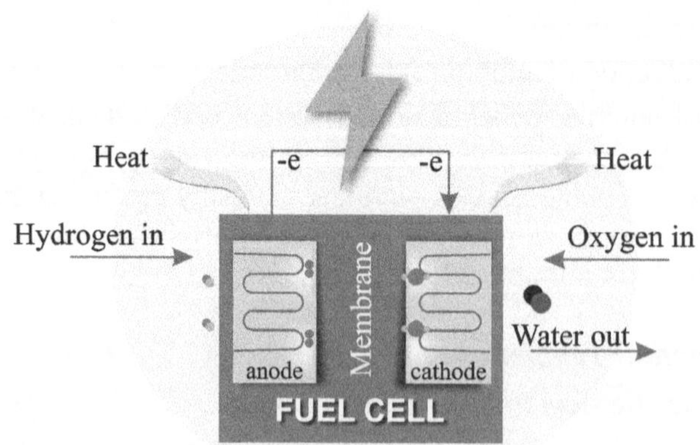

Figure 16: Fuel Cell: Hydrogen Production

How Fuel Cells Work

Fuel cells work like batteries, but they do not run down or need recharging as long as fuel is supplied.

A fuel cell consists of two electrodes—a negative electrode (or anode) and a positive electrode (or cathode)—sandwiched around an electrolyte. A fuel, such as hydrogen, is fed to the anode, and air is fed to the cathode.

In a hydrogen fuel cell, a catalyst at anode separates hydrogen molecules into protons and electrons, which take different paths to the cathode.

The electrons go through an external circuit, creating a flow of electricity. The protons migrate through the electrolyte to the cathode, where they unite with oxygen and the electrons to produce water and heat.

Reactions

Anode reaction: (oxidation reaction is loss of electrons)

$$2H_2 \text{ (hydrogen)} \rightarrow 4H+ \text{ (protons)} + 4e^- \text{ (electrons)}$$

Cathode reaction: (reduction reaction is gain of electrons)

$$O_2 \text{ (oxygen)} + 4H+ \text{ (protons)} + 4e^- \text{ (electrons)} \rightarrow 2H_2O \text{ (water)}$$

Overall Reaction

$$2H_2 + O_2 \rightarrow 2H_2O + \text{heat}$$

What these reactions mean?

The first thing it means is that heat and water management are important.

In a fuel cell, hydrogen is oxidized at the anode to produce protons and electrons. The protons then migrate through an electrolyte to the cathode, while the electrons flow through an external circuit, creating an electrical current.

At the cathode, oxygen is reduced by accepting protons and electrons to produce water as the by product.

This overall reaction generate electricity and water.

Thermodynamics

Cell Voltage

The cell voltage is a key parameter in fuel cell thermodynamics. It represents electrical potential generated by electrochemical reactions within the cell. The cell voltage is determined by standard reduction potentials of reactants and products involved in the anode and cathode half-reactions. The difference in standard reduction potentials between the anode and cathode reactions dictates the overall cell voltage. The Nernst equation is commonly used to calculate the cell voltage.

Gibbs Free Energy Change

The Gibbs free energy change (ΔG) of the overall electrochemical reaction in a fuel cell is a measure of the thermodynamic feasibility of the process. A negative ΔG value indicates that the reaction is spontaneous and can proceed to generate electrical energy. The standard Gibbs free energy change ($\Delta G°$) is related to the cell voltage through the equation $\Delta G° = -nF\Delta E°$, where n is the number of moles of electrons transferred in the reaction, F is the Faraday constant, and $\Delta E°$ is the standard cell potential.

Thermodynamic Efficiency

The thermodynamic efficiency of a fuel cell quantifies the ratio of the electrical work output to the heat input (enthalpy change) from the reactants.

Second Law Efficiency

In the context of thermodynamics, the second law efficiency of a fuel cell refers to the ratio of the maximum available work output to the total enthalpy change of the reactants. This efficiency accounts for the irreversibilities.

Thermodynamic Issues of Fuel Cells

Heat Management

Fuel cells operate at high temperatures, which can cause issues with heat management and thermal regulation. If the temperature is not controlled properly, it can affect the performance and longevity of the fuel cell.

Energy Balance

Fuel cells need to maintain a delicate balance between providing enough energy to power the system and preventing overheating. If the energy balance is not carefully managed, it can lead to inefficiencies and reduced performance.

Waste Heat Recovery

Fuel cells produce waste heat as a by product of their operation, and it is important to effectively recover and utilize this heat to improve overall efficiency. However, capturing and utilizing waste heat can be challenging and may require additional equipment and infrastructure.

Corrosion and Degradation

Fuel cells are susceptible to corrosion and degradation over time, due to the high temperatures, chemical reactions, and operating conditions they are subject to. This can lead to reduced performance, decreased efficiency, and potentially premature failure of the fuel cell system.

Water Management

Fuel cells require a controlled supply of water to function properly, as it is a key component in the electrochemical reactions that occur within the cell. Managing the water supply, distribution, and removal can be a complex process that requires careful monitoring and control to prevent issues such as flooding or drying out of the fuel cell.

Specific Issues

Cell Voltage Losses

Fuel cells have theoretical voltage outputs based on the Gibbs free energy change of the reaction occurring in the cell. However, in practice, the actual cell voltage is lower than the theoretical value due to various losses including activation overpotential, concentration overpotential, and resistance losses in the cell components.

Gibbs Free Energy Related Issues

The Gibbs free energy change of the electrochemical reactions in fuel cells determines the maximum thermodynamic efficiency of the cell. If the reactants and products are not at standard conditions (e.g., different concentrations, temperatures), the Gibbs free energy change and thus the cell voltage may deviate from the ideal values.

Fuel Cell Efficiency: Complexity of Reality

Reference: Galen J. Suppes, Truman S. Storvick, in Science Direct.

An ideal fuel cell running on pure oxygen and hydrogen produces 1.229 volts of electromotive force to push electrons through the electrical devices when there is no load on the circuit (the switch is open). Since the chemical reactions of a fuel cell provide the same flow of electrons independent of the fuel cell efficiency, this voltage is used to compute the efficiency.

For example, when the fuel cell is connected to an external load, the "open circuit" voltage of the fuel cell is split between the resistance to anion flow in the cell and the resistance to electron flow in the external load.

If the fuel cell produces 0.615 volts when connected to a light bulb, the efficiency of the fuel cell with that load is 50% (0.6151.229 × 100).

The conversion efficiency improves when the external load is decreased. It takes careful design of the fuel cell to handle the variable power requirements of commuter car travel and maintain high efficiency.

At the start of the twenty-first century, the efficiency of a good hydrogen fuel cell running with atmospheric air rather than pure oxygen varied from about 45% at maximum design load to about 65% when operating at low power output corresponding to the "most efficient" load

If electrical power is used to make the hydrogen, and this electric power is generated with the efficiency varying from 30% to 55%, this gives an overall fuel cell efficiency of 13.5–36%.

If the fuel cell operates on hydrogen produced by converting gasoline to hydrogen on the vehicle, the conversion of energy from gasoline to hydrogen is about 70–80% which gives an overall efficiency of 31.5–52%.

The technology to use gasoline to power a fuel cell car requires the technology development to reform the gasoline to hydrogen. The gasoline reformer and fuel cell cars must be road tested before they come to the dealer's showroom.

The reality for fuel cells in vehicles today is that they are less efficient than the best diesel engine. The current goal of an overall fuel cell efficiency of 70% is not realistic neglecting the energy loss during hydrogen production.

Operation at maximum efficiency requires a large fuel cell to keep the electrical current produced per unit area of the cell low.

A fuel cell does not respond to the quick starts and acceleration required for urban driving. The fuel cell power output can be "leveled out" by using a battery pack that provides surge power and is recharged during cruise and stop phases of the trip.

The gasoline powered hybrid cars on the market today provide an ideal test for an electric powered car. Design a fuel cell to replace the gasoline engine, the electricity generator, and battery in the hybrid and you have a fuel cell car.

WATER SPLITTING: HYDROGEN PRODUCTION

Green hydrogen, also known as renewable hydrogen, is produced through the process of electrolysis, which involves splitting water (H_2O) into hydrogen (H_2) and oxygen (O_2) using electricity generated from renewable sources such as wind or solar. This process is considered to be environmentally friendly and sustainable as it does not produce any greenhouse gas emissions.

One of the main challenges in the production of green hydrogen is the efficiency of the electrolysis process. Traditional electrolysis technologies, such as alkaline electrolysis and proton exchange membrane (PEM) electrolysis, are relatively energy-intensive and have low efficiencies. This results in higher operating costs and limits the scalability of green hydrogen production.

Another challenge is the thermodynamics of water splitting. The Gibbs free energy change for the electrolysis of water is quite high needing considerable electrical energy for the reaction to move forward and drive the reaction. This means that the cost of producing green hydrogen can be quite high, especially when renewable energy sources are not readily available or affordable.

In addition to these challenges, the cost of electrolyzer technology and infrastructure is also a significant barrier to the widespread adoption of green hydrogen. Electrolyzers are currently quite expensive to manufacture and operate, making it difficult for industry to invest in large-scale green hydrogen production facilities.

Despite these challenges, the production of green hydrogen is gaining momentum around the world. Several countries, including Germany, Japan, and South Korea, have announced ambitious goals to increase their green hydrogen production capacity in the coming years. The European Union has also set a target of producing 10 million tons of green hydrogen by 2030.

To overcome the challenges in technology, thermodynamics, and cost, research and development efforts are focused on improving the efficiency of electrolysis processes, developing new electrolyzer technologies, and reducing the overall cost of green hydrogen production. Advances in materials science, catalyst development, and process optimization are expected to drive down the cost of green hydrogen and make it more competitive with traditional fossil fuels.

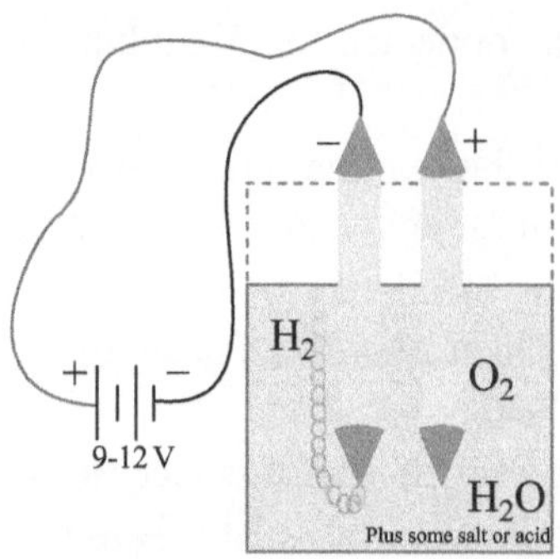

Figure 17: Electrolysis of Water

General

From thermodynamics angle water splitting is not a favourable reaction because the Gibbs free energy is + 237 KJ/mole. The single reason that unleashes many thermodynamic issues is the fact that water splitting is an endothermic reaction.

This reaction is endothermic, meaning it requires an input of energy in order to proceed. The main challenge with water splitting is the high energy input required to drive the reaction. This energy input adds to the cost of producing hydrogen, making it less economically viable compared to other sources of hydrogen production such as fossil fuels.

The standard thermodynamic potential for splitting water into hydrogen and oxygen is high, requiring an energy input of about 1.23 volts at room temperature. This voltage is known as the thermodynamic minimum voltage required for water electrolysis. In practice, however, additional voltage is needed to overcome various losses in the system, such as resistance in the electrolysis cell and overpotentials at the electrodes. These factors increase the overall energy input required for water splitting, leading to higher costs.

Electricity is the flow of electrons through a conductive path like a wire. This path is called a circuit. The electrical potential difference between the anode and the cathode makes the electrons to flow from the anode to cathode. Anode has more electrons and unstable because of crowding of electrons. The electrons want to rearrange themselves to get rid of this difference. Electrons repel each other and try to go to a place with fewer electrons. That place is cathode. Water splitting is slow red-ox reaction because pure water does not conduct electricity. Therefore, water splitting is not an easy reaction. Electrolytes are used to speed up the ionization of water. How an electrolyte is selected is a complex chemistry. This is not a part of this post.

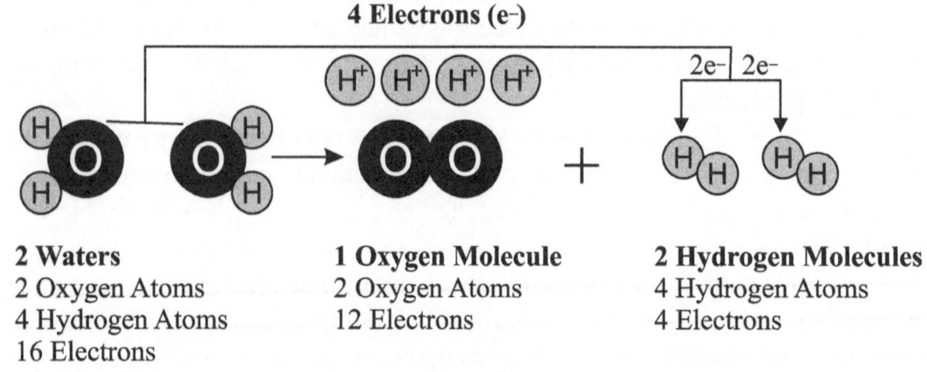

Figure 18

In the electrolyser there is one cathode and one anode connected to power source. Electrons always flow from anode to cathode. Anode is always a supplier of electrons and the cathode is the receiver of electrons. Oxidation happens at anode and reduction happens at cathode.

Cathode (reduction): $2H_2O(\mathrm{l}) + 2e- \rightarrow H_2(g) + 2OH-(aq)$ [gain of electrons]

Anode (oxidation): $2OH-(aq) \rightarrow \frac{1}{2}O_2(g) + H_2O(\mathrm{l}) + 2e-$ [loss of electrons]

Combination of these reactions produces:

$$2H_2O(l) \rightarrow 2H_2(g) + O_2(g)$$

H_2 is produced at the cathode and O_2 at the anode.

Process

A DC electrical power supply is connected to two electrodes, or plates, in the water (usually made of an inert metal like platinum or iridium). At the cathode (where electrons enter the water), hydrogen will appear, whereas oxygen will appear at the anode. Assuming perfect faradaic efficiency, the amount of hydrogen produced is twice that of oxygen, and both are proportional to the total electrical charge carried by the solution. However, competing side reactions occur in multiple cells, resulting in diverse products and a faradaic efficiency loss.

To overcome numerous activation obstacles, electrolysis of pure water necessitates additional energy in the form of overpotential. Electrolysis of pure water occurs very slowly or not at all without additional energy.

Thermodynamic and Kinetic Barriers to Water Splitting

Water splitting is a kinetically slow reaction because the water oxidation step is the rate-determining step. This is because oxygen is evolved more than four times slower than hydrogen.

The thermodynamic potential for water splitting is 1.23 volts at 25°C and 1 atm. However, the actual potential required to split water in a practical electrolysis cell is higher than this due to a kinetic barrier. This is because the reaction is kinetically controlled, meaning that additional potential is required to overcome factors such as activation energy, ion mobility, and surface hindrance. The amount of additional potential required is called the overpotential.

Water splitting is an endothermic process, meaning it requires energy to occur. The energy required is equivalent to $\Delta G = 237$ kJ/mol or a potential value of 1.23 eV per electron.

Difficult Oxidation Step

The oxidation step in water splitting is the reaction that generates oxygen,

$$2H_2O \rightarrow O_2 + 4H^+ + 4e-$$

This reaction is part of a redox reaction that also includes a reduction step that generates hydrogen,

$$4H^+ + 4e^- \rightarrow 2H_2$$

The oxidation step is the more difficult of the two because it requires the formation of an oxygen-oxygen bond and the coupling of four electron and proton transfers. In fact, oxygen evolves more than four times more slowly than hydrogen.

Large Kinetic Barrier

Water splitting requires a larger experimental potential than the thermodynamic potential.

What does it mean?

The large kinetic barrier in water splitting is the obstacle that must be overcome in the oxygen evolution reaction (OER), which is a thermodynamically unfavorable process.

Explanation

The OER (oxygen evolution reaction) is the primary reaction at the anode in many electrochemical energy conversion processes. The kinetic barrier for the OER is a significant obstacle in water-splitting systems because it requires a greater potential than the thermodynamic potential to overcome.

Thermodynamics

The endothermic nature of water splitting is one of the key reasons for its high cost. This process requires a significant amount of energy input in order to break the bonds of water molecules and separate them into hydrogen and oxygen gas. This energy input often comes from electricity, which can be expensive depending on the source. Additionally, the equipment and infrastructure needed for water splitting can also contribute to the high cost of the process.

The energy required to split water is 237 kilojoules per mole (kJ/mol) or 1.23 electron volts (eV) per electron. This is the standard Gibbs free energy of formation of water at 25 degc.

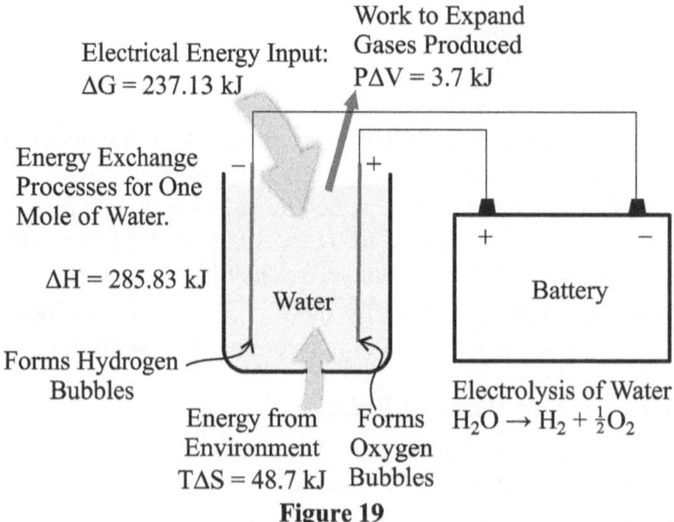

Figure 19

This is the minimum amount of electrical energy required to dissociate one mole of water through electrolysis.

Practical Energy

In practice, more than 1.8 volts is required to overcome the reaction's activation barrier. This is due to the slow kinetics of the anodic oxidation reaction and the fast kinetics of the cathode reduction reaction.

In summary, the enthalpy change for water splitting is 285.85 kilojoules (kJ) per mole of hydrogen gas generated. The internal energy change is 237.2 Kilojoules/mole. The thermodynamic work for expansion of H_2 and O_2 gases is 3.7 Kilojoules/mole. The entropy generation in water splitting is positive. TdS in Gibbs equation, $dG = dH - TdS$ is 48.7 Kilojoules/mole The electrical energy provides the internal energy [237.2 kilojoule/mole]. Thermal energy provides TdS equivalent energy 48.7 kilojoules/mole.

The thermochemical process of splitting water (H_2O) into hydrogen (H_2) and oxygen (O_2) involves overcoming the bond energies of the molecules and transitioning to a lower energy state for the products. The standard enthalpy changes for the water splitting reaction is 237 kJ/mole, which corresponds to the energy required to break the bonds in water and separate the hydrogen and oxygen atoms.

In terms of electrical energy, the theoretical minimum energy required to split water is 1.23 eV per electron, which corresponds to the standard potential for the water-splitting half-reaction at standard conditions. However, in practice, the energy requirements are higher due to various losses and inefficiencies in the system.

In addition to the electrical energy input, thermal energy is also required to account for entropy generation and the PV work associated with the expansion of the hydrogen and oxygen gases once they are formed. The exact amount of thermal energy required will depend on the temperature and pressure conditions of the system.

Overall, the energy requirements for water splitting are significant and represent a major challenge for the widespread adoption of green hydrogen production. Research and development efforts are focused on developing more efficient and cost-effective water splitting technologies to overcome these challenges and drive down the overall energy costs. Advances in materials science, electrochemistry, and process engineering are expected to play a key role in realizing the full potential of green hydrogen as a clean energy carrier.

What Makes Water Splitting Kinetically Slow?

Background

The cost is the major concern in green hydrogen making. It takes around 50–55 kilowatt-hours (kWh) of electricity to produce 1 kg of green hydrogen. This is because most commercial devices have inefficiencies, even though an electrolysis system with no inefficiencies would require 39.4 kWh.

The levelized cost of production (LCOH) for green hydrogen is generally between US\$3–6 per kilogram. However, the cost can vary depending on the capital cost of electrolyzers, the capacity of the electrolyzers, and the cost of renewable electricity.

Sluggish Kinetics in Water Splitting

The reason is that the activation energy of two water molecules combining to form O_2 at anode is much higher than H combining into H_2 by reduction reaction at cathode. The OER or oxygen evolution at anode is the slowest step and also the rate determine step of water splitting. The OER is four times slower than HER (hydrogen evolution rate).

In oxidation process, water molecules are oxidized at the anode by releasing electrons, which then travel through the external circuit to reduce protons to hydrogen gas at the cathode.

Anode half-cell reaction: $2H_2O \rightarrow O_2 + 4H^+ + 4e^-$

Cathode half-cell reaction: $4H^+ + 4e^- \rightarrow 2H_2$

The anodic oxidation of water involves breaking down water molecules into oxygen gas, protons, and electrons. This process requires overcoming the activation energy barrier for water molecules to lose electrons and form oxygen gas. This barrier is higher compared to the activation energy required for the reduction of protons to hydrogen gas at the cathode.

Additionally, the water molecule itself is stable and requires energy input to be broken apart, making the anodic oxidation of water a slower process compared to the reduction of protons at the cathode.

Thus if you analyse the water splitting challenges, two things are easy to conclude that impact cost [1] endothermic nature of reaction and (2) sluggish kinetics.

WATER-GAS SHIFT REACTION [WGSR]: STEAM-METHANE REFORMING [SMR] HYDROGEN PRODUCTION

Summary

Steam methane reforming followed by the water-gas shift (WGS) reaction is an interplay of endothermic (steam methane reforming) and exothermic (WGS) reactions. In steam methane reforming, methane is reacted with steam to produce hydrogen gas and carbon monoxide. This reaction is endothermic, meaning it requires heat input to proceed. In the WGS reaction, carbon monoxide is reacted with steam to produce carbon dioxide and more hydrogen gas. This reaction is exothermic, meaning it releases heat. The combination of these two reactions allows for the production of hydrogen gas with a net energy input.

The water-gas shift (WGS) reaction is a key step in the steam methane reforming (SMR) process, which is used to produce hydrogen.

Role

The WGS reaction uses a catalyst to react carbon monoxide and steam to produce carbon dioxide and more hydrogen. This reaction is exothermic, releasing energy.

Importance

The WGS reaction is essential for increasing the yield of hydrogen. It's also used in the production of ammonia, methanol, and hydrocarbons.

Process

The SMR process is endothermic, meaning it requires heat to proceed. The product gas from the SMR reformer is cooled and then fed into the WGS membrane reactor. The WGS reaction produces a syngas containing hydrogen, carbon monoxide, carbon dioxide, water, and natural gas.

Details

Water-Gas shift reaction is a wonderful example of balancing rate and conversion of a reaction by shifting equilibrium constant by the application of temperature. Steam methane reforming (SMR) is a process that uses steam to produce hydrogen from natural gas or other low-boiling hydrocarbons. Natural gas is the most common feedstock in steam reformers; naphtha and refinery off-gas may also be used. The process is called "reforming" because it involves rearranging the molecular structure of the hydrocarbons to change their properties. In the subsequent step called 'Water Gas Shift reactions' pure hydrogen is separated from impurities.

Steam-Methane Reforming [SMR]

Most hydrogen produced today in the United States is made via steam-methane reforming, a mature production process in which high-temperature steam (700°C–1,000°C) is used to produce hydrogen from a methane source, such as natural gas. In steam-methane reforming, methane reacts with steam under 3–25 bar pressure (1 bar = 14.5 psi) in the presence of a catalyst to produce hydrogen, carbon monoxide, and a relatively small amount of carbon dioxide. Steam reforming is endothermic—that is, heat must be supplied to the process for the reaction to proceed.

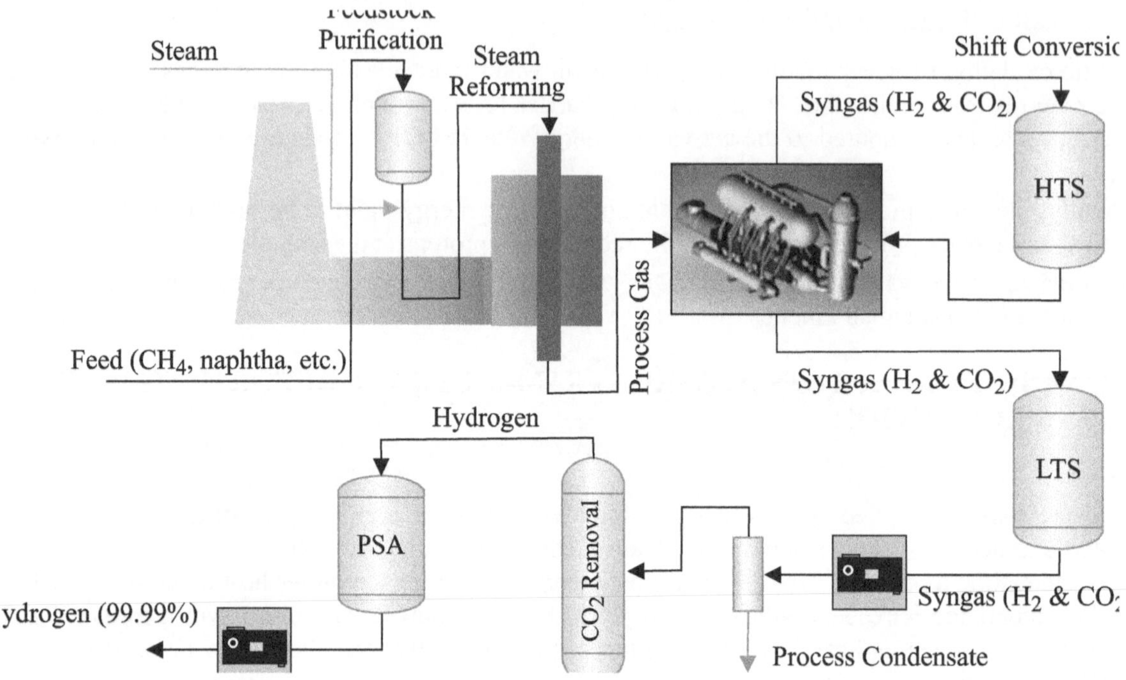

Figure 20: Hydrogen Production

Subsequently, in what is called the **"water-gas shift reaction,"** the carbon monoxide and steam are reacted using a catalyst to produce carbon dioxide and more hydrogen. In a final process step called "pressure-swing adsorption," carbon dioxide and other impurities are removed from the gas stream, leaving essentially pure hydrogen. Steam reforming can also be used to produce hydrogen from other fuels, such as ethanol, propane, or even gasoline.

Steam-methane reforming reaction

$$CH_4 + H_2O \text{ (+ heat)} \rightarrow CO + 3H_2$$

Water-gas Shift Reaction

$$CO + H_2O \rightarrow CO_2 + H_2 \text{ (+ small amount of heat)}$$

The water gas shift (WGS) reaction is a chemical process that converts carbon monoxide (CO) and water (H_2O) into carbon dioxide (CO_2) and hydrogen (H_2). In industrial applications, the WGS reaction is carried out in two stages: a high temperature shift (HTS) and a low temperature shift (LTS). The high-temperature shift (HTS) and low-temperature shift (LTS) are two stages in the water-gas shift (WGS) reaction that are used to take advantage of the reaction's kinetics and thermodynamics. The WGS reaction is a series of two adiabatic reactors that's used to adjust the H/CO molar ratio.

The combination of HTS and LTS allows for the adjustment of the final gas composition with few thermodynamic limitations. The currently used method to achieve high purity (99.999%) hydrogen, after the water gas shift reaction, is pressure swing adsorption (PSA).

HTS

This stage uses an iron-based catalyst and operates at a higher temperature (350–550°C) to take advantage of the reaction's high rates. However, the conversion of carbon monoxide is incomplete.

LTS

This stage uses a copper-based catalyst and operates at a lower temperature (200–300°C) to lower the carbon monoxide content to less than 1%.

HTS

The first stage of the WGS reaction, which takes place at a higher temperature to quickly convert the majority of carbon monoxide. HTS catalysts are typically made of iron oxide and chromium oxide, and operate at temperatures of 350–550°C.

LTS

The second stage of the WGS reaction, which takes place at a lower temperature to complete the remaining hydrogen generation. LTS catalysts are typically made of copper and zinc, and operate at temperatures of 200–300°C.

The WGS reaction is typically performed in two adiabatic reactors, with interstage cooling between the two stages. The catalysts used in each stage are different.

HTS

Uses an iron oxide–chromium oxide catalyst, such as Fe_2O_3/Cr_2O_3, which operates at temperatures of 350–550°C. The chromium oxide stabilizes the iron oxide.

LTS

Uses a copper-based catalyst, such as ZnO/CuO, which operates at temperatures of 200–300°C. The copper catalyst is sensitive to sulfur poisoning, so a guard bed is used to remove sulfur compounds before the LTS reactor.

The water gas shift reaction (WGSR) is a chemical reaction that converts carbon monoxide (CO) and water (H_2O) into carbon dioxide (CO_2) and hydrogen (H_2).

$$CO + H_2O \rightleftharpoons CO_2 + H_2$$

More about WGS

Reaction Type

The WGS reaction is moderately exothermic, meaning it releases heat. This makes it thermodynamically unfavorable at higher temperatures.

Equilibrium

The WGS reaction is operated close to thermodynamic equilibrium. To shift the equilibrium toward hydrogen production, the reaction is carried out in multiple adiabatic stages.

- **High Temperature Shift (HTS)**: This stage takes advantage of high reaction rates, but the thermodynamic limitations result in incomplete carbon monoxide conversion.
- **Low Temperature Shift (LTS)**: This stage produces a carbon monoxide exit composition of less than 1%. The LTS reactor is significantly lower in temperature than the HTS reactor.

Catalysts

Different catalysts are used at each stage to ensure optimal activity. The HTS catalyst is usually iron oxide–chromium oxide, while the LTS catalyst is usually copper-based.

Hydrogen Selective Membranes

Integrating hydrogen selective membranes can enhance WGS reactor performance by removing hydrogen from the reaction environment.

Kinetics of WGS Reaction

Here are some details about the kinetics of the water-gas shift (WGS) reaction:

Reaction equation.

The WGS reaction is $CO + H_2O \leftrightarrow H_2 + CO_2$

Reaction Rate

The reaction rate increases with higher concentrations of CO or H_2O, but decreases with higher concentrations of H_2. The reaction rate is nearly independent of CO_2 concentration.

Equilibrium

The WGS reaction is equilibrium-controlled at high temperatures and kinetically limited at lower temperatures.

Catalyst

A catalyst is required to drive the reaction forward at lower temperatures. The $CuO–ZnO/Al_2O_3$ catalyst is commonly used in industry.

Temperature

CO conversion increases with temperature. For example, one study found that a maximum CO conversion of 94% was achieved for temperatures greater than 300°C.

Role of Le Chatelier's Principle in WGS Reaction

Le Chatelier's principle can be applied to the water-gas shift reaction to predict how changes in temperature and pressure will affect the reaction's equilibrium.

Le Chatelier's principle states that when a reaction at equilibrium is disturbed by altering concentration, pressure, or temperature, the system will shift its equilibrium position to counteract the change.

Temperature

In an exothermic reaction like the water-gas shift reaction, increasing the temperature shifts the equilibrium towards the reactants, reducing hydrogen yield. Decreasing the temperature shifts the equilibrium towards the products, increasing the yield of hydrogen.

Pressure

In the water-gas shift reaction, pressure changes do not affect the equilibrium yield of hydrogen because the reaction involves equal moles of gas on both sides.

Concentration

The presence of CO_2 in the feed gas shifts the equilibrium to the right, favoring the formation of CO_2 and hydrogen.

Sequence of HTS and LTS in WGS Reaction

The sequence of the water-gas shift (WGS) reaction is high-temperature shift (HTS) followed by low-temperature shift (LTS).

HTS

The initial stage of the WGS reaction, which takes place at a higher temperature to convert carbon monoxide at a faster rate. HTS typically operates at 300–350°C and uses an iron-based catalyst.

LTS

The final stage of the WGS reaction, which takes place at a lower temperature to complete the hydrogen generation. LTS typically operates at 200–250°C and uses a copper-based catalyst.

The WGS reaction is a series of two adiabatic reactors that's used to adjust the *H/CO* molar ratio. The combination of HTS and LTS allows for the adjustment of the final gas composition with few thermodynamic limitations.

The currently used method to achieve high purity (99.999%) hydrogen, after the water gas shift reaction, is pressure swing adsorption (PSA).

In summary, the water-gas shift (WGS) reaction is a key step in the steam methane reforming (SMR) process, which is used to produce hydrogen.

Role

The WGS reaction uses a catalyst to react carbon monoxide and steam to produce carbon dioxide and more hydrogen. This reaction is exothermic, releasing energy.

Importance

The WGS reaction is essential for increasing the yield of hydrogen. It's also used in the production of ammonia, methanol, and hydrocarbons.

VAPOR COMPRESSION REFRIGERATION

Know Your Fridge

A refrigerator works on the principle of vapor compression refrigeration.
A refrigeration cycle typically has four parts.

[1] compressor [2] condenser [3] expansion valve and [4] evaporator. These four components keep the refrigerant moving in a closed loop called the refrigeration cycle.

To summarize Compressor works adiabatically, the condenser works isobarically, the throttling valve works adiabatically, Evaporator works isothermally/isobarically.

The throttling valve does partial vaporization of the refrigerant. In the evaporator, the liquid refrigerant does expansion work, W by extracting heat from the surroundings which are compensated by the compression by the compressor in the following step. In the subsequent steps, the compressed vapor refrigerant cools in the condenser and becomes the feed to the throttling valve.

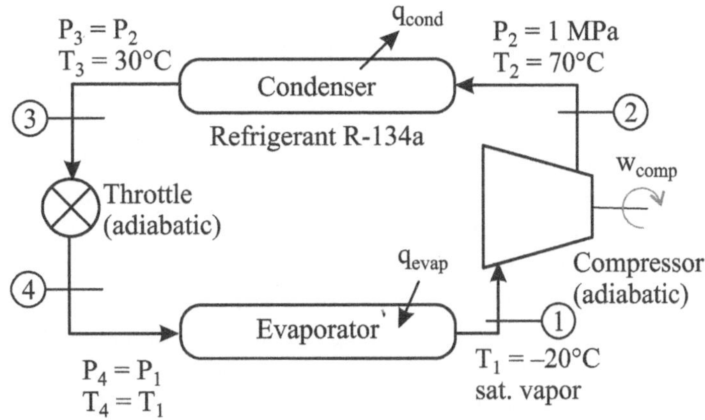

Figure 21: Vapor Compression Cycle

The above is a typical refrigeration cycle of refrigerant 134a.

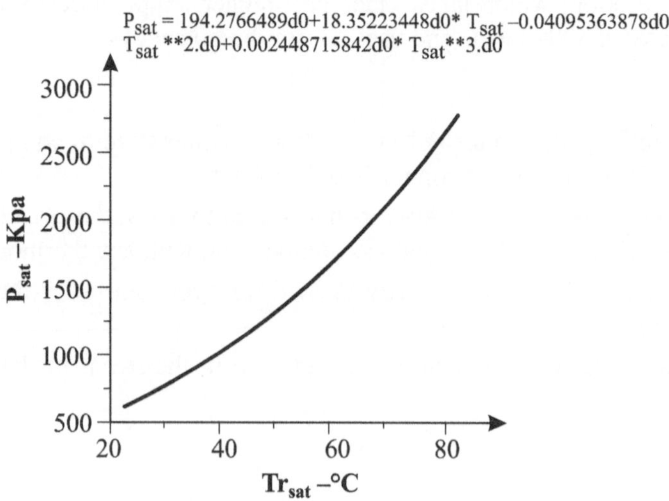

Figure 22: Saturation Properties of Refrigerant R134a

The above graph shows the saturation properties of refrigerant 134a.

Point by Point (refer to the refrigeration cycle)

Low-pressure refrigerant vapor at point.

[1] Enters the compressor at –20 degc. The compressor compresses the low-temperature refrigerant vapor to 1 Mpa and 70 degc. At 1 Mpa, the saturation temperature of 134a is equal to 46.02°C, therefore, the compressed refrigerant carries about [70–46] = 24 degc superheat. See the saturation properties of 134a in the graph above.

[2] The compressor works adiabatically on refrigerant vapor, $H = U + PV$, and adds internal energy U to refrigerant vapor.

[3] Superheated refrigerant vapor from the compressor enters and cools at constant pressure [isobaric] to liquid refrigerant by releasing its sensible and latent heat to water circulating inside the condenser to 30 degc at constant pressure 1Mpa. There is no expansion or compression of refrigerant inside the condenser. It is an isobaric phase change that occurs within the condenser. It is not an isothermal process. In the condenser, the temperature cools down from 70 degc at the exit of the compressor to 30 degc while pressure remains constant. The saturation temperature here is 46.02 degc. The liquid refrigerant sub-cools in the condenser by about 16 degc.

[4] Next, the liquid refrigerant enters the expansion valve and expands adiabatically, and cools isenthalpically. The cooled partially vaporized refrigerant enters the evaporator.

Evaporation is an isobaric/isothermal process. At constant temperature and pressure at –20 degc, the refrigerant vapor extracts heat from the surroundings hot space. The saturation temperature of 134a varies between –20 to –40 degc depending on the pressure. The refrigerant vapor exits the evaporator with superheat ensuring no liquid refrigerant in the feed to the compressor.

The most interesting point in the refrigeration cycle is the evaporator where both isothermal and isobaric conditions work. If you see the diagram both at the inlet and outlet of the evaporator, temperature and pressure are constant. The only difference is at the inlet of the evaporator the refrigerant is saturated liquid and at the exit, it is a superheated vapor while the temperature remains constant. At constant temperature and pressure, the heat extracted in the evaporator supplies the latent heat of the vaporization of the refrigerant. The latent heat is internal energy + work. In the subsequent compression process the compressor replenishes this enthalpy, internal energy + work + adds frictional heat. A very interesting thing is at constant temperature there is no change in the internal energy in the evaporator. Only the work PdV at constant pressure reduces. So, at evaporator $H = -PdV$. This enthalpy H which is the total energy (work) extracted from the hot space finally is rejected in the condenser to surroundings as – W which is made by the compressor by doing + W work. Finally, this work $H = W$, extracted from hot space along with frictional heat added by the compressor and is rejected by the condenser. Therefore, vapor compression refrigeration is a process of work transfer by and large.

Essentially, the compressor *COP* = Work energy rejected by condenser/Work energy added by compressor × 100.

Remember while work as heat rejection by condenser is the thermal energy, work addition to compressor is electrical energy. Therefore, the performance of refrigeration process is measured by *COP* (Coefficient of performance) and not in percent terms.

How it is Happening?

At constant pressure when the refrigerant is converting to vapor by taking latent heat from its surroundings [hot space] it is expanding. At constant P when V expands by breaking its intermolecular bonds, this needs energy.

The supply of energy to break bonds for phase change comes from the hot space [surroundings]. So, the process remains isothermal. The work of expansion of liquid refrigerant to vapor $- W$ is made up by the compressor by doing $a + W$ work on the refrigerant.

To Repeat

Compressor works adiabatically

Condenser works isobarically

The throttling valve works adiabatically

The evaporator works isothermally/isobarically

The throttling valve does partial vaporization of the refringent. It transfers its expansion work $[-W]$ to the evaporator.

Explanation

A fridge works by enthalpy transfer between the evaporator and the condenser. Briefly the expansion (throttling) valve sends saturated liquid refrigerant to the evaporator at a constant temperature and pressure, see the image.

The Job of Throttling the Valve

The throttling valve regulates the flow of the refrigerant and maintains a constant flow and pressure in the evaporator. It may further be noted as the throttling valve cools the liquid refrigerant its latent heat increases significantly. This increases the heat extraction capacity of the refrigerant in the evaporator. It is an adiabatic process. The liquid refrigerant cools isenthalpically keeping the total enthalpy constant across the expansion valve, $H_1 = H_2$ leaving behind, U [internal energy], and W [work], to compensate each other across the throttling valve $U_1 + W_1 = U_2 + W_2$, with no loss of energy.

The partially vaporized refrigerant then moves to the evaporator in the next station at the prevailing pressure in the evaporator [see the image].

The Job of the Evaporator

The liquid refrigerant takes heat from the hot space in the evaporator to change phase. The hot space in the fridge cools. The heat goes to break intermolecular bonds of the liquid refrigerant and converts the liquid to gas thereby increasing the internal energy of the refrigerant gas. A part of the heat extracted from the hot space expands the vapor and drops the pressure. Here, heat converts to work. The expansion of the refrigerant to vapor in the evaporator is – W, negative work by the refrigerant.

The Job of the Compressor

In the next station, the compressor restores the pressure by compression and converts the work of expansion back to heat. In other words, the compressor replenishes the $- W$ [expansion work] by the refrigerant in the evaporator by doing $+ W$ [compression work] on the refrigerant vapor. It is always necessary to have superheated vapor leaving the compressor. If it wasn't superheated, it would be an indication there is liquid refrigerant in the compressor. For the same reason, the inlet of the compressor should have superheated gases to it. No liquid.

The Job of the Condenser

The compressed refrigerant gas in the condenser (work to heat conversion) compresses into liquid by releasing heat (exothermic). The condenser squeezes out the heat added by the compressor at constant pressure, see the image and convert the gas refrigerant to a sub-cooled liquid [explained above] at the prevailing pressure in the condenser.

In summary, the first point is vapor compression refrigeration is a work transfer process.

To, repeat, the most interesting point in the entire refrigeration cycle is the evaporator where both isothermal and isobaric conditions are working. If you see the diagram both at the inlet and outlet of the evaporator, temperature and pressure are constant. The only difference is at the inlet of the evaporator the refrigerant is saturated liquid and at the exit, it is a superheated vapor while the temperature remains constant.

At constant temperature and pressure, the heat extracted in the evaporator supplies the latent heat of the vaporization of the refrigerant. The latent heat is internal energy + work. In the subsequent compression process the compressor replenishes this enthalpy = internal energy + work. A very interesting thing is at constant temperature there is no change in the internal energy in the evaporator. Only the work PdV at constant pressure reduces. So, at evaporator $H = -PdV$. This enthalpy H which is the total energy (work) extracted from the hot space finally is rejected in the condenser to surroundings as $- W$ which is made up by the compressor by doing $+ W$ work.

Finally, this work $H = W$, extracted from hot space is rejected by the condenser.

Therefore, vapor compression refrigeration is a process of work transfer.

The benefit of constant temperature and pressure in the evaporator is that there is no change in the internal energy of the refrigerant in the evaporator. dH (enthalpy change) = PdV.

Essentially, the compressor COP = Work energy rejected by condenser/Work energy added by compressor × 100. Remember while work as heat rejection by the condenser is the thermal energy, work addition to the compressor is electrical energy. Therefore, the performance of the refrigeration process is measured by COP (Coefficient of performance) and not in percent terms.

Here You May have a Couple of Questions

The question can arise as to why the condenser of a refrigerator feels so hot when the source of heat is heat inside the fridge which is near 0. The answer is the compressor by compressing, increases the saturation temperature of the refrigerant gas several times more than the saturation temperature of the liquid refrigerant in the evaporator. That makes the condensation temperature of the refrigerant gas in the condenser several times higher than the evaporator.

The obvious question that comes next is how refrigerant gas moves from near 0 degc at the evaporator to the condenser at ambient temperature. The answer is latent heat or the enthalpy of vaporization of the liquid refrigerant at nearly 0 degc in the evaporator is much higher than the latent heat of liquid refrigerant at ambient temperature in the condenser. The reason is simple. There are comparatively fewer intermolecular bonds at 30 degc to break than at 0 degc. This speaks of the role of the compressor.

The next station is the expansion valve [already explained.

The cycle repeats.

VAPOR ABSORPTION REFRIGERATION

An absorption refrigerator uses a heat source to provide the energy needed to drive the cooling process. The advantage of a vapor absorption system is that, unlike a conventional vapor compression system, it doesn't need mechanical work input to operate. Instead, it can be powered by heat sources like solar energy, waste heat, or fuel combustion. Because of this, it is a good option for applications that have access to waste heat or renewable energy sources.

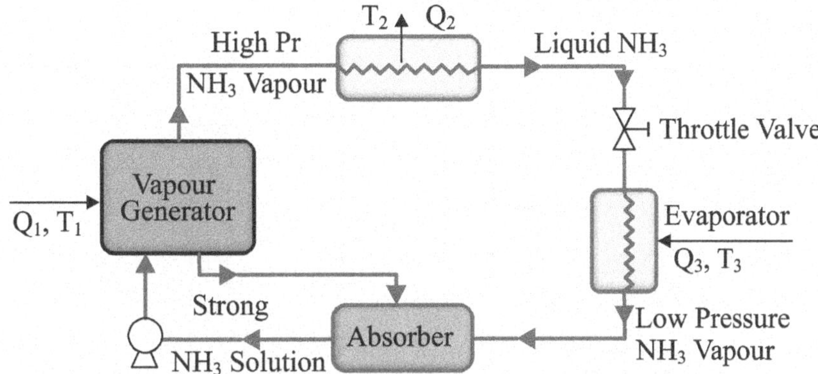

Figure 23: Simple Vapour Absorption System

Working Principle

An absorbent solution absorbs refrigerant vapor, which is the basis for the operation of a vapor absorption refrigeration system. The following are the essential actions in running a vapor absorption refrigeration system:

- **Refrigerant Evaporation**: Heat from the area being cooled is absorbed by the refrigerant, which is usually lithium bromide or ammonia. This causes the refrigerant to evaporate and turn into a low-pressure vapour.
- **Refrigerant Absorption**: Next, a strong absorbent solution made primarily of water and an absorbent such as lithium bromide or ammonia is added to absorb the refrigerant vapour. Heat is produced by this process of absorption.
- **Refrigerant Desorption**: To separate the refrigerant from the absorbent, the heat produced during absorption is extracted and the absorbent solution is heated. High-pressure refrigerant vapour is produced during this desorption process.
- **Refrigerant Condensation**: The latent heat of vaporization is released by cooling and condensing the high-pressure refrigerant vapor. Usually, this heat is released into the surroundings.
- **Refrigerant Expansion**: The cycle is then repeated by the liquid refrigerant entering the evaporator after passing through an expansion valve that lowers its temperature and pressure.

Detail

[Let us assume that the refrigerant is ammonia and water is the absorbent]

The image above stands for a typical vapor absorption refrigeration process.

It may be noted that in a vapor absorption cycle.

The process occurs in the following steps

Absorber: Ammonia absorption

It may be observed that low-pressure ammonia gas from the evaporator enters the absorber. The absorber contains an ammonia solution in water. The dissolution of ammonia gas in water is an exothermic process. To keep the ammonia gas dissolution moving forward the cold water is circulated in the absorber.

Ammonia Regeneration: Ammonia Desorption

The resulting strong solution of ammonia from the absorber is pumped into the ammonia gas generator in the next step as you may see in the image above. Here the strong solution of ammonia is heated under pressure for the desorption of the ammonia gas. The dilute solution of ammonia post-stripping of ammonia gas in the regenerator is fed back into the absorber.

Ammonia Condensation

In the next step, the high-pressure ammonia gas is fed to a condenser where the ammonia gas condenses into liquid ammonia.

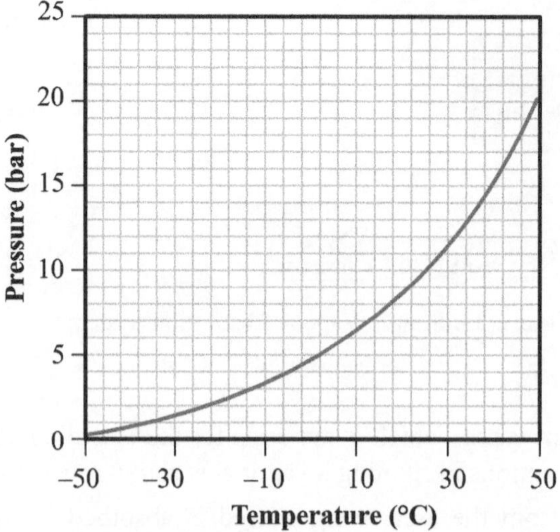

Figure 24: Ammonia Vapor Pressure Curve

In the condenser at 7–8 bar pressure the ammonia gas converts to liquid at 20–25 degc. Please refer to the vapor pressure curve of ammonia above.

The latent heat of vaporization is released by cooling and condensing the high-pressure refrigerant vapor. Usually, this heat is released into the surroundings.

Ammonia Expansion

The liquid ammonia from the condenser is cooled by using a throttling valve. The low-pressure liquid ammonia is then fed into the evaporator as you may see in the image of the vapor refrigeration cycle above. You may also notice in the vapor pressure curve of ammonia that you just need about 2 bar pressure for –18 to –20 degc saturation temperature of liquid ammonia in the evaporator. The cold liquid ammonia pulls the latent heat from the surroundings and cools the hot space while converting itself into gas.

The cycle repeats.

HEAT PUMP

Heat pump works on the principle of refrigeration.

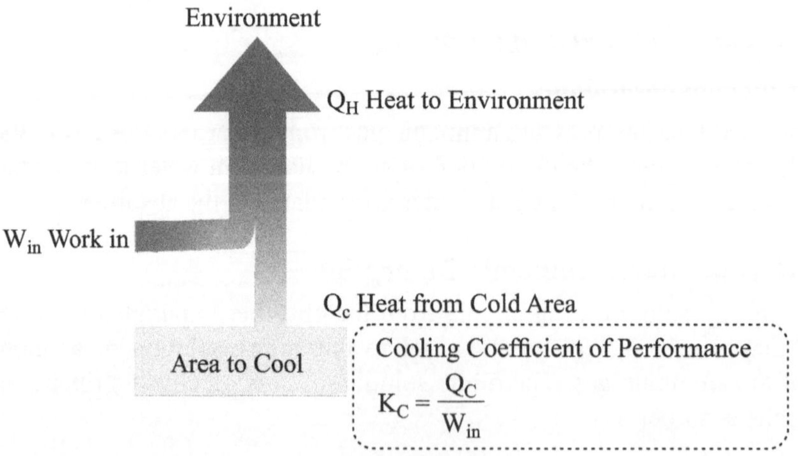

Figure 25: Environment of Heat Pump

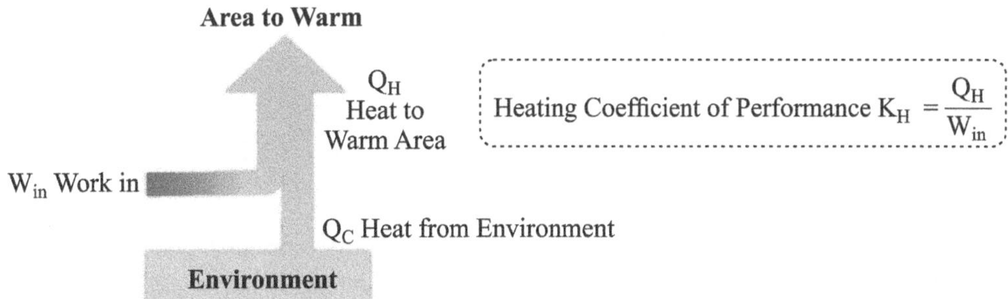

Figure 26: Heat Pump Area to Warm

Refrigerator/Air conditioning: Refrigerators extract heat from the interior and expel it to the external environment, maintaining a temperature below room temperature.

The cooling coefficient of performance of a system (refrigerator, air conditioning) is the ratio of heat out of the cold reservoir to the work put into the system.

Heat pump [above image]: Heat pumps, on the other hand, transfer heat from a colder area to a warmer area in heating mode and remove heat from space in cooling mode.

The heating coefficient of performance of a system (heat pump) is the ratio of heat delivered to the hot reservoir to the work put into the system.

PERFORMANCE RATING

Heat Engines

For a heat engine working between two heat reservoirs, a hot reservoir at high temperature T_H and a cold temperature at low temperature T_L, the maximal achievable efficiency by the heat engine is expressed as Carnot efficiency

$(nC = 1 - T_L/T_H)$

Heat Pump

The performance of a heat pump is expressed by *COP*. The Coefficient of Performance (*COP*) is a performance rating that tells us how effective a heat pump or air conditioner is transferring heat versus the amount of electrical power it consumes.

Why *COP* is expressed as a number: By convention *COP* is expressed as a dimensionless number.

COP = Heat extracted/Electrical energy in.

What is the convention? If the required output has to be in the form of work, we calculate efficiency. If the required output is in the form of heat, we calculate Coefficient of Performance (*COP*). In Refrigeration and Air Conditioning, the desired output is always in the form of heat. Hence the *COP*.

Coefficient of performance demonstrates instantaneous performance without any factor of time involved. That is to say, the *COP* tells us exactly how much heat a piece of HVAC equipment transfers when given a specific amount of electrical input power at a specific temperature, and therefore, at a specific and instantaneous moment of time, instead of over a period of time.

***COP* = Output watt [power]/Input watt [power]** [both numerator and denominator are expressed as watt]

So, when we speak in terms of the coefficient of performance, we are speaking in terms of instantaneous performance, which means we speak in terms of power and not energy (Energy is power over time) and more importantly, output power *vs* input power. Which means, this output power to input power can be expressed as a ratio, and this is exactly what the *COP* is. *COP* then, tells us how much heat output power an air conditioner or heat pump transfers versus its electrical input power. For example: A Heat Pump with a *COP* of 4:1, or 4, means that for every 1 unit of electrical input power put into the Heat Pump, it provides 4 units of heat output power.

There are Two COPS

- *COP* of cooling [refrigeration/air conditioning]
- *COP* of heating [heat pump]
- *COP* of cooling = *Qc*/*W* [*Qc* is cold space heat and *W* is work by compressor]
- *COP* of heating = [*W* + *Qc*]/*W* by compressor

The *COP* for heating and cooling are different because the heat reservoir of interest is different. When one is interested in how well a machine cools, the *COP* is the ratio of the heat taken up from the cold reservoir to input work. However, for heating, the *COP* is the ratio of the magnitude of the heat given off to the hot reservoir (which is the heat taken up from the cold reservoir plus the input work) to the input work:

COP of cooling in details: How COP cooling changes: Vapor compression vs vapor absorption

COP of cooling depends on the ratio of Qc and work input, W

Vapor absorption refrigeration COP < 1

The *COP* usually exceeds 1, especially in heat pumps [to be discussed later], because instead of just converting work to heat (which, if 100% efficient, would be a *COP* of 1), it pumps additional heat from a heat source to where the heat is required. Most air conditioners have a *COP* of 2.3 to 3.5. Less work is required to move heat than for conversion into heat [like vapor absorption], and because of this, heat pumps, air conditioners and refrigeration systems can have a coefficient of performance greater than one.

In the case of vapor compression system, it is the mechanical or electrical work, which is the Input to the machine. Since, "work" is considered as High-Grade energy, the conversion of "work" into other forms of energy, the process is more efficient, and the *COP* is mostly greater than 1. **In the case of vapor absorption system, the input is invariably "heat", which is a low-grade energy, its conversion into other forms is not efficient and the *COP* is less than 1.**

For a refrigerator, *Th* is the ambient temperature, and *Tc* is the temperature in the refrigerator for, a refrigerator, *COP* max = *Th*/*Th* – *Tc* [this comes from the inverse relation of *COP* with Carnot [maximum] efficiency.

COP of heating is always > 1

Fundamentals

Heat Pump Equation

Q h = *Q c* + *W*. The purpose of a heat pump is to transfer energy by heat to a warm environment, such as a home in the winter. A heat pump supplies *Qh* = *Qc* + *W*

COP of heating = [*W* + *QC*]/*W* = 1 + *Qc*/*W* = 1 + *COP*

It may be noted that while *COP* of cooling is *QC*/*W*, the *COP* of heating is [*QC* + *W*//*W*].

It may be observed in *COP* of heating the numerator is larger than *COP* of cooling. Note that the *COP* of a heat pump depends on its direction. The heat rejected to the hot sink is greater than the heat absorbed from the cold source, so the heating *COP* is greater by one than the cooling *COP*. *Qh* = *Qc* + *W*.

Adsorption has the maximum COP

COP = *Q* compression + *Q* latent heat of condensation/*W*

Q compression adds internal energy. That makes *COP* for adsorption very high.

COP and Carnot efficiency

Refrigeration, air conditioning and heat pumps follow inverse Carnot principle

Maximum Carnot efficiency = *Th*– *Tc*/*Th*

Maximum *COP* cooling = *Tc*/*Th* – *Tc*

Maximum *COP* heating = *Th*/*Th* – *Tc*

Part–II

Fundamentals of Thermodynamics with Technical Note for Engineers

Technical Notes

Welcome to the second part of this book, which is a compilation of technical notes on various thermodynamics subjects and related technical matters. Each note in this section is designed to be complete in itself, providing a detailed exploration of specific topics within the realm of thermodynamics.

You may find some repetition of certain matters from the first part of the book. This repetition is intentional and serves to reinforce key principles and ideas covered earlier, offering additional clarity and a deeper understanding of the concepts discussed.

The technical notes in this second part aim to provide a more detailed and in-depth analysis of complex thermodynamics concepts and related technical matters. By delving deeper into specific topics, we hope to offer a comprehensive resource for readers looking to enhance their knowledge and understanding of thermodynamics.

We invite you to explore the technical notes in this section and delve into the nuances of thermodynamics in a structured and organized manner. We hope that this compilation will serve as a valuable reference for students, researchers, and professionals in the field of thermodynamics.

Let's dive in and continue our exploration of the fascinating world of thermodynamics and technical matters.

CHAPTER 30

Temperature

Temperature is an important factor in determining the energy of particles in a system, which directly affects the overall behaviour and properties of the system in thermodynamics.

In thermodynamics, temperature plays a key role in defining the direction and magnitude of heat transfer between different components of a system. The measurement and control of temperature are essential in ensuring the efficiency and effectiveness of thermodynamic processes, such as heat engines and refrigeration systems. Changes in temperature can lead to phase transitions, such as melting or boiling, which are fundamental processes in thermodynamics.

WHAT IS TEMPERATURE?

The temperature gradient is a vector, it has both magnitude and direction In the Kelvin scale the temperature represents the average kinetic energy. The equation $KE = 1/2$ kBT is based on the law of equipartition of energy, which states that the average energy of each degree of freedom in thermal equilibrium is 1/KBT. KB is the Boltzmann constant and T is the temperature in Kelvin.

The velocity of molecules or the motion of molecules in a mass arises from the repulsion of electrons (electron cloud) around the nucleus. This is directly proportional to the temperature T. The molecular collisions generate temperature that is proportional to kinetic energy (KE).

Kinetic Molecular Theory

The kinetic molecular theory of gases states that the average kinetic energy of gas particles is directly proportional to the absolute temperature of the gas.

Temperature and KE

The temperature of a system increases if and only if the average kinetic energy of its particles increases.

What Follows?

The molecules in motion within a mass generate collisions. The collisions generate vibrations. The vibrations generate momentum. The momentum transfers from one molecule to another by collisions. The linear motions transfer momentum in a linear translational manner and the angular momentum generates rotational motions of molecules. The temperature is an intensive property of a substance linked to the atomic structure of the substance. Temperature is a fundamental property of a substance.

This is how three types of motions are created within a molecule (1) translational (2) rotational and (3) vibrational. These motions in x, y, and z-axes in the space store heat in a molecule that gives the heat capacity of a substance.

The storage space of heat in a molecule depends on the number of atoms in a molecule and how they store heat in different types of motions. This is called the degree of freedom". For instance, a single-atom substance has three degrees of freedom, all translational. A di-atomic molecule has five degrees of freedom. It can hold more heat than a single atom molecule. A tri-atomic molecule has seven degrees of freedom and hence it can hold more energy than a di-atomic molecule.

What Limits Temperature Rise?

The heat capacity which takes into account the storage space of heat in a substance limits the rise of temperature in a substance.

Example

At saturation temperature, a substance reaches its heat capacity and therefore, that stops any temperature rise at a given pressure.

HEAT AND TEMPERATURE

Temperature is an intensive property of substance. Temperature is a fundamental property of a substance. Heat is not a thermodynamic property. Heat is a scalar quantity. Heat is a thermal energy that always looks to reach an equilibrium with substances at a lower energy level and shares its enthalpy. Therefore, heat is always in transit. Heat has magnitude but no direction. Heat looks for a stable situation where Gibbs free energy change is zero. While doing so it interacts with the surroundings and performs thermodynamic work and while converting heat to work a portion of heat is lost as entropy. Heat shares its internal energy with its surroundings as it moves between two points.

It may be remembered that while heat shares its internal energy, as it moves between two points, within the surroundings, the heat has no control over the heat capacity of surroundings. The final call is taken by the specific heat of a substance that determines how much heat a substance can absorb.

Heat *vs.* Temperature at the Molecular Level

Heat

"Heat is the transfer of thermal energy between objects due to a temperature difference. It is described by the first law of thermodynamics ($Q = U + W$), where Q is the sum of the internal energy (a state function) and work (a path function). Since heat depends on its path for travel between two points, it is a property in transit that cannot be contained or stored. It can only be transferred from one object to another or converted to other forms of energy."

Temperature

"Heat is the internal energy of a substance, which includes both kinetic and potential energy, while temperature is a measure of the average kinetic energy of the particles in a substance. It's important to note that temperature alone does not provide a complete picture of the energy content of a substance, as it does not account for potential energy stored in chemical bonds. This distinction is crucial in understanding the behavior of materials and the transfer of energy in various physical processes. Another name of potential energy is latent heat. Latent heat is a form of potential energy. Heat or total thermal energy has both sensible and latent heat. Temperature has only sensible heat or temperature is the average kinetic energy.

A classic example is phase change. In a phase change, the potential energy of the molecules remains constant. The temperature remains constant. The increase in kinetic energy causes the molecules to move more freely, therefore changing their phase. In solid-liquid or liquid-gas phase transitions, energy is used to break the bonds between molecules (potential energy), rather than increasing their kinetic energy. During these transitions, the temperature remains constant as the energy added is used in overcoming the forces holding the molecules together, rather than increasing their kinetic energy."

HOW TEMPERATURE CAN BE EQUATED TO WORK?

Temperature and work can be equated through the concept of thermal energy. When work is done on a system or by a system, its internal energy changes, causing a change in temperature. This change in temperature can be quantified using the first law of thermodynamics, which states that the change in internal energy of a system is equal to the heat added to the system minus the work done by the system:

$$\Delta U = Q - W$$

Where:

ΔU = change in internal energy

Q = heat added to the system

W = work done by the system

From this equation, we can see that when work is done by a system, it loses internal energy, causing a decrease in temperature. Conversely, when work is done on a system, it gains internal energy, causing an increase in temperature. Therefore, temperature and work can be equated through the transfer of thermal energy associated with changes in internal energy.

TEMPERATURE'S IMPACT ON PROPERTIES OF MATERIALS

Temperature can have a significant impact on both physical and chemical properties of substances.

Physical Properties

- **Volume:** In general, most substances expand when heated and contract when cooled. This can lead to changes in volume, which is a key property in many physical processes.
- **Density:** The density of a substance is affected by temperature, as changes in temperature can alter the spacing between particles in a substance. As a result, the density of a substance may increase or decrease with changes in temperature.
- **Viscosity:** Viscosity, or the resistance of a fluid to flow, is often temperature-dependent. In general, fluids become less viscous (more fluid) at higher temperatures and more viscous (thicker) at lower temperatures.
- **Surface tension:** As temperature increases, surface tension decreases. This is because the kinetic energy of the molecules increases, which decreases the effectiveness of the intermolecular attraction. One typical example is superheated steam. Due to high temperature superheated steam exhibits near zero surface tension which is the reason for its film type condensation and low heat transfer coefficient.
- **Electrical conductivity:** The ability of a substance to conduct electricity can be influenced by temperature. In some cases, the conductivity of a material may increase with temperature, while in other cases it may decrease.
- **Thermal conductivity:** Thermal conductivity, which is a material's ability to conduct heat, has a complex relationship with temperature:

Metals

Thermal conductivity in metals generally decreases as temperature increases. This is because the increased vibrations within the lattice structure, called phonon scattering, reduces the mean free path of electrons. However, the thermal conductivity of alloys increases with temperature because the change in electrical conductivity is usually smaller.

Nonmetals

Thermal conductivity in nonmetals is approximately constant at high temperatures. At low temperatures, thermal conductivity decreases due to carrier scattering from defects.

Liquids

Thermal conductivity in liquids decreases as temperature increases because the liquid expands and the molecules move apart.

- **Specific heat capacity**: Specific heat capacity is a measure of the amount of heat required to raise the temperature of a substance by a certain amount. Specific heat capacity, or specific heat, is directly proportional to the amount of heat required to raise the temperature of a substance, and inversely proportional to the change in temperature.
- **State changes**: Changes in temperature can result in phase transitions, such as melting, freezing, evaporation, or condensation. These transitions are critical for understanding the behavior of substances at different temperature ranges.

Chemical Properties

- **Reaction rate**: Higher temperatures generally increase the rate of chemical reactions as the molecules move faster and collide more frequently, leading to increased reaction rates. Conversely, lower temperatures can slow down or even inhibit reactions.
- **Equilibrium constants**: Temperature can also affect the position of chemical equilibria. In general, increasing temperature favors endothermic reactions (reactions that absorb heat) and decreases the favorability of exothermic reactions (reactions that release heat).
- **Solubility**: Temperature can influence the solubility of substances in a solvent. In many cases, increasing temperature can increase the solubility of solids in liquids, as the increased thermal energy can break down intermolecular forces holding the solute together.
- **Vapor pressure**: The temperature can also affect the vapor pressure of a substance, which is the pressure exerted by the vapor phase above a liquid or solid. Higher temperatures generally lead to higher vapor pressures as more molecules have enough energy to escape from the liquid or solid phase.
- **Reaction selectivity**: In some cases, temperature can influence the selectivity of a chemical reaction, leading to the formation of different products depending on the temperature conditions.

WHY INTERMOLECULAR BONDS BREAK AT SATURATION TEMPERATURE?

This is a fundamental question in thermodynamics which is never explained because it is never questioned.

In summary

Phase change occurs when liquid-vapor phases reach thermodynamic equilibrium. This requires the Gibbs free energy becoming minimum and the entropy change becoming maximum or maximum disorder at phase change point. In order to meet this demand of the system the intermolecular bonds break at saturation point and generate maximum entropy.

If you look at internet everybody talks about the role of temperature and none talks about the critical role of entropy change at saturation point.

The first thing is it is interesting to note that no matter what the saturation temperature is the intermolecular bonds break at saturation temperature that is the phase transition temperature. The temperature is not important for intermolecular bond breaking at saturation temperature it is how Gibbs free energy changes at saturation point takes the call

Why?

One line answer. At saturation point vap and liquid reach thermodynamic equilibrium. This makes the Gibbs free energy minimum and maximum entropy (maximum disorder).at the saturation point. This is the reason why intermolecular bonds break at saturation point and generate maximum entropy change.

Explanation

Gibbs Free Energy

At constant temperature and pressure, a system's Gibbs free energy is minimized when it reaches equilibrium. The Gibbs free energy of a system is the enthalpy of the system minus the product of the temperature and the entropy of the system.

The Gibbs free energy reaches minimum for phase transition to occur. This necessitates maximum entropy change at phase transition point.

This requirement for phase change is met by intermolecular bonds breaking down at phase change point.

Entropy

The second law of thermodynamics states that the entropy of the universe always increases for a spontaneous process. Entropy is a scientific concept that is associated with a state of disorder, randomness, or uncertainty.

The entropy change in a liquid-vapor system is maximum at phase change point.

In order to achieve maximum disorder at phase change point the intermolecular bonds break at this point. Correspondingly the Gibbs free energy (G) change $G = H - TdS$ reaches near zero and that provides a state of thermodynamic equilibrium for phase transition.

HOW THE ATMOSPHERE COOLS ITSELF: THE BALANCING ACT OF TEMPERATURE AND PRESSURE

The temperature profile of the atmosphere is a result of an interaction between thermal conduction, thermal radiation, and natural convection. Sunlight hits the matter on the surface of the earth, land, and sea, and heats them. They then heat the air above the surface. If radiation were the only way to transfer energy from the ground to space, the greenhouse effect of gases in the atmosphere would keep the ground at roughly 333 K (60°C; 140°F). However, the atmosphere absorbs a major chunk of the sunlight before it hits the ground and keeps the average surface temperature on Earth is approximately 59 degrees Fahrenheit (15 degrees Celsius), according to NASA. Thermodynamic expansion and contraction of air in the atmosphere plays a key role in cooling the atmosphere.

The Balancing Act of Temperature and Pressure

In response to the compression or expansion associated with elevation change, the temperature of the atmosphere changes. The assumption is that the process is adiabatic i.e., no heat exchange occurs between the given air [parcel] and its surroundings. The rate at which the temperature of air changes as it rises above is called the adiabatic lapse rate.

Explanation

As altitude increases, temperature decreases. Various factors are responsible for this, including air pressure and water vapor content. With every 100 meters, the temperature drops by an average of 0.65°C. Where the air is very dry, such as in an area of high pressure, the air can cool by almost 1°C per 100 meters.

When air is hot, it tends to expand, which lowers its density. Thus, hot air tends to rise and carry internal energy upward. This is the process of convection. Vertical convective motion stops when a parcel of air at a given altitude has the same density as the other air at the same elevation.

When a parcel of air expands, it pushes the air around it, doing thermodynamic work. The expansion or contraction of the rising air parcel is considered to be adiabatic, without inward or outward heat transfer.

Air has low thermal conductivity, and the bodies of air involved are very large volumes, so the transfer of heat by conduction is negligibly small. Also, in such expansion and contraction, intra-atmospheric radiative heat transfer is relatively slow and so negligible.

As said above, since the upward-moving air expands adiabatically by doing thermodynamic work it does so at the expense of the internal energy which decreases the air temperature. Air cools.

Effect of Water in the Atmosphere

The dry adiabatic lapse rate is characteristic of dry air. Water in the air will condense or evaporate; in doing so it will release or absorb heat, respectively, complicating calculations of the lapse rate and atmospheric stability. In general, as a parcel of air rises, the water vapor in it condenses and heat is released. The rising air will therefore cool more slowly as it rises; the wet adiabatic lapse rate will in general be less negative than the dry adiabatic lapse rate. The wet adiabatic lapse rate has been observed to vary between −6.5°C/km and −3.5°C/km.

The dry adiabatic lapse rate may be calculated from basic physical principles, $dt/dz = -9.8$ degc/km, where T = temperature and z = altitude.

The actual measured rate at which air cools as it rises is called the ambient or prevailing lapse rate. The relationships between the ambient lapse rate and the dry adiabatic lapse rate essentially determine the stability of the air. When the ambient lapse rate is the same as the dry adiabatic lapse rate, the atmosphere has neutral stability. Superadiabatic conditions prevail when the air temperature drops more than 9.8°C/km (1°C/100 m). Subadiabatic conditions prevail when the air temperature drops at a rate of less than 9.8°C/km.

CHAPTER 31

Pressure and Pressure Drop

Pressure is a state variable. Pressure is a measure of the force exerted on a given area, often resulting from the collisions of gas molecules in a closed system. In thermodynamics, pressure plays an important role in determining the work done by a system as it undergoes changes in volume and temperature. The relationship between pressure, volume, and temperature is described by the ideal gas law, which helps in understanding the behavior of gases under different conditions.

Pressure can have a significant impact on the physical properties of substances, affecting their boiling points, melting points, and even their states of matter.

WHAT IS PRESSURE?

Pressure

Pressure is a type of potential energy per unit volume. For example, in a fluid, pressure can be used as a measure of energy density, precisely, kinetic energy density and potential energy density.

Pressure is a state variable. Pressure is a measure of the force exerted on a given area, often resulting from the collisions of gas molecules in a closed system. In thermodynamics, pressure plays an important role in determining the work done by a system as it undergoes changes in volume and temperature. The relationship between pressure, volume, and temperature is described by the ideal gas law, which helps in understanding the behavior of gases under different conditions.

Pressure can have a significant impact on the physical properties of substances, affecting their boiling points, melting points, and even their states of matter.

Kinetic and Potential Energy in Fluids

The molecule in motion has kinetic energy, but it can also have potential energy:

Kinetic Energy

Molecules in motion have kinetic energy, which is energy that any moving object possesses. The kinetic energy of a molecule is constantly changing as it changes speed, and the particles in a sample of matter don't all have the same energy. The kinetic energy of an object is directly proportional to its mass and the square of its velocity.

Potential Energy

Potential energy is the stored energy in molecules in motion. There are two sub-forms of potential energy.

Sub-forms of Potential Energy

The two most common types of potential energy are gravitational potential energy and elastic potential energy.

Gravitational Potential Energy

This is the energy an object has due to its mass, height above the ground, and the strength of the gravitational field. The equation for gravitational potential energy is $E P = m g h$.

Elastic Potential Energy

This is the energy stored in materials that can be stretched or extended, such as rubber bands.

According to kinetic theory, gases are made up of tiny particles that collide with each other and the container's wall at random, resulting in elastic collisions. At constant temperature, there is no kinetic energy loss. Pressure is the impact of these collisions perpendicular to the container's wall per unit area. Moving molecules have the same kinetic energy at a constant temperature, though their speed may differ if the molecules' masses differ.

Equation:

$p = 1/3\rho\langle u^2\rangle$, equals one-third rho and u squared. Rho is the density and u is the velocity.

Explanation: The kinetic theory of gases states that gas molecules are in constant, random motion and collide with each other and the walls of their container. The force exerted by the molecules on the walls per unit area is what generates gas pressure.

The Pressure is a State Function

The kinetic energy per unit volume (*KE/V*) varies with the state of the substance and can be a factor in determining pressure. *V* is volume. *KE* is kinetic energy.

The change in *KE/V* as a substance transitions from one state to another, such as from a gas to a solid, can affect the pressure exhibited by the substance regardless of the path of transition from one state to another.

Ultimately, pressure is a state function because it is determined by the properties of the system in a given state and is independent of the path followed to reach that state of the matter.

In gases, the pressure is determined by the number of gas molecules in a given volume and their average kinetic energy.

In liquids, the pressure is influenced by the density and arrangement of molecules, as well as external forces acting on the liquid.

In solids, the pressure is determined by the arrangement and bonding of atoms or molecules within the material.

Static Pressure

Static pressure exerted by a fluid that is not moving. It's caused by the weight of the fluid above it, and is measured perpendicular to the fluid surface. It generally applies to liquid.

Dynamic Pressure

Dynamic pressure is the kinetic energy per unit volume. It is one of the terms of Bernoulli's equation, which can be derived from the conservation of energy for a fluid in motion. Remember the pressure as such is not an energy.

Pressure and Kinetic Energy Relationship

Kinetic energy of a gas is described by the equation $PV = 2/3\ KE$,

where:

P : is the pressure
V : is the volume
KE : is the average kinetic energy per unit volume

Pressure is a measure of energy per unit volume because its units are the same as the units of energy per unit volume:

- **Pressure:** Pressure is defined as force per unit area, or $P = F/A$.
- **Energy per unit volume:** The expression for energy per unit volume is $E/V = \text{kg m/m}^2\ \text{s}^2$.
- **Units:** The units of pressure and energy per unit volume are both kg $\text{m/m}^2\ \text{s}^2$.

Pressure is a type of potential energy per unit volume. For example, in a fluid, pressure can be used as a measure of energy density, along with kinetic energy density and potential energy density.

IS PRESSURE ENERGY?

Pressure is not energy. Pressure when multiplied by volume becomes 'work energy'.

Explanation

What is Pressure?

Definition of Pressure

There are three ways how pressure can be described.

1. The term 'stress' (mechanical stress) is used as a synonym for pressure (SI). Pressure is known in physics as mechanical pressure, which is force per unit area, $p = F{\cdot}A{-}1$ [Pa].
2. According to kinetic theory, gases are made up of tiny particles colliding with each other and the container's wall randomly, resulting in elastic collisions. At constant temperature, there is no kinetic energy loss. Pressure is the impact of these collisions perpendicular to the container's wall per unit area. Moving molecules have the same kinetic energy at a constant temperature, though their speed may differ if the molecules' masses differ. In other words, pressure is the average kinetic energy of molecules.
3. Kinetic energy signifies temperature, average $KE = 1/2\ KT$. [K is a constant and T is temperature] This comes from the fact that for monoatomic particles with three degrees of freedom, each mode of motion carries 1/2 KT energy.

The temperature which is the translational motion of atoms carry 1/2 KT energy. Therefore, at constant volume, the pressure of a gas is directly proportional to temperature.

Is Pressure Energy?

The SI unit of pressure is Pascal [Pa]. The SI unit of energy is Joule. Therefore, pressure is not energy.

Pressure is a fundamental quantity that expresses energy per unit volume. The *SI* unit of pressure is generally pascal [*Pa*] = [$J{\cdot}m^{-3}$]. Therefore, 1 Joule = $Pa \times m^3$ or Pascal is Joule per unit volume. Therefore, pressure is energy per unit volume.

In summary, pressure is force per unit area, $P = F{\cdot}A^{-1}$ [Pa]. The SI unit of pressure is Pascal [Pa]. The dimensional formula of pressure is $M^1\ L^{-1}\ T^{-2}$. It is evident that pressure and energy have different dimensional formulas, so pressure is not the same as energy. But when pressure is multiplied by volume, PV becomes 'work energy' expressed as Joule, $PV = M^1\ L^{-1}\ T^{-2} \times L^3 = M^1\ L^2\ T^{-2}$ = Joule.

WHY ATMOSPHERIC PRESSURE IS 101.325 KPA?

This is an example how one can calculate atmospheric pressure without much complication. The calculation shows the theoretical value is about 98.6% of the observed pressure.

Atmospheric pressure = Mass of atmosphere/Surface Area of earth

Calculation

Mass of atmosphere (M^2) = 5.15×10^{18} kg
Acceleration due to gravity (g) = 9.81 m/s^2
Average radius of Earth = 6,371 km = 6.371×10^6 meters
Surface Area of Earth = $4 \times \pi \times (6.371 \times 10^6)^2$

[Surface Area of Sphere = $4\pi r^2$, where r is the radius of sphere. The earth is a sphere though not perfect sphere. The Earth is an irregularly shaped ellipsoid. The North and South Poles are slightly flattened due to the force of the Earth's rotation.]

Now, let's plug in these values into the formula:
Pressure = $(M^2 \times g)/(4 \times \pi \times (6.371 \times 10^6)^2)$

Calculating

Surface Area = $4 \times \pi \times (6.371 \times 10^6)^2 = 4 \times 3.14159 \times (4.045401 \times 10^{13}) \approx 5.09 \times 10^{14}$ m^2

Pressure = $(5.15 \times 10^{18} \times 9.81)/(5.09 \times 10^{14}) = 5.05 \times 10^{19}/5.09 \times 10^{14} \approx 99{,}225$ Pa

Converting Pa to hPa (hectopascal) by dividing by 100, we get:

Theoretical pressure $\approx$ 992.25 hPa (this is 98.6% of observed 101.325 kilopascal)

Therefore, the calculated atmospheric pressure at sea level based on the mass of Earth and the atmosphere is approximately 992.25 hPa, which is close to the observed value of 1013.25 hPa or 101.325 kilopascal.

PRESSURE-DROP: INTERPLAY OF POTENTIAL AND KINETIC ENERGY

In summary when there is a pressure drop in a fluid flow system, there is a conversion of potential energy (static pressure) to kinetic energy (dynamic pressure). As the fluid moves through the pipe and encounters obstacles or changes in the flow path, the pressure decreases and kinetic energy increases.

This conversion of energy is important for the movement of the fluid through the system. However, it is important to note that along with this conversion, there are also frictional losses that occur. These losses manifest as heat due to the interaction between the fluid and the walls of the pipe, as well as turbulence within the flow.

Therefore, while the pressure drop results in an increase in kinetic energy, it also leads to energy losses in the form of heat.

Managing pressure drop and minimizing friction within the system can help optimize energy efficiency and performance. It is essential to strike a balance between converting potential energy to kinetic energy and minimizing energy losses to ensure the smooth operation of fluid systems.

What is Pressure Drop?

The fundamental idea is that a moving fluid's potential energy and kinetic energy interact to produce a pressure drop. The static pressure is the potential energy at a constant altitude, and the dynamic pressure is the kinetic energy. The two together always remain the same as a sum. An excellent illustration of how potential and kinetic energies alternate in a fluid flow is provided by Bernoulli's equation.

The viscosity of the fluid, the roughness of the pipe's surface, gravity if elevation changes and the length of the pipes are the major factors for a circular cross-section pipe. Each makes their unique contribution. We'll talk about how each of these things affected the pressure drop. We will not go into equations as they are available in books.

The difference in total pressure between two points of a fluid in a pipe is known as pressure drop, which is frequently denoted by the abbreviations "*dP*" or "ΔP." When a fluid is flowing through a channel, pipe, or tube, frictional forces acting on the fluid brought on by the flow resistance cause a pressure drop. Some of the fluid's hydraulic energy is converted to thermal energy (internal energy) by this friction. The fluid loses pressure because the thermal energy cannot be transformed back into hydraulic energy, which is necessary for energy conservation.

The velocity of the fluid through the pipe and the viscosity of the fluid are the main factors that affect resistance to fluid flow.

According to the frictional shear forces present in the piping network, pressure drop rises proportionately.

The pressure drop will be influenced by the piping network's high relative roughness rating, numerous pipe fittings and joints, tube convergence, divergence, turns, surface roughness, and other physical characteristics. A pipe section, valve, or elbow joint, experiences a greater pressure drop when the flow rate or viscosity of the fluid is high. There will be little to no pressure drop at low velocities.

Fundamentals

When a single-phase fluid flows through a piping system that typically consists of bends, tanks heat exchangers, reducers etc., the fluid pressure changes. The fluid pressure changes as a result of the conversion of kinetic energy to potential energy and vice versa as well as an irreversible conversion of some of the mechanical energy of the fluid. Actually, fluid flow can be viewed as one-dimensional and is defined in each reference cross-section by the mean parameters, whereas static pressure is typically constant over the flow cross-section and only changes longitudinally.

In addition to the potential and kinetic energy of the fluid, there are four key players that contribute to the pressure drop of a fluid in a pipe:

- **Viscosity of the fluid:** Viscosity is a measure of a fluid's resistance to flow. A higher-viscosity fluid will experience a higher pressure drop as it flows through a pipe due to its internal friction. This is seen in the form of a higher-pressure gradient along the pipe length.
- **Surface roughness of the pipe**: The roughness of the inner surface of a pipe can cause additional frictional resistance to the flow of fluid, resulting in a pressure drop. Rough surfaces create more turbulence and disrupt the smooth flow of the fluid, leading to increased energy losses.
- **Length of the pipe:** The longer the pipe, the greater the pressure drop. This is because the fluid encounters more resistance and frictional losses along the length of the pipe, resulting in a decrease in pressure.
- **Gravity:** In systems where there are changes in elevation, gravity plays a role in the pressure drop. If the fluid is flowing against gravity, it will experience an increase in pressure drop due to the additional work required to lift the fluid to a higher elevation.

These factors, along with other considerations such as flow rate, pipe diameter, and fluid properties, contribute to the total pressure drop in a fluid flow system.

THERMODYNAMICS OF PRESSURE DROP

Thermodynamic properties such as enthalpy, entropy, free energy, and specific heat do not have a direct role in causing pressure drop in a fluid flow system. However, they can indirectly influence pressure drop by affecting other fluid properties such as temperature, density, and viscosity, which in turn can impact the pressure drop.

Examples

- **Enthalpy:** Enthalpy is the heat energy of a system plus the pressure-volume work done on or by the system. It represents the total energy content of a fluid. While enthalpy is not directly related to the pressure drop, it can affect the fluid's temperature and density, which can, in turn, impact the pressure drop. For example, changes in enthalpy due to heat transfer can cause the fluid to expand or contract, thereby affecting the pressure drop.
- **Entropy**: Entropy is a measure of the disorder or randomness of a system. It quantifies the distribution of energy within a fluid. While entropy is not directly linked to pressure drop, it can influence the changes in temperature and pressure within a fluid flow system. Changes in entropy can affect the efficiency of heat transfer and energy dissipation, which can impact the pressure drop indirectly.
- **Free energy:** Free energy measures the maximum amount of thermal energy in a system that is available for useful work. It is a combination of enthalpy and entropy. Although free energy is not directly related to pressure drop, it can indirectly influence it by affecting the fluid's thermodynamic state and behaviour, which in turn can impact pressure drop.
- **Specific heat**: Specific heat is the amount of heat energy required to raise the temperature of a substance by a certain amount. It represents the ability of a substance to store thermal energy. While specific heat is not directly related to pressure drop, it can influence the temperature changes within a fluid flow system. Changes in temperature can affect the fluid's density, viscosity, and other properties, which can subsequently influence the pressure drop.

In summary, enthalpy, entropy, free energy, and specific heat are important thermodynamic properties that can indirectly impact the pressure drop in a fluid flow system by influencing the fluid's temperature, density, viscosity, and other properties. These properties play a role in determining the overall behaviour of the fluid, which in turn affects the pressure drop.

HOW DOES THE PRESSURE DROP IN GAS DIFFER FROM THE LIQUID?

The pressure drop in gas flow is different from that in liquid flow due to several factors.

- **Compressibility:** Gases are highly compressible compared to liquids. This means that changes in pressure will result in significant changes in gas volume, whereas liquids are considered incompressible. As a result, the pressure drop in a gas is more affected by changes in gas density and volume compared to liquids.

- **Viscosity:** Gases generally have much lower viscosity than liquids. Viscosity is a measure of a fluid's resistance to flow, and higher viscosity results in higher resistance and pressure drop. Due to their lower viscosity, gases generally experience lower pressure drop compared to liquids.
- **Flow velocity**: Gases tend to have higher flow velocities compared to liquids for the same volumetric flow rate. This higher velocity can lead to higher frictional losses and pressure drop in gas flow.
- **Density**: Gases have lower densities than liquids. This lower density, combined with the compressibility of gases, leads to lower pressure drop as the gas can expand and fill the available space more easily.

In summary, the main differences in pressure drop between gases and liquids are due to the compressibility, viscosity, flow velocity, and density of the respective fluids.

HOW DOES REYNOLDS NUMBER AFFECT THE PRESSURE DROP OF FLUIDS?

The Reynolds number, denoted as Re, is a dimensionless quantity used to characterize the flow regime of a fluid. It is defined as the ratio of inertial forces to viscous forces within the fluid.

In terms of pressure drop, the Reynolds number has a significant impact on the flow behaviour and the associated pressure loss. The Reynolds number affects the flow regime, which can be classified into three main categories:

- **Laminar flow (Re < 2000):** In laminar flow, fluid particles move in smooth, parallel layers without much mixing. In this regime, the pressure drop is generally linear and directly proportional to the fluid flow rate. The pressure drop in laminar flow can be calculated using the Hagen-Poiseuille equation.
- **Transitional flow (2000 < Re < 4000)**: Transitional flow occurs when the Reynolds number is within a transitional range between laminar and turbulent flow. In this range, the flow behaviour is less predictable, and the pressure drop can exhibit irregular patterns.
- **Turbulent flow (Re > 4000):** In turbulent flow, fluid particles exhibit chaotic and irregular motion, leading to increased mixing and enhanced momentum transfer. Turbulent flow has higher pressure drops compared to laminar flow for the same flow rate. The pressure drop in turbulent flow can be calculated using various empirical correlations, **such as the Darcy-Weisbach equation or the Colebrook-White equation.**

In summary, the Reynolds number affects the flow regime and, consequently, the pressure drop of fluids. Lower Reynolds numbers are associated with laminar flow and lower pressure drops, while higher Reynolds numbers indicate turbulent flow and higher pressure drops.

PRESSURE AND STRESS DISTRIBUTION: EFFECT OF SHAPE

The spheres have the minimum surface area *vs.* volume ratio. The first point is spheres have minimum surface energy. This means it requires minimum energy to create a spherical surface. Horton sphere relates to the shape of a surface *vs.* pressure distribution. The two benefits of a spherical shape of Horton sphere arise from its minimum surface area, [1] the reduced surface area helps to minimize pressure buildup inside the sphere and ensures a more even distribution of pressure and [2] the smaller surface area means less heat transfer to the surroundings, which can be advantageous in certain applications.

A Horton sphere is a type of spherical tank that is specifically designed to store liquid or gas at high pressures. These tanks are typically used for storing liquefied gases or hazardous materials, as they can withstand high levels of pressure.

Horton sphere relates to Pascal's Law, which states that when there is an increase in pressure at any point in a confined fluid, there is an equal increase in pressure at every other point in the fluid. This ensures that the sphere can safely contain the high-pressure gas without experiencing uneven stress or pressure points that could lead to structural failure.

The Horton sphere works by allowing the stored material to expand and contract within the sphere as needed, without causing the tank to rupture or fail. *The spherical shape of the tank helps to evenly distribute the pressure of the stored material, making it a safe and reliable storage option for high-pressure applications.*

The great advantage of a Horton sphere is its ability to efficiently store large amounts of compressed gases or liquids at high pressure in a relatively small footprint. The smaller surface area of *the* Horton sphere also helps to reduce the amount of material and welding required for construction, making it a cost-effective and reliable storage solution for various industries.

Additionally, the spherical shape of the Horton sphere allows for easier maintenance and inspection, as well as better resistance to external forces such as earthquakes or high winds.

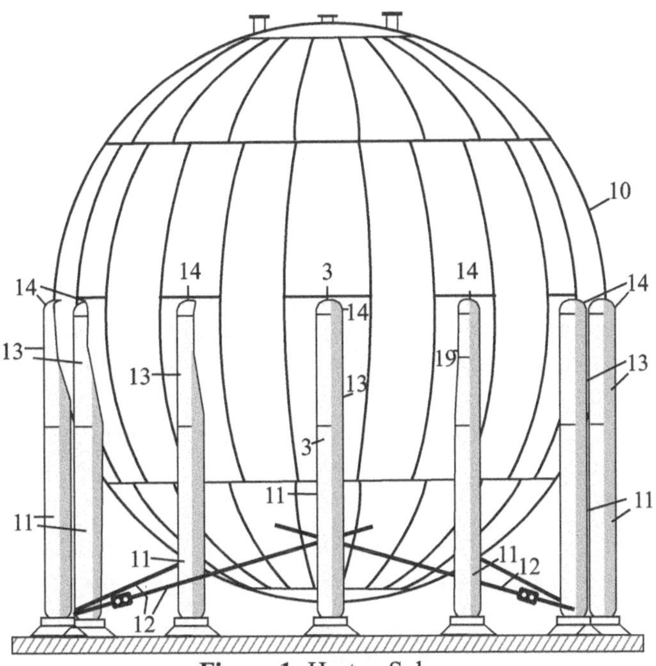

Figure 1: Horton Sphere

How Does the Sphere Work?

A sphere has the minimum surface area per unit volume of any three-dimensional shape. This means that for a given volume, a sphere will have the smallest surface area compared to any other shape, such as a cylinder, cube, or rectangular tank.

The reduced surface area of a sphere means that there is less area through which heat can transfer between the contents of the sphere and its surroundings. This results in lower heat transfer potential, making a spherical tank more efficient for storing substances that need to be kept at a specific temperature or protected from external heat sources.

Additionally, the reduced surface area minimizes the area through which pressure is exerted on the tank, helping to evenly distribute the pressure and reduce stress concentrations as discussed earlier. These properties make a spherical tank ideal for applications where heat transfer and pressure distribution are critical factors to consider.

CHAPTER 32

Specific Heat and Latent Heat

ROLE IN THERMODYNAMICS

Specific heat refers to the amount of heat required to raise the temperature of a unit mass of a substance by one degree Celsius, while latent heat is the amount of heat absorbed or released during a phase change at constant temperature. Both specific heat and latent heat are important concepts in thermodynamics, as they affect the heat transfer and energy changes in a system.

Specific heat is used to calculate the amount of heat energy needed to raise the temperature of a substance, while latent heat is used to determine the energy required for phase changes such as melting or vaporization. The specific heat of a substance varies depending on its physical state and composition, while the latent heat remains constant for a specific substance at a given temperature and pressure.

Gases have two specific heats, *Cp* and *Cv*, *Cp* signifies specific heat at constant pressure and *Cv* at constant volume.

WHAT IS SPECIFIC HEAT?

Specific heat, often denoted by the symbols *Cp* (specific heat at constant pressure) and *Cv* (specific heat at constant volume), is a fundamental property of materials that describes their ability to store and release heat energy.

At the molecular level, specific heat signifies the amount of heat required to raise the temperature of a unit mass of a substance by one degree Celsius.

The concept of specific heat is closely related to the equipartition theorem in statistical mechanics, which states that each degree of freedom of a molecule contributes an average energy of $kT/2$, where k is the Boltzmann constant and T is the temperature.

The degrees of freedom of a molecule include translational, rotational, and vibrational modes, and the specific heat of a substance depends on how these modes contribute to its internal energy change.

The specific heat at constant pressure, *Cp*, includes the heat required to increase the temperature of a substance along with the work done during expansion. This is because, at constant pressure, the substance may expand, leading to work being done in addition to heat being absorbed. *Cp* signifies sum of internal energy and work energy. This is also the total energy called enthalpy.

On the other hand, the specific heat at constant volume, *Cv*, only considers the heat required to change the temperature of the substance at constant volume, without any work being done. *Cv* signifies the internal energy.

The relationship between *Cp* and *Cv* can be expressed by the equation $Cp = Cv + R$, where R is the gas constant. This relationship is important in thermodynamics, as it helps determine the heat capacity of a material under various conditions. $Cp - Cv = R$ signifies the work energy in an ideal gas.

Cp/Cv

Cp/Cv refers to the ratio of specific heat capacities at constant pressure (Cp) and constant volume (Cv) for a substance. This ratio is denoted by the symbol γ (gamma) and is a critical parameter in analyzing the thermodynamic behavior of gases, under adibatic conditions. Under adiabatic condions the only source of energy is internal energy change or *Cv*.

For an ideal monatomic gas (e.g., helium, argon), the *Cp/Cv* ratio is approximately 5/3 (1.67), while for diatomic gases (e.g., nitrogen, oxygen), the ratio is around 7/5 (1.4).

The expression to relate '*y*' with degree of freedom (DOF) is $y = 1 + DOF/2$

For monoatomic gas

Degree of freedom = 3

$y = 1 + 2/3 = 1.66$

For diatomic gas

Degree of freedom = 5

$y = 1 + 2/5 = 1.4$

Degree of freedom is modes of energy storage in a molecule.

What These Numbers Tell You

These numbers tell you the amount of internal energy *Cv* available in a gas for adiabatic work.

In general, a smaller value of the ratio of specific heats (γ) implies a larger value for the specific heat at constant volume (Cv) for a given gas. This is because the specific heat at constant volume (Cv) is inversely related to the ratio of specific heats by the equation:

$$\gamma = Cp/Cv$$

Therefore, if γ is small, it means that *Cv* is relatively larger compared to *Cp*. This implies that for a gas with a smaller γ, there is more internal energy stored per unit mass at constant volume, making it more efficient for adiabatic work where only internal energy transfer is involved.

So, in the context of adiabatic work done using internal energy, a gas with a smaller γ will have a larger specific heat at constant volume (Cv), providing more internal energy per unit mass available for performing work.

SPECIFIC HEAT AND TEMPERATURE

We define specific heat as the quantity of heat that raises the temperature of 1 gram of a substance by 1 degc at constant pressure.

Unlike specific heat, the latent heat, signifies heat change during a phase transition at a constant temperature and at a given pressure in other words at the saturation temperature. The specific heat changes at every temperature at a given pressure. The latent heat is expressed as kJ/kg while specific heat is expressed as kJ/kg.K.

Explanation

The latent heat represents the stored energy in a substance at a molecular level held in the intermolecular bonds. The specific heat represents the stored energy in a substance at the atomic level held in the motions of atoms. This thermal energy change can be related to the random motion of molecules as they vibrate, rotate, and translate in space.

Translational Kinetic Energy

$Ek = 1/2\ mv^2$ is the energy associated with the movement of a chemical entity's center of mass, where m is the mass of the chemical entity (molecule, atom, or ion) and v is the velocity of its center of mass. This is essentially linear motion. In the fluid phases (gas and liquids) which causes collisions.

Vibrational Kinetic Energy

$Ek = 1/2\ kx^2$ is the energy associated with the atom's vibrational motion, which can be modelled as an oscillating spring with *k* being Hooke's law constant and × is the displacement from the equilibrium location.

Vibrational kinetic energy is the energy an object has due to its vibrational motion. For example, a ringing cell phone or a drum that's been hit by a hammer both have vibrational kinetic energy.

In a simple harmonic oscillator, the total energy is the sum of the kinetic energy and the elastic potential energy, and it remains constant. The formula for the total energy is

$$E_{Total} = 1/2\ mv^2 + 1/2\ kx^2$$

Rotational Kinetic Energy

$E\omega = 1/2\ I\omega^2$ is the energy of a chemical entity associated with rotation, where ω is the rotational frequency (how fast the molecule spins) and I is the moment of inertia (in a sense, its resistance to spinning. The formula shows that rotational kinetic energy is directly proportional to the moment of inertia and the square of the angular velocity.

This means that:

- If the moment of inertia increases, the rotational kinetic energy increases as well, assuming the angular velocity remains constant.
- If the angular velocity increases, the rotational kinetic energy increases as well, assuming the moment of inertia remains constant.

EQUIPARTITION THEOREM

The name "equipartition" means "equal division. According to the equipartition theorem, in thermal equilibrium energy is shared equally among all of its various forms of kinetic energy; for example, the average kinetic energy per degree of freedom in the translational motion of a molecule should equal that in rotational motion. For example, it predicts that every atom in a monoatomic ideal gas has an average kinetic energy of 3/2 *kBT* in thermal equilibrium, where k is the Boltzmann constant and T is the thermodynamic temperature. A monoatomic gas has three degrees of freedom. All of them store translational kinetic energy. Each shares an equal amount of energy 1/2 *KBT*.

What Does it Mean?

Among the above three types of kinetic energy, translational or the linear motion of molecules motion raises the temperature by collisions. Vibrational motion involves atoms within a molecule moving relative to each other, while rotational motion refers to the spinning of molecules around their center. Although these motions contribute to the overall energy of a molecule, they are not as directly related to temperature changes as translational motion.

Maxwell Boltzmann Distribution

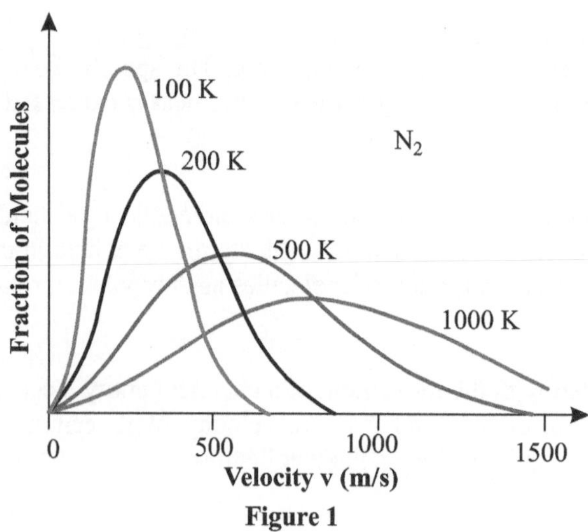

Figure 1

The image above is Maxwell Boltzmann's velocity distribution of nitrogen gas at different temperatures.

It may be noted that at every temperature the velocity of N2 molecules is different. This implies that at every temperature the KE is different and therefore the temperature is different.

Precisely, when we define specific heat as the quantity of heat that raises the temperature of 1 gram gas by 1 degc, it implies that 1 degc rise refers to the rise of temperature over its existing temperature and not delta *T*. That's how the unit of specific heat is J/kg/K.

SPECIFIC HEAT AND THERMODYNAMICS EQUATIONS

- The equation for specific heat capacity (c) is: $q = m\ c\ dt$, where q is the heat energy transferred, m is the mass of the substance, c is the specific heat capacity, and dt is the temperature change.
- The equation for heat capacity (C): $Q = Ct$, where Q is the heat energy transferred and t is the temperature change. Heat capacity is the amount of heat energy required to raise the temperature of an object by a given amount.

- The relationship between specific heat capacities at constant pressure (Cp) and constant volume for ideal gases (Cv): $Cp - Cv = R$, where R is the specific gas constant. R represents the work done per unit Kelvin per unit mole.
- The equation for enthalpy change (dH): $dH = n\ Cp\ dt$, where ΔH is the enthalpy change, n is the number of moles of the substance, Cp is the molar specific heat capacity at constant pressure, and dt is the temperature change.
- The equation for internal energy change (dU): $dU = n\ Cv\ dt$, where dU is the internal energy change, n is the number of moles of the substance, Cv is the molar specific heat capacity at constant volume, and dt is the temperature change.

MEANING OF CP AND CV

Cp is a ratio of enthalpy change which is the total energy change per unit temperature, dH/T. The total energy of a gas consists of internal energy and work at constant pressure, $H = U + PdV$. Therefore, $Cp = dH/T$

Cv is a ratio of internal energy change and temperature dU/T. $Cv = dU/T$. *Cv* does not take into account the work. Therefore, *Cp* is always more than *Cv*.

Cp/Cv signifies available internal energy in a gas when it undergoes adiabatic expansion.

SPECIFIC HEAT OF REAL GASES *VS.* IDEAL GASES

The specific heat of real gases is slightly higher than that of ideal gases, and the specific heat of real gases varies with temperature, while the specific heat of ideal gases does not.

SPECIFIC HEAT OF NOBLE GASES *VS.* IDEAL GASES

The specific heat of noble gases is slightly different from ideal gases because noble gases have filled valence shells, making them more stable and less likely to interact with other molecules. The specific heat of noble gases is generally close to the ideal value for monoatomic ideal gases this is because no work is done against external pressure when the temperature of a monoatomic gas increases.Additionally, noble gases have negligible intermolecular forces compared to other gases, so their specific heat remains relatively constant over a wide range of temperatures and pressures.

SPECIFIC HEAT AND LATENT HEAT

Specific heat and latent heat are two important thermodynamic properties in heat transfer. Specific heat is the energy demand of a substance at the atomic level to provide energy for different modes of atomic motions, e.g., translational, rotational, and vibrational motions. At the atomic level, these motions store kinetic energy. According to the Equipartition Theorem heat applied to a substance gets equally shared among these three degrees of freedom each receiving 1/2 KbT heat, Kb is a constant, and T is the temperature. It is also possible to think of specific heat as a resistance to heat transfer. Greater specific heat corresponds to larger heat storage and smaller heat transfer.

Specific heat depends on the complexity of the atoms in a molecule. A mono-atomic gas has three degrees of freedom, one each translational, rotational, and vibrational to store energy. A mono-atomic molecule can store 3/2KbT energy while a diatomic molecule with five degrees of freedom can store 5/2KbT energy and a triatomic molecule can store 7/2 KbT energy. As the degrees of freedom or storage space within a molecule increases the specific heat increases.

While specific heat and latent heat look like different phenomena, one interesting thing is that latent heat comes into action when there is no more heat capacity for sensible heat to increase.

Sensible heat is the heat that can be sensed driven by the heat capacity of a substance while latent heat is the hidden heat of a substance. It comes into play when there is no more heat capacity for sensible temperature increase. Any heat supply at this point is then absorbed latently to break the intermolecular bonds and phase change

The latent heat is the stored hidden potential energy in a substance held in the intermolecular bonds. Which the substance gains when heated or loses when the substance is cooled.

Solids, Liquids, and Gases: Specific and Latent Heat

Specific Heat

Solids have the highest specific heat because their molecules are closely packed and densest among all the states of matter. They can easily pass on heat energy through vibrations. Gases have a lower specific heat because their molecules are further apart and heat energy is transferred through collisions.

Latent Heat

The solids have generally high latent heat due to tightly held molecules and larger intermolecular forces. Water is an exception due to *H* bonds.

Temperature and Pressure

Specific Heat

The specific heat increases with temperature. With increased kinetic energy and molecular motions, the energy demand increases which increases specific heat.

Latent Heat

The latent heat decreases with temperature. With fewer intermolecular bonds available as temperature increases the bond-breaking energy reduces, the entropy generation reduces and it continues to reduce until it becomes zero at critical point.

WHAT FOLLOWS THE SUPPLY OF HEAT BEYOND THE SATURATION POINT?

Three things happen by heat supply beyond saturation point. [1] The saturation point marks the of the phase transition from liquid to vapor [2] heat attacks the weakest points which are the intermolecular bonds and begins breaking them into vapor and [3] vapor expands and increases entropy.

Why and how does liquid turn into vapor when intermolecular bonds break?

The vapor pressure of a liquid depends on its intermolecular forces. The relation between vapor pressure and intermolecular forces is quite simple. Vapor pressure is inversely proportional to the intermolecular forces. The stronger the intermolecular forces lower the vapor pressure and vice versa. As the intermolecular bonds begin breaking the intermolecular forces become weaker with the supply of heat and raise the vapor pressure. The rise in vapor pressure increases the evaporation of liquid to vapor.

What is the relation between specific heat and latent heat

At the point where the specific heat capacity ends, the sensible heat goes hiding. At this point the heat becomes latent.

Why does the temperature get fixed at the phase transition point at a given pressure?

At phase change, the two phases reach thermal equilibrium. The heat transfer between phases stops.

APPLICATION OF LATENT HEAT IN PHASE CHANGE

A substance changes its phase when its sensible heat has reached the saturation point or the saturation temperature. The end of sensible temperature rise marks the beginning of phase change at constant temperature and constant pressure. At phase, the liquid and vapor phases reach the thermodynamic equilibrium. Gibbs free energy reaches minimum and the entropy change reaches maximum as the vapor phase expands. At thermodynamic equilibrium both liquid and vapor phases reach a thermal equilibrium. That explains why at phase change temperature stays constant.

Why constant pressure is necessary at phase change?

The answer is simple. The reason is that phase change can't occur at constant volume. The Gibbs free energy can't reach a minimum for phase change to occur if volume stays constant.

Why?

At constant volume, a phase change between liquid and vapor states is not possible because the densities of liquid and vapor are different.

In a constant volume process, the volume of the substance remains constant, meaning that the substance cannot expand or contract to accommodate the change in density that occurs during a phase change.

Phase changes between liquid and vapor states typically occur at constant pressure, where the substance can expand or contract to accommodate the change in density.

At constant pressure, the substance can absorb or release heat energy to enable the phase change while its volume changes to maintain the constant external pressure.

Therefore, constant pressure is essential for phase changes between liquid and vapor states, as it allows for the adjustment of volume to accommodate the change in density that occurs during the phase change process.

SPECIFIC HEAT: ROLE OF MOLECULAR STRUCTURE

The specific heat of a substance is the amount of heat required to raise the temperature of a unit mass of the substance by one degree Celsius. The molecular structure of a substance plays a significant role in determining its specific heat.

The specific heat of a substance is influenced by the intermolecular forces between its molecules, as well as the degree of freedom of movement of the molecules. Substances with strong intermolecular forces, such as hydrogen bonds or metallic bonds, tend to have higher specific heats because more energy is required to break these bonds and raise the temperature of the substance.

In addition, substances with larger and more complex molecules tend to have higher specific heats because they have more degrees of freedom of movement, allowing them to absorb and store more heat energy before their temperature increases.

Overall, the molecular structure of a substance affects its specific heat by influencing the energy required to change its temperature and the ability to store heat energy.

WHY DOES WATER HAVE A GREATER SPECIFIC HEAT THAN ICE?

The specific heat of water is higher than the specific heat of ice:

- **Ice:** The specific heat of ice is 2.108 kilojoules per kilogram per kelvin (kJ/kgK).
- **Water:** The specific heat of water is 4.187 kJ/kgK.

The answer is water in solid form has mostly rotational degree of freedom while water as liquid at 0 degc have both translation and rotational degrees of freedom. Thus, water at 0 degc can store more heat than ice.

DEGREES OF FREEDOM: STORAGE AND DISTRIBUTION OF THERMAL ENERGY

Summary

In simple one line the internal energy consisting of kinetic and potential energy in a substance is its thermal energy storage capacity.

Thermal degrees of freedom refer to the different ways in which heat energy can be stored in a system. In other words, they represent the different forms of kinetic and potential energy associated with the motion and interactions of particles within the system. The concept of degrees of freedom is related to the molecular structure of substances, and it plays an important role in determining specific heat capacities. Remember, the specific heat capacity is the heat storage capacity of a substance.

Degrees of freedom represents the number of independent ways in which the molecules of a substance can store energy (kinetic and potential). Monoatomic molecules, such as noble gases like helium or argon, have only three translational degrees of freedom because they can move in three perpendicular directions (x, y, z) in space. This limitation means they can only store energy in terms of translational motion.

On the other hand, diatomic molecules, such as oxygen (O_2) or nitrogen (N_2), have an additional two rotational degrees of freedom, in addition to their translational degrees. This is because these molecules can also rotate along two independent axes, providing additional ways to store energy compared to monoatomic molecules.

The presence of additional degrees of freedom in diatomic molecules allows them to store more heat energy compared to monoatomic molecules. This means that diatomic molecules have higher specific heat capacities because they require more energy to increase their temperature by one degree compared to monoatomic molecules. This principle is in line with the Equipartition theorem, which states that each degree of freedom in a molecule contributes 1/2 kT of energy to the total internal energy of the molecule at thermal equilibrium.

In conclusion, the differences in degrees of freedom between monoatomic and diatomic molecules contribute to their respective specific heat capacities, with diatomic molecules having a higher specific heat capacity due to their additional rotational degrees of freedom. Triatomic water with seven degrees of freedom has even higher specific heat.

The specific heat capacity of a substance is related to the thermal degrees of freedom as it measures the amount of heat required to raise the temperature of a unit mass of the substance by one degree Celsius. The specific heat capacity is influenced by the number of degrees of freedom available for the substance to store energy and thus plays a crucial role in determining its thermal properties.

Substances with higher specific heat capacities require more heat energy to increase their temperature compared to substances with lower specific heat capacities. This means that substances with higher specific heat capacities can store more heat energy per unit mass, and can thus absorb and release more heat without experiencing significant temperature changes. This property is particularly important in applications such as regulating temperature changes in systems, maintaining thermal stability, and buffering against fluctuations in heat energy.

Degrees of Freedom in Thermodynamics

Different types of molecules in models will experience different amounts of temperature change when the same amount of kinetic energy flows into them. The less the temperature changes, the more places (or degrees of freedom) there are to store the energy.

Degrees of freedom enable us to estimate different thermodynamic variables for heat capacity, entropy, temperature, etc.

Translational Degrees of Freedom

Gas molecules' freedom of motion in space gives rise to translational degrees of freedom. A molecule in a Cartesian coordinate system can move in the *x, y,* and *z* directions. A particle is said to be in translational motion along the *x, y,* and *z* axes when its centre of mass shifts from its starting position to a new location. Thus, the gas molecule's translational motion is associated with three degrees of freedom. Since any molecule can move freely in any direction in three dimensions, this holds true for all gas molecules, whether they are polyatomic, diatomic, or monatomic. A molecule or an atom can move in three dimensions. As a result, the translational motion of the centre of mass concerning the *X, Y,* and *Z* axes gives any atom or molecule three degrees of freedom.

In three dimensions, the three translational degrees of freedom produce 3/2 kBT of energy.

Rotational Degrees of Freedom

From a classical perspective, the fact that an atom's mass is concentrated in its nucleus means that it has very little rotational energy. Atoms have a very small rotational moment of inertia due to the radius of a nucleus, which is approximately 10^{-15} m. However, molecules differ in that they possess the ability to have kinetic energy during rotation. A linear molecule can rotate about two axes that are perpendicular to the molecular axis, giving it two rotational degrees of freedom. Because the mass is concentrated so close to the axis of rotation, the rotational energy of the rotations along the molecular axis is negligible.

A nonlinear molecule such as water, whose atoms are not arranged along a single axis. has three rotational degrees of freedom This is because the molecule can rotate around any one of three perpendicular axes.

The rotational degrees of freedom contribute kBT to the energy of linear molecules and 3/2 kBT to the energy of non-linear molecules.

Vibrational Degrees of Freedom

A molecule has vibrations as well. A diatomic molecule has a single molecular vibration mode in which the chemical bond between the two atoms acts as a spring to cause the oscillations. More complex molecular vibrational modes are found in molecules with N atoms: 3N − 5 vibrational modes for linear molecules and 3N − 6 modes for nonlinear molecules.

The translational motions contribute maximum to temperature increase. The average kinetic energy of the molecules rises as the material gets hotter. There is enough energy imparted by the collisions for rotation to happen. Next, rotation increases the specific heat and adds to the internal energy. The vibrational energy states are farther apart than the rotation energy states. Vibration does not jump out of the ground state until high temperatures (1000 k).

WHAT IS LATENT HEAT?

Phase Change: Evaporation *vs.* Boiling

Latent heat is the stored energy in the intermolecular bonds of a substance. Latent heat is part of the potential energy of a substance.

The definition of latent heat [Wikipedia] says that it is the latent energy or heat of transformation released or absorbed by a body or a thermodynamic system during a constant-temperature process phase transition.

The definition does not specify that latent heat is only available at the boiling point.

Mark two keywords in the definition: [1] It is a constant temperature process.[2] It's a phase transition temperature.

What is phase change? A phase of a thermodynamic system and states of matter have similar physical properties. During a phase transition of a given medium, certain properties of the medium change due to change in external conditions.

When external interventions cause the system to become unstable, it seeks a new temperature and pressure and switches to a different state of matter to achieve stability.

As previously stated, two conditions are necessary for latent heat to be available.[1] Constant temperature, [2] phase transition. While constant temperature is a derived requirement, the driving force behind making latent heat available is phase change.

Let's look at two common examples, [1] Evaporation: sweating at body temperature.[2] Water boiling at 100 degrees Celsius. In this case, evaporation occurs at 37 degrees Celsius and boiling occurs at 100 degrees Celsius. The two conditions are completely different; evaporation occurs at 37 degrees Celsius, which is our body temperature, whereas boiling occurs at 100 degrees Celsius. Both involve the phase transition of water to vapor by rejecting heat called 'Latent heat'.

How does this Happen?

It is frequently questioned how a similar liquid-to-vapor phase change can occur at such low temperatures, such as 37 degrees Celsius when boiling requires 100 degrees Celsius.

What Causes the Phase Change?

Evaporation and Boiling

Definition

Evaporation is the process by which a liquid turns into a gas at temperatures below its boiling point, primarily occurring at the surface of the liquid. Boiling, on the other hand, is the rapid transformation of a liquid into a gas at its boiling point, happening throughout the entire volume of the liquid.

How They Occur

Evaporation occurs when individual molecules of a liquid gain enough energy to escape into the surrounding air, forming a gas. This process can happen at any temperature and is influenced by factors such as surface area, temperature, and humidity. Boiling, on the other hand, requires the application of heat to raise the liquid's temperature to its boiling point. At this temperature, the liquid undergoes a rapid phase change, leading to the formation of bubbles within the liquid that rise to the surface.

Difference

One key difference between evaporation and boiling lies in the temperature at which they occur. Evaporation can happen at any temperature below the boiling point, while boiling specifically occurs at the boiling point of the liquid. Additionally, evaporation is a slower, surface-level process involving individual molecules escaping into the air, while boiling is a rapid process that involves the entire volume of the liquid forming bubbles and turning into gas within a short period.

Thermodynamics

From a thermodynamic perspective, evaporation is an endothermic process, meaning it absorbs heat from the surroundings to provide the energy needed for molecules to escape the liquid and turn into gas. This leads to a cooling effect on the remaining liquid, making evaporation a common method for cooling systems such as sweating in humans or refrigeration. In contrast, boiling is also an endothermic process that requires heat input to raise the liquid's temperature to the boiling point, after which the energy is used to break intermolecular bonds and overcome the forces holding the liquid together. The heat required for boiling is significantly higher than for evaporation due to the involvement of the entire liquid volume.

Is Latent Heat a form of Work Energy?

Latent heat is a type of energy that arises from the work required to overcome the forces that hold together atoms or molecules in a material. Latent heat is the energy that is absorbed or released when a substance changes state without changing its temperature or pressure. This can happen during a first-order phase transition, such as melting or condensation.

Latent heat is different from sensible heat, which is energy that is transferred as heat and causes a change in temperature. Latent heat is only observed when a substance is undergoing a phase change.

Latent heat is a form of internal or potential energy stored by evaporated or melted water. As ice melts or liquid water evaporates, the molecules change state—from a solid to a liquid, from a liquid to a gas, or from a solid directly to a gas.

Details

The phase change of a substance, such as from solid to liquid or liquid to gas, does involve breaking or forming intermolecular bonds. In a solid, molecules are closely packed together and held in place by strong intermolecular forces. As heat is added to the substance, the molecules gain energy and eventually have enough energy to overcome these forces, leading to the breaking of intermolecular bonds and the transition to a different phase. Also in a liquid-vapor system there is PV work that expands the vapor phase.

Entropy is a measure of the randomness or disorder of a system, and during a phase change, the entropy of the system can increase as the molecules become more disordered. This is because the molecules have more freedom to move and rearrange themselves in a different phase.

Gibbs free energy is a measure of the potential energy of a system that is available to do work. During a phase change, the Gibbs free energy of the system tends to minimize to reach equilibrium. This is because the system is trying to minimize its free energy to achieve stability.

Therefore, the work done during phase change is precursor to entropy increase and the thermodynamic equilibrium.

How and Why?

When a liquid and vapor reach thermodynamic equilibrium, it means that the system is at a point where the chemical potential and other thermodynamic properties are equal in both phases, and there is no net transfer of molecules between the liquid and vapor phases. This represents a stable state.

Therefore, in the context of latent heat, when a substance transitions from a liquid to a vapor phase, energy is required to break intermolecular bonds and increase the entropy of the system. In addition to this, work is also done in expanding the vapor as it transitions to a more gaseous state. This expansion work is the result of the vapor pushing against its surroundings as it expands, which can also be seen as a form of work energy being utilized during the phase change process.

Summary

All type 1 phase changes are work transfer. Take a step back and go to fundamentals. Phase change occurs when a system seeks to achieve stability or minimum free energy that is only possible when the entropy reaches maximum. In order to reach maximum entropy the system expands to the maximum extent possible by PV work. This expansion or stretching breaks intermolecular bonds. This explains intermolecular bonds break not because of temperature but to facilitate a thermodynamic equilibrium to maximize stability when a phase reaches saturation and has no capacity absorb any more energy.

Therefore, in a sense, latent heat can be considered a form of work energy.

LATENT HEAT THERMODYNAMICS

Latent heat means the heat which remains latent with no visual effect during phase change.

Let us expand this statement. Latent heat is the heat absorbed or released by a substance during a phase change without a temperature change. It is called latent heat because the heat remains "hidden" or "latent" as it is not reflected in a temperature change, even though the molecules within the substance are gaining or losing energy.

When a substance undergoes a phase change, such as melting or vaporization, the added or removed heat energy is used to break or form intermolecular bonds, rather than increasing the average kinetic energy of the molecules, which would result in a temperature change.

For example, when water is heated, it eventually reaches its boiling point and begins to vaporize. During this phase change, the added heat energy is used to break the hydrogen bonds holding the water molecules together, allowing them to escape into the gaseous phase. This added heat energy is the latent heat of vaporization.

Why and when the Phase Change Occurs?

Every substance wants to stay in a state with minimum free energy and in a maximum stable condition. A phase change occurs when the substance is thermodynamically unstable and far from its internal thermodynamic equilibrium or minimum free energy. When the conditions in which the substance exists (e.g. temperature, pressure) result in the current phase being less stable than a different phase, a phase change will occur to move the substance towards a lower energy state or equilibrium. This can lead to transitions from solid to liquid, liquid to gas, or other phase changes depending on the specific conditions and properties of the substance.

Gibbs free energy change in phase transition

Gibbs free energy change, often denoted as ΔG, is a thermodynamic quantity that represents the maximum amount of reversible work that can be performed by a thermodynamic system at constant temperature and pressure.

Phase change occurs when the Gibbs free energy reaches minimum and entropy reaches maximum leading to a thermodynamic equilibrium between liquid and vapor phase.

Why phase change occurs at a constant temperature at a given pressure?

Gibbs phase rule

You can predict what will stay constant and when by the Gibbs phase rule.

$$F = C - P + 2$$

which relates the degrees of freedom F (meaning intensive thermodynamic variables you can change) to the number of components C and number of phases P. When there is a single component ($C = 1$) phase change (meaning $P = 2$), there is only one degree of freedom. If you fix the pressure that completely sets the state and you have no freedom to change the temperature.

Pressure–Enthalpy [P-H] diagram is a powerful tool

A *P-H* diagram is a graphical representation of the thermodynamic properties of a substance. It plots pressure on the y-axis and enthalpy on the x-axis and can be used to visualize phase changes and other thermodynamic processes. When it comes to latent heat, the *P-H* diagram is particularly useful because it shows the heat transfer that occurs during a phase change. For example, when a substance undergoes a phase change from a liquid to a vapor, the latent heat of vaporization is the amount of heat that must be added to the substance at constant pressure for the phase change to occur. By looking at a P-H diagram, one can easily see the change in enthalpy that occurs during this phase change, as well as how the pressure and temperature of the substance change. This makes it a powerful tool for understanding latent heat and other phase-change properties of substances.

Explanation

On the left of the dome, it is a water region. On the outline on the left side, it is the saturated water region or line. It has only sensible heat. At 1 bar/100 degc it is 419 KJ/kg.

Let us consider 1 kg water at 0 degc the sensible heat in saturated water with dryness fraction = 0.

Q at 1 bar = $m \times Cp \times dt$ $m = 1$ kg $Cp = 4.187$ KJ/kg/degc $Q = 1 \times 4.19 \times 100$ KJ/kg = 419 KJ/kg.

This is the enthalpy of saturated water at 100 degc when the dryness fraction of water is = 0.

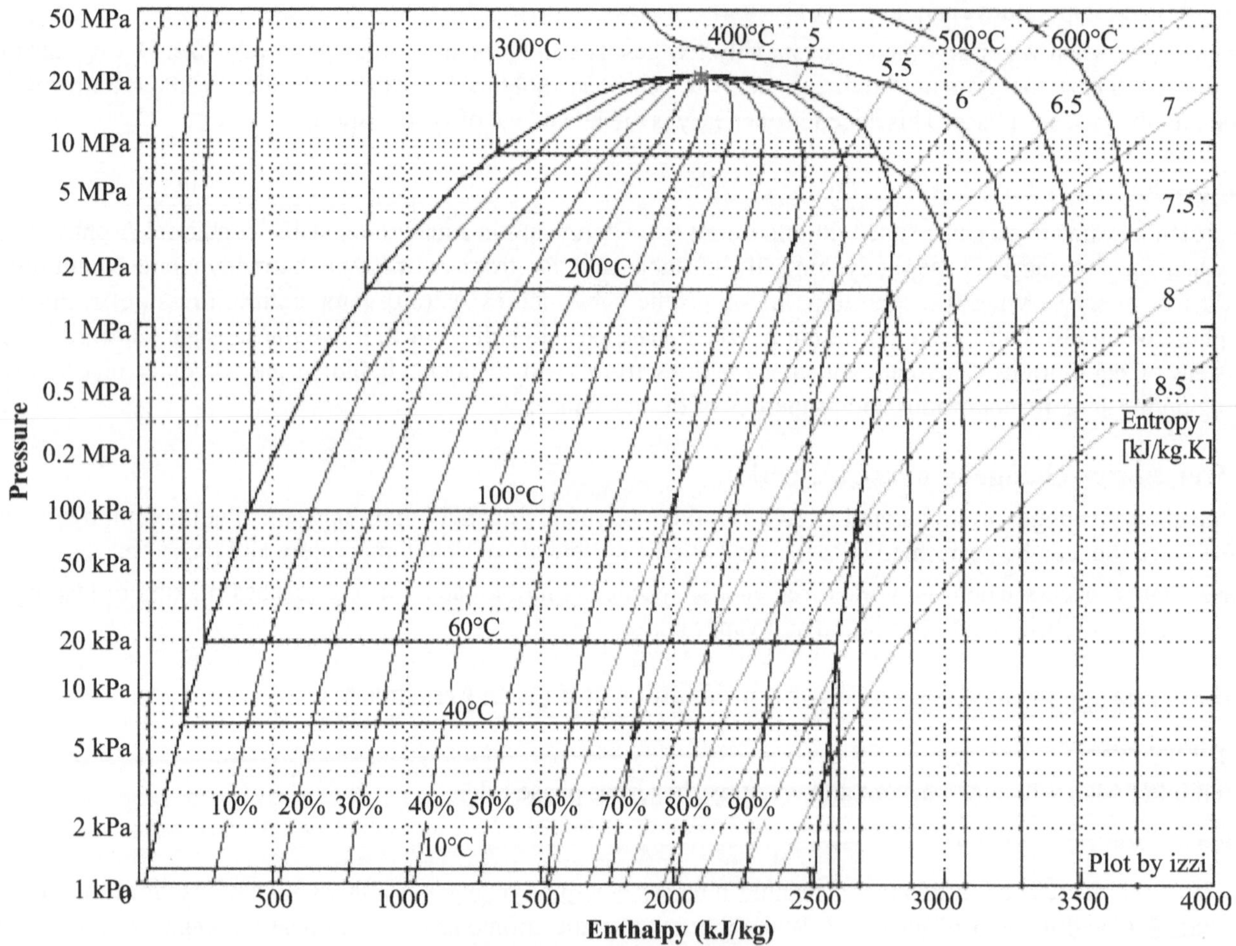

Figure 2: *P-h* Diagram of Water

You will find the same value on the P-H diagram. Inside the dome, it is the saturation region. Water at 0 dryness fraction converts to steam at constant temperature and pressure by breaking intermolecular bonds.

On the right outline, it is the saturated steam region. Here the dryness fraction has increased to 1. The steam is a dry saturated steam on this line. The width of the dome at 1 bar/100 degc is the heat of vaporization, 2257 KJ/kg. This is the latent heat. The effect of heat is not visible here. Please refer to the P-H diagram. On the right-side outline which is the saturated steam line, the total energy is 419+ 2257 = 2676 KJ/kg. This is the total enthalpy of saturated steam at 1 bar. On the x-axis, you can find all enthalpy values.

CP/CV'S ROLE IN THE ANALYSIS OF CONTROL VALVES AND COMPRESSORS

In one line control valves and compressors involve adiabatic processes. Adiabatic processes with no heat coming or exiting into the surroundings consume the internal energy of the gas to perform the job. *Cv* measures the internal energy of a gas.

Background

Specific heat constant pressure: Cp

Cp symbolizes total energy or enthalpy

Cp is the amount of heat energy released or absorbed by the unit mass of the substance with the change in temperature at constant pressure. In other words, it is the heat energy transfer at a constant pressure between a system and its surroundings.

Specific Heat Constant Volume: Cv

Cv symbolizes the internal energy

During a small change in the temperature of a substance, *Cv* is the amount of heat energy released or absorbed by the unit mass of the substance with the change in temperature at constant volume. In other words, it is the heat energy transfer between a system and surroundings when there is no change in volume.

$Cp/Cv = y =$ Adiabatic index

The ratio *Cp/Cv* signifies the internal energy available in an adiabatic process as a proportion to the total energy *Cp dT* at constant pressure. This ratio is also known as the adiabatic index for an ideal gas, and it is a measure of the amount by which the pressure and density of a gas vary in an adiabatic process. For diatomic gases, the adiabatic index is typically 1.4, meaning the internal energy available in an adiabatic process is $Cp = 1.4x\ Cv$ or $Cv = Cp/1.4$.

Role of *Cp/Cv* in Control Valves and Compressors

The ratio of specific heats, *Cp/Cv,* plays an important role in the adiabatic processes involved in control valves and compressors, which operate using internal energy represented by *Cv*.

In control valves, the adiabatic process relies on the internal energy (*Cv*) of the fluid to determine the flow capacity of the valve. The *Cp/Cv* ratio is used to calculate the compressibility factor and predict the pressure drop and flow rate through the valve. By understanding this ratio, engineers can design control valves that efficiently regulate flow and pressure in a system, ensuring optimal performance.

Similarly, in compressors, the adiabatic process is fundamental for efficient compression, utilizing the internal energy (*Cv*) of the working fluid. The *Cp/Cv* value is crucial in optimizing compressor design and operating conditions to achieve the ideal adiabatic compression process.

Therefore, *Cp/Cv* serves as a key parameter in the adiabatic processes of control valves and compressors to know the availability of internal energy in the gas, where internal energy (*Cv*) is utilized to drive efficient operation and performance in industrial systems and processes.

CHAPTER 33

Energy

Energy in thermodynamics is the driving force behind all physical and chemical processes. It is the ability to do work or produce heat. The first law of thermodynamics states that energy cannot be created or destroyed, only transferred or converted from one form to another. The second law of thermodynamics explains how energy tends to disperse and become more disordered over time. The study of thermodynamics helps us understand the behavior of systems and predict how they will change under various conditions.

JOULE: A UNIVERSAL UNIT OF ENERGY

The joule is a special unit. It stands for various types of energy. The International System of Units uses the joule as its unit of energy (SI). It is named after the English physicist James Prescott Joule. The work performed when a force of one Newton acts over a distance of one meter is equal to one Joule.

Mathematically, the relationship between energy (E), work (W), force (F), and distance (d) is given by:

$$E = W = F \times d$$

Why is the Joule a Special Unit?

The ability to represent multiple types of energy makes the joule a special kind of unit. Heat, work, kinetic energy, potential energy, chemical energy, and mechanical energy can all be measured with this flexible unit. Because of its adaptability, it is a basic and frequently used unit in physics and engineering for calculating various Joule is a unique unit.

The dimensional units of the Joule are distinct ($M \cdot L^2 \cdot T^{-2}$). The dimensional units of mass (M), length (L), and time (T) all fit with the dimensional unit of Joule in each example, indicating the versatility of the Joule as a unit of measurement for energy. Indicating the versatility of the Joule as a unit of measurement for energy. Because a joule is a measure of work capacity or heat transfer, it is a universal unit of measurement for energy.

How Joule Represents Various Forms of Energy

Heat

In the context of heat, one joule is the energy transferred when the heat produced by one watt of power is dissipated for one second.

Work

In the context of work, one joule is the work done when a force of one Newton moves its point of application one meter in the direction of the force.

Other Forms of Energy

When considering kinetic energy, potential energy, chemical energy, and mechanical energy, the dimensional unit of Joule ($M \cdot L^2 \cdot T^{-2}$) can be applied as follows:

- **Kinetic energy**: 1 Joule = 1/2 mass velocity^2 ($M \cdot L^2 \cdot T^{-2}$)
- **Potential energy**: 1 Joule = mass gravitational acceleration height ($M \cdot L^2 \cdot T^{-2}$)-
- **Chemical energy**: 1 Joule = amount of energy released by a chemical reaction ($M \cdot L^2 \cdot T^{-2}$)
- **Mechanical energy**: 1 Joule = work done by a force of one newton acting over a distance of one meter ($M \cdot L^2 \cdot T^{-2}$)

In each case, the dimensional units of mass (M), length (L), and time (T) fit with the dimensional unit of Joule, demonstrating how Joule can represent various forms of energy.

HOW IS ENERGY BORN IN A SUBSTANCE?

What is Energy?

Energy is most commonly defined as the capacity for work. Put differently, all entities possess energy. Every time work is done, energy is either transferred or transformed, for example, for a gas, $H = U + PV$ [H is the enthalpy, or the total energy, U is the internal energy consisting of the kinetic and potential energy, P is pressure and V is the volume. PV symbolizes the 'work' energy.

How is energy born in a substance?

Energy is born in a substance at the atomic level when electrons become excited. When an atom absorbs energy, the orbiting electrons are raised to higher energy levels. This process is called excitation, and the energy added to the system is called excitation energy.

Photon Emission

When the excited electrons relax and return to their original shells, the atom emits energy as photon'.

The energy is born in a substance at the atomic level consisting of electrons, protons, and neutrons. Atoms possess huge amounts of energy to overcome attractions and repulsions of oppositely charged particles and prevent an imbalance of positive and negative charges while keeping an atom neutral.

This is the beginning of the creation of energy within a substance. This is the creation of stored energy within a substance. This is the beginning of the birth of potential energy. Atoms use a fraction of this energy to supply the kinetic energy of motions. Rest remains as stored energy.

The kinetic energy is a function of temperature, $KE = \frac{1}{2} kT$. The average KE of a gas particle has ½ kT per energy mode (k is Boltzmann's constant, and T is its absolute temperature in Kelvin).

What does this Mean?

It means that at every temperature, there is untapped potential energy with a magnitude that depends on the size of the atom and a limited fixed amount of kinetic energy that the temperature allows the substance to possess. The kinetic energy is limited to performing work because of its temperature dependence. At any temperature, the kinetic energy can only perform the amount of work that the heat capacity of the substance allows.

Creation of Molecules by Potential Energy

As atoms begin the creation of molecules things start becoming complex. Every two dissimilar atoms have their atomic structure and polarity, making the molecular polar like in H_2O where H has a positive and O has a negative charge. Consequently, this produces intermolecular attraction and repulsion and generates intermolecular bonds. Potential energy expands its storage space for more energy as molecules form. However, the role of the potential energy remains as an energy reservoir without doing any 'work'. In a true sense, potential energy is still not fully converted energy.

Potential Energy: The Energy Supplier

The potential energy supplies energy only when the location of the potential energy carrier changes concerning a reference point. Potential energy changes only by external intervention.

Conversion of Potential Energy to Kinetic Energy and Reverse

This is a demand-supply situation. When the kinetic energy runs out of, the substance becomes saturated with energy, the further demand of energy is met by the potential energy.

When a gas is compressed against intermolecular attraction, the molecules change their locations and come closer generating larger collisions. In the process, the kinetic energy adds energy to potential energy. The gas becomes hot.

Similarly, when a gas expands, the molecules go away from one another, they change their locations, and the potential energy supplies energy to kinetic energy, and the gas cools.

However, every substance has a limited capacity to create kinetic energy which is fixed by the substance's temperature. This is what we call the heat capacity of the substances. As said above, when a substance exhausts it is kinetic energy it draws potential energy at a constant temperature. Potential energy represents latent heat, whereas kinetic energy is a manifestation of sensible heat.

In Summary

- Potential energy changes only when the location of molecules with respect to a reference point in an energy field change.
- Potential energy changes when a molecule is under stress.
- In simple words when Coulomb forces between molecules increase as the molecules come closer, the potential energy increases. The opposite happens when the Coulomb forces between molecules decrease with molecules going away from one another.
- When the work is + W, the surroundings add potential energy to the system. When, the work is – W, the system supplies potential energy which converts to kinetic energy to perform the work of expansion. The sum of potential + kinetic energy remains constant.
- At every temperature, there is a fixed amount of kinetic energy available which is the heat capacity of the substance. When the kinetic energy reaches the limiting point the temperature increase stops. This marks the beginning of phase change with potential energy coming into dominance.

WHAT DOES THE WORD "SATURATION" MEAN?

This is another term that confuses all of us and books don't explain.

"Saturation" means saturation with energy. Saturated water means water is saturated with energy which occurs at 1 bar/100 degc. At saturation, water has a maximum energy of 419 kJ/kg. Water at saturation has no heat capacity to add any more energy.

Saturated steam has a maximum energy of 2675 kJ/kg.

Beyond the saturation point, any supply of energy makes the phase to go into a new phase like saturated water becomes steam.

In psychrometry, saturated air has the maximum enthalpy. Any supply of heat beyond the saturation point makes moisture in the saturated air change phase to water.

The driving force here is the heat capacity. The moist air heat capacity 1.8 kJ/kg does not allow it to absorb heat beyond the saturation point. The moisture turns into a new phase of water at this point which has a much bigger heat capacity of 4.19 kJ/kg-k.

To summarize, the stage of saturation for a phase marks the beginning of phase change.

At saturation the entropy reaches maximum and the Gibbs free energy reaches minimum.

QUALITY OF ENERGY

While all forms of energy may be equal, some are more than equal, highlighting the fact that all forms of energy do not have the same ability to perform work, the yardstick to map the quality of an energy is to determine its ability to deliver work. High grade energy is defined as the form of energy that can fully convert to work and the opposite.

The ability of an energy to perform work is its quality. Quantity of energy is easier to measure than quality because not all energy directed towards work reaches its destination; some energy dissipates along the way, a process known as entropy loss. This dissipation, or entropy energy loss, distinguishes one type of energy from another. Although all energy is, by definition, the same and is when a force of one Newton displaces a body one meter in the direction of the force applied. But when it comes to quality, some forms of energy are more equal than others.

This note aims to examine various forms of energy and perform a quality rating in order to determine how and why they differ from each other. The discussion will primarily centre around four types of energy: thermal energy, work energy, electrical energy, and chemical energy.

Before I go into the specifics, allow me to briefly discuss the basics of low-grade energy using the example of thermal energy. Thermal energy is defined as follows: it moves in the direction of its heat capacity, and if it encounters anything that has a higher specific heat or can absorb more heat than it can, its path becomes dispersed, in other words all kinetic energy that was directed to its destination do not reach and therefore, it is unable to perform the same amount of work that it was assigned.

Details

Different forms of energy are equivalent, according to the first law of thermodynamics, and a suitable conservation equation can be formulated. Nevertheless, the intrinsic losses in an energy conversion process are missed by the first law analysis. The second law of thermodynamics, however, makes up for this weakness in the first law.

Every time energy is converted from one form to another, energy is lost. The intrinsic losses that occur during energy conversion can be substantial in some circumstances. Different types of energy are divided into two categories based on these intrinsic losses: high-quality energy (like work and electricity) and low-quality energy (like heat and wind). With a very high conversion efficiency, high-quality energy can be converted into low- or high-quality energy. However, even in theory, low-quality energy cannot be entirely transformed into high-quality energy. The second law of thermodynamics is the reason behind these restrictions on energy conversion efficiencies.

Let us look at a few examples of energy transformations. Whenever heat (low-quality energy) is converted to work (high-quality energy), Carnot's efficiency is important. The second law of thermodynamics states that a heat engine that converts heat to work cannot be 100% efficient, and there has to be some heat rejection (Kelvin–Plank statement of the second law of thermodynamics). The maximum efficiency of a heat engine is known as the Carnot efficiency. The Betz limit directs the maximum work that can be produced whenever wind energy is converted to work in a wind turbine. Similarly, the Shockley–Queisser limit is crucial when producing electricity from solar radiation through a photovoltaic device.

To assess the quality of any energy form, we should concentrate on the fundamental components of the energy carrier. High-quality energy is defined as basic constituents carrying the same amount of energy in the same direction; otherwise, the form of energy is low-quality. Thermal energy is an expression of internal particle motions; each particle moves at a different speed in different directions, and only their combined effect manifests heat, so heat is low-quality energy.

Because different air molecules move at different speeds, the net kinetic energy of these molecules reflects the energy content in the wind. Wind energy is a low-quality energy source.

Electricity

Since all electrons have the same charge and move in the same direction, electricity has a higher quality than other forms of energy, according to the second law of thermodynamics.

In summary, low-grade energy

According to thermodynamic concepts, energy sources can be classified as high-grade or low-grade based on how easily they can be transformed into other forms. For example, electrical energy is classified as high-grade because it can be easily converted almost entirely into thermal energy (for example, by using an electrical heater). On the other hand, thermal energy cannot be completely converted into electrical energy (thermal power plants typically have efficiencies of about 30%), so thermal energy is classified as low-grade. Consequently, high-grade energy sources are more costly than low-grade energy sources.

EXERGY: A SIMPLIFIED NOTE

Exergy is a thermodynamic property that represents the maximum potential work that can be extracted from a system as it comes into equilibrium with its surroundings. Exergy is often referred to as "available energy" or "useful energy," as it measures the useful potential of a system to perform work.

In summary, exergy provides a more precise measure of the useful energy content of a system compared to total energy.

Exergy losses occur due to irreversibilities within a system, such as heat transfer across temperature gradients, friction losses, mixing processes, and combustion inefficiencies, which reduce the potential work that can be extracted from a system

Energy *vs* Exergy

Energy Definition

Energy is the ability to do work or produce heat. It is a fundamental property of a system that can exist in various forms such as mechanical, thermal, chemical, and nuclear.

Exergy Definition

Exergy, on the other hand, is a measure of the potential work that can be extracted from a system when it reaches thermodynamic equilibrium with its surroundings. Exergy accounts for both the quantity and quality of energy within a system.

Measure of Usefulness

Energy

Energy represents the total amount of work and heat contained within a system, but not all of this energy can be converted into useful work.

Exergy

Exergy, however, quantifies the portion of energy that is available to perform useful work. It provides a more accurate representation of the useful energy content of a system.

Causes of Exergy Losses

Exergy losses occur due to irreversibilities within a system. Irreversibilities are processes that prevent a system from returning to its initial state, leading to a decrease in exergy.

Some common causes of exergy losses include:

- **Heat transfer across a finite temperature difference:** Heat transfer between a system and its surroundings at different temperatures leads to exergy losses due to the irreversibility of the process.
- **Friction and fluid flow resistance:** Mechanical or fluid flow losses due to friction and resistance result in exergy losses as some of the energy is converted into thermal energy and cannot be used for work.
- **Mixing of different substances:** Mixing of substances at different temperatures or pressures leads to exergy losses as the potential for work extraction decreases.
- **Combustion inefficiencies:** In combustion processes, not all of the chemical energy is converted into useful work, resulting in exergy losses.

Components of Exergy

In exergy calculations, there are four contributing factors. These are kinetic energy, potential energy, chemical exergy, and physical exergy. These components make up the total available energy of a system.

- **Kinetic Energy:** Kinetic energy is the energy associated with the motion of an object. In the context of exergy calculations, kinetic energy may be considered as a form of energy that can potentially be converted into useful work. The kinetic energy of a system can be included in the total energy content when calculating exergy.
- **Potential Energy**: Potential energy is the energy associated with the position of an object relative to a reference point. In the context of exergy calculations, potential energy can also be considered as a form of energy that can be converted into useful work. The potential energy of a system can be included in the total energy content when calculating exergy.

Chemical Exergy

Chemical exergy represents the maximum useful work that can be obtained from a substance through chemical reactions. It takes into account the chemical composition of the substance and the potential work that can be extracted by reacting it with other substances.

Chemical exergy is an important component in the exergy analysis of systems involving chemical reactions.

Physical Exergy

Physical exergy, also known as thermodynamic or mechanical exergy, represents the maximum useful work that can be obtained from a system through physical processes such as expansion, compression, heat transfer, etc. Physical exergy accounts for the thermodynamic properties of the system, such as pressure, temperature, and entropy, and assesses the quality of the energy available for conversion into work.

In exergy calculations, these various components of energy are considered based on the specific characteristics of the system under study. The total exergy of a system is then calculated by accounting for all relevant forms of energy and their potential for useful work conversion.

Exergy Calculation

Equation

The equation for calculating exergy is:

$$\text{Exergy} = \text{Energy} - (T_0 \times S)$$

Where: Energy is the total energy content of the system or substance, T_0 is the reference temperature (in Kelvin), S is the entropy of the system or substance

What does this equation mean?

As explained above, the four components of exergy are: internal energy, kinetic energy, potential energy, and flow energy. These components make up the total available energy of a system. T_0S represents the irreversibilities in a system due to temperature differences. So, in a sense, the four components of exergy contribute to the total energy available in a system, while T_0S represents the irreversibilities present.

Next Step

Calculate the total energy content of the system or substance. This can be calculated using the specific energy formula

$$\text{Energy} = \text{Specific energy} \times \text{Mass}$$

Where:

Specific energy is the energy content per unit mass of the substance (in Joules/kg)

Mass is the total mass of the substance

In the equation Exergy = Energy – $T_0 \times S$, the energy component typically includes the internal energy, potential energy, kinetic energy, and chemical energy of a system.

The calculation of energy is dependent on the specific components and forms of energy present in the system. Here's how we can calculate energy using the components of exergy:

Internal Energy (U)

The internal energy of a system is the sum of the kinetic and potential energies of the molecules within the system. It can be calculated using the equation:

$$U = U\_\text{kinetic} + U\_\text{potential}$$

Where U_kinetic is the kinetic energy of the molecules and U_potential is the potential energy of the molecules relative to some reference point.

Kinetic Energy

The kinetic energy of a system is given by the formula:

$$\text{Kinetic Energy} = 0.5 \times m \times v^2$$

Where m is the mass of the system and v is the velocity.

Potential Energy

The potential energy of a system is given by the formula:

$$\text{Potential Energy} = m \times g \times h$$

Where m is the mass of the system, g is the acceleration due to gravity, and h is the height of the system relative to a reference point.

Chemical Energy

Chemical energy is the energy stored in the chemical bonds of molecules within the system. The calculation of chemical energy depends on the specific chemical reactions and compounds involved.

Entropy (S)

Entropy is a measure of the disorder or randomness in a system. It can be calculated using thermodynamic principles and the system's temperature, pressure, and volume. The calculation of entropy involves considering the system's state variables and their changes under reversible processes.

By accounting for internal energy, kinetic energy, potential energy, and chemical energy, we can determine the total energy content of the system and then use the formula

$\text{Exergy} = \text{Energy} - T_0 \times S$ to calculate the exergy content, where S is the entropy of the system.

Calculation of Specific Energy

Specific energy in the context of exergy calculation refers to the amount of energy per unit mass of a substance. It is typically expressed in terms of energy per unit mass (such as joules per kilogram or BTUs per pound).

To calculate specific energy in exergy calculations, we would first determine the total energy content of a substance (considering internal energy, kinetic energy, potential energy, and chemical energy as mentioned earlier) and then divide this total energy by the mass of the substance.

The formula for specific energy can be written as:

$$\text{Specific Energy} = \text{Total Energy/Mass}$$

Where:

Specific Energy is the energy content per unit mass (e.g., J/kg, BTU/lb)

Total Energy is the total energy content of the substance, including internal energy, kinetic energy, potential energy, and chemical energy

Mass is the mass of the substance

Once we have calculated the specific energy for a substance, you can use this value in exergy calculations to determine the exergy content and potential work that can be extracted from the substance.

How the four components of exergy contribute?

Contribution of the four components of exergy to total exergy.

Kinetic Energy Contribution

Calculate the kinetic energy of the system using the formula:

$$\text{Kinetic Energy} = 0.5 \times m \times v^2$$

where m is the mass of the system and v is the velocity.

Potential Energy Contribution

Calculate the potential energy of the system using the formula:

$$\text{Potential Energy} = m \times g \times h$$

where m is the mass of the system, g is the acceleration due to gravity, and h is the height of the system relative to a reference point.

Chemical Exergy Contribution

Calculate the chemical exergy of the system by considering the chemical composition and balancing of the chemical reactions involved.

$$\text{Chemical Exergy} = \Sigma n\_i \times \mu\_i$$

where n_i is the number of moles of each component i, and μ_i is the chemical potential of component i.

Physical Exergy Contribution

Calculate the physical exergy of the system by considering the thermodynamic properties of the system, such as pressure, temperature, and entropy.

$$\text{Physical Exergy} = E - T_0S + P_0V$$

where E is the total energy content of the system, T_0 is the reference temperature, S is the entropy of the system, P_0 is the reference pressure, and V is the volume of the system.

Total Exergy Contribution

Calculate the total exergy of the system by summing up the contributions from kinetic energy, potential energy, chemical exergy, and physical exergy.

$$\text{Total Exergy} = \text{Kinetic Energy} + \text{Potential Energy} + \text{Chemical Exergy} + \text{Physical Exergy}$$

By following these steps and calculating the contribution of each component one by one, we can determine the total exergy content of the system.

ENERGY: 'DISORDERED' AND 'DEGRADED' ENERGY

Ordered and Disordered Energy

Solids have the most ordered energy because their particles are closely packed and have minimal movement. Liquids have slightly less ordered energy because their particles have more freedom to move but are still close together. Gases have the most disordered energy because their particles are far apart and have the most freedom to move. Monoatomic gases have the most disordered energy within gases because their particles are not connected to each other and move freely in all directions. As we transition from solids to liquids to gases, the degree of order and organization of the molecules decreases, leading to an increase in entropy. This increase in entropy represents the measure of disorder or randomness in a system. The reduction in potential energy as we move from solids to gases also corresponds to a decrease in the amount of free energy available to do work. Additionally, this decrease in potential energy is reflected in the decrease in specific heat, as the energy required to change the temperature of a substance decreases with the reduction of potential energy in the system.

Disordered and Degraded Energy

Disordered energy is a state of energy that is spread out and less useful, while degraded energy is energy that has become less available for work:

Disordered Energy

Also known as energy with high entropy, disordered energy is a state of energy that is spread out and less useful. For example, when an airplane flies through the air, some of its energy is lost as heat energy due to friction with the air.

Degraded Energy

Degraded energy is energy that has become less available for work. This happens when energy transforms from a useful or ordered state to a disordered state. For example, when a battery is used to power a device, some energy is lost as heat due to the battery's internal resistance.

Electrical energy is a more ordered form of energy since it flows along a conductor. Unlike heat energy electrical energy does not disperse on its way to a destination.

WHY IS PRESSURE POTENTIAL ENERGY?

Background

Potential energy since not visible and not felt like kinetic energy of motion is not easily understood.

In simple language, the potential energy represents the energy that is stored in a system and has the potential to do work but has not yet been utilized. When the system is triggered or pushed in some way, the potential energy can be converted into kinetic energy or other forms of energy to perform work.

In a system, the total energy is the sum of the kinetic energy (KE) and potential energy (PE) of all the particles within the system. Kinetic energy is associated with the motion of the particles, while potential energy is associated with the interactions between particles and their positions within the system. These two forms of energy are constantly interacting within the system and can be converted from one form to another, leading to the conservation of total energy in the system. This principle is known as the conservation of energy, which states that energy cannot be created or destroyed, only transferred or converted from one form to another.

Essentially, potential energy exists and is tied up in the electron-electron repulsion, electron-proton attraction, proton-proton repulsion, and intermolecular attraction in a molecule.

Thus, to summarize, the potential energy of an electron in an atom is determined by its distance from the nucleus and interactions with other electrons and the positively charged protons in the nucleus. The ground state energy represents the lowest possible energy level that an electron can occupy in an atom. Energy never becomes zero.

This interplay of repulsions and attractions determines the stability and structure of atoms and molecules.

At a molecular level, in molecular systems, **potential energy** is determined by the interactions between molecules, which are influenced by the nature of **intermolecular forces**. These forces include hydrogen bonding, dipole-dipole interactions, van der Waals forces, and ion-ion interactions. The arrangement and orientation of molecules relative to each other can greatly impact the potential energy of the system.

The potential energy of a molecular system changes based on the distance [location] and orientation between molecules. For example, when molecules come closer together, they may experience attractive forces that lower the potential energy, leading to stability. Conversely, if molecules are forced too close together, repulsive forces may dominate, resulting in a rise in potential energy.

Is pressure a form of Potential Energy

Pressure can be considered a form of energy density, or energy per unit volume, and can be related to potential energy in a number of ways.

Pressure Energy

The work required to introduce a fluid into a system without changing its volume. Pressure energy per unit mass of fluid is calculated by dividing PV by m (P is pressure, V is volume and m is mass)

Bernoulli Equation

In the Bernoulli equation, pressure represents the flow energy per unit volume, while the term rho g z (rho is density, g is gravitational constant and z is elevation) represents the gravitational potential energy per unit volume.

Energy Density

Energy densities from other causes can be expressed as an effective "pressure". For example, the energy density of solvent molecules that leads to osmosis is expressed as osmotic pressure.

Gravitational Potential Energy

In a fluid at rest, gravitational potential energy density decreases as you move vertically downward, so the pressure must increase by the same amount

SPECIFIC HEAT IS KINETIC ENERGY AND LATENT HEAT IS A FORM OF POTENTIAL ENERGY

Specific heat is a measure of the amount of heat energy required to raise the temperature of a unit mass of a substance by one degree Celsius. It is related to the kinetic energy of the molecules within a substance, as heat energy is absorbed by the substance and causes the molecules to move faster, translating into an increase in temperature.

On the other hand, latent heat refers to the amount of heat energy required to change the phase of a substance, such as from solid to liquid or liquid to gas, without a temperature change. This heat energy is used to overcome the intermolecular forces between molecules in the substance and thus is related to the potential energy associated with these interactions. Therefore, while specific heat is indeed related to the demand for kinetic energy in a molecule, latent heat is more related to the demand for potential energy in terms of the rearrangement of molecules relative to each other during a phase change.

Energy Transfer

Kinetic to Potential and Reverse

The surroundings can influence the direction of energy transfer or conversion within a system. For example, in a thermodynamic process, energy can flow from the system to the surroundings (exothermic) or from the surroundings to the system (endothermic) depending on the conditions and interactions between the system and surroundings.

In a system where there is a transfer of energy (such as heat) from the surroundings to the system, this can lead to an increase in the internal energy of the system, causing changes in the kinetic and potential energy of the particles within the system.

Conversely, if energy is transferred from the system to the surroundings, it can lead to a decrease in the internal energy of the system.

Therefore, the interactions and exchanges of energy between the system and surroundings can influence the direction of energy transfer and the conversion between kinetic and potential energy within the system.

Typical Example

Bond-Breaking and Bond-Making Energy

During a bond-breaking process, energy is required to break the bonds holding atoms together, resulting in the conversion of potential energy stored in the bonds to kinetic energy. This kinetic energy is responsible for the movement of atoms and molecules as the bond breaks.

On the other hand, during a bond-making process, energy is released as new bonds form between atoms. This energy is in the form of kinetic energy, which is then converted back to potential energy as the new bonds are established. The potential energy is then stored in the newly formed bonds until it is needed for another chemical reaction. Energy transfer always takes place from a higher to a lower level which gives the thermodynamic stability to a molecule.

Overall, the breaking and making of chemical bonds involve the conversion of energy between potential and kinetic forms, allowing for the rearrangement of atoms and molecules in chemical reactions.

KINETIC ENERGY IS A STATE FUNCTION

KE connects all three key state properties. *P*, *T* and *V*, *KE/V* is a unique number for a state of matter independent of path. In summary, *KE/V* distinguishes one state from another.

Kinetic energy is an important property that can connect mass, velocity, pressure, and temperature. This is because kinetic energy is directly related to an object's mass and velocity, and it can also be influenced by pressure and temperature in certain systems.

How?

Relation with Temperature

Average kinetic energy is proportional to temperature, $KE = 1/2 \text{ Kb} \times T$. Kb is Boltzmann constant and T is temperature.

Relation with pressure and volume

$$PV = 3/2\ KE$$

Relation with mass and velocity

$$KE = 1/2\ mv^2$$

At constant temperature *KE* is constant. At constant temperature for the same mass, *KE* is proportional to velocity squared. This explains why H_2 gas among all gases at any given temperature has the highest velocity. This also explains why H_2 gas requires a higher amount of energy for compression.

Since *KE* connects all key three-state properties. *P, T,* and *V*, *KE*/*V* is a unique number for a state of matter independent of path.

In summary, KE/V is a useful number to distinguish one state from another.

The path functions such as work and heat are of course easy to conceive. Heat has several options to travel between two points like isothermal, adiabatic, etc generating different amounts of work.

HEAT *VS.* ELECTRIC ENERGY: WHY ELECTRICITY IS A CLEAN FORM OF ENERGY?

Electrical energy is a form of work energy. Work energy is the capacity of a system to do work, and electrical energy is the energy associated with the flow of electric charge [electrons] through a conductor. When electrical energy is converted into other forms of energy, such as mechanical energy or heat, work is performed. Since electrical energy is typically a form of work energy, electrical energy is considered as 'pure' energy. Electrical energy is a more ordered form of energy since it flows along a conductor. Unlike heat energy electrical energy does not disperse on its way to a destination. Electrical energy is not a path function, as it is a form of energy dependent on the potential difference between two points in an electric circuit. The potential difference determines the flow of electrons through a circuit, or voltage, between the points, rather than the specific path the electrons take.

Detail

Let me expand on this point. This is the key to this post. Work energy is a "purer" form of energy in the sense that it is not associated with the generation of entropy, as it does not involve heat transfer. Work energy involves the transfer of energy through mechanical means, such as pushing, pulling, or lifting an object, and it is a form of energy that can be used to perform useful work without necessarily leading to the generation of waste heat or an increase in entropy. In contrast, energy transfer through heat involves random molecular motion and is associated with an increase in entropy. This is a fundamental aspect of the second law of thermodynamics, which states that in any energy transfer or conversion process, some energy will be lost as waste heat, leading to an increase in entropy in the system. Therefore, work energy can be considered a "cleaner" form of energy compared to heat energy in terms of entropy generation. The energy of electrical energy is typically expressed as 100% since it is the initial form of energy input. When electrical energy is converted into other forms of energy such as heat, light, or mechanical energy, the total amount of energy remains the same, but the efficiency of the conversion process determines how much of that energy is successfully transformed into the desired output form. The entropy Q/T [Q is heat and T is temperature] is generated as a result of the conversion process. This is due to the inherent tendency of energy to disperse and become more disordered as it is transformed from one form to another.

Electrical energy is work energy: Electrical energy work, W [Work] = VQ, and V is the potential difference between the two points. Q is the charge, Q = It [Ohm's law]

$W = V$ It, Therefore, to conclude, in the case of electrical energy the conversion to work depends only on the potential difference V which is in volts, and current which is in ampere. There is no entropy or heat reservoir concept like in heat energy.

How electrical energy generates heat: Electrical energy can generate heat through the process of resistive heating, which is described by the formula $I^2 R T$. In this formula: – I represent the current flowing through a resistor, – R represents the resistance of the material through which the current is flowing, – T represents the time duration for which the current flows. When an electric current flows through a resistor, such as a heating element in an appliance, the resistance of the material causes some of the electrical energy to be converted into heat energy. This heat generation is a result of the collisions between electrons and atoms in the material as the current flows through it, and it is what powers devices like electric heaters, stoves, and toasters.

In summary, the source of electricity is typically work, which is a pure form of energy, while the source of heat is often the transfer of energy through infrared radiation or other forms of energy transfer. Work, in the context of electricity, involves the movement of charged particles, such as electrons, through a conductor to generate electrical energy. One way to understand why electricity is considered purer than heat is by looking at the nature of the energy transfer involved. When electricity is generated and transmitted through a circuit, the flow of electrons is a direct and controlled process that can be harnessed to perform specific tasks, such as powering devices or lighting up bulbs. This transfer of electrical energy is efficient and can be easily converted into other forms of energy, such as mechanical or light energy, with minimal loss.

On the other hand, heat energy is often the result of the random motion or vibration of particles within a system, which makes it more difficult to control and harness for specific purposes. Heat transfer is typically a less efficient process compared to the flow of electricity, as heat tends to disperse and dissipate more readily, leading to energy loss and decreased efficiency in energy conversions. Therefore, the relative purity of electricity compared to heat stems from the direct and controlled nature of electrical energy transfer, as well as its efficiency in performing work and powering devices. This distinction highlights the importance of understanding the different forms of energy and their characteristics in various applications and systems.

In final words, electrical energy is a more ordered form of energy since it flows along a conductor. Unlike heat energy electrical energy does not disperse on its way to a destination. Electrical energy is not a path function, as it is a form of energy that is dependent on the potential difference between two points in an electric circuit. The potential difference determines the flow of electrons through a circuit, or voltage, between the points, rather than the specific path the electrons take.

CHAPTER 34

Thermodynamics

Thermodynamics is the branch of physics that deals with the relationships between heat and other forms of energy. It governs the behavior of systems at the macroscopic level, describing how energy is transferred and transformed within these systems. The laws of thermodynamics establish fundamental principles that dictate the limits and possibilities of energy conversion. Engineers and scientists use thermodynamics to design efficient systems, such as engines or refrigerators, and to study the behavior of complex systems like the Earth's climate.

THERMODYNAMIC DIAGRAMS

Diagrams called thermodynamic diagrams are used to show the thermodynamic states of a material, usually a fluid, as well as the effects of manipulating the material. A temperature-entropy diagram (T-s diagram), for example, can be used to show how a compressor changes a fluid's behaviour.

The equivalency of area and energy in thermodynamic diagrams is their main characteristic. Air undergoes temperature and pressure changes during a process, which results in a closed curve in the diagram. The area enclosed by this curve relates to the energy that the air has gained or released.

1. PV Diagram

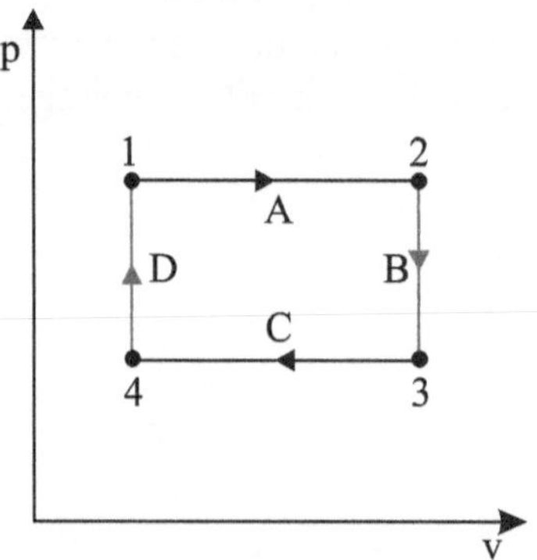

Figure 1: PV Diagram

Plotting the change in pressure *P* in relation to volume *V* for a process or processes is done using a *PV* diagram. In thermodynamics, a sequence of events usually culminates in a cycle, meaning that at the end of the cycle, the system's state remains unchanged; that is, the pressure and volume of the device return to their initial values.

An idealised *PV* diagram's features are depicted in the figure. A list of the numbered states is displayed (1 through 4). Each state is reached by a process (A through D) that modifies the system's volume or pressure, or both.

One important aspect of the diagram is that the area enclosed by the four lines represents the net amount of work, which means that the amount of energy received or used by the system as work can be measured. The net amount of work in the figure is the difference between the two processes, [1] compression and [2] expansion represented by the area inside the rectangle.

What does it mean?

In a PV cycle diagram, the forward path (compression stroke) does more work than the return path (expansion stroke) because work is done on the system during compression and work is done by the system during expansion.

During the compression stroke, work is done on the gas by an external source in order to compress it. This work increases the internal energy of the gas and increases the pressure and temperature. In contrast, during the expansion stroke, the gas does work on the surroundings as it expands, pushing against a piston or other external resistance. This work decreases the internal energy of the gas and decreases the pressure and temperature. Therefore, the work done during the compression stroke is greater than the work done during the expansion stroke in a PV cycle diagram. This is why the forward path does more work than the return path.

2. Mollier (H–S) Diagram

A Mollier diagram, also known as the *h-s* (enthalpy-entropy) diagram, is a graphical representation of the thermodynamic properties of a substance. It is commonly used in the field of thermodynamics to analyze and visualize the behavior of substances in various processes such as heating, cooling, and phase change.

In Mollier's diagram, the x-axis is entropy and the y-axis is enthalpy.

The diagram allows for the visualization of the relationship between enthalpy, entropy, temperature, and pressure for a substance. By plotting the properties of a substance in the Mollier diagram, it becomes possible to predict changes in enthalpy and entropy as the substance undergoes various thermodynamic processes.

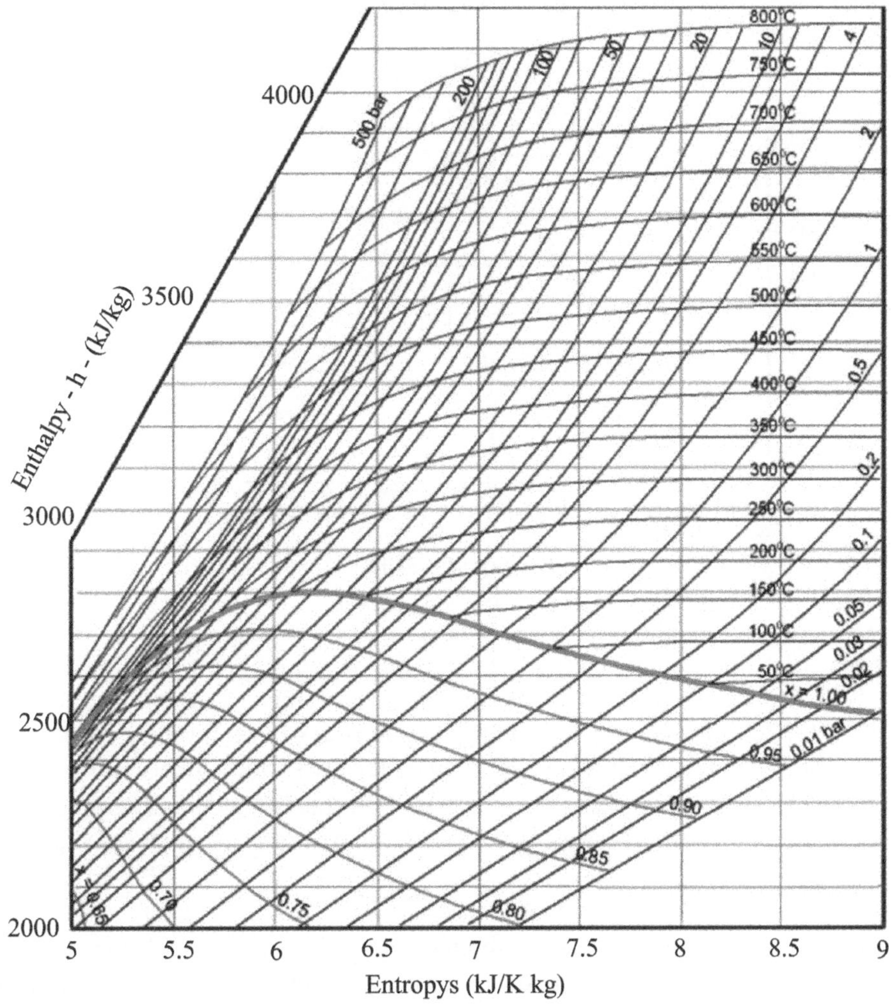

Figure 2: Mollier Diagram Water-Steam (*h-s* diagram)

In the context of predicting changes in enthalpy and entropy, the Mollier diagram facilitates the analysis of processes such as compression, expansion, condensation, and other phase changes. The diagram provides insights into how these processes will impact the state variables of the substance, such as temperature, pressure, and specific volume, among others.

Mollier Diagram is a Useful Tool to Study Phase Change

During a phase change, such as the transition from liquid to vapor, the enthalpy of the system progressively increases as vapor forms. This increase in enthalpy is accompanied by an increase in internal energy and PV work, as the molecules transition from a more ordered liquid state to a more disordered vapor state. Simultaneously, the entropy of the system also increases as the vapor expands. This increase in entropy indicates a higher degree of disorder in the system as the molecules move from a more structured arrangement in the liquid phase to a more chaotic arrangement in the vapor phase. These changes in enthalpy, internal energy, and entropy during a phase change can be observed on a Mollier diagram, where the vapor curve shows a sharp increase in enthalpy and entropy during the phase change. Overall, the phase change process is characterized by an increase in enthalpy, internal energy, and entropy, reflecting the energy and thermodynamic changes occurring within the system.

T–S Diagram

It's a temperature *vs.* entropy diagram.

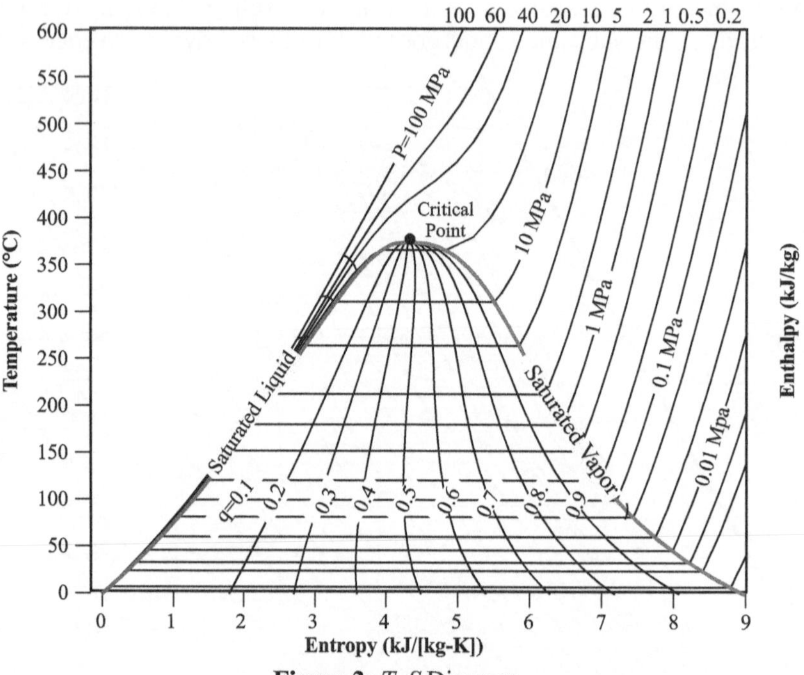

Figure 3: *T–S* Diagram

A temperature–entropy (*T–S*) diagram is a type of thermodynamic diagram used in thermodynamics to show changes in temperature (*T*) and particular entropy (s) as a curve graph during a thermodynamic process or cycle. It is a common and helpful tool, especially for helping to visualise the transfer of heat during a process. The heat transferred to the system during a reversible (ideal) process is represented by the area under the process's *T–S* curve.

A common classification scheme for working fluids is based on the *T–S* diagram's shape. On a *T–S* diagram, an isentropic process is represented by a vertical line, while an isothermal process is represented by a horizontal line.

More Details

A TS diagram is a graphical representation of the thermodynamic properties of a substance. It plots temperature on the y-axis and specific entropy on the x-axis. This allows for a visual representation of the changes in a substance's temperature and entropy during a thermodynamic process.

TS diagrams are commonly used in the study of thermodynamics to analyze and understand the behavior of substances such as steam, refrigerants, and ideal gases. By looking at the shape and slope of the lines on a *TS* diagram, it is possible to determine important properties such as enthalpy, specific volume, and quality. It's a common tool for visualizing heat transfer during a process. For example, an isentropic process appears as a vertical line on a *T–S* diagram, while an isothermal process appears as a horizontal.

The main difference between a temperature-entropy (*T–S*) diagram and an enthalpy-entropy (*H–S*) diagram is that a *T–S* diagram only shows temperature changes, while an *H–S* diagram shows both enthalpy and entropy.

4. P–H Diagram

This is a pressure-enthalpy diagram of water.

It's a pressure *vs.* enthalpy diagram. Refrigeration uses this diagram. The relationship between a specific refrigerant's pressure and enthalpy is depicted in the diagram. It is best to go through the vapour compression cycle on a P–H diagram in order to fully comprehend this diagram. Enthalpy is shown on the x-axis and pressure is shown on the y-axis of the P-H diagram. Pressure is commonly measured in pounds per square inch (psi), while enthalpy is typically measured in Btu/lb. The diagram's inverted U shape indicates the locations where the refrigerant undergoes phase changes. The saturated liquid curve is shown by the left vertical curve, and the saturated vapour curve is shown by the right vertical curve.

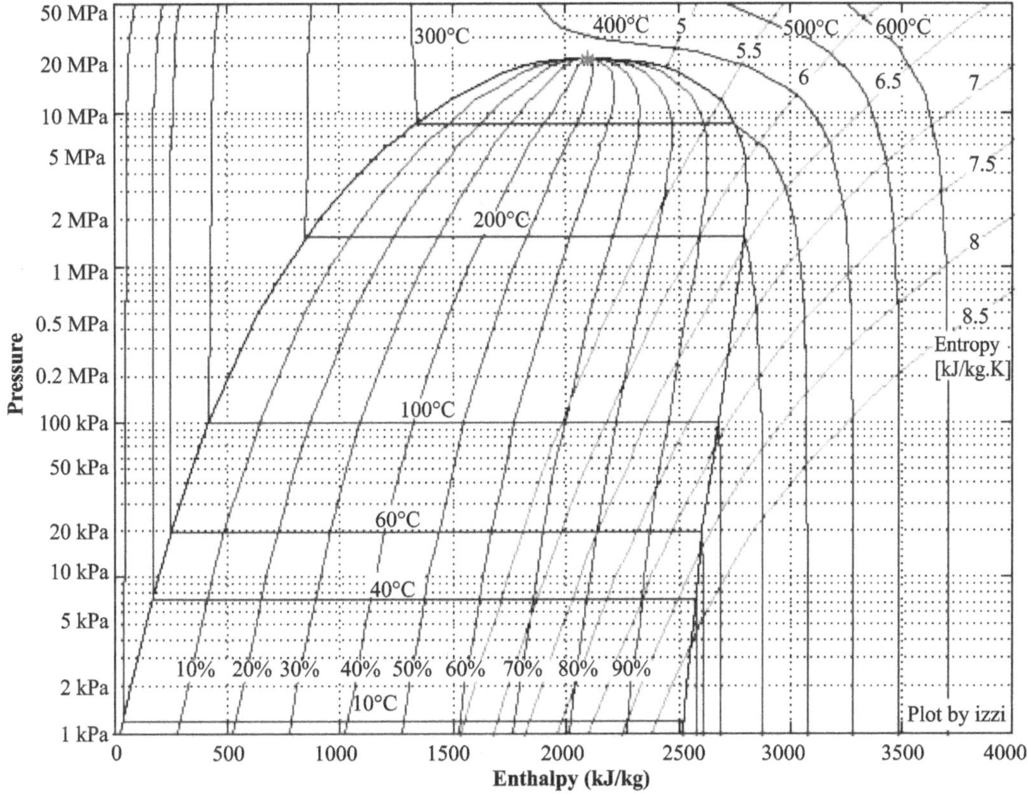

Figure 4: *T–h* Diagram to Water

A mixture of liquid and vapor-containing refrigerant states are represented by the area between the two curves. The refrigerant is in liquid form when it is located to the left of the saturated liquid curve and in vapour form when it is located to the right of the saturated vapour curve. The critical point is the location where the two curves converge. This point is crucial because it indicates that no amount of pressure higher than this will turn the vapour into a liquid.

Summary of all three thermodynamic diagrams, HS, TS and PH diagrams

- **HS (Enthalpy-Entropy) Diagram**: This diagram plots specific enthalpy against specific entropy. It is useful in determining the state of a substance, as well as analyzing processes such as refrigeration cycles and power cycles. It helps in visually representing the energy changes and heat transfer that occur during a thermodynamic process.
- **TS (Temperature-Entropy) Diagram**: This diagram plots specific entropy against temperature. It is useful in analyzing reversible and irreversible processes, as well as determining the efficiency of systems like heat engines and refrigeration cycles. It can also help in identifying the direction of heat flow and the availability of energy in a system.

- **PH (Pressure-Enthalpy) Diagram:** This diagram plots specific enthalpy against pressure. It is useful in analyzing processes involving gases and understanding the behavior of refrigerant fluids in heating and cooling systems. It can also help in determining the state of a substance, as well as identifying phase changes such as condensation and evaporation.

Overall, these thermodynamic diagrams are essential tools in the field of thermodynamics, as they provide a visual representation of the relationships between key properties of a substance and help in analyzing and understanding various thermodynamic processes.

5. Psychrometric Chart

This chart is used to analyze the properties of moist air, such as temperature, humidity, and enthalpy, and is crucial for applications related to heating, ventilation, and air conditioning (HVAC) systems.

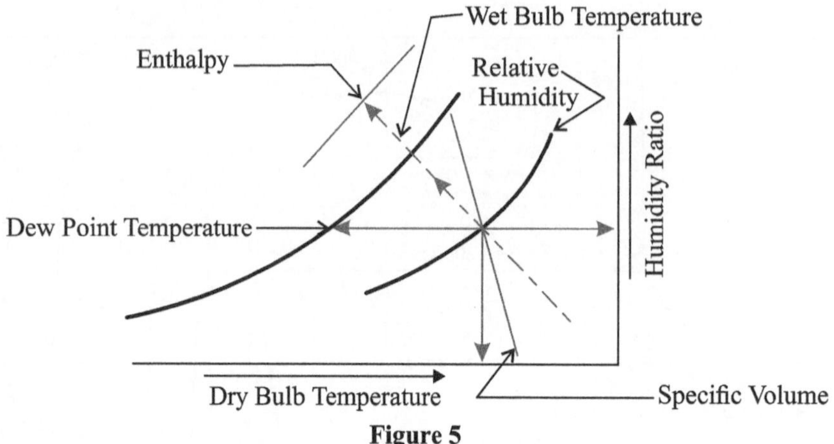

Figure 5

A psychrometric chart is a graphical representation of the psychrometric processes of air at constant pressure. Psychrometric processes include physical and thermodynamic properties such as dry bulb temperature, wet bulb temperature, humidity, dew point, enthalpy, and air.

6. P and T Phase Diagram

Phase diagrams depict the equilibrium phases of a substance or mixture under various conditions of temperature and pressure. They are essential for understanding phase transitions and predicting the behavior of substances under different thermodynamic conditions.

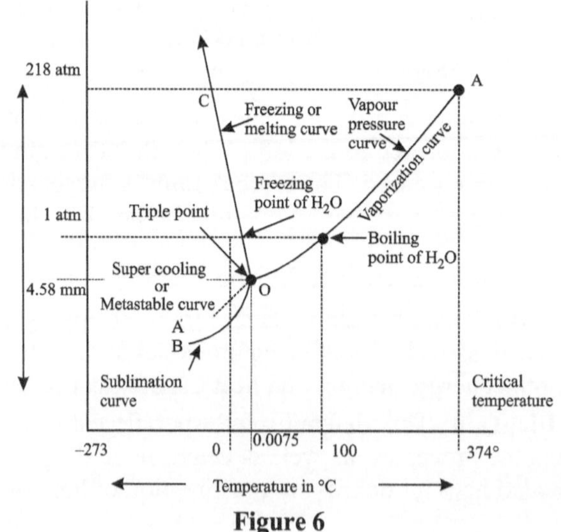

Figure 6

7. Compressibility Chart

It's Z *vs.* reduced pressure chart.

A generalized compressibility chart is a graphical representation of compressibility factor Z as a function of reduced pressure Pr and reduced temperature Tr for a given substance. The compressibility factor is a measure of how much a real gas deviates from ideal gas behavior.

In the generalized compressibility chart, the compressibility factor Z is plotted on the y-axis, while the reduced pressure Pr (ratio of actual pressure to critical pressure) is plotted on the x-axis. The reduced temperature Tr (ratio of actual temperature to critical temperature) is typically held constant for a given chart.

The chart is typically used in thermodynamics and fluid dynamics to estimate the compressibility factor of a gas at various conditions, without the need for complex calculations.

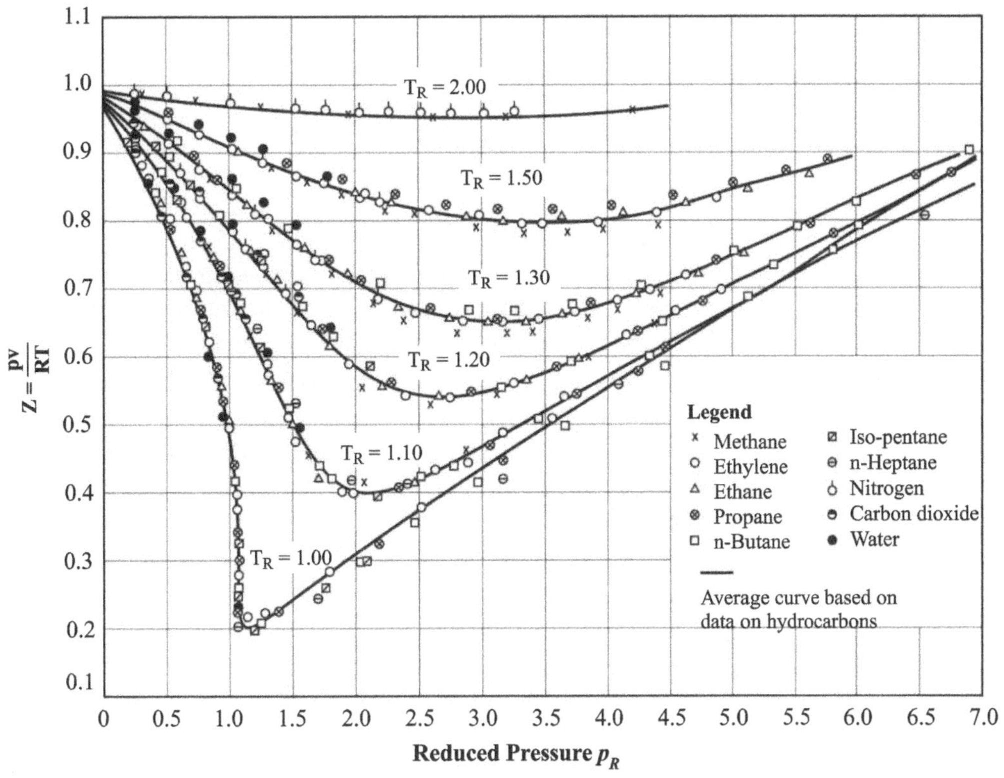

Figure 7: Z *vs.* Reduced Pressure Chart

Background

Theorem of Corresponding States

The principle of Corresponding States (PCS) was stated by van der Waals and reads: "Substances behave alike at the same reduced states."

Reduced Properties

Reduced properties are used to define corresponding states. Reduced properties provide a measure of the "departure" of the conditions of the substance from its own critical conditions and are defined as follows:

$Pr = P/Pc$, $Tr = T/Tc$, and $Vr = V/Vc$ P, T, and V are temperature, pressure, and volume. Subscript c stands for critical point and r stands for reduced temperature and pressure. If $Pr = Tr = Vr = 1$, the substance is at its critical condition. If we are beyond critical conditions, $Tr > 1$, $Pr > 1$ and $Vr > 1$. By the same token, if all the conditions are subcritical, $Tr < 1$, $Pr < 1$ and $Vr < 1$. Critical conditions become the scaling factor by which substances can be compared among each other in terms of their "departure from criticality" or reduced properties.

8. Saturation Vapor Curve

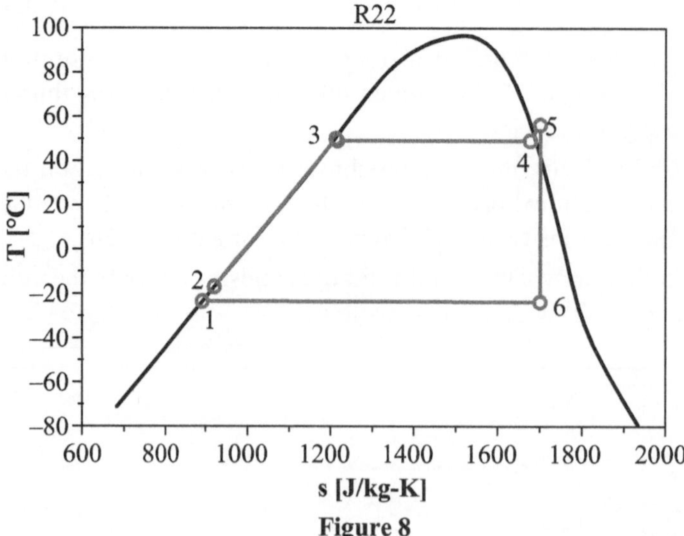

Figure 8

The curve in the temperature–entropy diagram (*T–S* diagram) that divides the two-phase state from the superheated vapour state is known as the saturation vapour curve in thermodynamics. By looking at this curve you get an idea of expansion of liquid-vapor in the saturated region [two phase regions] between the saturated liquid and saturated vapor curves and hence the compressibility of the fluid and the thermodynamic work potential.

In the *T–S* diagram, the curve that divides the subcooled liquid state from the two-phase state is known as the saturated liquid curve.

The fluid expansion in a power cycle is highly dependent on the characteristics of this saturation curve: A fluid that is "wet" exhibits a negative saturation vapour curve. At the end of the expansion, a two-phase state is obtained if overheating occurs before the expansion is limited.

There is a vertical saturation vapour curve in a "isentropic" fluid. In the event of a hypothetical isentropic expansion, it stays extremely near to the saturated vapour state.

By looking at this curve you get an idea of expansion of liquid-vapor in the saturated region [two phase regions] between the saturated liquid and saturated vapor curves and hence the compressibility of the compressibility of the fluid and the thermodynamic work potential.

THERMODYNAMIC KEY CONCEPTS

A quick look at some of the key definitions of thermodynamics.

Enthalpy

Enthalpy is a thermodynamic property that is defined as the sum of a system's internal energy and the product of its pressure and volume. It is not possible to measure enthalpy directly, but the change in enthalpy is usually easier to measure. This is because the change in enthalpy can be related to heat transfer, which is easier to measure. Enthalpy is a state function. Enthalpy is expressed as *Kj*/mole in SI unit. Enthalpy is expressed as $H = U + PV$, where H is enthalpy, U is internal energy, P is pressure and V is volume. Enthalpy is an extensive property.

Heat

Heat is form of energy. Heat is a path function. Heat is an energy in transit. SI unit of heat is Joule, Heat is expressed by equation $Q = U + W$, where Q is heat, U is internal energy and W is work.

Internal Energy

Internal energy is sum of kinetic and potential energy. The *SI* unit for internal energy is Joule. Internal energy is a state function. Internal energy is expressed by $U = Q - W$.

Work

Work is mechanical energy. Work is a path function. SI unit for work is Joule. When work is done on a system the work is given a positive sign, + *W*. When work is done by the system work is given a negative sign, – *W*.

Entropy

A measure of the randomness or disorder of molecules in a system. Entropy tells us how far a system is from equilibrium. Entropy is measured in Joules/grams/Kelvin. The Second Law of Thermodynamics states that the entropy of the universe increases in a spontaneous process. Entropy is a state function. Formula for entropy, $dS = dQ/T$. The change in entropy dS is the inverse of the temperature integrated over the change in heat transfer.

Free Energy

Free energy is the amount of energy in a system that can be used to perform useful work. It's a state function or energy-like property of a system that's in thermodynamic equilibrium. Gibbs free energy is expressed as $dG = dH - TdS$, where G is free energy, H Is enthalpy, T is temperature in Kelvin and S is entropy. Free energy is calculated by subtracting the product of a system's temperature and entropy from its enthalpy. It's used to determine how systems change and how much work they can produce. *Si* Unit of free energy is Joule.

Sensible Heat

Sensible heat is heat that can be sensed. Sensible heat is primarily the kinetic energy. causes a temperature change, but the body does not change phase.

Latent Heat

Latent heat is the heat that cannot be sensed. Latent heat is the stored potential energy in the chemical bonds. Latent heat plays important role during phase change by either absorbing heat to provide energy for intermolecular bonds breaking or releasing heat to provide energy for exothermic bond making at constant temperature and pressure.

Adiabatic Process

An adiabatic process is one in which there is no heat-in or heat-out. In adiabatic cooling of gases, the gas expands by using its internal energy.

From the 1st law of thermodynamics, $U = Q - W$, U is internal energy, Q is heat and W is work since in an adiabatic process. $Q = 0$, in adiabatic expansion, $U = -W$. In other words, the cooling in an adiabatic process occurs at the expense of the expansion work – W.

Isenthalpic Process

An isenthalpic process is also an adiabatic expansion process but it occurs at constant enthalpy across the throttling valve.

Let us take two points across a throttling valve 1 (upstream) and 2 (downstream).

Isenthalpic process cooling across the throttling valve can be expressed as: $H_1 = H_2$ (H stands for enthalpy,) $U_1 + W_1 = U_2 + W_2$. There is no change in total energy. In the isenthalpic process when the gas expands the internal energy and the work energy compensate each other while the gas cools by keeping the total energy constant.

Remember U is a state function and W is a path function. They are two different forms of energy and act independently to compensate each other while keeping the total energy constant in a constant enthalpy process.

The internal energy '*U*' is the sum of kinetic and potential energy. Work '*W*' is mechanical energy.

A point to note is that an adiabatic process does not necessarily have to be either an isentropic (constant entropy) or isenthalpic process (constant enthalpy) because of some thermodynamic restrictions.

Isentropic Process

An isentropic (constant entropy) process is adiabatic + an internally reversible process e.g., the power plant turbine. In contrast, an isenthalpic (constant enthalpy) process is adiabatic + constant enthalpy process. Joule Thomson effect is considered to be an isenthalpic process. A turbine is idealised as a isentropic process.

What is Friction

Friction at a molecular level involves rubbing of atoms when two surfaces slide over one another. During friction, atoms tend to repel (essentially electron-electron repulsion) each other. So, when something touches something else the atoms in both objects push on each other, which creates a small amount of lost energy as heat. Frictional losses since occur at atomic can't be reversed.

COMMON THERMODYNAMIC EQUATIONS

The First Law of Thermodynamics: $\Delta U = Q - W$

- This equation states that the change in internal energy of a system is equal to the heat added to the system minus the work done by the system.

The Second Law of Thermodynamics (Clausius statement): $\oint(\delta Q/T) \leq 0$

- This equation states that for any thermodynamic process, the integral of heat transfers over the temperature must be less than or equal to zero.

The Third Law of Thermodynamics: $S = 0$ ***as*** $T \rightarrow 0$

- This equation states that the entropy of a perfect crystal approaches zero as the temperature approaches absolute zero.

Ideal Gas Law: $PV = nRT$

- This equation relates the pressure (P), volume (V), amount of substance (n), temperature (T), and the universal gas constant (R) for an ideal gas.

Enthalpy Equation: $H = U + PV$

- This equation defines enthalpy (H) as the internal energy (U) of a system plus the product of pressure (P) and volume (V).

Gibbs Free Energy Equation: $G = H - TS$

- This equation defines the Gibbs free energy (G) of a system in terms of enthalpy (H), temperature (T), and entropy (S).

Helmholtz Free Energy Equation: $A = U - TS$

- This equation defines the Helmholtz free energy (A) of a system in terms of internal energy (U), temperature (T), and entropy (S).

Maxwell Relations

These are a set of equations that describe the relationships between the partial derivatives of thermodynamic state functions, such as those involving entropy, enthalpy, internal energy, and Gibbs free energy.

Clapeyron Equation: $dP/dT = \Delta S/\Delta V$

This equation relates the slope of the phase boundary between two phases in a phase diagram to the entropy change (ΔS) and volume change (ΔV) of the phase transition.

Van't Hoff Equation: $\ln(K_2/K_1) = -\Delta H/R\ (1/T_2 - 1/T_1)$

This equation relates the equilibrium constant (K) of a reaction to the enthalpy change (ΔH) and the temperature change (ΔT) of the reaction.

- ***Carnot Efficiency:*** The Carnot efficiency is the maximum theoretical efficiency that any heat engine can achieve. It is determined by the temperatures of the heat source and heat sink in the engine and is given by the equation $\eta = 1 - (Tc/Th)$, where η is the efficiency, Tc is the temperature of the cold reservoir, and Th is the temperature of the hot reservoir.
- **van der Waals Equation:** The van der Waals equation of state is a modification of the ideal gas law that accounts for the volume occupied by gas molecules and the attractive forces between them. The equation is $(P + a(n/V)^2)(V - nb) = nRT$, where P is pressure, V is volume, n is number of moles, a and b are van der Waals constants, R is the ideal gas constant, and T is temperature.

Cp and Cv equations

The specific heat at constant pressure (Cp) and specific heat at constant volume (Cv) are related by following equations:

1. $Cp - Cv = R$, where R is the gas constant
2. $Cp = Cv + R$

   $Cp/Cv = \gamma$

   where γ is the adiabatic index,

   For an ideal diatomic gas, $\gamma = 1.4$

INTERPLAY OF THERMODYNAMICS AND KINETICS

Thermodynamics and kinetics are both important factors in understanding and predicting chemical reactions. Thermodynamics predicts the 'feasibility' of ['free energy'] and activation energy predicts the 'rate of a reaction'.

Thermodynamics deals with the energy changes and equilibrium of a chemical reaction. It helps determine whether a reaction is thermodynamically favorable or unfavorable by providing information about the stability of the products and reactants. Thermodynamics gives insights into the direction and extent of a reaction, as well as the equilibrium constant (K). It provides information about the feasibility and spontaneity of a reaction, but it does not provide information about the rate at which the reaction will occur.

On the other hand, kinetics focuses on the rate at which a reaction proceeds. It provides information about the mechanism of the reaction, the factors that affect the reaction rate (such as temperature, concentration, and catalysts), and the rate constant (k). Kinetics explains the "how" of a reaction, including the specific steps involved, the activation energy required, and the reaction rate expression.

The rate equation or rate law derived from kinetic studies can be influenced by thermodynamic parameters, such as temperature, concentration, and activation energy.

The most commonly used equation that relates kinetics to thermodynamics is the Arrhenius equation:

$$k = A \times e^{(-Ea/RT)}$$

where:

k	is the rate constant of the reaction
A	is the pre-exponential factor or frequency factor
Ea	is the activation energy of the reaction
R	is the gas constant
T	is the temperature in Kelvin

The Arrhenius equation shows how the rate constant of a reaction is affected by the temperature and the activation energy. It helps establish the relationship between kinetics (rate constant) and thermodynamics (temperature and activation energy).

In the Arrhenius equation, the units of the pre-exponential factor A are identical to those of the rate constant and will vary depending on the order of the reaction. If the reaction is first order it has the units: s^{-1}.

K is the number of collisions that result in a reaction per second, A is the number of collisions (leading to a reaction or not) per second occurring with the proper orientation to react, and $e^{-Ea/(RT)}$ is the probability that any given collision will result in a reaction.

Additionally, the concept of equilibrium constant (K) from thermodynamics can be related to the rate constant (k) through the relationship known as the Gibbs free energy (ΔG) change:

$$\Delta G = -RT^{*}\ln(K)$$

By considering the rate constants at equilibrium, thermodynamics can provide insight into the relationship between the equilibrium constant and the rate constant.

While there isn't a single equation that combines thermodynamics and kinetics, these equations illustrate how thermodynamic parameters can influence the rate of a chemical reaction.

KINETICS AND THERMODYNAMICS: PUZZLING QUESTIONS

Two Questions

Would a reaction occur if

1. *Ea* is not met but *dG* is negative
2. *Ea* is met but *dG* is positive

Case1

Can a reaction proceed if dG is negative but Ea is not met?

dG = Free energy change and *Ea* = activation energy

The answer is no

Free energy change is a thermodynamic parameter that tells us if a reaction can occur spontaneously in the given conditions. When *dG* is negative, it tells that the reaction is thermodynamically favorable and can proceed spontaneously. However, the activation energy (*Ea*) is the minimum energy that a reaction must possess to convert the reactants into products.

If the activation energy is not met, the reaction will not proceed, even if *dG* is negative. This is because, despite the reaction being energetically favorable or thermodynamically meeting the conditions for a reaction to occur, the molecules must still acquire sufficient energy to overcome the activation barrier.

Therefore, the following are important points.

Negative dG

The reaction is thermodynamically favorable. Activation energy not met.

The reaction will not happen. Therefore, for a reaction to occur, both thermodynamic and kinetics conditions must be satisfied: a negative *dG* and sufficient energy to overcome the activation energy barrier.

Case 2

Can a reaction occur if Ea is met and dG is positive?

The answer is yes.

On the other hand, if Ea is met but the free energy change *dG* is not favorable like in the following equation *dG* is not negative the reaction can still be made to occur by increasing the term *TdS* in the equation below to make *dG* negative. $dG = dH - TdS$.

We find such cases typically in endothermic reactions in which the enthalpy change dH is positive. Such reactions generally require high temperatures to increase the *TdS* term.

Example

Splitting of Water

The splitting of water is not a thermodynamically favorable reaction at standard conditions. It's an endothermic reaction.

The reaction for water splitting can be represented as:

$$2H_2O(l) \rightarrow 2H_2(g) + O_2(g)$$

At standard conditions (STP: 25°C, 1 atm pressure), the standard enthalpy change ($\Delta H°$) for this reaction is +483.6 kJ/mol, indicating that the reaction is endothermic.

The standard entropy change ($\Delta S°$) for the reaction is +163.4 J/mol*K, meaning that the products have higher entropy than the reactants.

Using the equation: $\Delta G° = \Delta H° - T\Delta S°$

where T is the temperature in Kelvin, we can calculate the standard Gibbs free energy change ($\Delta G°$) at standard conditions: $\Delta G°$ = +483.6 kJ/mol – (298 K)*(+163.4 J/mol*K)$\Delta G°$ = +483.6 kJ/mol – 48.682 kJ/mol$\Delta G°$ = +434.9 kJ/mol

Since $\Delta G°$ is positive at standard conditions, the reaction for water splitting is not thermodynamically favorable under these conditions. It would require an input of energy to drive the reaction in the forward direction.

This added energy cost makes H_2 gas-making by splitting water a cost-intensive process.

WHAT ARE STATE VARIABLES?

A state variable is an independent variable of a state function in thermodynamics. Internal energy, enthalpy, temperature, pressure, volume, and entropy are a few examples. Heat and work are process functions, not state functions.

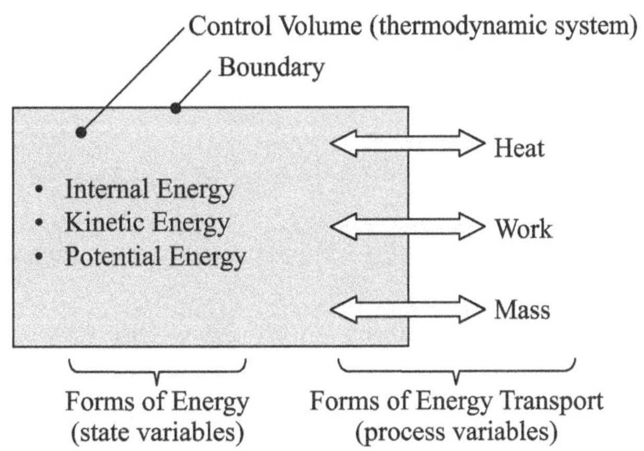

Figure 9

State variables are path independent, which means that no matter which path you take between two states, the change in the value of the state variable will be the same. This does not apply to the work W or the heat Q.

A state variable is one of a collection of variables used to describe the mathematical "state" of a dynamical system. The state of a system intuitively describes enough about the system to predict its future behavior in the absence f any external forces affecting the system.

Explanation

Energy Transport Variables are Not State Variables

Mass

Mass is a quantity that does not change with respect to time and hence, it does not define the present state of the system. Thus, mass is not a thermodynamic state variable.

Work and Heat

Heat and work are not state functions. Work can't be a state function because it is proportional to the distance an object is moved, which depends on the path used to go from the initial to the final state. If work isn't a state function, then heat can't be a state function either.

State Dependent Forms of Energy

How state variables control internal energy, enthalpy and entropy are linked with each other?

Internal energy, enthalpy and entropy internal link

Fundamentally, the internal energy is the heat energy of the system at constant volume while the enthalpy is the heat energy under constant pressure. Enthalpy is the amount of internal energy contained in a compound whereas entropy is the amount of intrinsic disorder within the compound. Enthalpy is defined as the sum of the system's internal energy and the product of its pressure and volume, $dH = dU + PdV$. H is enthalpy, U is internal energy, P is pressure and V is volume. The total enthalpy of a system cannot be measured directly, because the internal energy contains components that are unknown. In practice, a change in enthalpy (dH) is the preferred expression for measurements at constant pressure, because it simplifies the description of energy transfer. The U term can be interpreted as the energy required to create the system and the PV term as the work that would be required to "make room" for the system if the pressure of the environment remained constant.

WHAT IS CONTROL VOLUME CONCEPT?

The control volume concept is applied to systems that are not stationary.

Control volume analysis is commonly used to study dynamic flow systems where the flow of fluids is changing over time.

By defining a control volume that moves with the fluid flow, we can analyze how properties such as velocity, pressure, and temperature vary within the system as it evolves.

Control volumes are commonly used in the analysis of fluid flow, heat transfer, and other engineering systems.

What are the Assumptions?

1. **Conservation of mass:** The total mass within the control volume remains constant, with no mass entering or leaving the control volume.
2. **Conservation of energy:** Energy cannot be created or destroyed within the control volume, only transferred in or out as work or heat.
3. **Steady-state conditions:** The properties of the fluid within the control volume do not change with time.
4. **Uniform properties:** Properties such as density, temperature, and velocity remain constant throughout the control volume.
5. **No internal heat generation:** Heat is not generated within the control volume by internal sources.
6. **Negligible kinetic and potential energy changes:** Changes in kinetic and potential energy of the fluid within the control volume are small and can be neglected.
7. **No leaks or losses:** There are no leaks in the control volume and no losses of mass or energy through its boundaries.
8. **No chemical reactions:** Chemical reactions do not occur within the control volume.

These assumptions help simplify the analysis of fluid flow and thermodynamic processes within the control volume and determine important parameters such as mass flow rate, energy transfer, and work done by the fluid.

Details

Control volume is a concept commonly used in fluid mechanics and thermodynamics to analyze mass and energy flow through a defined region in a system. It involves setting a boundary around a specific area of interest, through which mass and energy can enter or leave.

The control volume can be stationary or moving, depending on the nature of the system being analyzed. It can also be open or closed, with an open control volume allowing mass and energy to flow in and out, while a closed control volume does not permit this exchange.

Examples of control volume include a pump in a hydraulic system, where the pump serves as the boundary for fluid flow entering and leaving the system. Another example would be a steam turbine in a power plant, with the turbine acting as the control volume for steam entering the system and power leaving in the form of electricity.

The main purpose of using the control volume concept is to apply the principles of conservation of mass and energy to analyze how mass and energy are transferred within a system. By defining a specific region as the control volume, engineers and scientists can accurately calculate properties such as flow rate, pressure, temperature, and energy conversion efficiency.

Overall, the control volume concept is a fundamental tool in the study of fluid dynamics and thermodynamics, helping to model and predict the behavior of complex systems in various engineering applications.

WHEN IS CONTROL VOLUME USED?

A control volume is a fictitious region of space that's either fixed or moves with a constant flow velocity. The closed surface that encloses the region is called the control surface.

A control volume is a mathematical abstraction used in continuum mechanics and thermodynamics to create mathematical models of physical processes. It's used when analyzing complex systems by breaking them down into manageable volumes such as.

Computational fluid dynamics and computational heat transfer

The control volume method is the most widely used numerical discretization method in these fields.

Thermodynamics

Control volumes are used to apply the first law of thermodynamics to components of a system, such as a compressor or evaporator.

Control volumes can be used for both open and closed systems. In a closed system, a fixed amount of mass can be chosen.

WHAT IS STEADY STATE?

A steady state is a condition where all state variables remain constant, even though there are processes that are trying to change them. This happens when the rate of input of a state variable is equal to the rate of output of that variable.

Here are some examples of steady state:

- **Bathtub**: When the water flowing into a bathtub is equal to the water flowing out, the water level stabilizes and the system is at steady stat
- **Flow chemistry**: When the process temperature and reagent feed flow rates are constant.

Some other characteristics of a steady state include:

- There is no accumulation of mass or energy over time.
- The same mass flow rate remains constant in the flow.
- All conditions in the process remain constant with time, such as temperature, pressure, mass of material, and flow rate.

CHEMICAL EQUILIBRIUM AND STEADY STATE

Chemical equilibrium in chemical engineering is a state where the rates of the forward and reverse reactions in a reversible chemical reaction are equal, resulting in no net change in the amounts of reactants and products.

The main difference between steady state and equilibrium is that steady state is maintained by fluxes into and out of a system, while equilibrium is a state where the forward and backward reactions occur at the same rate.

Steady State

A state where all variables remain constant over time, even though there is a flow through the system. For example, a bathtub with the tap running but the drain unplugged is in a steady state because the water level stabilizes after a while.

Equilibrium

A state where the concentration of reactants and products are constant over time. This often occurs when the forward and backward reactions occur at the same rate. For example, a metal bar at room temperature is in thermal equilibrium because there is no flow of heat and both systems are at the same temperature.

Equilibrium is a special case of steady state.

WHEN STEADY STATE IS NOT APPLIED?

Steady state does not apply in situations where:

There are significant changes

In the case of temperature, a system is unlikely to reach a state of unchanging distribution over time. This is known as non-steady-state conduction.

The system is unstable and not in thermal equilibrium

An unstable steady state is when small disturbances cause the system to diverge over time.

A nonequilibrium steady state is when the system's statistical physical parameters are constant over time, even though it's not in thermal equilibrium.

The steady state approximation can be used in chemistry when the concentration of an intermediate is low and more or less constant.

ROLE OF ENTROPY IN STEADY STATE

Entropy is a measure of disorder or randomness in a system. Entropy is a measure of how disorganized a system is, and how much energy is unavailable to do work. In a steady state, the entropy change of a system is zero if the state of the system does not change during the process.

Here are some other things to know about entropy and steady state:

Entropy Production Rate

At steady state, the time-averaged contributions of entropy production rate are equal in magnitude but opposite in sign.

Entropy Generation

In a steady-flow device, the specific entropy generation is never the same as the change in specific entropy.

Entropy and Energy

The more disordered a system is, the higher its entropy and the less energy is available to do work.

ROLE OF ENTROPY IN EQUILIBRIUM

Entropy plays a key role in equilibrium, as described by the Second Law of Thermodynamics:

In a chemical reaction, equilibrium is when the concentrations of both reactants and products have no further tendency to change over time.

Entropy increases in spontaneous changes

The Second Law of Thermodynamics states that the entropy of the universe increases in spontaneous changes.

Entropy stays the same at equilibrium

In an isolated system, the entropy of the system reaches a maximum at equilibrium and stays there. Any further change would reduce entropy.

Entropy change is zero at equilibrium

The total entropy change is zero at equilibrium.

Entropy is cancelled out in a chemical reaction

Any entropy gained or lost by the forward reaction is canceled out by the opposite occurring in the reverse reaction.

Entropy reaches a maximum value in a chemical reaction

When a chemical reaction reaches equilibrium, the entropy of the system reaches a maximum value.

DARCY'S CONTRIBUTION TO FLUID MECHANICS

Darcy's Law

A quantitative relationship between the flow rate of a fluid through a porous medium and the hydraulic gradients within it. This law is used to describe the flow of water through aquifers, oil and gas through petroleum reservoirs, and more.

Darcy-Weisbach Equation

Describes the relationship between pressure drop, pipe friction, and flow rate in pipes of different sizes and materials.

Darcy's Law

What drives groundwater flow?

- Water flows from high elevation to low elevation and from high pressure to low pressure, gradients in potential energy drive groundwater flow.
- Groundwater flows from high to low head.

Darcy's work focused on how to measure the potential or head in a well. The elevation of the water level in the well is a measure of the potential energy at the opening of the well in 1856, a French hydraulic engineer named Henry Darcy published an equation for flow through a porous medium that today bears his name.

$Q = KA\ (h_1 - h_2)/L$ or $q = Q/A = -K\ dh/dl$, h: hydraulic head, $h = p/rg + z$

thought experiment: hydraulic head distribution in a lake:

$q = Q/A$ is the specific discharge $[L/T]$, dh/dl is the hydraulic gradient.

K is the hydraulic conductivity $[L/T]$.

The law is very similar to *Ohm's law* for electrical curcuits $I = 1/R \times U$ (current = voltage divided by resistance).

Darcy-Weisbach Equation

Darcy Weisbach equation is an empirical equation in fluid mechanics named after Henry Darcy and Julius Weisbach. The Darcy Weisbach Equation relates the loss of pressure or head loss due to friction along the given length of pipe to the average velocity of the fluid flow for an incompressible fluid.

The Darcy–Weisbach equation contains a dimensionless friction factor, known as the **Darcy friction factor**. This is also variously called the Darcy–Weisbach friction factor, friction factor, resistance coefficient, or flow coefficient.

Pressure-Loss Equation

In a cylindrical pipe of uniform diameter D, flowing full, the pressure loss due to viscous effects Δp is proportional to length L and can be characterized by the Darcy–Weisbach equation:

$$\Delta p/L = fd \times rho/2 \times v^2/DH$$

where the pressure loss per unit length $\Delta p/L$ (*SI units:* Pa/m) is a function of the density of the fluid (kg/m^3);

DH is the hydraulic diameter of the pipe (for a pipe of circular section, this equals D; otherwise, $DH = 4A/P$ for a pipe of cross-sectional area A and perimeter P) (m).

v is the mean flow velocity, experimentally measured as the volumetric flow rate Q per unit cross-sectional wetted area (m/s);

fd is the Darcy friction factor (also called flow coefficient A).

Darcy Friction Factor

The Darcy friction factor is a dimensionless quantity that measures the resistance to flow in a pipe and is used to calculate friction losses.

Definition

The Darcy friction factor, also known as the Darcy-Weisbach friction factor or resistance coefficient, is a theoretical equation that predicts frictional energy loss in a pipe.

Calculation

The Darcy friction factor is calculated using the Darcy-Weisbach equation, which is a combination of geometric factors such as the Reynolds number, relative roughness of the pipe, and π.

Value

The Darcy friction factor is usually between 0.01 and 0.03.

Comparison to Fanning Friction Factor

The Darcy friction factor is four times larger than the Fanning friction factor.

Moody Diagram

The Darcy friction factor is usually selected from a chart called the Moody diagram, which plots the friction factor against the Reynolds number.

Hazen-Williams and Darcy-Weisbach Equations

Hazen-Williams and Darcy-Weisbach equations are both used to calculate head loss due to friction in fluid flow. The Hazen-Williams equation is commonly used for water flow through pipes, while the Darcy-Weisbach equation is a more versatile and accurate method for calculating head loss due to friction in fluid flow.

The Hazen-Williams equation is easier to use than the Darcy-Weisbach equation, as it doesn't require iterative calculations. It estimates friction loss in a pipeline with just a few parameters. However, this equation is only applicable to water flow and cannot be used for other fluids. It also is less accurate for larger pipes and higher flow velocities.

On the other hand, the Darcy-Weisbach equation can be used for various fluids, not just water. It provides more accurate results for a wide range of pipe sizes and flow velocities. However, the equation is more complex, requiring iterative calculations and additional parameters.

Values of the friction loss coefficient, C, used in the Hazen-Williams equation are accurate only if the flow velocity is close to that at which the value of C was measured.

Manning and Fanning's Formulas

Manning's formula is a mathematical formula used to calculate the flow of water in open channels, such as rivers and streams. It is typically used in hydraulic engineering to determine the discharge of water based on characteristics of the channel, such as slope, roughness, and cross-sectional area.

Fanning's formula is a similar formula to Manning's formula, but it is used for pipe flow calculations. Fanning's formula is used to determine the flow rate of water through a closed conduit, such as a pipe, based on the characteristics of the pipe, such as diameter, length, and roughness.

Darcy-Weisbach equation is used to calculate the head loss due to friction in a pipe flow. It is similar to Fanning's formula in that it is also used for pipe flow calculations, but it takes into account additional factors such as Reynolds number and relative roughness of the pipe. Darcy-Weisbach equation is commonly used in fluid mechanics and hydraulic engineering applications.

The main difference between Darcy-Weisbach and Fanning's formula lies in the calculations of head loss: Darcy-Weisbach equation takes into account the Reynolds number, which is a measure of the flow regime in the pipe, while

Fanning's formula does not consider the Reynolds number. Darcy-Weisbach equation is generally considered more accurate for a wide range of flow conditions, while Manning's formula is specifically for open channels and Fanning's formula is for closed conduits like pipes.

PRESSURE-DROP, ENTROPY, FRICTION AND DARCY'S EQUATION

What is the Pressure Drop?

At constant elevation, it is simply the difference in static pressure between two points. It is a trade-off for velocity in any heat exchange process. If you want velocity, you must accept pressure drop.

$$V1^2/2g + P1/rho = V2^2/2g + P2/rho$$

Pressure drop = $[P1 - P2]/rho = [V2^2 - V1^2]/2g$

Pressure drop is simply the difference in the kinetic energy at two points while the total energy is conserved.

What is Kinetic Energy?

Kinetic energy is the energy of translational motions of molecules. Kinetic energy generates thermal energy.

Can in a moving liquid at two points there be different thermal energy? Would not they try to equalize?

The answer is no. A moving liquid cannot have two levels of thermal energy at two points. Two thermal energies will equalize.

How thermal energies will equalize?

Entropy will equalize the thermal energy either by adding excess energy to internal energy if the system is adiabatic or by generating irreversible entropy in the surroundings.

This Explains How Pressure Drop and Entropy are Related

How?

What are the consequences of entropy generation by the system? The consequence is turbulence. The consequence is the wear and tear of the surface. This is friction.

This explains how entropy and friction are related.

Does friction cause loss of energy? Friction generates thermal energy when two molecules are pressed and rubbed. This causes the stretching of intermolecular bonds.

What happens when intermolecular bonds are stretched? Who supplies energy? The frictional thermal energy stretches intermolecular bonds and adds to internal energy. The energy remains within the system. Heat converts to temperature here depending on the specific heat of the substance. Q/T is lost as an irreversible entropy in the surroundings. If the system is adiabatic thermal energy produced by friction adds up to internal energy. Such systems are called internally reversible or 'isentropic'.

Darcy's Equation

Darcy has done an excellent job of equating pressure drop and entropy.

Darcy's Equation of Friction [fluids]

$$fD = \Delta p/L \times 2/\rho \times DH/[v]^2$$

ρ is the density of the fluid (kg/m^3); DH, the hydraulic diameter of the pipe (for a pipe of circular section, this equals D. v is the mean flow velocity; f is the Darcy friction factor (also called flow coefficient. The friction factor is therefore, for a fixed diameter, directly proportional to the pressure drop/length, inversely proportional to density, and inversely proportional to the square of velocity.

What does it mean?

- The friction factor directly proportional to pressure drop accounts for kinetic energy at two points and **that is entropy.**
- The friction factor is inversely proportional to the density or directly proportional to the specific volume **that is entropy**.
- The friction factor inversely proportional to the square of the velocity signifies less kinetic energy and **hence less entropy.**

WATER'S JOURNEY TO CRITICAL TEMPERATURE AND BEYOND

The critical point is the temperature and pressure at which a pure material can exist in vapor/liquid equilibrium at the highest possible temperature and pressure. Phase boundaries vanish at the critical point, defined by a critical temperature *Tc* and a critical pressure *pc*. The substance cannot exist as a liquid at temperatures above the critical temperature, regardless of pressure. At temperatures and pressures above the critical point, the substance is classified as a fluid, neither a gas nor a liquid. The substance is considered a gas at pressures lower than the critical pressure (but at higher temperatures). Near the critical point, many physical properties of pure fluids exhibit unusual behavior.

Two-phase system behavior

A phase is defined as "a quantity of matter that is homogeneous throughout." A liquid and its vapor are two common examples of multiphase systems.

There is a one-to-one correspondence between the temperature at which vaporization occurs and the pressure for a pure substance, as shown in the figure below. These values are known as saturation pressure and temperature.

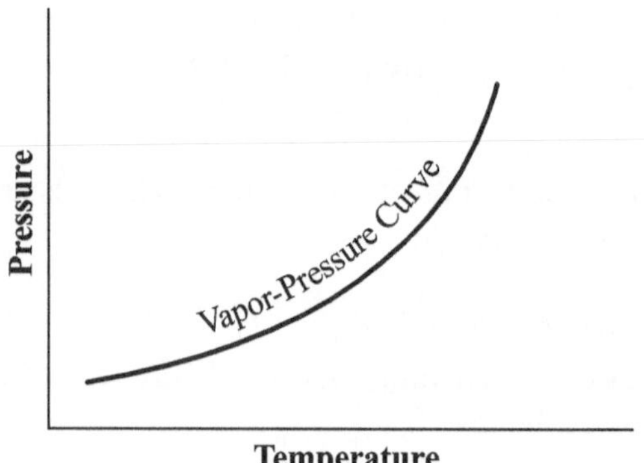

Figure 10: *P–T* Relation for a Liquid-Vapor System

This means there is an additional constraint for a liquid-vapor mixture, in addition to the equation of state. The consequence is that we only need to specify one variable to determine the state of the system. For example, if we specify *T* then *P* is set. In summary, for two phases in equilibrium, $P = P(T)$. If both phases are present, any quasi-static process at constant *T* is also at constant *P*.

Phase Behavior of a Two-Phase System

The figure below is a *TS* diagram of steam. In the *TS* diagram, the y-axis is enthalpy and the x-axis is entropy.

The left side of the bell curve is the liquid water region. Inside the bell is a saturated region where water and steam co-exist [mixed phase]. The right side of the bell is a superheated region. The topmost point is the critical point where saturated water and saturated dry steam two outlines of the bell curve intersect. Red lines are enthalpy, black lines are pressure and green lines are dryness lines.

We will study the saturation region in the bell curve starting from 100 degc to 373.95 degc which is the critical temperature of the water with the help of a steam table. For this purpose, we have divided the saturation region in the bell curve into four segments.

100–150 degc
150–250 degc
250–350 degc
350 to the critical point.

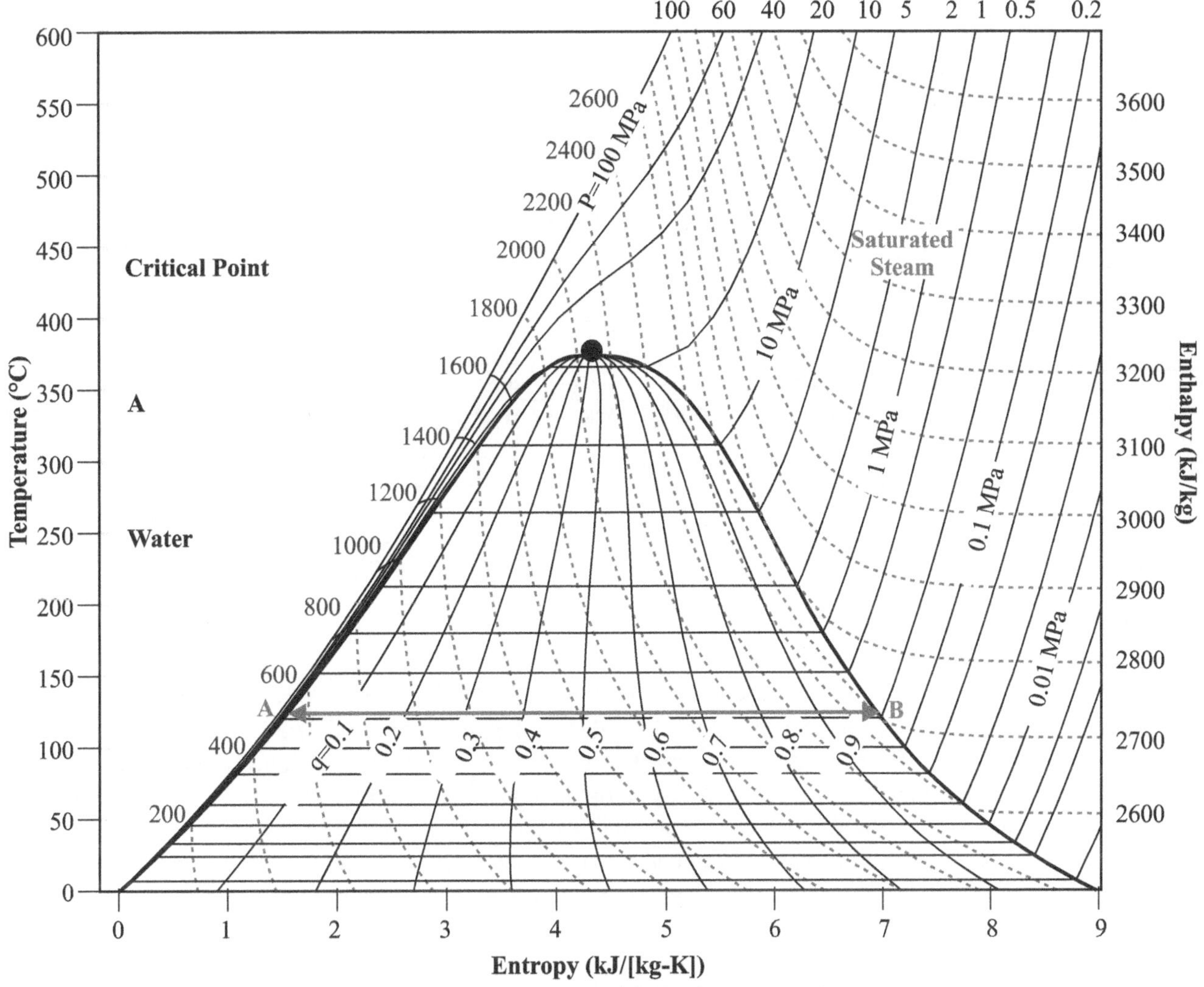

Figure 11: T-S Diagram of Water

We will study some of the segments and look at each of the four key thermodynamic properties to understand how two-phase water-steam systems behave during their journey to the critical point.

The thermodynamic properties that we would look into are as follows:

- Volume (m^3/kg)
- Energy (kJ/kg)
- Enthalpy (kJ/kg)
- Entropy (kJ/kg.K

Table 1: Saturation Properties for Steam (Temperature (0.01°C–150°C))

Temp. (°C)	*Pressure* *MPa*	*Volume* (m^3/kg)		*Energy* (kJ/kg)		*Enthalpy* (kJ/kg)			*Entropy* (kJ/kg.K)		
		vf	*vg*	*uf*	*ug*	*hf*	*hfg*	*hg*	*sf*	*sfg*	*sg*
0.01	0.00061	0.00100	205.99	0	2374.9	0.001	2500.9	2500.9	0	9.1555	9.1555
5	0.00087	0.00100	147.01	21.02	2381.8	21.0	2489.1	2510.1	0.0763	8.9485	9.0248
10	0.00123	0.00100	106.30	42.02	2388.6	42.0	2477.2	2519.2	0.1511	8.7487	8.8998
15	0.00171	0.00100	77.875	62.98	2395.5	63.0	2465.3	2528.3	0.2245	8.5558	8.7803
20	0.00234	0.00100	57.757	83.91	2402.3	83.9	2453.5	2537.4	0.2965	8.3695	8.6660
25	0.00317	0.00100	43.337	104.83	2409.1	104.8	2441.7	2546.5	0.3672	8.1894	8.5566
30	0.00425	0.00100	32.878	125.73	2415.9	125.7	2429.8	2555,5	0.4368	8.0152	8.4520
35	0.00563	0.00101	25.205	146.63	2422.7	146.6	2417.9	2564.5	0.5051	7.8466	8.3517
40	0.00739	0.00101	19.515	167.53	2429.4	167.5	2406.0	2573.5	0.5724	7.6831	8.2555
45	0.00960	0.00101	15.252	188.43	2436.1	188.4	2394.0	2582.4	0.6386	7.5247	8.1633
50	0.01235	0.00101	12.027	209.33	2442.7	209.3	2382.0	2591.3	0.7038	7.3710	8.0748
55	0.01576	0.00102	9.5643	230.24	2449.3	230.3	2369.8	2600.1	0.7680	7.2218	7.9898
60	0.01995	0.00102	7.6672	251.16	2455.9	251.2	2357.6	2608.8	0.8313	7.0768	7.9081
65	0.02504	0.00102	6.1935	272.09	2462.4	272.1	2345.4	2617.5	0.8937	6.9359	7.8296
70	0.03120	0.00102	5.0395	293.03	2468.9	293.2	2333.0	2626.1	0.9551	6.7989	7.7540
75	0.03860	0.00103	4.1289	313.99	2475.2	314.0	2320.6	2634.6	1.0158	6.6654	7.6812
80	0.04741	0.00103	3.4052	334.96	2481.6	335.0	2308.0	2643.0	1.0756	6.5355	7.6111
85	0.05787	0.00103	2.8258	355.95	2487.8	356.0	2295.3	2651.3	1.1346	6.4088	7.5434
90	0.07018	0.00104	2.3591	376.97	2494.0	377.0	2282.5	2659.5	1.1929	6.2852	7.4781
95	0.08461	0.00104	1.9806	398.00	2500.0	398.1	2269.5	2667.6	1.2504	6.1647	7.4151
100	**0.10142**	**0.00104**	**1.6718**	**419.06**	**2506.0**	**419.2**	**2256.4**	**2675.6**	**1.3072**	**6.0469**	**7.3541**
110	0.14338	0.00105	1.2093	461.26	2517.7	461.4	2229.7	2691.1	1.4188	5.8193	7.2381
120	0.19867	0.00106	0.8912	503.60	2528.9	503.8	2202.1	2705.9	1.5279	5.6012	7.1291
130	0.27028	0.00107	0.66800	546.09	2539.5	546.4	2173.7	2720.1	1.6346	5.3918	7.0264
140	0.36154	0.00108	0.50845	588.77	2549.6	589.2	2144.2	2733.4	1.7392	5.1901	6.9293
150	0.47616	0.00109	0.39245	631.66	2559.1	632.2	2113.7	2745.9	1.8418	4.9953	6.8371

100–150 degc

Let us look at the *AB* line [x-axis] on the *TS* diagram which is the most fundamental line of the phase diagram. A is the point on the saturated water line on the extreme left and B is the point on the saturated steam line on the bell curve.

Simultaneously look at the steam table and track how thermodynamic properties are behaving. On line *AB* as we go from left to right, saturated water to saturated steam that is 0 dryness to 100 dryness entropy increases as the steam-making process starts and entrained water in steam begins reduction and it continues to do so at a constant temperature until point *B* has reached, all entrained water has evaporated. At *B* it is dry saturated steam.

Water evaporates isothermally as heat is applied at a constant temperature 100 degc and 0.10142 mpa [1 atmosphere] while water and steam remain in thermal equilibrium. The H-bonds of water break at a saturation temperature 100 degc thus making water free to expand to become steam. Steam contains entrained water and such steam is called wet steam usually expressed as a dryness fraction of steam. If steam contains 10% water by mass, it's said to be 90% dry or have a dryness fraction of 0.9. Saturated water, which has no latent heat and therefore 0% dryness, will only contain sensible heat. As saturated water at 100 degc converts to saturated steam, the steam becomes dry and free from water. At atmospheric pressure, the saturation temperature is 100°C.

As we go up from 100 degc line to 150 degc line we notice the following:

Specific volume: The specific volume of water increases. Water expands as more and more H-bonds break with increased temperature. van der Waals intermolecular forces turn into repulsive force as pressure increases and molecules come closer.

van der Waals forces may be attractive or repulsive, depending on the distance between the molecules involved. At extremely high pressures the molecules are so close to each other that their atoms' positive nuclei will repel, and this repulsive force will exceed the attractive force from the dipole, dispersion, or H-bond attraction. The net result as temperature and pressure increase there is more repulsion between water molecules than attraction and that makes the water expand.

As water expands more and more, the specific volume of vapor is compressed when pressure increases. In other words, the vapor gets hammered as the water expands in the saturated region. Water pushes the vapor molecules closer causing the density of saturated vapor to increase. Since the specific volume of saturated vapor is equal to the inverse of the density of saturated vapor, it follows that the specific volume of saturated vapor decreases.

Internal Energy: The internal energy of both liquid and vapor increases. Heat, once absorbed as energy, contributes to the overall internal energy of the object. One form of this internal energy is kinetic energy; the particles begin to move faster, resulting in greater kinetic energy. This more vigorous motion of particles is reflected by a temperature increase. This increases the internal energy of both liquid and steam as temperature increases.

Enthalpy and Entropy: Let us look at 150 degc saturation line in the TS diagram and simultaneously look at the enthalpy and entropy values on the steam table. On this line, you can see the sensible heat of the water has increased. Pressure has simultaneously increased to 0.47616 mpa. Correspondingly the entrainment of water in steam has reduced and therefore correspondingly the entropy generation on the saturation line at 150 degc/0.47616 mpa in the mixed-phase region has reduced from 6.01 to 4.99 kj/kg-k. Consequent to the reduction of entropy generation on this line the enthalpy of vaporization in the mixed-phase region has reduced from 2256.4 kj/kg to 2113.7. The saturation line has become shorter than what it was at 100 degc/0.10142 mpa.

Steam table 2: 150 degc to the critical point.

Other segments of TS diagram/steam table

- 150–250 degc
- 250–350 degc
- 350 to the critical point

Table 2: Saturation Properties for Steam (Temperature (150°C–373.95°C))

Temp	*Pressure*	*Volume (m^3/kg)*		*Energy (kJ/kg)*		*Enthalpy (kJ/kg)*			*Entropy (kJ/kg.K)*		
(°C)	*MPa*	*vf*	*vg*	*uf*	*ug*	*hf*	*hfg*	*hg*	*sf*	*sfg*	*sg*
150	0.4762	0.00109	0.39245	631.66	2559.1	632.2	2113.7	2745.9	1.8418	4.9953	6.8371
160	0.6182	0.00110	0.30678	674.79	2567.8	675.5	2081.9	2757.4	1.9426	4.8065	6.7491
170	0.7922	0.00111	0.24259	718.20	2575.7	719.1	2048.8	2767.9	2.0417	4.6233	6.6650
180	1.0028	0.00113	0.19384	761.92	2582.8	763.1	2014.1	2777.2	2.1392	4.4448	6.5840
190	1.2552	0.00114	0.15636	806.00	2589.0	807.4	1977.9	2785.3	2.2355	4.2704	6.5059
200	1.5549	0.00116	0.12721	850.47	2594.2	852.3	1939.7	2792.0	2.3305	4.0997	6.4302
210	1.9077	0.00117	0.10429	895.39	2598.3	897.6	1899.7	2797.3	2.4245	3.9318	6.3563
220	2.3196	0.00119	0.08609	940.82	2601.2	943.6	1857.3	2800.9	2.5177	3.7663	6.2840
230	2.7971	0.00121	0.07150	986.81	2602.9	990.2	1812.7	2802.9	2.6101	3.6027	6.2128
240	3.3469	0.00123	0.05971	1033.4	2603.1	1037.6	1765.4	2803.0	2.7020	3.4403	6.1423
250	3.9762	0.00125	0.05008	1080.8	2601.8	1085.8	1715.1	2800.9	2.7935	3.2786	6.0721
260	4.6923	0.00128	0.04217	1129.0	2598.7	1135.0	1661.6	2796.6	2.8849	3.1167	6.0016
270	5.5030	0.00130	0.03562	1178.1	2593.7	1185.3	1604.4	2789.7	2.9765	2.9539	5.9304
280	6.4166	0.00133	0.03015	1228.3	2586.4	1236.9	1543.0	2779.9	3.0685	2.7894	5.8579
290	7.4418	0.00137	0.02556	1279.9	2576.5	1290.0	1476.7	2766.7	3.1612	2.6222	5.7834
300	8.5879	0.00140	0.02166	1332.9	2563.6	1345.0	1404.6	2749.6	3.2552	2.4507	5.7059
310	9.8651	0.00145	0.01834	1387.9	2547.1	1402.2	1325.7	2727.9	3.3510	2.2734	5.6244
320	11.284	0.00150	0.01547	1445.3	2526.0	1462.2	1238.4	2700.6	3.4494	2.0878	5.5372
330	12.858	0.00156	0.01298	1505.8	2499.2	1525.9	1140.1	2666.0	3.5518	1.8904	5.4422
340	14.601	0.00164	0.01078	1570.6	2464.4	1594.5	1027.3	2621.8	3.6601	1.6755	5.3356
350	16.529	0.00174	0.00880	1642.1	2418.1	1670.9	892.7	2563.6	3.7784	1.4326	5.2110
360	18.666	0.00190	0.00695	1726.3	2351.8	1761.7	719.8	2481.5	3.9167	1.1369	5.0536
370	21.044	0.00222	0.00495	1844.1	2230.3	1890.7	443.8	2334.5	4.1112	0.6900	4.8012
373.95	22.064	0.00311	0.00311	2015.7	2015.7	2084.3	0	2084.3	4.4070	0	4.4070

The same repeats as explained above as we go up the bell curve. The specific volume of steam continues to increase while the specific volume of vapor continues to decrease. Water's sensible heat continues to increase with fewer H bonds to break. van der Waals repulsive forces get stronger. Correspondingly entropy continues to decrease and so is the enthalpy of vaporization in the mixed-phase region.

At Critical Point

The density of liquid finally reaches the same value as that of vapor. Both reach the same specific volume of 0.00311 m^3/kg. They become a single phase. The enthalpy of vaporization line continues to be shorter finally becoming a point at the critical point. Entropy changes for the mixed-phase become zero as there is no more phase transition at the critical point. The entropy at the critical point is the contribution of water which stands at 4.4070 kj/kg-k.

Beyond Critical Point

Supercritical Water [SCW]: A Few Notable Properties

Thermal Properties: Specific heat *Cp* increases become infinitely large at high temperatures as repulsive forces of water begin to dominate as does thermal conductivity, whereas thermal diffusivity becomes zero.

Viscosity Decreases: Viscosity is resistance to flow. For liquids, typically the larger the intermolecular forces (IMF) the higher the viscosity. The absence of intermolecular attraction forces and dominance of repulsion forces reduces viscosity.

Surface Tension is Zero: At liquid-gas interfaces, surface tension results from the greater attraction of liquid molecules to each other (due to cohesion) than to the molecules in the air (due to adhesion). The net effect is an inward force at its surface that causes the liquid to behave as if its surface were covered with a stretched elastic membrane. **The absence of intermolecular forces at critical point makes surface tension zero.**

SCW water is a poor solvent for electrolytes

An excellent solvent for non-polar molecules

Auto-dissociation of water increases.

CRITICAL POINT A TO Z: WHY DO AT CRITICAL POINT THE LIQUID-VAPOR EQUILIBRIUM BREAK?

Summary: The thermodynamics of water-steam changes dramatically near the critical point. The direct relationship between pressure and temperature [*P* and *T*] of saturated water and steam ceases to exist near the critical point. When water expands due to the breaking of *H* bonds, the resulting repulsion begins to compress the vapour phase. There comes a point when the vapour can no longer withstand further compression. The vapour sub-cools. The volume of the vapour decreases as it cools. The density of the vapour phase increases as it contracts. This process is repeated until water and vapour have the same density. The two phases become indistinguishable.

This is the Critical Point

Further Clarification

The critical temperature is the temperature at which a substance transitions from a gas to a liquid as pressure is increased, while the critical pressure is the pressure at which this transition occurs at the critical temperature.

The critical pressure measures the force required to compress the gas into a liquid at the critical temperature. At critical point the compression of the vapor phase exceeds the vapor pressure of the liquid phase and the liquid-vapor equilibrium break.

Detail

Above a certain temperature called the critical temperature, liquid and vapour combine to form a single phase known as a supercritical fluid. Near the critical point, the physical properties of the liquid and vapour change dramatically, with both phases becoming increasingly similar. Liquid water is nearly incompressible under normal conditions, has a low thermal expansion coefficient, and a high dielectric constant, and is an excellent electrolyte solvent. All of these properties reverse near the critical point: water becomes compressible, expandable, a poor dielectric, a poor electrolyte solvent, and prefers to mix with nonpolar gases and organic molecules.

The following fundamental knowledge is required to comprehend critical temperature

The phase diagram of water [PT diagram]: Pressure and temperature are represented by the axes. The phase diagram depicts the lines of equilibrium or phase boundaries between the three pressure-temperature phases of solid, liquid, and gas. On the phase diagram, the curves represent the points that separate the two phases. Along this line, the two phases are in equilibrium, with Gibbs free energy = 0.

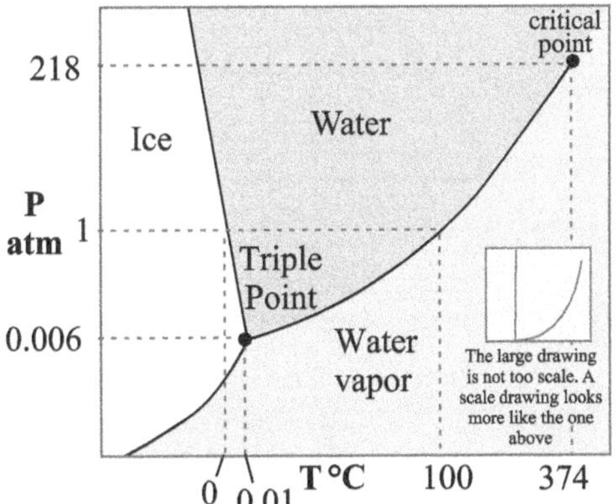

Figure 12: Phase Diagram of Water

The phase boundary between liquid and gas does not continue indefinitely in the diagram on the left. Instead, it comes to an end at a point on the phase diagram known as the critical point. This reflects the fact that, at extremely high temperatures and pressures, the liquid and gaseous phases of a supercritical fluid become indistinguishable. *Tc* = 647.096 K (373.946°C), pc = 22.064 MPa (217.75 atm) are the critical points in the water.

Critical point: The critical point defines the boundary where the water-steam [subcritical] region ends on the phase diagram and the supercritical phase region begins. The critical point (or critical state) is the end point of a phase equilibrium curve.

Phase boundaries vanish at the critical point, which is defined by a critical temperature Tc and a critical pressure pc. At temperatures above the critical temperature, the substance cannot exist as a liquid, regardless of pressure. At temperatures and pressures above the critical point, the substance is classified as a fluid, which is neither a gas nor a liquid. The substance is considered a gas at pressures lower than the critical pressure (but at higher temperatures). Near the critical point, many physical properties of pure fluids exhibit unusual behaviour.

Below the critical point, the region is called a subcritical region. Above the critical point, the region is known as the supercritical region.

WHAT HAPPENS WHEN WATER IS HEATED?

Part 1: Water: Liquids are incompressible because there is comparatively less free space between molecules. Gases are compressible because gases have plenty of free space between molecules.

Please follow the attached TS diagram of the water-steam

As you heat water and as long as the water is saturated with steam that is water and steam are inside the bell curve, they are in equilibrium. Inside the bell curve, the temperature and pressure of both phases increase at constant volume. **P and T follow a direct relation inside the bell curve. The topmost point of the bell is the critical point.**

At critical point the compression of the vapor phase exceeds the vapor pressure of the liquid phase and the liquid-vapor equilibrium break.

P and *T* direct relations are lost near the critical point. The standard EOS is not valid at the critical point. At the critical point, it is different thermodynamics. Near critical points water and steam, behaviour changes dramatically. Water breaks its H-bonds leaving the positively charged *H* ions of *H–O–H* to repel each other. Water expands at critical pressure and its density reduces. The density of water reduces to 0.322 g/cm^3.

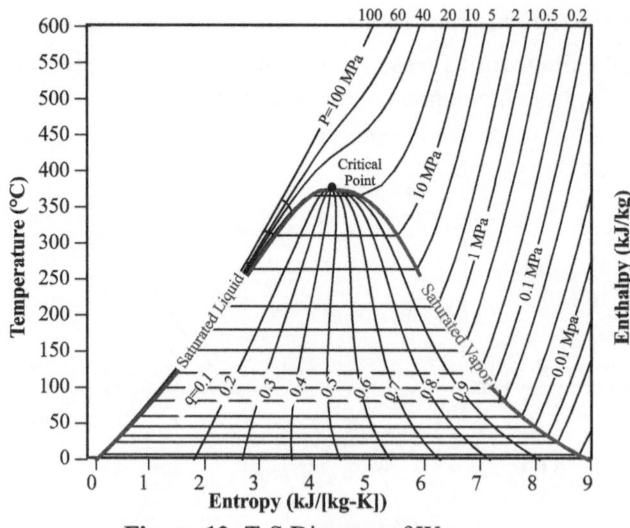

Figure 13: T-S Diagram of Water

At temperatures and pressures above the critical point, the substance is classified as a fluid, which is neither a gas nor a liquid. The liquid and gas phases combine into a single phase.

Part 2: Vapor

When water expands due to the breaking of *H* bonds, the resulting repulsion, begins to compress the vapour phase. There comes a point when the vapour can no longer withstand further compression. The vapour sub-cools. The volume of the vapour decreases as it cools. The density of the vapour phase increases as it contracts. This process is repeated until water and vapour have the same density.

THERMODYNAMICS OF AIR *vs.* STEAM: WHICH HAS MORE ENERGY?

Summary

This post compares two aspects of air and steam [1] the ability to store energy and,[2] the ability to perform work.

Which has more energy is determined by the *KE* + *PE* of air and water. The molecular structure influences *KE* and *PE*. Superheated steam has more than 1.4 times the internal energy of air, according to a conservative estimate. This explains why superheated steam has significantly more energy than air. This also explains why, under similar conditions, superheated steam has much higher specific heat than air. At 100 bar and 500 degrees Celsius, air has a specific heat of about 1.108 kj/kg-k, whereas superheated steam has a specific heat of about 2.589 KJ/kg-k, or about 2.5 times.

Explanation

Water molecules [mol weight = 18 against air mol weight 28 g/mol] are much smaller. That makes superheated steam more compressible than air [steam's more empty space/unit volume than air makes steam more compressible].

Details

To understand the distinction between air and water, you must first comprehend their molecular structure and energy distribution.

Energy Analysis

Ability to Store Energy: Internal Energy

Air

Air is made up of the gases N_2 and O_2. Both gases are diatomic in nature and non-polar. All diatomic molecules are linear and have only one parameter: the bond length or distance between the two atoms. Diatomic nitrogen has three bonds, while diatomic oxygen has two. They are only held together by a weak van der Waals intermolecular attraction.

A diatomic molecule has five degrees of freedom [three translational and two rotational]. Each of these modes receives an equal share of 1/2 kBT's energy. [K denotes the Boltzmann constant, and T denotes the temperature. As a result, the total internal energy of a diatomic molecule equals 5/2KBT. A molecule's atoms can store energy in vibrations, rotations as well as translations. Each way energy can be stored in the molecule is called a degree of freedom. The compressibility factor for air at high temperatures is > 1. That is there is a positive deviation from ideal gas laws.

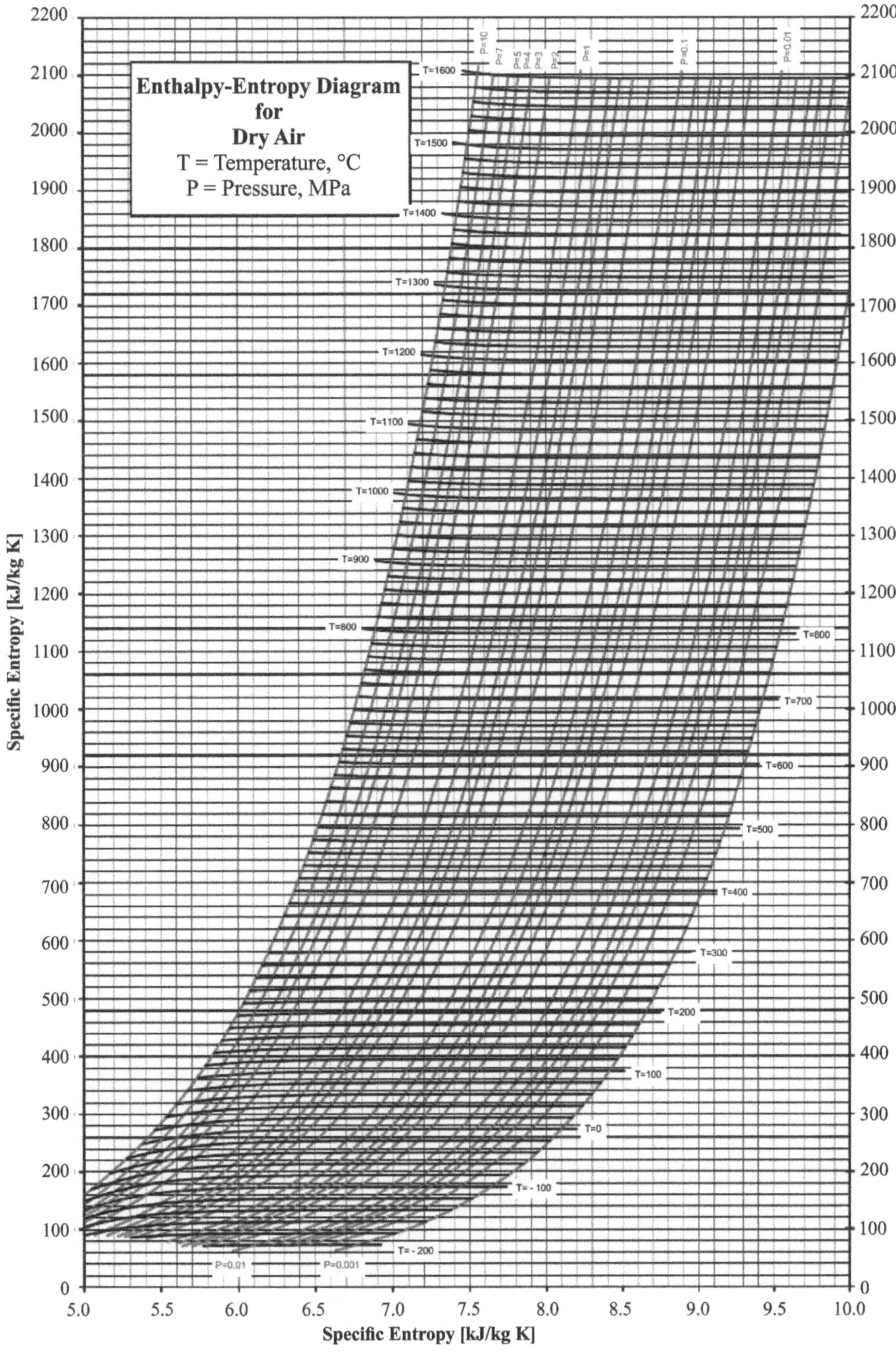

Figure 14: Enthalpy – Entropy of Air

At the initial condition (50 bar a/300°C): $h_g = 2.927$ kJ/kg

and $s_g = 6.212$ kJ/kg°C

For dry saturated steam 0.04 bar a: $s_f = 0.422$ kJ/kg°C

$s_{fg} = 8.051$ kJ/kg°C

and $s_g = 8.473$ kJ/kg°C

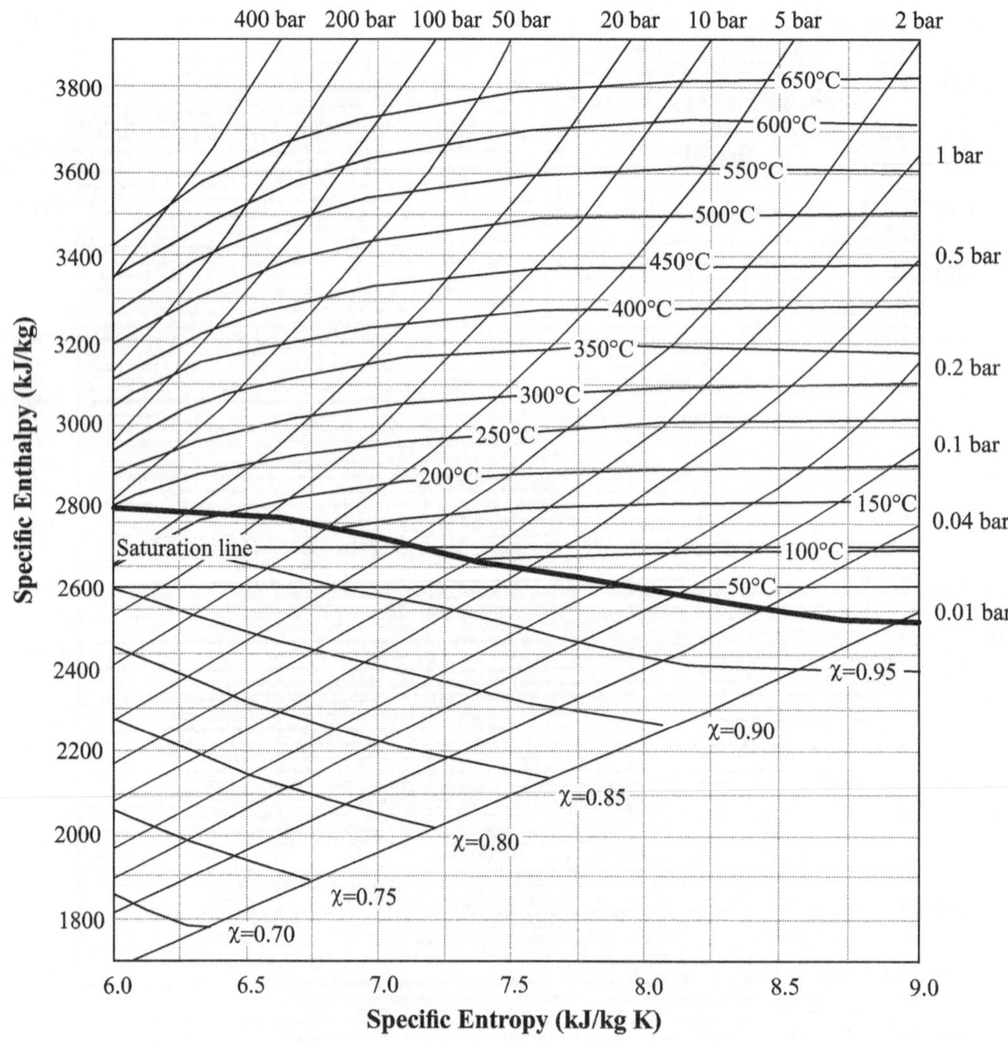

Figure 15: Enthalpy – Entropy or Mollier Chart for Steam – Example

Superheated Steam: It is a product of water which is a triatomic polar molecule. Water's H-bonds break at 100 degc that leaves van der Waals forces to stay as steam gets superheated. Water molecules [mol weight = 18 against air mol weight 28 g/mol are much smaller. That makes superheated steam more compressible than air [emptier space/unit volume]. A non-linear triatomic molecule like superheated steam [water] has a total, of nine possible degrees of freedom (three translational, three rotational, and three vibrational), but only seven are accessible at lower temperatures. The internal energy of a superheated steam molecule could be anything from 7/2 KBT to 9/2 KBT. Much more than air. A very conservative estimate is superheated steam has internal energy of more than 1.4 times of air. That explains why superheated steam has much more energy than air. This also explains why superheated steam has much more specific heat than air under similar conditions. At 100 bar, 500 degc, the air has about 1.108 kj/kg-k specific heat while superheated steam has about 2.589 kj/kg-k specific heat which is about 2.5 times. Consequently, for the same hot reservoir and cold reservoir *dT*, the air has less specific enthalpy dh. ($dh = Cp\ [T\text{ hot} - T\text{ cold}]$). This needs a gas turbine *o* have much higher *dT* across hot and cold reservoirs to compensate its smaller enthalpy.

Enthalpy-Entropy Diagram for Air

Y-axis is specific enthalpy in kj/kg. The X-axis is entropy in kj/kg-k. Red lines are pressure lines in Mpa. The extreme right-side red line stands for 0.001 Mpa and the extreme left-side red line stands for 10 Mpa. Black bold almost horizontal lines are temperature lines. There are two important observations emerging from this diagram [1] at a constant temperature as pressure increases, enthalpy remains practically constant while entropy reduces, and [2] at constant pressure as temperature increases, the enthalpy and entropy both increase.

The most important observation is specific enthalpy of air 100 bar/500 degc is only = 795.9 kj/kg.

Enthalpy-Entropy Diagram for Superheated Steam

Y-axis is enthalpy expressed as kj/kg. The X-axis is entropy expressed as kj/kg-k. There are two sets of curved lines in the image. These are well explained what they stand for. The curves rounded upwards are temperature lines expressed as degc. The curves rounded downwards are pressure lines. The image suggests the following: [1] at a constant temperature when pressure increases the enthalpy reduces and also entropy reduces [2] At constant pressure when the temperature is increased enthalpy and entropy both increase. The most important observation is the specific enthalpy of superheated steam at 100 bar/500 degc = 3373.81 kj/kg. There is a huge difference in the enthalpy between superheated steam and air. Superheated steam has enthalpy at 100 bar/500 degc more than four times that of air. Air cannot achieve this superheated steam's enthalpy even at 1700 degc and 100 bar. So, air's enthalpy is a big limitation while delivering mechanical work. The question is why? This is explained below through the compressibility graphs of air and superheated steam.

Air *vs.* Steam

Ability to Perform work: Compressibility

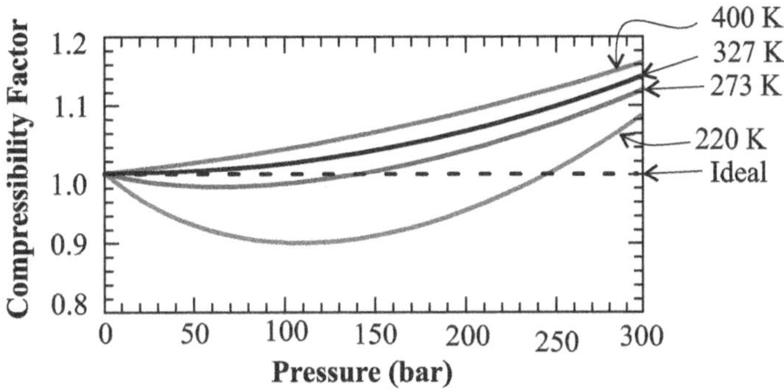

Figure 16: Compressibility of N2

Water molecules [mol weight = 18 against air mol weight 28 g/mol are much smaller. That makes superheated steam more compressible than air [emptier space/unit volume].

The images suggest compressibility of air $z > 1$ and compressibility of superheated steam < 1. It means steam is more compressible than an ideal gas. Air with $z > 1$ is less compressible than an ideal gas.

In other words, under similar conditions, superheated steam is more compressible. Steam is the most compressible gas among all gases because water's stable form is liquid. Steam always wants to go back to liquid.

The specific volume of air at 100 bar/500 dec is 0.023073 m^3/kg while superheated steam specific volume is 0.032453 m^3/kg which is one and half times more than air. This suggests, that under similar temperatures and pressure, superheated steam can perform more work than air.

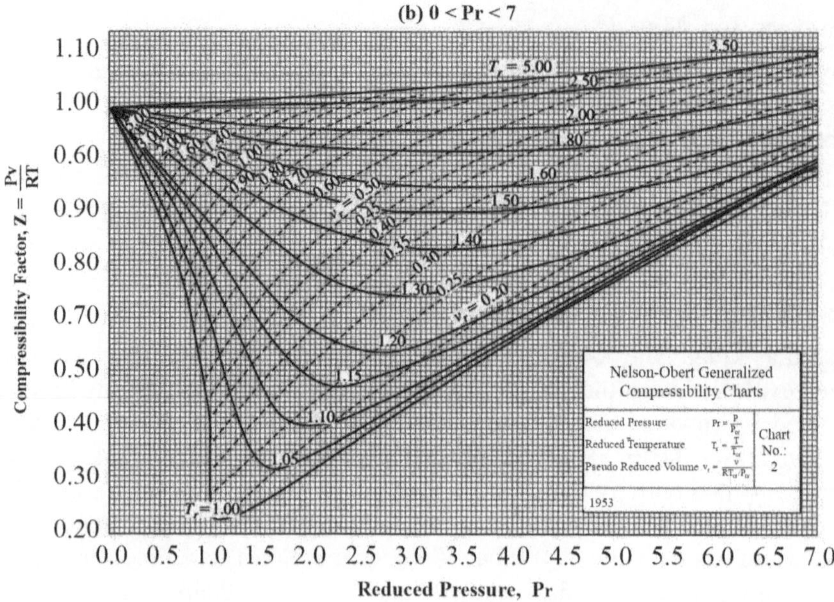

Figure 17: Compressibility of Superheated Steam

Summary: Air *vs.* Superheated Steam Comparison

At the Molecular Level

Table 3: Air *vs* Superheated Steam Comparison

Air Property (100 bar/500 degc)	*Value*	*Unit*
Pressure	100	(bar)
Temperature	500	(Celsius)
Specific volume	0.023073	(m^3/kg)
Specific Enthalpy	795.9	(kJ/kg)
Specific isobar heat capacity: Cp	1.108	(kJ/kg K)
Compressibility, Z	>1	

Superheated steam Property (100 bar 500 degc)	*Value*	*Unit*
Pressure	100	(bar)
Temperature	500	(Celsius)
Specific volume	0.032453	(m^3/kg)
Specific Enthalpy	3373.81	(kJ/kg)
Specific isobar heat capacity: Cp	2.589	(kJ/kg K)
Compressibility, Z	<1	

All important points have been explained in the note. There are two key points for air compared to superheated steam [1] air has less internal energy and [2] it is less compressible. Therefore, under similar pressure temperatures, the air has less energy or enthalpy than steam.

THERMODYNAMICS *vs.* AERODYNAMICS

In simple terms, thermodynamics deals with the study of the flow and transfer of thermal energy, while aerodynamics deals with the study of the flow and dynamics of air around objects in space. Overall, while aerodynamics focuses on the behavior of air and the forces acting on objects in motion, thermodynamics provides the fundamental principles to explain energy transfer, heat transfer, and energy conversion within the aerodynamic system. By combining both disciplines, engineers and researchers can fully understand the complex interactions between fluid flow, heat transfer, and energy conversion in various aerodynamic applications.

Space technology encompasses a wide range of complex technologies that are designed to overcome the challenges of operating in outer space. One of the fundamental challenges is the vast difference in gravitational forces between the Moon, Earth, and other celestial bodies. Spacecraft must navigate this gravitational tug of war to achieve a stable equilibrium. Achieving equilibrium requires a deep understanding of physics and mathematics, including concepts such as Newton's laws of motion, orbital mechanics, and the gravitational forces between objects. Additionally, material science plays a crucial role in space technology as engineers must develop materials that can withstand the extreme conditions of space, such as temperature variations, radiation, and vacuum.

In More Detail

Thermodynamics, the science of energy transformation and heat transfer, also plays a vital role in space technology. As space missions involve propulsion systems, thermal management, and power generation, an understanding of thermodynamics is necessary to optimize energy efficiency, ensure proper heat dissipation, and maintain stable operating conditions.

Aerodynamics is a branch of physics that studies the motion of air and how it interacts with objects moving through it, primarily focusing on the flow of gases around solid bodies, such as airplanes, cars, or spacecraft.

In the context of aerodynamics, thermodynamics plays a crucial role in understanding the energy transfer and conversion that occur during the flow of air around an object. Some of the key relationships between aerodynamics and thermodynamics are explained below:

Three key points are as follows:

Energy Conversion: In aerodynamics, the interaction of air with an object can cause the conversion of potential energy into kinetic energy. Thermodynamics provides insights into the mechanisms of energy conversion, such as the conversion of heat into work or the interchange between kinetic energy and internal energy.

Heat Transfer: Thermodynamics studies the transfer of heat energy between systems. In the context of aerodynamics, understanding heat transfer is crucial for analyzing the behavior of objects when exposed to high-temperature environments, such as supersonic flight or re-entry into Earth's atmosphere.

Pressure and Temperature: Both aerodynamics and thermodynamics deal with the concepts of pressure and temperature. In aerodynamics, the pressure and temperature distribution around an object affects its aerodynamic performance and stability. Thermodynamics provides the principles explaining the relationship between pressure, temperature, and fluid properties and behavior, helping to analyze and predict aerodynamic phenomena.

Complexities

Earth is rotating around its axis also it is moving around the sun. The moon is moving in an elliptical orbit around Earth. There is a relative velocity between the moon and the earth. Moon's distance from Earth is changing constantly. The air sandwiched between the moon and earth in an asymetric manner moving around each planet with constantly changing temperature, pressure, density, and compressibility and this is changing the orbital velocity of spacecraft continuously.

Material science, physics, and thermodynamics are crucial subjects in space technology. Here's why each of these fields is important:

- **Material Science**: Space technology requires the use of advanced materials that can withstand the extreme conditions of space. Materials need to have properties like high strength, structural integrity, resistance to extreme temperatures, radiation, and fatigue. Material scientists work on developing and testing materials that can meet these requirements, ensuring the safety and reliability of spacecraft and equipment.
- **Physics**: Physics forms the foundation of space technology. Understanding principles like Newton's laws of motion, gravitational forces, electromagnetism, optics, and thermodynamics is essential for designing spacecraft, calculating trajectories, launching vehicles, and conducting experiments in space. Physics also plays a significant role in areas such as astrodynamics (the study of spacecraft motion) and the behavior of various celestial bodies.
- **Thermodynamics**: Thermodynamics is crucial in space technology for understanding heat transfer, energy conversion, and the behavior of fluids in space. It helps engineers design thermal control systems to regulate temperatures inside spacecraft, manage heat generated by equipment, and optimize energy usage. Thermodynamic principles are applied to propulsion systems, life support systems, and in the design of thermal protection systems.

These subjects, along with many others, provide the theoretical knowledge and principles needed to address the unique challenges of space exploration. They enable engineers and scientists to develop innovative solutions and make informed decisions in the design, construction, and operation of space technologies.

WHY IN THE FIRST LAW OF THERMODYNAMICS; DELTA U = DELTA Q – DELTA W?

We frequently overlook a lovely aspect of the first law of thermodynamics that combines two non-additive thermodynamic functions. One internal energy state function and one work function.

Fundamentally, the first law of thermodynamics states that energy cannot be created or destroyed in a closed system, only transferred or transformed. In other words, the total energy of a closed system remains constant over time. This fundamental principle of energy conservation is often expressed as:

delta U = delta Q – delta W.

This equation illustrates that the change in internal energy is equal to the heat added to the system minus the work done by the system.

The first law of thermodynamics in a rearranged form says:

delta Q = delta U + delta W

It may be noted in this equation:

Heat is not a sum of two path functions. Heat, Q is a sum of one state function, the internal energy, U, and one work function, work, W. State and work functions are not additive.

Explanation

The rearranged form of the 1st law of thermodynamics delta Q = delta U + delta W, clearly states that heat Q is a sum of internal energy that depends only on the current state of the system and not on the path taken to reach that state. This means that the change in internal energy from point A to point B is the same, regardless of how the system got there.

On the other hand, W is a path function, meaning it does depend on the path taken. The work done by the system on its surroundings can vary depending on the process by which the system reaches its final state.

The work done by a system can indeed vary at each point between points A and B due to variations in pressure and volume, as well as other factors such as the specific heat capacity. Since the specific heat is different at each point between A and B, the temperature is different. Therefore, P and V at each point between A and B are different. So, the equation $Q = U + W$ helps to clarify that while U is solely determined by the initial and final states of the system, W can vary at each point along the path between those states.

This differentiation is crucial for understanding the transfer of energy in thermodynamic processes involving a state and work function.

IMPORTANT 1ST LAW OF THERMODYNAMICS EQUATIONS DERIVATIONS AT A GLANCE

You just need to remember one equation dU = delta Q - delta W, U is internal energy, Q is heat and W is work.

This equation represents the first law of thermodynamics, which states that the change in internal energy of a system is equal to the net heat transfer minus the net work done on the system.

Details

At Constant Temperature

The 1st law of thermodynamics equation is:

$$U = Q - W$$

Where:

U = Internal energy of the system

Q = Heat added to the system

W = Work done by the system

At constant temperature, the change in internal energy (ΔU) of the system is equal to the heat added to the system (Q) minus the work done by the system (W). This equation essentially states that the change in internal energy of a system is equal to the sum of the heat added to the system and the work done by the system.

For ideal gases the internal energy change is zero. As internal energy is sum of kinetic and potential energy and at constant temperature the kinetic energy change is zero and potential energy does not exist in ideal gases in absence intermolecular attractions therefore, the change in internal energy is zero.

Therefore, for ideal gases at constant temperature $Q = W$.

At constant pressure

$U = Q - W$

$dU =$ delta $Q - (PdV + VdP)$

at constant pressure $VdP = 0$

$dU + PdV =$ delta Q

since $dU + PdV = dH$, H is enthalpy

At constant pressure delta $Q = dH$

Under adiabatic conditions delta Q = 0

1st law equation reduces to $dU =$ delta work.

Adiabatic processes use internal energy U to perform work.

Concept of *Cp* and *Cv*

Specific heat vs. work relationship

Cp is specific heat at constant pressure. This involves both changes in internal energy and work per unit temperature

Cv is specific heat at constant volume. This involves only change in internal energy per unit temperature.

Enthalpy change (*dH*)- Internal energy change (*dU*,) = Work.

$dH = CpdT,\ dU = Cv\ dT$

$Cp - Cv$ signifies change in work per unit change in temperature.

$Cp = dH/dT\ Cv = dU/dT$

$CpdT - CvdT = R\ dT$

(From ideal gas equation for 1 mole $PV = RdT$

$PV = Work = RdT$)

Therefore, $Cp - Cv = R$ (applies to ideal gases)

Adiabatic index Y (gamma)

$Y = Cp/Cv$

This is an important index for adiabatic processes. It tells us the availability of internal energy for any adiabatic process.

There are many industrial processes that work adiabatically therefore, it is important that thermodynamics of these processes is understood correctly.

WHAT IS FREE ENERGY?

The free energy of a system is one that makes it available to do work. A system in equilibrium is at its least state of energy and has no free energy.

Free energy is the ability of a system to do work. Everything on earth wants to stay at minimum energy and as a consequence everything wants to stay in a state of maximum entropy.

Remember, at equilibrium the total energy and its entropy at the given temperature always stay balanced, $H = TS$, H is enthalpy, T is Kelvin and S is entropy.

When you break the equilibrium by taking away some energy from a system and disturb the equilibrium the system restores the equilibrium by moving forward. This motivates the system to move forward.

In exothermic chemical reactions this is how forward reactions work. The rise in temperature makes the reaction quicker.

Similarly if there is excess energy in a system, the system restores equilibrium by going reverse and slowing down the reaction and by stopping energy generation.

In endothermic chemical reactions this is why the reactions reverses.

When free energy change is negative the system moves forward. When the free energy change is positive the system reverses.

Any consumption of energy, generates new energy called free energy to restore equilibrium. Free energy generation is always led by the consumption of energy.

Let's look at Gibbs free energy equation $dG = dH - TdS$ (G is free energy, H Is total energy or enthalpy, T is temperature and S is entropy.

When a system is in equilibrium there is no free energy in the system $dG = 0$ or $dH = TdS$. But as you consume some energy, the process moves forward and generate energy by reducing entropy.

Let's take an example. Imagine a glass of water on a table. The glass and water are in thermal equilibrium with the surroundings. The glass of water has no free energy. At equilibrium the glass of water is already in the most stable state and it is not at all keen to disturb itself.

However, if you make the glass of water do work and consume its energy, it will move forward and generate energy by increasing its entropy to compensate for the loss of energy.

This is how any spontaneous thermodynamic process creates energy and works.

How the Signs of Gibbs Free Energy are Important?

$$dG = 0$$

When $dG = 0$, it means $dH - TdS = 0$ or $dH = TdS$.

This means all enthalpy produced by the reaction is consumed by the generation of entropy during the reaction leaving no free energy to perform any 'Work'. When $dG = 0$, the reaction goes into equilibrium.

$$dG > 1$$

When dG is greater than one, $dH > TdS$. The reaction wants to reach $dH = TdS$ to achieve equilibrium, in such a situation the free energy change dG goes reverse from positive to zero. This explains why $dG > 1$ symbolizes a reverse reaction.

$$dG < 1$$

When dG is less than one or $dH < TdS$, the reaction wants to reach $dH = TdS$ to achieve equilibrium. The enthalpy change dH moves forward from negative to zero towards equilibrium. This explains when $dG < 1$ why the reaction moves forward spontaneously in the quest for an equilibrium.

WHAT IS ENTROPY?

Entropy is a disorder. Anything that contributes to the creation of disorder generates entropy. In many cases, you would find a close relation between thermodynamics and human behavior. A person in a situation where he feels completely relaxed, totally composed, and sure of himself or in control of himself always has more capability to work in any form. He is more motivated to do work. Molecules behave exactly like this. A relaxed molecule relatively can focus more on work than when there is too much chaos that distracts the molecule from doing work. It loses focus on work and engages itself to cope with chaos/disorder. The same applies to entropy. Entropy is always increasing in the universe in everything. Because everything in nature wants to go to a stable state of equilibrium from high energy to low energy by dissipating its energy by creating disorder. This disorder is entropy.

Definitions of Entropy

- Entropy is a disorder of molecules - Entropy is a loss of energy when heat converts to work.
- Entropy is energy loss in Joule/mole-K expressed as $\Delta S = Q/T$.
- Entropy is the spread of energy.
- Entropy measures how far a system is from equilibrium.
- Entropy changes at equilibrium is zero. When $\Delta H = T\Delta S$, no enthalpy is available to do work at equilibrium. At equilibrium, $\Delta S = Q$ rev/T.

How does Entropy Change?

Some Examples

- Entropy changes when the heat does work by compression/expansion.
- Entropy changes when a process is irreversible' Any heat that goes to surroundings cannot be brought back. All such systems are called irreversible systems. When the sum of entropy generation is a system and surroundings is > 1, the system is called irreversible.

  S_1 [system] = Q/T_1,

  S_2 [surroundings] = Q/T_2,

  Q is the heat. T_1 is the system temperature. T_2 is the temperature of the surroundings.

  When $S_1 = S_2$ there is no generation of entropy. The system is an equilibrium When $S_2 > S_1$ or in other words $T_2 < T_1$, there is a generation of entropy. S (generation) = $S_2 - S_{13}$.
- Entropy changes with an increase in the concentration of molecules.
- Entropy reduces with an increase in pressure. Increased pressure brings the molecules closer and reduces disorder.
- Entropy increases with the increasing number of molecules or more spread of molecules. More spread of molecules means more disorder.
- Entropy increases with the size of a molecule.
- Entropy increases with an increase in volume.
- Entropy changes with specific heat. For a change in temperature T k the entropy changes by Q/T, if Q is the heat. The same Q/T is the specific heat of the substance if the mass is constant. The higher the specific heat capacity, the smaller the temperature change and the colder the substance for the same amount of heat applied to the same mass of substance, therefore, the larger the entropy increase.

VASTNESS OF ENTROPY

What is Entropy?

I'm not sure Rudolf Clausius knew he discovered a powerful demon who rules the universe when he invented entropy.

Entropy is the arrow of time. Entropy is increasing with time.

Entropy was born with the creation of the universe and dies with its demise. Without entropy, the universe cannot exist.

Just like in finance, every transaction comes with a tax. Entropy is a tax at source for the conversion of heat to work. The universe is run by the entropy paid by heat.

In many cases, you would find a close relation between thermodynamics and human behavior. A person in a situation where he feels completely relaxed, totally composed, and is sure of himself or in control of himself always has more capability to work in any form. He is more motivated to do work. Molecules behave exactly like this. A relaxed molecule relatively can focus more on work than when there is too much chaos that distracts the molecule to do work. It loses focus on work and engages itself to cope up with chaos/disorder. The same applies to entropy. Entropy is always increasing in the universe in everything. Because everything in nature wants to go to a stable state of equilibrium from high energy to a low energy by dissipating its energy by creating disorder. This disorder is entropy.

Definitions of entropy

- **Entropy** is a disorder of molecules.
- **Entropy** is a loss of energy when heat converts to work.
- **Entropy** is energy loss in Joule/mole-K expressed as $\Delta S = Q/T$.
- **Entropy** is the spread of energy.
- **Entropy** measures how far a system is from equilibrium.
- **Entropy** at equilibrium $\Delta H = T\Delta S$, no enthalpy available to do work at equilibrium.
- **Entropy** at equilibrium is zero. No entropy at equilibrium expressed as $\Delta S = Q$ rev/T Joule/mole-K.
- **Entropy** is an arrow of time: Entropy is an arrow of time. With time entropy is increasing and free energy is reducing. Time is not reversible. Time is only moving forward. The time arrow points toward the future. If the entropy reduces arrow of time will point toward the past.

How and When Entropy Changes?

- **Entropy changes** when the heat does work by compression/expansion.
- **Entropy does not change** if heat flows between two points like a heat exchanger unless there is a phase change
- **Entropy changes** when a process is irreversible
- **Entropy changes** with temperature.

At constant Q the entropy S can only increase if there is a reduction in temperature. This reduction of energy provides kinetic energy to the molecules to increase motions and hence the temperature.

With changing Q entropy increases with rise in temperature.

- **Entropy changes** with an increase in the concentration of molecules.
- **Entropy reduces** with an increase in pressure. Increased pressure brings the molecules closer and reduces disorder.
- **Entropy increases** with the increasing number of molecules or more spread of molecules. More spread of molecules means more disorder.
- **Entropy increases** with the size of a molecule.
- **Entropy increases** with an increase in volume.
- **Entropy changes** with specific heat**.** For a change in temperature **T** k the entropy changes by Q/T, if Q is the heat. The same Q/T is the specific heat of the substance if the mass is constant. The higher the specific heat capacity, the smaller the temperature change for the same amount of heat applied to the same mass of substance, therefore smaller the entropy increase. Entropy and specific heat are two sides of the same coin. Both have the same units, J/mole-k. The main distinction between heat capacity and entropy is that heat capacity is dependent on the material or object, such as measuring the change in temperature when the material absorbs energy, whereas entropy is independent of any object.
- **Entropy changes** with latent heat. The more latent heat the more breakdown of molecular clusters held by intermolecular forces. The more disorder. **Example**: The entropy increase when saturated water converts to saturated steam.
- **Entropy of a pure** substance at a given temperature is the sum of all the entropy it would acquire on warming from absolute zero (where $S = 0$) to the particular temperature.
- **Entropy does not change** for ideal gases because there is no work.
- **Entropy change does not alter the** temperature. Entropy gets generated only when the heat does PV work. Every time a real gas molecule does PV work its actual volume change is not V it is

  $(V - b)$. b is the van der Waals constant for real gases. 'b' is the volume of the molecule.

  Therefore, every time gas molecules perform work it is less by van der Waals constant 'b'.

  Every time a gas molecule displaces another the displacement is less by the "van der Waals constant." This is what entropy loss fundamentally means. Entropy is the loss of work. It has nothing to do with temperature change. This explains why entropy increases with molecular weight and size. Entropy loss in an ideal gas is zero because the "van der Waals constant" does not figure in an ideal gas Therefore, strictly speaking, the 1st law of thermodynamics $H = U + PV$ applies to only ideal gases. It does not account for entropy. If you just let the heat flow between two points and if there is no work, there is no entropy loss.

- **Entropy increases** in the isobaric process.
- **Entropy does not change** in the iso-volumetric process.
- **Entropy does not increase** in isothermal systems with the system and surrounds being at the same temperature.
- **Entropy can increase** in adiabatic systems if it is an internally irreversible system. An internally reversible adiabatic system is isentropic.
- **Entropy increases** in isolated systems.

  An isolated system is a closed system that can neither transfer nor receive heat, work, or mass. In other words, the path for energy transfer into or out of a closed system is closed. An isolated system's only source of energy is its own internal energy. Because its temperature is constant, so is its total kinetic energy. However, molecular collisions do occur at the given temperature based on their kinetic energy. When one molecule collides with another, one slows down while the other accelerates, but the total momentum remains constant because momentum is conserved. Consider what happens when the velocity of some molecules increases while the total energy remains constant. This increases the molecule's spread continuously within a confined space. Imagine 10^{23} molecules/mole repeating this every moment. This is big chaos happening in the system. This lowers the density of molecules per unit volume, resulting in more microstates and increased entropy.
- **Entropy increases** with more microstates.

Entropy and Probability

Microstates: Explanation

A microstate is a term used to describe the number of different possible arrangements of molecular position and kinetic energy at a particular thermodynamic state. A process that gives an increase in the number of microstates, therefore, increases the entropy.

The interpretation of entropy is the measure of uncertainty, which remains about a system after its observable macroscopic properties, such as temperature, pressure, and volume, have been taken into account. For a given set of macroscopic variables, the entropy measures the degree to which the probability of the system is spread out over different possible microstates.

In contrast to the macrostate, which characterizes plainly observable average quantities (temperature, for example), a microstate specifies all molecular details about the system, including the position and velocity of every molecule. With more available microstates, the entropy of a system increases. This is the basis of an alternative (and more fundamental) definition of entropy:

$$S = k \ln w$$

In this, k is the Boltzmann constant (the gas constant per molecule, 1.38×10^{-23} J/K) and w is the number of microstates that correspond to a given macrostate of the system. The more such microstates, the greater the probability of the system being in the corresponding macrostate. For any physically realizable macrostate, the quantity w is an unimaginably large number.

Entropy and Microstates

As explained, a microstate (W) is a specific configuration of the locations and energies of the atoms or molecules that comprise a system like the following:

$$S = k \ln W$$

As for other state functions, the change in entropy for a process is the difference between its final (Sf) and initial (Si) values:

$$\Delta S = Sf - Si = [k \ln Wf - k \ln Wi] = k \ln Wf/Wi$$

For processes involving an increase in the number of microstates, $Wf > Wi$, the entropy of the system increases, $\Delta S > 0$. Conversely, processes that reduce the number of microstates, $Wf < Wi$, yield a decrease in system entropy, $\Delta S < 0$.

What does no entropy mean?

- **No chemical reactions**: In the absence of entropy, the universe will have extremely high free energy. Both reactants and products will have the same free energy. There would be no chemical reactions or chemical processes.
- **All processes will be extremely slow**: The irreversibility of a process causes entropy. Without entropy, the universe would be a reversible slow process in equilibrium with its surroundings.
- **No heat transfer from high to low temperature**: Heat transfer takes place in the direction of entropy increase from high to low temperature.
- **Heat will no longer be a path function**: Heat is always in transit because it is being chased by entropy.
- **No phase change**: Entropy is a disorder. Because entropy always increases, there is a phase change. When water is boiled, it converts to steam because it can no longer accommodate entropy. To share its entropy, water moves to a higher energy state known as vapor.

  That means there will be no steam and only low-energy hot water to do any mechanical work. So, no mechanical work.
- **No volume expansion**: Without entropy, there would be no new microstate. The temperature has no effect on volume. Nothing will be compressible.
- **There is no thermodynamic work and everything is incompressible**: Entropy is created when heat is converted to work. There would be no thermodynamic work if there was no entropy.
- **No equilibrium**: There will be no equilibrium: all processes will be unidirectional. The human body is in balance with its surroundings. Entropy regulates our body temperature by making us sweat and cooling us down. In the absence of a place to dump the entropy, humans will experience significant increases in body fluctuations. temperature.

WHY IS ENTROPY CHANGE dS = dQ/T?

Rudolf Clausius made a statement in 1854 that heat can never go from a colder temperature to a warmer temperature without some other change. In 1865, Clausius gave the first mathematical version of the concept of entropy and also gave it its name. Clausius chose the word because the meaning (from Greek ἐν en "in" and τροπή tropē "transformation") is "content transformative" or "transformation content". [Wikipedia]. The direction of heat movement and the concept of entropy became the 2nd law of thermodynamics.

Clausius proposed an equation called the equation of inequality, The generalized equation of Clausius's inequality *dS* (surr) is > or = *dQ*/*T* (sys) *dS* (sys) is equal to or greater than delta *Q*/*T* (surr). Clusius implied that the entropy change in an irreversible process can increases or remain stationary but it never decreases per cycle.

Explanation of Clausius's Law of Inequality

In the simplest language, let us consider the following. Let us imagine a system at temperature > the surrounding temperature. Heat goes from high to low temperature. If Q heat is supplied to system, then the entropy generation in the system is $S_1 = Q/T$(sys).

In a heat permeable system, the same heat Q goes to surroundings, the entropy generation is $S_2 - Q/T$(surr) As a consequence the system the entropy losses stand at $[S_2 - S_1] = [Q/T$ (surr) $- Q/T$(sys)] as kJ/kg-k.

$[S_2 - S_1]$, the entropy loss is a function of the differences in the heat capacities between surroundings and systems which sets the direction of heat flow. This process would continue until $S_2 = S_1$ and the system has no heat available to supply and the surroundings have no heat capacity to absorb heat. This saturation on both sides would occur when the entropy has reached the maximum value and eaten up all enthalpy, $dH = TdS$, $dG = 0$. At this stage, the system – and surroundings both have established equilibrium.

Therefore, one single equation of Clausius's inequality dS (surr) is > or = *dQ*/*T* (sys) explains both requirements of the 2nd law of thermodynamics. When the surroundings are at lower temperatures the entropy generation in the surroundings would be more for the same Q since the surroundings are at lower temperature, T surr < T sys and entropy will always increase.

Only at equilibrium when $S_2 = S_1$, the heat Q is reversible. At aquarium $dS = Q$ rev/T.

In a nutshell, the irreversibility of a system is a function of the difference in the heat capacities Q/T of the system and surroundings. Therefore, entropy change and heat capacity are both expressed by Q/T, kJ/kg -K.

When S (surr) > S (sys) it's an irreversible system. When S (surr) = S (sys), the system is a reversible system in equilibrium.

ENTROPY BEYOND HEAT AND WORK

This is an overview. This is a collection of information from various sources on the internet including AI.

The objective of this article is to highlight the "Reach of Entropy in the different spectrums of science"

You'll discover that the review provides astounding information that challenges our perception that entropy is solely applicable to thermodynamics. It reveals the astounding application of entropy to information technology, particularly data science.

The concept of disorder, or entropy, extends far beyond just heat transfer and thermodynamics, permeating the entire spectrum of science and influencing diverse fields such as information theory, statistical mechanics, quantum mechanics, and biology.

Entropy is a measure of the randomness or disorder in a system, and its understanding has profound implications in various disciplines.

Application of Entropy in Various Disciplines

- **In information theory**, entropy is used to quantify the amount of information or uncertainty in a message. The concept was introduced by Claude Shannon in the 1940s, who defined entropy as the average amount of information produced by a random variable. This idea revolutionized communication and laid the foundation for modern encryption, data compression, and error correction techniques.
- **In statistical mechanics**, entropy plays a role in understanding the behavior of systems with a large number of particles. The Boltzmann entropy formula relates the microscopic configurations of particles to their macroscopic properties, providing a bridge between the microscopic world of atoms and molecules and the macroscopic world we observe. This has implications for diverse phenomena such as phase transitions, diffusion, and the behavior of complex fluids.
- **In quantum mechanics**, the concept of entropy is used to characterize the entanglement and information content of quantum states. Entropy measures such as von Neumann entropy or Rényi entropy provide insights into the quantum correlations and complexity of multipartite systems, with applications in quantum computing, cryptography, and quantum information theory.
- **In biology**, entropy is used to describe the diversity, complexity, and organization of living systems. The concept of negentropy, introduced by Erwin Schrödinger in his book "What is Life," highlights how living organisms can reduce their internal entropy by capturing and utilizing energy from the environment to maintain order and sustain life.

Entropy considerations are also relevant in understanding biological processes such as protein folding, gene regulation, and evolutionary dynamics.

Overall, the concept of disorder, as quantified by entropy, transcends traditional disciplinary boundaries and offers a unifying framework to study complex phenomena across science and nature. By accounting for entropy beyond heat transfer, it has been possible to uncover deep connections between different fields, leading to new insights, discoveries, and technological advancements.

A Few Important Equations

1. **The Boltzmann entropy equation:** This equation quantifies the entropy of a system and is given by $S = k \ln \Omega$, where S is the entropy, k is the Boltzmann constant, and Ω is the number of microstates corresponding to a

particular macrostate. This equation is used in statistical mechanics to understand the randomness and disorder in a system.

2. **Gibbs entropy equation:** This equation accounts for the entropy change in a thermodynamic system and is defined as $\Delta S = \Delta H/T$, where ΔS is the change in entropy, ΔH is the change in enthalpy, and T is the temperature. This equation is used in chemical thermodynamics to determine the spontaneity and direction of chemical reactions.
3. **Shannon entropy equation:** This equation is used in information theory to quantify the amount of uncertainty or randomness in a communication system. It is given by $H = -\Sigma\ P(x)\ \log P(x)$, where H is the entropy, $P(x)$ is the probability of a particular message x, and the sum is taken over all possible messages. This equation helps in optimizing communication systems and coding schemes.
4. **Von Neumann entropy equation:** This equation is used in quantum mechanics to quantify the entropy of a quantum system. It is defined as $S = -Tr(\rho \ln \rho)$, where S is the entropy, ρ is the density matrix of the system, and Tr denotes the trace operation. This equation helps in understanding the information content and uncertainty in quantum systems.
5. **Black hole entropy equation:** This equation, given by $S = A/4G$, relates the entropy of a black hole to its surface area A, and the gravitational constant G. This equation, proposed by Stephen Hawking, provides insight into the thermodynamic properties of black holes and their connection to entropy and information theory.
6. **Landauer's principle:** This equation accounts for the minimum amount of energy required to erase one bit of information in a computing system. It is given by $\Delta E = kT \ln(2)$, where ΔE is the energy required, k is the Boltzmann constant, T is the temperature, and ln(2) represents the conversion of one bit of information to energy. This equation is important in understanding the energetic costs of computation and has implications for the design of energy-efficient computing systems.
7. **Shannon's channel capacity equation:** This equation determines the maximum rate at which information can be reliably transmitted through a communication channel. It is given by $C = B \log_2(1 + S/N)$, where C is the channel capacity, B is the bandwidth, S is the signal power, N is the noise power, and $\log_2(1 + S/N)$ represents the channel capacity in bits per second per hertz. This equation helps in optimizing communication systems by maximizing the data transmission rate while minimizing errors due to noise.
8. **Boltzmann's equation for entropy production:** This equation describes the rate at which entropy is produced in a system due to irreversible processes. It is given by $dS/dt \geq 0$, where dS/dt represents the rate of change of entropy with respect to time. This equation highlights the second law of thermodynamics, which states that the total entropy of an isolated system will always increase over time due to irreversible processes.
9. **Kolmogorov-Sinai entropy equation:** This equation measures the rate at which information is generated by a dynamical system. It is given by $h = \lim(t \to \infty)\ (1/t)\ H(\varepsilon, t)$, where h is the Kolmogorov-Sinai entropy, t is the time interval, and H(ε, t) is the metric entropy at scale ε. This equation helps in quantifying the complexity and unpredictability of chaotic systems by measuring the flow of information over time.
10. **Shannon-McMillan-Breiman theorem:** This theorem relates the entropy of a stochastic process to the typical behavior of its sample paths. It states that the entropy rate of a stationary and ergodic process is equal to the mean entropy per symbol generated by the process. This theorem is used in information theory to understand the long-term behavior and predictability of random processes.

WHY DOES THE ENTROPY OF AN ISOLATED SYSTEM ALWAYS INCREASE?

Entropy is a scientific concept commonly associated with disorder, randomness, or uncertainty.

Definition: Wikipedia

'The second law may be formulated by the observation that the entropy of isolated systems left to spontaneous evolution cannot decrease, as they always tend toward a state of thermodynamic equilibrium where the entropy is highest at the given internal energy'.

This post aims to find the meaning of the above statement.

Background reading

What is entropy?

Entropy symbolizes disorder randomness or dispersed energy. Entropy is a loss of useful energy.

What is useful energy?

An energy is termed useful energy when it can be converted fully to work.

How and why entropy causes 'work' losses?

One reason is that entropy is directly related to the 'work losses' is that not all heat transfer can be converted into work. More precisely. all heat can't reach its destination to perform the assigned task. Some heat gets dispersed on the way. A good example is thermal energy. If one directs Q amount of heat from point A to B to perform work all Q will not reach B. Anything that comes on the path from A to B having a higher heat capacity or a lower temperature than B will get a share of the heat Q. It may be noted that while the entropy is a state function, the heat and work are path functions. Both heat and work have multiple options to travel between two points choosing different paths with each path having its unique potential to reach heat and work to the destination.

Expression of Entropy

The second law states that there exists a useful state variable called entropy. The change in entropy (delta S) is equal to the heat transfer (delta Q) divided by the temperature (T). The second law further states that if the physical process is irreversible, the entropy of the system and the environment must increase; the final entropy must be greater than the initial entropy.

The SI unit of entropy is *KJ/kg/K*. It is expressed as a ratio of heat/temperature or Q/T [said above]. This expression symbolizes the average kinetic energy per unit temperature. The larger the change in kinetic energy change per unit temperature the larger the disorder or larger the randomness or larger the spread of energy and the smaller the available useful energy to perform work. For a reversible system, the entropy change is expressed as $[S_1–S_2] = dS = Q$ rev/T, here S_1 and S_2 represent the system and surroundings. For irreversible systems, $[S_1 - S_2] = dS$ is equal or more than Q/T. It simply means that in a reversible system, the heat can be reversed from the surroundings while heat that once goes to the surroundings can't be reversed and it is a permanent 'work' loss.

Friction is Not Entropy

Friction does not involve 'work'. Frictional losses are irreversible. Friction is the force resisting the relative motion of solid surfaces, fluid layers, and material elements sliding against each other. The process is irreversible because energy loss in the form of heat generated due to friction cannot be recovered. Friction is a non-conservative force, meaning energy is transferred to new forms not useful to the system. Friction loses kinetic energy, which is dissipated as thermal energy. At a molecular level, when two molecules rub against each other heat is generated which goes partly to increase the internal energy of molecules and partly get dissipated.

Concept of System and Surroundings

A system is defined as a part of a process that is under study and possesses fixed boundaries. These boundaries can be real or imaginary. Surroundings are the part outside the system.

Isolated System

An isolated system is a thermodynamic system that neither exchanges energy nor matter (mass) with its surroundings. An isolated system differs from a closed system from the standpoint of the transfer of energy. Closed systems are only closed to material transfer, energy can be exchanged across the system's boundaries. An isolated system has no surroundings and it cannot transfer energy to surroundings to share heat and entropy with surroundings. The entropy has two options in an isolated system, [1] to stay constant or [2] increase.

Thermal energy boxed-up in an isolated system overtime increases until a thermodynamic equilibrium is reached and the Gibbs free energy change dG becomes zero, since by Gibbs equation $dG = dH - TdS$, reaching equilibrium by an isolated system means $dH = TdS$. In other words, the entropy generation dS, completely neutralizes the available total energy change dH [enthalpy change] in the system.

Discussion

In thermodynamics, we deal with two types of equilibriums [1] thermal equilibrium and [2] thermodynamic equilibrium. Thermal equilibrium is reached when all parts of a system have the same temperature. That means there is no net heat flow anywhere. Thermodynamic equilibrium means that there is no change in free energy. In other words, thermodynamic equilibrium is one when $dH = TdS$, which means that the entropy generation neutralizes the total energy of a system. Free energy stays at minimum in thermodynamic equilibrium and entropy stays at the maximum.

Interplay of H [total energy] and S [entropy] change in an isolated system

Let us take an example. Imagine a room in your apartment that has no surroundings. You have many items in this room. While these items are in thermal equilibrium at the same temperature which means they are in thermal equilibrium they are not in thermodynamic equilibrium. The reason is they do not have the same internal energy. While they have the same average kinetic energy since they are at the same temperature, they have different potential energy due to their different molecular construction and degrees of freedom to hold energy. In short, they have different heat capacities. They have different capacities to absorb heat.

In summary, the molecular structure of molecules in every item is different with different internal energy. Since $Q = 0$ in isolated systems, and from the 1st law of thermodynamics, $U = Q - W$, every item in the room internally has a different temperature and draws energy from the item's internal energy to perform 'work' and generate entropy.

In other words, as the time proceeds the internal energy depletes and the entropy increases. This continues till all items have reached zero internal energy which means zero work, and maximum entropy. This means the terms dH at constant pressure is equal to TdS in the Gibbs equation, $dG - dH - TdS$. and that makes $dG = 0$ when a system reaches equilibrium.

ROLE OF ENTROPY IN RUSTING

What is Rust?

Rust is hydrated iron oxide, $Fe_2O_3.nH_2O$. During the course of reaction of iron with oxygen several hydroxides and oxides form which are bigger in volume than the volume of the metal. As a consequence, they can't provide a secured cover to metal surface. The oxide coating chip off exposing the metal that follows initiation of self-supporting rusting process on the surface.

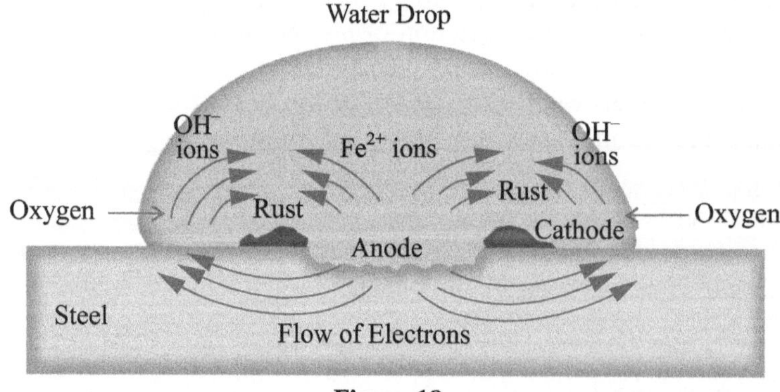

Figure 18

How Rusting Starts?

A drop of stationary water on iron surface is sufficient for the initiation process of rusting. The drop on metal becomes barrier for [air] O_2 on metal surface. Differential O_2 concentration between O_2 starved point on metal and other normal part of metal sets up electrolytic corrosion cells consisting of anodes/cathodes because of potential difference. Oxygen starved point becomes anode. Corrosion occurs at the anode, where metal oxidizes and dissolves. At cathode reduction takes place. Current flows [electron] through the metal from the anode to the cathode.

Reaction at anode

$Fe = Fe^{2+} + 2e^{-}$

Reaction at cathode

O_2 gets reduced to OH^{-} ions at cathode

$\frac{1}{2} O_2 + H_2O + 2e^{-} = 2OH^{-}$

Fe^{++} and OH^{-} ions combine to produce hydroxides which dehydrate to $Fe_2O_3.nH_2O$ which is rust.

What makes rusting a slow but self-supporting spontaneous chemical reaction?

Rusting reaction on metal surface:

$4Fe\ (s) + 3O_2(g) \longrightarrow 2Fe_2O_3(s)$: $\Delta r\ H = -1648$ kJ mol–1, ΔS system $= -549.4\ JK^{-1}\ mol^{-1}$ at 298 K. where, $\Delta r\ H$ is enthalpy change of reaction and ΔS is entropy change for the reaction.

The first question that comes to our mind who provides energy for this chemical reaction? How iron gets activation energy to cross the hill at ambient temperature? Image below, represents the activation energy barrier for rusting.

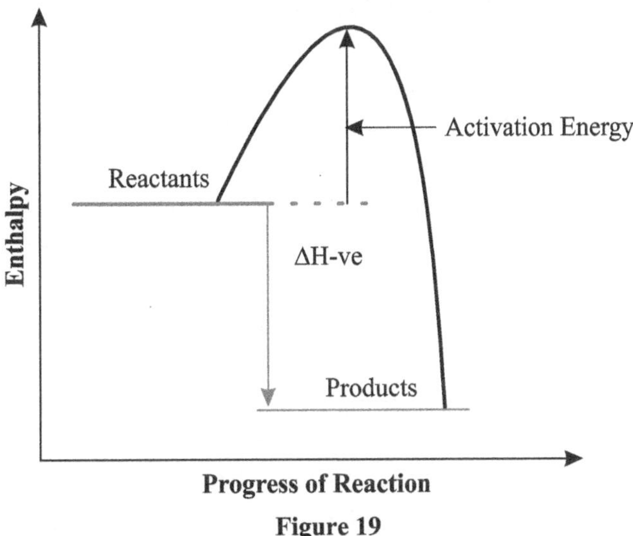

Figure 19

The answer is for iron plus oxygen to form iron oxide or rust, the reactants, iron and oxygen don't have to be at a high temperature to have energy localized within them. Iron atoms (as –Fe–Fe–Fe–) plus oxygen molecules of the air (O–O) have more energy localized within their bonds than does the product of their reaction, iron rust. The rust formation reaction is exothermic. In other words, the exothermic reaction between iron and oxygen provides energy for iron's rusting. To explain it further, the bond energy of Fe–Fe–Fe–Fe atoms + O = O atoms > heat of formation of Fe_2O_3 [rust]. This surplus stored energy in Fe and O_2, makes the corrosion reaction exothermic and it has huge implication in making rusting a spontaneous reaction.

Why rusting initiation is a slow process?

The reason is, even though iron has significant affinity for oxygen, O_2 at normal room temperature around 298 K, has only few atoms moving exceptionally fast which can hit the Fe just right so that Fe–Fe and an O–O bonds break and Fe–O bond forms. Essentially, kinetically the combination of iron and oxygen at ambient temperature to form iron oxide is a slow reaction.

Rusting reaction really takes off and becomes a self-supporting reaction when the surplus energy 1648 kJ mol^{-1}, localized in Fe and O_2 [exothermic heat] becomes dispersed to the surroundings as 'heat' and raises the entropy of surrounding. The entropy increase of the surroundings makes the moist air molecules in the surroundings move faster and make more hard-hitting collisions to speed up rusting.

Entropy and free energy change in rusting reaction:

$\Delta r\ H = -1648$ kJ/mol.

ΔS system = –549.4 J/k/mol

ΔS surrounding = Q/T = 1648/298.15 [T = 298.15 k ambient temperature]

= 5.53 kj/mole = 5530 Jk mol–1

ΔS total = ΔS system + ΔS surrounding = [5530 + (–549.4]) Jmol–1K

= 4980.6 J/K/mole or 4.98 kj/mole/k

Huge gain in entropy by surrounding makes the rusting reaction spontaneous.

$\Delta G = \Delta H - T\ \Delta S = [-1648 - 298.15*4.98]$, where ΔG = Gibbs free energy

$\Delta G = -2840$ kj/mole, negative ΔG makes the reaction spontaneous.

The total entropy increases by the rusting favours spontaneity by making the free energy change negative.

HIGH ENTROPY NEXT GENERATION ALLOYS [HEA]

In one line, the challenge in alloying lies in finding alloying elements that contribute to the desired properties without causing a large enthalpy change that cannot be compensated by the entropy change, limiting the feasibility of the process. Obtaining a sufficient negative free energy change for a spontaneous process is a challenge in commercial alloying. HEA addresses this challenge by adding entropy.

Perhaps you're wondering how alloys and entropy are related. Even though entropy is a powerful concept, it is generally regarded negatively. One of the wonders of entropy is HEA. Remember, "a system becomes more stable when its energy is spread out in a more disordered state", this is precisely what entropy does to high entropy alloys.

This note aims to provide a brief overview of HEAs and explain how entropy, or thermodynamics, contributes to HEAs' amazing properties.

Several investigations have demonstrated that a high entropy strategy holds great promise for creating new materials with properties that go beyond those of conventional materials based on one principal element or component.

High-entropy alloys (HEAs) have garnered significant interest because of their superior tribological characteristics, thermodynamic stability, mechanical performance, and resistance to corrosion. High-entropy alloys (HEAs) are presently of great research interest in materials science and engineering. Unlike conventional alloys, which contain one and rarely two base elements, HEAs comprise multiple principal elements, with the possible number of HEA compositions extending considerably more than conventional alloys. High-entropy alloys are a novel class of materials. Research indicates that some HEAs have considerably better strength-to-weight ratios, with a higher degree of fracture resistance, tensile strength, and corrosion and oxidation resistance than conventional alloys as said above.

Let me start with where entropy fits in alloys.

Thermodynamic challenges in alloys

Obtaining a sufficient free energy change for a spontaneous process is a challenge in commercial alloying. This limitation affects the choice of alloying elements by considering that the enthalpy change should not exceed the product of temperature (T) and entropy change (ΔS), which is denoted as $T\Delta S$.

The challenge in alloying lies in finding alloying elements that contribute to the desired properties without causing a large enthalpy change that cannot be compensated by the entropy change, limiting the feasibility of the process.

Here comes the role of entropy.

Details

What is an alloy?

Basics

A chemical mixture of elements with at least one metal is called an alloy. The most popular and traditional method of alloying involves melting the base metal above its melting point, and then dissolving the solute into the molten liquid. The internal stresses produced by differences in the size of atoms in the metal crystal lattice improve metal properties.

There are two mechanisms for an element to insert itself into the molten base metal. The mechanisms depend on the relative size of atoms.

Atom Exchange Mechanism

The atom exchange method typically takes place when the atoms are fairly similar in size.

Bronze and brass are examples of atom exchange alloys.

Interstitial Mechanism

In the interstitial mechanism, one type of atom is typically much smaller than the other and cannot effectively replace the other atom in the base metal. Instead, the smaller atoms get stuck in the spaces between the crystal's atoms, or interstitial sites. Steel is an example of an interstitial alloy.

What is Entropy?

Free energy ΔG, enthalpy ΔH, and entropy change ΔS are all state functions.

$$\Delta G = \Delta H - T\Delta S \text{ [T is temperature]}$$

ΔG, signifies the available energy to perform work in a process. ΔH signifies the total energy in a process. $T\Delta S$ is the product of temperature and entropy changes in a process. When a process occurs spontaneously, there is the generation of entropy that consumes a part of the total energy ΔH. When $T\Delta S > \Delta H$, there is the consumption of free energy ΔG or ΔG becomes negative. This shortfall of ΔG is made up of the energy from the surroundings to keep the reaction moving forward spontaneously. This goes on until the process reaches equilibrium with the surroundings and the surroundings stop the supply of energy.

To summarize, for any process to occur spontaneously the process must have enough free energy.

Thermodynamics of Alloys

In the context of thermodynamics, the process of metal alloying involves mixing two or more metals to form a new alloy. This process can be described in terms of enthalpy (H), entropy (S), and free energy (G) changes.

- **Enthalpy (ΔH):** The enthalpy change during alloying refers to the heat absorbed or released during the mixing process. It depends on the specific metals being alloyed and their respective enthalpies of mixing. The enthalpy change can be positive (endothermic) or negative (exothermic), depending on whether heat is absorbed or released during alloying.
- **Entropy (ΔS):** The entropy change reflects the disorder or randomness in the alloying process. Upon mixing metals, the randomness generally increases, leading to a positive entropy change. This increase in entropy contributes to the thermodynamic driving force for alloy formation.
- **Free energy (ΔG):** The free energy change is a measure of the overall spontaneity of the alloying process. The Gibbs free energy change (ΔG) takes into account both the enthalpy and entropy changes. A negative ΔG indicates that the alloying process is favorable and spontaneous; energy is released and the disorder increases.

What is HEA?

HEAs are solid solutions composed of at least five equimolar or near-equimolar alloying elements. The term "high-entropy alloys" was created because the entropy increase of mixing is substantially higher when there is a larger number of elements in the mix, and their proportions are more nearly equal.

How do HEAs help?

High entropy alloys (HEAs) can mitigate the issue of obtaining enough free energy for alloy formation to some extent by incorporating a large entropy contribution, $T\Delta S$. HEAs are multicomponent alloys consisting of equal or nearly equal atomic percentages of multiple elements. The high configurational entropy, resulting from the random distribution of multiple elements in their crystal lattice, contributes to a significant entropy change during alloy formation.

The high entropy contribution increases the overall entropy change (ΔS) in the free energy equation ($\Delta G = \Delta H - T\Delta S$), allowing for a larger enthalpy change (ΔH) without making the free energy change positive. This enables the formation of alloys that may not have been possible with traditional alloying approaches, where the enthalpy change alone may have exceeded the product of temperature and entropy change (TdS).

By adjusting the composition of HEAs, it is possible to achieve desirable properties while still maintaining a negative free energy change. This flexibility in composition and the increased entropy contribution in HEAs expands the range of alloys that can be formed and potentially overcomes the limitations associated with obtaining enough free energy change for spontaneous alloying.

HEA Thermodynamics

In a metal or in a binary or ternary alloy the molecular configuration is far more ordered with less entropy and less spread-out energy and therefore, less stable than a multicomponent alloy. HEA creation is just a step of creating more entropy in the alloy by mixing different elements of different atomic sizes to generate more entropy in the alloy and thereby making the energy more spread out in the alloy and reaching a more stable state with less internal stress.

The faster-moving particles have more energy; the slower ones have less. High entropy increases the random distribution of the energy. In essence . . . "a system becomes more stable when its energy is spread out in a more disordered state" free from any internal stress. A system that is more disordered in space will tend to have more disorder in the way the energy is arranged as well. ... The entropy has increased in terms of the more random distribution of the energy. In essence . . . "a system becomes more stable when its energy is spread out in a more disordered state".

The Future Potential of HEA

High entropy alloys (HEAs) have shown promising potential in various applications due to their exceptional mechanical, thermal, and chemical properties. The use of HEAs in addressing the challenges of H_2 gas transport can be explored by tailoring the alloy composition to optimize the desired properties.

One of the main challenges in H_2 gas transport is hydrogen embrittlement, which is the degradation of mechanical properties in materials exposed to hydrogen. HEAs, with their unique composition and crystal structure, have the potential to mitigate hydrogen embrittlement. By carefully selecting appropriate elements in an alloy, the mechanical properties can be enhanced, and the resistance to hydrogen embrittlement can be improved.

Furthermore, HEAs can also offer high ductility and toughness, which are crucial in H2 gas transport where materials need to withstand high-pressure conditions. Additionally, the thermal stability of HEAs allows them to maintain their properties under varying temperature conditions, making them suitable for H2 gas transport applications.

WHY WATER EVAPORATES AT AMBIENT TEMPERATURE? ROLE OF ENTROPY AND FREE ENERGY

The one line answer is that the free energy change of water at 298.15 K and 1 atm is 8.6 kJ/mol.

This means water has a positive free energy. Water in your glass at ambient temperature is not in a state of thermodynamic equilibrium of minimum free energy and maximum entropy change.

Gibbs free energy is expressed by equation

$dG = dH - TdS$, here G is free energy, H Is enthalpy, T is Kelvin and S is entropy.

In the case of liquid water $dG = 8.6$ KJ/mol at 298.15 k and 1 atm pressure.

Water while trying to reach a stable thermodynamic equilibrium state always try to minimize it's free energy to go towards $dG = 0$ by increasing entropy, please refer back to Gibbs equation.

How?

Evaporation of water and phase change increase entropy because they increase the randomness of the system's particles.

Explanation

Entropy is a measure of disorder in a system. When water evaporates, it changes from a liquid to a gas, which increases the randomness of the water molecules and the entropy of the system.

This balancing act of water of reaching a stable thermodynamic state by generating entropy is therefore leads water to evaporate even when it is in subcooled state at 25 degc or 298.15 K.

So what evaporates from your glass of water at ambient temperature is in fact the irreversible entropy losses to surroundings that water rejects in its attempt to reach stability.

WHAT IS ENTHALPY?

Enthalpy is the total energy at constant pressure. Change of enthalpy per unit temperature is the specific heat of a gas at constant pressure, dH/dT = Cp

The energy has two components (1) heat and (2) work.

These two components are inseparable. When heat travels from one point to another in a gas it changes the volume of the gas and does thermodynamic work.

In incompressible substances like liquids or solids, there is practically no change in volume so when heat travels through these substances heat only changes the internal energy which is the sum of kinetic and potential energy.

Internal Energy

Internal energy is the sum of kinetic and potential energy.

For ideal gases, the average kinetic energy is directly proportional to temperature, $KE = 1/2$ kb T, kb is constant. This is also an expression to convert Joule to Kelvin.

Potential energy is the stored energy in molecules in the form of intermolecular forces.

Coming back to enthalpy, from the 1st law of thermodynamics at constant pressure, $dH = dU + PdV$. here, H is enthalpy, U is internal energy, P is pressure and V is volume.

In enthalpy, the internal energy and work are separately accounted for while both represent energy.

The reason is internal energy is a state variable. Water, in liquid form, has much smaller internal energy than when water becomes steam.

Enthalpy does not depend on the path chosen by water to become steam. The values of internal energy of liquid water and steam remain constant regardless of the path liquid followed to become steam.

Work on the other hand is a path function. Between two points there could be uncountable options for work to travel from one point to another by doing different amounts of work, PdV work

The work and internal energy are always estimated separately because their behavior in a thermodynamic system can't be truly predicted unless each is separately tracked and accounted for.

Why is energy expressed as enthalpy?

In any energy calculation between two points the enthalpy gives an account of the total energy which is conserved.

Example: Steam

The process starts with incompressible water. For water, the enthalpy is its internal energy m $Cp\ dt$. H water = $m\ Cp\ dt$.

When water becomes a compressible fluid called steam, the enthalpy of heat of vaporization gets added which is the sum of steam's internal energy and the thermodynamic PV work of expansion of water to steam. H steam = internal energy of steam + PV work.

Therefore, at every point in a process, the total energy can be tracked and estimated by enthalpy.

ENTHALPY REJECTION AT CONSTANT PRESSURE

The simplest example is a constant pressure condenser. The steam enters the condenser at a constant pressure and its corresponding saturated temperature. The steam goes from a higher energy level to a lower energy level by doing a phase change to saturated water at the temperature and pressure of the condenser by releasing its enthalpy which is primarily the latent heat.

This latent heat is then carried by the coolant to a heat rejection device. The density of steam undergoes an increase as steam condenses from gas to liquid. This amounts to a compression process that holds the pressure constant and works for enthalpy transfer.

The advantage of such a process is facilitating enthalpy $U + PV$ transfer or complete energy transfer at constant pressure.

From, dU = delta Q – delta W, in an isothermal condenser, $dU = 0$. Therefore, delta Q = delta W.

In a constant pressure condenser $dH = dU$ + delta W (at constant pressure the enthalpy is the sum of internal energy and work).

$dU = dH$ – delta W

Therefore, dH – delta W = delta Q – delta W

Therefore, delta $Q = dH$

To summarise, In an isothermal condenser delta Q = delta W. In a constant pressure condenser delta Q = delta H

This is the advantage of a constant pressure condenser over an isothermal condenser.

Steam heat transfer is an example of constant pressure enthalpy or total energy transfer when steam gives up its latent heat at constant pressure.

Rankine cycle condenser is another example, of constant pressure enthalpy/total energy transfer.

ISENTHALPIC PROCESS

To put it simply, an isenthalpic process is a type of thermodynamic process where the system's enthalpy stays constant, $dH = 0$ while an isentropic process is one in which the generation of entropy [internal entropy and external entropy both] remain zero, $dS = 0$. An isentropic process is simple to understand, but the working principle of an isenthalpic process can frequently be difficult to understand. In an isenthalpic process the enthalpy H_1 upstream equals the enthalpy H2 down stream meaning $U_1 + W_1$ upstream equals $U_2 + W_2$ down stream.

An Example of an Isenthalpic Process is the Throttling Process

Joule Thompson process is a typical isenthalpic process.

Throttling (Isenthalpic) process: When a fluid suddenly runs into a restriction in a passage, the process of throttling takes place.

There could be a valve that is nearly entirely closed, a sudden and significant drop in flow area, etc., causing the restriction.

The fluid's pressure suddenly drops as a result of this restriction because the fluid must force itself through it. This highly irreversible process lowers the refrigerant's temperature and pressure in a refrigeration system.

Both isenthalpic and isentropic processes are adiabatic, meaning that there is no heat exchange with the surroundings. However, there is an important distinction between the two in terms of reversibility.

In an isentropic process, the entropy remains constant, and the process is internally reversible, meaning there are no irreversibilities within the system. On the other hand, in an isenthalpic process, the entropy (S) can increase due to internal irreversibilities.

WHY AND WHEN A PROCESS BECOMES ISENTHALPIC?

This is a difficult concept. You don't find any explanation easily.

An isenthalpic process is one where there is no change in total energy or enthalpy.

A porous plug and throttling valve consists of pores/tiny openings in the range of 5 to 125 micrometers through which high pressure gas at a given temperature is forced suddenly into a large space.

Joule-Thomson cooling formula

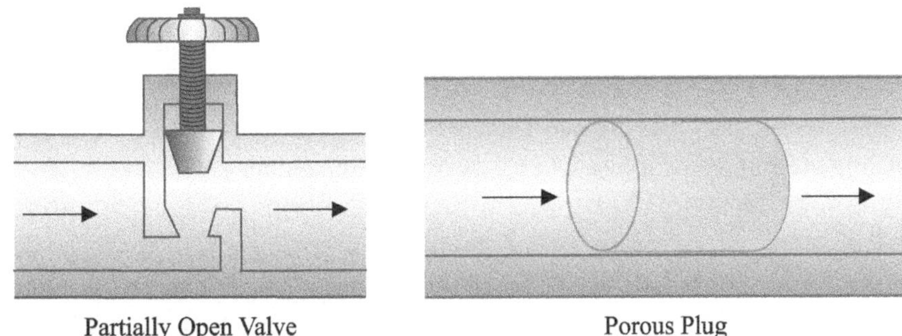

Figure 20

In Steps

You know initial pressure of gas Pi in bar. Find out the final pressure of the gas (Pf) in bar.

Next, determine the Joule Thomson coefficient (μ) in Kelvin per bar.

Use the formula from above

$= Tf = Ti + \mu \times (Pi - Pf).$

Finally, calculate the final temperature (*Tf*) in Kelvin.

How JT does Isenthalpic Cooling?

Joule-Thomson cooling porous plug or throttling valve can have pores/tiny openings in the range from 5 to 125 micrometers. The upstream pressure and temperature remain constant. As the gas leaks through these micro pores there is significant drop in kinetic energy downstream that does cooling called JT cooling. The gas leaks into a large volume effortlessly without doing any work. Therefore, we always say in JT cooling there is no "work" involved. But never understand, why?

To sum up (1) large reduction in kinetic energy across the throttling valve gives the cooling at no cost of energy because of effort less expansion of small volume of gas into a much larger space suddenly involves "no work". This explains why *JT* cooling is "isenthalpic".

Explanation

An isenthalpic process occurs when suddenly a real gas is made to expand into a large space which precisely defines what follows a throttling process.

Since the gas expands into a large space it does so without much pushing the surroundings and practically no work.

Thus the "Work "component of enthalpy $H = U + W$ remains constant. H is enthalpy U is internal energy, W is work.

The second thing that follows the expansion of gas in large space is dramatic reduction of molecular collisions that reduces the temperature or internal energy.

Overall in equation, $H = U + W$ the sum remains constant except the value of each changes.

Summary

To summarize, in an isenthalpic process such as the Joule-Thomson expansion, the gas undergoes a sudden expansion into a large space through a throttling valve.

This expansion occurs without the gas doing much work on the surroundings, leading to a constant enthalpy.

The decrease in pressure and temperature during the expansion results in a reduction of molecular collisions, leading to a decrease in internal energy.

Despite these changes in internal energy and work, the total enthalpy of the system remains constant.

EXPANSION *vs.* THROTTLING: ARE THESE SAME?

The answer is no.

A porous plug and throttling valve consists of pores/tiny openings in the range of 5 to 100 micrometers through which high pressure gas at a given temperature is forced suddenly into a large space.

The key difference is throttling is isenthapic and expansion can be isentropic.

Expansion and throttling are related processes, but they are not the same.

Expansion (turbine)

In a turbine, expansion of a fluid can produce useful work output and be close to a reversible process. In expansion it is more gradual than a sudden expansion by a throttling valve with considerable frictional losses.

Throttling (refrigeration)

Throttling is an expansion process that is highly irreversible. In throttling, a fluid's pressure and temperature are reduced suddenly when it passes through a throttling valve. This is due to the fluid expanding into a larger space and frictional effects.

Explanation

Throttling valves are used in many systems, including refrigeration and air conditioning.

For example, in a refrigerator, a throttling valve reduces the pressure and temperature of the refrigerant so it can cool the evaporator.

In gas-expansion processes, the gas expands in an expansion machine after being compressed and rejecting heat. This method has a higher cooling intensity than throttling

Gas expansion is isentropic that is more efficient cooling process than throttling which is isenthalpic.

ROLE OF THROTTLING IN VAPOR COMPRESSION REFRIGERATION

What is a Throttling Valve?

Throttling valve is a device that causes high pressure fluid at upstream to suddenly expand through 100–125 micropore into a large volume effortlessly without doing any work and without loss of internal energy, $H = U$ (H is enthalpy and U is internal energy. The cooing across a throttling valve is just a conversion of kinetic energy to potential energy with no change in total internal energy.

Further Clarification

Explanation

In an isenthalpic process, enthalpy of a system remains constant, even though there may be significant changes in pressure and temperature. This is because the net sum of the associated terms in the energy balance $H = U + W$ is zero. $W = 0$ as explained above, therefore, $H = U$.

Throttling Process

A throttling valve is a device that reduces the pressure of a fluid passing through it, causing a simultaneous decrease in temperature. This process is an isoenthalpic process because the enthalpy remains constant.

Throttling valve does three things in vapor compression refrigeration

1. ***Maintaining Pressure** in the Isobaric Evaporator:* The throttling valve helps to maintain a constant pressure in the evaporator by regulating the flow of refrigerant into this component of the system.
2. ***Cooling the Refrigerant** from the condenser and Increasing Latent Heat of Vaporization in the Evaporator:* As the refrigerant passes through the throttling valve, it undergoes a rapid expansion resulting in a drop in pressure

and temperature. This causes the refrigerant to cool down significantly, which is essential for absorbing heat from the surrounding environment in the evaporator. The increase in latent heat of vaporization allows the refrigerant to efficiently absorb heat energy during the phase change from liquid to vapor in the evaporator.

3. ***Isenthalpic Expansion:*** The expansion process in the throttling valve is considered isenthalpic, meaning that it occurs at constant enthalpy. This implies that the expansion and cooling of the refrigerant in the throttling valve occur without any external work being done on or by the refrigerant. The reduction in temperature during this process is solely attributed to the decrease in pressure and temperature as the refrigerant moves through the valve by a conversion of kinetic to potential energy.

Overall, the throttling valve plays an important role in the refrigeration cycle by regulating pressure, cooling the refrigerant, increasing the latent heat of vaporization, and facilitating isenthalpic expansion.

INTERPLAY OF 'INTERNAL ENERGY' AND 'WORK' IN THERMODYNAMIC PROCESSES

Thermodynamics is the study of the relationship between heat, work, and energy. In thermodynamic processes, both internal energy and work play a crucial role. **Internal energy, also known as thermal energy, is the storehouse of energy within a system, while work is the vehicle for the transport of energy between systems.** There are four types of thermodynamic processes: isothermal, isobaric, adiabatic, and constant volume (or isochoric) processes. In each process, internal energy and work play different roles.

Isothermal Process

In an isothermal process, the temperature of the system remains constant. The equation for the work done in an isothermal process is given by $W = nRT \ln(V_2/V_1)$, where n is the number of moles of gas, R is the gas constant, T is the temperature, and V_1 and V_2 are the initial and final volumes, respectively.

Isobaric Process

In an isobaric process, the pressure of the system remains constant. The equation for the work done in an isobaric process is given by $W = P(V_2 - V_1)$, where P is the constant pressure, and V_1 and V_2 are the initial and final volumes, respectively.

Let me expand on the concept of the isobaric process because it often creates confusion.

In an isobaric process, additionally, as the pressure is held constant, the system also transfers internal energy, U as it changes volume, and this transfer of internal energy takes place due to the constant pressure from the surroundings. **Therefore, for an isobaric process, $W = dU + P(V_2 - V_1)$** This transfer of internal energy, U is a key aspect of the isobaric process, as the system interacts with its surroundings while maintaining a constant pressure. The addition of internal energy from the surroundings, along with the work done by the system, highlights the interplay of internal energy and work in the context of the isobaric process.

Adiabatic Process

In an adiabatic process, there is no heat exchange between the system and its surroundings. The equation for the work done in an adiabatic process is given by $W = -(\gamma/\gamma^{-1})(P_2V_2 - P_1V_1)$, where γ is the ratio of specific heats, P_1 and V_1 are the initial pressure and volume, and P_2 and V_2 are the final pressure and volume, respectively.

Isochoric Process

In an isochoric process, the volume of the system remains constant. In this process, no work is done, as the volume does not change. Instead, the heat added to the system is equal to the increase in internal energy, given by $Q = \Delta U$.

In each of these processes, the interplay of internal energy and work is evident.

The transfer of energy through work and the change in internal energy are essential components of understanding thermodynamic processes. In addition to the mentioned points, it is important to note that the first law of thermodynamics, also known as the law of conservation of energy, states that the change in the internal energy of a system is equal to the heat added to the system minus the work done by the system.

This fundamental principle further emphasizes the significance of internal energy and work in thermodynamic processes. In conclusion, the interplay of internal energy and work in thermodynamic processes is essential for understanding the transfer and transformation of energy within systems. Each type of thermodynamic process has its equation to represent the work done, highlighting the different roles of internal energy and work in these processes.

CLAUSIUS AND BOLTZMANN'S APPROACH TO ENTROPY

Clausius and Boltzmann approached entropy differently.

Clausius' concept of entropy is related to the energy flow in a system and its surroundings. According to Clausius, the entropy of a system and its surroundings always increases in any irreversible process. This aligns with the understanding that entropy is a measure of the system's dispersal of energy or its tendency to move toward equilibrium. In an isolated system, where there is no energy exchange with the surroundings, the entropy will naturally increase as the system moves towards equilibrium. This aligns with the concept that entropy always increases in isolated systems.

Example

The second law states that there exists a useful state variable called entropy. The change in entropy (dS) is equal to the heat transfer (delta Q) divided by the temperature (T), dS = delta Q/T. For a given physical process, the entropy of the system and the environment will remain constant if the process can be reversed. If we denote the initial and final states of the system by "*i*" and "*f*", then: $Sf = Si$ [reversible process]. The law also says that heat moves from a high to a low energy level.

Assume for the moment that heat is transferred from object 1 (which is hot) at temperature T_1 to object 2 (which is cold) at temperature T_2. We will refer to the ultimate equilibrium temperature for both objects as T_f, and the quantity of heat transferred as Q. As the heat is dispersed from the hot object, its temperature fluctuates. We will refer to the hot object's average temperature during the process as Th, which is equal to the average of T_1 and T_f.

$$Th = [T_1 + T_f]/2$$

Similarly, for the cold object, the final temperature is T_f and the average temperature during the process is T_c which is the average of T_f and T_2.

$$T_c = [T_2 + T_f]/2$$

Th will always be greater than T_c, because T_1 is greater than T_2.

$$Th > T_C$$

The entropy change for the hot object will be ($-Q/Th$), with the minus sign applied because the heat is transferred away from the object.

$$dS\,[h] = -Q/Th$$

For the cold object, the entropy change is (Q/T_c), positive because the heat is transferred into the object.

$$dS\,[c] = Q/T_c$$

So, the total entropy change for the whole system would be given by the equation

$$S_f - Si = -Q/Th + Q/Tc$$

with Si and S_f being the final and initial values of the entropy. The term (Q/T_c) will always be greater than ($-Q/Th$) because Th is greater than Tc. Therefore, Sf will be greater than Si, as the second law predicted by Clausius.

Boltzmann's Approach

On the other hand, Boltzmann's concept of entropy is related to the microscopic level and the number of possible microstates of a system.

According to Boltzmann, entropy (S) is proportional to the natural logarithm (ln) of the number of microstates (W) available to a system in a particular macrostate.

This aligns with the idea that entropy is a measure of the system's disorder or randomness. As the number of microstates increases, the system becomes more disordered, leading to an increase in entropy.'

Boltzmann's Entropy Formula

$S = kb \ln W$, where S represents entropy, kb is the Boltzmann constant, and W is the number of microstates or possible arrangements of a system.

As the number of microstates (W) increases, the entropy (S) also increases. This means that as disorder or randomness in a system increases, the number of possible ways the system can be arranged also increases. Consequently, there is a greater opportunity for energy or heat to be dispersed and transferred within the system, leading to a higher likelihood of energy loss and reduced efficiency in performing assigned work.

This means that the increased disorder or entropy can result in a higher chance for energy dissipation, reducing the capacity for heat to be effectively utilized for doing assigned work.

WHY EVERY THERMODYNAMIC PROCESS HAS ENERGY VENT?

The answer in one line is if you don't let the energy to exit you convert the process into an isolated process where entropy eats up the energy over a period of time and force the system to go to equilibrium minimizing the Gibbs free energy.

Details

In thermodynamics, venting out energy is critical to ensure that the Gibbs free energy of a system remains negative. This is essential in preventing the system from reaching equilibrium and increasing entropy.

By continually allowing energy to be vented out of the system, the Gibbs free energy remains negative, indicating that the system is in a state of non-equilibrium and can perform work.

In the absence of venting out energy, the system's entropy would continue to increase, eventually leading towards equilibrium and the system becomes an isolated system.

As the entropy increases, the Gibbs free energy approaches zero, and the system reaches a state of equilibrium where no further work can be done.

In various thermodynamic processes, energy can be vented out in different ways to maintain a negative Gibbs free energy and prevent the system from reaching equilibrium.

Below are examples of how energy is vented out in isothermal, isobaric, and adiabatic processes:

Isothermal Process

In an isothermal process, the temperature of the system remains constant. Energy can be vented out by transferring heat to the surroundings while maintaining the temperature constant. For example, in a gas expansion process within a piston-cylinder system, heat can be continuously transferred to the surroundings to maintain the temperature of the gas. This process allows the system to perform work while keeping the Gibbs free energy negative.

Isobaric Process

In an isobaric process, the pressure of the system remains constant. Energy can be vented out by allowing the system to expand against a constant pressure, performing work in the process. For instance, in a chemical reaction that occurs at a constant pressure, the system can vent out energy by expanding or contracting to maintain the pressure while allowing work to be done.

Adiabatic Process

In an adiabatic process, there is no heat transfer between the system and its surroundings ($Q = 0$). Despite this, energy can still be vented out in the form of work transfer. For example, in an adiabatic expansion of a gas, the gas performs work on the surroundings as it expands against a piston, venting out energy in the process as work

This work transfer allows the system to maintain a negative Gibbs free energy even without heat transfer.

In each of these processes, energy is vented out in different ways to keep the Gibbs free energy negative and prevent the system from reaching equilibrium and not letting entropy neutralize all energy.

WORK TO HEAT' TRANSFER

From the 1st law of thermodynamics, dU = delta Q – delta W.

When $dU = 0$, delta Q = delta W. Therefore, at constant temperature and pressure when $dU = 0$, the work converts to heat.

This typically happens when an adiabatic process transfers work isothermally to a cooler at constant pressure. At constant temperature, $dU = 0$, the work transfers to heat in the cooler which the cooler rejects and sends to the surroundings.

This is how adiabatic processes in compressor and power plant turbine transfers work to surroundings.

Work is a mechanical energy consisting of potential and kinetic energy. Total energy is always conserved in mechanical energy if you ignore drag by air. Energy never leaves the system like a swinging pendulum.

ADIABATIC PROCESS: WORK TRANSFER

The adiabatic process is not permeable to heat but it is permeable to "Work". Generally, all adiabatic processes are followed by a "Work sink" You may be familiar with the word "Work sink" in heat engines.

The adiabatic process is a thermodynamically closed process that occurs by using its internal energy. Adiabatic processes generally occur so rapidly that the process does not have the time to pull energy from the surroundings or transfer energy to the surroundings.

Expansion and compression processes are generally examples of an adiabatic process.

While an adiabatic process is not permeable to heat transfer it is permeable to work transfer. A typical example is a compressor. A compressor compresses air at constant volume by increasing temperature using PV compression work on the air. The compression adds work energy to the air in the compression. The temperature increase in compression is a hazard and unless controlled it can damage the compressor and also it becomes a safety concern. To remove this issue the compressors are provided with isothermal coolers. Isothermal coolers operating at ambient temperature and pressure extract the work energy generated in the compressor and maintain the compressor temperature nearly at ambient temperature. These interstage coolers act as "Work sink".

In a power plant turbine, the condenser acts as a "Work sink".

SIGN CONVENTION OF THERMODYNAMIC 'WORK'

Often, the sign convention of thermodynamic work becomes confusing. The sign convention for work (positive or negative) depends on the perspective of the system and its surroundings. This can indeed be confusing without a clear understanding of the direction of work.

When the system does work on the surroundings, the work is considered to be negative, indicating a loss of energy from the system to the surroundings. Similarly, when the surroundings do work on the system, the work is considered to be positive, indicating a gain of energy by the system.

An easy way to remember this convention is to consider the energy flow. If the energy is flowing from the system to the surroundings, the work is considered to be negative. Conversely, if the energy is flowing from the surroundings to the system, the work is considered to be positive.

It's also important to keep track of the direction of the force and the resulting displacement to correctly determine the sign of work according to the perspective of the system and surroundings.

Keeping this perspective in mind can help clarify the sign of work, making it easier to understand whether it is adding to or extracting energy from the system.

Details

The fundamentals of thermodynamic work come from the 1st law of thermodynamics. The relationship between internal energy (U), heat (Q), and work (W) is governed by the first law of thermodynamics, which states that the change in internal energy of a system is equal to the heat added to the system minus the work done by the system on its surroundings. This relationship is expressed by the equation $U = Q - W$.

As stated above, when work (W) is positive, it signifies that the surroundings are doing work on the system, such as in the case of compressing gas molecules. This results in the addition of internal energy to the system. For example, when gas is compressed, the work done by the surroundings on the system increases the internal energy of the gas. Conversely, when work (W) is negative, it indicates that the system is doing work on its surroundings, as in the case of a gas expanding and pushing against its surroundings. In this scenario, the system extracts internal energy by performing work against intermolecular forces.

In compressors, work is done on the gas to compress it, the positive work leads to an increase in internal energy. This is because the compression of the gas increases the kinetic energy of its particles, increasing temperature and internal energy. The compressor works as an adiabatic process. On the other hand, during adiabatic expansion of air from a leaking tire, the process is adiabatic, meaning no heat is exchanged with the surroundings. Consequently, the only way for the internal energy of the air to decrease is for work to be done as the air expands, causing it to lose energy. This explains why the internal energy of the air decreases as it expands during the leak in process.

PDV *vs.* VdP THERMODYNAMIC WORK

References: Research Gate/Wikipedia

In one line, Pdv is used when there is “dv”, ie change in volume while P is a function of volume. This typically happens in a non-flow process. On the other hand, VdP is used when there is “dP”, ie change in pressure. This happens during a flow process (since the difference in head causes flow).

A typical example of VdP work is the gas turbine

For the gas or steam to enter inside the turbine to spin the shaft gas or steam to maintain a steady flow of steam, the steam has to push the fluid already inside and therefore, needs to perform work. This net work is the shaft work that spins an isentropic turbine, this is Vdp.

What is thermodynamic work?

Thermodynamic work can be categorized into work at constant pressure (PdV) and work at constant volume (VdP).

In the case of work at constant pressure, the system changes volume while the pressure remains constant. This type of work is often associated with processes occurring in closed systems, such as a piston-cylinder arrangement. On the other hand, work at constant volume involves a change in pressure while the volume remains constant. This type of work is commonly associated with processes occurring in open systems. Turbines are considered an open system.

Pdv is used when there is “dv”, ie change in volume while P is a function of volume. This typically happens in a non-flow process. On the other hand VdP is used when there is “dP”, ie change in pressure. This happens during a flow process (since difference in head causes flow).

Perhaps the most significant thing here is that ∫ PdV is the specific work for a closed system, as in a piston and cylinder, while ∫ VdP is the specific work for an open system, as in a steam or gas turbine.

There is also a relationship between these two and PV, which arises from the integration of parts.

In a typical steam power plant with three separate machines (high-pressure, intermediate-pressure, and low-pressure turbines), the LPT produces about 45% of the total shaft power. It’s also by far the largest in size. People think that because the pressure is much higher (say 200 atm) entering the HPT and much lower (say 3 atm) entering the LPT that the HPT should put out much more power, but it doesn’t. This is because steam (and gas) turbines are volumetric flow devices, described by ∫ VdP, not pistons described by ∫ PdV. Pdv is used when there is “dv”, ie change in volume while P is a function of volume. This typically happens in a non flow process. On the other hand, VdP is used when there is “dP”, ie change in pressure. This happens during a flow process (since the difference in head causes flow).

THERMODYNAMIC PROCESSES AND POLYTROPIC INDEX

In a nutshell, thermodynamic processes involve different types of heat and work transfer. Typically, there are four thermodynamic processes. [1] isobaric or constant pressure process [2] isothermal or constant temperature process [3] adiabatic process; [4] isochoric or constant volume process. The constant pressure process uses heat from outside to keep the pressure constant. Isothermal processes maintain a constant temperature by absorbing heat from the surroundings and returning it to the surroundings, whereas adiabatic processes use their internal energy in the process. Internal energy increases in the isobaric process, it remains constant in the isothermal process and decreases in the adiabatic process. No work in the isochoric process.

The change in internal energy of a system is the sum of the heat transferred and the work done.

Thermodynamic processes follow the equation $PV^n = C$. In this case, P is pressure, V is volume, n is the polytropic index, and C is a constant.

What Does the Polytropic Equation Signify?

Polytropic index 'n' signifies the PV work. The polytropic process equation can be used to describe a variety of expansion and compression processes, including heat transfer.

Polytropic Work Transfer

$n = 0$, constant pressure or isobaric process, $W = P\,[V_2 - V_1]$
$n = 1$, constant temperature or isothermal process, $W = Q = RT \ln V_2/V_1$
$n = y$, reversible adiabatic process, $W = P_1V_1 - P_2V_2/Y - 1$
n = infinity, constant volume or isochoric process, $W = 0$

Isobaric *vs.* Isothermal Process

Between constant pressure or isobaric process and isothermal process, in the isobaric process pressure is maintained constant by supplying additional heat that increases the internal energy of the gas. Therefore, the work done by an isobaric process is purely a function of pressure and not limited to atmospheric pressure as in the case of the isothermal process. In the isothermal process, the internal energy stays constant.

Isothermal *vs.* Adiabatic Process

An infinite amount of heat is available from the surroundings to maintain a constant temperature. This means that the system can exchange as much heat as needed to keep the temperature constant, allowing for unlimited work to be done. On the other hand, in an adiabatic process, no heat is exchanged with the surroundings. The work done in an adiabatic process is limited by the available internal energy or the specific heats (*Cp* and *Cv*) of the gas. This limitation on the available energy means that the adiabatic process can do only as much work as the internal energy allows, making it less than the theoretical maximum for the isothermal process.

Isochoric Process

As explained, there is no work in the isochoric process.

Finally, the work done by Isobaric process > Isothermal process > Adiabatic process.

ROLE OF POLYTROPIC INDEX *n* = *CP/CV* IN COMPRESSION OF GASES

Polytropic Index

A polytropic process is a thermodynamic process that obeys the relation:

$$PV^n = C$$

where P is the pressure, V is volume, n is the polytropic index, and C is a constant. The polytropic process equation can describe multiple expansion and compression processes which include heat transfer. The term "polytropic" was originally created to describe any reversible process on any open or closed system of gas or vapor which involves both heat and work transfer, such that a specified combination of properties were maintained constant throughout the process. In such a process, the expression relating the properties of the system throughout the process is called the polytropic path. There are an infinite number of reversible polytropic paths between two given states; the most commonly used polytropic path *is* $TdS/dT = C$, which is a constant and is equal to zero for an adiabatic process. T is Temperature, S is Entropy.

The value of 'n' is different in different thermodynamic processes. Some examples are given below. The polytropic index is a measure of the work done by the system.

- **Word of caution**: Many gas compression and expansion processes may be approximated by a polytropic process. In each case, the polytropic coefficient must be determined experimentally by measurement of the heat and work transfer and the initial and final states.

$n = 0$, constant pressure process $\quad w = P(v_2 - v_1)$

$n = 1$, constant temperature process $\quad w = q = RT \ln \frac{v_2}{v_1}$

$n = \gamma$, reversible adiabatic process $\quad w = \frac{P_1 v_1 - P_2 v_2}{\gamma - 1}$

$n = \infty$, constant volume process $\quad w = 0$

In isentropic expansion and compression $n = Y$ [Gamma] $= Cp/Cv$

Cp and Cv

The *Cp* and *Cv* are the specific heats of an ideal gas at constant pressure and at constant volume. These indicate the quantity of heat that can increase the temperature of unit mass by 1°C.

Cp > Cv

According to the 1st law of thermodynamics, delta H = delta U + delta W [H is enthalpy, U is internal energy and W is work.

At constant pressure, the heat absorbed is not only increasing internal energy but also goes for doing work. At constant volume since delta-$V = 0$ heat goes to increase only internal energy. delta $W = 0$. Hence $Cp > Cv$.

Cp/Cv ratio

Thermodynamic significance of *Cp/Cv*.

For ideal gases PV = constant. *Cp/Cv* is a measure of how much internal energy is getting converted to work in any compression/expansion process. Every change in internal energy impacts entropy changes and brings irreversibility in the process. Thus, it can also be said that the heat capacity ratio is the ratio between the enthalpy to the internal energy: $y = Cp/Cv = H/U$ [Enthalpy divided by internal energy]

The value of 'n' can vary from zero to infinity. n = 0 for an isobaric process, in which the pressure of the system stays constant: $\Delta P = 0$. n = infinity for an isochoric process, when volume stays constant i.e., $\Delta V = 0$.

Some typical values of Cp/Cv

Table 4

Gas	*Ratio of Specific Heats*
Acetylene	1.3
Air, Standard	1.4
Ammonia	1.32
Carbon Dioxide	1.28
Carbon Monoxide	1.4
Chlorine	1.33
Ethane	1.18
Helium	1.66
Hydrogen	1.41
Methane	1.32
Natural Gas (Methane)	1.32
Nitrogen	1.4
Oxygen	1.4
Propane	1.12
Steam	1.28
Suplhur dioxide	1.26

Relation with Degrees of Freedom

The heat capacity ratio (γ) for an ideal gas can be related to the degrees of freedom (f) of a molecule by,

Gamma = $1 + 2/f$

Thus, we observe that for a monatomic gas, with 3 degrees of freedom: $y = 5/3 = 1.66$ like helium.

While for a diatomic gas, with 5 degrees of freedom (at room temperature: 3 translational and 2 rotational degrees of freedom; the vibrational degree of freedom is not involved, except at high temperatures): $Y = 7/5 = 1, 4$ like nitrogen gas.

Significance of Polytropic Index *n* in Compression and Expansion

The heat capacity ratio is important for its applications in thermodynamically reversible processes, especially involving ideal gases.

If you have a value for n, then you can determine the heat of compression by Eq. below. It is common to look at the measured suction and discharge temperatures and compression ratios and determine the polytropic index.

$$Cp \ln (T_2/T_1) = R \ln (p_2/p_1)$$

We divide both sides by "*cp*" and take the exponential function of both sides (this "un-does" the logarithms).

$$T_2/T_1 = (p_2/p_1)^{(R/Cp)}$$

where the symbol "^" denotes an exponent. Now we substitute the expression for "*R*/*Cp*" to obtain:

$$T_2/T_1 = (p_2/p_1) \wedge [(\text{gamma} - 1)/\text{gamma}]$$

where the numbers 1 and 2 denote the states at the beginning and end of the compression process.
where gamma = n = polytropic exponent.

Typical Example

To understand this relation, consider the following thought experiment. A closed pneumatic cylinder contains air [diatomic]. The piston is locked. The pressure inside is equal to atmospheric pressure. This cylinder is heated to a certain target temperature. Since the piston cannot move, the volume is constant. The temperature and pressure will rise. When the target temperature is reached, the heating is stopped. The amount of energy added equals $CV\,\Delta T$, with ΔT representing the change in temperature. The piston is now freed and moves outwards, stopping as the pressure inside the chamber reaches atmospheric pressure. We assume the expansion occurs without the exchange of heat (adiabatic expansion). Doing this work, the air inside the cylinder will cool to below the target temperature. To return to the target temperature (still with a free piston), the air must be heated, but is no longer under constant volume, since the piston is free to move as the gas is reheated. This extra heat amounts to about 40% more than the previous amount added. In this example, the amount of heat added with a locked piston is proportional to CV, whereas the total amount of heat added is proportional to CP. Therefore, the heat capacity ratio in this example is 1.4.

WHAT IS THE ENTHALPY AND ENTROPY OF MIXING?

Enthalpy of Mixing

In a nutshell, what constitutes an ideal mixture is the first thing to understand before learning about enthalpy of mixing. Any mixture that has the same arithmetic mean for the two pure substances as the final mixture (mole fraction) is considered ideal. This indicates that there is no enthalpy of mixing, among other significant thermodynamic simplifications:

$$\text{Mix } \Delta H \text{ ideal} = 0.$$

It is reasonable to assume that any gas that obeys the ideal gas law will mix perfectly, as will hydrocarbons and liquids with comparable molecular interactions and characteristics. The primary cause of changes in a mixture's enthalpy is intermolecular forces.

Details

Enthalpy of mixing is the change in energy that occurs when particles in a mixture interact with each other, compared to when they are in their pure states. It can also be referred to as the heat of mixing or excess enthalpy. A positive enthalpy of mixing indicates an endothermic mixing, while a negative enthalpy of mixing indicates an exothermic mixing. In ideal mixtures, the enthalpy of mixing is zero, meaning no heat is released or absorbed during mixing. In non-ideal mixtures, the thermodynamic activity of each component is different from its concentration. When two substances are mixed together without initiating a chemical reaction, heats of mixing are released or consumed. If the mixing process takes place at constant pressure, the heat of mixing is equal to the enthalpy of mixing.

When two substances are mixed the resulting enthalpy change is not an addition of the pure component enthalpies, unless the substances form an ideal mixture. The interactions between each set of molecules determines the final change in enthalpy. For example, when compound "*x*" has a strong attractive interaction with compound "*y*" the resulting enthalpy is exothermic. In the case of alcohol and its interactions with a hydrocarbon, the alcohol molecule participates in hydrogen bonding with other alcohol molecules, and these hydrogen bonding interactions are much stronger than alcohol-hydrocarbon interactions, which results in an endothermic heat of mixing.

Ideal Mixtures

Ideal mixture is any in which the arithmetic mean (with respect to mole fraction) of the two pure substances is the same as that of the final mixture.

Among other important thermodynamic simplifications, this means that enthalpy of mixing is zero. Any gas that follows the ideal gas law can be assumed to mix ideally, as can hydrocarbons and liquids with similar molecular interactions and properties.

Entropy of Mixing

In summary there could two reasons. Mixing of two substances always increase randomness and disorder simply because there are more microstates and hence the entropy. The second possibility is when two substancse are mixed and if they are nonideal materials, the total final volume may be different from the sum of the separate initial volumes, and there may be transfer of work or heat, to or from the surroundings; also, there may be a departure of the entropy of mixing from that of the corresponding ideal case.

The entropy of mixing is the increase in the total entropy when several initially separate systems of different composition, each in a thermodynamic state of internal equilibrium, are mixed without chemical reaction by the thermodynamic operation of removal of impermeable partition(s) between them, followed by a time for establishment of a new thermodynamic state of internal equilibrium in the new unpartitioned closed system.

In the special case of mixing ideal materials, the final common volume is in fact the sum of the initial separate compartment volumes. There is no heat transfer and no work is done. The entropy of mixing is entirely accounted for by the diffusive expansion of each material into a final volume not initially accessible to it.

In the general case of mixing non-ideal materials, however, the total final common volume may be different from the sum of the separate initial volumes, and there may occur transfer of work or heat, to or from the surroundings; also, there may be a departure of the entropy of mixing from that of the corresponding ideal case. That departure is the main reason for interest in entropy of mixing. These energy and entropy variables and their temperature dependences provide valuable information about the properties of the materials.

On a molecular level, the entropy of mixing is of interest because it is a macroscopic variable that provides information about constitutive molecular properties. In ideal materials, intermolecular forces are the same between every pair of molecules. In non-ideal materials, there may be differences of intermolecular forces or specific molecular effects between different species, even though they are chemically non-reacting.

CONFUSING TERMS: SOLUTION, ABSORPTION, AND ADSORPTION

Solution

A solution is a type of homogeneous mixture composed of two or more substances in chemistry. A solute in such a mixture is a substance that has been dissolved in another substance known as a solvent. If the attractive forces between the solvent and solute particles are stronger than the attractive forces holding the solute particles together, the solvent particles separate and surround the solute particles. These encircled solute particles then move away from the solid solute and into the solution.

How Does it Work

LHS image: Exothermic solution
RHS image: Endothermic solution

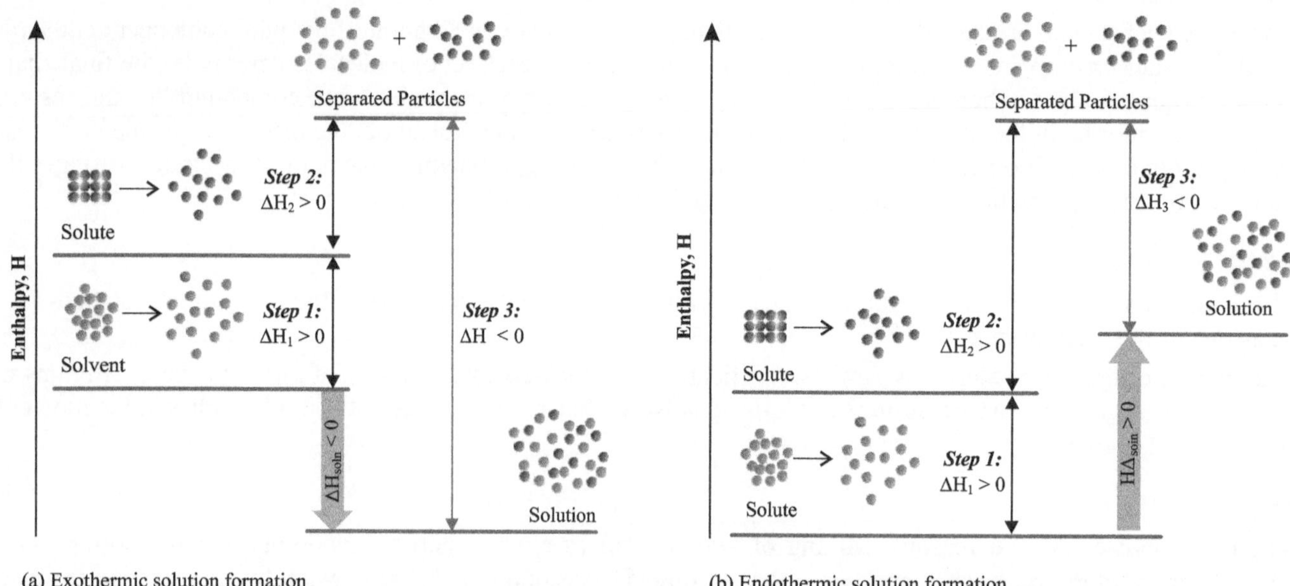

(a) Exothermic solution formation (b) Endothermic solution formation

Figure 21

A solution forms when there is a difference of solute + solvent energy and the energy in the solution. A solution is a property of a substance. A solution can be both exothermic and endothermic. When a solution has lesser enthalpy than the solute + solvent. The solution releases heat and it is exothermic and the opposite is endothermic. Solubility is expressed generally as gram of solute/100-gram solvent at a given temperature. As said solubility entirely depends on the solvation capacity of a solvent for a solute. It is specific to a solute.

Absorption

Absorption could be a chemical reaction or purely a physical process. Chemical absorption is exothermic. Reactions between acid gases and amines are exothermic,

Physical absorption is a type of solution that is always endothermic. In absorption, there are not enough intermolecular forces that can influence the making or breaking of intermolecular bonds.

Chemical absorption [not a physical process]

An example of chemical absorption is the purification of natural gas by passing the natural gas through an aqueous solution of an ethanolamine in which any acid gases, such as hydrogen sulfide (H_2S) and carbon dioxide] (CO_2), are removed from the natural gas by reacting with the ethanolamine. This is purely a chemical reaction.

Physical absorption: Absorption is a process that occurs when atoms, molecules, or ions enter a bulk phase – liquid or solid material. Molecules undergoing absorption are taken up by the volume rather than the surface (as is the case with adsorption). The absorption process occurs when one substance absorbs another.

A substance absorbs and transforms energy through the absorption process. An absorbent distributes the material it captures throughout the entire structure.

How does it work?

Physical absorption is a type of solution. The absorbing and the absorbed substances do not have enough intermolecular energy to break or make intermolecular bonds.

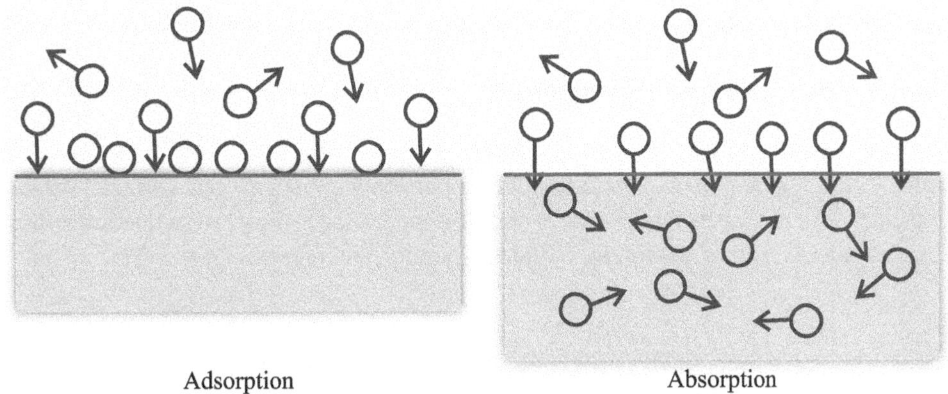

Figure 22

What is Adsorption? Why is Adsorption Exothermic and Absorption Endothermic?

Adsorption is a surface phenomenon where particles adhere to the top layer of a substance. The nature of the adsorbent, temperature, pressure, and the activation of the solid adsorbent all affect adsorption.

In adsorption, the substance accumulates on the surface, it doesn't enter the bulk of the solid or liquid. In many applications, activated carbon removes impurities from fluids, vapours, or gas through adsorption.

What is Absorption *vs.* Adsorption

Absorption is when a substance is dissolved by another substance.

Absorption involves the entire volume of the absorbing substance, while adsorption is a surface-based process.

In absorption, atoms pass through or enter the absorbing substance. In adsorption, molecules are loosely held on the surface of the absorbing substance.

Adsorption is Exothermic

Adsorption is an exothermic process because the surface energy of the adsorbent decreases during adsorption, which appears as heat.

What is surface energy and why it decreases during adsorption?

Why and how?

Explanation

Surfaces try to minimize their energy, (to become more stable) and adsorption does this by replacing exposed surface atoms with lower energy atoms or molecules. This process decreases the number of exposed atoms with high surface energy.

Heat

The decrease in surface energy is released as heat, making the adsorbent surface hot.

Entropy

The entropy of the system also decreases during adsorption.

Absorption is endothermic

Absorption is an endothermic process.

Why and how?

Explanation

Endothermic reactions absorb energy from their surroundings to break the bonds in reactants, which allows new bonds to form in products. This process requires more energy than is released, so the reaction absorbs energy and feels cold.

Thermodynamics of Adsorption *vs.* Absorption

Adsorption

This process occurs on the surface of a substance, where particles adhere to the surface. Adsorption is exothermic, meaning it produces heat. As the adsorbate particles stick on the surface, the system becomes more stable and releases excess energy. The area of adsorption decreases as the temperature increases.

Absorption

This process involves the mass transfer of particles into another material. In absorption, a substance's molecules are equally dispersed throughout the body.

Kinetics of Adsorption *vs.* Absorption

What they Measure

The kinetics of adsorption measure the rate at which a solute is adsorbed onto a surface, and how long the adsorbates remain there. The kinetics of absorption measure the reaction rate and activity of reactants over time.

What Affects the Rate

The rate of adsorption is affected by factors such as pH, contact time, temperature, adsorbate concentration, and adsorbent dose.

ADSORPTION ISOTHERM

An adsorption isotherm is a graph that shows the relationship between the amount of a substance adsorbed and the pressure or concentration at a constant temperature.

It's a useful tool for predicting how much a substance can adsorb, and it's used in designing adsorption systems.

The slope of the graph gives the rate of adsorption.

How it Works

Adsorption isotherms are created by injecting a small amount of gas into a sample and measuring how much gas is adsorbed at a constant temperature. The amount adsorbed is calculated from the change in gas pressure.

What it Shows

Adsorption isotherms can show the surface area and pore size distribution of a solid. They can also show the interactions between the adsorbate and adsorbent.

How Adsorption Isotherms are Interpreted?

Adsorption isotherms are graphs that show the relationship between the amount of adsorbate and pressure at a constant temperature. They are used to understand how adsorbates interact with adsorbent surfaces, and to predict the performance of adsorption systems.

Equilibrium

Adsorption occurs until the concentration of material on the adsorbent surface reaches equilibrium with the concentration of material in the solution.

Saturation Pressure

Once the saturation pressure is reached, the amount of adsorbate on the adsorbent stops increasing because the surface area is full.

Isotherm Models

There are many different isotherm models, including the Langmuir, Freundlich, and BET models. Each model has its own assumptions about the adsorption process.

Where it's Used

Adsorption isotherms are used in many applications, including:

- **Gas masks:** Poisonous gases are adsorbed onto the surface of gas masks, preventing them from coming into contact with the wearer.
- **Vacuum creation:** Charcoal is used to adsorb traces of air, which are then removed from devices being evacuated.
- **Humidity control:** Silica gel pellets are used to adsorb moisture in medicines and plastic bottles.
- **Removing coloring:** Animal charcoal is used to remove coloring agents from cane juice.
- **Catalysts:** Reactants adhere to the surface of a catalyst, which speeds up the reaction.

ADSORPTION PROCESS APPLICATIONS

Waste Water Treatment

The adsorption process is a common method for treating wastewater and can be used for a variety of applications, including:

- **Removing contaminants**: Adsorption can remove a wide range of contaminants from wastewater, including dyes, heavy metals, phenolic compounds, and odor and taste.
- **Treating industrial wastewater**: Adsorption is often used to treat industrial wastewater from organic and inorganic pollutants.
- **Removing organic matter**: Adsorption can be used to remove low concentrations of non-degradable organic compounds from groundwater.

Adsorption is a process that involves the binding of molecules in a liquid to the surface of a solid substance. The solid substance, known as an adsorbent, has a high internal surface area that allows for adsorption.

HAVAC industries

Adsorption Chillers

These systems use solid sorption materials, like silica gel, as adsorbents to cool chilled water. They can be powered by low-grade waste heat and are considered a more environmentally friendly alternative to conventional refrigeration technologies.

Air Quality Improvement

Adsorption is an effective and economical way to remove chemical pollution from HVAC systems. Activated carbon is a popular adsorbent for this purpose.

Adsorption systems have several environmental advantages over conventional systems:

- Lower greenhouse gas emissions.
- Adsorption systems use natural refrigerants or environmentally friendly alternatives, like water, which have a lower global warming potential.
- Higher energy efficiency.
- Adsorption systems can use waste heat to reduce energy consumption and reliance on fossil fuels.

- Reduced resource depletion.
- Adsorbent materials that can be regenerated and reused.

WHAT IS PRESSURE SWING ADSORPTION (PSA)

Pressure Swing Adsorption (PSA) is a technology used for separating gases in a mixture through an adsorption process. It relies on the differential adsorption behavior of gases on a solid adsorbent material under varying pressure conditions.

Some typical adsorbents used in air separation include

- **Activated carbon**: Activated carbon is a highly porous material with a large surface area. It can adsorb a wide range of gases, including nitrogen, oxygen, and carbon dioxide.
- **Molecular sieves:** Molecular sieves are crystalline aluminosilicates with small uniform pore sizes that can selectively adsorb certain molecules based on their size and polarity. They are often used in air separation processes to remove water and other impurities from the air.
- **Zeolites:** Zeolites are a type of aluminosilicate mineral with a porous structure that allows them to adsorb gases and liquids. They are commonly used in air separation applications to remove impurities and separate nitrogen and oxygen.
- **Silica gel**: Silica gel is a porous form of silicon dioxide that is often used as a desiccant to remove moisture from the air. It can also adsorb other gases and organic compounds.
- **Activated alumina**: Activated alumina is a porous material made from aluminum oxide that can adsorb a wide range of gases, including water vapor and carbon dioxide. It is commonly used in air separation processes to remove impurities and moisture from the air.

Details

- The gas mixture is fed into a vessel containing the solid adsorbent material, often in the form of beads or pellets.
- Under high pressure, the adsorbent material selectively adsorbs the target gas component while allowing the remaining gases to pass through.
- Once the adsorbent material is saturated with the target gas, the pressure in the vessel is reduced, causing the captured gas to desorb and release into a separate stream.
- The vessel is then purged with a small amount of waste gas to remove any remaining traces of captured gas, and cycle is repeated.

PSA technology is used in a variety of processes. Some common applications include:

For example

- **Air separation:** PSA systems are used to separate O2 and N2 from air for industrial processes and medical applications.
- **Hydrogen purification:** PSA is used to purify hydrogen gas for fuel cells and industrial processes.
- **Natural gas processing:** PSA systems can separate methane, ethane, and other gases from natural gas streams.

Challenges of PSA technology include

Energy consumption:

PSA systems require significant energy input to operate, especially in high-pressure applications.

Adsorbent degradation:

Over time, adsorbent material in the system may degrade due to repeated cycling, leading to reduced efficiency.

Performance limitations:

PSA systems may not be suitable for separating gases with very similar adsorption characteristics, limiting their applicability in certain gas separation processes.

PRESSURE SWING ADSORPTION: ROLE OF KINETIC DIAMETER

References: Wikipedia and Science Direct.

The pressure swing adsorption (PSA) process is based on the phenomenon that under high pressure, gases tend to be trapped onto solid surfaces, i.e. to be "adsorbed". The higher the pressure, more gas is adsorbed. When pressure is dropped, gas is released, or desorbed.

PSA can be used to separate gases in a mixture because different gases are adsorbed onto a given solid surface more or less strongly. For example, if a gas mixture such as air is passed under pressure through a vessel containing an adsorbent bed of zeolite that attracts nitrogen more strongly than oxygen, a fraction of nitrogen will stay in the bed, and the gas exiting vessel will be richer in oxygen than mixture entering. When bed reaches the limit of its capacity to adsorb nitrogen, it can be regenerated by decreasing the pressure, thus releasing the adsorbed nitrogen. It is then ready for another cycle of producing oxygen-enriched air.

Using two adsorbent vessels allows for near-continuous production of the target gas. It also allows a pressure equalisation, where the gas leaving the vessel being depressurised is used to partially pressurise the second vessel. This results in significant energy savings, and is a common industrial practice.

Role of kinetic diameter of adsorbate and adsorbent.

The kinetic diameter of an adsorbate and adsorbent plays a role in the following ways:

What is Kinetic Diameter?

Kinetic diameter is a measure of the size of a molecule and its likelihood of colliding with other molecules. It's different from atomic diameter, which is the size of an atom's electron shell

Pore Size

The kinetic diameter of an adsorbate affects the limiting pore volume of an adsorbent. If the adsorbate's kinetic diameter is larger than the pore width, it won't be able to access the pore.

Diffusion

The kinetic diameter of an adsorbate affects how quickly it diffuses into the adsorbent's pores. For example, oxygen molecules diffuse faster into carbon molecular sieve than nitrogen molecules because of their different kinetic diameters.

Adsorption Capacity

The kinetic diameter of an adsorbate affects the accuracy of the Dubinin–Radushkevich (D–R) equation in predicting adsorption capacity.

What is Dubinin—Radushkevich (D-R) Equation?

Dubinin showed the defining characteristic of a microporous material is its micropore volume rather than its surface area. The DR equation is widely used for linearizing the gas adsorption isotherms, and the micropore volumes can be obtained from the intercept of the DR plots.

CALCULATION OF AIR FLOW TO PRODUCE N_2 GAS BY PSA

To calculate the volume of nitrogen N_2 produced by a pressure swing adsorption (PSA) system, you can consider the following parameters:

Air: N_2 ratio

The ratio of air to nitrogen.

Flow rate:

The core parameter of the nitrogen generator, which is usually measured in volumetric flow in m^3/h or L/min.

Standard State

The volume flow is usually referred to the standard state, which is a temperature of 20°C, pressure of 0.101 MPa, and relative humidity of 65%.

Example

A N2 source produces 30.9 m^3/h at a 6 bar(g) air inlet and the air: N2 ratio is 3.4:1, then the air inlet flow requirement to the generator is 30.9 × 3.4 = 105 m^3/h.

(An air-to-nitrogen ratio of 3.4:1 is a ratio that can be used in the sizing and selection of nitrogen gas generators.

For example, if a nitrogen gas generator has an air-to-nitrogen ratio of 3.4:1 and produces 30.9 m^3/h, then the air inlet flow requirement to the generator is 105 m^3/h).

Steps

- Volume calculation.
- To express the volume in standard conditions use $P_1 V_1 = P_2 V_2$.
- Pressure to flow relationship.

Standard Cubic Metres per hour (Sm^3/h means air at 20 degrees C and a barometric pressure of 101.3 kPa absolute (dry air).

From this relationship one convert pressure into flow.

SUPERCRITICAL *vs.* SUPERHEATED STEAM

These two terms are often confused.

Supercritical State

Supercritical steam is a supercritical fluid, which is a substance with properties of both a gas and a liquid.

Supercritical steam is more efficient than superheated steam because it has a higher energy content per unit mass of water and can be used in more efficient steam-generating units.

Super critical steam is not superheated steam because they are two different thermodynamic states of steam.

Steam is typically produced by boiling water in a boiler, which generates saturated steam at a specific temperature and pressure. When steam is at its saturation temperature and pressure, it is known as saturated steam. If more heat energy is added to saturated steam, it becomes superheated steam.

Superheated steam has higher energy content than saturated steam.

More specifically.

Superheated Steam

Steam is created when a liquid evaporates and the remaining gas reaches a certain temperature. Superheated steam is created at pressure conditions below critical point.

Supercritical Steam

On the other hand, supercritical steam refers to steam that is at a critical pressure and temperature, where it exhibits properties of both a gas and a liquid. At this point, the distinction between liquid and vapor phases becomes less clear, and the properties of the steam become unique.

More Specifically

Supercritical Steam

Water is heated to a state where it's indistinguishable from a liquid or gas, and there's no clear distinction between the two. This state is above the critical point, which for water is 3206 psi and 706°F. At the critical point, the latent heat of steam is zero and its specific volume is the same whether it's liquid or gas.

In summary, superheated steam and supercritical steam are distinct thermodynamic states of steam, with superheated steam being steam that has been heated beyond its saturation temperature, and supercritical steam being steam that is at a critical pressure and temperature.

Thermodynamics of Supercritical *vs.* Superheated Steam

Supercritical steam has more energy than superheated steam because of elevated pressure.

Supercritical steam is denser than superheated steam. This tells a lot about their thermodynamics.

Preference of Supercritical Steam over Superheated Steam

The choice between supercritical steam and superheated steam depends on the intended application.

Supercritical Steam

Used for power generation and has several advantages, including:

- **Lower fuel consumption:** Supercritical steam can consume less fuel than other systems for a given output.
- **Lower carbon emissions:** Supercritical steam can produce lower carbon emissions than less efficient systems.

Superheated Steam

Higher temperature: Superheated steam has a higher temperature and lower density than other types of steam.

ENERGY CONTENT OF SUPERHEATED STEAM *vs.* SUPERCRITICAL STEAM

Supercritical steam is more efficient than superheated steam.

Supercritical Steam

Supercritical steam plants can achieve efficiencies of over 45%, which is higher than the 41–42% efficiency of conventional thermal electric power plants. Supercritical steam plants also require less coal per megawatt-hour, which leads to lower emissions and fuel costs.

Why and How?

This question was crossing my mind for last couple of days. I couldn't find any clear views from internet.

There are Two Thoughts

The first one is that because of very high temperature and lager temperature drop behot and cold reservoirs the thermodynamic efficiency (Carnot efficiency) of supercritical steam is certainly more than superheated steam.

The second thought is related to the thermodynamic properties of supercritical fluids and I quote the "surface tension of supercritical steam is zero" This gels well with fact that supercritical fluids are non-polar.

It means supercritical fluids have no intermolecular forces and there is the dominance of intermolecular repulsion in supercritical steam.

These strong repulsive forces between molecules, cause the supercritical steam molecules push against each other with greater force.

Explanation

Supercritical steam typically has a lower density compared to superheated steam. At supercritical conditions, steam exists as a single phase with properties of both liquid and gas, exhibiting properties such as a higher specific volume and lower density than superheated steam.

As a general comparison, the density of supercritical steam (at supercritical conditions of approximately 220.6 bar pressure and 374°C temperature) can be around 5–10 times lower than the density of superheated steam at corresponding temperature and pressure conditions.

For reference, the density of superheated steam at 220.6 bar pressure and 374°C temperature is approximately 5–10 times higher than the density of supercritical steam at the same conditions.

What Does This Tell Us?

This tells us supercritical steam has more specific volume than superheated steam?

What Does this Mean?

The dominance of repulsive forces present in supercritical steam provides more specific volume and that can lead to increased expansion and more PV work done compared to superheated steam.

This is the second reason why supercritical steam is more efficient for a power plant than superheated steam.

In summary therefore, supercritical steam has two advantages (1) because of high temperature and high temperatures difference between hot and cold reservoir, supercritical steam produces higher thermodynamic (Carnot efficiency) than superheated steam and (2) because of its larger specific volume supercritical steam produces more thermodynamic PV work. These two factors make supercritical steam more efficient than superheated steam.

UNIQUE PROPERTIES OF SUPERCRITICAL FLUIDS

References: Research Digest, Wikipedia and many.

Supercritical fluids (SCFs) have become of increasing significance in the past two decades for several reasons. However, it should be recognized that scf's have been a subject of research for some time.

Supercritical carbon dioxide and supercritical water have been investigated extensively as media for application in solvent-free chemistry processes and, therefore, suggesting to some people, green chemistry.

They have also been used in environmental remediation. SCFs offer opportunities to carry out novel chemistry and have been shown to be media for study of organic, inorganic, and organometallic reactions.

Supercritical fluid is a substance at a temperature and pressure above its critical temperature and pressure where it does not condense or evaporate to form a liquid or gas.

Supercritical fluids have properties between those of a liquid and gas.

Supercritical fluids exhibit unique properties due to their proximity to the critical point of their phase diagram.

At the critical point, the distinction between liquid and gas phases disappears, leading to unique thermodynamic properties that are distinct from both liquids and gases.

One important aspect is the density of supercritical fluids, which can be tuned by adjusting the temperature and pressure. At the critical point, the density of the supercritical fluid is close to that of a liquid, while its viscosity is closer to that of a gas. This makes supercritical fluids highly compressible, allowing for changes in density and solvation properties with small variations in pressure.

Additionally, supercritical fluids display enhanced mass transfer properties compared to liquids or gases.

Their high diffusivity and low viscosity enable efficient mass transfer, making them ideal for extraction processes and chemical reactions.

The unique thermodynamic properties of supercritical fluids at the critical point make them versatile solvents for various industrial applications.

They don't have surface tension.

More specifically, the properties of supercritical fluid include:

- **Density:** Lower density than liquids.
- **Viscosity:** Lower viscosity than liquids.
- **Diffusion:** Higher coefficient of diffusion than liquids.
- **Dissolution:** Can dissolve materials that are not normally soluble in a liquid or gas.

Mass and heat transfer: Faster rates of mass and heat transfer.

SUPERCRITICAL CO_2: A POTENTIAL SOURCE OF ENERGY FOR POWER GENERATION

Supercritical CO_2 has a unique low critical temperature and pressure. This gives supercritical CO_2 a huge advantage of low specific volume at critical point compared to steam and a much smaller footprint of the turbine. The high density and volumetric heat capacity of CO_2 concerning other working fluids make it more energy-dense, meaning that the size of most system components such as turbine and pump can be considerably reduced, which leads to a smaller plant footprint and possibly lower capital costs.

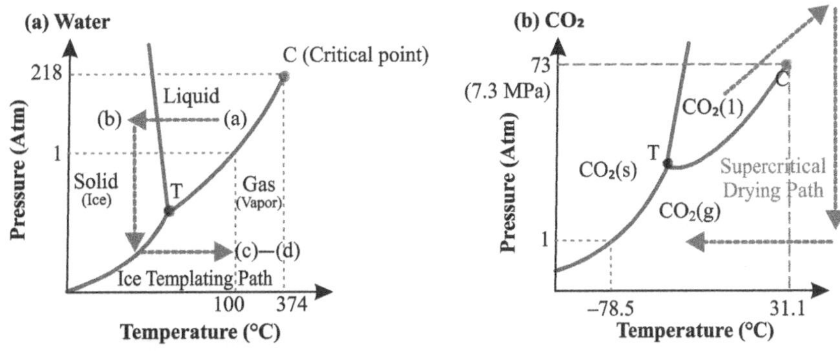

Figure 23: Phase Diagram of Water and Carbon-dioxide

The left-side image is the phase diagram of water and the right-side image is the phase diagram of CO_2.

Definitions

- **Critical point**: It is the endpoint of a phase equilibrium curve. It is the endpoint of the pressure-temperature curve in the phase diagram when a liquid and its vapor can coexist. At the critical point, defined by a critical temperature and a critical pressure, phase boundaries disappear. At higher temperatures than the critical temperature, the gas cannot be liquefied by pressure alone.
- **Supercritical fluid**: A supercritical fluid (SCF) is any substance at a temperature and pressure above its critical point, where distinct liquid and gas phases do not exist.

At the Molecular Lever Level

At STP water [molar mass 18 g/mol] is liquid and CO_2 is gas [STP is a temperature of 298.15 K (25°C, 77°F) and an absolute pressure of exactly 1 atm (101,325 Pa, 1.01325 bar)] Although carbon dioxide has a greater molecular mass than water, water is a liquid at room temperature while carbon dioxide is gas because of the strong intermolecular forces that hold water molecules together. Primarily, it is water's hydrogen bonds that keep it liquid at room temperature.

Boiling Point

The normal boiling point of a liquid is the temperature at which its equilibrium vapor pressure, *P*, equals 1 atm. For water, this is 100°C. For CO_2, the liquid phase does not exist at 1 atm, so instead of boiling, solid carbon dioxide sublimes at a temperature of –78.5 degc.

At Critical Point

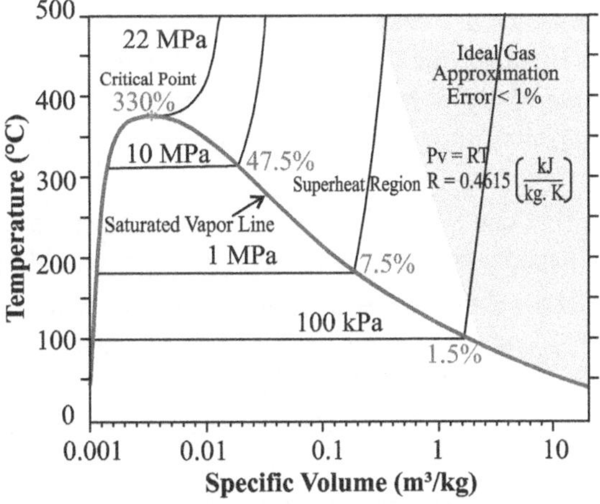

Figure 24: P-V Diagram for Steam

Gas densities of interest are marked with dotted green lines. The critical temperature is marked with a solid yellow line.

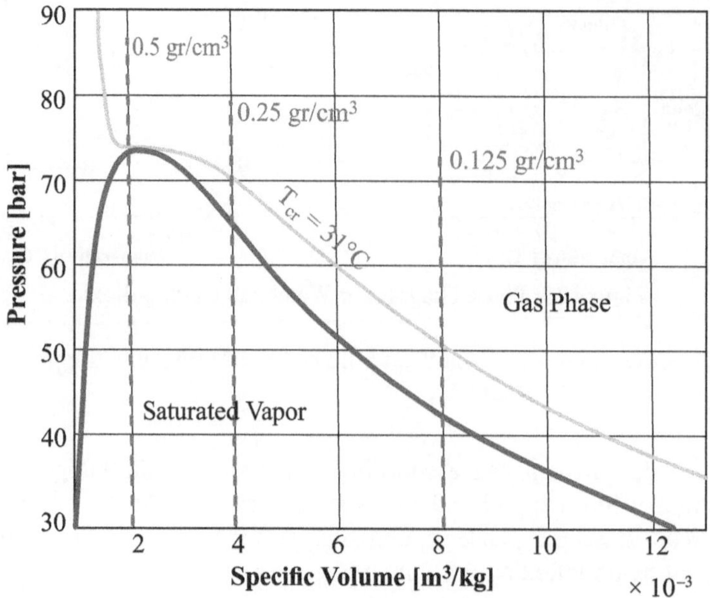

Figure 25: P-V Diagram of Carbon-dioxide

Specific volume at the critical point

Water [374°C, 22.1 MPa] is 0.003155 m^3/kg.

CO_2 [(31.0°C, 7.3773 MPa] is 0.002 m^3/kg.

At the critical point, CO_2 is denser than water by 63%.

Why CO_2 is denser than steam at the critical point?

From a thermodynamic perspective at its supercritical state, carbon dioxide is nearly twice [actually 63%] as dense as steam, resulting in a very high- power density. Supercritical carbon dioxide is easier to compress and requires less work energy than supercritical steam since it is more compressible than supercritical steam. A denser supercritical fluid has larger compressibility.

Why a denser supercritical fluid has larger compressibility?

At critical point, the density of a gas reaches its maximum value. At this point, the gas molecules are packed closely together, making it easier to compress them further. This is because the gas molecules are already in close proximity to each other, so additional pressure can easily reduce the volume of the gas even further. In other words, at critical point, the gas is already at its maximum density, so any additional pressure applied to it will have a more significant impact on its volume.

The net result is a simpler turbine that can be 10 times smaller than its steam equivalent. A steam turbine usually has between 10 and 15 rotor stages. A supercritical CO_2 turbine equivalent would have four.

Supercritical turbines would also be an attractive upgrade from steam systems aboard ships and submarines, producing the same power while occupying less space. Because they use carbon dioxide instead of water as their process fluid, these turbines would also work well in drought-stricken areas.

Supercritical CO_2 Power Cycle

In recent years, the use of CO_2 power cycles for waste heat recovery has gained attention. Such cycles are usually operated under trans-critical or supercritical conditions, due to the relatively low critical point of carbon dioxide. Its main advantages can be summarized as follows. First of all, the CO_2 cycles are suitable for the recovery of waste heat sources with a wide range of temperatures, while the steam-based Rankine cycle is only efficient in recovery of waste heat at high temperatures. In addition, the CO_2 power cycles occupy a smaller space as compared with the Steam Rankine Cycle, which has a large working fluid steam volume and therefore shows great potential for system downsizing and lightweight.

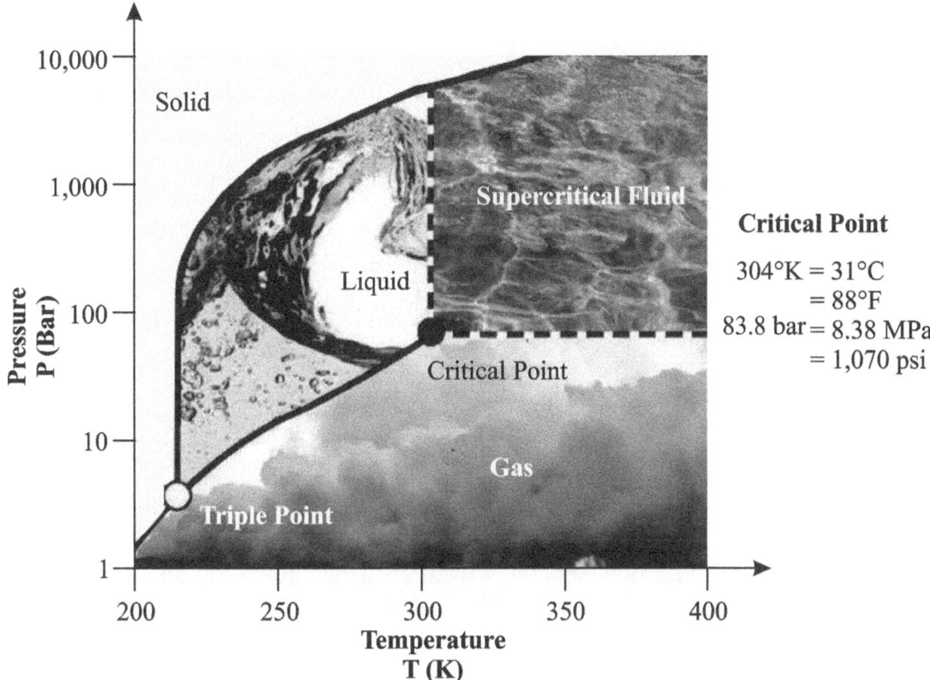

Figure 26: Phase Diagram of Carbon-dioxide

Just to repeat, CO_2 has a relatively low critical pressure of 7.4 megapascals (MPa) and a critical temperature of 31C, a consequence of this is that it can be compressed directly to supercritical pressures and readily heated to a supercritical state before expansion. The high density and volumetric heat capacity of CO_2 concerning other working fluids make it more energy-dense, meaning that the size of most system components such as turbine and pump can be considerably reduced, which leads to a smaller plant footprint and possibly lower capital costs.

COMPRESSOR: SURGE, STALL AND STONEWALLING MEANING AND WHY THEY OCCUR?

Surge, stall, and stonewalling are all undesirable conditions that can occur in compressors, leading to reduced efficiency, increased wear and tear, and potential damage to the equipment.

Surge: Surge is typically caused by operating the compressor at too low of a flow rate or by sudden changes in the system load.

Compressor surge is a form of aerodynamic instability in axial compressors or centrifugal compressors. The term describes violent air flow oscillating in the axial direction of a compressor, which indicates the axial component of fluid velocity varies periodically and may even become negative and can cause reverse flow.

This instability can lead to a backflow of compressed air, causing a decrease in pressure, and potentially damaging the compressor. This phenomenon can occur due to various factors, such as a sudden increase in the downstream pressure or a sudden decrease in the flow rate.

The term "low reverse pressure in surge" could refer to the situation where the pressure at the outlet of the compressor drops significantly during the surge event. This drop in pressure occurs as a result of the reverse flow of the compressed gas, which disrupts the normal operation of

Stall: Stall occurs when the airflow over the compressor blades becomes disrupted, leading to a decrease in the compressor's efficiency and capacity. This can result in a decrease in pressure and flow rate, as well as an increase in noise and vibration. Stall is often caused by operating the compressor at too high of a flow rate or by poor maintenance practices.

In thermodynamics, “stall” typically refers to a condition where the flow of air over an airfoil (such as a turbine blade, compressor blade, becomes separated from the surface of the airfoil, leading to a loss of efficiency.

Stall in a compressor or turbine typically occurs when the flow of air separates from the blades due to high angles of attack or other flow disruptions.

The separation of flow reduces the pressure difference across the airfoil, causing a decrease in lift (in the case of aircraft wings) or a decrease in pressure rise (in the case of compressors.

Stonewalling: Stonewalling occurs when the compressor is unable to achieve the desired pressure and flow rate, despite operating at full capacity. This can be caused by a variety of factors, including damage to the compressor blades, worn-out seals, or a malfunctioning control system. Stonewalling can lead to reduced efficiency, increased energy consumption, and potential damage to the compressor.

Stonewalling typically refers to the phenomenon where the flow separation becomes persistent and does not recover easily. This can happen in situations where the angle of attack or the flow conditions are such that the airfoil is unable to reattach the flow.

WHAT IS FLOW SEPARATION IN AXIAL COMPRESSOR?

Flow separation is a phenomenon that occurs when a fluid’s flow detaches from a surface, which can cause issues in compressors and other machinery:

Definition

Flow separation happens when a fluid’s flow separates from a surface due to an adverse pressure gradient. This occurs when pressure increases in the direction of the fluid flow, causing the fluid to decelerate and potentially change direction.

Is it boundary layer separation?

Yes, it is boundary layer separation.

The flow separation is a type of boundary layer separation that can occur at high Reynolds numbers.

It occurs when a boundary layer detaches from a surface and forms a wake. This can happen when a boundary layer travels far enough into an adverse pressure gradient, where the boundary layer’s speed relative to the surface reverses direction.

When it occurs

Flow separation can occur when flow slows down and pressure increases, such as when passing through a widening passage or the thickest part of a streamline body.

What it causes

Flow separation can cause an unsteady area with high turbulence and recirculation zones. It can also lead to the formation of vortical structures, such as the Kelvin-Helmholtz and von Karman, instabilities.

When it can happen

Flow separation can begin at a Reynolds 24.

Effects

In compressors, flow separation can cause stalling and vibrations in machinery blading, and increased losses.

COMPRESSOR: AERODYNAMIC ISSUES

Typical Doubts

Are surge, stall and stonewalling all caused by flow separation?

Surge, stall, and stonewalling are all aerodynamic issues that can occur in a compressor, but they do not all necessarily happen due to flow separation.

Surge is a condition where airflow through the compressor becomes unstable, leading to a rapid drop in pressure and flow rate. This can occur due to a disruption in the airflow, such as a sudden increase in back pressure or a change in operating conditions. While surge can sometimes be caused by flow separation.

Stall is another aerodynamic issue in compressors, where airflow separation occurs near the compressor blades, leading to a loss of efficiency and performance. This can happen when the compressor operates at high angles of attack or during rapid changes in operating conditions. Flow separation is a common cause of stall, but it can also be influenced by factors such as blade design and inlet conditions.

Stonewalling is a term used to describe a condition similar to surge, where airflow becomes unstable and pressure fluctuations occur. This can happen in compressors that are operating near their maximum performance limits, leading to a loss of efficiency and potential damage to the compressor blades. Stonewalling can be influenced by flow separation, but it can also be triggered by other factors such as inadequate cooling or inadequate inlet conditions.

Overall, while surge, stall, and stonewalling can all be aerodynamic issues in compressors, they are not always directly caused by flow separation. Other factors such as inlet conditions, operating conditions, and compressor design can also play a significant role in triggering these issues.

PRESSURE RATIO OF COMPRESSOR AND CONSEQUENCES

Pressure ratio is the ratio of the air total pressure (pt) exiting the compressor to the air pressure entering the compressor. This number is always greater than 1.0.

Pressure ratio and flow are important in compressors because they affect the compressor's operating range, efficiency, and stability:

What is Pressure Ratio?

The ratio of the discharge pressure to the inlet pressure is a key parameter in compressor design and selection.

Consequences of High-Pressure Ratio

A higher pressure ratio requires more horsepower, and can cause the gas to heat up to undesirable temperatures. It can also make the compressor more sensitive to changes in pressure, which can limit its operating range.

Operating Range

The operating range of a compressor is the space between the surge point and the choke point.

The surge point is when the flow reverses, which can destroy the compressor. The choke point is when the flow reaches Mach 1, and no more flow can pass through.

Stability

If the downstream pressure is increased beyond the maximum possible, the compressor will stall and become unstable.

Efficiency

Choke conditions cause a large drop in efficiency.

Details

Pressure ratio is a key factor in the efficiency and performance of a compressor.

Volumetric Efficiency

The pressure ratio, along with clearance volume, determines the amount of re-expansion that happens when the piston draws gas into the cylinder. This affects the compressor's volumetric efficiency, which is the ratio of the volume of gas entering the compressor to the compressor's displacement.

Discharge Temperature

Most gases increase in temperature when compressed, so the final outlet temperature is a concern. A high discharge temperature can cause internal components to fail due to thermal expansion or material degradation.

Horsepower

The required horsepower for a stage increases as the compression ratio increases.

Sensitivity to Pressure Changes

Higher compression ratios make the compressor more sensitive to changes in discharge and suction pressure.

Pressure Ratio and Surge

Compressor surge is directly related to the pressure ratio of a compressor. When the pressure ratio becomes too high, particularly at low flow rates, the compressor can become unstable and surge can occur. This is because the compressor is unable to maintain the necessary pressure difference between the inlet and outlet, causing the airflow to reverse.

Pressure Ratio and Stall

Stall in a compressor occurs when the airflow becomes separated from the blades of the compressor, resulting in reduced or disrupted airflow and decreased efficiency.

The pressure ratio of a compressor is a key factor in determining its operating limits and performance characteristics. As the pressure ratio increases, the compressor must work harder to pressurize the air, which can increase the likelihood of stall occurring.

WHY AND HOW GASES DEVIATE FROM IDEAL GAS BEHAVIOUR NEAR CRITICAL POINT?

Explanation

Near the critical point the assumptions of the ideal gas law are no longer valid: (1), Gas molecules have negligible volume and (2) Intermolecular interactions are negligible.

As a gas approaches its liquefaction point, the deviation from ideal behavior increases. This is because the volume will not be zero, even if the pressure increases, because the molecules occupy some space that cannot be compressed.

What do we know?

Ideal gas behavior:

Gases generally behave like ideal gases at high temperatures and low pressures.

Deviations at High Pressures

At high pressures, the finite particle volumes lower the actual volume available to the gas particles, which results in a higher pressure than the ideal gas value.

Compression and Cooling

Compression brings gas molecules closer together, and reduced temperature slows their random movement. This combination of compression and cooling leads to intermolecular interactions, which cause the molecules to move closer together and change state.

Application of van Der Waals Equation at Critical Point

The van der Waals equation applies to the critical point of a gas by defining the critical point as the point where the first and second derivatives of pressure with respect to volume are zero: $(\partial p/\partial v)\ T = 0$ and $(\partial_2 p/\partial V_2)\ T = 0$.

The critical point is the boundary between two different behaviors of the equation:

Above the critical point: The curves are monotonic.

Below the critical point: The curves exhibit maxima and minima.

The van der Waals equation of state is: $(P + aV_2)(V-b) = RT$.

The critical constants are related to the van der Waals constants as follows: $Vc = 3\ b$, $P\ c = a\ 27\ b\ 2$, and $T\ c = 8\ a\ 27\ R\ b$.

In summary, gases deviate from ideal gas behavior near the critical point due to the increased intermolecular forces that are present at high pressures and temperatures.

As a gas is compressed and heated towards its critical point, the molecules are forced closer together and begin to interact more strongly with each other. This causes deviations from the ideal gas law, leading to non-ideal behavior.

Near the critical point, the gas molecules are no longer able to move freely and independently of each other, as they do in ideal gases. Instead, they begin to exhibit more complex behavior, such as clustering and forming intermolecular bonds. This results in the gas behaving more like a liquid than a typical gas.

Additionally, at the critical point, the distinction between the liquid and gas phases becomes blurred, as the properties of both phases merge together. This can lead to significant deviations from the ideal gas law, as the gas no longer conforms to simple ideal gas behavior.

Overall, gases deviate from gas laws near the critical point due to the increased intermolecular forces and complex interactions that occur at high pressure and temperature conditions.

COMPRESSIBILITY FACTOR Z AT CRITICAL POINT

What Do We Know

The compressibility factor is the ratio of a gas's actual molar volume to its molar volume if it were an ideal gas at the same temperature and pressure.

The compressibility factor (Z) of a gas at its critical point is:

For an ideal gas: $Z = 1$

For a real gas: Z is between 0.20 and 0.30

For a van der Waals gas: $Z = 0.375$

How a Real Gas Different from van der Waals Gas?

Real Gases

Real gases are made up of molecules that occupy space and interact with each other. They don't follow the ideal gas law under all conditions, but only at high temperatures and low pressures. Real gases have mass, volume, and velocity. The volume of a gas is the volume of the container it's in. Real gas molecules have intermolecular interactions, called van der Waals interactions Real gas is one that does not obey gas laws under all conventional pressure and temperature conditions. Real gases obey gas laws only at high temperatures.

van der Waals Gases

The van der Waals equation of state is used to describe van der Waals gases. The equation includes two constants: a corrects for intermolecular forces, and b corrects for the finite size of molecules. The van der Waals equation approaches the ideal gas law as the values of these constants approach zero.

At the critical point, the gas and liquid are the same. As the temperature and pressure increase along the coexistence curve, the gas becomes more like a liquid and the liquid becomes more like a gas.

Just above the critical point, there is a range of pressure for which Z drops quite rapidly. At higher temperatures, the process is entirely gradual.

Derivation of Z at Critical Point

The van der Waals equation of state is given by:

$$(P + a/V^2)(V - b) = RT$$

where:

P = pressure

V = molar volume

T = temperature

R = gas constant

a and b are van der Waals constants dependent on the gas.

At the critical point, the first and second derivatives of Gibbs free energy with respect to pressure and volume vanish, i.e. $(\partial G/\partial P)\text{–}T = (\partial^2 G/\partial P^2)\_T = 0$.

From van der Waals equation, we have:

$$\partial G = V\,dP - S\,dT$$

$$(\partial G/\partial P)\_T = V = (RT/(P + a/V^2)) - S \times (\partial V/\partial P)\_T$$

Since $(\partial G/\partial P)\_T = 0$ at the critical point, $V = Vc$. Also, at the critical point, $(\partial V/\partial P)\_T = 0$. And $(\partial V/\partial P)\_T = -(RT/(P + a/V^2)^2)(2a/V^3) = 0$, which simplifies to $Vc = 3b$.

Plugging Vc = 3b into the van der Waals equation, we get:

$$Pc = a/(27b^2)$$

Substitute Pc and Vc back into the compressibility factor equation $Z = PVc/RTc$, we obtain:

$$Z = PcVc/(RTc) = (a/(27b^2))(3b)/[(8a/27b)Tc] = 3/8$$

Therefore, $Z = 0.375$ at the critical point. Factor Z at critical point.

What Does it Mean?

This indicates that there is a significant amount of work being done against intermolecular forces at the critical point. This work is necessary to overcome strong intermolecular attractions present in the substance and allow it to exist in a state where the distinction between gas and liquid phases becomes blurred to reach equal density.

Compressibility and compressibility factor are two close terms and often become confusing.

When discussing the properties of gases, it is important to understand the concepts of compressibility and compressible factor.

Compressibility refers to the ability of a gas to be compressed or reduced in volume under pressure. Gases that are easily compressed have high compressibility, while gases that are difficult to compress have low compressibility. The compressibility of a gas can be affected by factors such as temperature, pressure, and the type of gas molecules present.

On the other hand, the compressibility factor is a dimensionless quantity that accounts for the deviation of a real gas from ideal behavior. The compressibility factor is defined as the ratio of the molar volume of a real gas to the molar volume of an ideal gas at the same temperature and pressure. A compressibility factor of 1 indicates ideal gas behavior, while values greater or less than 1 indicate either compressible or non-compressible behavior, respectively.

For example, at high pressures and low temperatures, gases tend to deviate from ideal behavior and exhibit high compressibility. This can be seen in the case of carbon dioxide (CO_2) which has a compressibility factor greater than 1 at high pressures. On the other hand, at low pressures and high temperatures, gases tend to behave more like ideal gases and have low compressibility. For instance, helium (He) has a compressibility factor close to 1 at low pressures and high temperatures.

CHOKED FLOW

In simple language, choked flow occurs when the flow of a gas reaches a critical point where the velocity is so high that it begins to interact with sound waves in the fluid.

As the gas velocity increases and approaches the speed of sound (Mach 1), the sound waves become asymmetric and act as a barrier, limiting the flow from accelerating any further.

This phenomenon is known as flow choking. Essentially, it means that the flow is restricted and cannot increase beyond a certain point due to the interference of sound waves.

Why flow chokes at Mach 1?

Explanation

Why and when choked flow occurs?

As a gas accelerates from subsonic toward supersonic speed in the air, different types of wave phenomena occur. To illustrate these changes, at a stationary point ($M = 0$) sound waves are symmetric. The speed of sound is the same in all directions in a uniform fluid, so these waves are simply concentric spheres.

As the sound-generating point begins to accelerate, the sound waves “bunch up” in the direction of motion and "stretch out" in the opposite direction.

When the gas reaches sonic speed ($M = 1$), it travels at the same speed as the sound waves. At this stage, an infinite number of sound waves “pile up” ahead of the gas preventing any further increase in gas velocity or flow rate.

Role of Cp/Cv in choked flow

A gas with a small *Cp/Cv* meaning with relatively larger *Cv* (internal energy) can tolerate more pressure drop for choked flow.

Cv is the source of energy for the expansion of gas at the constriction:

Cv, or the internal energy of the fluid, provides the energy for fluid’ adiabatic expansion at the nozzle/orifice, and thus *Cv* is important in determining flow across a nozzle/orifice and choked pressure drop.

There is a wonderful relationship between *Cv* and choked flow. The more the *Cv*, the less *Cp/Cv*. This means more internal energy. The more the internal energy of a fluid the more it does adiabatic work or expand and therefore more prone to choking. A triatomic gas with gamma [*Cp/Cv*] = 1.33 has more chance to choke than a diatomic gas with gamma = 1.4.

Example

O_2 is a diatomic gas. It’s *Cp/Cv* is 1.4. The ratio of *Pu/Pd* for choked flow is 1.89 at 20 degc.

Pu is upstream pressure

Pd is downstream pressure

He gas is a monoatomic gas. It’s *Cp/Cv* is 1.67. The ratio of *Pu/Pd* for choked flow is 2.049 at 20 degc.

What does it mean?

It means He gas has smaller *Cv* or internal energy than O_2. This further means O_2 has higher internal energy than He gas to expand at a constriction. The *Pu/Pd* ratios show He can tolerate a larger pressure drop than O_2 for choked flow. This tells that O_2 is more prone to choked flow than He.

Role of pressure ratio at choked flow

The relation is Q is proportional to sqareroot of differential pressure at throat, *dP*.

He with a higher-pressure ratio than O_2 can tolerate more *dP* at choked flow.

THERMODYNAMIC ISSUES OF FUEL CELLS

Heat Management

Fuel cells operate at high temperatures, which can cause issues with heat management and thermal regulation. If the temperature is not controlled properly, it can affect the performance and longevity of the fuel cell.

Energy Balance

Fuel cells need to maintain a delicate balance between providing enough energy to power the system and preventing overheating. If the energy balance is not carefully managed, it can lead to inefficiencies and reduced performance.

Waste Heat Recovery

Fuel cells produce waste heat as a byproduct of their operation, and it is important to effectively recover and utilize this heat to improve overall efficiency. However, capturing and utilizing waste heat can be challenging and may require additional equipment and infrastructure.

Corrosion and Degradation

Fuel cells are susceptible to corrosion and degradation over time, due to the high temperatures, chemical reactions, and operating conditions they are subject to. This can lead to reduced performance, decreased efficiency, and potentially premature failure of the fuel cell system.

Water Management

Fuel cells require a controlled supply of water to function properly, as it is a key component in the electrochemical reactions that occur within the cell. Managing the water supply, distribution, and removal can be a complex process that requires careful monitoring and control to prevent issues such as flooding or drying out of the fuel cell.

Specific Issues

Cell Voltage Losses

Fuel cells have theoretical voltage outputs based on the Gibbs free energy change of the reaction occurring in the cell. However, in practice, the actual cell voltage is lower than the theoretical value due to various losses including activation overpotential, concentration overpotential, and resistance losses in the cell components.

Gibbs Free Energy Related Issues

The Gibbs free energy change of the electrochemical reactions in fuel cells determines the maximum thermodynamic efficiency of the cell. If the reactants and products are not at standard conditions (e.g., different concentrations, temperatures), the Gibbs free energy change and thus the cell voltage may deviate from the ideal values.

FUEL CELL EFFICIENCY: COMPLEXITY OF REALITY

Reference: Galen J. Suppes, Truman S. Storvick, in Science Direct.

An ideal fuel cell running on pure oxygen and hydrogen produces 1.229 volts of electromotive force to push electrons through the electrical devices when there is no load on the circuit (the switch is open). Since the chemical reactions of a fuel cell provide the same flow of electrons independent of the fuel cell efficiency, this voltage is used to compute the efficiency.

For example, when the fuel cell is connected to an external load, the "open circuit" voltage of the fuel cell is split between the resistance to anion flow in the cell and the resistance to electron flow in the external load.

If the fuel cell produces 0.615 volts when connected to a light bulb, the efficiency of the fuel cell with that load is 50% (0.6151.229×100).

The conversion efficiency improves when the external load is decreased. It takes careful design of the fuel cell to handle the variable power requirements of commuter car travel and maintain high efficiency.

At the start of the twenty-first century, the efficiency of a good hydrogen fuel cell running with atmospheric air rather than pure oxygen varied from about 45% at maximum design load to about 65% when operating at low power output corresponding to the "most efficient" load.

If electrical power is used to make the hydrogen, and this electric power is generated with the efficiency varying from 30% to 55%, this gives an overall fuel cell efficiency of 13.5–36%.

If the fuel cell operates on hydrogen produced by converting gasoline to hydrogen on the vehicle, the conversion of energy from gasoline to hydrogen is about 70–80% which gives an overall efficiency of 31.5–52%.

The technology to use gasoline to power a fuel cell car requires the technology development to reform the gasoline to hydrogen. The gasoline reformer and fuel cell cars must be road tested before they come to the dealer's showroom.

The reality for fuel cells in vehicles today is that they are less efficient than the best diesel engine. The current goal of an overall fuel cell efficiency of 70% is not realistic neglecting the energy loss during hydrogen production.

Operation at maximum efficiency requires a large fuel cell to keep the electrical current produced per unit area of the cell low.

A fuel cell does not respond to the quick starts and acceleration required for urban driving. The fuel cell power output can be "leveled out" by using a battery pack that provides surge power and is recharged during cruise and stop phases of the trip.

The gasoline powered hybrid cars on the market today provide an ideal test for an electric powered car. Design a fuel cell to replace the gasoline engine, the electricity generator, and battery in the hybrid and you have a fuel cell car.

FUGACITY: A SIMPLIFIED NOTE

Introduction

What is fugacity? In simple words, fugacity is a thermodynamic concept that takes into account the escaping tendency of a gas from a system or its deviation from ideal gas behavior. It represents an effective measure of the escaping tendency of a gas at a specific condition, accounting for both the pressure and non-ideal behavior of the gas. Fugacity is represented by the symbol "f" and is often used in thermodynamic calculations to describe the behavior of gases under non-ideal conditions. Unlike pressure, fugacity considers the effects of intermolecular forces, molecular size, and other factors that influence the escaping tendency of a gas. In the context of escaping tendency, fugacity provides a more accurate representation of the actual behavior of gases in a system, especially when dealing with non-ideal conditions.

Details

Fugacity is used in place of partial pressure in non-ideal gas calculations, offering a better representation of the escaping tendency or tendency to deviate from ideal behavior in real gases.

The fugacity of a gas is defined as the hypothetical pressure the gas would exert if it were ideal. In other words, it is a correction factor that accounts for the deviations of real gases from ideal behavior. The fugacity of a gas is denoted by the symbol f and is related to the partial pressure of the gas by the equation:

$$f = P \times \varphi$$

where P is the partial pressure of the gas and φ is the fugacity coefficient. The fugacity coefficient takes into account the non-ideal behavior of the gas and is typically less than 1 for real gases.

In other words, the relationship between fugacity (f) and pressure (P) in non-ideal gases is given by the above equation.

The fugacity coefficient (Φ) is a correction factor that accounts for the deviations from ideal gas behavior. It is a function of temperature, pressure, and the composition of the gas mixture.

The fugacity of a gas can also be related to the chemical potential of the gas by the equation:

$$\mu = \mu\text{^}o + RT\ln(f/P\text{^}o)$$

where μ is the chemical potential of the gas, μ^o is the standard chemical potential of the gas, R is the gas constant, T is the temperature, and P^o is the reference pressure. This equation shows how the fugacity of a gas is related to its chemical potential, which is a measure of the energy required to add a component to a system such as gas molecule, while keeping other components constant.

The chemical potential represents the potential energy of a substance in a system, and the fugacity is used to account for non-ideality in gases or liquids. The relationship between fugacity and chemical potential allows for the determination of the escaping tendency or the tendency of a substance to leave a mixture or phase based on its thermodynamic properties.

What is escaping tendency of real gases?

The escaping tendency of a gas is its tendency to expand or escape, which is related to its partial pressure. It's also known as fugacity. It's a property of a substance that describes its tendency to leave its thermodynamic state through a chemical or physical process

The escaping tendency of a gas is closely related to the decrease of the Gibbs energy of the system. The escaping tendency of gas molecules increases when pressure increases and the molecules almost touch each other. This is because the repulsive interactions between the molecules increase

When a real gas escapes, it can undergo a throttling process, which is characterized by a significant drop in pressure. The temperature of the gas may increase, decrease, or remain constant, depending on the fluid and its initial conditions.

What does fugacity tell us?

fugacity tells	
Phase	The phase with the lowest fugacity is the most favorable
Chemical potential	Fugacity is a measure of chemical potential
Equilibrium	When fugacities are equal, there is no net transfer of a chemical between compartments
Vaporization	Fugacity measures a component's tendency to vaporize from a liquid mixture
Absorption	Fugacity provides information on how easily a gas can be absorbed into materials

Fugacity and compressibility factor Z relationship

Fugacity can be determined from measurements of volume as a function of pressure at a constant temperature. The compressibility factor is directly related to the fugacity coefficient.

The compressibility factor is also directly proportional to volume. The strength of intermolecular forces changes the volume of a gas compared to an ideal gas, which affects the compressibility factor.

CHEMICAL POTENTIAL: A SIMPLIFIED NOTE

Introduction

Chemical potential is a form of energy. Its unit is Jule. It is expressed by symbol **μ.**

$$\mu = \partial G/\partial n.$$

μ is the partial derivative of the free energy G with respect to the amount of the species n.

According to **Wikipedia** the **chemical potential** of a species is the energy that can be absorbed or released due to a change of the particle number of the given species, e.g. in a chemical reaction or phase. The chemical potential of a species in a mixture is defined as the rate of change of free energy of a thermodynamic system with respect to the change in the number of atoms or molecules of the species that are added to the system. Thus, it is the partial derivative of the free energy with respect to the amount of the species, all other species' concentrations in the mixture remaining constant. When both temperature and pressure are held constant, and the number of particles is expressed in moles, the chemical potential is the **partial molar Gibbs free energy.**

Discussion

Chemical potential measures the change in thermodynamic potentials like Gibbs free energy with respect to the change in the number of atoms or molecules of the species that are added to the system. In a sense it measures potentials like internal energy change, enthalpy change, Gibbs free energy change and Helmowtz free energy change with respect to the change in the number of atoms or molecules of the species that are added to the system

The chemical potential of a species in a system is the rate of change in the free energy of the system when the number of molecules of that species changes. It's the partial derivative of the free energy with respect to the amount of the species.

Gibbs free energy

The Gibbs free energy is the maximum amount of work that can be extracted from a closed system at a fixed temperature and pressure. The chemical potential of a substance is the Gibbs free energy per mole of that substance

In a sense, chemical potential measures the energy change that happens when a system adds a molecule.

What does chemical potential do

Measures energy change

Chemical potential measures the change in energy when a particle is added to a system while keeping the entropy and volume constant.

It measures the energy a component contributes to a mixture when it moves from one phase to another.

It's a key concept in chemistry and thermodynamics that helps determine the stability of substances and their tendency to react chemically.

Determines stability

Chemical potential determines the stability of substances and their tendency to react chemically, transform into new physical states, or move to a different location.

Related to Gibbs free energy

Chemical potential is the slope of the free energy of a system with respect to a change in the number of moles of a specific component.

Central to thermodynamics of materials

Chemical potential is central to thermodynamics of materials because it can be used to obtain a material's thermodynamic properties at a given temperature and pressure.

Chemical potential is often considered the "potential" for a chemical reaction to take place. It's closely related to concepts like entropy, energy, and chemical equilibrium.

What is partial derivative?

A partial derivative is a derivative of a function that has multiple variables, where it measures how the function changes when only one of the variables changes:

Explanation

The partial derivative of a function with respect to a variable gives the rate of change of the function as that variable changes. To calculate a partial derivative, we treat all other variables as constants and take the first derivative of the function with respect to the variable of interest.

Symbol for partial derivative

The symbol for a partial derivative is the lowercase Greek letter delta, ∂

The Gibbs free energy equation has two variables: delta H and delta S. Delta H represents the enthalpy of a system, and delta S represents its entropy.

The Gibbs free energy equation is $\Delta G = \Delta G° + RT \ln Q$,

where:

ΔG : The free energy at any moment.

ΔGo : The standard-state free energy.

R : The ideal gas constant, which is 8.314 J/mol-K.

T : The absolute temperature in Kelvin.

lnQ : The natural logarithm of the reaction quotient.

The Gibbs free energy is a state function, which means it's defined in terms of thermodynamic properties that are also state functions.

It's a thermodynamic potential that's useful for phase transitions. The Gibbs free energy indicates how stable a system is, and provides insight into atomic bonding strength and alloy stability.

What role chemical potential play in fugacity?

Chemical potential plays a central role in fugacity, which is a measure of the difference between a gas's chemical potential in a system and its chemical potential in a hypothetical ideal-gas standard state.

Fugacity is a measure of chemical potential.

Fugacity is a convenient way to relate chemical potential to the composition of a system.

Fugacity is used to predict equilibrium.

When the fugacities of a chemical in two compartments are equal, the chemical is in equilibrium and there is no net transfer.

When the fugacities are unequal, the chemical will move from the higher fugacity compartment to the lower fugacity compartment.

Fugacity is a measure of chemical potential that's expressed as an "adjusted pressure". It's a convenient way to relate chemical potential to the composition of a system.

Fugacity is also a measure of a substance's tendency to escape from one compartment to another.

Chemical potential is a type of energy, while fugacity is a measure of chemical potential.

Chemical potential is the energy stored in the chemical bonds of a substance. For example, gasoline contains a large amount of chemical potential energy that is released when it is burned in a car engine.

Fugacity is a measure of how likely a substance is to escape from one compartment to another. Fugacity is proportional to concentration, while chemical potential is

Fugacity is related to chemical potential in the following ways.

Fugacity as a measure of chemical potential

Fugacity is a measure of the difference between a substance's chemical potential in a system and its chemical potential in its standard state.

Fugacity and equilibrium

When a system is at equilibrium, the chemical potential of each chemical species in the solution is the same as its chemical potential in the gas phase.

Fugacity is a function of the difference between a substance's chemical potential in a system and its chemical potential in its ideal gas standard state. The concept of fugacity was originally used to explain the non-ideal behavior of real gases, but it has since been generalized to calculate phase equilibrium.

Chemical potential can apply to all substances. For example, in a condensed phase (liquid or solid), the chemical potentials of the two phases are equal if the condensed phase is saturated.

Fugacity is a measure of how likely a substance is to move from one compartment to another. It's proportional to concentration and has units of pressure. Fugacity is the difference between a substance's chemical potential in a system and its chemical potential in an ideal-gas state at the same temperature.

Chemical potential is related to Gibbs free energy

The change in Gibbs free energy when the elements that make up one mole of a substance move from their standard states into a system as one mole of that substance. Chemical potential is related to the Gibbs energy for a mole of a pure substance.

Common between chemical potential and fugacity

Chemical potential and fugacity are both related to a substance's chemical potential and how it relates to a system's composition.

Relationship

Fugacity is a convenient way to relate chemical potential to a system's composition. Fugacity can be used to express the difference in chemical potential between two states of a substance.

Application

Fugacity can be used to predict how a chemical will redistribute between two compartments that are in contact. For example, if the fugacities of a chemical in two compartments are different, the chemical will move from the compartment with the higher fugacity to the compartment with the lower fugacity.

Effective pressure

The effective pressure of a real gas is lower than the pressure of an ideal gas because of intermolecular forces.

Real gases don't follow the laws of gases at all temperatures and pressures, and they have some volume. The particles of real gases attract each other, which creates a backward force that prevents them from colliding with the container's wall as efficiently as they would if there was no attraction.

Units

Fugacity is a thermodynamic property that's used to describe the effective pressure of a real gas, and is expressed in bars. Chemical potential is energy measured in Joules.

WHAT IS PARTIAL DERIVATIVE?

In thermodynamics many variables are multivariable functions therefore it is an important subject.

A partial derivative is a derivative of a multivariable function with respect to one of its variables while holding all other variables constant. It measures how the function changes with respect to one variable as the others remain fixed. The symbol of partial derivative is ∂.

In thermodynamics, there are several quantities that depend on multiple variables, such as temperature, pressure, volume, and entropy. Here are ten examples of partial derivatives in thermodynamics with explanations:

- $\partial T/\partial V$: This represents the partial derivative of temperature with respect to volume. It measures how temperature changes with respect to volume while keeping pressure constant. This is known as the coefficient of thermal expansion.
- $\partial P/\partial T$: This represents the partial derivative of pressure with respect to temperature. It measures how pressure changes with respect to temperature while keeping volume constant. This is known as the coefficient of thermal pressure.
- $\partial H/\partial T$: This represents the partial derivative of enthalpy with respect to temperature. It measures how enthalpy changes with respect to temperature while keeping pressure and volume constant.
- $\partial G/\partial T$: This represents the partial derivative of Gibbs free energy with respect to temperature. It measures how Gibbs free energy changes with respect to temperature while keeping pressure and entropy constant.
- $\partial S/\partial V$: This represents the partial derivative of entropy with respect to volume. It measures how entropy changes with respect to volume while keeping temperature constant.
- $\partial U/\partial V$: This represents the partial derivative of internal energy with respect to volume. It measures how internal energy changes with respect to volume while keeping temperature constant.
- $\partial P/\partial S$: This represents the partial derivative of pressure with respect to entropy. It measures how pressure changes with respect to entropy while keeping temperature constant.
- $\partial H/\partial P$: This represents the partial derivative of enthalpy with respect to pressure. It measures how enthalpy changes with respect to pressure while keeping temperature and volume constant.
- $\partial T/\partial S$: This represents the partial derivative of temperature with respect to entropy. It measures how temperature changes with respect to entropy while keeping pressure constant.
- $\partial G/\partial P$: This represents the partial derivative of Gibbs free energy with respect to pressure. It measures how Gibbs free energy changes with respect to pressure while keeping temperature and entropy constant.

THERMODYNAMIC POTENTIAL

A thermodynamic potential is a scalar quantity that represents the state of a thermodynamic system and has the capacity to do work. Thermodynamic potentials are also known as fundamental functions.

Definition

Thermodynamic potentials are a type of potential energy, which is the capacity to do work.

The four most important thermodynamic potentials are:

- **Internal energy:** The energy of a system's configuration, which is based on the kinetic and potential energy of its molecules.
- **Enthalpy**: The heat content of a system, which is a function of pressure and entropy.
- **Helmholtz free energy**: A thermodynamic potential.
- **Gibbs free energy**: A thermodynamic potential.

Thermodynamic potentials can be expressed using the variables pressure, volume, temperature, and entropy. They can't be measured directly, but computational chemistry can be used to predict them.

Purpose of thermodynamic potentials

Here are some purposes of thermodynamic potentials:

Calculate equilibrium results

Thermodynamic potentials are used to calculate the equilibrium results of chemical reactions that take place under constraints like constant pressure.

Design thermal systems

Thermodynamic potentials are used to analyze, design, and optimize thermal systems like engines and refrigerators.

Predict spontaneous changes

Thermodynamic potentials can be used to predict the direction of spontaneous changes.

Compute equilibrium compositions

Thermodynamic potentials can be used to compute equilibrium compositions.

How do thermodynamic potentials work?

Examples

The four fundamental thermodynamic potentials are internal energy U, enthalpy H, Helmholtz free energy F and Gibbs free energy G.

Equations

Internal Energy (U):

$$U = E + PV$$

Where U is the internal energy, E is the total energy of the system, P is the pressure, and V is the volume of the system. Internal energy represents the total energy stored in the system, including both its kinetic and potential energy.

Enthalpy (H):

$$H = U + PV$$

Where H is the enthalpy, U is the internal energy, P is the pressure, and V is the volume of the system. Enthalpy represents the total heat content of the system at constant pressure.

Helmholtz Free Energy (F):

$$F = U - TS$$

Where F is the Helmholtz free energy, U is the internal energy, T is the temperature of the system, and S is the entropy of the system. Helmholtz free energy represents the maximum reversible work that can be extracted from the system at constant temperature and volume.

Gibbs Free Energy (G):

$$G = H - TS$$

Where G is the Gibbs free energy, H is the enthalpy, T is the temperature of the system, and S is the entropy of the system. Gibbs free energy represents the maximum reversible work that can be extracted from the system at constant temperature and pressure.

Each of these potentials plays a key role in thermodynamics by providing a unique perspective on the energy and work capabilities of a system under different conditions.

CHEMICAL POTENTIAL AND FUGACITY

Chemical potential and Fugacity are two confusing and difficult concepts to grasp. An attempt has been made in this post to explain them in simple words as much as possible. Complex equations have been avoided as much as could be done. Equations have been explained whenever used.

The post covers [1] What does chemical potential mean? Why it is important? [2] Gibbs free energy and chemical potential [3] $\mu = G/N$ [4] Fugacity [5] Fugacity and chemical potential and [6] $\mu = \mu_0 + RTln\, f/P_0$.

A most simple definition of chemical potential

In simple words, the chemical potential of a chemical substance, μ, can be simply defined as the chemical energy (Uc) possessed by 1 mol of the substance. The chemical energy Uc is the same as the Gibbs free energy G. The unit of chemical potential is Gibbs [G].

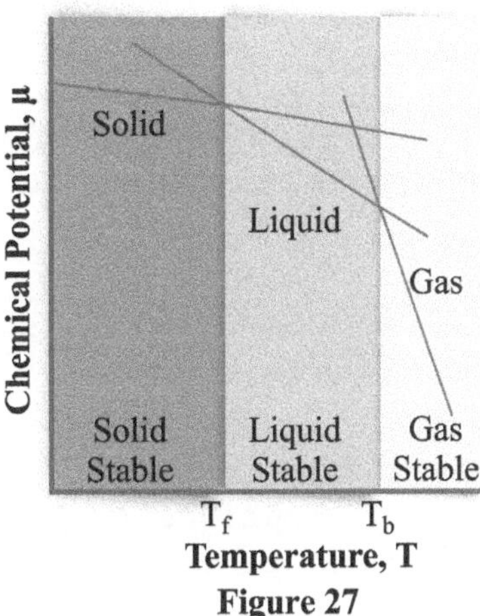

Figure 27

How do the chemical potentials compare for the vapour and liquid phases at, below, and above the boiling point in phase change?

Assume that you have a container of water.

Above 100°C, a water molecule that is in the vapour phase has a lower chemical potential than one that is in the liquid phase.

Water molecules in the liquid phase convert to gas, where their chemical potential is lower, so the water evaporates.

Below 100°C, the molecules in the liquid phase have a lower chemical potential, so the vapour condenses to liquid.

At 100°C the chemical potentials in liquid and vapour are the same, and the system is at equilibrium.

Chemical potential is a central concept in the thermodynamics of materials because all of the thermodynamic properties of a material at a given temperature and pressure can be obtained from knowledge of its chemical potential. Under the most common thermodynamic condition of constant temperature and pressure, chemical potential determines the stability of substances, such as chemical species, compounds, and solutions, and their tendency to chemically react to form new substances, to transform to new physical states, or to migrate from one spatial location to another.

A Most Simple Definition of Fugacity

Fugacity is a thermodynamic property of a real gas which if substituted for the pressure or partial pressure in the equations for an ideal gas gives equations applicable to the real gas.

In thermodynamics, the fugacity of a real gas is an effective partial pressure that replaces the mechanical partial pressure. It is equal to the pressure of an ideal gas which has the same temperature and molar Gibbs free energy as the real gas. Fugacities are determined experimentally or estimated from various models such as a van der Waals gas that are closer to reality than an ideal gas. The real gas pressure and fugacity are related through the dimensionless fugacity coefficient φ.

$\Phi = f/P$ [f is fugacity and P is pressure]

For an ideal gas, fugacity and pressure are equal and so $\varphi = 1$. Taken at the same temperature and pressure, the difference between the molar Gibbs free energies of a real gas and the corresponding ideal gas is equal to $RT \ln \varphi$. The fugacity is closely related to the thermodynamic activity. For a gas, the activity is simply the fugacity divided by a reference pressure to give a dimensionless quantity. This reference pressure is called the standard state and normally chosen as 1 atmosphere or 1 bar.

Chemical Potential

Let us begin with the basics: What is potential?

Definition of Potentials

A potential in physics is defined as the energy stored per unit of matter (i.e., a potential describes the corresponding potential energy intensity). Potential is an intensive property independent of system size.

Electrical Potential

For example, the familiar electric potential, φ, is the electrostatic potential energy, $U\ E$, with one unit (e.g., one Coulomb) of charge, q. In its simplest mathematical form, the electrical potential is defined as,

$\varphi = UE/q$

Therefore, electrical potential represents the electrical energy intensity. Conceptually, it is fundamentally different from electrical energy, which is proportional to the system size and is an extensive quantity.

Gravitational Potential

Another familiar example for potential is the gravitational potential or gravitational energy intensity, which is the gravitational potential energy per unit mass. Similarly, temperature, T, which can be considered the thermal potential or thermal energy intensity, is the thermal potential energy, UT, possessed by one unit of thermal matter, or one unit of entropy S.

$T = UT/S$

Pressure p can be considered as the mechanical potential. It is a measure of mechanical potential energy, Um, stored per unit volume V, or the mechanical energy intensity.

$P = -Um/V$

What is Chemical Potential [μ]?

The chemical potential of a species in a mixture is defined as the rate of change of free energy of a thermodynamic system with respect to the change in the number of atoms or molecules of the species that are added to the system.

What Does Chemical Potential Mean? Why it is Important?

In thermodynamics, chemical potential of a species is energy that can be absorbed or released due to a change of the particle number of the given species, e.g., in a chemical reaction or phase transition. The greater μ, the more active or "driven" the substance. The smaller μ, the more passive it is. Chemical potential [μ] tells us the tendency of a substance to [1] to react with other substance [2] to undergo a phase or physical change and [3] to reorganize or rearrange in space.

The extent of the above tendencies, that is the numerical value of μ - is determined by the nature of the substance, as well as - by its system environment, but neither by the nature of reaction associates nor the resulting products.

A reaction, transformation, redistribution, etc. can only proceed spontaneously if the tendency for the process is more pronounced in the initial state in other words the process has more energy in the initial state than in the final state.

The chemical potential μ is a measure of the magnitude of this tendency. We write μA or $\mu(A)$ to signify the potential of substance A. The greater μ, the more active or "driven" the substance. The smaller μ, the more passive it is.

Gibbs Free Energy and Chemical Potential

At equilibrium, the sum of the chemical potentials of the reactants is equal to the sum of the chemical potentials of the products. The Gibbs free energy change for the reactions, ΔrG, is equal to the difference between these sums and therefore, at equilibrium, is equal to zero.

Particles tend to move from higher chemical potential to lower chemical potential. In diffusion, molecules tend to move from areas with high concentration to low concentration, until eventually, the concentration is the same everywhere. The microscopic explanation for this is based on kinetic theory and the random motion of molecules. However, it is simpler to describe the process in terms of chemical potentials: For a given temperature, a molecule has a higher chemical potential in a higher concentration area and a lower chemical potential in a low concentration area. The movement of molecules from higher chemical potential to lower chemical potential is accompanied by a release of free energy. Therefore, it is a spontaneous process.

In direct analogy to electrical potential, gravitational potential, and mechanical potential, the chemical potential of a chemical substance, μ, can be simply defined as the chemical energy (Uc) possessed by 1 mol of the substance,

$\mu = Uc/N$, where N is the number of moles of the substance (i.e., the chemical potential of a chemical substance represents its chemical energy intensity in a given homogeneous system or at a given location of an inhomogeneous system).

The chemical energy Uc is the same as the Gibbs free energy G, and equation $\mu = Uc/N$ holds for each chemical component of a homogeneous multicomponent solution as well as for the entire homogeneous solution. The definition of chemical potential based on the above equation should be significantly easier to follow for most people.

Explanation

Molar Gibbs Free Energy and Chemical Potential

We consider the total internal energy, U, of a simple system by adding up the thermal, mechanical, and chemical energy from Equations $T = UT/S$, $P = -Um/V$, and $\mu = Uc/N$

We can write, $U = UT + UM + UC = TS–PV + \mu N$

This can be written as $U–TS + PV = \mu N$

$$U–TS + PV = G$$

Therefore, $G = \mu N$

Gibbs defined a simple system as a system without considering the surface, electric, magnetic, and non-hydrostatic mechanical energy contributions.

$$\mu = G/N$$

Therefore, the chemical potential of a substance introduced by Gibbs is simply the Gibbs free energy or chemical energy per mole of that substance (i.e., the molar Gibbs free energy is precisely the chemical potential).

Fugacity

Fugacity is the state function of matter at a fixed temperature. It only becomes useful when dealing with substances other than an ideal gas. For an ideal gas, fugacity is equal to the pressure. In the real world, though under low pressures and high temperatures some substances approach ideal behaviour, no substance is truly ideal, so we use fugacity not only to describe non-ideal gases but liquids and solids as well. The fugacity coefficient is defined as the ratio of fugacity/pressure. For gases at low pressures (where the ideal gas law is a good approximation), fugacity is roughly equal to the pressure. Thus, for an ideal gas, the ratio $\varphi = f/P$ between fugacity f and pressure P (the fugacity coefficient) is equal to 1. This ratio can be thought of as 'how closely the substance behaves like an ideal gas,' based on how far it is from 1. For a given temperature T, the fugacity f satisfies the following differential relation:

$$d\ln[f/f_0] = dG/RT = VdP/RT$$

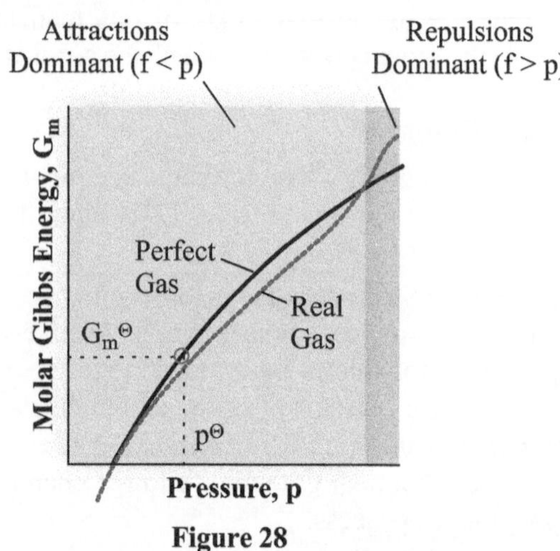

Figure 28

where G is the Gibbs free energy, R is the gas constant, is the fluid's molar volume, and f_0 is a reference fugacity which is generally taken as that of an ideal gas at 1 bar. For an ideal gas, when $f = P$, this equation reduces to the ideal gas law. Thus, for any two mutually isothermal physical states, represented by subscripts 1 and 2, the ratio of the two fugacities is as follows:

$$f_2/f_1 = \exp\left[1/RT \int (G_1 \text{ to } G_2)\, dG\right] = \exp\left[1/RT \int (P_1 \text{ to } P_2)\, VdP\right]$$

LHS Image: The chemical potential of a real gas. As p approaches 0, m coincides with the value for a perfect gas (shown by the pale line). When attractive forces are dominant (at intermediate pressures), the chemical potential is less than that of perfect gas and the molecules have a lower 'escaping tendency'. At high pressures, when repulsive forces are dominant, the chemical potential of a real gas is greater than that of a perfect gas. Then the `escaping tendency' is increased.

Important Concepts

In order to have a definition of chemical potential similar to the ideal gas case, we define $\mu\, v\, (T, P) = \mu\, 0\, (T) + RT \ln (f v\, (T, P))$, where $f v$ is called the fugacity which depends on the temperature and pressure.

Fugacity (fi) is the pressure value needed at a given temperature to make the properties of a non-ideal gas satisfy the equation for an ideal gas, i.e.,

$$fi = Yi\, Pi$$

where Yi is the fugacity coefficient and Pi is the partial pressure for component i of the gas. For an ideal gas, $Yi = 1$

$dG = VdP – SdT$ (The temperature is contstant so $SdT = 0$)

$Gpf = Gpi + \int VdP$ (integrate from pi top f)

For an ideal gas

$$Gpf = Gpi + RT\int (1/P)dP$$

$$Gpf = Gpi + RT{*}\ln (pf/pi)$$

If $pi = p^o$

$G(p) = G^o + RT^* \ln (P/P^o)$ Gibbs Energy for ideal gas

$G = G^o + RT^* \ln(f/P^o)$

$f = øP^o$

Fugacity and Chemical Potential

Refer to the following important equations

We have, $d\mu = VdP$

$F = U - TS, H = U + PV,$

$G = H - TS = F + PV$

Therefore $dG = -SdT + VdP + \mu dN$

The chemical potential for one species is the specific Gibbs free energy!

$\mu = G(T, P, N)/N$ we find that:

$d\mu = -sdT + vdP$

For every pure substance, we have the relation $dG = -SdT + VdP$ for Gibbs free energy and we can integrate this expression remembering the chemical potential is a function of T and P. We must also set a reference state. In this case, for an ideal gas, the only reference state will be the pressure, and we set $P = 1$ bar.

$\int (\mu_0 \text{ to } \mu)\, d\mu = \int (P_0 \text{ to } P)\, VdP],$

Now, for the ideal gas $V = RT/P$

$\mu - \mu_0 = \int (P_0 \text{ to } P)\, RT/P\, dP = RTLn\, P/P_0$

Reordering we get, $\mu = \mu_0 + RTLn\, P/P_0$

This gives the chemical potential for an ideal gas in an isothermal process, where the reference state is $P = 1$ bar.

For a real gas, we cannot calculate $\int(P_0 \text{ to } P)\, VdP$ because we do not have a simple expression for a real gas' molar volume. We do not know molecular interactions. We would like the expression for a real gas' chemical potential to be similar to the one for an ideal gas. We can define a magnitude, called fugacity so that the chemical potential for a real gas becomes $\mu = \mu_0 + RT\ln f/f_0$ with a given reference state. We can see that for an ideal gas, it must be $f = P$.

But for $P \rightarrow 0$, every gas is an ideal gas. Therefore, fugacity must obey the limit equation.

Limit $\rightarrow P = 0, f/P = 1$

We determine f by defining a function

$\Phi = [PV - RT]/P$, φ is fugacity coefficient

We can obtain values for Φ experimentally easily by measuring V, T, and P. (note that for an ideal gas, $\Phi = 1$

From the expression above we have, $V = RT/P + \Phi$

We can then write, $\int (\mu_0 \text{ to } \mu)\, d\mu = \int (P_0 \text{ to } P)\, VdP] = \int (P_0 \text{ to } P)\, RT/P\, dP + \int (P_0 \text{ to } P)\, \Phi dP$

Since the expression for an ideal gas was chosen to be $\mu = \mu_0 + RT\ln f/f_0$, we must have,

$\mu_0 + RT\ln f/f_0 = RT\ln P/P0 + \int (P_0 \text{ to } P)\, \Phi dP$

The fugacity coefficient will then verify

$\text{Ln}\, \Phi = 1/RT + \int (0 \text{ to } P)\, \Phi\, dP$

The integral can be evaluated via graphical integration if we measure experimentally values for φ while varying P. We can then find the fugacity coefficient of a gas at a given pressure P and calculate, $f = \varphi P$.

The reference state for the expression of a real gas' chemical potential is taken to be "ideal gas, at $P = 1$ bar and work T". Since in the reference state the gas is considered to be ideal.

(it is a hypothetical reference state), we can write that for the real gas

$$\mu = \mu_0 + RT\ln f/P_0$$

HOW COMPRESSIBILITY RELATES TO THE FUGACITY OF A GAS?

Fugacity is the escaping tendency of a real gas into its condensed form. Fugacity can be defined as a measure of the tendency or willingness of a substance to escape or leave a particular phase. It takes into account factors such as the pressure, temperature, and composition of the substance as well as the characteristics of the phase it is in. It is related to compressibility, entropy, free energy, and inversion temperature in that these factors can affect the fugacity of a substance.

Explanation

Real gas is more compressible than ideal gas.

Real gases have a non-zero volume due to the size and intermolecular forces between the gas molecules. This can cause them to occupy less space and hence be more compressible than ideal gases.

Real gases have compressibility factor $Z < 1$ [generally]

Real gases have a compressibility factor (Z) that is generally less than 1, which means they are more compressible than ideal gases. This is because the intermolecular forces between real gas molecules allow them to be squeezed together more easily.

Real gases have higher entropy

Real gases tend to have higher entropy than ideal gases due to the additional degrees of freedom associated with intermolecular interactions and the non-zero volume of the molecules.

Real gases and free energy

At the same temperature and pressure, real gases have less free energy compared to ideal gases because extra work is needed to overcome intermolecular attractions. The intermolecular attractions in real gases cause the gas molecules to be held closer together than in ideal gases. This extra attraction means more work is required to break the bonds between the gas molecules when the gas is compressed or expanded. This results in a decrease in the free energy of the real gas.

Phase change tendency

The escaping tendency of gas from the gas phase to the liquid phase depends on the difference in free energy between the gas and liquid phases. Real gases typically have less escaping tendency than ideal gases because they have lower free energy and higher entropy.

Fugacity effects

The difference between real and ideal gas-free energy is related to the fugacity of the gas, which is a correction factor that accounts for the deviation of the gas from ideal behavior. The fugacity reflects the escaping tendency of a real gas and is related to the difference in free energy between the gas and liquid phases.

In general, the fugacity of a real gas is lower than that of an ideal gas due to the attractive forces between gas molecules.

Still, You Might Ask for More Explanation

Let us take two gases N_2 and NH_3 and explain the role of compressibility, enthalpy, entropy, free energy and relate them to fugacity.

Compressibility

NH_3 has stronger intermolecular attractions than N_2 due to its polarity, hence requires more work against intermolecular forces to compress. Therefore, at the same temperature and pressure, N_2 is expected to be more compressible than NH_3.

Entropy

NH_3 has more entropy due to its complex molecular structure than N_2.

Enthalpy

NH_3 being polar has more potential energy of enthalpy than N_2.

Free Energy

NH_3 has more free energy than N2 at the same temperature despite more entropy because it is a polar molecule with a lot of potential energy.

Fugacity

Therefore, NH_3 has a greater capacity to work and escape from gas to liquid compared to N_2 [same temperature and pressure]. This means that, under these conditions, NH_3 would have a higher fugacity than N_2.

JOULE THOMSON EXPANSION (REFRIGERATION) *vs.* ISENTROPIC EXPANSION (TURBINES)

Joule Thomson Expansion

In Joule Thomson expansion, the process is characterized by sudden expansion through tiny pores of the porous plug leading to a significant reduction in kinetic energy and conversion to potential energy. The process is adiabatic.

In an isenthalpic process, such as when high-pressure gas is forced through small pores downstream, very little work is involved, as the focus is on the conversion of kinetic energy to potential energy. This is what gives JT cooling at constant enthalpy

Explanation

From 1st law of thermodynamics dU = delta Q - delta W. $Q = 0$ in adiabatic process therefore, dU = : delta W.

Practically, *JT* is a switch between *KE* and *PE*. So practically, dU (change in internal energy) is near zero in *JT* effect and that makes near zero work (W) in *JT*. This eventually, makes *JT* isenthalpic process.

Due to high frictional losses, as the gas is forced through tiny pores of about 100–150 micrometres of porous plug, Joule Thomson expansion is not internally reversible or isentropic.

The focus in Joule Thomson expansion is on the enthalpy of the gas, with no change in enthalpy during the process, as it is isenthalpic.

Total energy is kept constant in Joule Thomson expansion through the conversion of kinetic energy to potential energy.

Finally, one can conclude that the sum of internal energy and work in isenthalpic expansion across the porous plug up and downstream remains constant, $U_1 + W_1 = U_2 + W_2$ or $H_1 = H_2$ (H is enthalpy).

Isentropic Expansion: (Turbine)

Isentropic expansion involves a gradual expansion of a high-pressure gas.

The process is adiabatic and idealized as internally reversible, with no heat transfer and no change in entropy.

In isentropic expansion, the change in enthalpy dH does the mechanical work.

In isentropic process since the expansion of gas is gradual it is idealized as a frictionless expansion and no energy loss. An expander is designed such that the expansion of gas in the expander does not generate friction. This is why an isentropic process is idealized as internally reversible. It is an assumption that an isentropic process doesn't generate entropy either externally or internally.

In summary, while both Joule Thomson expansion and isentropic expansion involve adiabatic processes, there are key differences in their characteristics, reversibility, and focus on enthalpy or entropy.

ADIABATIC PROCESS *vs.* ISENTHALPIC PROCESS *vs.* ISENTROPIC PROCESS

These three terms may often be confusing.

Here is an explanation

Equation for adiabatic process is PV^n = Constant. n is adiabatic index expressed as Cp/Cv

Equation of isenthalpic process is $H_1 = H_2$, $U_1 + W_1 = U_2 + W_2$

Equation for isentropic process is $dS = 0$.

Adiabatic Process

It's a process that is heat impermeable, $dQ = O$. No heat can enter or exit the process.

Adiabatic process uses its own internal energy to perform work since they have no external supply of energy.

Adiabatic work can be both – W which is expansion or + W which is compression. Adiabatic processes are externally reversible since they can't transfer heat to surroundings.

Isenthalpic Process

It is an adiabatic process, in addition, it is a constant enthalpy process. We can define isenthalpic process as a special type of adiabatic process in which a flowing real gas is suddenly throttled. The sudden reduction in kinetic energy produces cooling at constant enthalpy. Such processes have very little thermodynamic work involved in the process.

Isenthalpic process is an irreversible process. The gas when forced through tiny openings during throttling generate friction. The frictional heat at molecular level, when two molecules rub against each other partly get dissipated and partly goes in the excitation of electrons that can't be reversed.

Isentropic Process

This is also a special type of adiabatic process where a gas expands without generating any friction. The expander is so designed that the expanding gas in the expander does not produce friction. Such processes are called internally reversible since they don't produce entropy internally in addition to being externally reversible because they are also adiabatic process in nature.

TURBOEXPANDER [TE] AND JOULE THOMSON [JT] EXPANSION

In one line, TE is idealized as an adiabatic reversible process while JT is approximated as an irreversible adiabatic process. In JT skin friction between the flowing gas and the non-moving surface of the valve, pores of the porous plug generate irreversible heat which cannot be reversed. . This heat, more precisely, goes to excite the molecules of the valve or the material of the porous plug causing the excitation of electrons that cannot be reversed. In the case of TE, its design eliminates friction losses. This makes TE an internally reversible adiabatic isentropic [no entropy] process while the irreversibility in the JT due to frictional loss of heat makes JT just an adiabatic process with an overall much less efficient cooling process than TE. Both TE and JT are approximated as adiabatic process.

Turbo-expander [TE]

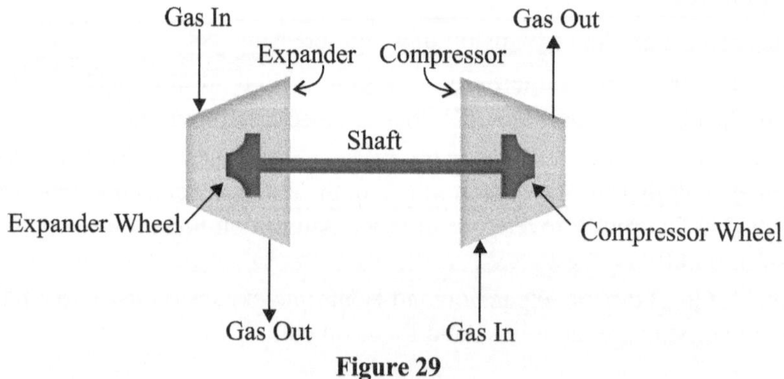

Figure 29

Turbo-expander is a centrifugal or axial-flow turbine, through which high-pressure gas is expanded to produce work that is often used to drive a compressor or generator. Because work is extracted from the expanding high-pressure gas, the expansion is approximated by an isentropic process (i.e., a constant-entropy process), and the low-pressure exhaust gas from the turbine is at a very low temperature, −150°C or less, depending upon the operating pressure and gas properties.

Turbo-expanders handle a wide range of services, from those involving cold gases (say, –270°C) to hot gases (above 350°C). Typically, the use of a turbo-expander makes sense only for relatively large gas flows.

Typical applications of Turboexpander

Turboexpander for Power Generation

As a very rough indication, units currently in operation range in power ratings from about 25 kW to around 25 MW. The work produced by the gas expanding through the turboexpander is available at room temperature. This work may be absorbed as heat into a cooling loop. The heat is then recovered for use elsewhere. In some large cryo-plants, the turboexpanders may be on the same shaft as compressors elsewhere in the cycle. The work produced by the gas expanding through the turboexpanders in this case then helps drive the compressors.

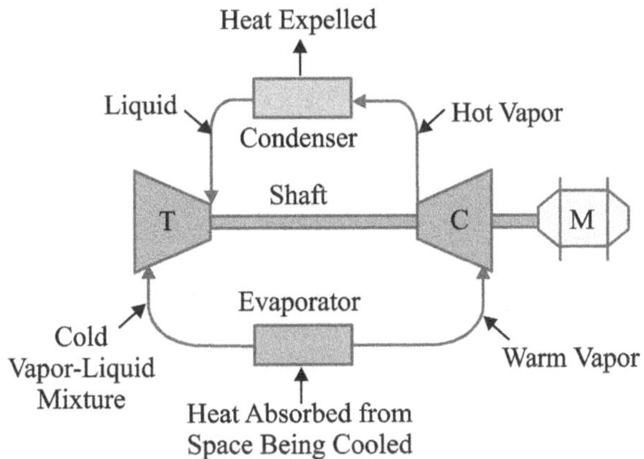

T = Expansion Turbine, C = Compressor, M = Motor

Figure 30

The use of turboexpanders in cryogenic plants has been the result of several technological advances. The development of sophisticated computational fluid dynamic models coupled with computer-aided design and manufacturing has allowed the design and construction of efficient turboexpanders operating at cryogenic temperatures. The development of reliable rotating bearing systems that function at high speeds and cryogenic temperatures has also been necessary for the successful use of turboexpanders.

Hot Gas Applications

Lots of plants have streams of high-pressure hot processes or waste gases that require cooling before further processing or disposal. Not infrequently, sites opt for turbo-expanders to recover useful work from these streams.

Cold Gas Services

Many turboexpanders find use in low-temperature, refrigeration, and cryogenic services. Such turbo-expanders primarily serve to efficiently reduce the temperature in a high-pressure gas stream. The expansion causes the gas to cool dramatically while providing mechanical energy to rotate equipment to do useful work. Some configurations couple the turbo-expander to a compressor, with the generated work used for the compression of the gas in the process. Sometimes, the turbo-expander and compressor are packaged in a single unit on a single shaft.

Joule Thomson expansion: JTE describes the temperature change of a real gas or liquid (as differentiated from an ideal gas) when it is forced through a valve or porous plug while keeping it insulated so that no heat is exchanged with the environment. This procedure is called a throttling process or Joule–Thomson process. JTE is considered to be an isenthalpic process. When the stream of a real gas is forced through a valve or porous plug while keeping it insulated [image above] so that no heat and work are exchanged with the environment, there is a change in the temperature of the stream of gas. An isenthalpic process works at constant *U* + *W* [*U* is internal energy and *W* is work] with no change in enthalpy. At room temperature, all gases except hydrogen, helium, and neon cool upon expansion. These three gases experience the same effect but only at lower temperatures.

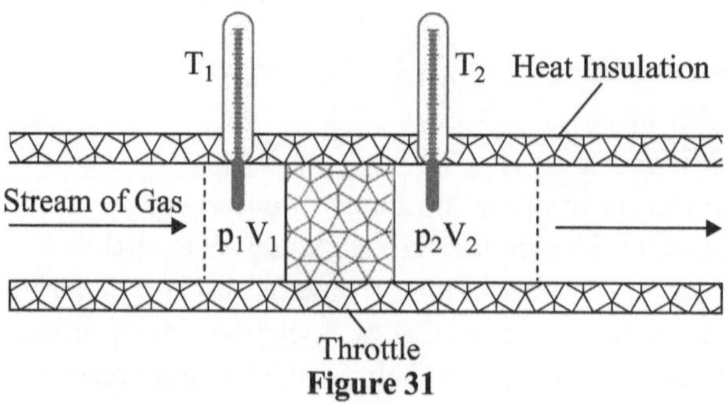

Figure 31

The simple answer is that Turboexpander does not produce entropy, whereas *JT* does. *TE* is an adiabatic process that is internally reversible. As a result, TE has more energy to expand and perform mechanical work. *TE* produces more cooling as a result.

Fundamentals

Gases are a phase of matter where all the atoms or molecules are moving independently of each other all the time. All these molecules move at different speeds but if we take the average speed of all the molecules; we get the temperature of the gas. *KE* [average] = 1/2Kb *T* [Kb is Boltzmann's constant and *T* is temperature]. If a particular gas has a large number of fast molecules, the gas is hot. On the other hand, if a particular gas has mostly slow molecules, then the gas is cold.

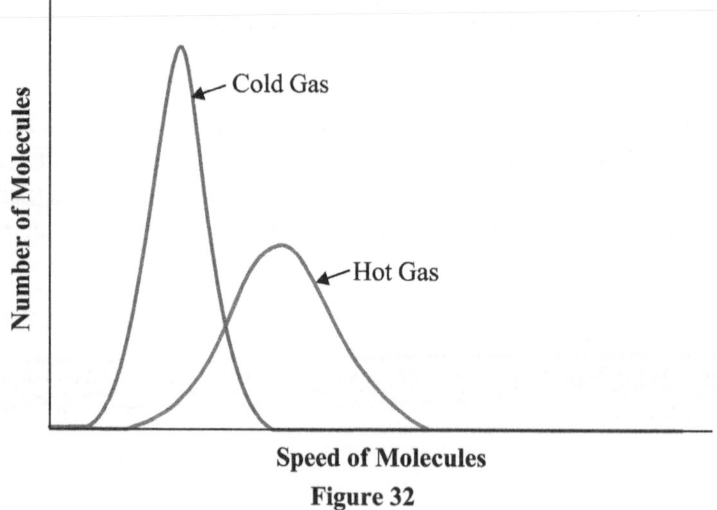

Figure 32

The average speed of a gas is also called the kinetic energy of the gas. When you combine the average speed for each molecule with all the molecules in the gas you get heat. Heat is the average speed of a single gas molecule multiplied by all the molecules in the gas. Heat can be seen as the total amount of energy, *H* [enthalpy]. $dH/dT = Cp$, [*Cp* is the heat capacity of the gas] of all the molecules in a certain gas.

At the fundamental level, every gas has some kinetic energy, according to kinetic theory, "the average kinetic energy of gas molecules depends upon the absolute temperature. When a gas expands, the gas works to overcome the intermolecular forces of attraction (it implies that the gas is spending its energy). The change in the sum of internal energy and work, U+W across the porous plug up and downstream with one compensating the other keeping the sum total which is enthalpy produce isenthalpic cooling in JT process.

Key points of distinction between TE and JT

[1] Both JT and TE [turboexpander] are adiabatic processes.

[2] JT is an irreversible process due to huge frictional irreversible energy losses as heat converts to work when the gas is throttled.

[3] TE is also an adiabatic process.

[4] TE does not generate entropy; which means TE is an internally reversible process.

[5] Because there are no entropy losses in TE, there is more energy available for gas to work and expand, resulting in more cooling.

[6] With the same base value, the output pressure in TE is much smaller because it does more work and produces more cooling.

[7] JT is more economically viable at a low gas flow rate. A normal throttling pressure-reducing valve does not have to withstand the stresses and temperature differences imposed on a cryogenic valve.

[8] TE offers fewer economic advantages below 10 MMScfd and lose efficiency below 5 MMScfd for ethane recoveries of 10 to 30%. JT expansion may be sufficient for this job.

[9] JT is insensitive to variable flow rates while the turboexpander loses efficiency when operating off the design rates. Isentropic expansion in turboexpander makes it much more effective than JT. JT is less costly than turboexpander. Turboexpander is very sensitive to feed gas quality while JT is flexible regarding feed gas quality.

Cooling curve of JTE *vs.* turboexpander

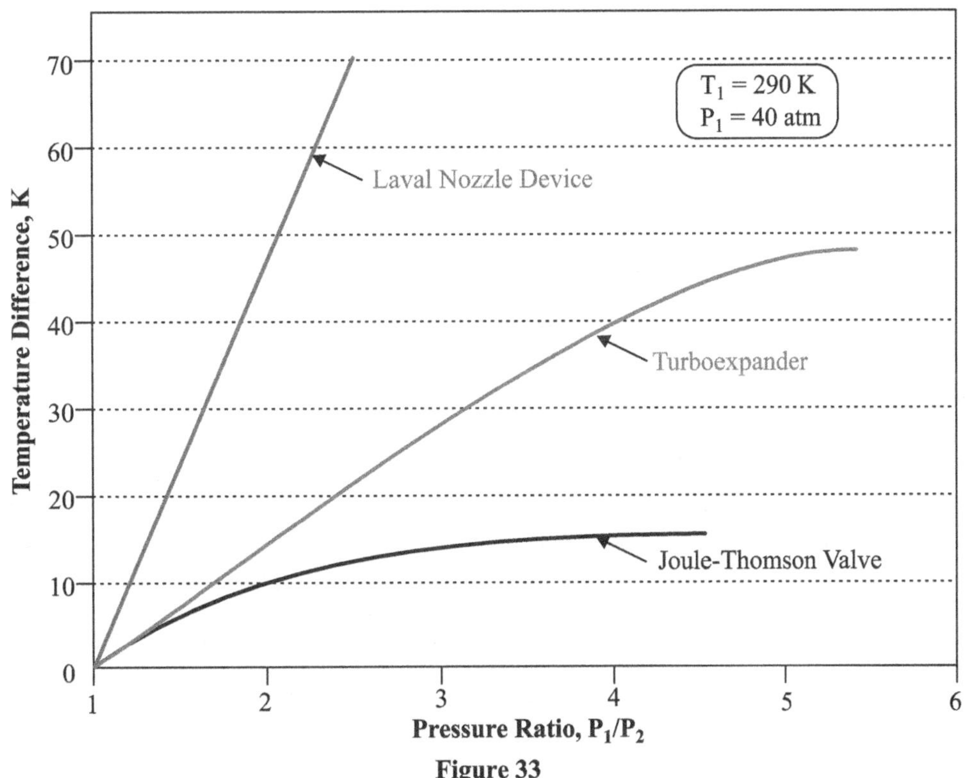

Figure 33

Summary

A turboexpander can generate low-temperature gas far more efficiently than options such as a "Joule-Thomson" (JT) valve or others in many refrigeration, cryogenic, and low-temperature gas services. Given a certain pressure reduction, the almost isentropic expansion in a turbo-expander allows for a lower temperature of the expanded gas than an isenthalpic expansion using a throttling valve or other devices. Indeed, the application of a cold gas turbo-expander instead of low-efficiency, traditional methods (such as a JT valve) can significantly improve the cooling capacity, performance, efficiency, and operational costs of such a processing plant. The lower temperature considerably increases the overall cold gas or refrigeration cycle efficiency.

DIFFERENCE BETWEEN ADIABATIC COOLING AND COOLING BY THROTTLING

Some thermodynamic concepts often become confusing.

Adiabatic Process

An adiabatic process is one in which there is no heat-in or heat-out. In adiabatic cooling of gases, the gas expands by using its internal energy.

From the 1st law of thermodynamics, $U = Q - W$, U is internal energy, Q is heat and W is work since in an adiabatic process. $Q = 0$, in adiabatic expansion, $U = -W$. In other words, the cooling in an adiabatic process occurs at the expense of the expansion work $-W$.

Isenthalpic Process

An isenthalpic process is also an adiabatic expansion process but it occurs at constant enthalpy across the throttling valve.

Let us take two points across a throttling valve 1 (upstream) and 2 (downstream) isenthalpic cooling across the throttling valve can be expressed as: $H_1 = H_2$ (H stands for enthalpy,) $U_1 + W_1 = U_2 + W_2$. There is no change in total energy. In the isenthalpic process when the gas expands the internal energy and the work energy compensate each other while the gas cools by keeping the total energy constant by a switch between kinetic and potential energy constant.

Remember U is a state function and W is a path function. They are two different forms of energy and act independently to compensate each other while keeping the total energy constant in a constant enthalpy process.

The internal energy 'U' is the sum of kinetic and potential energy. Work 'W' is mechanical energy.

A point to note is that an adiabatic process does not necessarily have to be either an isentropic (constant entropy) or isenthalpic process (constant enthalpy) because of some thermodynamic restrictions.

Adiabatic *vs.* Isenthalpic Process

The main difference between an adiabatic process and isenthalpic process is that an adiabatic process cools or heats up by adiabatic expansion or compression while doing $-W$ or + Work either by consuming or by adding internal energy. An isenthalpic process occurs when a real gas flow is throttled suddenly causing a large reduction in the kinetic energy that cools the gas. In isenthalpic process no thermodynamic work is involved. The cooling occurs by converting the kinetic energy to potential energy when the gas expands.

THERMODYNAMICS OF STATES OF MATTER: ENTROPY AND FREE ENERGY'S ROLES

What makes a solid? What makes a liquid? What makes a gas?

The key point is that every substance at a given temperature and pressure with the given intermolecular complexities seeks a stable state. Let's take the example of water when water is cooled it finds the ice as the most stable place to migrate by lowering its entropy. On the other hand, when water is heated and entropy increases water finds the vapor phase as the most suitable place to migrate to off-load its entropy. This is a universal phenomenon happening all the time.

Matter can exist in different states, such as solid, liquid, or gas, due to the principles of thermodynamics. Every substance seeks a stable state where Gibbs free energy is minimized, taking into account the constraints of intermolecular bonds and molecular structure. Despite water being liquid it has higher specific heat compared to its solid state (ice) due to a larger density than ice. So, these phase changes are complex.

Throughout these state changes, entropy plays a crucial role. Entropy seeks to maximize to a vapor state, while Gibbs free energy strives to minimize. The outcome is determined by the factor between free energy and entropy, that wins the battle, based on the given temperature, pressure, and intermolecular forces at play. Overall, the existence of matter in different states is intricately tied to the interplay between free energy, entropy, and the physical properties of the substance.

Further Explanation

A matter chooses the most stable state to exist: Thermodynamics

Two things decide the state of a matter [1] Gibbs free energy and [2] entropy. In this tug of war, the one who wins, decides the state of a matter. Entropy plays a critical role. Ice has a lower entropy than water. Ice forms by external intervention when the molecules are pushed to clump together. A matter chooses a state to exist where it is stable in a given temperature and pressure. When water is cooled water achieves stability in the formation of ice by going into a lower entropy state. When water is cooled it chooses to turn into solid ice with lower entropy. When water evaporates due to a large kinetic energy of molecules and entropy generation water chooses to go to a higher energy state which is vapor where it can off-load its entropy and relax. While the water-to-vapor transition is a spontaneous process when water maximizes its entropy, the formation of ice is not a spontaneous process. Formation of ice requires the external intervention of cooling and reducing the kinetic energy of molecules and bringing them closer to packing into a solid. Therefore, ice forms by external intervention like cooling of water which takes water to a state of a lower entropy.

While cooling the water, it contracts and brings molecules closer to facilitate a large number of H bonds getting active to form many clusters of water molecules, and that forces the water to reduce its entropy and turn into solid ice,

CHAPTER 35

Mass, Momentum and Energy

In thermodynamics, mass is typically considered to be conserved, meaning it cannot be created or destroyed but only transferred between systems. Momentum plays a crucial role in thermodynamics, as it describes the motion of particles within a system and how their interactions affect the overall behavior. Energy is another fundamental concept in thermodynamics, representing the ability of a system to do work or produce heat. In thermodynamic processes, energy can be transferred between systems as heat or work, leading to changes in the system's internal energy. The conservation of mass and energy, along with the principle of momentum, form the basis of many thermodynamic laws and equations used to analyze and predict the behavior of physical systems.

WHAT IS MASS? MASS – MOMENTUM- ENERGY INTERACTION

A body's mass is an inherent quality. Prior to the discovery of the atom, it was widely believed to be related to the amount of matter in a physical body. It was discovered that, despite having the same amount of matter in theory, different atoms and elementary particles have different masses. It is possible for two substances to have the same mass if their product of the volume of density is the same. M = Density × Volume. Therefore, strictly, it is not the amount of matter in a substance is its mass. **Mass is something that resists its change of inertia.**

What is Inertia?

In simple words, the inertia of a substance is "laziness to change its motion.". Inertia is a property of matter that describes its resistance to changes in motion, and it is caused by the mass of the object.

Newton's first law of motion states that "An object at rest stays at rest and an object in motion stays in motion with the same speed and in the same direction unless acted upon by an unbalanced force." Objects tend to "keep on doing what they're doing." In fact, it is the natural tendency of objects to resist changes in their state of motion. This tendency to resist changes in their state of motion is described as inertia.

The resistance of the body to acceleration (change of velocity) in the presence of a net force can be measured experimentally as mass. The gravitational pull an object has on other bodies is also influenced by its mass. The kilogram (kg) is the SI's fundamental unit of both mass and inertia. A heavier mass is more difficult to move than a lighter mass because the more gravitational force acting on the heavier object produces more weight [mg]. Mass is directly proportional to its inertia.

- **When a mass is at rest?** When a mass is at rest, it means that the net force acting on it is zero. In the case of drag and viscous forces, if the drag force is greater than the viscous force when the mass is at rest, it means that there is an external force exerted on the mass opposing its motion, causing it to remain stationary. This external force could be a result of friction or other restraining factors. In simple words, drag force in overcoming viscosity to resist a body's forward motion.
- **When a mass is in motion?** When a mass is in motion, its viscous forces are more dominant than drag forces. This concept forms the basis of the Reynolds number, which is used to determine the type of flow (laminar or turbulent) based on the relative magnitudes of viscous and inertial forces.
- **When a mass is in a steady state motion?** A mass is in a steady state of motion when the drag forces equal viscous forces.

Mass, Momentum, and Energy Interaction

Mass, momentum, and energy interact in various ways depending on the physical system and the specific interactions involved. However, there are some general principles that can help understand their interactions:

- **Mass-energy equivalence**: According to Einstein's famous equation $E = mc^2$, mass and energy are interchangeable. This means that mass can be converted into energy and vice versa. This is evident in nuclear reactions, where a small amount of mass is converted into a large amount of energy.
- **Conservation laws**: In isolated systems, the total mass, momentum, and energy are conserved. This means that these quantities cannot be created or destroyed but can only be transferred or transformed between different forms. For example, in a collision between two objects, the total momentum before and after the collision remains constant.
- **Energy and momentum transfer**: In interactions between objects, the transfer of energy and momentum can occur. For example, in elastic collisions, objects bounce off each other, transferring both momentum and kinetic energy. In inelastic collisions, objects stick together, and there is a partial or complete loss of kinetic energy.
- **Kinetic energy and momentum**: Kinetic energy is directly related to momentum. The kinetic energy of an object can be expressed as $K.E. = \frac{1}{2}mv^2$, where m is the mass and v is the velocity. Momentum is the product of mass and velocity, given by $p = mv$. Therefore, changes in momentum can lead to changes in kinetic energy.
- **Work and energy**: Work is the transfer of energy from one system to another. When a force is applied to an object and causes it to move, work is done on the object, transferring energy to it. This change in energy can result in a change in momentum, as momentum is closely related to the object's energy.

Examples

- **Collisions**: When two objects collide, they interact through the exchange of momentum and kinetic energy. For example, in a car crash, the momentum of the vehicles is exchanged, resulting in deformation and the dissipation of kinetic energy.
- **Projectile motion**: When an object is launched into the air, such as a ball thrown or a bullet fired, it experiences both momentum and energy interactions. The object gains momentum due to the propelling force, and its energy is converted between potential and kinetic as it travels through the air.
- **Fluid flow**: In fluid dynamics, mass, momentum, and energy interactions occur as fluids (liquids or gases) move through pipes and channels. The flow of fluids is characterized by the exchange of mass (fluid entering and leaving a system), momentum (changes in fluid velocity), and energy (pressure differences, heat transfer, etc.).
- **Elastic collisions**: In elastic collisions, objects interact through the exchange of momentum and kinetic energy while maintaining their original shape and without any energy loss. For example, billiard balls colliding on a table demonstrates elastic collisions where momentum and energy are conserved.

These are just a few examples of how mass, momentum, and energy interact in various physical phenomena.

MOMENTUM: A SHORT NOTE

Contents

- Newton's second law of motion
- Four aspects of the momentum
- Momentum *vs.* pressure: Newton's third law
- Conservation of momentum
- Momentum equation
- Momentum transfer

What is momentum? Mass multiplied by velocity is the one-word answer.

The term "momentum" refers to "mass in motion." Because all objects have mass, if they are moving, they have momentum – their mass is moving. The amount of momentum that an object possesses is determined by two factors: the amount of material moving and the speed at which it is travelling. The factors mass and velocity influence momentum.

Momentum is a vector quantity in the sense that it has both a magnitude and a direction similar to velocity.

According to Newton's second equation of motion, the particle's time-based rate of change of momentum is equal to the force acting on it.

The units for momentum would be mass units' times velocity units. The standard SI unit of momentum is the kg × m/s.

Newton's Second Law of Motion

The second law of Newton is a quantitative description of the effects that a force can have on a body's motion. It asserts that the force imposed on a body equals the time rate of change of its momentum in both magnitude and direction. The product of a body's mass and velocity determines its momentum. A force acting on a body can affect the magnitude, direction, or both of the momentum's components. It can be stated in the form $F = ma$ for a body with a constant mass m. If a body has a net force acting on it, it is accelerated in accordance with the equation. Conversely, if a body is not accelerated, there is no net force acting on it.

Detail

Momentum

Linear and angular momentum are the two different types of momentum. The angular momentum of a spinning object is different from the linear momentum of a moving object. We'll concentrate on linear momentum, referring to it simply as momentum.

There are four aspects of momentum to consider

Here are four key points to remember when it comes to momentum.

[1] The first is the definition of momentum, which is the product of mass and velocity: Momentum: $p = mv$.

[2] The second point to keep in mind is that momentum is a vector, and it has the same direction as velocity.

[3] The link between motion and force is the third point to consider. The two concepts of force and motion are inextricably linked. A change in momentum is caused by a force acting for a specific amount of time (known as an impulse).

$$\text{Impulse} = F\,\Delta t = \Delta p$$

Because this is a vector equation, the change in momentum follows the force in the same direction.

[4] The fourth and most significant aspect of momentum is that it is conserved; an isolated system's total momentum remains constant. The term "isolated" refers to the fact that the system, which consists of interacting items, is not subjected to any external forces. If a net force is operating on a system, the momentum varies according to the impulse equation.

Momentum *vs.* Pressure: Newton's Third Law

A molecule experiences a change in momentum when it collides with a container wall; during the collision, an impulse is imparted by the wall to the molecule that is equal and opposite to the impulse imparted by the molecule to the wall. This is required by Newton's third law. The sum of the impulses imparted by all the molecules to the wall is, in effect, the pressure. The gas molecules have a mass m and some average velocity v so they have a momentum mv. When one molecule rebounds from the wall the change of momentum is 2mv, but the rate of change of momentum is just the force exerted on the wall. So, if the number of collisions with the wall per second per unit area is N, then the pressure on the wall will be $P = 2Nmv$ (not quite, because not all collisions are at right angles, but let's skip over this).

Conservation of Momentum

The law of conservation of momentum asserts that unless an external force is introduced, the total momentum of two or more bodies acting on each other in an isolated system remains constant. As a result, neither the creation nor the destruction of momentum is possible.

Newton's rules of motion state that momentum can only be modified by the action of forces. Because momentum is a vector quantity with both a magnitude and a direction, it is more complex to deal with than mass and energy. At the same time, momentum is conserved in all three physical directions. When dealing with a gas, it's even more complicated because forces in one direction might alter momentum in another due to the collisions of many molecules.

Equation of Momentum

In open channel flow issues, the momentum equation is utilised to determine unknown forces (F) acting on the walls or bed in a control volume. The momentum equation, in contrast to the energy equation, which deals with scalar quantities like mass (m), pressure (P), and velocity magnitude (V), deals with vector quantities like velocity vector and forces (F). As a result, writing a momentum equation in a given direction and using the component of the forces within the defined direction is crucial. The momentum equation can be expressed as follows, using Newton's second rule of motion, which correlates the rate of change of momentum ($M = mV$) with the algebraic total of all external forces:

$$dM/dt = \Sigma F$$

Momentum Transfer

The amount of momentum transferred from one particle to another is known as momentum transfer. The only way to transfer momentum is by having one object apply a force to another. The force is the rate of change of momentum. when two bodies collide, they transfer only momentum. Because, when being precise, the word "collision" is used only for impacts that transfer only momentum. This implies that the transfer of matter or of energy other than kinetic is so small that it will be ignored.

Typical Examples of Momentum Transfer

Momentum transfer in liquids

Momentum is readily transferred from one region of the flow to another in high viscosity fluids. The molecules in a fluid move momentum from one region of the fluid to another by bouncing around at random. The viscosity of a fluid is assessed in a continuum model to determine its capacity to transfer momentum.

Momentum transfer in gases

In gases, molecular collisions transfer momentum between fluid layers. As slower molecules collide with faster molecules, the slower molecules speed up and the faster molecules slow down.

NAVIER STOKES EQUATIONS

In one line, Navier Stokes equations relate mass, momentum, and energy, the fundamental conserved quantities governing fluid flow.

The Navier-Stokes equations are a set of differential equations that describe the motion of fluid substances. These equations are derived from the conservation principles of mass, momentum, and energy. They represent the conservation of mass, the conservation of momentum (Newton's second law applied to fluid motion), and the conservation of energy in fluid flow. The Navier-Stokes equations are fundamental in fluid mechanics and are used to model a wide range of flows.

The equations take into account factors such as the pressure of the fluid, its viscosity (how sticky it is), and how the fluid responds to changes in temperature and velocity.

Mass, momentum, and energy are studied together in the Navier-Stokes equations because they are interconnected and influence each other in fluid flow. In a fluid flow system, mass conservation ensures that the total mass of the fluid remains constant, momentum conservation relates to the forces and velocities of the fluid particles, and energy conservation accounts for the internal energy of the fluid and its interactions with its surroundings.

In the Navier-Stokes equations, these three quantities are coupled through various physical phenomena such as pressure, viscosity, and thermal effects. By studying these equations together, one can gain a comprehensive understanding of the fluid flow behavior and make accurate predictions about its dynamics and properties. Additionally, solving the Navier-Stokes equations allows engineers to design and optimize fluid flow systems for various applications, including hydrodynamics, and heat transfer.

Detail

The Navier-Stokes equations can be broken down into two main components.

The continuity equation:

This part of the equation deals with the principle of conservation of mass. In simple terms, it states that the amount of fluid flowing into a given region must be equal to the amount flowing out. This helps us understand how the fluid is distributed and moving within a system.

The momentum equation:

This part of the equation describes how the velocity of the fluid changes over time in response to forces acting on it. It takes into account factors like pressure, viscosity, and external forces. This helps us predict how the fluid will flow and behave in different situations.

The Navier-Stokes equations in their full form are a set of partial differential equations as said above, that describe the motion of fluid substances. They can be written in vector form as:

Continuity equation:

$$\partial \rho/\partial t + \nabla \cdot (\rho v) = 0$$

where:

$\partial \rho/\partial t$ represents the rate of change of density over time.

ρ is the density of the fluid at a given point.

$\nabla \cdot (\rho v)$ is the divergence of the density times the velocity vector, representing how the fluid is moving in different directions.

This equation essentially tells us how the density of the fluid changes over time in response to the flow of the fluid.

Momentum equation:

$$\partial v/\partial t + v \cdot \nabla v = -1/\rho \, \nabla P + \nu \nabla^2 v + f$$

where:

$\partial v/\partial t$ represents the rate of change of velocity over time.

v is the velocity vector of the fluid at a given point.

∇v is the gradient of the velocity vector, representing the change in velocity in different directions.

P is the pressure of the fluid.

ρ is the density of the fluid.

ν is the kinematic viscosity of the fluid.

∇^{2v} is the Laplacian of the velocity vector, representing the rate of change of the velocity gradient.

f represents any external forces acting on the fluid.

This equation describes how the velocity of the fluid changes over time in response to pressure, viscosity, and external forces. It helps us predict how the fluid will flow and behave in a given system.

Though, the basic form of the Navier-Stokes equations consists of the continuity equation and the momentum equation. Any additional terms or equations that are added to the system depend on the specific characteristics of the flow being studied that is the energy equation.

The energy equation in the Navier-Stokes equations describes the conservation of energy within a fluid flow. This equation accounts for the transfer of thermal energy, work done by pressure forces, and any sources or sinks of energy within the flow. The energy equation is important for analyzing the temperature distribution in a fluid flow and understanding how energy is transferred within the system.

The energy equation is not always included in the standard form of the Navier-Stokes equations, depending on the particular analysis being conducted. It is typically used in situations where heat transfer and temperature effects are important, such as in flows involving significant temperature gradients or in situations where the internal energy of the fluid needs to be accurately modeled.

Overall, the energy equation serves as a valuable tool in fluid dynamics for understanding the thermodynamic behavior of fluids and can provide additional insights into the overall flow behavior.

Energy Equation

The energy equation in the Navier-Stokes equations describes the conservation of energy within a fluid flow. It takes into account the transfer of thermal energy, work done by pressure forces, and any sources or sinks of energy within the flow.

The general form of the energy equation in the Navier-Stokes equations is:

$$\partial(\rho E)/\partial t + \nabla\bullet(\rho E\, u) = -\nabla\bullet(p\, u) + \nabla\bullet(\tau\, u) + \nabla\bullet(q) + S$$

where:

ρ is the fluid density

E is the total energy per unit mass of the fluid

t is time

u is the fluid velocity

p is the pressure

τ is the stress tensor

q represents the heat flux

S represents any sources or sinks of energy

The terms in the equation describe the change in energy with respect to time, the convective transport of energy by the fluid flow, the work done by pressure forces, the work done by viscous stresses, the heat transfer via conduction, and any additional sources or sinks of energy.

In summary, in simpler terms, the Navier-Stokes equations are a set of fundamental equations in fluid dynamics that describe how fluids like air and water move and interact. They are used to predict how fluids flow and how forces like pressure, viscosity, and external influences affect their behavior.

The Navier-Stokes equations connect the properties of a fluid, such as its velocity, pressure, and density, to the forces acting on it. By solving these equations, one can analyze and understand complex fluid flow phenomena, such as turbulence, drag, and vortices.

In essence, the Navier-Stokes equations provide a mathematical framework for studying and predicting the behavior of fluids in various practical applications, and industrial processes. They help researchers and engineers design more efficient and effective systems by providing insights into how fluids behave in different scenarios.

NAVIER STOKES EQUATIONS: MEANING

In a closed system, the sum of mass, momentum, and energy remains constant. This is known as the principle of conservation of mass, momentum, and energy. The interconnection between mass, momentum, and energy in fluid flow is closely related to fundamental thermodynamic state properties such as pressure, temperature, viscosity, and density in each case. Navier Stokes equations are sets of equations which connect mass, momentum and energy of a fluid. An engineer should be concerned with mass, momentum, and energy in fluid dynamics because these properties are essential for understanding and analyzing the behavior of fluids in various engineering applications.

Mass is important because it determines the amount of substance that is present in a fluid and helps in calculating density, which in turn affects the flow and pressure of the fluid.

Momentum, being the product of mass and velocity, plays a crucial role in determining the direction and speed of fluid flow. Conservation of momentum is a fundamental principle in fluid dynamics and is crucial for predicting the behavior of fluids in different scenarios.

Energy is vital because it is transferred and converted during the flow of fluids, affecting the temperature and pressure within the system. Understanding the energy balance in fluid flow helps engineers optimize the efficiency and performance of fluid systems.

The Navier-Stokes equations are considered to be one of the most challenging subjects in fluid dynamics due to their complex nature and the use of differential equations. Many individuals often shy away from delving deeper into understanding these equations, but it is essential to grasp the fundamentals behind them in order to truly appreciate their significance in the field of fluid dynamics. In this note, I will attempt to write my take in simple words from Navier Stokes equations.

Detail

The Navier-Stokes equations are a set of differential equations that describe the motion of fluid substances. These equations are derived from the conservation principles of mass, momentum, and energy. They represent the conservation of mass, the conservation of momentum (Newton's second law applied to fluid motion), and the conservation of energy in fluid flow. The Navier-Stokes equations are fundamental in fluid mechanics and are used to model a wide range of flows

The equations take into account factors such as the pressure of the fluid, its viscosity (how sticky it is), and how the fluid responds to changes in temperature and velocity.

All you need to know is how to solve differential equations.

Mass, momentum, and energy are studied together in the Navier-Stokes equations because they are interconnected and influence each other in fluid flow. In a fluid flow system, mass conservation ensures that the total mass of the fluid remains constant, momentum conservation relates to the forces and velocities of the fluid particles, and energy conservation accounts for the internal energy of the fluid and its interactions with its surroundings.

The Navier-Stokes equations can be broken down into three components:

The continuity equation:

This part of the equation deals with the principle of conservation of mass. In simple terms, it states that the amount of fluid flowing into a given region must be equal to the amount flowing out. This helps us understand how the fluid is distributed and moving within a system.

The momentum equation:

This part of the equation describes how the velocity of the fluid changes over time in response to forces acting on it. It takes into account factors like pressure, viscosity, and external forces. This helps us predict how the fluid will flow and behave in different situations

The energy equations

Energy equation

The energy equation in the Navier-Stokes equations describes the conservation of energy within a fluid flow. It takes into account the transfer of thermal energy, work done by pressure forces, and any sources or sinks of energy within the flow.

Explanation

Continuity equation

$$\partial\rho/\partial t + \nabla \cdot (\rho v) = 0$$

where:

$\partial\rho/\partial t$ represents the rate of change of density over time.

ρ is the density of the fluid at a given point.

$\nabla \cdot (\rho v)$ is the divergence of the density times the velocity vector, representing how the fluid is moving in different directions.

This equation essentially tells us how the density of the fluid changes over time in response to the flow of the fluid.

Momentum equation:

$$\partial v/\partial t + v \cdot \nabla v = -1/\rho \, \nabla P + v\nabla^{2v} + f$$

where:

$\partial v/\partial t$ represents the rate of change of velocity over time.

v is the velocity vector of the fluid at a given point.

∇v is the gradient of the velocity vector, representing the change in velocity in different directions.

P is the pressure of the fluid.

ρ is the density of the fluid.

ν is the kinematic viscosity of the fluid.

∇^{2v} is the Laplacian of the velocity vector, representing the rate of change of the velocity gradient.

f represents any external forces acting on the fluid.

This equation describes how the velocity of the fluid changes over time in response to pressure, viscosity, and external forces. It helps us predict how the fluid will flow and behave in a given system.

Energy equations

The general form of the energy equation in the Navier-Stokes equations is:

$$\partial(\rho E)/\partial t + \nabla \bullet (\rho E\, u) = -\nabla \bullet (p\, u) + \nabla \bullet (\tau\, u) + \nabla \bullet (q) + S$$

where:

ρ is the fluid density

E is the total energy per unit mass of the fluid

t is time

u is the fluid velocity

p is the pressure

τ is the stress tensor

q represents the heat flux

S represents any sources or sinks of energy

The terms in the equation describe the change in energy with respect to time, the convective transport of energy by the fluid flow, the work done by pressure forces, the work done by viscous stresses, the heat transfer via conduction, and any additional sources or sinks of energy.

What do these equations mean?

In a closed system, the sum of mass, momentum, and energy remains constant. This is known as the principle of conservation of mass, momentum, and energy.

For a fluid system, the conservation laws can be expressed in terms of continuity equation (for mass), Navier-Stokes equations (for momentum), and energy conservation equations. These equations describe how mass, momentum, and energy are conserved within a fluid system and how they change as the fluid flows through the system.

It is important to note that these conservation laws may not always apply if the system is open or if external forces are acting on the system. In such cases, the conservation laws need to be modified to account for the additional factors influencing the system.

The interconnection between mass, momentum, and energy in fluid flow is closely related to fundamental thermodynamic state properties such as pressure, temperature, viscosity, and density in each case.

How and why?

- **Mass**: The density of a fluid is directly related to mass, as density is defined as mass per unit volume. The mass of a fluid flowing through a system affects its density, which in turn influences the fluid's behavior in terms of flow rate, pressure, and temperature. The density of a fluid can also be affected by changes in temperature, pressure, and composition.

- **Momentum:** Momentum is related to both density and velocity in fluid flow. The momentum of a fluid element is determined by its mass and velocity, and changes in momentum can be caused by forces acting on the fluid, such as pressure gradients or viscous forces. The viscosity of the fluid plays a critical role in momentum transfer, as it affects the resistance of the fluid to flow and the generation of shear stresses within the fluid.
- **Energy:** Energy in fluid flow is related to temperature, pressure, and internal energy of the fluid. The energy of a fluid element includes its kinetic energy (related to velocity), potential energy (related to elevation), and internal energy (related to temperature). The pressure of the fluid is a measure of the internal energy per unit volume, and changes in pressure can affect energy transfer within the fluid system. The temperature of the fluid also plays an important role in energy transfer and can impact the density, viscosity, and thermal properties of the fluid.

Dimensions of Navier Stokes equations

In final words, the interconnection between mass, momentum, and energy in fluid flow is closely linked to three fundamental laws of physics: Fick's Law, Fourier's Law, and Newton's Second Law.

- **Fick's Law**: Fick's Law describes the diffusion of mass in a fluid system. It states that the rate of mass transfer (diffusion) is directly proportional to the concentration gradient of the species being transferred. In fluid flow, Fick's Law can be applied to processes such as the diffusion of solutes in a liquid or gas phase, where mass transfer occurs due to concentration differences. The interconnection between mass, momentum, and energy influences the diffusion process, as changes in density, viscosity, and temperature can impact the rate of mass transfer.
- **Fourier's Law**: Fourier's Law describes the conduction of heat in a fluid system. It states that the rate of heat transfer is directly proportional to the temperature gradient in the system. In fluid flow, Fourier's Law can be applied to processes such as the conduction of heat through a pipe or boundary layer, where energy transfer occurs due to temperature differences. The interconnection between mass, momentum, and energy influences the heat transfer process, as changes in temperature, viscosity, and thermal properties of the fluid can affect the rate of heat conduction.
- **Newton's Second Law**: Newton's Second Law describes the relationship between the net force acting on an object and its acceleration. In fluid flow, Newton's Second Law can be applied to the momentum balance equation, which relates the net forces acting on a fluid element to changes in momentum. The interconnection between mass, momentum, and energy plays a crucial role in the application of Newton's Second Law, as changes in mass, velocity, pressure, and viscosity can affect the forces acting on the fluid system and the resulting acceleration.

Overall, the interconnection between mass, momentum, and energy in fluid flow is intricately linked to the fundamental laws of physics, including Fick's Law, Fourier's Law, and Newton's Second Law. Understanding and applying these laws are essential for analyzing and predicting the behavior of fluid flow systems in various engineering and scientific applications.

APPLICATION OF MASS, MOMENTUM AND ENERGY IN A CHEMICAL PROCESS

The mass, momentum, and energy are interconnected and are often considered together in the study of various phenomena.

Changes in one of these quantities can result in changes in the other two, demonstrating their interconnectedness.

In every process mass, momentum and energy work as a function of state variables such as pressure, temperature, viscosity and density

In two-phase processes such as boiling, the difference in momentum between the two phases (liquid and vapor) plays a critical role in various aspects of the process. Momentum transfer is essential for the efficient transfer of mass, energy, and momentum between the phases, affecting heat and mass transfer rates, phase distribution, and overall process performance.

Let's take two familiar examples

Bernoulli's equation

While Bernoulli's equation itself is an energy conservation equation, it can be related to momentum conservation in fluid dynamics. The equation can be used to analyze the balance between pressure, velocity, and height, providing a way to understand the distribution of forces and momentum within a flowing fluid. In this sense, Bernoulli's equation can be viewed as a useful tool for examining the conservation principles at play in fluid flow.

Boiling

There are four points

In the context of boiling, momentum transfer between the liquid and vapor phases is important for several reasons:

1. **Bubble formation:** The momentum of the vapor phase influences bubble formation during boiling. As heat is applied to the liquid, vapor bubbles form at nucleation sites and grow as they rise through the liquid. The momentum of the vapor phase affects the size and velocity of the bubbles.
2. **Heat transfer:** The momentum of the vapor phase plays a critical role in enhancing heat transfer during boiling. The movement of vapor bubbles away from the heating surface helps to remove the latent heat of vaporization and maintain a high heat transfer.
3. **Phase distribution:** The momentum difference between the liquid and vapor phases affects the distribution of phases within the boiling system. Proper phase distribution is essential for ensuring uniform heat transfer and boiling across the surface.
4. **Flow dynamics:** Momentum transfer between the phases influences the flow dynamics within the boiling system. The movement of vapor bubbles and liquid flow is determined by the momentum difference between the phases.

Proper flow dynamics are important for efficient heat and mass transfer, as well as for preventing issues such as vapor lock or flow instabilities.

Overall, the key in a boiling processes, is the momentum transfer between the liquid and vapor phases.

Understanding and optimizing momentum transfer in boiling systems is essential for enhancing efficiency, improving heat transfer rates, and achieving optimal process outcomes.

Why mass, momentum and energy do not decay?

The sum total of mass, momentum and energy in any unit process (closed system) always stay constant.

It is imperative that you cannot design any process that involves transport of a fluid without accounting for the mass, momentum and energy changes as a function of state properties such as pressure, temperature, viscosity and density.

The application of Navier Stokes equations can predict these three conserved properties of fluid in any unit process.

In the case of mass, it remains constant because the number of protons and neutrons in an object typically do not change (except in nuclear reactions).

Momentum is conserved through collisions between molecules, which are often elastic and do not result in a loss of momentum.

And energy, a combination of kinetic and potential energy, can change forms (such as when work is done), but the total energy in a closed system remains constant.

These explanations help to illustrate how the conservation of mass, momentum, and energy is maintained in the physical world.

DOES MOMENTUM RULE MOTION ON EARTH?

Mathematically, momentum is defined as the product of an object's mass and velocity. Mass is a scaler quantity while velocity is a vector. Momentum is a vector.

Momentum plays a fundamental role in controlling motion on Earth, as well as in many other physical systems. The product of mass and velocity gives an object its momentum, which is a measure of its motion and the resistance it has to change in that motion.

According to Newton's second law of motion, the rate of change of momentum (or the force) acting on an object is equal to the mass of the object multiplied by its acceleration. This force is what governs the motion of objects on Earth and determines how they respond to external influences.

In simpler terms, momentum helps us understand how objects move and how they interact with other objects or forces. It is a crucial concept in physics and has wide-ranging applications in various scientific fields.

Momentum is Conserved

This is a great concept based on Newton's third law.

Momentum is a conserved quantity, meaning that it does not decay or disappear under normal circumstances. According to the law of conservation of momentum, the total momentum of an isolated system remains constant as long as no external forces act on it.

This principle can be understood by looking at Newton's third law of motion, which states that for every action, there is an equal and opposite reaction. When two objects interact, the total momentum of the system before and after the interaction remains the same.

For example, if two objects collide, the total momentum of the system before the collision is equal to the total momentum after the collision. Momentum may be transferred between the objects depending on the type of collision (elastic or inelastic), but the total momentum remains constant.

While external forces can change the momentum of an object or a system, the total amount of momentum within the system remains the same, demonstrating that momentum is conserved.

How Mass, Momentum, Velocity, and Force are Related?

Mass, velocity, momentum, and force

Inertia refers to an object's resistance to change in motion, and when an object with mass is in motion (i.e., has velocity), it possesses momentum. Mathematically, momentum is defined as the product of an object's mass and velocity. The rate of change of momentum, often referred to as force, is indeed responsible for governing the motion of objects on Earth.

Newton's second law of motion states that the force acting on an object is equal to the rate of change of its momentum. This law describes how forces influence the acceleration and movement of objects.

So, overall, mass's inertia is given velocity, it becomes momentum, and the rate of change of momentum (force) acts as the governing factor for the motion of objects on Earth.

Specific Roles of Momentum

Kinetic and potential energy.

Potential Energy

Momentum can play a role in potential energy in certain situations. The potential energy associated with an object's position or configuration can indeed be influenced by its mass and velocity.

For example, in the case of gravitational potential energy, the potential energy of an object near the Earth's surface is given by the equation $U = m \times g \times h$, where m is the mass of the object, g is the acceleration due to gravity, and h is the height or vertical position of the object relative to a reference point.

If the object is in motion, its potential energy formula still includes the mass, which is related to its momentum through the equation $p = m \times v$, where p is the momentum and v is the velocity of the object.

So, while momentum may not directly affect the potential energy itself, it is related to the object's mass and velocity, which in turn can influence the potential energy of the system.

Kinetic Energy

Momentum transfer through collisions is indeed a fundamental mechanism through which energy is transferred. However, it is more accurate to say that energy is the fundamental quantity that governs the behavior of systems, including the Earth, rather than momentum alone.

Energy encompasses various forms, such as kinetic energy (associated with motion), potential energy (associated with position or configuration), and thermal energy (associated with the random motion of particles). Energy can be transformed from one form to another, and it is conserved in isolated systems according to the principle of conservation of energy.

Gravity

Gravity is a force that exists between two objects with mass, and it is responsible for the attraction between them.

Momentum can indirectly influence gravity in certain situations. According to Newton's laws of motion, an object in motion will tend to stay in motion unless acted upon by an external force. So, if an object is moving toward another object due to its momentum, gravity can further accelerate it toward the second object. On the other hand, if an object has a large momentum in a direction away from another object, gravity can slow it down and potentially pull it back toward the second object.

In the context of celestial bodies, the momentum of a planet or satellite plays a role in determining its trajectory under the influence of gravity. The momentum of a planet's motion around the sun affects its orbit, shape, and speed. In this way, momentum indirectly influences the gravitational interactions between celestial bodies.

Force

The rate of change of momentum is force.

Momentum and Bernoulli's Principle: Fluid Dynamics

Momentum and Bernoulli's principle are connected through the concept of fluid dynamics.

Momentum is the product of an object's mass and its velocity, and it is a measure of the object's motion. In fluid dynamics, momentum is conserved in a closed system, meaning the total momentum remains constant unless acted upon by external forces.

Bernoulli's principle, on the other hand, states that in a fluid flow, the sum of the pressure energy, kinetic energy, and potential energy per unit volume is constant along a streamline. This principle describes the relationship between pressure, velocity, and height in a fluid, and it is often used to analyze the flow of fluids in various applications.

The connection between momentum and Bernoulli's principle lies in the conservation of energy. As fluid flows, changes in velocity occur, which affect the fluid's kinetic energy. According to Bernoulli's principle, as the fluid's velocity increases, its pressure decreases. This pressure difference can create a force, known as the Bernoulli force, that acts on objects in the fluid flow.

The momentum of the fluid, as well as the objects within it, can be affected by the Bernoulli force. For example, when fluid flows around an object, the higher velocity on one side of the object results in lower pressure, creating a pressure gradient. This pressure gradient exerts a force on the object, causing it to experience a change in momentum.

In summary, momentum and Bernoulli's principle are connected through the conservation of energy in fluid flow. Changes in fluid velocity, described by Bernoulli's principle, can result in pressure differences that affect the momentum of the fluid and objects within it.

CHAPTER 36

Dimensionless Numbers

Dimensionless numbers play a crucial role in thermodynamics by providing a way to compare different systems without the influence of specific units of measurement. The most commonly used dimensionless numbers in thermodynamics include the Reynolds number, Prandtl number, and Nusselt number, which help to characterize fluid flow, heat transfer, and mass transfer processes. Dimensionless numbers in thermodynamics can also reveal important information about the behavior of a system, such as the level of turbulence, rate of heat transfer, or efficiency of a process. By using dimensionless numbers, engineers and researchers can better understand and predict the behavior of thermodynamic systems, leading to more efficient designs and processes. Overall, dimensionless numbers provide a valuable tool for analyzing and interpreting thermodynamic phenomena by simplifying complex equations and allowing for clearer comparisons between different systems.

DIMENSIONLESS NUMBERS' ROLE IN HEAT TRANSFER

The first question that arises in our mind is why the dimensionless numbers are always ratios? The answer is that Dimensionless numbers are comparison to the magnitude of properties in the denominator of the number. It's a comparison of the numerator with the denominator. It's relative to the denominator of the number.

For example, Re number [Renolds number] stands for "inertial forces" relative to "viscous forces".

Dimensionless numbers represent ratios between different physical quantities or properties, and they are often used to quantify the resistance or interaction between these quantities. By comparing the numerator to the denominator, we can understand the relative magnitude of these properties and how they influence the overall system.

In one-line, dimensionless numbers are a group of ratios of different functional parameters of heat transfer based on experimental data that represent the resistance to heat transfer.

Complexities

Heat transfer from water vapor in a beaker is a complex process, involving multiple mechanisms such as conduction, convection, and evaporation. The specific complexity depends on various factors such as the temperature difference between the vapor and the surroundings, the geometry and material of the container, and the presence of any additional heat sources or sinks.

The behavior of water vapor in a container can be described using different dimensionless numbers, such as the Nusselt number, the Reynolds number, and the Prandtl number, which help characterize the relative importance of different heat transfer mechanisms and provide insights into the overall heat transfer process.

Additionally, the laws of thermodynamics and heat transfer principles, such as Fourier's law of conduction, Newton's law of cooling, and the heat transfer equation, are used to analyze and understand the heat transfer events in a beaker.

What are Dimensionless Numbers in Heat Transfer?

Similarity Numbers	*Formulation*	*Physical Meaning*
Biot number	$Bi = \frac{hl}{\lambda_{SI}}$	$\frac{\text{Convective heat transfer in fluid}}{\text{Conductive heat transfer in solid}}$
Fourier number	$Fo = \frac{\lambda_{SI} t}{C_p, SI \rho_{SI} l^2}$	$\frac{\text{Diffusive transport rate}}{\text{Storage rate}}$
Grashof number	$Gr = \frac{g\beta\Delta T l^3}{\mu^2}$	$\frac{\text{Buoyancy forces}}{\text{Viscous forces}}$
Nusselt number	$Nu = \frac{hl}{\lambda_{Fl}}$	$\frac{\text{Convective heat transfer in fluid}}{\text{Conductive heat transfer in fluid}}$

Similarity Numbers	*Formulation*	*Physical Meaning*
Prandt number	$\mathrm{Pr} = \frac{\mu c_p}{\lambda_{FI}}$	$\frac{\text{Viscous diffusion rate}}{\text{Thermal diffusion rate}}$
Reynolds number	$\mathrm{Re} = \frac{\rho F I^{cl}}{\mu}$	$\frac{\text{Inertial forces}}{\text{Viscous forces}}$

In simple words, dimensionless numbers are ratios of fundamental properties of a process or system without any units. This captures the essence of their purpose and significance concisely. These are six typical dimensionless numbers used in heat transfer. Explained above.

No Single Dimensionless Number can Connect All Heat Transfer Activates

There is no single equation that connects all heat transfer dimensionless numbers, as each dimensionless number represents a different aspect of heat transfer and has its own significance and calculation method. However, some dimensionless numbers can be related to each other in specific situations or through empirical correlations. For example, in certain cases, the Nusselt number can be related to the Reynolds and Prandtl numbers through an empirical correlation called the Nusselt number correlation. Overall, the relationship between dimensionless numbers depends on the specific heat transfer problem and the underlying physics involved.

How to Choose a Dimensionless Number?

Choosing the right dimensionless number depends on the specific problem or phenomenon being studied. Different dimensionless numbers capture different aspects of the system's behaviour or characteristics. One way to select the appropriate dimensionless number is by considering the relevant physical phenomena or processes involved in the problem. For example, if fluid flow is a dominant factor, the Reynolds number may be relevant. If heat transfer is important, the Nusselt number or the Biot number may be more appropriate.

Furthermore, reviewing literature, consulting experts, can also provide insights into the most appropriate dimensionless numbers for a given situation.

Common Dimensionless Numbers in Heat Transfer

There are six commonly used ones along with their significance:

- **Biot Number (Bi):** This dimensionless number represents the ratio of internal thermal resistance to surface thermal resistance. It is important in problems involving heat conduction within a solid material and heat transfer at the solid-fluid interface.
- **Fourier number (Fo):** The dimensionless number associated with Fourier's law of heat conduction is called the Fourier number (Fo). The Fourier number represents the ratio of the rate of heat conduction within a material to the rate of thermal energy storage in that material.
- **Grashof Number (Gr):** This dimensionless number is vital in natural convection problems and represents the ratio of buoyancy forces to viscous forces. It helps determine the dominance of natural convection and its impact on heat transfer.
- **Nusselt Number (Nu):** The Nusselt number relates convective heat transfer to conductive heat transfer and is used to characterize heat transfer in convection-dominated problems. It provides a measure of the effectiveness of convective heat transfer compared to conduction.
- **Prandtl Number (Pr):** This number represents the ratio of momentum diffusivity to thermal diffusivity and is important in problems involving fluid flow and heat transfer. It helps quantify how effectively momentum and heat are transferred within a fluid.
- **Reynolds Number (Re):** It represents the ratio of inertial forces to viscous forces and is crucial for characterizing fluid flow and heat transfer in systems involving fluids. It helps determine flow patterns, heat transfer regimes, and fluid behaviour.

Dimensionless Numbers' Role in Heat Transfer from Water-vapor Life Cycle

Heating water in a beaker till it boils involves multiple heat transfer processes and various dimensionless numbers. Here's a brief explanation of each process and the associated dimensionless numbers:

- **Conduction on the beaker wall**: This process involves the transfer of heat through the solid material of the beaker wall. It can be characterized by Fourier's law of heat conduction and the dimensionless number associated with it is the Fourier number.
- **Conduction through the no-slip boundary layer**: As the heat is transferred from the beaker wall to the water, it involves conduction through a thin layer near the wall called the boundary layer. This process is characterized by the Biot number, which represents the ratio of internal resistance to external resistance.
- **Internal resistance *vs.* fluid heat transfer**: This refers to the heat transfer within the fluid itself, which can be influenced by factors such as fluid properties and flow conditions. The Biot number also reflects this aspect of heat transfer.
- **Convection:** Convection occurs when heat is transferred by fluid motion. In the case of boiling water, convection plays a significant role. The dimensionless numbers associated with convection heat transfer are the Reynolds number (Re), which represents the ratio of inertial forces to viscous forces, and the Prandtl number (Pr), which represents the ratio of momentum diffusivity to thermal diffusivity.
- **Boiling**: Boiling is a phase change process that involves the vaporization of liquid. It can be characterized by the heat flux and the associated dimensionless numbers such as the Boiling number, Leidenfrost number, or Bond number.
- **Vapor:** The behavior of the vapor phase can be characterized using Dalton's law, which relates the pressure of a mixture of gases to the partial pressures of its constituent gases.

These are just some of the many dimensionless numbers and processes involved in heat transfer during the boiling of water in a beaker. Each process and the dimensionless number represent a different aspect of heat transfer and helps to analyze and understand specific phenomena.

PRANDTL NUMBER: A SHORT REVIEW

Summary

Prandtl Number

The Prandtl number (Pr) is a dimensionless number defined as the ratio of momentum diffusivity to thermal diffusivity. It is a property of a fluid.

The Prandtl number is given as:

$$Pr = Cp \times \mu/k$$

Cp is specific heat, μ is dynamic viscosity and k is thermal conductivity.

Prandtl number is also a ratio of velocity or hydrodynamic and thermal boundary layers.

Prandtl number plays an important role in heat transfer as it is connected to heat transfer coefficient '*h*'. *L* is the characteristic length.

$Nu = h \times L/k = 0.023 \times Re^{4/5} \times Pr^n$, Nu is Nusselt number.

What is the Boundary Layer?

A boundary layer is a thin layer of viscous fluid close to the solid surface of a wall in contact with a moving stream. Viscous stresses are very prominent within this layer in which (within its thickness) the flow velocity varies from zero at the wall (where the flow "sticks" to the wall because of its viscosity) and then it moves up to U_0 at the boundary. This layer is called the boundary layer.

Shear andhear forces are confined to this part of the fluid. Outside the boundary layer, potential flow survives.

In the region, very near to the surface (in the thin layer), there is friction in the flow which signifies that the fluid is retarded until it adheres to the surface (no-slip condition).

The transition of the mainstream velocity from zero at the surface (with respect to the surface) to full magnitude takes place across the boundary layer.

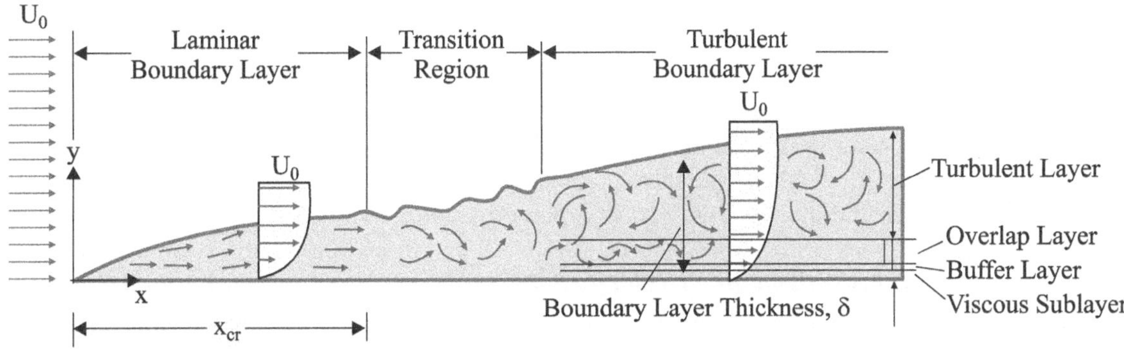

Figure 1

Important Concepts

What is Shear and Shear Stress?

The term shear flow refers to a type of fluid flow that is caused by forces. In a shearing flow, adjacent layers of fluid move parallel to each other in a stratified flow with different speeds. Viscous fluids resist this shearing motion.

What is Shear Stress?

Shear stress is defined as a force per unit area, acting parallel to an infinitesimal surface element. Shear stress is primarily caused by friction between fluid particles, due to fluid viscosity.

What is the Shear Rate?

Shear rate is the rate at which a fluid is sheared or "worked" during flow. In more technical terms, it is the rate at which fluid layers or laminae move past each other. Shear rate is determined by both the geometry and speed of the flow.

What is Potential Flow?

Except for fluids moving at low velocities or with high viscosities, the effect of solid boundary wall is confined to a layer of the fluid immediately adjacent to the solid wall. As explained, this layer is called the boundary layer. Shear and shear forces are confined to this part of the fluid. Outside the boundary layer, potential flow survives.

Prandtl Number

As explained above, the Prandtl number (Pr) is a dimensionless number defined as the ratio of momentum diffusivity to thermal diffusivity.

The Prandtl number is given as:

$$Pr = Cp x \mu / k$$

Cp is specific heat, μ is dynamic viscosity and *k* is thermal conductivity. *Pr* is the ratio of momentum diffusivity to thermal diffusivity.

Note that whereas the Reynolds number and Grashof number are subscripted with a scale variable, the Prandtl number contains no such length scale and is dependent only on the fluid and the fluid state.

Prandtl number is a ratio of the measure of velocity and thermal boundary layers.

What is a Hydrodynamic or Velocity Boundary Layer?

The velocity boundary layer is a thin layer of viscous fluid close to the solid surface of a wall in contact with a moving stream in which (within its thickness δ) the flow velocity varies from zero at the wall (where the flow "sticks" to the wall because of its viscosity) up to at the boundary, which approximately (within 1% error) corresponds to the free stream velocity. The region of the flow over the surface bounded by viscous shearing forces caused by fluid viscosity is called the velocity boundary layer or hydrodynamic boundary layer. The mode of heat transfer in the velocity boundary layer is predominantly convection.

What is a Thermal Boundary Layer?

The fluid particles in the adjacent layer to the solid surface get the same temperature as that of the surface. The particles exchange heat energy with particles in adjoining fluid layers and so on. But as we move away from the surface the temperature decreases hence after a particular thickness the temperature will be the same as 99% free stream temperature. Thus, the layer up to which temperature gradient exists significantly is called the thermal boundary layer. The mode of heat transfer in the thermal boundary layer is predominantly conduction.

Prandtl number is a ratio of hydrodynamic or velocity boundary layer and thermal boundary layer.

The relationship between the hydrodynamic boundary layer and thermal boundary layer is given by the Prandtl number.

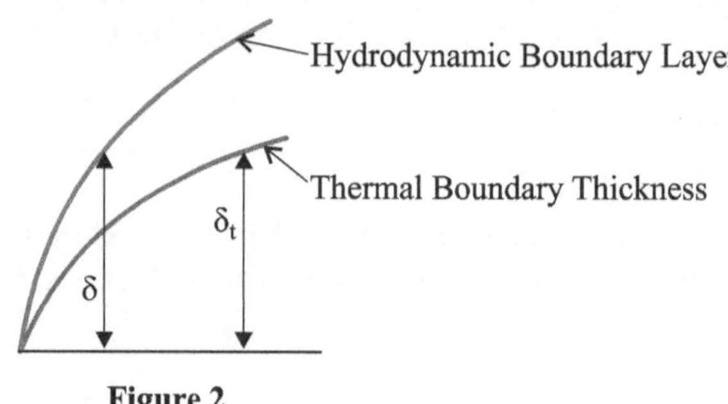

Figure 2

The relation between the two is given by

$$\delta/\delta t = Pr^{1/3}$$

is the thickness of the hydrodynamic layer, the region of flow where the velocity is 99% of the free stream velocity. In other words, the hydrodynamic boundary or the momentum boundary while begins from the boundary surface of the substrate it ends when the flow achieves 99% of the momentum that exists in the free stream

δt is the thickness of the thermal layer, the region of flow where the temperature y is 99% of the free stream velocity. In other words, the thermal boundary begins from the boundary surface of the substrate it ends when the flow achieves 99% of the temperature that exists in the free stream.

I repeat again, we define the thickness of the momentum boundary layer as the distance from the wall to the point where the velocity is 99% of the "free stream" velocity and thickness of the thermal boundary layer as the distance from the wall to the point where the temperature is 99% of the "free stream" temperature.

Free stream is the layer of fluid where pressure, temperature, and relative velocity are unaffected and are free of any external influence like the boundary layer.

In simple and short, the boundary layer thickness, δ is the distance normal to the wall to a point where the flow velocity has essentially reached the 'asymptotic' velocity.

- **When $Pr > 1$**, the momentum or hydrodynamic boundary layer will increase more than the thermal boundary layer
- **When $Pr < 1$**, the momentum or hydrodynamic boundary layer will decrease more than the thermal boundary layer
- **When $Pr = 1$**, the hydrodynamic boundary layer and thermal boundary layer would increase at the same rate

How *Pr* Number Plays Important Role in Heat Transfer

Prandtl number is a measure of the ratio of the dominance of velocity or hydrodynamic boundary layer over the thermal boundary layer.

Relationship between Prandtl number and Heat Transfer Coefficient

We use the following relationship to find out heat transfer coefficient 'h' using Nusselt number.

In fluid dynamics, the Nusselt number (Nu) is the ratio of convective to conductive heat transfer at a boundary in a fluid.

$$Nu = h \times L/k$$

where h is the convective heat transfer coefficient of the flow, L is the characteristic length, k is the thermal conductivity of the fluid. Nusselt number connects to heat transfer coefficient.

Dittus-Boelter Equation

The Dittus-Boelter Equation Connects Nusselt Number to Prandtl Number

$$Nu = 0.023 \times Re^{4/5} \times Pr^{n}$$

$n = 0.4$ for the fluid being heated, and, $n = 0.3$ for the fluid being cooled.

That is how Pr number can influence heat transfer coefficient. Heat transfer coefficient increases with an increase in Pr number.

To Summarize

The temperature decreases with increasing Pr. An increase in the Prandtl number reduces the thermal boundary layer thickness. The Prandtl number is the ratio of momentum diffusivity to thermal diffusivity. In heat transfer problems, Pr controls the relative thickening of the momentum and thermal boundary layers. When Pr is small, heat diffuses quickly because there is less heat compared to the velocity (momentum), which means that for liquid metals, the thickness of the thermal boundary layer is much larger than that of the momentum boundary layer. Fluids with lower Prandtl numbers have higher thermal conductivities (and thicker thermal boundary layer structures) so that heat can diffuse from the sheet faster than that for higher-Pr fluids (thinner boundary layers). Hence the Prandtl number can be used to determine the rate of cooling in conducting flows. It is well known that if the Prandtl number is greater than 1, the thermal boundary layer is thinner than the velocity boundary layer. If the Prandtl number is less than 1, which is the case for air under standard conditions, the thermal

The Prandtl number for water is around 10. So viscous effects cause the velocity boundary layer to develop about ten times faster than the thermal boundary layer.

For oils, the value is much larger than for water, so this difference is even more pronounced.

For Mercury, it goes the other way around. Mercury, being a metal, conducts heat very well. So in mercury, the thermal boundary layer would develop faster than the viscous boundary layer.

For many gases, the value is lower than 1, so then the thermal boundary layer grows faster than the velocity boundary layer.

REYNOLDS NUMBER DUrho/μ: A SHORT NOTE

In one line, the Reynolds number is the ratio of the inertial force that propels a fluid flow to the viscous forces that resist it. The Reynolds number is a dimensionless number used to categorize the fluids systems in which the effect of viscosity is important in controlling the velocities or the flow pattern of a fluid.

$Re = F$ inertia/F viscous = [kg/m^3 × m/s × m]/[$Pa \times s$] = F/F **[Dimensionless number]**

The SI unit for dynamic viscosity is the Pascal-second

SI unit for pascal *Pa* is kg·m^{-1}·s^{-2}

where ρ (kg/m^3) is the density of the fluid, V (m/s) is the characteristic velocity of the flow, and L (m) is the characteristic length scale of flow.

Background Reading

Fluid Forces and Energy

Fluid flows, like any mechanical system, obey the three basic conservation rules of mass (the continuity principle), momentum (Newton's second law), and energy (Newton's third law) (the first law of thermodynamics). These can be expressed in words as follows:

- **Conservation of Mass:** The amount of fluid entering a region of space in unit time is equal to the amount leaving plus the amount stored inside the region due to density variations. The latter is, of course, zero for incompressible flows.
- **Conservation of Momentum:** Conservation of momentum can be expressed in one of two ways, either directly as the amount of fluid momentum leaving a region of space minus the amount entering in unit time is equal to the sum of the fluid forces on the boundaries of that space or indirectly as the mass of the fluid element multiplied by the total (i.e., spatial and temporal) acceleration being equal to the sum of the forces.

There are three types of forces that act on a fluid element:

[1] Pressure forces that act perpendicular to the element's faces; [2] friction forces that act parallel to the element's faces; and [3] body forces such as gravity. In low-speed atmospheric fluxes, the latter are typically small.

Conservation of Energy

Energy conservation can be expressed as the heat transfer minus the work done by the fluid [zero for incompressible flows] equals the energy leaving the region of space minus the energy entering the unit of space.

The forces imposed on the surfaces over which the fluid passes predominantly are pressure forces and friction forces. However, it is often difficult to distinguish between these two types of force in any one situation, and very often it is only possible to specify the overall forces.

Reynolds Number

The Reynolds number is a dimensionless number used to categorize the fluids systems in which the effect of viscosity is important in controlling the velocities or the flow pattern of a fluid. The Reynolds number is the ratio of inertial forces to viscous forces.

What is an Inertial Force and Viscous Force?

Newton's First Law: Inertia

An object at rest remains at rest, and an object in motion remains in motion at a constant speed and in a straight line unless acted on by an unbalanced force.

Newton's first law states that every object will remain at rest or in uniform motion in a straight line unless compelled to change its state by the action of an external force. This tendency to resist changes in a state of motion is inertia. If all the external forces cancel each other out, then there is no net force acting on the object. If there is no net force acting on the object, then the object will maintain a constant velocity. If that velocity is zero, then the object remains at rest. If an external force acts on an object, the velocity will change because of the force.

Inertial Force

Inertial force is the force due to the momentum of the fluid. This is usually expressed in the momentum equation by the term $\rho(du/\mathrm{d}t)$ or $(\rho v)\ v$. So, the denser the fluid is, and the higher its velocity, the more momentum (inertia) it has. According to Newton's 2nd Law, "The Rate of change of momentum of a body is equal to the resultant force acting on" Force = Rate of change of momentum = $m\ [u^2 - u^1] = Q\ \rho\{u^2 - u^1]$ [m and Q are mass and volume and u^1 and u^2 are velocities at two points of the fluid and ρ is the density.] Momentum forces are inertial forces. Inertia is a property of matter as explained above by which it continues in its existing state of rest or uniform motion in a straight line unless that state is changed by an external force. The force that keeps fluid moving against viscous [viscosity] forces is the inertial force. The inertial forces are characterized by the product of the density rho times the velocity V times the gradient of the velocity dV/dx.

Viscous Force

It is a measure of a fluid's resistance to flow. Viscous forces are shear forces. The viscous forces are the forces due to the friction between the layers of any real fluid. Since the fluid particles are very closely packed so necessarily there is friction between layers of fluid. When a body moves in a fluid, it can experience two types of forces. Forces normal to the surface are called pressure forces and forces tangential to the surface are called shear forces. Pressure forces change linear momentum and shear forces are described by 'viscosity effects' and the forces become viscous forces that oppose the flow

Inertial force (in the direction of flow) and viscous force (in the reverse direction of flow) act in opposite direction.

The viscous force counteracts or counterbalances the inertial force. In the case of fluid flow, this is represented by Newton's law, $\tau x = \mu \, dv$ dy. This is only dependent on the viscosity and gradient of velocity.

Reynolds Number is Inertial force/Viscous Force

The derivation of Re = DV ρ/μ

The Reynolds number is the ratio of the inertial force that propels a fluid flow to the viscous forces that resist it.

$$\text{Re} = \text{Inertial force/Viscous force} = DV\,\rho/\mu \qquad \ldots (1)$$

Equation [1] can also be written as

$$\text{Re} = DV/\nu \qquad \ldots (2)$$

Here, μ is the dynamic viscosity of the fluid, and ρ is the density of the fluid. The ratio $\nu = \mu/\rho$ = is termed the kinematic viscosity.

If we multiply the equation [1] both the numerator and the denominator by the average velocity V, we get

$$Re = \rho\, V^2/\mu\, [V/D] \qquad \ldots (3)$$

In equation [3], we see that the denominator represents characteristic shear stress in the flow because it is the product of the viscosity of the fluid and a characteristic velocity gradient obtained by dividing the average velocity by the diameter of the tube. The numerator, in contrast, describes inertial stress; recall that the larger the density, the more massive material is, and mass is a measure of inertia.

This is how Re is a ratio of Inertial stress/Viscous stress.

Stress is force per unit area. Therefore, it is common to express the physical significance of the Reynolds number as Re = Inertia Force/Viscous fore.

$$Re = F\text{ inertia}/F\text{ viscous} = [\text{kg/m}^3 \times \text{m/s} \times \text{m}]/[Pa \times s] = F/F$$

The SI unit for the dynamic viscosity is the Pascal-second

SI unit for pascal *Pa* is $\text{kg}\cdot\text{m}^{-1}\cdot\text{s}^{-2}$

where ρ (kg/m^3) is the density of the fluid, V (m/s) is the characteristic velocity of the flow, and L (m) is the characteristic length scale of the flow.

PHYSICAL SIGNIFICANCE OF REYNOLDS NUMBER

At large Reynolds numbers, the inertial forces, are large relative to the viscous forces, and thus the viscous forces cannot prevent the random and rapid fluctuations of the fluid. At small or moderate Reynolds numbers, however, the

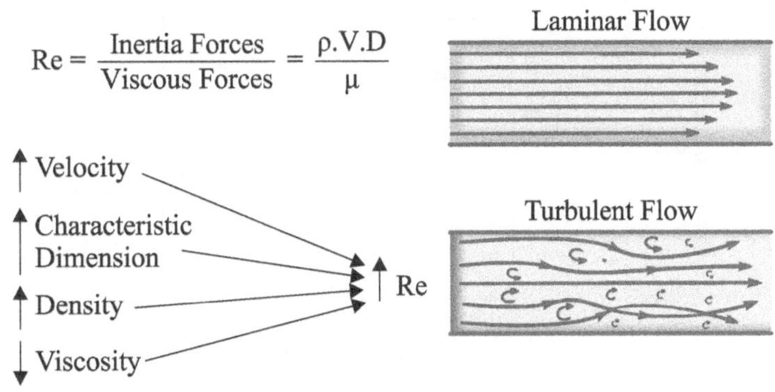

Figure 3

viscous forces are large enough to suppress these fluctuations and to keep the fluid "in line." Thus, the flow is turbulent in the first case and laminar in the second. In a circular pipe when the Reynolds number is < 2300 the flow will be always viscous. When Reynolds number > 4000, the flow will be always turbulent. The transition from laminar to turbulent flow does not occur suddenly; rather, it occurs over some region in which the flow fluctuates between laminar and turbulent before it becomes fully turbulent. The intense mixing of the fluid in turbulent flow as a result of rapid fluctuations enhances momentum transfer between fluid particles, which increases the friction force on the surface and thus the required pumping power. The friction factor reaches a maximum when the flow becomes fully turbulent.

Laminar to Turbulent Flow Transition

It is certainly desirable to have precise values of Reynolds numbers for laminar, transitional, and turbulent flows, but this is not the case in practice. It turns out that the transition from laminar to turbulent flow also depends on the degree of disturbance of the flow by surface roughness, pipe vibrations, and fluctuations in the flow. Under most practical conditions, the flow in a circular pipe is laminar for *Re* < or equal 2300, turbulent for *Re* > or equal 4000, and transitional in between.

Flow Regimes

For low Reynolds numbers, the behavior of a fluid depends mostly on its viscosity and the flow is steady, smooth, viscous, or laminar.

For high Reynolds numbers, the momentum of the fluid determines its behavior more than the viscosity and the flow is unsteady, churning, roiling, or turbulent.

For intermediate Reynolds numbers, the flow is transitional—partly laminar and partly turbulent as said above.

Turbulent Flow in Pipes

When a flow is turbulent, it comprises eddying motions of various sizes, and a significant portion of the mechanical energy in the flow is used to generate these eddies, which eventually lose their energy as heat. As a result, the drag of a turbulent flow is greater than the drag of a laminar flow at a given Reynolds number. Furthermore, surface roughness affects turbulent flow, so increasing roughness increases drag.

Fluid particles flow in an ordered manner along route lines in laminar flow. Molecular diffusion transports momentum and energy through streamlines. In turbulent flow, swirling eddies move mass, momentum, and energy far faster than molecular diffusion, greatly enhancing mass, momentum, and heat transfer. As a result, turbulent flow is associated with much higher values of friction, heat transfer, and mass transfer coefficients.

Pipe Friction

Pressure drops seen for the fully developed flow of fluids through pipes can be predicted using the Moody diagram which plots the Darcy-Weisbach friction factor f against Reynolds number Re and relative roughness ε/D. The diagram clearly shows the laminar, transition, and turbulent flow regimes as the Reynolds number increases. The nature of pipe flow is strongly dependent on whether the flow is laminar or turbulent.

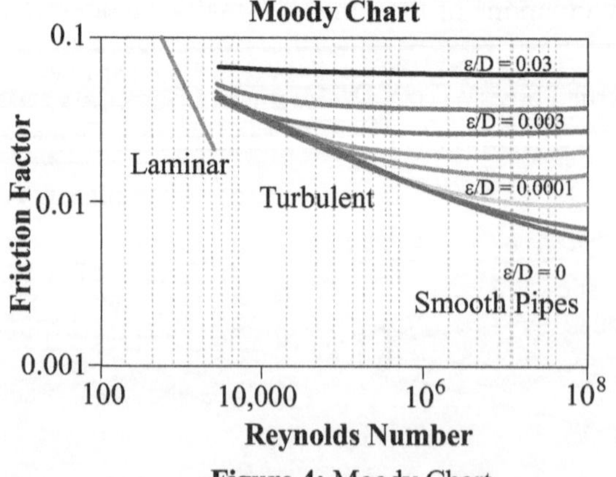

Figure 4: Moody Chart

NUSSELT NUMBER: ROLE IN HEAT TRANSFER

In summary, the Nusselt number (*Nu*) is an important factor in evaluating the heat transfer coefficient (*h*), as it represents the ratio of convective to conductive heat transfer, $Nu = h\ L/K$. [*h* is heat transfer coefficient, *L* is the length and *k* is the thermal conductivity]. However, the relationship between Nu, Reynolds number (Re), and Prandtl number (Pr) is more complex.

In general, *Nu* is a function of *Re* and *Pr*, with different correlations for different flow regimes (such as laminar or turbulent) and heating conditions. As the temperature increases, the momentum diffusivity typically decreases and the thermal diffusivity increases, resulting in a decrease in Prandtl number (*Pr*).

As the temperature increases, therefore, the Reynolds number (which is a measure of the flow velocity) typically increases, while the Prandtl number (which is a measure of the relative importance of momentum and thermal diffusivities) decreases. This change in *Re* and *Pr* can affect the overall heat transfer process, transitioning from a dominant momentum diffusivity to a dominant thermal diffusivity.

The Nusselt number is a dimensionless number that relates the convective heat transfer coefficient to the thermal conductivity of a fluid. It plays an important role in heat transfer analysis and design processes for various engineering applications.

The Nusselt number is used to characterize the heat transfer behavior of a fluid flow by correlating the heat transfer rate to the fluid properties and flow characteristics. It helps in predicting the convective heat transfer coefficient for different fluid flow conditions, such as laminar or turbulent flow regimes.

Remember you can't arbitrarily fix Reynolds number in heat transfer. A high Reynolds number is a compromise to momentum transfer and it pushes you to conductive heat transfer from convective heat transfer limiting the heat transfer to the thermal conductivity of the fluid. This is all about this article.

Nusselt number is a ratio of total heat transfer relative to conductive heat transfer.

It is a dimensionless number. Dimensionless numbers reduce the number of variables that describe a system, lowering the amount of experimental data required to make physical phenomenon correlations to scalable systems. Dimensionless numbers are useful for determining fluid flow characteristics.

Nusselt [Nu]number is the ratio of total heat transfer to conductive heat transfer at a boundary in a fluid. Total heat transfer combines conduction and convection. Convection includes both advection (fluid motion) and diffusion (conduction). The convection and conduction heat flows are parallel to each other and the surface normal of the boundary surface.

A Nusselt number of order one represents heat transfer by pure conduction. A value between one and 10 is characteristic of slug flow or laminar. A larger Nusselt number corresponds to more active convection, with turbulent flow typically in the 100–1000 range.

$$\text{Nu} = \text{Total heat transfer/Conductive heat transfer} = h \times L/K$$

where *h* is the convective heat transfer coefficient of the flow or the heat transfer coefficient, *L* is the characteristic length and *k* is the thermal conductivity of the fluid.

The formula to connect Nu with the heat transfer coefficient, *h* is

$$Nu = hL/k,$$

where *h* is the heat transfer coefficient, *L* is the characteristic length, and *k* is the thermal conductivity of the fluid.

The Dittus-Boelter equation is used (for turbulent flow) to calculate the Nusselt number.

The Dittus-Boelter equation is a heat transfer correlation used to calculate the Nusselt number for turbulent flow in smooth pipes:

$$Nu = 0.023\ Re^{0.8}\ Pr^{n}$$

$n = 0.4$ for heating and $n = 0.3$ for cooling

The Dittus-Boelter equation is valid when: $0.7 \le \text{Pr} \le 160$, $\text{Re} \le 10{,}000$, and $\text{L/D} \le 10$.

Nusselt number is a function of Reynold's and Prandtl's.

ROLE OF REYNOLDS AND PRANDTL'S NUMBERS IN HEAT TRANSFER

Reynolds Number

The **Reynolds** number represents the **ratio of inertial forces to viscous forces** within a fluid subjected to relative internal movement due to different fluid velocities. A region where these forces change behavior is known as a boundary layer, such as the bounding surface in the interior of a pipe.

$$Re = Du\rho/\mu,$$

where ρ is the fluid density, u is the average fluid velocity, D is the system's cross-sectional length and is μ the dynamic fluid viscosity.

Inertial Forces

The force that keeps fluid moving against viscous [viscosity] forces is the inertial force. The inertial forces are characterized by the product of the density rho times the velocity V times the velocity dV/dx gradient.

Prandtl Number

The prandtl number is a dimensionless number. It is a ratio of momentum heat transfer and thermal [conduction] heat transfer.

Prandtl number: Role in heat transfer

Prandtl number is a ratio of the measure of velocity and thermal boundary layers.

The Prandtl number is written as:

$$Pr = Cp \times \mu/k$$

Cp is specific heat, μ is dynamic viscosity and k is thermal conductivity.

How do *Re* and *Pr* work together? Does any compromise occur?

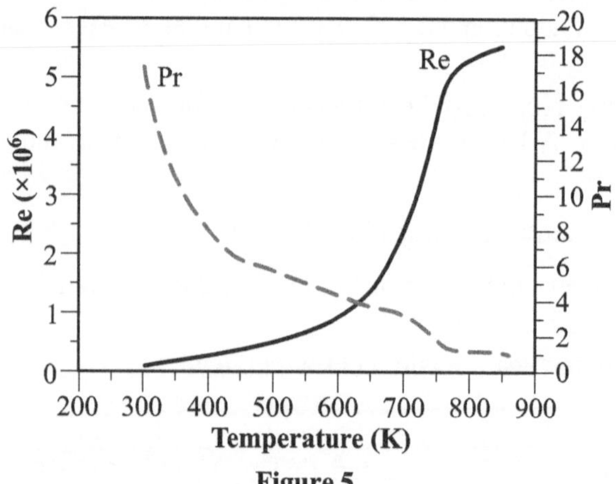

Figure 5

Explanation

Reynolds number and momentum transfer

At large Reynolds numbers, the inertial forces, are large relative to the viscous forces, and thus the viscous forces cannot prevent the random and rapid fluctuations of the fluid. At small or moderate Reynolds numbers, however, the viscous forces are large enough to suppress these fluctuations and to keep the fluid "in line." Thus, the flow is turbulent in the first case and laminar in the second. In a circular pipe when the Reynolds number is < 2300 the flow will be always vicious. When Reynolds number > 4000, the flow will be always turbulent. The transition from laminar to turbulent flow does not occur suddenly; rather, it occurs over some regions in which the flow fluctuates between laminar and turbulent before it becomes fully turbulent. The intense mixing of the fluid in turbulent flow as a result of rapid fluctuations enhances momentum transfer between fluid particles, which increases the friction force on the surface and thus the required pumping power.

Viscosity aids the momentum transfer and adds to viscous forces

One of the causes of viscosity is the interchange of momentum between layers of fluid moving with different velocities due to the chaotic molecular motion. The motion of molecules in a flowing fluid is the superposition of the motion of the fluid media and of the chaotic motion of molecules due to thermal agitation. Molecules are transferred from one layer to another. Thus, molecules with higher momentum move into layers with lower momentum. There they mix with slower molecules and increase the overall momentum of that layer. Similarly, molecules with lower momentum moving into the higher momentum layer reduce the overall momentum of that layer. Both processes tend to reduce the relative velocity between the layers. This momentum exchange generates an effective shear force between the two layers.

Role of Reynolds Number on Heat Transfer

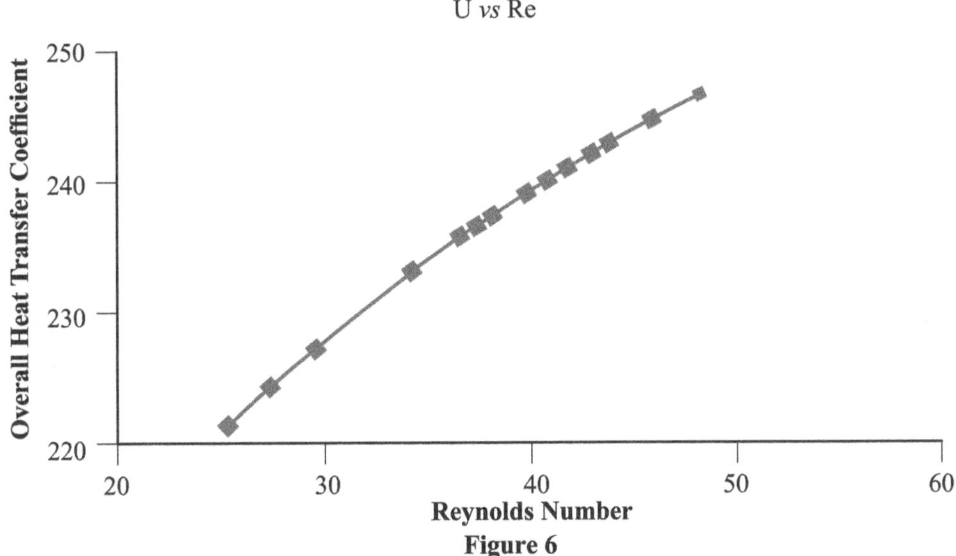

Figure 6

This is the Reynolds number *vs.* the overall heat transfer coefficient graph. The graph shows the variation of the overall heat transfer coefficient with the Reynolds number. It may be observed that the overall heat transfer coefficient increases from 220 to 250 W/m 2-K with the increase of Reynolds number.

Therefore, the overall heat transfer coefficient which depends on the convective heat transfer coefficient increases with Reynolds number. To summarize, while the Reynolds number overcomes the viscous forces it also compromises with the momentum transfer at high Re number due to the reduction of velocity gradient.

This is an important point in this discussion. Let me repeat it. A high Reynolds number overcomes viscous forces but may compromise momentum transfer due to a reduction in velocity gradients. This is because at high Reynolds numbers, the inertia of the flowing fluid dominates over viscous forces, leading to increased momentum transfer. However, the reduction in velocity gradients can decrease the efficiency of momentum transfer in certain cases.

CHAPTER 37

Heat Engines

Heat engine thermodynamics involves the study of the conversion of heat into mechanical work. The efficiency of a heat engine is determined by the ratio of the work output to the heat input. The Carnot cycle is a theoretical model that represents the maximum efficiency for a heat engine operating between two temperature reservoirs. Real heat engines are less efficient due to factors such as friction and heat loss. Thermodynamics principles are used to optimize the performance of heat engines and to improve energy conversion processes in various applications.

CARNOT *vs.* RANKINE CYCLE

In brief, the Carnot heat engine is theoretical. It was conceived to deliver maximum efficiency to become a benchmark for efficiency that a heat engine can produce. Rankine cycle is a modified version of the Carnot cycle that addresses some of the weak links of the Carnot cycle and makes it practical to use on the ground. The Rankine cycle is extensively used in power generation.

The Carnot cycle is a constant internal energy cycle at constant temperature while in the Rankine cycle the change in the enthalpy between the hot and cold reservoirs at constant pressure produces the mechanical work. In Carnot cycle the heat converts to work at constant internal energy.

In the Carnot cycle, the working fluid moves at a constant temperature through its four arms and reaches the same pressure and temperature at the end of the cycle from the point the cycle starts its journey. Thus, the net internal energy change, $dU = 0$. This means from the 1st law of thermodynamics, $dU =$ delta $Q -$ delta $W = 0$. In other words, delta $Q =$ delta W remains throughout the cycle. This makes the Carnot cycle a reversible cycle with $Q = W$ and no entropy generation. However, this makes the Carnot cycle an infinitesimally slow cycle.

The Carnot cycle has one isothermal expansion hot reservoir which adds internal energy to saturated water by expanding and breaking its H bonds releasing the huge potential energy held in intermolecular bonds and converting saturated water to saturated steam at constant temperature.

The next arm of the Carnot cycle is an adiabatic expansion, ($-W$ work) of steam that pulls internal energy to spin the turbine and cools down the steam to saturated steam temperature. The next is an isothermal compression arm ($+ W$ work) that adds back the internal energy consumed in stage 2 by compressing the fluid from stage 2 at constant temperature by reducing volume or increasing the density of water making water release of latent heat of condensation. This stage is called heat sink.

The fourth arm is adiabatic compression. This arm's (pump) job is to raise the internal energy to the same level that is stage 1 from where the working fluid (steam) starts the journey and completes the cycle. Rankine realized isothermal heat addition and heat rejection is impossible for a practical heat engine on the ground.

Rankine modified the Carnot cycle by replacing the two isothermal arms by constant pressure arms and focussed on the total energy, enthalpy at constant pressure, $dH = dU + PdV$. This opened up the option to use superheated steam in the Rankine cycle.

Essentially, Rankine tracked the total energy as it moved round the cycle while Carnot tracked the internal energy. The Rankine cycle has also two adiabatic arms like the Carnot cycle. Their jobs are similar. In stage 2 the adiabatic arm extracts enthalpy to run the turbine and in stage 4 the adiabatic arm adds back the enthalpy. In the Rankine cycle, the heat reservoirs are not reversible. They do generate entropy as the working fluid expands and gets compressed.

CARNOT AND RANKINE'S APPROACH TO HEAT ENGINE

Carnot cycle is an interplay of heat and temperature in a reversible process, $dS = Q\ \text{rev}/T$.

Carnot's idea was to conceive an ideal reversible heat engine with maximum efficiency that could become a benchmark for heat engines.

He conceived a cyclic machine that had two isothermal energy reservoirs that did not generate entropy. One is an isothermal expansion hot reservoir [typically a boiler] to supply energy isothermally and the other is a cold reservoir with isothermal compression that also acts reversibly to act as an energy sink.

Energy moves in a clockwise direction from the hot reservoir first to an adiabatic expansion station to perform work reversibly. Being an adiabatic process there is no entropy generation.

Next, is a cold reservoir where the machine deposits the residual energy after performing mechanical work.

The next station is an adiabatic compression station which fills up the energy consumed by the machine while doing mechanical work and restores the initial energy before feeding back the gas to a hot reservoir.

The entropy generation in the hot reservoir is negated in the cold reservoir as the volume of expansion of gas in the hot reservoir equals the volume of compression of gas in the cold reservoir.

This completes a Carnot cycle.

Further explanation

Please refer to the image below.

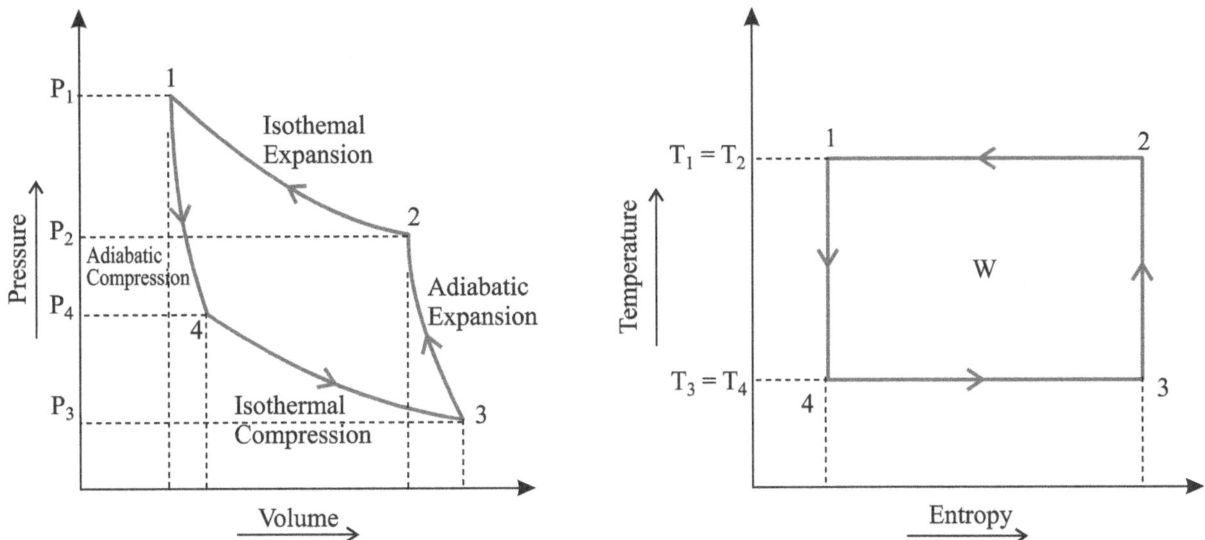

Figure 1: *P-v* and *T-s* Diagrams of a Carnot

Point 1-2: Isothermal Expansion. Heat is transferred reversibly from the high-temperature reservoir at constant temperature T_1 (isothermal heat addition or absorption). During this step the gas is allowed to expand, doing work on the surroundings. Although the pressure drops from points 1 to 2 the temperature of the gas does not change during the process because it is in thermal contact with the hot reservoir at *T*1, and thus the expansion is isothermal. Heat energy Q1 is absorbed from the high-temperature reservoir. Since heat is absorbed reversibly there is no entropy generation.

Point 2-3: Isentropic (reversible adiabatic) expansion of the gas (isentropic work output). For this step, the gas in the engine is thermally insulated from both the hot and cold reservoirs. Thus, they neither gain nor lose heat, an 'adiabatic' process. The gas continues to expand by reducing pressure, doing work on the surroundings, equal to the mechanical work done. The gas expansion without heat input causes it to cool to the "cold" temperature, T_3. The entropy remains unchanged in the adiabatic process.

Point 3-4: Isothermal Compression. Heat is transferred reversibly into the low-temperature reservoir at constant temperature T_3. (Isothermal heat rejection)

Now the gas in the engine is in thermal contact with the cold reservoir at temperature T_3. Cooling does compression work. This causes an amount of heat energy Q_2 to leave the system in the cold reservoir. There is no entropy generation since it's reversible process.

Point 4-1: Adiabatic reversible compression. Once again, the gas in the engine is thermally insulated from the hot and cold reservoirs, and the engine is assumed to be frictionless, hence reversible. During this step, the surroundings do work on the gas. compressing it, and causing its temperature to rise back to T_1 due solely to the work added to the system. No entropy generation in the reversible adiabatic compression process.

At this point, the gas is back to the same state as at the start of step 1.

This completes a Carnot cycle.

The notable point is Carnot chose isothermal reversible heat reservoirs. The question is why he chose isothermal reversible heat reservoirs despite fully realizing that for any isothermal process, the process must be very slow to be a reversible process.

The answer is he had no options other than isothermal heat reservoirs.

He did not consider constant-pressure heat reservoirs because these reservoirs are not reversible in nature. A constant pressure process has external intervention. In constant pressure process the pressure is maintained constant by supplying heat externally. A constant pressure process is not a reversible process because such processes do not remain in equilibrium with surroundings as they require external heat supply to maintain constant pressure.

That is the reason Carnot went with isothermal reservoirs.

It is only the cyclic movement of internal energy that takes place in a reversible Carnot cycle with zero change in internal energy $dU = 0$.

Rankine had a more Practical Approach

He conceived a real-life cycle with similar four arms but not constrained to deliver a non-entropy reversible process. Rankine's focus was more on a workable practical machine. Rankine's cycle represents the movement of enthalpy which is total energy unlike the movement of internal energy in a cycle in Carnot's heat engine. Rankine took a practical approach and made a departure from the Carnot cycle by replacing slow isothermal heat reservoirs which are impossible to work in a real on the ground situation by constant pressure heat reservoirs for maximum delivery of enthalpy. It may be noted that constant pressure processes are not reversible. A constant pressure is not reversible because it receives heat from its surroundings and is not in equilibrium because the system and surroundings are not in equilibrium during the expansion process.

This is one of the key differences between Carnot and Rankine cycle. Carnot cycle is a reversible cycle while Rankine cycle is not.

dH = delta Q, that is the total thermal energy. In Rankine cycle the enthalpy change at constant pressure between the hot and the cold reservoir produce the mechanical work.

The Rankine cycle is followed in most power plants.

The issue is its low efficiency. It is about 30–35%. The efficiency is limited by the cooling tower water temperature that carries the enthalpy of compression (condensation) of vapor in the constant pressure [vacuum] condenser for heat rejection.

WHY THE CARNOT EFFICIENCY EXPRESSED IN TERMS OF TEMPERATURE AND THE RANKINE EFFICIENCY IN TERMS OF ENTHALPY CHANGE?

Why Carnot Chose Temperature to Express Efficiency?

Temperature is a fundamental property of matter. Temperature is an intensive property.

Carnot efficiency is expressed in terms of temperature because Carnot theorem says that all heat engines operating between the same two thermal or heat reservoirs cannot have efficiencies greater than a reversible heat engine operating between the same. Therefore, the origin of expressing Carnot efficiency in terms of temperature lies Carnot theorem.

It means that when an ideal reversible cycle operates between two temperature reservoirs (a hot reservoir at temperature Th and a cold reservoir at temperature Tc) the efficiency of the ideal (Carnot) cycle only depends on the temperatures of the reservoirs and is given by the formula.

While temperature signifies average kinetic energy, enthalpy accounts for total energy which is internal energy and work.

Another notable thing is during an isothermal process since $dU = 0$, in a Carnot cycle, delta Q = delta W.

In Rankine cycle $dH = dU + PdV$ In constant pressure heat reservoirs.

Carnot theorem (Wikipedia)

"Carnot's theorem states that all heat engines operating between the same two thermal or heat reservoirs cannot have efficiencies greater than a reversible heat engine operating between the same reservoirs. A corollary of this theorem is that every reversible heat engine operating between a pair of heat reservoirs is equally efficient, regardless of the working substance employed or the operation details. Since a Carnot heat engine is also a reversible engine, the efficiency of all the reversible heat engines is determined as the efficiency of the Carnot heat engine that depends solely on the temperatures of its hot and cold reservoirs."

Carnot Efficiency Formula

$$\eta\_\text{Carnot} = 1 - (Tc/Th)$$

This formula highlights importance of temperature in determining the efficiency of a heat engine operating on the Carnot cycle.

Rankine Cycle

The Rankine efficiency is expressed in terms of enthalpy because it is more relevant to consider the heat and work transfers in the cycle, which are determined by changes in enthalpy of the fluid. The Rankine efficiency is given by the formula:

$$\eta R = w/Q_{\text{in}} = (h_{\text{in}} - h_{\text{out}})/h_{\text{in}}$$

Where

ηR is the Rankine efficiency

w is the work output

Q_{in} is the heat input

h_{in} is the enthalpy of the steam entering the turbine

h_{out} is the enthalpy of the steam leaving the turbine

It may be noted that Rankine cycle works between contant pressure hot and cold reservoirs that involves both internal energy and work transfer which is called enthalpy transfer ($H_{\text{in}} - H_{\text{out}}$) does the mechanical work W.

Rankine cycle is not a reversible cycle therefore, Carnot's theory does not apply to Rankine cycle. While the temperature signifies average kinetic energy, enthalpy accounts for total energy which is internal energy and work.

WHY PHASE CHANGES IN RANKINE CYCLE BUT NOT IN CARNOT?

Two questions often arise (1) why there is phase change in Rankine cycle but not in Carnot cycle and (2) how an adiabatic expansion In turbine $dQ = 0$, communicate with cold reservoir.

Discussion

The fundamental point is Carnot cycle works at constant temperature $dU = 0$, $dH = PdV$. Carnot cycle is essentially a reversible work transfer process.

Work doesn't generate entropy.

Rankine cycle is a constant pressure process $dH = dU + PdV$. Rankine cycle involves enthalpy tranfer.

Question 1

In the development of heat engine cycles, Carnot and Rankine approached the problem of energy conversion in different ways.

Carnot's idealized cycle, characterized by reversible adiabatic and isothermal processes, aimed to maximize efficiency by eliminating internal energy losses and maintaining constant internal energy.

However, the Carnot cycle does not incorporate phase changes and may not accurately reflect real-world conditions. The reason is, a phase change involves entropy generation and not a reversible process.

On the other hand, Rankine introduced a more practical approach to power generation with Rankine cycle.

By choosing constant pressure heat addition to better simulate real-world heat transfer processes, Rankine was able to utilize superheated steam and include a phase change through condensation.

This allowed for extraction of latent heat and maximization of heat to work transfer, making the Rankine cycle a more efficient and suitable real life option for power generation applications.

Overall, while the Carnot cycle provides a theoretical framework for ideal energy conversion, the Rankine cycle's practical considerations and two-phase operation make it a realistic choice for heat engine design in practical applications.

Question 2

Work transfer

It's a very interesting question as to how an adiabatic process in a turbine with heat and mass impermeable barrier. Q = 0, communicate with the cold reservoir.

Remember every thermodynamic process needs a vent. Without a vent the system becomes an isolated system. with increase of entropy bringing the system to a dead state equilibrium. In heat engines the cold reservoir acts as a vent to transfer energy in the form of Work since an adiabatic process can't transfer heat.

Here is the answer

In the adiabatic expansion process of a turbine, energy is transferred to the cold reservoir in the form of work and not heat.

Adiabatic processes, by definition, do not involve heat transfer (Q = 0), but they can still transfer energy in the form of work.

In the turbine, the expansion of the high-pressure steam (or gas) drives the turbine blades, converting the energy of the expanding fluid into mechanical work. This work output can then be used to drive other components in a power plant or machinery, ultimately aiding in the overall energy transfer and conversion process without the need for heat exchange with a cold reservoir.

BACK-PRESSURE TURBINES: APPLICATION OF SATURATED STEAM

While superheated steam is the preferred choice for a power plant with condensing turbines, saturated steam is also used in the industry and is generally known as combined heat and power (CHP) systems. In a cogeneration system, a back-pressure turbine is used to generate electricity while also capturing and utilizing the waste heat produced in the process.

Back-pressure turbines are commonly used when saturated steam is utilized for power generation. In this context, the back pressure turbine is designed to generate electricity by expanding the high-pressure saturated steam and extracting useful work from it. Unlike condensing turbines, which are primarily used with superheated steam and operate at lower pressures, back-pressure turbines are specifically designed to accommodate the relatively high pressure of saturated steam. When the saturated steam is exhausted from the back pressure turbine, it retains a significant amount of energy in the form of heat. This can be utilized for various industrial processes, heating applications, or as part of a cogeneration system to maximize the overall efficiency of the power generation process.

Condensing *vs.* Non-condensing Turbines

Condensing Turbines

These turbines condense steam at a pressure below atmospheric to extract the maximum amount of energy from it. They exhaust directly to condensers that maintain a vacuum at the discharge.

Non-condensing Turbines

Also known as back-pressure turbines, these turbines exhaust steam at a pressure closer to the process heat requirements of the facility. They maintain a higher pressure than condensing turbines.

Back-pressure Turbine Details

A back-pressure turbine is a type of steam turbine that is designed to operate with a certain amount of back pressure in the steam exhaust system. The back pressure is maintained to meet the heating or process requirements of a facility.

How a back pressure turbine works:

1. Steam is produced in a boiler and then directed to the back pressure turbine.
2. The steam enters the turbine at high pressure and high temperature, causing the turbine rotor to spin.
3. As the steam flows through the turbine blades, it loses pressure and temperature, while simultaneously transferring some of its energy to the turbine rotor. This energy transfer causes the rotor to rotate at a high speed.
4. The rotating rotor is connected to a generator, which converts the mechanical energy of the rotor into electrical energy.
5. After passing through the turbine, the steam is exhausted at a certain pressure, referred to as the back pressure, and is then used for various heating or process applications within the facility.
6. The exhaust steam may also be condensed and returned to the boiler to be used again in the steam cycle.

CHAPTER 38

Heat Pumps

The thermodynamics of a heat pump involves the transfer of heat from a low-temperature source to a high-temperature sink using a refrigeration cycle. The first law of thermodynamics states that the energy input to a heat pump must be greater than the heat it transfers to the higher temperature reservoir. The second law of thermodynamics dictates that the efficiency of a heat pump is limited by the temperature difference between the heat source and sink. The Carnot efficiency of a heat pump is the maximum possible efficiency and is determined by the temperatures of the heat reservoirs.

COP OF HEAT PUMPS

Heat pumps are a common HVAC system used for both heating and cooling purposes. When operating on cooling duty, a heat pump works by transferring heat from inside a building to the outside, effectively cooling the indoor space. This process involves extracting heat from the air inside the building, transferring it to a refrigerant, and then releasing it into the outdoor air.

On the other hand, when a heat pump is operating on heating duty, it reverses the direction of the heat transfer process. In this mode, the heat pump extracts heat from the outdoor air (even in cold temperatures) and releases it inside the building to provide warmth. The refrigerant absorbs heat from the outside air, compresses it to increase its temperature, and then transfers it to the indoor space.

The key difference between cooling and heating duties of a heat pump lies in the direction of heat transfer. While cooling duty involves transferring heat from inside to outside, heating duty involves transferring heat from outside to inside. This dual functionality makes heat pumps an energy-efficient option for maintaining a comfortable indoor environment year-round.

Coefficient of Performance [COP]

The coefficient of performance or COP of a heat pump, refrigerator, or air conditioning system is a ratio of useful heating or cooling to work (energy) required. Higher COPs signify higher efficiency, lower energy (power) consumption and thus lower operating costs.

Heat Pump

A heat pump is a device that uses work to move heat using a refrigeration cycle. In winter, a heat pump moves heat from the outdoors to warm your house. During the summer, heat pumps move heat from your house into the outdoors. Air conditioners and refrigerators work on the same principle.

COP is Not Efficiency

The Coefficient of Performance (COP) is performance rating. COP tells us how effective a heat pump or air conditioner is in transferring heat versus the amount of electrical power it consumes.

Why COP is Expressed as a Number

By convention, COP is expressed as a dimensionless number.

How Efficiency is Expressed?

$$\text{Efficiency} = \text{Work out/Work in} \times 100$$

Efficiency is derived by dividing two quantities with the same units. Energy in versus Energy out, for example.

Refrigeration requires electrical energy in and heat flow out (BTUs per hour, for example). These are incompatible units. We could talk about "BTUs per hour per kilowatt-hour" but that would be a bit awkward.

COP = Heat extracted/Electrical energy in

What is the Convention?

If the required output has to be in the form of work, we calculate efficiency. If the required output is in the form of heat, we calculate the coefficient of Performance (COP). In Refrigeration and Air Conditioning, the desired output is always in the form of heat. Hence the COP.

Difference between Efficiency and COP

Efficiency ratings, precisely, have in common with each other a factor of time. For example: SEER shows us energy efficiency over the course of a cooling season, [SEER stands for Seasonal Energy Efficiency Ratio or "Rating." This is the ratio of the cooling output of an air conditioner and heat pump over a typical cooling season, divided by the energy it uses in Watt-Hours. HSPF [Heating Seasonal Performance Factor] shows us energy efficiency over the course of a heating season, and EER [Energy Efficiency Ratio] could be used to show us energy efficiency over any period of time that we define.

Coefficient of performance demonstrates instantaneous performance without any factor of time involved. That is to say, the COP tells us exactly how much heat a piece of HVAC equipment transfers when given a specific amount of electrical input power at a specific temperature, and therefore, at a specific and instantaneous moment, instead of over a while.

How does a Heat Pump Work?

A heat pump is a device that extracts energy from the air to heat or cool a space. We call this procedure "space conditioning." Since heat pumps use electricity as an input to move heat from a cold place to a warm one, they function similarly to a heat engine in reverse. Heat pumps follow the reverse Carnot cycle. In a heat pump, the Second Law of Thermodynamics would appear to be violated, but the reason that it doesn't is heat transfer is not spontaneous in a heat pump; energy must be input for it to occur.

A heat pump is used to transfer warm air from the outside into the home by extracting and heating it further. The heat pump reverses this process for home cooling, and heat is extracted from the indoor air and expelled to the outside, just like a refrigerator or air conditioner, thereby making the inside air cooler.

Heating Cycle

A heat pump's heating cycle entails drawing heat from the outside air, increasing its temperature, and then utilizing that heated air to heat the air inside. It does so through the following procedure:

- The liquid refrigerant turns into a gas in the "evaporator" when it absorbs heat from the outside air. A "compressor" is used to pass the refrigerant through, raising the gas's pressure and temperature. The hot gas enters the area that needs to be heated and passes through "condenser coils" inside. Because the hot gas is hotter than the area, it condenses [exothermic] and transfers heat to the room.
- To cool down, the liquid eventually returns through a valve that lowers the pressure and repeats the cycle.

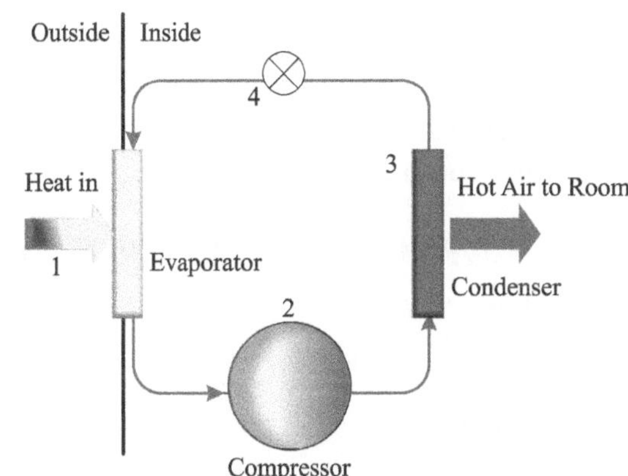

Figure 1: Heat Pump Heating Cycle

Cooling Cycle

A heat pump's cooling cycle removes heat from a space and directs it somewhere else. This is accomplished by reversing the refrigerant flow and switching the functions of the "evaporator" and "condenser coils":

The room cools because the cold refrigerant in the evaporator absorbs heat from the hotter room. After that, the compressor is used to raise the temperature. This heat is transferred to the outside air as it travels through the condenser coils. The cycle is then repeated when it expands to release pressure and cool down to a temperature below that of the room.

Performance rating [COP]: As said, the performance of a heat pump is expressed by the ratio of heat output to the work needed to be input.

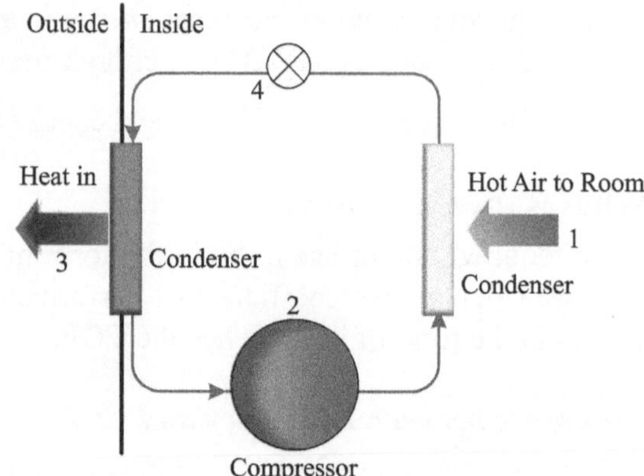

Figure 2: Heat Pump Cooling Cycle

The equation is *Q/W*

Q is the useful heat supplied or removed by the considered system.

W is the is the net work put into the system in one cycle.

The COP for heating and cooling are different because the heat reservoir of is different. For heating, the COP is the ratio of the magnitude of the heat given off to the hot reservoir (which is the heat taken up from the cold reservoir plus the input work) to the input work.

COP of cooling = Qc/W

COP heating = $Qc + W/W$

This difference should be noted. In cooling the COP is heat extracted from the hot space Qc divided by work, W. In heating, the COP is the heat extracted from the surroundings, Qc plus the work input by the compressor.

Therefore, COP for heating = $W + Qc/W = 1$ + COP for cooling.

Therefore, the COP of cooling is always less than the COP of heating.

WHY COP IS GREATER THAN ONE?

The answer is COP is a ratio of heat energy extraction and electrical energy addition.

Heat is known as second grade energy and electricity is considered as first grade energy.

For this reason 1 watt electricity = 3. 4 Btu heat.

Except in vapor absorption refrigeration the COP for all other processes > 1.

Vapor absorption refrigeration is an exception because it's COP is a ratio of heat extraction divided by heat absorption (two second grade energy).

In summary, in all types of refrigeration or heat pump processes where the input energy is electricity the COP is > 1 simply because 1 watt = 3.4 Btu. Inherently electricity is a clean energy than thermal energy.

WHY COPS OF HEAT PUMPS CAN BE VERY HIGH?

Heat pump equation

$$Qh = Qc + W$$

The purpose of a heat pump is to transfer energy by heat to a warm environment, such as a home in the winter. A heat pump supplies.

$$Qh = Qc + W$$

COP of heating = $[W + Qc]/W = Qh/W = 1 + Qc/W = 1 + COP$

It may be noted that while the COP of cooling is Qc/W, the COP of heating is $[Qc + W/W]$.

ISOTHERMAL-ISOBARIC REFRIGERATION EVAPORATOR

First point: Isothermal + isobaric process extracts "work" $dH = pdV$ since, in the equation $dH = dU - d$ (PV), ($dU =$ and $VdP = 0$).

How does it Work?

The key is refrigeration is a work transfer process. It is not a heat transfer process. This may surprise many viewers.

An isothermal process is a thermodynamic process in which the temperature of a system remains constant throughout the process. This means that the internal energy of the system remains constant.

An isobaric process is a thermodynamic process in which the pressure of a system remains constant throughout the process. This means that the system is allowed to exchange heat and work with its surroundings, but the pressure remains the same.

In summary, an isothermal process has constant temperature, while an isobaric process has constant pressure.

What Happens when Both Work Together

Example: Refrigeration Evaporator

In thermodynamics, isothermal and isobaric processes are important concepts in understanding the behavior of systems undergoing changes in pressure, temperature, and volume.

An isothermal process typically involves keeping the temperature of a system constant, which means that the heat transfer into or out of the system is maintained at a constant temperature. This process is represented by a horizontal line on a pressure-volume (PV) diagram, indicating that the temperature remains constant as volume and pressure change.

An isobaric process, on the other hand, involves keeping the pressure of a system constant, while allowing temperature and volume to change. This can be achieved by allowing the system to exchange heat with the surroundings while keeping the pressure constant. On a PV diagram, an isobaric process is typically shown as a vertical line, indicating that the pressure remains constant as volume and temperature change.

The meaning of isothermal and isobaric process is work extraction..

Refrigeration Evaporator

An example is refrigeration evaporator.

In the evaporator of a refrigeration system, the refrigerant undergoes an isothermal and isobaric process

The change in enthalpy is equal to:

$$dH = dU + d(PV)$$

At constant temperature dU (internal energy change) = 0

At constant pressure $VdP = 0$

Therefore, enthalpy (Total energy) change in evaporator = PdV

$$dH = PdV$$

This formula represents the heat absorbed at a constant temperature in an isothermal and isobaric process. In the context of a refrigeration system, this would correspond to the heat extraction in the evaporator where the refrigerant absorbs heat to evaporate at a constant temperature. Work in evaporator = $-pdV$ (expansion) in the subsequent step the compressor adds back $+pdV$ and converts work back to heat. This is the heat extracted from hot space. This is "Work in" as watt by compressor.

This heat finally gets rejected in condenser at constant pressure. This is "heat out" as Btu.

COP = W out (Heat out in condenser)/W in (power input by compressor or Btu/Watt.

HOW A VAPOR COMPRESSION REFRIGERATION EVAPORATOR AND CONDENSER WORK?

It is a vapor compression process the aim is to raise the refrigerant pressure, as it flows from an evaporator.

The vapor compression refrigeration cycle involves four components: compressor, condenser, expansion valve/throttle valve and evaporator.

The Focus is on Evaporator and Condenser

Evaporator is an isothermal and isobaric process. Its pressure is maintained by the throttling valve and temperature is fixed by the saturation temperature of the refrigerant.

Before proceeding further let's clarify following terms.

Isothermal Process

In an isothermal process, the temperature of a system remains constant while the pressure and volume change. When PV = constant, temperature automatically gets fixed. This is achieved by simultaneously and oppositely changing both the pressure and volume.

Isobaric Process

In an isobaric process, the pressure of a system remains constant while the volume and temperature change. This is achieved by adding heat, which increases the gas's internal energy and does work as its volume changes.

A refrigeration evaporator is both an isobaric and isothermal process:

In the Context of Refrigeration

The pressure of the refrigerant remains constant as it flows through the evaporator.

The temperature inside the evaporator remains constant fixed by saturation temperature. The internal energy remains constant and also at constant pressure the VdP component of work remains zero.

While absorbing heat, refringent expands and does – PdV work while converting to vapor. Both sensible and latent heat transfer occurs in evaporator

Compressor

The vapor from evaporator moves to adiabatic compressor which compresses the gas and raises pressure and add internal energy. The heat generated during compression is equal to heat the refrigerant absorbed in the hot space and the frictional heat by the compressor. The work done by compressor $Is + PdV$.

Condenser

The condenser in a vapor compression cycle is an isobaric process, which means that heat is rejected at a constant pressure.

The compressed gas from compressor gets condensed by a process of exothermic compression process in condenser to reject the heat carried from the hot space. In addition, the condenser also extracts heat of compression – frictional heat from compressor.

Summary

Evaporator

The evaporator is an expansion process that allows refrigerant to boil and expand in a controlled environment. The refrigerant absorbs heat from the surrounding area and turns from a liquid to a vapor.

Condenser

The condenser is a compression process that cools and condenses the refrigerant gas by rejecting the heat. The refrigerant vapor enters the condenser and is condensed at a constant.

The heat rejected in the condenser is always greater than the heat absorbed in the evaporator. Since it includes the heat produced by friction. This also reflects in COP = Btu/Watt.

CHAPTER 39

Fluid Dynamics

Fluid dynamics involves the study of the behavior of fluids in motion, while thermodynamics deals with the transfer of heat and energy within a system. The principles of thermodynamics play a important role in understanding the behavior of fluids, as changes in temperature can impact fluid properties such as viscosity and density. Fluid flow processes are often governed by thermodynamic laws, such as the conservation of mass and energy, as well as the second law of thermodynamics which dictates the direction of heat transfer. Both disciplines are interconnected when studying systems such as heat engines or refrigeration cycles, where fluid dynamics and thermodynamics must work together to optimize efficiency.

FLUID FLOW FROM LAMINAR TO TURBULENT COMPLEXITIES

Bernoulli's equation does not apply during flow transition. Bernoulli's equation is only applicable to steady, incompressible, and frictionless flows. It cannot be used when the flow is rotational, turbulent, or unsteady. Bernoulli's equation is no longer applicable when the flow becomes excessively viscous or compressible and when the Reynolds number is large enough that the flow transitions to turbulence. There are laws which don't strictly apply under the conditions we often use. They introduce errors. Darcy's law is an example. Darcy's law is invalid for turbulent flow, which can occur in very coarse-grained sediments or in pipes with a Reynolds number above 4000.

Following changes occur during transition from laminar to turbulent flow.

- **Wall shear stress:** The frictional force exerted by the fluid on a surface change. In laminar flow, wall shear stress is directly proportional to the velocity gradient near the surface. In turbulent flow, wall shear stress is more complex and varies with turbulent fluctuations.
- **Mixing:** In laminar flow, there is little mixing between adjacent layers. In turbulent flow, mixing is highly

Flow Separation

Turbulent Flow is Less Likely to Separate

Turbulent flow is more resistant to separation than laminar flow. This is because turbulent flow has better wall-normal momentum transport. Flow separation can occur when flow transitions from laminar to turbulent:

- **Transitional separation:** This type of separation occurs when the flow transitions from laminar to turbulent, and then back to laminar. It's characterized by a more abrupt separation of the boundary layer from the surface than laminar separation
- **Laminar separation bubble:** This occurs when the flow transitions from laminar to turbulent over a surface due to an adverse pressure gradient. The region between the laminar separation and turbulent reattachment is called a laminar separation bubble
- **Increase in flow velocity:** The transition from laminar to turbulent flow is typically triggered by an increase in flow velocity. As the velocity of the fluid increases, the flow patterns become more chaotic and irregular, ultimately leading to turbulent flow.
- **Formation of eddies and vortices**: In laminar flow, the fluid moves in smooth, orderly layers with little to no mixing between them. In turbulent flow, however, the fluid experiences non-linear, chaotic motion that results in the formation of eddies and vortices. These structures cause significant mixing of fluid particles and can enhance heat and mass transfer.
- **Increase in pressure drop**: Turbulent flow is associated with higher pressure drops compared to laminar flow. This occurs because of the increased mixing and turbulence in the flow, which results in higher frictional losses.

- **Enhancement of heat transfer**: Turbulent flow is more effective at transferring heat compared to laminar flow. The chaotic motion of fluid particles in turbulent flow results in better mixing and increased contact between the fluid and the heat transfer surface, leading to higher heat transfer coefficients.
- **Increase in energy dissipation**: Turbulent flow is characterized by higher levels of energy dissipation compared to laminar flow. This is due to the formation of eddies and vortices, which require additional energy to maintain the chaotic motion of the fluid.

Overall, the transition from laminar to turbulent flow is accompanied by significant changes in flow behavior, including increased mixing, flow separation, pressure drop, heat transfer, and energy dissipation. These changes have important implications for a wide range of industrial and natural processes involving fluid flows.

Complexity

The transition from laminar to turbulent flow significantly increases the complexity and unpredictability of fluid behavior due to the following reasons:

- **Chaotic motion**: Turbulent flow is characterized by chaotic, irregular motion of fluid particles. This chaotic motion leads to the formation of vortices, eddies, and swirling patterns, which constantly evolve and change in a seemingly random manner. The unpredictable nature of these flow structures makes it difficult to predict the behavior of the fluid at any point in time.
- **Increased mixing**: Turbulent flow is highly efficient at mixing different fluid layers and particles, leading to rapid and thorough mixing of fluid components. This results in a loss of distinct layers and boundaries within the fluid, making it challenging to track individual fluid particles or predict their trajectories.
- **Energy dissipation**: Turbulent flow is associated with higher levels of energy dissipation compared to laminar flow. The constant generation and dissipation of turbulent kinetic energy in the form of eddies and vortices contribute to the complexity and unpredictability of fluid behavior. This energy dissipation leads to fluctuations in flow properties such as velocity, pressure, and temperature.
- **Non-linear dynamics**: Turbulent flow exhibits non-linear behavior, meaning that small changes in initial conditions can lead to significant changes in the flow patterns and behavior over time. This sensitivity to initial conditions, known as the butterfly effect, makes it challenging to accurately predict the behavior of turbulent flows over longer time scales.
- **Lack of universal laws**: Unlike laminar flow, which can be described by simplified mathematical equations such as the Navier-Stokes equations, turbulent flow does not have universally applicable laws that can accurately predict its behavior under all conditions. The complexity and unpredictability of turbulent flow often require the use of computational fluid dynamics (CFD) simulations and empirical models to understand and predict fluid behavior.

Overall, the transition from laminar to turbulent flow introduces a high degree of complexity, variability, and unpredictability in fluid behavior, making it challenging to accurately predict and control turbulent flows in engineering and natural systems.

WHAT IS A LIQUID? HOW LIQUID IS DIFFERENT FROM SOLID?

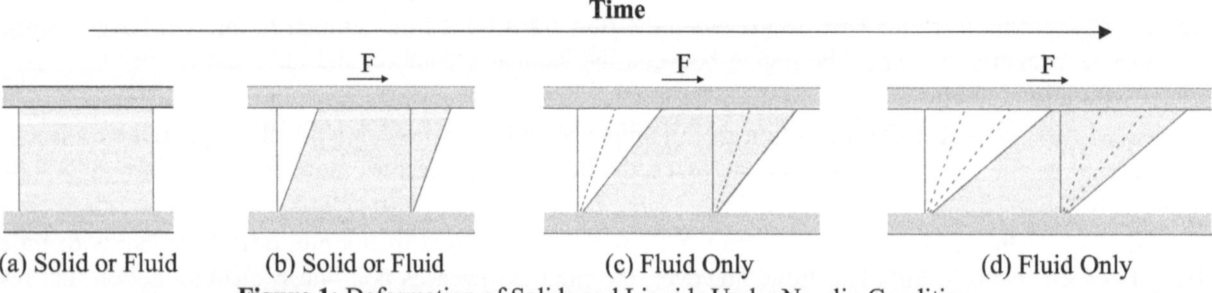

Figure 1: Deformation of Solids and Liquids Under No-slip Condition

Because of the no-slip condition, a fluid in contact with a solid surface does not slip-it has the same velocity as the surface. The amount of deformation of the solid is determined by its modulus of rigidity. The rate of fluid deformation is determined by the fluid's viscosity. Solids are described as elastic, while fluids are described as viscous.

Table 1

Solid	*Fluid*
• More Compact Structure • Attractive Forces between the molecules are larger therefore more closely packed • Solids can resist tangential stresses in static condition • Whenever a solid is subjected to shear stress (a) It undergoes a definite deformation α or breaks (b) α is proportional to shear stress upto some limiting condition • Solid may regain partly or fully its original shape when the tangential stress is removed	• Less Compact Structure • Attractive Forces between the molecules are smaller therefore more loosely packed • Fluids cannot resist tangential stresses in static condition. • Whenever a fluid is subjected to shear stress (a) No fixed deformation (b) Continious deformation takes place until the shear stress is applied • A fluid can never regain its original shape, once it has been distorded by the shear stress

Source: Fundamentals of fluid mechanics by Fox and Mcdonald.

What is Tangential or Shearing Force?

Explanation

Definition of stress: Stress is defined as a force per unit area, acting on an infinitesimal surface element.

Stresses have both magnitude (force per unit area) and direction, and the direction is relative to the surface on which the stress acts. There are normal stresses and tangential stresses.

Pressure is an example of normal stress, and acts inward, toward the surface, and perpendicular to the surface.

A shear stress is an example of tangential stress, i.e., it acts along the surface, parallel to the surface. Pressure is an example of normal stress, and acts inward, toward the surface, and perpendicular to the surface. Friction due to fluid viscosity is the primary source of shear stresses in a fluid.

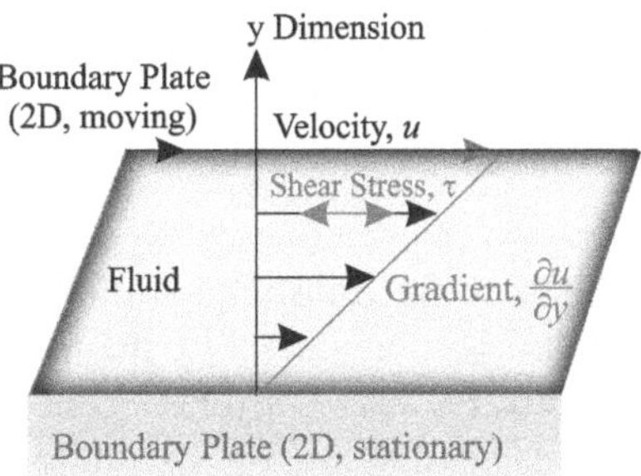

Figure 2

When a fluid flows, it is assumed to do so in layers, as shown in the image above. According to the 'No Slip Condition,' the layer of fluid in contact with the surface over which it is moving (the bottom-most layer) has zero velocity. The velocity is present at the free surface of the flowing fluid (the topmost layer). As a result, each layer drags the layer below it. This dragging force will be a shear force on a horizontal plane that is perpendicular to the plane of the image. This means that each layer exerts a shear force on the layer beneath it. Newton's Law of Viscosity describes the relationship between shear stress and velocity gradient (change in velocity with respect to distance).

Fluid at Rest

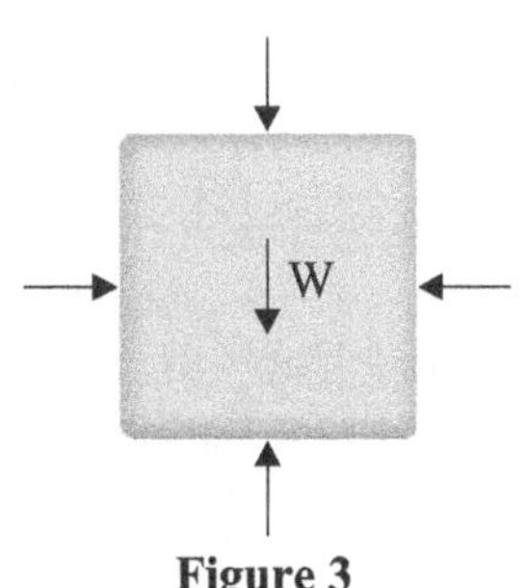

Figure 3

Consider a tiny fluid element in a case where the fluid is at rest. A fluid at rest can have only normal stresses since a fluid at rest cannot resist shear stress. In this case, the sum of all the forces must balance the weight of the fluid element. This condition is known as hydrostatics. Here, pressure is the only normal stress that exists. Pressure always acts positively inward.

Obviously, the pressure at the bottom of the fluid element must be slightly larger than that at the top, in order for the total pressure force to balance the weight of the element. Meanwhile, the pressure at the right face must be equal to that on the left face, so that the sum of forces in the horizontal direction is zero. This diagram is two-dimensional, but an actual fluid element is three-dimensional. Hence, the pressure on the front face must also balance that on the back face.

In a fluid at rest, all frictional and inertial stresses vanish and the state of stress in the system is called hydrostatic. When this condition of $V = 0$ is applied to the Navier–Stokes equations, the gradient of pressure becomes a function of body forces only. In this part, we only consider fluids at permanent rest without any macroscopic local or global motion. Microscopically, fluids like all kinds of matter are of course never completely at rest because of random molecular motion (heat), and internal atomic motion.

Fluid in Motion

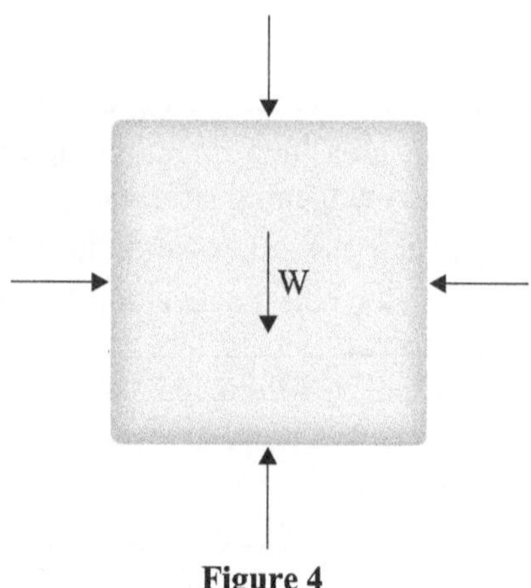

Figure 4

Consider a tiny fluid element (a very small chunk of the fluid) that is moving around in some flow field. Since the fluid is in motion, it can have both normal and shear stresses, as shown by the free-body diagram. The vector sum of all forces acting on the fluid element must equal the mass of the element times its acceleration (Newton's second law). Likewise, the net moment about the center of the body can be obtained by summing the forces due to each shear stress times its moment arm. As the size of the fluid element shrinks to "zero," i.e., negligibly small, the shear stress acting on one face of the element must be the same magnitude as those acting on the other faces. Otherwise, there would be a net moment, causing the fluid element to spin rapidly!

[*Note:* To obtain force, one must multiply each stress by the surface area on which it acts, since stress is defined as force per unit area.]

If shear stress is applied to the surface of a solid, the solid will deform a little, and then remain at rest (in its new distorted shape). One can say that the solid (at rest) is able to resist the shear stress.

Now consider a fluid (in a container). When shear stress is applied to the surface of the fluid, the fluid will continuously deform, i.e., it will set up some kind of flow pattern inside the container. In other words, one can say that the fluid (at rest) is unable to resist the shear stress. That is to say, it cannot remain at rest under the application of shear stress.

Summary: Fluid is any liquid or gas or generally any material that cannot sustain a tangential, or shearing, force and undergoes a continuous change in shape when subjected to such stress. This continuous and irrecoverable change of position of one part of the material relative to another part when under shear stress constitutes flow, a characteristic property of fluids. In contrast, the shearing forces within an elastic solid, held in a twisted or flexed position, are maintained; the solid undergoes no flow and the spring comes back to its original shape. Compressed fluids can spring back to their original shape, too.

In nutshell, when the vector sum of all forces acting on a fluid element becomes equal to the mass of the element times its acceleration (Newton's second law), the fluid becomes unstable. The fluid begins to deform in the direction of stress on the surface [tangential stress] with little push. In such unstable conditions, a liquid layer just needs a small push to generate shear stress on the layer below it to make the liquid deform and flow. The liquid will continue to deform until the inertial forces have reached a point when the viscous forces have started dominating over the inertial forces. To reiterate, fluid does not deform by itself, it needs a push as said above when all vectors sum = mass × acceleration.

WHAT CAUSES THE LIQUID TO FLOW?

Imagine a liquid flow in a pipeline. The Liquid's bottom surface in contact with the pipe surface is the no-slip layer. Here the velocity is zero. Therefore, the liquid can only deform in layers and because each layer is rubbing against the layer below it there is stress generation. This is shearing stress. Each layer is shearing the layer below it. This is also the viscous force that is resisting the liquid's flow. There are two types of stresses acting on a moving liquid element [1] normal stress, acting perpendicular downward toward gravity [2] the applied stress acting parallel to the surface. This is the stress which is pushing the liquid forward when a liquid is at rest there is no deformation. The shear stress is zero. Only the normal force is acting on the liquid. In a moving liquid, both normal and tangential stresses are working. Stress has both magnitude and direction. When the sum vectors of all stresses acting on the liquid become equal to its mass time acceleration due to gravity, the liquid is ready to roll. With little push, the liquid layers start rolling one over the other.

$N\,Re = D \times u \times$ rho/Viscosity

The denominator is viscous force. The numerator is an inertial force. When the inertial forces exceed the viscous force, liquids begin to roll one layer over the other.

When the vector sum of all forces acting on a fluid element becomes equal to the mass of the element times its acceleration (Newton's second law), the fluid becomes unstable. The fluid begins to deform in the direction of stress on the surface [tangential stress] with little push. In such unstable conditions, a liquid layer just needs a small push to generate shear stress on the layer below it to make the liquid deform and flow. The liquid will continue to deform until the inertial forces have reached a point when the viscous forces have started dominating over the inertial forces. To reiterate, fluid does not deform by itself, it needs a push as said above when all vectors sum = mass × acceleration. This is fundamental to how a fluid generates viscosity.

A liquid is defined as a substance that continuously deforms and changes its shape by tangential shear forces. Liquids and gases cannot resist deformation forces; they move or flow as a result of the force. As long as the force is applied, the fluid's shape will change continuously. We usually distinguish three states of matter: solid, liquid, and gas. However, liquid and gas are both fluids: unlike solids, they cannot resist deformation. Because fluid cannot resist deformation force, it moves or flows as a result of it. As long as the force is applied, its shape will change continuously. In a fluid flow, mass, momentum, and energy are all transferred. The Navier-Stokes equations are conservation equations for these quantities that describe fluid motion. A fluid is unconcerned about its shape no matter how much it is deformed. A fluid is concerned with how quickly it is deformed; its resistance to motion is proportional to the rate of deformation. The dynamic viscosity of a fluid is a property that indicates how well it resists the rate of deformation.

While at rest, a solid can withstand a deformation force. While a force can cause displacement, the solid cannot move indefinitely.

Almost everything around, us is either a solid, a liquid, or a gas, they are all made up of tiny particles that connect to one another differently. These particles are tightly bound together in solids. These particles can flow over or alongside one another in liquids and gases. Because liquids and gases can flow, they are also referred to as fluids. This flow can be smooth, chaotic, or somewhere in the middle. Laminar flows are smooth, while turbulent flows are turbulent. Fluids can deform their shape because their molecular structure is elastic rather than rigid like solids. Solid layers cannot move one over the other layer because of tightly held intermolecular forces.

BERNOULLI'S PRINCIPLE: LIMITATIONS IN THERMODYNAMICS

The original idea behind Daniel Bernoulli's principle was to conceive of an ideal flowing fluid that conserves energy as it moves through a system. Bernoulli's principle is based on the concept of conservation of energy along a streamline in an ideal, non-viscous fluid flow. Bernoulli's principle states that in a flowing fluid, an increase in the speed of the fluid occurs simultaneously with a decrease in pressure, and vice versa. This principle is derived from the conservation of energy, where the total energy of the fluid remains constant along a streamline – comprising the fluid's potential energy, kinetic energy, and pressure energy.

Bernoulli's equation

$$\frac{v^2}{2}+\frac{p}{p}+gz = \text{constant}$$

Where:

v is fluid velocity

p is fluid pressure

g is fluid density

g is gravitational constant

z is elevation

The Bernoulli equation states that for an ideal fluid (that is, zero viscosity, constant density, and steady flow), the sum of its kinetic, potential, and thermal energy must not change. This constraint gives rise to a predictable relationship between the velocity (speed) of the fluid, its pressure, and its elevation (relative height).

What does Bernoulli's equation mean?

Pressure has two parts [1] static and [2] dynamic. In incompressible fluids dynamic pressure 'q' sometimes called velocity pressure is the quantity defined

$q = 1/2$ rho V^2

q is dynamic pressure in pascals (i.e., $kg{\cdot}m^{-1}{\cdot}s^{-2}$),

rho is fluid mass density (e.g., in kg/m^3, in *SI* units),

V is flow speed in m/s.

The dynamic pressure can be thought of as the fluid's kinetic energy per unit volume. For incompressible flow, the dynamic pressure of a fluid is the difference between its total pressure and static pressure. From Bernoulli's law, dynamic pressure is given by

$p_0 - ps = ½$ rho V^2

p_0 is total pressure and ps is static pressure.

$p_0 - ps$ always remains content and it is independent of total pressure. An increase or reduction of total pressure is just a switch between dynamic and static pressure.

The kinetic energy can increase only if there is an increase in temperature. For incompressible liquids, at constant density, the kinetic energy can't change. Hence velocity remains always constant in a steady-state flow for incompressible liquids. This is what Bernoulli's equation of energy conservation means within all limitations.

Pressure = Static pressure + dynamic pressure

Bernoulli's principle says, $p_0 - ps = ½$ rho V^2.

p_0 is total pressure and ps is static pressure. $p_0 - ps$ is the dynamic pressure.

This means Bernoulli's principle says that an increase in the speed of a fluid occurs simultaneously with a decrease in static pressure or a decrease in the fluid's potential energy. This further means if there is an increase in kinetic energy there will be an increase in dynamic pressure. This is the fundamental point of Bernoulli's principle,

Meaning of assumptions: By assuming that the fluid is incompressible, inviscid, and follows a steady flow path with no external work being done on or by the fluid, Bernoulli's principle simplifies the complex dynamics of fluid flow into a more manageable form. This idealized representation allows for the prediction and analysis of fluid behavior in various engineering applications, such as in aerodynamics, hydraulics, and fluid mechanics. While Bernoulli's principle provides valuable insights into fluid behavior and has various practical applications, it is important to recognize its limitations and the idealized assumptions on which it is based. In real-world scenarios with compressible fluids or systems involving thermodynamic work, more sophisticated models and analyses are required to accurately capture the complexities of fluid flow and energy interactions.

Explanation of Each Limitation from Thermodynamics' Point

Inviscid flow assumption: The assumption of inviscid flow in Bernoulli's principle implies that the fluid is not experiencing any viscous effects, such as frictional losses. This assumption neglects the effects of viscosity on state variables such as temperature, pressure, and density. In reality, viscous effects can cause changes in these state variables, leading to deviations from the idealized predictions of Bernoulli's principle.

Incompressibility assumption: The assumption of incompressible flow in Bernoulli's principle implies that the fluid density remains constant along the flow path. This assumption neglects the effects of compressibility on state variables such as density and pressure. In compressible flow, changes in density can result in significant variations in pressure and temperature, which are not accounted for in Bernoulli's principle.

Heat transfer and real gas effects: The neglect of heat transfer effects and real gas behavior in Bernoulli's principle means that changes in state variables such as temperature and entropy are not considered. In real-world situations with heat transfer or when dealing with real gas behavior, these changes in state variables can have a significant impact on the accuracy of predictions based on Bernoulli's principle.

Turbulence effects: The assumption of laminar flow in Bernoulli's principle implies that the fluid particles follow smooth, parallel paths. In reality, however, flow conditions can be turbulent, with chaotic and irregular paths. Turbulent flow can lead to changes in velocity, pressure, and other path variables, which are not accounted for in the idealized assumptions of Bernoulli's principle.

Irreversible processes: The assumption of reversible flow in Bernoulli's principle implies that the flow process is fully reversible, with no irreversible processes such as friction or heat transfer present. In real-world applications, however, irreversible processes can lead to changes in path variables such as work, heat transfer, and entropy production. These irreversible processes can result in deviations from the idealized predictions of Bernoulli's principle.

No thermodynamic work: One of the major limitations of Bernoulli's principle is that it is derived based on assumptions that apply to ideal, inviscid, and incompressible fluid flow. In this idealized scenario, the principle assumes that there is no thermodynamic work done on or by the fluid in the system. In real-world situations, however, fluids are often compressible and may experience changes in temperature, pressure, and density along the flow path. These changes can result in variations in thermodynamic work within the system, which are not accounted for in the basic form of Bernoulli's equation. As a result, when dealing with compressible fluids or systems where thermodynamic work is significant.

Implications of the Thermodynamic Limitations

The limitations in thermodynamics terms predict as follows:

$PdV = 0$

$TdS = 0$ [S = Entropy]

This means, $dH = dU$ [H = Enthalpy and U = Internal energy]

$dG = dU$ [G = Gibbs free energy]

As the pressure exerted on the gas increases, so do these molecular collisions. As a result, both the kinetic energy and the internal energy of the gas see a consequential rise. In contrast, when you elevate the volume under a constant number of gas molecules, the average distance between the molecules expands. Therefore, a change in internal energy signifies a change in pressure.

Change of Pressure/Volume

Pressure in a fluid can be seen to be a measure of energy per unit volume through the definition of 'work'.

dU is, therefore, pressure change/volume.

That makes *dG* only dependent on pressure.

Free Energy and Pressure Relationship

When a system changes from an initial state to a final state, the Gibbs free energy (*dG*) equals the work exchanged by the system with its surroundings, minus the work of the pressure force, *dG* = [Net change of work between two points] – Pressure.

Since Bernoulli's principle assumes no 'thermodynamic work'.

dG = – Pressure.

Therefore, Bernoulli's principle assumes that there is only one state variable, pressure that controls the transport of a fluid between two points.

CAVITATION: INTERPLAY OF STATIC AND DYNAMIC PRESSURE

The term "cavitation" usually refers to the phenomenon where a liquid's static pressure falls to a level lower than its vapor pressure, causing the liquid to form tiny cavities filled with vapor. The so-called "bubbles" or "voids" collapse under increased pressure and can produce shock waves that harm machinery.

NPSHA

The available NPSH (NPSHA) is an indicator of the fluid's proximity to flashing and, consequently, cavitation at a specific point. In technical terms, it is the liquid's vapor pressure less the absolute pressure head.

NPSHR

The required NPSH (NPSHR) is the head value at the suction side (such as the pump's inlet) needed to prevent the fluid from cavitating. This is provided by the manufacturer.

NPSH is especially important inside turbines and centrifugal pumps.

Cavitation occurs in a pump when.

$$\text{NPSHA} = [Pi/\text{rho } g + Vi^2/2g] - Pv/\text{rho g}$$

Note that NPSHA and NPSHR are in absolute units and usually expressed in "m" or "ft,"

Here, *Pi* is the absolute pressure at the inlet, *Vi* is the average velocity at the inlet, rho is the fluid density, g is the acceleration of gravity and *Pv* is the vapor pressure of the fluid. NPSH is equivalent to the sum of both the static and dynamic heads. Which is the stagnation head minus the equilibrium vapor pressure head. This is called a "net positive suction head".

How *P* = rho *g h* is derived?

(rho)g h represents the hydrostatic pressure due to the gravity force on a fluid. This is derived from the mass of the fluid times the acceleration of gravity or $m \times g$, mass can be written as rho × *V*. This gives, *P* = rho × *V* × *g*. The volume of fluid is simply the height *h* of the fluid column. The force per unit area *P* thus becomes equal to rho**g***h*.

Bernoulli's Equation

The simplified form of Bernoulli's equation is written as follows:

Static pressure + Dynamic pressure = Total pressure

The changes in the 'rho g *h*' term between two points in Bernoulli's equation is very small and therefore, ignored. *h* is elevation and rho is density.

Static pressure = P

Dynamic pressure = ½ rho × V^2

Therefore, cavitation occurs when the sum of static pressure and dynamic pressure becomes equal to the vapor pressure of the fluid.

Static pressure

Static pressure is the pressure that a fluid exerts when it is not moving, and it's caused by the weight of the fluid above it. It is the pressure exerted by molecules hitting the walls. As the molecules collide with the walls of the pipe the molecules impart momentum to the walls, producing a force perpendicular to the wall. The sum of the forces of all the molecules striking the wall divided by the area of the wall is defined to be the pressure.

Dynamic pressure

It symbolizes the kinetic energy of the fluid at the eye of the impellor.

TYPES OF FLOW IN FLUID DYNAMICS

Among all key types of flow, the Coriolis flow is very interesting. The Coriolis force is a product of the Earth's rotation that influences the Coriolis flow. The Coriolis force is responsible for the deflection of moving objects in fluids or gases on rotating platforms such as the Earth.

The Coriolis force is directly proportional to the angular velocity of the Earth's rotation and the velocity of the moving object. The magnitude and direction of the Coriolis force change with latitude. Near the equator, the Coriolis force is negligible, while it becomes stronger as you move toward the poles.

In the Northern Hemisphere, the Coriolis force causes moving fluids or objects to deflect to the right, while in the Southern Hemisphere, it causes deflection to the left. This deflection due to the Coriolis force is known as the Coriolis effect and has significant impacts on large-scale phenomena such as ocean currents, atmospheric circulation patterns, and weather systems.

So, the location of a planet on Earth affects the Coriolis flow by determining the strength and direction of the Coriolis force experienced by moving fluids or objects on that planet.

Why there are different types of flow in fluid dynamics?

In fluid dynamics, the choice of flow type depends on several factors, including the specific requirements and constraints of the situation. While minimizing resistance is often desirable, it is not always the sole determining factor in selecting a particular flow type. Different flow types have their own characteristics and advantages, and the specific application or problem at hand may require considerations beyond just resistance minimization.

For example, laminar flow is characterized by smooth and orderly fluid motion with low resistance, making it suitable for situations where minimal energy losses are desired. Turbulent flow, on the other hand, involves complex and chaotic fluid motion with higher resistance but increased mixing capabilities. Turbulent flow might be preferred in situations where efficient heat transfer or mixing is required, even if it comes at the expense of increased resistance.

Other factors that can influence the choice of flow type include the Reynolds number (which determines whether flow is laminar or turbulent), the presence of obstacles or boundaries, the nature of fluid properties (such as viscosity and density), flow stability considerations, and specific engineering requirements.

In summary, while minimizing resistance is often an important consideration in fluid dynamics, the choice of flow type depends on multiple factors and can include considerations beyond resistance reduction. The diverse range of flow types in fluid dynamics reflects the various requirements and constraints encountered in different applications and scenarios.

Why Knowing Different Types of Flow Important?

Four Key Points

Designing Systems and Structures: Fluid dynamics plays a key role in the design of many systems and structures, such as aircraft wings, pipelines, heat exchangers, and turbines. Different types of flow have distinct effects on the performance and efficiency of these systems, so understanding and predicting flow patterns is essential for their optimal design.

Energy Transfer and Mixing: Fluid flow types affect the transfer of energy within a fluid system. For example, in laminar flow, there is minimal kinetic energy transfer between layers, while in turbulent flow, energy is rapidly transferred and mixed across the fluid. Harnessing these flow characteristics is crucial for processes such as heat transfer, stirring, and mixing in various industrial applications.

Dynamic Loads and Fluid-Structure Interaction: Different types of flow generate various dynamic loads on objects immersed in the fluid. For instance, turbulent flow can create high-pressure fluctuations and vortices that exert significant forces on structures. Understanding the flow behavior allows engineers to design structures that can withstand these loads and optimize their performance.

Transport Phenomena: Fluid dynamics is involved in the study of transport phenomena, such as mass transfer, momentum transfer, and heat transfer. Different types of flow influence these transport processes, and understanding flow types assist in analyzing and modelling these phenomena accurately.

There are Various Types of Flow in Fluid Dynamics, Including

Laminar Flow: In this type of flow, the fluid particles move in parallel layers without any significant mixing between them. It is characterized by smooth and orderly flow.

Turbulent Flow: This is a type of flow with irregular, chaotic, and unpredictable behavior. The fluid particles move in a random fashion, and there are significant mixing and eddy formations.

Compressible Flow: it is a flow in which the density and pressure changes occur due to compressibility effects. This occurs when the fluid velocity is comparable to the speed of sound.

Incompressible Flow: In contrast to compressible flow, incompressible flow assumes the fluid to be incompressible with constant density. This is often applicable to liquids or when the fluid velocity is significantly lower compared to the speed of sound.

Steady Flow: In this type of flow, fluid properties such as velocity, pressure, and density do not change with time at any given point. It remains constant over time.

Unsteady Flow: Here, the fluid properties change with time at a specific point. The velocities, pressures, and other properties vary with time.

Channel Flow: It refers to the flow of fluids (usually water) in channels with a free surface exposed to the atmosphere. Examples include rivers, canals, and open pipes.

Closed Conduit Flow: This type of flow occurs in pipes or conduits where the fluid is completely enclosed. It is characteristic of pipe systems used for water supply, oil/gas pipelines, etc.

Poiseuille Flow: It is a type of laminar flow in a cylindrical conduit where there is viscous shearing in the radial direction. This flow is often used to analyze fluid flow through capillaries and small tubes.

Couette Flow: It refers to the flow between two parallel plates where one plate is stationary and the other moves with a constant velocity. It is used to study fluid flow in lubrication and shear-induced flow in thin film applications.

Karman Vortex Street: It is a phenomenon characterized by the formation of alternating rotating vortices behind an obstacle in a flowing fluid. This phenomenon is often observed in the wake of buildings, around cylinders, and other bluff bodies.

Coriolis Flow: This flow occurs due to the Coriolis effect, which is a result of the rotation of the Earth. It causes a deflection of fluid flow in large-scale atmospheric and oceanic circulations.

These represent just a few examples of different flow types encountered in fluid dynamics, and there are many other flow regimes and phenomena studied in this field.

WHAT MAKES A FLOW TURBULENT?

In simple words, the turbulence sets in when the inertial forces in a flowing fluid begin dominating over the viscous forces. What are the inertial forces in a flowing fluid and what are the viscous forces in a flowing fluid? Let us answer this question first.

Fundamentals

Inertial Forces in Moving Fluid

Turbulence, also known as turbulent flow, is a type of fluid motion in fluid dynamics defined by erratic flow velocity and pressure variations. On the other hand, a laminar flow happens when a fluid flows in parallel layers without interfering with one another.

Excessive kinetic energy or the inertial forces in certain areas overcome the viscous forces or the viscosity and do some kind of damping effect on the viscosity and that's the beginning of the turbulence.

Turbulence is therefore frequently observed in low-viscosity fluids. In general, turbulent flow is characterized by the appearance of unstable vortices of various sizes that interact with one another, increasing drag due to frictional forces. A region in a fluid where the flow revolves around an axis line is called a vortex. Linear momentum is a property of objects changing their position concerning a reference point. Angular momentum in a vortices is a property of objects that are changing the angle of their position vector concerning a reference point. Angular momentum depends on the rotational velocity of an object, but also its rotational inertia. When an object changes its shape (rotational inertia), its angular velocity will also change if there is no external torque. In summary, in a vortex, the velocity of the spinning particles along with the change in the shape of vortices is continuously changing and creating a chaotic situation in some pockets of the flow spreading turbulence all across.

The energy required to pump fluid through a pipe rises as a result.

Let us answer a couple of questions here:

- What are the inertial forces?
- What are the viscous forces?
- What are the vortices and drag?

One by one

Inertial forces?

Inertia is the property of an object by which it cannot change its state of rest along a straight line on its own unless acted upon by an external force. Inertia increases with an increase in the mass of the body and vice-versa. Inertial forces keep fluid moving against viscous [viscosity]. The inertial forces are characterized by the product of the density rho, velocity, U and pipe diameter, *D*.

Viscous forces

In simple words, viscosity generates viscous forces which oppose the flow of the fluid. In liquids, viscosity originates with the cohesive forces between molecules in adjacent layers. The viscous force results because the attractive forces between molecules tend to prevent the relative motion of molecules. In low-density gases cohesive forces make only a minor contribution to the viscosity. The molecules are widely separated and exert very weak attractive forces

Eddies and Vortices

In the context of fluid dynamics, an eddy is the circular movement of a fluid that is usually seen behind an obstruction or in the wake of a moving object. Eddies are caused by the separation of the fluid flow from a surface, creating a region of recirculating flow, and they can have a significant effect on fluid dynamics. On the other hand, the vortices are there is actually no difference between a vortex and an eddy. However, the words are used in different contexts.

In turbulence, we usually speak of eddies and these eddies decay to smaller eddies and so forth which forms a turbulent cascade. This has nothing to do with the type of fluid. Eddies are formed in fluids, gas, and plasma.

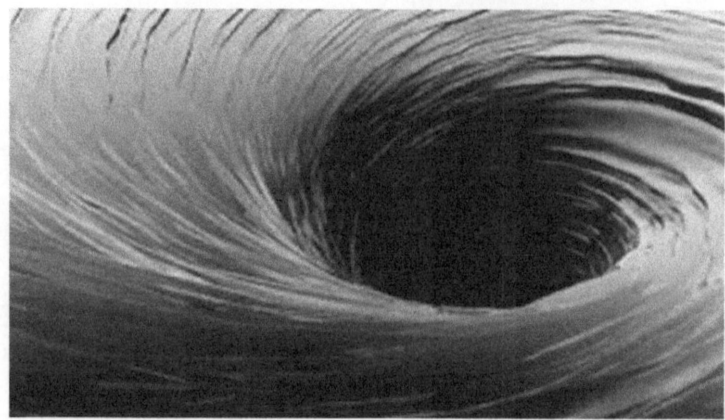

Figure 5: Vortex

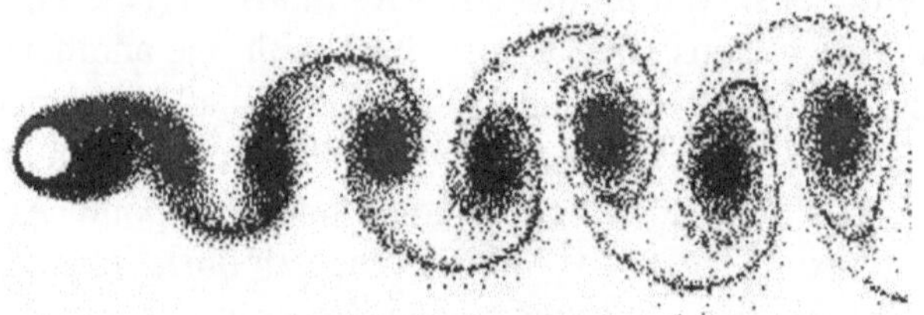

Figure 6: Eddy

Vortices on the other hand are more stable structures with a fixed axis circular motion. They do not decay like changing axis eddies.

What is Drag?

Drag, also known as fluid resistance. Drag is a force that acts against object's relative motion. when it moves concerning a surrounding.

It can occur between two solid surfaces, between two fluid layers, or between a fluid and a solid surface.

Detail

There are two factors [1] inertial forces and [2] viscous forces. The inertial forces in a flow constitute the [1] pipe diameter [2] density an [3] the velocity of the fluid. The viscous forces arise from the viscosity of the fluid that opposes the flow. When the inertial forces exceed the viscous forces beyond a certain point which is expressed by the Reynold's number the turbulence sets-in.

Key Factors that Cause Turbulence

- **Velocity and viscosity** are related. In a turbulent flow, the velocity of the flow fluid will be much higher. This can be related to viscosity by stating that lower viscosity makes the fluid velocity higher. Hence, in turbulence fluids, the velocity of the flow will be much higher which is due to lower viscosity.
- **Velocity and turbulence**: The drag between adjacent layers of fluid and between the fluid and its surroundings can form swirls and eddies if the velocity is great enough. Essentially the high velocity generates eddies which has been explained above and causes turbulence.
- **Pipe diameter and turbulence:** When the mass flow rate and viscosity are fixed, increasing the pipe diameter will increase the Reynolds number and turbulence.
- **Velocity and density**: Higher density causes higher velocity gradients in the flow, causing flow instabilities, causing turbulence.
- **Surface roughness of the pipe**: The roughness of the inner surface of a pipe can cause additional frictional resistance to the flow of fluid and more turbulence and disrupt the smooth flow of the fluid.

HOW DOES LIQUID FLOW THROUGH A PIPE AT CONSTANT ENERGY?

In one line the flow of liquid through a pipe is interplay of potential and kinetic energy and the sum total of potential and kinetic energy remains constant.

The flow of a liquid in a pipe when total energy remains constant is due to the pressure difference between two points along the pipe. When there is a pressure difference, the liquid will flow from the higher pressure point to the lower pressure point in order to equalize the pressure. This flow is driven by the potential energy of the liquid, which is converted into kinetic energy as it moves through the pipe.

The flow rate of the liquid will depend on the magnitude of the pressure difference, the viscosity of the liquid, the diameter of the pipe, and any obstructions or bends in the pipe that may impede the flow. As long as the total energy of the system remains constant, the flow of the liquid will continue as long as there is a pressure difference driving it.

Potential energy is the energy associated with the elevation of the fluid in the pipe, or the height difference between two points in the pipe. This potential energy is converted into kinetic energy as the fluid flows downhill or from a higher pressure to a lower pressure point in the pipe.

Kinetic energy is the energy associated with the motion of the fluid particles as they flow through the pipe. The velocity of the fluid particles is directly related to the kinetic energy of the system.

Pressure energy, also known as the flow energy, is the energy associated with the pressure difference between two points in the pipe. This pressure energy is responsible for driving the flow of the fluid from high-pressure areas to low-pressure areas.

Therefore, fluid flow is a dynamic process where potential energy, kinetic energy, and pressure energy work together to maintain the flow of the fluid in a pipe.

When Elevation Head is Zero

When the elevation head is zero, meaning there is no height difference between two points in the pipe, the potential energy component of the total energy in the system becomes zero. In this case, the flow of the fluid in the pipe is solely driven by the pressure energy and kinetic energy.

The pressure energy is still responsible for driving the flow of the fluid from high-pressure areas to low-pressure areas, while the kinetic energy represents the energy associated with the motion of the fluid particles as they flow through the pipe.

Without the potential energy component, the flow of the fluid in the pipe is primarily determined by the balance between the pressure and kinetic energies. The total energy in the system remains constant, with the pressure and kinetic energy interplaying to maintain the flow of the fluid in the absence of elevation head.

In fluid flow, pressure energy can convert to kinetic energy and vice versa as the fluid moves through a pipe or conduit. This conversion is governed by the principle of conservation of energy, specifically the Bernoulli's equation.

Bernoulli's equation states that the total mechanical energy of a fluid flowing in a steady, non-viscous, incompressible flow system remains constant along a streamline. The equation includes three components: pressure energy, kinetic energy, and potential energy.

The equation is expressed as:

$$P + 1/2\rho v^2 + \rho g h = \text{constant}$$

where:

- P is the pressure energy per unit mass of the fluid
- ρ is the density of the fluid
- v is the velocity of the fluid
- g is the acceleration due to gravity
- h is the height above a reference point

As the fluid flows through a pipe, changes in the pipe diameter, flow velocity, pipe roughness, and elevation result in changes in pressure, velocity, and height. When these changes occur, the total energy in the system remains constant, but the energy can be converted between pressure and kinetic energy. For example, if the fluid accelerates in a narrowing pipe, pressure energy can be converted into kinetic energy. Conversely, when the fluid decelerates or changes direction, kinetic energy can be converted back into pressure energy.

This conversion between pressure and kinetic energy plays a important role in determining the behavior of fluid flow in pipes and conduits.

HEAT TRANSFER TO LIQUID FLOWING THROUGH PIPE

[1] Eddy transport in turbulent flow [2] Laminar and turbulent heat transfer correlations [3] Hydrodynamic and Thermal entrance's role on fluid flow and temperature [4] Nusselt number [5] Prandtl number [6] Sieder-Tate equation [7] Dittus-Boelter correlation.

Heat transfer between the fluid and the pipe's surroundings is an important aspect in heat exchangers, boilers, condensers, evaporators, and many other process equipment. The calculation of heat transfer coefficient, typically by convection or phase transition between a fluid and a solid is one of the major exercises we do in our heat transfer calculations. We classify the flow of a fluid in a straight circular pipe into either laminar or turbulent flow. We assume fully developed incompressible, Newtonian, steady flow conditions.

The first question is how flow develops in a pipe? Why does the velocity profile of laminar flow is parabolic and turbulent flow flat?

A fluid travels a distance in a pipe before it becomes fully developed. The distance a fluid travels before becoming fully developed is called the entrance length.

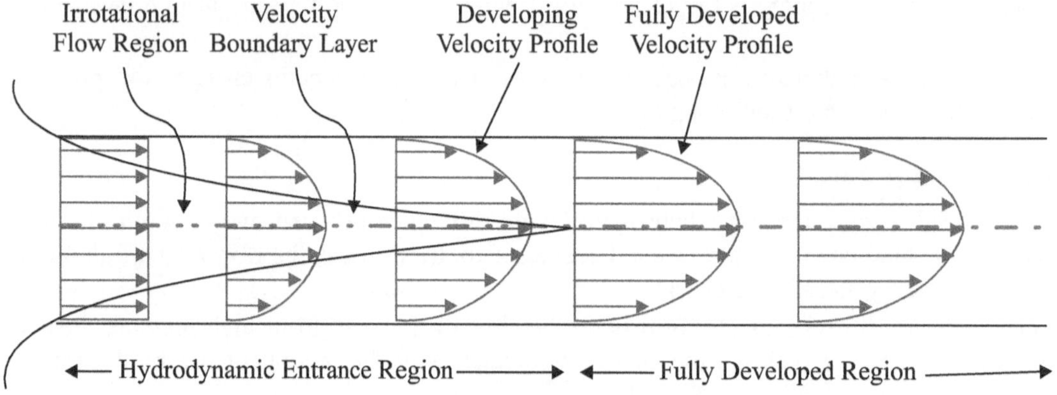

Figure 7: Velocity Profile of Laminar and Turbulent Flow

The area following the pipe entrance where effects originating from the interior wall of the pipe propagate into the flow as an expanding boundary layer. When the boundary layer expands to fill the entire pipe, the developing flow becomes a fully developed flow, where flow characteristics no longer change with increased distance along the pipe.

Why velocity profile of laminar flow is parabolic and flat for turbulent flow?

If the flow in a pipe is laminar, the velocity distribution at a cross-section will be parabolic in shape with the maximum velocity at the center being about twice the average velocity in the pipe. The velocity of layers of the fluid in the center of the pipe increases to compensate for the reduced velocities of the layers of fluid near the pipe.

In the case of turbulent flow, it gets a little flatter due to vigorous mixing in radial direction and eddy motion. surface.

How Flow Modifies Heat Transfer?

Hydrodynamic entrance length vs. thermal entrance length on fluid flow and temperature

Hydrodynamic entrance length

The hydrodynamic entrance region refers to the area of a pipe where fluid entering a pipe develops a velocity profile due to viscous forces propagating from the interior wall of a pipe. This region is characterized by a non-uniform flow. The fluid enters a pipe at a uniform velocity, then fluid particles in the layer in contact with the surface of the pipe come to a complete stop due to the no-slip condition. Due to viscous forces within the fluid, the layer in contact with the pipe surface resists the motion of adjacent layers and slows adjacent layers of fluid down gradually, forming a velocity profile. For the conservation of mass to hold true, the velocity of layers of the fluid in the center of the pipe increases to compensate for the reduced velocities of the layers of fluid near the pipe surface. This develops a velocity gradient across the cross section of the pipe.

The hydrodynamic entry length is the point at which the velocity profile is fully developed in the tube from the point of entry for the fluid. The thermal entry length is the point at which the temperature profile is fully developed from the point at which the tube wall is heated or cooled.

The hydrodynamic entry length is a function of Reynold's and Prandtl's numbers. For turbulent flow, entry length is approximately independent of the Reynolds Number.

Thermal entrance length

The thermal entrance length describes the distance for incoming flow in a pipe to form a temperature profile of the stable shape. The shape of the fully developed temperature profile is determined by temperature and heat flux conditions along the inside wall of the pipe, as well as fluid properties.

The Prandtl number modifies the hydrodynamic entrance length to determine thermal entrance length. The thermal entrance length for a fluid with a Prandtl number greater than one will be longer than the hydrodynamic entrance length, and shorter if the Prandtl number is less than one.

To summarize, the turbulent flow hydrodynamic and thermal entrance lengths are much shorter than those for laminar flow due to velocity fluctuations that enhance the heat transfer convection and transfer of heat and momentum between the coolant particles.

The short entrance length in turbulent flow contributes significantly to an improved heat transfer compared to laminar flow.

Concept of boundary layer

When the fluid just enters the pipe, the thickness of the boundary layer gradually increases from zero moving in the direction of fluid flow and eventually reaches the pipe center and fills the entire pipe. At the boundary layer, the shearing viscous forces are significant. This boundary layer is a hypothetical concept. It divides the flow in the pipe into two regions, boundary layer region—the region in which viscous effects and the velocity changes are significant and [2] the irrotational (core) flow region: The region in which viscous effects and velocity changes are negligible, also known as the inviscid core.

Reynolds Number

The Reynolds number is the ratio of inertial forces to viscous forces within a fluid that is subjected to relative internal movement due to different fluid velocities. A region where these forces change behavior is known as a boundary layer, such as the bounding surface in the interior of a pipe.

We define the Reynolds number as follows. Reynolds number, $Re = DV\rho/\mu$, D is the inside diameter of the pipe, V is the average velocity of the fluid, ρ is the density of the fluid, and μ is dynamic.

The value of the Reynolds number permits us to determine whether the flow is laminar or turbulent

Flow in a pipe is considered laminar when the Reynolds number is below 2,300. In the range of Re 2,300 to Re 4,000, the flow is considered in transition from laminar to turbulent. Above Re = 4000, the flow is considered turbulent. Results for heat transfer in the transition window are difficult to predict, and it is generally avoided when someone is designing a heat exchanger.

Turbulent flow is inherently unsteady, being characterized by time-dependent fluctuations in the velocity and pressure, but we usually average over these fluctuations and define time-smoothed or time-average velocity and pressure; these time-smoothed entities can be steady or time-dependent. In this post, we have considered only steady conditions.

Heat Transfer in Laminar *vs.* Turbulent Flow

The key difference is in the laminar flow it is the flow direction, and in the turbulent, the directions lie in a plane perpendicular to the tube axis. In the turbulent flow, the heat flow can be broken into radial and azimuthal components.

Azimuthal Components of Fluid Flow

In simple words, the vector from an observer (origin) to a point of interest is projected perpendicularly onto a reference plane; the angle between the projected vector and a reference vector on the reference plane is called the azimuth.

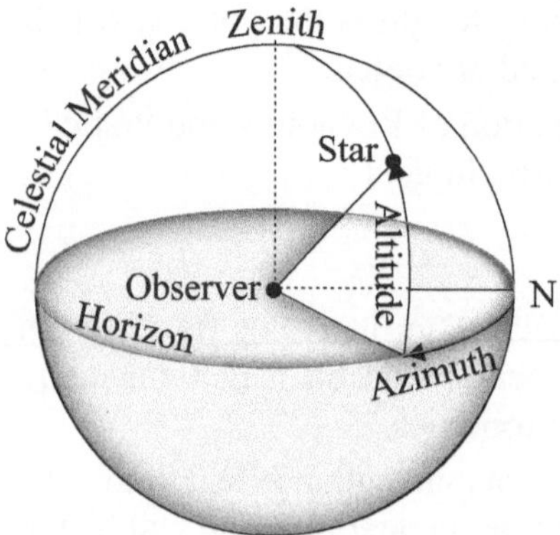

Figure 8

The important point is in the turbulent flow heat transfer takes place in the radial and azimuthal directions. This whole thing is commonly termed "eddy transport".

Eddy Transport

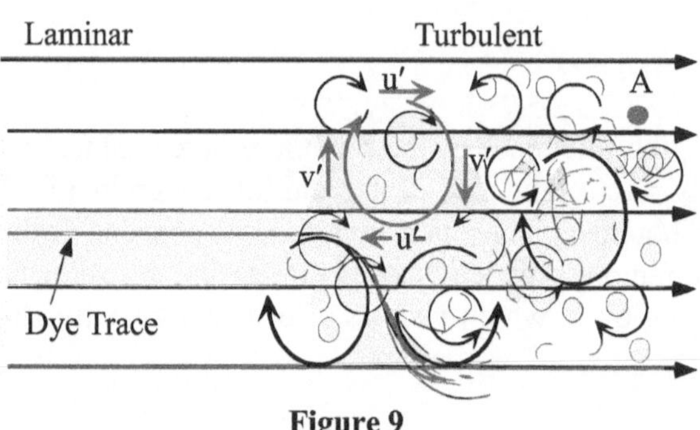

Figure 9

The eddy transport provides a much better transfer of energy across the flow at a given axial position in the turbulent flow than in laminar flow.

Wherein conduction the only mechanism that operates in the transverse directions made at right angles to the long axis of the body.

Another important point is the fluid is assumed to have a fully developed turbulent velocity profile throughout the length of the pipe. Local and fully developed Nusselt numbers are presented for fluids with Prandtl numbers ranging from 0.7 to 100 for Reynolds numbers between 50000 and 500000.

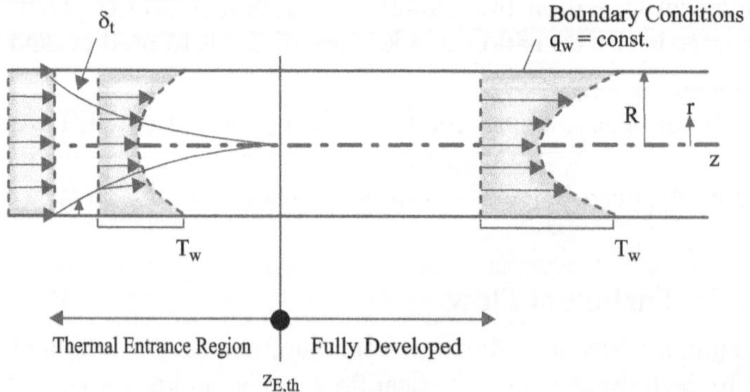

Figure 10

Laminar Heat Transfer Correlations

There is a number of dimensionless correlations which can predict heat transfer rates in laminar flow.

Nusselt Number

$Nu = \varphi\ (Re, Pr\ldots)$, where $Nu = hD/k =$ is the Nusselt number, φ is some function, and $Pr = \mu\ Cp/k = \nu/\alpha$ is the Prandtl number. Here, h is the heat transfer coefficient, k is the thermal conductivity of the fluid, and Cp is the specific heat of the fluid at constant pressure. The Prandtl number can also be written as the ratio of the kinematic viscosity ν to the thermal diffusivity of the fluid α. The heat transfer coefficient appearing in the Nusselt number is usually the average value over the heat transfer surface. You should assume this to be the case unless otherwise stated explicitly.

The Physical Significance of Prandtl Number,

$Pr = \nu/\alpha$

= [Ability of fluid to transport momentum by molecular means]/[Ability of that fluid to transport energy by molecular means].

Gases typically have Prandtl numbers in the range 0.7–1, while the Prandtl number for most liquids is much larger than unity. The Prandtl number for water ranges from 4–7, while that for oil might be of the order of 50–100. It is not uncommon to encounter Prandtl numbers for viscous liquids that are of the order of several thousand or even larger.

Prandtl Number of Turbulent *vs.* Laminar Flow

Prandtl number for turbulent flow is about 0.7 while for water it is around 7.56 (At 18°C).

Sieder and Tate's correlation for the Nusselt number for laminar flow heat transfer

$Nu = 1.86 \times Re^{1/3} \times Pr^{1/3} \times [D/L]^{1/3} \times [\mu b/\mu w]^{0.14}$, b stands for bulk and w stands for wall.

Turbulent Flow Heat Transfer

The distance required for flow to develop is called entrance length [Le]. For flow entering a circular pipe from a reservoir the Le can be calculated from the following equations

Le/D = 0.05 Re [Laminar flow, Re </ = 2000]

Le/D = 50 Re [Turbulent flow, Re >/ = 3000]

The entrance length of laminar vs. turbulent flows

The entrance lengths are much shorter for turbulent flows, because of the additional transport mechanism explained earlier, across the cross-section. Thus, typical hydrodynamic entrance lengths [the point at which the velocity profile is fully developed] in turbulent flow are 10–15 tube diameters, and the thermal entrance lengths are even smaller. Therefore, for most engineering situations we use correlations $L/D \geq 50$, for fully developed conditions.

Dittus-Boelter Correlation

The historic equation for use in turbulent pipe flow is the Dittus-Boelter Correlation.

A common form is for fluids with Prandtl number in the approximate range of 0.7–100, and tubes with $L/D > 50$

$Nu = 0.023\ Re^{0.8} \times Pr^{n}$, here, $n = 0.4$ if the fluid is being heated, that is, if the wall is at a higher temperature than the entering fluid, and n = 0.3 if the fluid is being cooled. All the physical properties used in the Dittus-Boelter correlation must be evaluated at the average bulk temperature of the fluid. This is the arithmetic average of the bulk average temperatures at the entrance and the exit. The usual recommendation is to use the Dittus-Boelter correlation for Re 10,000 >, but in practice, it is used even at lower Reynolds numbers so long as the flow is turbulent because it is a simple correlation to use.

NEWTONIAN AND NON-NEWTONIAN LIQUIDS: ROLE OF SURFACE TENSION AND VISCOSITY. WHAT IS THIXOTROPY?

Non-Newtonian fluids are extensively used for gas/oil separation in the well by a process called **'Thixotropy'** Overall, thixotropy is an extension of non-Newtonian fluid behavior. **Thixotropic fluids** can be used in systems where gas and liquids need to be separated. When the fluid is at rest, it forms a thick gel-like layer that effectively traps gas bubbles, preventing them from rising to the surface. However, when the system is agitated or sheared, the viscosity decreases, allowing the gas bubbles to escape and be separated from the liquid phase.

Role of Surface Tension

Surface tension is the tendency of a liquid to minimize its surface area, which plays a crucial role in the behavior of fluids. It is a force that acts in the plane of the surface of the liquid, causing liquids to form droplets and exhibit capillary action.

Role of Viscosity

Viscosity, on the other hand, is a measure of a fluid's resistance to deformation or flow. It determines how easily a fluid can flow and is influenced by factors such as temperature and pressure. Viscous fluids have a high resistance to flow, while non-viscous fluids flow easily.

In Newtonian fluids, viscosity is constant regardless of the shear rate, while in non-Newtonian fluids, viscosity changes with the shear rate. This means that the flow properties of non-Newtonian fluids are more complex and can exhibit behaviors such as shear-thinning, shear-thickening, or viscoelasticity.

Surface tension and viscosity both play important roles in determining the behavior of fluids, affecting characteristics such as flow rate, drag force, and stability. For example, high surface tension can lead to the formation of droplets or bubbles, while high viscosity can result in slower flow rates and higher resistance to deformation.

In non-Newtonian fluids, the relationship between surface tension, viscosity, and shear rate is more complex and can result in unique and varied flow behaviors. Understanding the role of surface tension and viscosity in both Newtonian and non-Newtonian fluids is crucial for various industrial processes and applications.

Detail

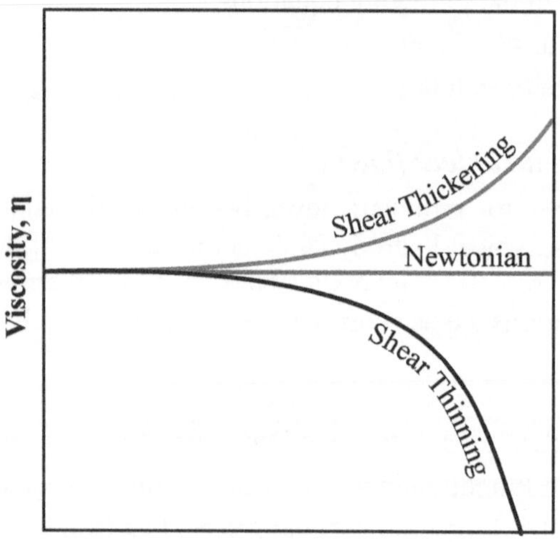

Figure 11: Viscosity of Newtonian, Shear Thinning and Shear Thickening Fluids as a Function of Shear Rate

Definition of three key words: Shear stress, strain, and viscosity

Shear stress is the force tangent to a material surface divided by the area on which the force acts. Shear strain is the angular deformation caused by shear force on the material. The shear strain increases with a fluid as long as the stress is applied. The viscosity of a fluid is the ratio of the shear stress, to the rate of change of shear strain.

Newtonian and Non-Newtonian Fluids

In Newtonian fluids, like water, the viscosity of fluids is only affected by temperature. The shear rate does not affect viscosity in Newtonian fluids.

In non-Newtonian fluids, the viscosity is affected by factors other than temperature. The viscosity of a non-Newtonian fluid will change due to agitation or pressure-technically known as shear stress.

Those fluids that become less viscous when shaken or stirred are known as shear-thinning fluids.

Shear Thinning Principle

Applying a shear force breaks hydrogen bonds (or other intermolecular forces in secondary structures) and allows the molecules to flow past each other more easily. When the force is removed, the hydrogen bonds between the molecules form again.

Shear Thickening Principle

Those fluids that become more viscous due to agitation or pressure-technically known as shear stress are shear thickening fluids. Applying a shear force can cause random coils of a polymer to unwind and become entangled with each other, raising the viscosity.

Examples

Shear-thinning fluids become less viscous under shear stress common examples include:

1. Ketchup 2. Shampoo 3. Toothpaste 4. Blood

Shear-thickening fluids, on the other hand, become more viscous under shear stress.

Some common examples include: 1. Cornstarch and water mixture 2. Certain clay suspensions 3. Custard 4. Silly Putty.

Thixotropy

Thixotropy is a unique property exhibited by certain non-Newtonian fluids, where the viscosity of the fluid decreases over time when subjected to shear stress, but gradually increases back to its original viscosity when the stress is removed. This behavior is in stark contrast to the typical Newtonian fluids, where the viscosity remains constant regardless of the applied stress.

Thixotropy is often observed in materials such as gels, pastes, and certain types of paint. When these materials are at rest, their viscosity is high, making them appear solid. However, when they are agitated or sheared, such as when being pumped through a pipeline, the viscosity decreases, making them flow more easily.

In gas-liquid separation processes, thixotropy can be leveraged to enhance efficiency. For example, in industries such as oil and gas, thixotropic fluids can be used in systems where gas and liquids need to be separated. When the fluid is at rest, it forms a thick gel-like layer that effectively traps gas bubbles, preventing them from rising to the surface.

However, when the system is agitated or sheared, the viscosity decreases, allowing the gas bubbles to escape and be separated from the liquid phase.

HOW PRESSURE-DROP AND FRICTION ARE RELATED? INTERPRETATION OF DARCY'S EQUATION

In summary, the continuous changes in kinetic energy (switch between kinetic and potential energy and reverse or switch between static and dynamic pressure) within a pipe can lead to the generation of turbulence and entropy. Turbulence is the chaotic, swirling motion of fluid flow characterized by rapid changes in velocity and pressure. These fluctuations in velocity and pressure are a result of the kinetic energy transformations occurring within the pipe.

Entropy is a measure of the disorder or randomness in a system. As the fluid flow inside the pipe experiences changes and fluctuations in kinetic energy, it can lead to an increase in entropy within the system. This increase in entropy is a result of the energy dissipation and mixing that occurs due to the turbulent flow conditions.

Overall, the continuous changes in kinetic energy within the pipe can generate turbulence and increase entropy, contributing to the overall complexity and randomness of the flow system.

Frictional forces are created when there is a pressure drop in a moving fluid between two points. Two points of pressure drop imply two kinetic energies in a moving fluid. Such an unstable situation in a fluid flow is not permitted by thermodynamics. The difference in kinetic energy would cause entropy, turbulence, and disorder in the moving fluid. This is friction.

Darcy's equation of friction [fluids]: It accounts for entropy

$$fD = \Delta p/L \times 2/\rho \times DH/[v]^2$$

ρ is the density of the fluid (kg/m^3); DH, the hydraulic diameter of the pipe (for a pipe of circular section, fD is the Darcy friction factor (also called flow coefficient], Δp is the pressure drop and L is the length of the pipe. The friction factor is therefore, for a fixed diameter, directly proportional to the pressure drop/length, inversely proportional to density, and inversely proportional to the square of velocity.

What does it mean?

In fluid dynamics, friction can occur between two points in a pipe when there is a change in kinetic energy between those points. This can lead to turbulence and mixing in the flow, increasing the entropy of the system. When there is a difference in flow speed or direction within the pipe, the molecules in the fluid will experience interactions and collisions that result in friction, generating heat and increasing the disorder and randomness (entropy) of the system. This process contributes to the overall loss of energy and efficiency in fluid flow systems.

Darcy's Equation is related to entropy

The friction factor directly proportional to pressure drop accounts for kinetic energy at two points and that is entropy.

The friction factor is inversely proportional to the density or directly proportional to the specific volume that is entropy.

The friction factor inversely proportional to the square of the velocity signifies less kinetic energy and hence less entropy.

DRAG IN FLUID DYNAMICS?

In one line the drag is a skin friction caused by the viscous forces. Viscosity generates drag at low opposite relative motion. At high velocity, the skin friction disappears as the boundary layer becomes turbulent with a dramatic decrease in drag. Turbulent drag is independent of viscosity.

The skin friction is a consequence of momentum transfer between the surfaces, due to the no-slip wall condition. As skin friction reduces with increased Reynolds number form-drag takes over. This continues to be within the domain of momentum transfer. Only the change that takes place is the fluid particles move in random directions. The transfer of momentum and energy is mainly through the interchange of fluid particles

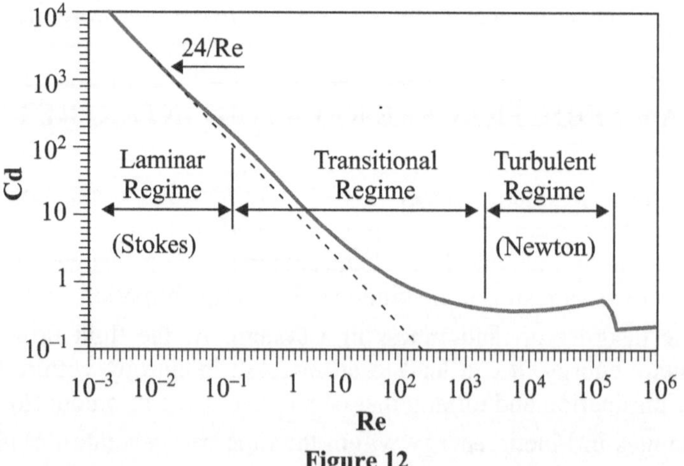

Figure 12

Drag is a type of friction or fluid resistance. It is a force acting opposite to the relative motion of any object moving with respect to a surrounding fluid. This can exist between two fluid layers (or surfaces) or a fluid and a solid surface. The dry friction is nearly independent of velocity. Drag force however, depends on velocity and can change significantly with the Reynolds number.

Are Friction and Drag the Same?

While dry friction is nearly independent of velocity, drag force depends on velocity. The frictional force is always parallel to the surface. An object that is stationary with respect to the fluid will certainly not experience any drag force. As an object starts moving in a fluid a resistive force will arise. Friction is a viscosity-controlled phenomenon. Drag is controlled by velocity.

Then What is a drag?

Drag is a type of friction. It is a skin friction. Skin friction drag is a resistant force exerted on an object moving in a fluid. Skin friction drag is caused by the viscosity of fluids and is developed from laminar drag to turbulent drag as a fluid moves on the surface. Skin friction drag is generally expressed in terms of the Reynolds number, which is the ratio between inertial force and viscous force. Laminar flow over a body occurs when layers of the fluid move smoothly past each other in parallel lines. In nature, this kind of flow is rare. As the fluid flows over an object, it applies frictional forces to the surface of the object which works to impede the forward movement of the object; the result is called skin friction drag. Skin friction drag is often the major component of drag on objects in a flow.

A fluid flowing through a pipe experiences the roughness of the contact surface. The pipe wall has surface roughness. The amount of roughness affects the drag on the fluid. Roughness is measured by the height of the projections sticking up from the pipe wall. In the valleys between projections the fluid moves slowly. Above the projections, it moves faster. The drag between layers tears, or shears, them apart and each layer moves at a different speed. The shear rate decreases as the distance from the wall increases. The velocity at the wall is zero and fastest at the center. This means the central core of the fluid exits the pipe first.

The drag force D exerted on a body traveling through a fluid is given by

$$D = \tfrac{1}{2} C \rho A v^2$$

C is the drag coefficient, which can vary along with the speed of the body. But typical values range from 0.4 to 1.0 for different fluids (such as air and water).

ρ is the density of the fluid through which the body is moving.

v is the speed of the body relative to the fluid.

A is the projected cross-sectional area of the body perpendicular to the flow direction (that is, perpendicular to v)

C is the drag coefficient—a dimensionless number.

The drag coefficient depends on the shape of the object and on the Reynolds number $Du\rho/\mu$. the ratio between inertial force and viscous force.

Drag *vs.* Reynolds Number

Drag force is proportional to the velocity for a laminar flow and the squared velocity for a turbulent flow. Even though the ultimate cause of a drag is viscous friction, the turbulent drag is independent of viscosity.

Reynolds number = Inertial force/Viscous force

- **Inertial force**: Inertia is the resistance of any physical object to any change in its velocity. This includes changes to the object's speed or direction of motion. An aspect of this property is the tendency of objects to keep moving in a straight line at a constant speed when no forces act upon them.
- **Viscous force**: The viscous force is the force between a body and a fluid (liquid or gas) moving past it, in a direction so as to oppose the flow of the fluid past the object.

Viscous forces are negligible compared with inertial forces when the Reynolds number is large

When the Reynolds numbers are small (1 and below) the flow behaves like a potential flow, the inertial forces are much smaller than the viscous forces. There is no separation. There is a no-form drag because the flow does not separate. The velocity increases with distance away from the surface. The pressure gradient pushes the flow round in the direction of flow. Therefore, there are no adverse pressure gradients and no separation of the boundary layer.

As the Reynolds number increases the drag decreases. At Reynolds numbers around 2–30, there is a separation of the boundary layer, but the region of disturbed flow is of a limited length. For Reynolds numbers close to 40–70, there is a periodic oscillation of the disturbance back and forth. For higher Reynolds numbers the eddies break off from the surface. As the Reynolds number increases further, the eddies are continuously shed from the surface and washed downstream.

In other words, as the Reynolds number is increased, the-form drag increases relative to the skin friction. Now the pressure form–drag contributes to almost 90% of the total drag. The value of coefficient of drag reaches a minimum of around 0.9 at a Reynolds number of around 2000. Increasing the Reynolds numbers further results in large angular velocities and degeneration of vortices into turbulence. In the Reynolds number range, 10^4 to 10^5 a wide region of disturbance is seen downstream. The pressure in the separated regions is almost constant. A dramatic change takes place when the Reynolds number is around 2×10^5 when the boundary layer becomes turbulent before separation [see the image above].

THE ROLE OF SPEED OF SOUND IN FLUID DYNAMICS

Summary

The speed of sound is associated with fluid dynamics because as a fluid travel, the invisible sound waves also travel alongside the fluid and interfere with the fluid flow.

When the extent of pressure drop of a fluid at a constriction by the interference of sound waves is such that the reduction of density or the expansion of the fluid stops further increase of the flow of fluid, this limiting flow is called choked flow.

This article will try to clarify "how" and "why" fluids' compressibility and sound speed are related. The article also aims to give a brief explanation of something that many of us frequently lack understanding of the causes and mechanisms of flow choking.

The answer is simple. Sound waves are in motion everywhere except vacuum. When sound waves travel through a fluid the waves generate vibrations in the fluid and a sort of chaos in the medium. The speed of sound in a medium depends on how quickly vibrational energy can be transferred through the medium.

While the sound waves move forward the medium starts moving back and forth. This compression and expansion of the fluid as long as to an extent balance each other [Mach number 0.2–0.3] and do not reduce the density of the fluid using overwhelming compression dominating over-expansion, the sound waves don't interfere with the flow of fluids. When the pressure drop of the traveling fluid arising from the interference of sound waves at a constriction is so much that the resulting decrease in the density of the fluid and expansion stops allowing further increase of flow of the fluid, this limiting flow is called choked flow. Typically choked flow occurs when the relative speed of the fluid and the speed of sound reaches zero. This is expressed as Mach 1.

Details

Sound is a mechanical wave produced by the back-and-forth vibration of the particles in the medium through which the sound wave travels. If a sound wave moves from left to right through the air, air particles will be displaced both rightward and leftward as the sound wave's energy passes through it. The particle motion is parallel (and anti-parallel) to the direction of energy transport. This is what distinguishes sound waves in the air as longitudinal waves. Because of the longitudinal motion of the air particles, there are areas in the air where the air particles are compressed together and areas where the air particles are spread apart. These areas are referred to as compressions and rarefactions, respectively. The compressions are high-pressure zones, whereas the rarefactions are low-pressure zones. A sound wave is called a longitudinal wave as said above because it is produced by compressions and rarefactions in the air. The air particles vibrate parallel to the direction of propagation.

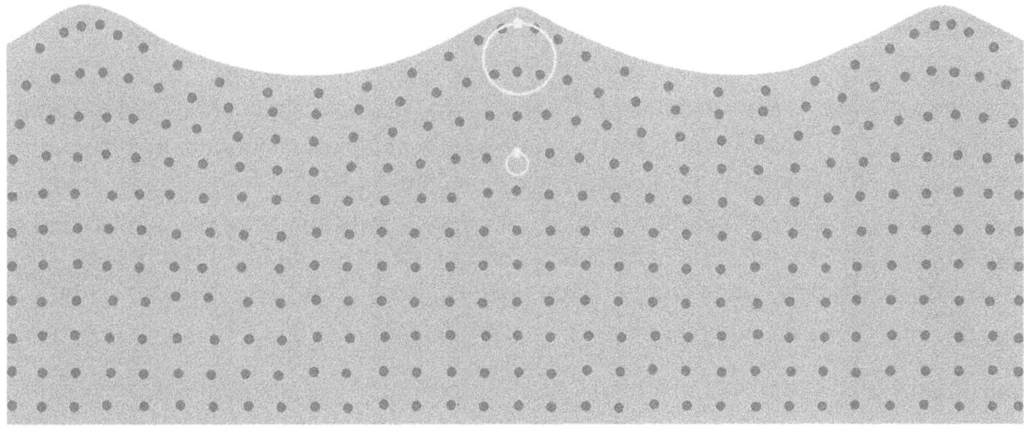

Figure 13

When fluid moves through the air, the air molecules near the fluid are disturbed and move around it as the fluid moves through the air. When the fluid travels at a low speed, typically less than 250 mph, the density of the air remains constant. At higher speeds, some of the fluid's energy is expended compressing the air and changing the density of the air locally. Because this force is proportional to air density, the compressibility effect modifies the amount of resulting force on the fluid. As the speed of fluid increases, the effect becomes more pronounced.

Explanation

Please refer to the images below which are self-explanatory. You will notice that there are two cases. Case 1 relates to Mach number < 0.2–0.3, here the fluid jetting through the constriction is acting as an incompressible fluid. The pressure drop at the constriction is a linear function of the fluid flow. There is no change in the density of the fluid at the nozzle. The sound waves are not able to compress the air. Case 2 relates to Mach 1, here the sound and the fluid have the same speed. Sound waves have completely compressed the fluid jetting through the constriction. As a result of the compression and expansion of the fluid at the constriction, there is a reduction of the density of the fluid. The relationship between the flow and pressure drop at the constriction is no longer linear. This erratic fluid flow behavior at the constriction is called choked flow. At Mach1, the relative speed between the sound waves and fluid reaches zero.

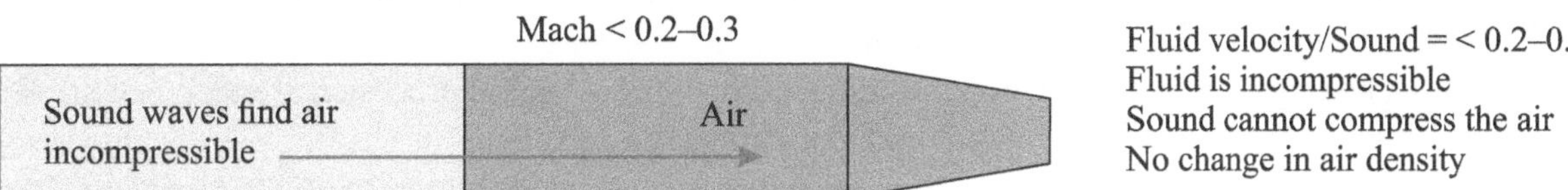

Fluid velocity/Sound = < 0.2–0.3
Fluid is incompressible
Sound cannot compress the air
No change in air density

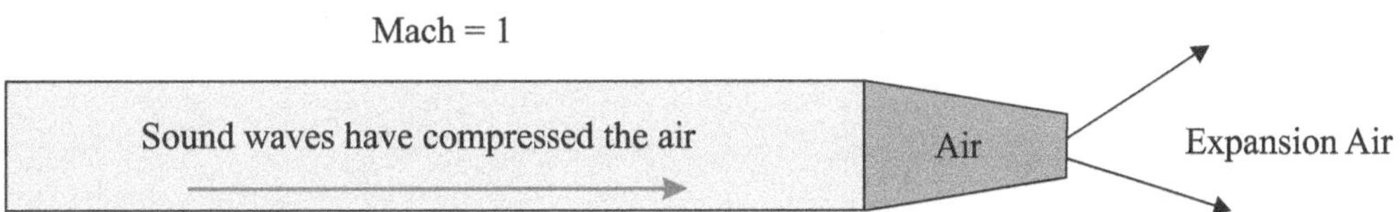

Fluid velocity/Sound speed = 1
Fluid is completely compressible
Sound has compressed the air
Air density reduces with expansion
Air flow chokes

Figure 14: Fluid Compressibility by Sound Waves

WHY FLOW DOES NOT CHOKE IN JOULE THOMSON COOLING?

JT doesn't involve 'work,' while flow choking does. In JT, gas density increases as gas cools. In flow choking, gas density decreases as the gas expands at the restriction. These are two opposite phenomena that don't occur together.

In simple one line, in Joule Thomson expansion the process is isenthalpic and does not involve external work. Sound waves at the sonic velocity of the medium can't compress the medium and do flow choking by doing work externally.

It may be noted that JT cooling at constant enthalpy occurs by switching between kinetic and potential energy without involving 'work'. In JT expansion a small amount of gas forces through a tiny opening or through 100–150 micrometre pores of the porous plug into a large volume downstream almost like a free expansion without involving any 'work'.

What is Choked Flow?

Choked flow occurs when a fluid flows through a restriction and reaches sonic velocity. As the flow of fluid increases, at restriction the specific volume of the fluid increases, when the fluid velocity reaches sonic velocity the sound waves compress the fluid and stop further expansion of fluid or increase of flow. This involves external work done by sound waves on the fluid. The final result is that the flow gets restricted even if the downstream pressure continues to decrease or the upstream pressure continues to increase. At choked flow, the flow velocity at the restriction reaches Mach number (the ratio of the flow velocity to the speed of sound) is equal to 1.

In conclusion, choked flow is a critical phenomenon at Mach = 1. At this point, the sound waves stop further expansion of fluid or further increase of fluid flow at restriction.

Explanation

The Joule-Thomson effect occurs when a pressurized fluid expands through a throttling valve, causing a decrease in temperature. This phenomenon is described by the Joule-Thomson coefficient, which determines whether the fluid will experience cooling or heating as it expands. In the case of flow through a throttling valve, the fluid does not choke because the flow is isenthalpic, meaning that the enthalpy remains constant during the process. In other words, the sum of internal energy and work energy always remains constant.

As said in the case of the Joule-Thomson effect, the flow is isenthalpic and does not involve any external work. This implies that no work is done on the fluid, so the sound waves present in the fluid are unable to compress or obstruct the flow as they would in a compressible flow situation. In other words, the sound waves are unable to choke the flow at the constriction by compression when the flow is isenthalpic.

This is why flow choking does not occur in the Joule-Thomson effect, as the process is isenthalpic and the sound waves are unable to impede the flow of the fluid through the throttling valve.

In summary, the Joule-Thomson expansion doesn't choke because the enthalpy remains constant during the expansion. This means that no external work is done, and the potential energy of the gas increases. This increase in potential energy causes the kinetic energy to decrease, which in turn causes the temperature of the gas to fall.

CHAPTER 40

Centrifugal Pumps

Pumps play a crucial role in thermodynamics as they are used to transfer fluids from one place to another, often against a pressure gradient. This is essential in many thermodynamic processes where fluids need to be circulated, compressed or delivered to different parts of a system.

In thermodynamics, pumps are often used to increase the pressure of a fluid, which can then be used to drive other processes such as generators, turbines, or heat exchangers. The work done by a pump to increase the pressure of a fluid is an important aspect of thermodynamic analysis, as it affects the overall efficiency of the system. Additionally, pumps are used in many thermodynamic systems to control the flow rate of fluids, which is critical for maintaining temperature, pressure, and other important parameters within the system. By controlling the flow rate, pumps help to ensure that the system operates efficiently and effectively.

Overall, pumps are important parts of thermodynamics because they play a key role in the transfer and circulation of fluids within thermodynamic systems, helping to maintain the desired conditions for efficient operation.

UNIT OF ENERGY IN BERNOULLI'S EQUATION

In one line Bernoulli's equation represents total energy per unit volume/mass because it's used to determine the volume of fluids in terms of pressure, velocity, and elevation.

Often there is a mix-up of 'energy/volume and energy/mass' in how we write Bernoulli's equation. While both are correct but their expressions are different.

Bernoulli's equation is just a convenient statement of conservation of energy for an incompressible fluid in the absence of friction.

Bernoulli's equation, also known as Bernoulli's theorem, is a fundamental principle in fluid dynamics. It states that the total energy per unit volume of a flowing fluid remains constant along its flow path, as long as there's no friction. This total energy is made up of the fluid's pressure energy, kinetic energy, and gravitational potential energy.

Details

Since $P = F/A$, its units are N/m^2. If we multiply these by m/m, we obtain N·m/m^3 = J/m^3, or energy per unit volume. F is force and A is area, N is Newton and m is meter.

In Bernoulli's equation, energy is expressed in terms of energy per unit volume:

- P: The pressure, which represents the flow energy per unit volume.
- 1/2 rho u^2: The kinetic energy per unit volume.
- rho gz,: The gravitational potential energy per unit volume.

The sum of these three terms is the available energy per unit volume. The equation for Bernoulli's principle is

P + 1/2 u^2 + rho g z = Constant

Bernoulli's equation is based on the law of conservation of energy. It states that the total mechanical energy of a moving fluid remains constant, even as the various forms of energy (kinetic, potential, and flow) are transformed into one another.

When Bernoulli's is equation is expressed as head

P/rho g + u^2/2g + z = Constant

P/rho g = pressure head or pressure energy per unit weight,

$u^2/2g$ = kinetic head or kinetic energy per unit weight,

z = potential head or potential energy by unit weight.

Where P refers to pressure, z refers to the height of the fluid, g refers to acceleration due to gravity, u refers to the velocity of flow.

Hence from the above equation, we can see that Bernoulli's equation represents energy per unit mass.

PRESSURE *vs.* HEAD IN FLUID DYNAMICS

The first point to note is that pressure is not energy but pressure head is energy.

The term "head" is used to represent the vertical height of a static column of liquid; it corresponds to the energy contained in the liquid per unit mass

The main difference between head and pressure is that head is fluid independent, while pressure is fluid dependent.

P is not energy but P/rho is energy per unit mass.

Derivation

The unit of pressure is Pascal (Pa), which is equal to 1 N/m^2 (Newton per square meter).

The unit of density ρ is kg/m^3 (kilogram per cubic meter).

Therefore, the unit of pressure head (P/ρ) can be derived as follows:

$$
\begin{aligned}
\text{Pressure head } (P/\rho) &= (Pa)/(\text{kg/m}^3) \\
&= (\text{N/m}^2)/(\text{kg/m}^3) \\
&= (\text{N/m}^2) \times (\text{m}^3/\text{kg}) \\
&= \text{Nm/kg (1 Nm = 1 Joule)} \\
&= \text{Joule (J)/kg}
\end{aligned}
$$

Hence, the unit of pressure head (P/ρ) is Joule (J)/kg, which is the energy per unit mass.

Head

The height a pump can lift a fluid, regardless of the fluid's density. Head is usually measured in meters of liquid column.

Pressure

The amount of force applied perpendicular to a surface per unit area. Pressure is affected by gravity and the density of the fluid. Pressure is measured using a manometer on the pump's suction and delivery line.

Pressure Head

It is simply the pressure value divided by the fluid density. In fluid mechanics, pressure head is the internal energy of a fluid due to the pressure exerted on its container.

Velocity Head

The kinetic energy of a liquid due to its velocity, represented by the formula $u^2/2$ g where u is the velocity and g is the acceleration due to gravity.

The unit of head is feet or meters.

Details

In fluid dynamics, the term "head" refers to the energy per unit weight of a fluid, typically expressed in meters (m). One meter of head is equivalent to the potential energy of a fluid at that height above a reference point. Head can also be expressed in other units, such as feet.

In the context of Bernoulli's equation, all terms are expressed in terms of head, which signifies the total energy of the fluid per unit weight. This includes the pressure head (pressure energy), velocity head (kinetic energy), and elevation head (potential energy). By expressing all terms in the equation as head, it allows for a consistent representation of the different forms of energy in the fluid flow.

CENTRIFUGAL PUMP: AN ENERGY CONSERVATION MACHINE

Summary

The impeller is offset from the casing in a centrifugal pump. The fluid channel between the impeller and the casing widens as they get closer to the discharge outlet. If the flow rate (Q) remains constant, increasing the cross-section area reduces velocity (kinetic energy). Because the total energy must remain constant, the fluid's pressure energy increases. From Bernoulli's equation [Total pressure – Static pressure] = Dynamic pressure = Kinetic energy/unit volume. When kinetic energy/unit volume reduces more pressure is generated. This is how the kinetic energy converts to pressure energy in the volute casing in a centrifugal pump to keep the total energy constant.

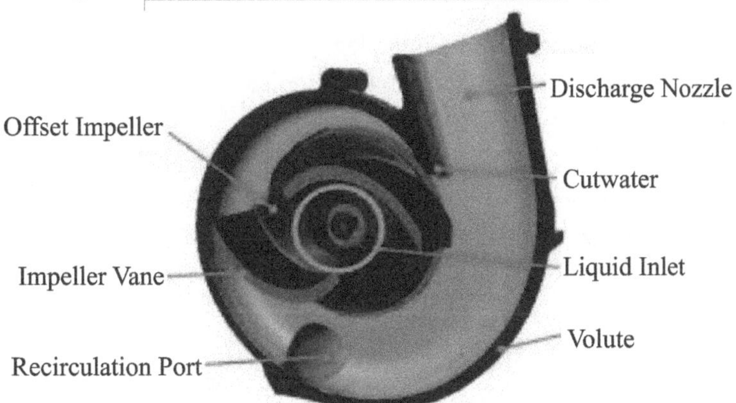

Figure 1: Centrifugal Pump

Any centrifugal pump's job is to move liquid, which necessitates the use of energy. The pump must provide an appropriate form of energy to the fluid in order to do work and accomplish this. What exactly do we mean when we say the most appropriate form of energy? Because for work to be done, we need the right type or form of energy to do the task.

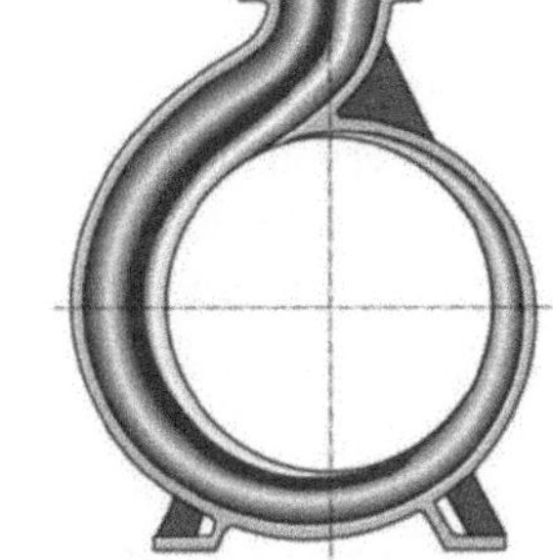

Figure 2: Sectional View of a Pump Volute

Which Type of Energy do We Look for?

We may have access to thermal energy, but heating water will not cause it to flow from point *A* to point *B*; instead, it will boil. However, in a power plant, the same thermal energy is transformed into electrical energy, which rotates the motor and pump shaft, converting the energy to kinetic and pressure energy, which is exactly what we need for fluid movement.

A Centrifugal Pump is an Energy Conservation Machine

A centrifugal pump is a machine that converts energy. By raising the velocity of the fluid leaving the impeller, the mechanical energy of the shaft is transferred to kinetic energy. In the pump volute or diffuser, the majority of this energy is transformed to pressure energy.

How? Simply put, for incompressible flow, according to Bernoulli's principle, any arbitrary position along the streamline has the same total energy. When one kind of energy increases or decreases, other forms of energy decrease or rise, but the overall energy remains the same. In our situation, total energy refers to the kinetic, potential, and pressure energies at any point in the pumping system (pumps, pipelines, valves, etc.).

CENTRIFUGAL PUMP: TORQUE AND ANGULAR MOMENTUM

Torque and angular momentum are inextricably linked. Angular momentum ('*l*') is the rotational analogue of linear momentum ('*p*'). It is a vector item. The particle's angular momentum is $l = r \times p$.

Summary

Step by step what goes on inside a centrifugal pump:

- The rotating shaft generates torque.
- The torque generates angular momentum of the impeller.
- The impeller transfers angular momentum to the rotating liquid.
- The momentum transfer increases the fluid's velocity.
- The kinetic energy of fluid increases.
- High-velocity fluid enters the volute casing.
- In a volute casing, a continuously increasing cross-sectional area converts the kinetic energy of the fluid into fluid pressure.

Given, that the torque on a particle is its angular momentum,

$$\tau = r \times F$$

where F is the force acting on the particle, τ is the torque and r is the moment arm as r, the moment arm of rotation increases in the offset volute, the angular momentum/radial forces increase. By using a concentric casing where the centerline of the impeller and the casing are congruent the magnitude of radial forces would be significantly less.

Finally, momentum converts to pressure or they are the same thing. Pressure is defined as force/area which is the same as momentum/area/time since $F = dp/dt$.

Details

A pump adds mechanical energy to a fluid by increasing its pressure. It is a device that uses mechanical action to move fluids (liquids or gases). A centrifugal pump converts driver energy to kinetic energy in a liquid by accelerating the fluid to the outer rim of an impeller. The amount of energy delivered to the liquid corresponds to the velocity at the impeller's edge or vane tip. The higher the velocity of the liquid at the vane tip and the more energy imparted to the liquid, the faster the impeller revolves or the larger the impeller.

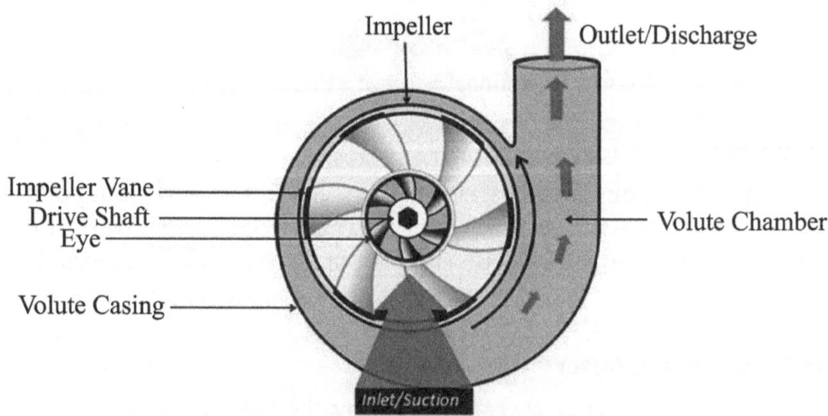

Figure 3: Volute Case Design

Centrifugal Pump Energy Conversion

When fluid enters a centrifugal pump, it is immediately directed to the low-pressure area at the impeller's center or eye. The rotating impeller and blades transfer momentum to the incoming fluid as they rotate. A momentum transfer to a moving fluid increases the fluid's velocity. The kinetic energy of a fluid increases as its velocity increases. High-kinetic-energy fluid is forced out of the impeller area and into the volute.

The volute is a region with a continuously increasing cross-sectional area that is designed to convert fluid kinetic energy into fluid pressure. This energy conversion mechanism is the same as that for subsonic flow through a diverging section of a nozzle.

Explanation

What is an Impeller?

The impeller is the most important component of a centrifugal pump. It is made up of a series of curved vanes. Normally, these are sandwiched between two discs (an enclosed impeller). An open or semi-open impeller (backed by a single disc) is preferred for fluids containing entrained solids. By accelerating the fluid radially outwards from the center of rotation, impellers transfer energy from the motor that drives the pump to the fluid being pumped.

Figure 4: Impeller Types (l to r): Open, Semi-Enclosed (or Semi-Open), Enclosed

Fluid enters the impeller through its axis (the 'eye') and exits through the circumference between the vanes. The impeller is connected to a motor via a drive shaft and rotates at high speed on the opposite side of the eye (typically 500–5000 rpm). The fluid is accelerated out of the pump casing by the rotational motion of the impeller vanes.

Pump casings are classified into two types: volute and diffuser. The goal of both designs is to convert fluid flow into a pressure-controlled discharge.

What is the Eye of the Impeller? Why it is a Low-Pressure Area?

Impeller's center is the eye of the impeller.

[1] **The intuitive explanation**: The rotating impeller imparts a centrifugal force on the fluid in between the impeller blades. This centrifugal force pushes the fluid radially outward leaving an empty void at the center. This void translates to low pressure at the eye of the impeller, which the surrounding fluid tries to equalize by rushing into that space.

[2] **The mathematical explanation**: Bernoulli's principle states that:

P/rho + $V2/2$ = constant along a streamline so long as no external work is done on the fluid.

P = pressure of the fluid at a point in the flow
rho = density of the fluid at that point
V = velocity of the fluid at that point

This equation means the higher the velocity, the lesser would be its pressure, and vice versa as long as no external work is done on the fluid.

At the suction end, the fluid is accelerated towards the eye of the impeller and a high velocity of the fluid at the eye means a lower pressure than the surrounding fluid.

Most Fundamental Point

How impeller transfers momentum to fluid in the pump?
First, what is momentum in a centrifugal pump?

Torque and angular momentum:

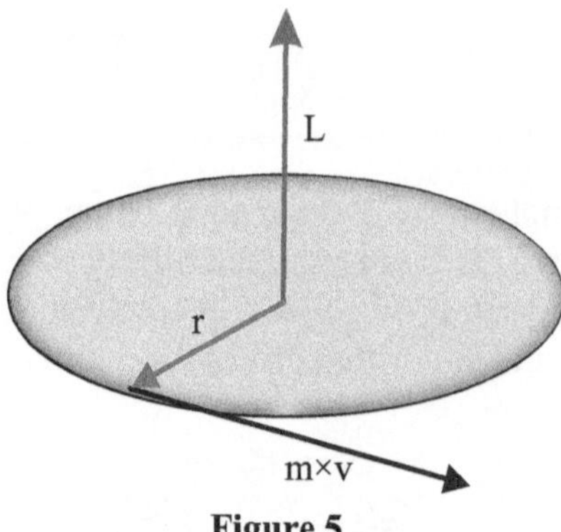

Figure 5

The turning effect of a force is called Torque. The rotating shaft generates torque. The torque generates angular momentum. Angular momentum is defined as the property of any rotating object given by moment of inertia times angular velocity. [what is the moment of inertia? moment of inertia also known as the angular mass or rotational inertia can be defined w.r.t. rotation axis, as a quantity that decides the amount of torque required for a desired angular acceleration or a property of a body due to which it resists angular acceleration]. The angular momentum of the body is defined as the moment of momentum about the axis of rotation. More generally, the torque on a point particle (which has the position r in some reference frame) can be defined as the cross product,

$$\tau = r \times F$$

where F is the force acting on the particle. The magnitude τ of the torque is given by

$$\tau = r \times F \text{ Sin } \theta$$

where F is the magnitude of the force applied, and θ is the angle between the position and force vectors.

A transfer of momentum to the moving fluid increases the fluid's velocity. As the fluid's velocity increases its kinetic energy increases. The fluid of high kinetic energy is forced out of the impeller area and enters the volute.

In the context of centrifugal pump: What is happening here?

Given, that the torque on a particle is its angular momentum,
where F is the force acting on the particle, τ is the torque and r is the moment arm as r, the moment arm of rotation increases in the offset volute, the angular momentum increases.

Finally, momentum converts to pressure or they are the same thing. Pressure is defined as force/area which is the same as momentum/area/time since $F = dp/dt$. Momentum flow would be the momentum passing through a unit area per unit time. So it has the same unit.

Bernoulli's Principle on the Kinetic Energy of Moving Fluids

The Bernoulli equation states that for an ideal fluid (that is, zero viscosity, constant density, and steady flow), the sum of its kinetic, potential, and thermal energy must not change. This constraint gives rise to a predictable relationship between the velocity (speed) of the fluid, its pressure, and its elevation (relative height). Specifically, given two points along a streamline (an imaginary line tangent to the direction of flow, as shown below.

Bernoulli's equation

$$\frac{v^2}{2}+\frac{p}{p}+gz = \text{constant}$$

Where:

v is fluid velocity
p is fluid pressure
g is fluid density
g is gravitational constant
z is elevation

Pressure has two parts [1] static and [2] dynamic. In incompressible fluids dynamic pressure 'q' sometimes called velocity pressure is the quantity defined,

$$q = 1/2 \text{ rho } V^2$$

q is dynamic pressure in pascals (i.e., $kg \cdot m^{-1} \cdot s^{-2}$),
rho is fluid mass density (e.g., in kg/m^3, in SI units),
V is flow speed in m/s.

The dynamic pressure can be thought of as the fluid's kinetic energy per unit volume. For incompressible flow, the dynamic pressure of a fluid is the difference between its total pressure and static pressure. From Bernoulli's law, dynamic pressure is given by,

$$p_0 - ps = ½ \text{ rho } V^2$$

p_0 is total pressure and ps is static pressure.

$p_0 - ps$ always remains constant. The static pressure and the dynamic pressure switch from one to another to keep the total pressure constant.

Fundamentals

The impeller of a centrifugal pump is offset to the casing see the sectional view of a pump volute.

As the impeller and casing come closer to the discharge port, the fluid path between them widens. The flow rate (Q) is equal to the velocity multiplied by the area ($Q = V \times A$). As a result, for a given constant flow rate (Q), since increasing the cross-section area reduces velocity (kinetic energy), for keeping the total energy constant the pressure energy increases. This is how the kinetic energy converts to pressure energy in a centrifugal pump.

In summary, the impeller is offset from the casing in a centrifugal pump. The fluid channel between the impeller and the casing widens as they get closer to the discharge outlet. If the flow rate (Q) remains constant, increasing the cross-section area reduces velocity (kinetic energy). Because the total energy must remain constant, the fluid's pressure energy increases. From Bernoulli's equation [Total pressure – Static pressure] = Dynamic pressure = Kinetic energy/unit volume. When kinetic energy/unit volume reduces more pressure is generated. This is how the kinetic energy converts to pressure energy in the volute casing in a centrifugal pump to keep the total energy constant.

AFFINITY LAWS OF PUMPS

Background

While dealing with centrifugal pumps, we are typically most concerned with the three primary performance characteristics: flow, head, and power. These three characteristics all vary in an orderly manner with changes in impeller speed, following very simple equations. These equations are known as the pump affinity laws and allow for the prediction of pump performance at varying speeds. Given flow, head, and power at one speed for a specific pump, one can determine new values for those parameters at a different speed.

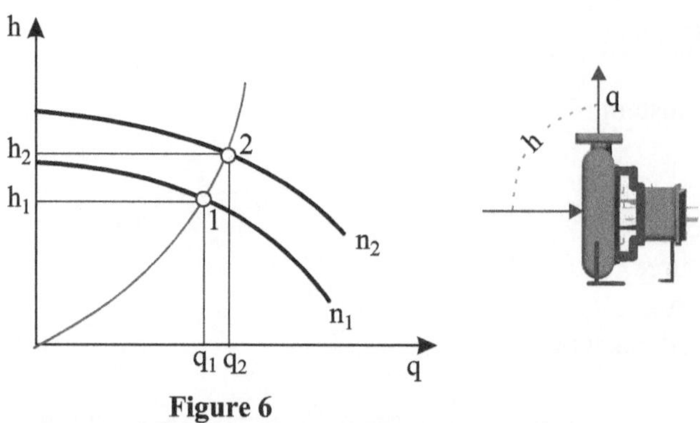

Figure 6

Image Credit: Google & NAPTEL

Total dynamic head (or total head), discharge and power of a pump are related to the speed of an impeller (i.e., pump speed). These relations can be developed by using fundamental equations of pumps. As we know, the peripheral velocity of the impeller (u) is given as:

u = [Pie × D × N]/60, u is proportional to N

Pump Affinity Laws 1

The first pump affinity law deals with flow or pump capacity. The flow varies in direct proportion with the impeller speed. If the impeller speed for a particular pump is increased by 10%, the flow from that pump increases by 10%. Likewise, if the impeller speed is reduced by 20%, the flow is reduced by 20%. The affinity law equation for flow is:

$$Q_1/Q_2 = N_1/N_2$$

where N_1 = Original speed, Q_1 = Original flow, N_2 = New speed, Q_2 = New flow. Rearranging we get, $Q_2 = Q_1 \times N_2/N_1$.

The second pump affinity law deals with fluid head. The head that a particular pump generates varies with the square of the proportional speed change. For the head, when the impeller speed is increased by 10%, the head is increased by 21%, and so on. The affinity law equation for the head is:

$$H_1/H_2 = [N_1/N_2]^2$$

where, N_1 = Original speed, H_1 = Original head, N_2 = New speed, H_2 = New head. Rearranging we get, $H_2 = H_1 \times [N_2/N_1]^2$.

The third pump affinity law deals with hydraulic power. The power that a particular pump generates varies with the cube of the proportional speed change, and thus is most impacted by speed adjustments. The affinity law equation for power is:

$$P_1/P_2 = [N_1/N_2]^3$$

where N_1 = Original speed, P_1 = Original power, N_2 = New speed, P_2 = New power. Rearranging we get, $P_2 = P_1$ X$[N_2/N_1]^3$.

Pump Affinity Laws 2

Thus, similar to the impact of change in pump speed on pump performance, we obtain three equations that relate the impact of change in impeller diameter to changes in pump performance. If D_1 and D_2 are the two impeller diameters, and Q_1, Q_2; H_1, H_2 and P_1, P_2 are respectively discharges, heads and powers corresponding to the impeller diameters D_1 and D_2, the three equations relating the discharge, head, and power with the impeller diameter are as follows:

$$Q_1/Q_2 = [D_1/D_2]$$

$$H_1/H_2 = [D_1/D_2]^2$$

$$P_1/P_2 = [D_1/D_2]^3$$

The above three equations are known as Affinity Law II, which suggests that the discharge varies linearly with a change in the impeller diameter, the head varies as the square of the ratio of the two impeller diameters, and the power varies as the cube of the ratio of the two impeller diameters.

Note that the applicability of the Affinity Law II is limited to less than 20% changes in the original outer diameter of the impeller.

Example

Problem: A centrifugal pump requires 5 kW power when it runs at 1450 rpm and delivers water against a head of 10 m. If the pump is operated at 1750 rpm, calculate the head developed and the power required by the pump.

Solution: From the question, $N_1 = 1450$ rpm, $N_2 = 1750$ rpm, $H_1 = 10$ m, and $P_1 = 5$ kW. The head developed by the pump (H_2) and the power required (P_2) at 1750 rpm can be calculated by using the Affinity Law I.
From the Affinity Law I, we have:

$$H_2 = H_1 \times [N_2/N_1]^2 = 10 \times [1750/1450] = 14.57 \text{ m}$$

From affinity law 1

$$P_2 = P_1 \times [N_2/N_1]^3 = 5\text{x } [1750/1450] = 8.79 \text{ kw}$$

WHAT IS PUMP CURVE?

A pump curve shown below describes the relationship between flow and head in a graph. The flow appears on the x-axis and head pressure on the y-axis. The curve begins at the point of zero flow, or shutoff head, and gradually descends until it reaches the pump runout point or maximum flow rate. Other variables covered in a pump curve are: the relationship between RPM, impeller size, power efficiency, and net positive suction head (NPSH) in centrifugal and positive displacement pumps.

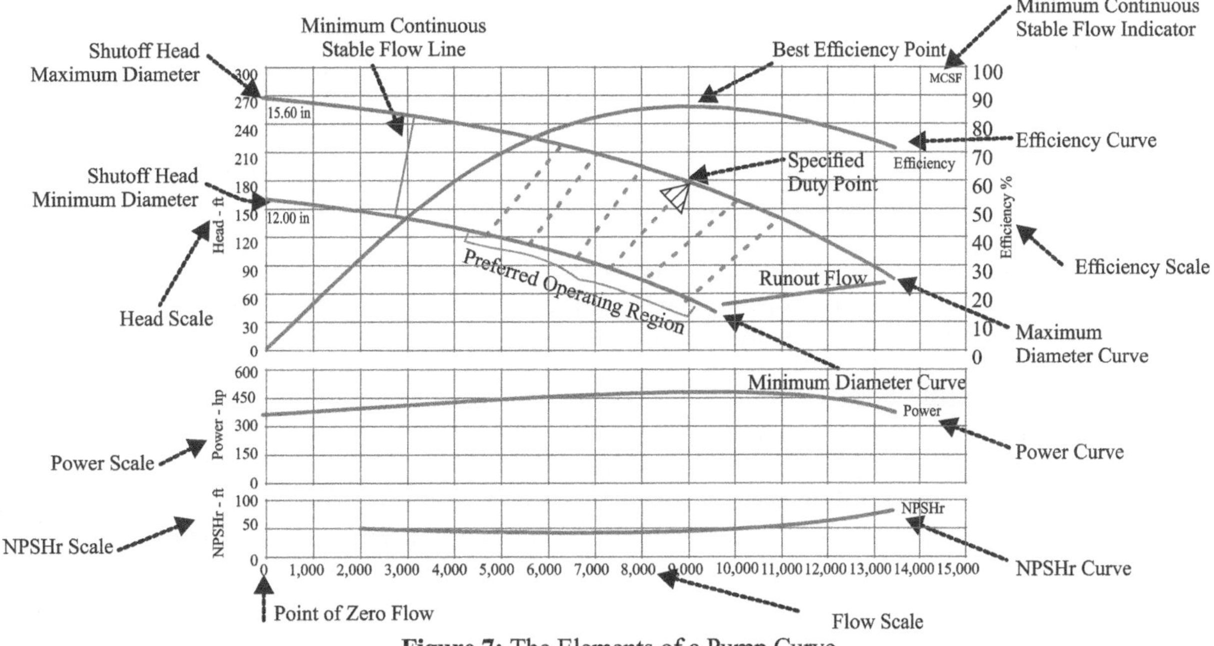

Figure 7: The Elements of a Pump Curve

Image Credit: Google & NAPTEL

Explanation

BEP

The optimal efficiency point (BEP), also known as the pump's operating "sweet spot," is often found close to the curve's centre. When a pump can operate close to its BEP, as specified by the manufacturer, it will operate most efficiently and last the longest. **The preferred operating region (POR) for a pump is typically defined as the region on the curve that falls between 70 and 120 percent of the BEP by blue dashed lines in the above image.** The impellor size plays a crucial role in determining the most stable area for the maximum efficiency called the preferred operating region (POR). This area falls between maximum diameter impeller size shutoff head and minimum diameter impeller size shutoff head.

What is Stable Flow and What is Unstable Flow in a Pump?

For stable head-flow characteristic the differential head –*H* progressively falls with increasing flow rate – *Q*.

Stable Flow

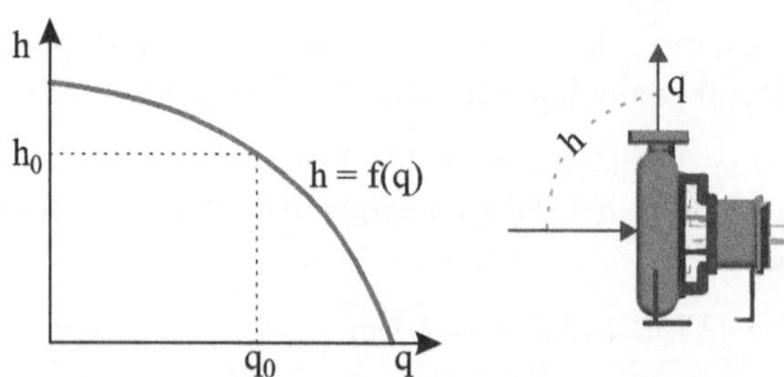

Figure 8: The Engineering Tool Box
Image Credit: Google & NAPTEL

For unstable head-flow characteristic the differential head –*H* rises to a maximum and then progressively falls with increasing flow rate - *Q*.

Unstable Flow

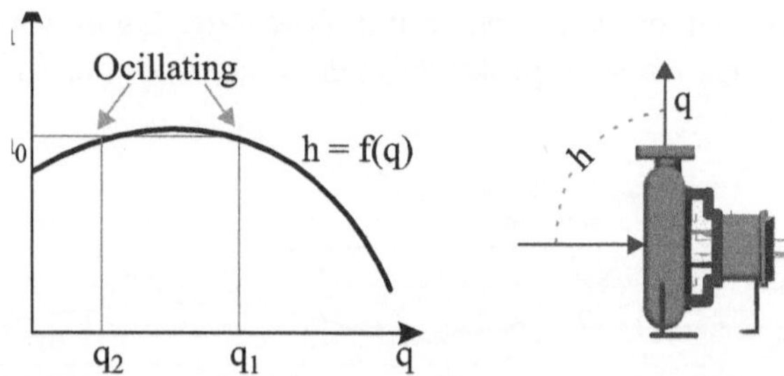

Figure 9: The Engineering Tool Box
Image Credit: Google & NAPTEL

In summary an unstable flow is an oscillating completely unpredictable flow. The impeller size plays a critical role. At both extreme ends, a too large diameter impeller (too much pressure drops) and a two small impeller diameter (if the suction starves) can generate an unstable flow by cavitation.

System Curve and Operating Point

Together with the pump curve, a second curve known as the system curve can be superimposed on the same graph. The system curve, which is produced by figuring out the system's static head and friction loss, represents the system head in your particular application at different flow rates. The energy needed to overcome flow resistance is referred to as the head (or pressure) loss due to friction. On a system curve, a rise in system head, or the pressure needed to make the liquid move, corresponds with an increase in flow rate. The graph below shows the location of operating point.

Operating Point on Pump Curve

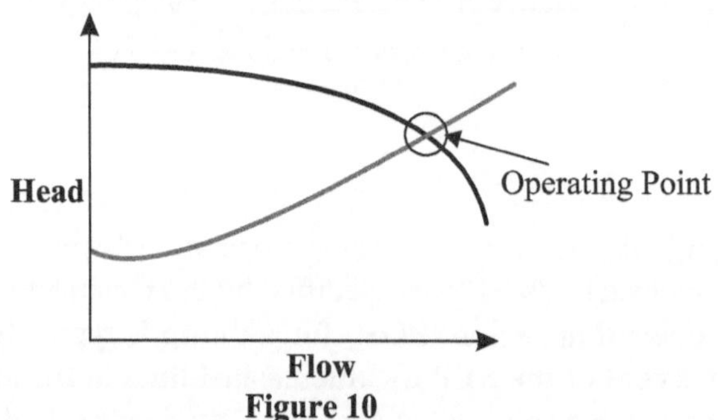

Figure 10
Image Credit: Google & NAPTEL

Pump curve is provided by the manufacturer of the pump and the system curve is generated at the site by using site data.

In a pump curve provided by the pump manufacture a system curve generated at the site with site data is overlaid.

When the system curve and the pump curve are superimposed, it becomes clear how the pump will function at a given flow rate and head pressure based on the impeller diameter and pump control valve position. The pump's actual operating point in that specific system is indicated by the graph's intersection of the pump and system curves.

If the impeller diameter is increased or decreased, the pump's performance can be depicted on the same graph. The diameter is given in inches as can be observed in the pump curve image. The system curve is unaffected by changes in impeller diameter and only shifts in response to changes in system head, such as a closed valve. The intersection points are where the pump will perform at each diameter.

As long as the pump performance remains within the POR, it is acceptable to alter the impeller diameter and system parameters.

Efficiency Curve

What is BEP?

The BEP, or best efficiency point, is the point at which a pump operates at its peak efficiency. At the BEP, flow enters and exits the pump with the least amount of flow separation, turbulence, and other losses. The closer a pump operates to its BEP, the less wear it will experience, increasing the pump's reliability. The BEP is located at the apex of the pump efficiency curve. The terms "BEP" and "preferred operating line" (POR) are synonymous. The POR is a range of flow rates to either side of the predicted BEP within which the pump's hydraulic efficiency and operational reliability are not significantly reduced. The pump's design service life is unaffected by flow-induced vibration or internal hydraulic loads within this region.

The efficiency of a pump over its whole operating range is represented by the pump efficiency curve. The curve graph's right side displays percentages representing efficiency. The peak of the efficiency curve represents the BEP, and efficiency decreases as the curve arcs away from the BEP to the right or left beyond the stable region of the pump as explained above when the flow becomes unstable. The horsepower needed for an application can also be calculated with the efficiency % in mind. The most important thing to remember from this discussion is that any modification of the impeller will change the BEP of the pump. Trimming the outside diameter (OD) of an impeller, replacing an impeller with one of different diameter or number of vanes or changing the rotating speed of the impeller will alter the BEP for the pump. Before modifying an impeller in any way, make sure that you determine how the change will impact the pump curve, the efficiency curve and the BEP.

Power Curve

The power curve aids in appropriate motor sizing by showing the load the pump places on the driver at a specific point on the pump curve. It is shown as a separate curve graph that rises gradually to its peak load, which for the majority of rotodynamic pump types is close to the BEP. Following that, it declines as it approaches the runout point.

Net Positive Suction Head Curve

The amount of force required to force liquid into the pump impeller's eye is indicated by the net positive suction head required (NPSHr). It is shown below the main pump curve graph in feet. Pump cavitation, vibration, and premature failure can be avoided by knowing the appropriate NPSHr.

Specific Points

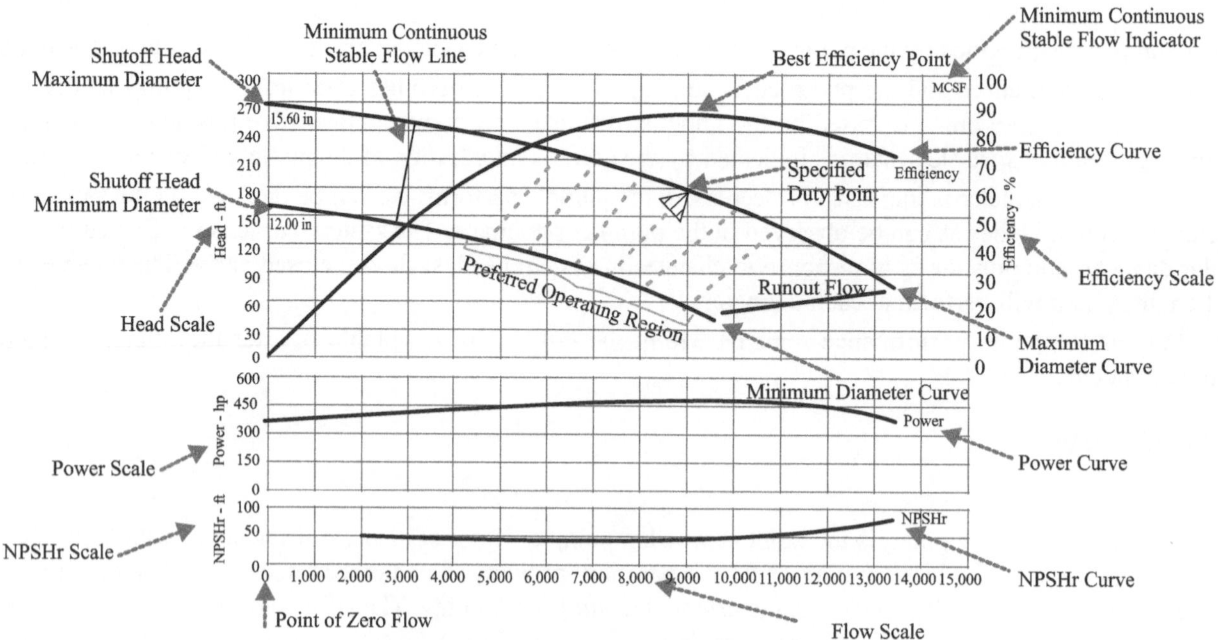

Figure 11: The Elements of a Pump Curve

Image Credit: Google & NAPTEL

Why NPSHr Tends to Increase Both at Very Low and Very High Flow Rate?

At both ends of very low and very high flow rates NPSHr increase. At very high flow rate the high velocity causes a pressure drop that can lead to cavitation. At very low flow rate if the suction starves there can be vaporization due to very low pressure and cavitation. NPSHr line generally remains flat between very low and very high flow rate. Note that the NPSHr curve is roughly a lazy "*U*," reaching a minimum value at a capacity *Q* about 40 percent of the best-efficiency capacity. Although most published curves do not show the increased NPSHr at low flows, all centrifugal pumps exhibit such a characteristic. The NPSHr always rises as the pump capacity approaches shut-off (zero flow rate). Because pump head is a function of the impeller diameter, D2, the traditional 3 percent-head-drop NPSH varies as the impeller diameter varies.

Head Drop *vs.* Impeller Diameter

With the same degree of cavitation in the impeller eye, a pump with a maximum-diameter impeller will exhibit a smaller percentage of head drop than the same pump with a reduced-diameter impeller. As the impeller diameter decreases, added clearance between the impeller and the fixed pump casing increases internal flow recirculation, causes head loss, and lowers pumping efficiency.

Why Brake Horse Power Increases and Near BEP Reaches Maximum?

The most important factor in determining a pump's BEP is the impeller design, which determines how efficiently power (brake horsepower or BHP) is transmitted to the liquid being pumped. A well-designed impeller optimizes flow while reducing turbulence and increasing efficiency. With increasing flow BHP increases and reaches the maximum stable efficiency at BEP. In summary, impeller design is the most significant factor for determining the BEP of a pump because it dictates how efficiently power (brake horsepower or BHP) is transmitted to the liquid being pumped. A properly designed impeller optimizes flow while minimizing turbulence. Points to remember before modifying an impeller in any way, make sure that you determine how the change will impact the pump curve, the efficiency curve and the BEP.

Brake horse power and best efficiency relation

Brake horse power of pumps = $Q \times P/(C \times n)$

Where:

BHP = Brake Horsepower in horsepower

Q = Flow Rate in gallons per minute

P = Pressure in pounds per square inch

n = Efficiency expressed as a decimal

C = A conversion factor depending on units of flow and pressure

The formula for calculating *NPSHa*, is

$$NPSHa = h_1 + hss - hf - hvp$$

Where:

h_1 is the absolute pressure over the liquid in feet

hf is the friction in the suction line in feet

hss is the static head in feet – this is negative if there is a suction lift

hvp is the vapor pressure of the liquid at the pump in feet

The change in your pump's performance when the impeller diameter is reduced, can be predicted using affinity laws: Pump flow rate (Q) varies directly with the diameter (D), i.e. $Q_1/Q_2 = D_1/D_2$. But the power P is related to impeller diameter by equation $P_1/P_2 = [D_1/D_2]^3$. In other words, for a given pump with a constant speed, the capacity will be directly proportional to the impeller diameter, the head will be directly proportional to the square of the impeller diameter, and the required power will be directly proportional to the cube of the impeller diameter. At BEP the BHP is most optimal. When you operate on the left side of BEP the BHP is even smaller.

Why impellers are often trimmed?

Trimming of impeller is often done to avoid replacement of impeller in oversized pumps. Oversized and throttled pumps produce excess pressure and increase power consumption. So, the trimming of impeller is done primarily to reduce energy cost. Trimming involves machining the impeller to reduce its diameter. Trimming should be limited to about 75% of a pump's maximum impeller diameter, because excessive trimming can result in a mismatched impeller and casing. Essentially, trimming reduces the impeller's tip speed, which in turn reduces the amount of energy imparted to the pumped fluid; as a result, the pump's flow rate and pressure both decrease. A smaller or trimmed impeller can thus be used efficiently in applications in which the current impeller is producing excessive head. This is the advantage part of impeller trimming. The disadvantage part is as the impeller diameter decreases, added clearance between the impeller and the fixed pump casing increases internal flow recirculation and turbulence and causes head loss, and lowers pumping efficiency.

References

Pumps and Systems

Wikipedia

Research Direct

Engineering too box

PUMP CURVE EXPLANATION

A pump's primary job is to exert force on a liquid. It has the ability to compress and move liquids. Pumps are divided into two categories: positive displacement pumps and centrifugal pumps, depending on how they work. Positive displacement pumps transfer fluid by altering the size of the pump's chamber, whereas centrifugal pumps do so by moving their blades. Positive displacement pumps are typically used with high viscosity fluids, and centrifugal pumps are typically used with low viscosity fluids.

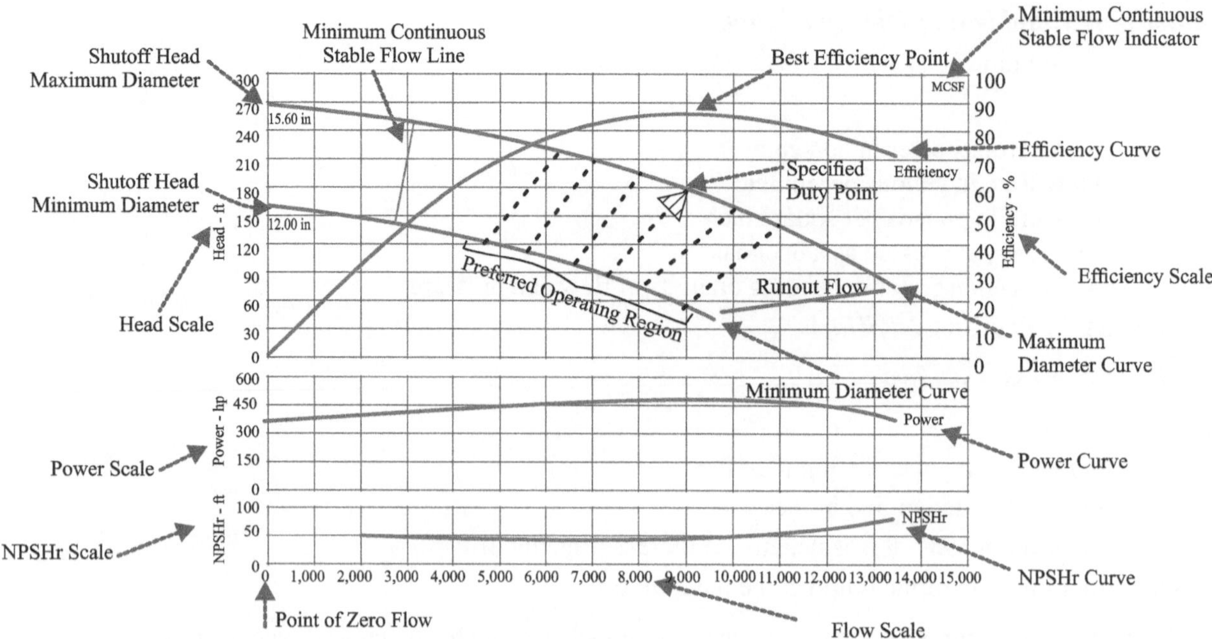

Figure 12: The Elements of a Pump Curve

Image Credit: Google & NAPTEL

Fundamental Understanding

The volume of fluid passing through the pump in a unit of time is indicated by the volume flow rate (Q). As a result, it provides the fluid flow rate through the pump. The volume flow rate can also be calculated using the mass flow rate (m), and the density of the operating fluid. The formula for calculating the relationship between the mass flow rate and volume flow rate, or capacity, is as follows: $m = rho \times Q$, where rho is the density of the operating fluid.

Discharge head

The net head that is obtained at a pump's outlet is known as the discharge head. With a centrifugal pump, the density of the fluid and the suction or inlet pressure also affect the discharge pressure. As a result, depending on the inlet pressure, we may have different discharge pressure values for the same fluid flow rate. As a result, the fluid's discharge head, the height it can reach after being pumped varies depending on the operating circumstances.

Total head

For a centrifugal pump, total head is the difference between the maximum height the fluid can reach at the inlet and the maximum height it can reach at the outlet. This is a widely used parameter for determining industrial requirements and is essential for choosing the right pump.

Total head = Discharge head – Suction head

The pump's energy (head) contribution to the fluid flow is all that remains after the inlet head is removed, eliminating the impact of the pressure that is supplied to the pump.

Total Dynamic Head (TDH)

The total dynamic head is the sum potential energy + pressure + kinetic energy. The total dynamic head (TDH) is the work that a pump needs to do, in height of fluid, to pump a fluid, considering the static lift + the pressure head + the velocity head (Kinetic energy) + friction loss.

SENSITIVITY OF OPERATING POINT ON A PUMP CURVE

Pump operating point is the intersection of pump characteristic [pump curve] and piping system characteristic [system curve. The pump will always operate where the system and pump curves intersect. The flow rate you can anticipate from the pump in this configuration can be found by plotting the pump and system curves on the same graph as shown below, and identifying the intersection of the lines. The operating point is this intersection.

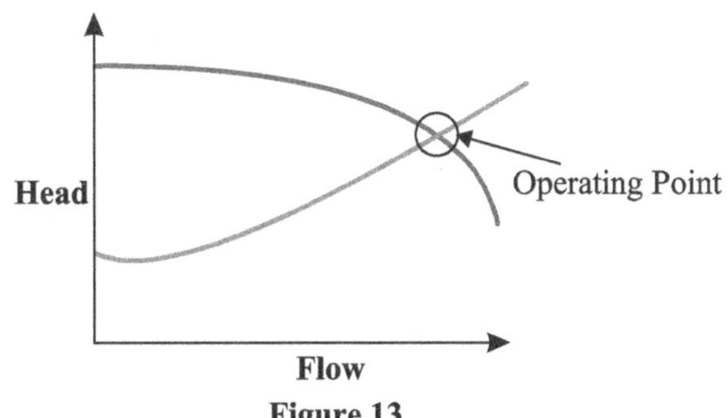

Figure 13

Image Credit: Google & NAPTEL

A pump's efficiency is displayed as a curve that spans the range of pressures and flows the pump can produce. The efficiency on the left side of the curve will increase from 0 to a maximum of about 85% efficiency, after which it will start to decline at the midpoint of the performance curve. The best practice is to be between 10% and 5% to the left of BEP. It goes without saying that choosing the best pump for your system is crucial, but it doesn't ensure that your pump will continue to function when it gets close to BEP. Systems alter over time in the real world.

Best Efficiency Point [explained earlier]

The point on a pump performance curve where efficiency is highest is called the best efficiency point (BEP). Stated differently, BEP represents the maximum pump efficiency. It is ideal for any pumping system to operate at or near the pump's BEP. The operating point would move on the pump performance curve depending on what happens in downstream and upstream, and if the situation is not managed properly, the operating point might be far from BEP. Running at an operating point far from BEP could be problematic. The desire is to keep the operating points at or around BEP regardless of disturbances in upstream and downstream. Operation at a flow rate lower or higher than the best efficiency point (BEP) flow rate is not efficient nor desirable. However, this might happen in transients and unexpected cases.

Ensuring that every pump in the system is running as close to its BEP as feasible is essential to achieving the best possible overall system performance. This not only provides you with the best energy return on investment, but it will also lower your maintenance expenses. When operated close to their BEP, pumps operate most smoothly and have the longest service lives.

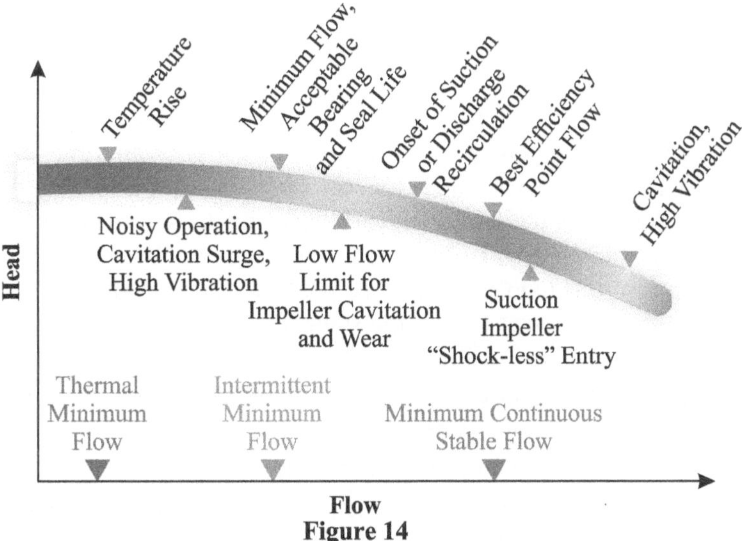

Figure 14

Image Credit: Google & NAPTEL

The illustration above and across the pump curve shows the Best Efficiency Point, which is the most efficient point at which the pump can be operated. It is usually located midway along the curve.

As demonstrated in the illustration below, operating the pump to the left may result in severe vibration, low bearing life, and mechanical seal failure.

If a pump is run too far left on its curve, no additional capacity is allowed in case the system pressures are calculated incorrectly.

If the pump is positioned too far to the right, there is a chance of cavitation, which can quickly damage the impeller and pump casing and raise the liquid's temperature and could be a safety concern in certain cases.

Good practice is to always have a safety margin maybe 10% towards the left of the point to ensure the pump can operate as required, as a pumps performance can always be reduced, but not increased.

Key Points

BEP and Sizing of the Pump

Gaining a thorough understanding of the upstream and downstream systems is the first step in pump selection and sizing. When choosing a pump, it is important to determine or compute the suction pressure, discharge pressure, needed head, flow rate, net positive suction head (NPSH), and other crucial factors.

Finding the kind and number of stages required to meet the specified set of operating conditions—typically the flow rate range, head, NPSH available (NPSHa), and others—is the first stage in the selection and sizing of a pump. In case the pump needs to function under multiple sets of operating conditions, it is recommended to select one set as the rated point. This point should be situated as near to the pump's BEP as feasible.

The chosen pump's acceptable operating range should encompass all other operating points. Stated differently, they ought to be sufficiently near BEP. It might take some trial and error to make the right choice. This indicates that the appropriate pump should be chosen based on its performance curve and details. Prior to determining the corresponding pump performance curves, operating range, and distances to BEP, one should first make a trial choice, perform a preliminary selection/sizing, and repeat these steps as needed.

BEP and Head

It's crucial that the head rise continuously from BEP to shutoff. The value of the head rise from the BEP/rated point to the shutoff point is another crucial factor. A slight rise in the required head would not move the operating point too far from BEP if the pump curve is steep enough and the head rise is substantial enough. In this manner, while the operating point is still reasonably close to BEP, the necessary head rise would be produced with only a minor adjustment. At least 8% should be the head rise from the BEP/rated point to shutoff. The head rise from the rated point/BEP to shutoff on some pumps, like the American Petroleum Institute (API) 610 pumps or pumps operating in parallel, should be greater than 10%.

BEP and Shut Off Head

Long periods of time spent running a pump at or close to its shutoff point (low flow, maximum head) are not recommended. Operations errors or disruptions, such as accidentally shutting a main valve downstream, can result in transient operation close to the shutoff point. This operation close to the shutoff point should be restricted to 15–40 seconds for certain pumps. A dedicated pressure instrument or switch may be configured in certain pumping systems to turn the pump off when the pressure is high (low or no flow). Complex and vital pumps should have more advanced safeguards and controls to prevent operation at or near the shutoff point.

The majority of pumps shouldn't be used at flow rates lower than a specific threshold. Usually, this is referred to as the minimum continuous stable flow. A recirculation line or other comparable device should be used if a pump is to be operated at a flow rate lower than 60% of the flow rate at BEP.

The following are some elements that influence minimum stable flow:

- increases in the liquid's temperature. One may assume that the temperature rise will not exceed 8 or 10 degrees Celsius. A pump may overheat if it runs at a very low flow rate.
- thrust on impellers.

This is a significant issue. This increase in force/load could occur even at flow rates as low as 60% of BEP flow rate. This is a significant issue. This increase in force and load could shorten the life of the bearing(s), even at flow rates as low as 60% of BEP flow rate. The high forces could cause impeller rubbing, excessive shaft deflection, seal failures, shaft breakage, and other issues.

- recirculated flow within the pump. Noise, vibration, cavitation, and mechanical damage could result from this.

A recirculation line or another capacity control technique, like a variable speed drive (VSD), should be used if a lower flow rate is required. While the pump is still operating at or close to BEP, this results in an overall low flow.

Challenges on the Right Side of BEP

High flow rate operation near the curve's end on the right side of the BEP may also present challenges. Near the curve's end, a pump would vibrate excessively and make noise. In this area, the pump would require a significant amount of power due to its low efficiency. The system is unreliable and inefficient.

Final Words

Many pumps need to operate at different pressures (heads). A pump will only deliver the flow rate (capacity) that corresponds to the intersection of the system head's curve and its performance curve. In order to vary the flow rate, something needs to be changed, preferably without shifting the operating point away from BEP. This can be done by using variable speed, control valves, and other control methods. The important thing is to have the desired head and flow rate while the pump is still operating close to its BEP.

COMBINING TWO PUMPS IN SERIES AND PARALLEL

One often decides to combine another pump when there is insufficient flow or head. Combining two pumps in parallel or series gives you two options. The advantages and restrictions of each of these choices are briefly discussed in this note.

Pumps can be linked in parallel for increased flow rate or in series for increased head; the linked pumps do not have to be of the same design.

Pumps in series

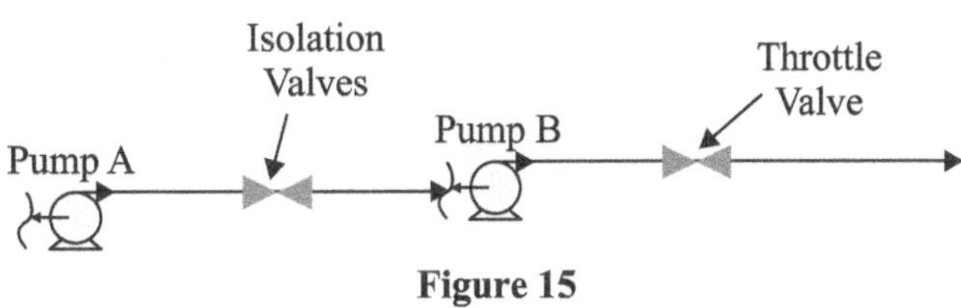

Figure 15

Image Credit: Google & NAPTEL

The above image shows a typical combination of two pumps *A* and *B* in series.

Operating Procedure

The important point to note is that the discharge pressure of Pump-A is equal to the suction pressure of Pump-B. Therefore, Pump-B must be rated for the higher suction and discharge pressures. Pump-B experiences a higher suction pressure than Pump-A, so care must be taken to choose a pump that can withstand the higher working pressure, even though Pump-B and Pump-A have the same differential pressure and flow rate. When using the B pump in series, it's also necessary to inspect the thrust bearings and mechanical seals after Pump-A, Pump-B, and the connected piping are primed full of the pumping liquid, the proper starting procedure is to start Pump-A first, which is the booster pump feeding Pump-B. Next, turn on Pump-B, making sure that its discharge throttle valve is opened to between 30 and 40% at first. Pump-A and Pump-B should then produce the desired flowrate.

The System Resistance Curve crosses the Series Curve as indicated below, not the individual Pump Curves, to determine the actual Rated flowrate of Pumps *A* and *B*.

Pump curve

The above is an image of two identical pumps combined in series, both having the same head, *H*. The head & flow curves for Pump-A and the combined curve, which is plotted by adding the heads of Pump-A and Pump-B at the same flowrates, are displayed in the schematic below. One can observe the shift of the single pump operating point to the right when two pumps are combined. Ideally the combined operating point should be on the left of BEP. It can also be noticed that the two heads add up to 2*H* while the there is no increase in the flow, Q on the x-axis.

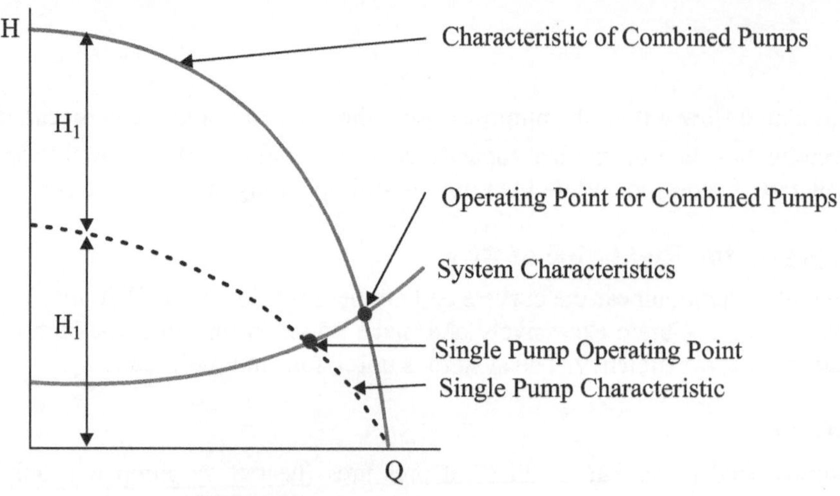

Figure 16

Image Credit: Google & NAPTEL

Pump-B may be a larger pump producing higher head when operating in series, or it may even be a multi-stage pump producing much higher head due to the size difference between the two pumps. But it is imperative that Pump-B and Pump-A have the same flowrates.

Pumps in Parallel

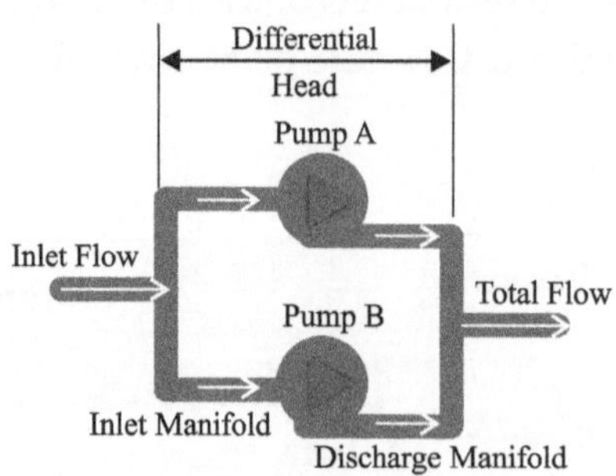

Both Pumps See the Same Differential Head
Total Flow = Pump A + Pump B (at the same differential head)

Figure 17

Image Credit: Google & NAPTEL

The image above shows two pumps *A* and *B* are combined in parallel.

Figure below represents the combined *H-Q* characteristic for two identical pumps connected in parallel.

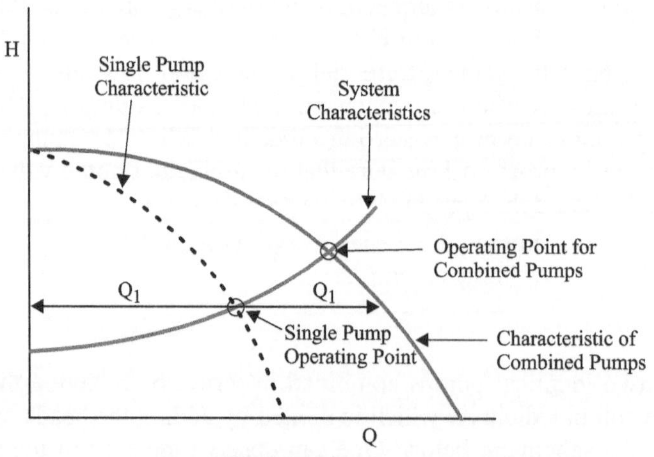

Figure 18

Image Credit: Google & NAPTEL

As it may be observed in the above image pumps running in parallel will increase flow (Q), but not head (H). Pumps are operated in parallel when two or more pumps are connected to a common discharge line and share the same suction conditions.

Limitation of Combining Pumps

When pumps are set to operate in series, the effects of a pump failure are more severe than when they are set to operate in parallel. If Pump-B fails, for example, Pump-A would experience a sudden and severe system resistance, which could be caused by the pump itself or by the stationary Pump-B. If Pump-A fails, on the other hand, the suction flow to Pump-B would be cut off, potentially leading to cavitation or even prime loss if Pump-B wasn't stopped right away. One way to safeguard both Pump-A and Pump-B against damage caused by a pump failure is to connect each motor driver with a PCM (Power Control Monitor). This would detect when a pump failed and would stop the motor.

In the case of parallel combination if one pump fails the liquid will go in the mode circulation with no damage to the running pump.

WHY DOES FLOW INCREASE WITH HEAD IN SYSTEM CURVE?

The pump and system curves both show head losses; the way they are interpreted differs, though. That is the one-line answer to query in the post.

While the system curve depicts the actual head losses during use at the location, the performance curve shows the intended head losses with increased flow supplied by the manufacturer. The pump's operating point is where these two head loss lines intersect.

A pump curve typically shows the pressure on the vertical axis (usually measured in bars or metres) and the flow rate on the horizontal axis (usually measured in litres per minute or m³/h). It illustrates the connection between a pump's head, efficiency, power consumption, and flow rate. A pump curve is very helpful in figuring out a pump's most efficient operating point because it gives a visual depiction of the pump's capacity to move fluid and the energy needed to do so. A straightforward head vs. flow pump curve is shown below. The upper line, coloured blue, represents the performance curve, and the lower line, coloured red, represents the system curve or system resistance curve. The operating point is where these two lines intersect. The system curve is created at the location where the pump is used, while the performance curve is supplied by the pump manufacturer.

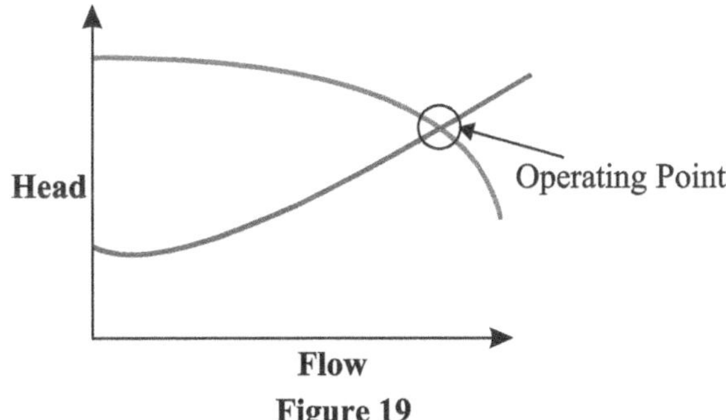

Figure 19

Image Credit: Google & NAPTEL

The pump curve and system curve depict different facets of a fluid system, they are in opposition to one another. In the pump curve it may be observed that the head, H drops on the y-axis as the flow Q increases. Just the opposite can be noticed in the system curve which shows that both the head and the flow are unidirectional. That is as the flow increases we can observe the head also increases simultaneously.

The relationship between the head (H) that a pump generates and the flow rate (Q) that it can deliver is depicted by the pump curve. It is a characteristic curve created by testing the pump in various flow conditions. It is unique to the pump that is being used.

The relationship between the fluid system's head loss and flow rate, on the other hand, is represented by the system curve. It is a curve that shows the overall head loss as a function of flow rate caused by friction, elevation, and other factors.

The system curve is unique to the given fluid system and all of its parts, including fittings, pipes, valves, and other constraints.

In order for a pump to provide the intended flow rate in a fluid system, it must be able to overcome the head losses caused by the system's components. The operating point of the pump in the system is found by combining the pump and system curves. The operating point, which is the intersection of the system and pump curves, indicates the head and flow rate that the pump provides to the fluid system when it is in operation.

The system curve is opposite to the pump curve because as the flow rate increases, the head loss in the system also increases, reducing the head available to the pump.

Therefore, the system curve is a decreasing function of flow rate, while the pump curve is an increasing function of flow rate.

To sum up, the pump curve shows the head produced by the pump, and the system curve shows the head losses in the fluid system. The system pump's operating point is found at the intersection of the two curves.

HOW CENTRIFUGAL PUMP DEVELOPS PRESSURE?

A centrifugal pump is driven by angular momentum:

What is Momentum?

Momentum is a fundamental concept that quantifies the motion possessed by an object. It is defined as the product of an object's mass and velocity. In other words, momentum is determined by multiplying the mass of an object by its velocity.

What is Angular Momentum?

Angular momentum is defined as: The property of any rotating object given by moment of inertia times angular velocity. It is the property of a rotating body given by the product of the moment of inertia and the angular velocity of the rotating.

Formula

$$L = m\,v\,r$$

L = angular momentum
m = mass
v = velocity
r = radius

What is Momentum of Inertia?

In the simplest of terms: the moment of inertia is a body's resistivity to angular acceleration around a certain axis.

How Centrifugal Pump Develop Pressure?

Angular momentum

A centrifugal pump uses a rotating impeller to transfer energy to a fluid by changing its angular momentum.

Kinetic energy

The pump increases the fluid's kinetic energy, which is its angular momentum.

Potential energy

The pump also increases the fluid's potential energy, which is its static pressure.

How Angular Momentum Converts to Pressure?

The angular momentum is converted to pressure through a series of steps that involve the transfer of energy and momentum to a fluid:

Impeller rotation

The impeller's rotation transfers momentum to the fluid, increasing its velocity and kinetic energy.

Fluid movement

The fluid moves through the impeller and into the volute, where it's forced to move into a region with a continuously increasing cross-sectional area.

Kinetic Energy Conversion

The volute's increasing cross-sectional area converts the fluid's kinetic energy into pressure.

This process is similar to subsonic flow through a nozzle's diverging section.

Pressure-increase

The fluid's velocity decreases and its pressure increases in volute casing.

The final important point

The flow rate is not determined by the rate of flow or pressure in the pipe, but by the pressure drop gradient along the pipe.

So, to work out the flow, length of the pipe and the differential pressure at both ends give the flow rate.

How to Calculate Flow Rate from Pressure Drop?

This is a familiar equation

Darcy-Weisbach equation

This equation calculates the pressure drop in a pipe: $\Delta Pf = f \times (L/D) \times (\rho v^2/2)$. In this equation, ΔPf is the frictional pressure drop, f is the friction factor, L is the pipe length, D is the pipe diameter, ρ is the fluid density, and v is the fluid velocity.

Volumetric flow rate

This equation describes the relationship between flow rate, pressure gradient, and pipe diameter: $Q = \pi\ (P_1 - P_2)r^4/8\mu L$. In this equation, Q is the volumetric flow rate, P_1 and P_2 are the pressures at both ends of the pipe, r is the radius of the pipe, μ is the viscosity of the fluid, and L is the length of the pipe.

HEAD AND SHUT-OFF HEAD OF THE CENTRIFUGAL PUMP

What is "head"?

Head is a concept in fluid dynamics that relates the energy in an incompressible fluid to the height of an equivalent static column of that fluid. According to Bernoulli's principle, the total energy at a given point in a fluid is the energy associated with the fluid's movement plus energy from the fluid's static pressure plus energy from the fluid's height relative to an arbitrary datum. Head is expressed in distance units such as metres or feet. In a gravitational field, the force per unit volume on a fluid is equal to g, where is the density of the fluid and g is the gravitational acceleration. On Earth, each additional foot of fresh water column height adds a static pressure of about 9.8 kPa per metre (0.098 bar/m) or 0.433 psi per foot of water column height.

A pump's static head is the maximum height (pressure) that it can deliver. The *Q-H* curve of the pump can be used to determine its capability at a given RPM (flow vs. height).

A common misconception is that the head equals the fluid's energy per unit weight, when in fact, the term with pressure does not represent any type of energy (in the Bernoulli equation for an incompressible fluid this term represents work of pressure forces). Because centrifugal pumps' pumping characteristics are generally independent of fluid density, head is useful in specifying them.

There are Generally Four Types of "Head"

1. **Velocity head** is due to the bulk motion of a fluid (kinetic energy):

   $hv = 1/2\ v^2/g$

2. **Elevation head** is due to the fluid's weight, the gravitational force acting on a column of fluid. The elevation head is simply the elevation (h) of the fluid above an arbitrarily designated zero point:

   $he = [rho \times g \times h]/[rho \times g]$

3. **Pressure head** is due to the static pressure, the internal molecular motion of a fluid that exerts a force on its container. It is equal to the pressure divided by the force/volume of the fluid in a gravitational field:

   $hp = p/rho \times g$

4. **Resistance head** (or friction head or Head Loss) is due to the frictional forces acting against a fluid's motion by the container. In a piped system head losses are described by the Hagen–Poiseuille equation and the Bernoulli Equation.

Pump head

The head of a pump is the maximum height that it can reach when pumping against gravity. Intuitively, a pump with higher pressure can pump water higher and produce a higher head. Also, because of the head exerted by the liquid in the suction tank, the higher the liquid in the tank, the higher the pump will be able to pump the water into the vertical discharge pipe.

Total head

The difference between the liquid level in the suction tank and the head in the vertical discharge pipe is a much more useful measure of the head. This is known as the "total head" that the pump can generate.

Increasing the level of liquid in the suction tank causes an increase in the head while decreasing the level causes a decrease in the head. Pump manufacturers and suppliers frequently refuse to tell you how much head a pump can generate because they cannot predict the height of the liquid in your suction tank. They will instead report the total head of the pump, which is the difference in height between the level of liquid in the suction tank and the height of a column of water that the pump can achieve. The total head is independent of the liquid level in the suction tank.

The total head *Ht* is calculated as $Ht = Hd - Hs$

The discharge head is denoted by *Hd*, and the suction head is denoted by *Hs*. Also, keep in mind that this equation applies whether the suction head is positive (the level of liquid in the suction tank is higher than the pump) or negative (level of the liquid in the suction tank is below the pump). The pump will still produce the same total head if the level is negative, but because the suction head is negative, the discharge head will be reduced by this amount, according to our equation.

Shut-off "head"

In the image below, a pump moves liquid from the suction tank into a vertical pipe, where it rises until it can't overcome gravity and stops rising. The flow of the pump is zero in this case. The pump is running, but gravity causes the water in the vertical discharge pipe to rise and the net flow to stop. This is referred to as the "shut-off head," and it is the amount of head that a pump can produce at zero flow.

WHY SHUT-OFF HEAD IS IMPORTANT?

To select the appropriate pump, you must first determine the total head and the required flow rate. These two quantities are, as one might expect, related. At a flow rate of zero, the maximum head (shut-off head) is achieved. As the liquid travels through the pipes from the suction tank to the pump and from the pump to the discharge pipe, the flow rate increases, introducing friction into the system. This friction reduces the total head that the pump can generate. In fact, as the flow increases, so does the friction, and the total head decreases. The amount of head lost due to friction is referred to as "friction head" or "friction loss."

Why is ''Head', Used to Determine a Pump's Ability to Pump Liquids?

Many pumps were historically used to pump water uphill to a higher level, such as into a storage tank at the top of a hill. If you have to pump water to a height of 60 metres to get it up the hill, using head (measured in metres) makes sense. If a pump does not have 60 metres of head, you know it is not suitable for your application. Another reason for using that head is that as long as the liquid being pumped has a viscosity similar to water, the head will be identical for different liquids.

When using pressure to define a pump's characteristics, this may or may not be the case. Although some pump manufacturers use pressure to describe their products, the vast majority of pumps are still defined by the total head they produce.

WHY DENSITY OF LIQUID DOES NOT FIGURE IN A PUMP CURVE?

In a pump curve, the y-axis stands for the head, H, and the x-axis for the flow rate, Q.

What is head?

The head is the height at which a pump can raise fluid. It is measured in meters or feet. The height is just a number independent of the density. We use 'head' when specifying centrifugal pumps to make sure that their pumping characteristics are independent of the fluid's specific gravity, often referred to as relative density. The head is liquid independent. The use of the head therefore eliminates the effect of density on the y-axis. See below for more explanations of how the head is independent of density.

We can see a spinoff advantage on the x-axis which represents flow for 'head 'being used on the y-axis. The flow Q is proportional to velocity or pump speed, typically expressed in revolutions per minute (RPM) and the head is proportional to the square of the velocity or pump speed [affinity laws]. Thus, the flow becomes independent of density. Therefore, both × and y axis of Q-H plot become independent of density. This explains why the density of the liquid does not matter in predicting the pump's performance on the Q-H curve.

However, the specific gravity or the relative density does have a role in power consumption by the pump. This is evident from the absorbed power equation for any pump. Considering the volume flow rate, head and efficiency remain unchanged an increase in density increases the absorbed power and a decrease in density decreases the absorbed power for the pump.

Absorbed power equation of pumps

$$kW \text{ Power} = (\text{Flow} \times \text{Pressure} \times SG)/K$$

here:

SG = Specific gravity = pump efficiency as decimal fraction k = constant,

Essentially, it means that if the fluid being pumped has a higher density, it may require more energy to move through the system, resulting in a lower flow rate and higher power consumption.

The higher density can also impact the pump selection, as higher-head pumps may be required to move denser fluids.

The pump efficiency depends on the pump's two parameters, head and flow rate, in addition to other factors such as properties of the fluid, impeller design, and motor speed selected.

Therefore, in summary for a given pump, the density does not impact the Q-H characteristics of the pump. While the power consumption of the pump gets impacted by density changes.

References

Wikipedia
Global pumps

WHY DOES THE PRESSURE HAVE NO ENERGY BUT THE HEAD DOES?

Pressure *vs.* Head

An object's energy is its capacity to produce a force that can do work and transfer heat. When one object acts upon another, altering its momentum, the former's energy (its capacity to perform work) diminishes and the latter's energy increases, through the transfer of heat.

Work is a process of energy transfer in which temperature is not considered during energy transfer. Unit [SI] of energy is always in Joules. It is also a result of force acting through the distance expressed as W = Force × Distance. The dimensional formula of energy is [M1L2T-2], here M is mass. L is time and T is time.

Pressure

Pressure is force per unit area, $P = F{\cdot}A\text{-}1$ [Pa]. The SI unit of pressure is Pascal [Pa]. The dimensional formula of pressure is $M1\ L\text{-}1\ T\text{-}2$. It is evident that pressure and energy have different dimensional formulas, so pressure is not the same as energy. But when pressure is multiplied by volume, PV becomes 'work energy' expressed as Joule, the dimensional formula for energy is $PV = M1\ L\text{-}1\ T\text{-}2 \times L3 = M1\ L2\ T\text{-}2$ = Joule.

In summary, the *SI* unit of pressure is Pascal [Pa]. The SI unit of energy is Joule. Therefore, pressure is not energy.

It is noteworthy to mention that the expression of the pressure term in Bernoulli's equation is frequently given incorrectly as energy per unit area. Energy/volume equals pressure always.

Head

The head is energy. Put simply, the head is the height at which a pump can raise fluid and is measured in metres or feet. The advantage of 'head 'over the pressure is that pressure is fluid-dependent and influenced by gravity, whereas the head is fluid-independent, meaning that a pump will raise the fluid to the same height regardless of its relative density, making it irrelevant whether the fluid is heavy sludge or water. The head depends only on the rpm and impeller size of the pump.

One can find the best explanation of what is head in Bernoulli's equation.

A simple form of Bernoulli's equation between two points 1 and 2 in a fluid flow can be written as follows:

$$P_1/rho\ g + V_1^2/2g + Z_1 = P_2/rho\ g + V_2^2/2g + Z_2$$

Z = height above reference (ft.)

V = average velocity (ft./sec.)

g = acceleration due to gravity (32.17 ft./sec.^{2})

rho = density

The first term represents the pressure head, the second term represents the velocity head and the third term is the elevation head.

The terms "heads" (pressure head, velocity head, and elevation head) refer to the units of distance for all the different forms of energy.

As said above, the term "head" refers to pressure; it is the height, usually in feet, of a column of water that a given pressure will support.

Head is a unit of measurement for all the energies possessed by a fluid; the elevation head represents the potential energy of a fluid due to its elevation above a reference level. The velocity head represents the kinetic energy of the fluid, and it is the height in feet that a flowing fluid would rise in a column. The total head of a fluid is the sum of its elevation head, velocity head, and pressure head.

Bernoulli's equation states that the total head of a fluid is constant. The velocity head is the height in feet that a flowing fluid would rise in a column if all of its kinetic energy were converted to potential energy. The pressure head is the flow energy of a column of fluid whose weight is equivalent to the pressure of the fluid.

PUMP THERMODYNAMICS

The note includes the following:

- Work done by the pump
- Fluid friction loss
- Equipment friction loss
- Control volume concept
- Total head and energy balance
- Head equation for single inlet and outlet.

What do Pumps Do?

In simple words, the pump compensates for the kinetic energy and potential energy change in the flow of a liquid between two points. In open systems, fluid particles move from the inlet to the outlet and leave the system. If the velocity varies between inlet and outlet, the kinetic energy varies. The pump is the energy source that compensates for the variation. When the liquid particles leave the system at a different elevation than they enter producing a variation in potential energy. The pump also compensates for this energy variation.

Potential Energy

The difference in the vertical position at which the fluid enters and leaves the system causes the potential energy of a fluid particle to increase or decrease. The potential energy increases when the fluid leaves at a higher elevation than it enters. The pump provides energy to compensate for the increase in potential energy. If the fluid leaves the system at a lower elevation than it enters, the potential energy decreases and it is possible to convert this energy to useful work. This is what happens in a hydroelectric dam. A channel behind the dam wall brings water to a turbine located beneath the dam structure. The flow of water under pressure turns a turbine that is coupled to a generator producing electricity. The difference in elevation or potential energy of the fluid is the source of energy for the generator.

Kinetic Energy

There is often a velocity difference between the inlet and outlet of a system. Usually. the velocity is higher at the outlet versus the inlet; this produces an increase in kinetic energy. The pump compensates for this difference in kinetic energy levels. The complete energy balance for an open system is:

$$QE - W = \Delta En + \Delta KE + \Delta PE \qquad \ldots (1)$$

Equation [1] has the thermodynamic properties on the right-hand side. The heat loss (QE) and the work done on the system (W), which are not thermodynamic properties, are on the left-hand side of the equation. To determine QE and W, it is necessary to know how the heat is produced (i.e., the path that the fluid must take between inlet and outlet) and how the work is done.

Work done by the pump

The flow of liquid is opposed by energy losses in a pipe due to friction, elevation, velocity, and pressure differences between two points. The role of a pump is to provide sufficient pressure to move fluid through the system at the required flow rate.

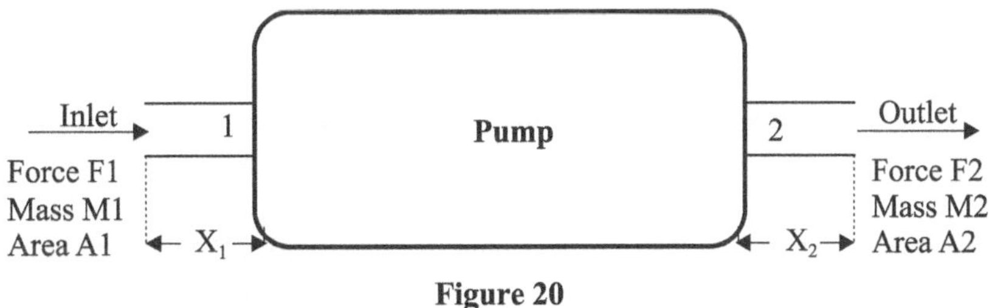

Figure 20

Work at point 1

$$F_1 \times X_1 = P_1 \times A_1 \times X_1 = P_1 \times V_1$$

Work at point 2

$$F_2 \times X_2 = P_2 \times A_2 \times X_2 = P_2 \times V_2$$

Net work W = Work at inlet – Work at outlet

$= F_1 \times X_1 - F_2 \times X_2$

$W = P_1V_1 - P_2V_2$ [V is volume which is control volume and constant]

$W = V[P_1 - P_2]$

$= \Delta P$[pump] $\times$ mass/density

Fluid and equipment friction loss

The total heat loss (QE) is the sum of fluid friction loss and equipment friction loss. $QE = QF + QE$.

Friction occurs within the fluid and between the fluid and the walls. Friction in fluids produces heat (QF) which dissipates to the environment. The temperature increase is negligible. If we can estimate the friction force, we know that we must supply an equal and opposite force to overcome it.

Fluid Friction Loss

Imagine a pipe between two points length d and cross section area A

$$F\text{ [fluid]} = F_1 - F_2$$

QF [fluid $= F$ [fluid] $\times d$

$= [F_1 - F_2] \times d$

$= [P_1 - P_2] \times A \times d$

$= \Delta P$[fluid] V [V is volume]

QF [fluid] $= \Delta P$[fluid] $\times$ mass of fluid/density

The energy corresponding to the heat loss (QF) must be supplied by the pump. The net force (FF) required to balance the friction force is $F_1 - F_2$. These forces are the result of the action of pressures p_1 and p_2. The difference between p_1 and p_2 is the pressure differential or pressure drop due to friction for a given length of pipe.

Equipment Friction Loss (QEQ)

There are many different types of equipment in typical industrial systems. Control valves, filters, etc., are examples of equipment that have an effect on the fluid. Because of the great variety of different equipment that can be installed, it is not possible to analyze every one of them. However, we can quantify the effect of the equipment by determining the pressure drop across the equipment much as we did with the pressure drop across the pump. Again, it is not necessary to know exactly what happens inside the equipment to be able to estimate this effect. Pressure drop (the difference between inlet and outlet pressure) is the outward manifestation of the effect of equipment. This pressure drop produces heat (QEQ) which is then lost through the fluid to the environment. The heat loss is calculated in the same manner as the pump work (W).

$$Q\text{ [equipment]} = \Delta P\text{ [equipment]} \times \text{mass of fluid/density}$$

Control Volume

A control volume is an imaginary boundary that surrounds a system and intersects all inlets and outlets. This makes it possible to apply the principle of conservation of mass. The boundary of the system is determined by appropriately locating the inlet and the outlet. The term "appropriate" refers to locating the boundary at points where the conditions (pressure, elevation, and velocity) are known. The control volume encompasses all internal and external energy sources affecting the system. This makes it possible to apply the principle of conservation of energy. How is the control volume positioned in real systems? Where exactly should we locate it, will any position do? Clearly no. Point 1 could be located anywhere on the feed line to the suction tank. The problem with locating point 1 on the inlet line is precisely that the location is indeterminate.

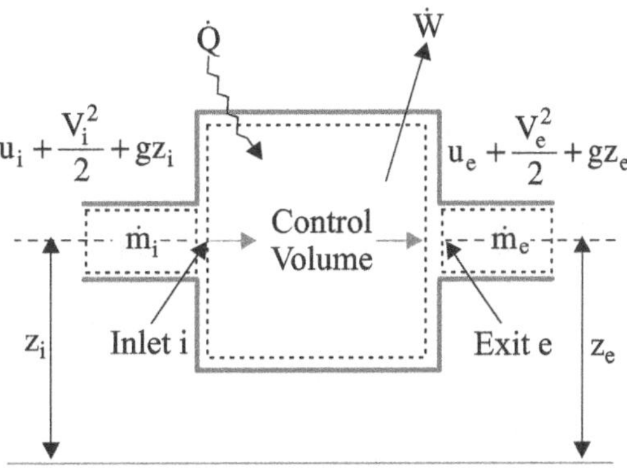

Figure 21

The pressure head at the inlet of the pump is proportional to the elevation of the suction tank fluid surface. We locate the inlet of the system (point 1) on the suction tank fluid surface since any change in the elevation of point 1 will affect the pump. The inlet pipe is the means by which we supply the pump with enough fluid to operate. As far as this system is concerned, the inlet feed pipe is irrelevant. This is because the pressure head at the inlet of the pump (or any point within the system) is dependent on the level of the suction fluid tank surface. The other reason that requires us to locate point 1 on the liquid surface of the suction tank is that the control volume must contain all the energy sources that affect the system and if the suction and discharge tanks are open to the atmosphere then this pressure will be present at points 1 and 2 (Head1 = Head 2 = 0 = atmospheric pressure).

A general system must take into account the possibility that the suction and discharge tanks are pressurized. The terms Head1 and Head2 represent respectively the suction and discharge pressure head at the fluid surfaces of these tanks. A general system must take into account the possibility that the system contains no tanks and that points 1 and 2 are connections to pipelines of other systems. The only difference between a tank and a pipe is that a tank can contain a much greater volume of fluid. Therefore, whether the system has tanks or not is immaterial. However, the conditions (i.e., pressure, velocity, and elevation) of the inlet and outlet of the system must be known. It will always be possible to disconnect our system from another if the pressure head, velocity, and elevation (H, v, and z) at the connection point are known.

Total Head and Energy Balance

Step by Step

We now have all the background information necessary to determine the work required of the pump and therefore the Total Head. First, we calculate the energy lost in friction due to fluid movement through pipes (QF) and equipment (QEQ.). Next, calculate the energy variations due to conditions at the inlet and outlet.

1. Potential energy variation (ΔPE) is associated with the difference in height of a fluid element between the outlet and inlet of the system.
2. Kinetic energy variation (ΔKE) is associated with the variation in the velocity of a fluid element between the outlet and inlet of the system.
3. Enthalpy variation (ΔEn) consists of two components $\Delta En = \Delta U + \Delta(pV)$. The first is the internal energy (ΔU) of the fluid. In most common fluid transfer situations, there is no significant temperature change and therefore no significant internal energy change ($\Delta U = U_1 - U_2 = 0$). The other component of enthalpy $\Delta(pV)$ is the energy associated with the difference in pressure at the inlet and outlet of the system. The difference between the inlet and outlet energies added to the total friction loss (QE) gives the resultant work W that must be supplied by the pump.

Head equation for single inlet and outlet

The simplest case we can consider is a system with one inlet and one outlet. The first law of thermodynamics or the principle of conservation of energy for open systems states.

$$Q\,E - W = \Delta En + \Delta KE + \Delta PE$$

Above expresses the rate of variation of the energy terms instead of the energy variation. This allows us to balance mass flow rates instead of mass between the inlet and outlet of the system. EQ represents the effect of all the equipment present in the line between points 1 and 2. Head1 and Head 2 are the pressure heads present at the suction tank and the discharge tank fluid surface respectively. z_1 and v_1 are respectively the elevations and the velocity of a fluid particle at the inlet of the system, and z_2 and v_2 are the same variables for a fluid particle at the outlet.

TYPICAL ENERGY LOSSES IN CENTRIFUGAL PUMPS

Energy losses in a pump can be caused by a number of factors.

An important point. Some amount of cavitation in a pump is unavoidable due to the fact that liquids have vapor pressure at all temperatures. This means that there will always be a small amount of dissolved gases in the liquid that can vaporize and form bubbles and there will be some head loss.

2nd important point

There is always some aerodynamic unavoidable loss in pumps due to factors such as turbulence, friction, and flow separating.

3rd point

When impeller does work on liquid there is some internal energy loss as entropy. Though Bernoulli's equation ignores it, this is unavoidable.

Factors of energy loss

Hydraulic loss: The largest source of energy loss in a pump, hydraulic loss is affected by flow channel's geometry, structure, flow field distribution, and size.

Examples

- **Friction**: As fluid moves through the pump, it encounters resistance from the pump's internal components, such as impellers and volutes.
- **Volumetric loss**: In a reciprocating pump, volumetric loss is caused by slippage through valves, the ratio of liquid chamber volume at the end of the stroke, and liquid compressibility
- **Mechanical loss**: A source of energy loss in a pump.
- **Leakage:** A source of energy loss in a centrifugal pump.

The path through valves and fittings also affects the amount of energy loss. A more circuitous path results in greater energy losses

Details of some typical sources

- **Friction**: Friction between the moving parts of the pump, such as bearings, seals, and impeller, can result in energy loss.
- **Viscous losses**: Viscous losses occur when the fluid being pumped exerts resistance to flow, which requires additional energy input from the pump.
- **Mechanical losses**: Mechanical losses, such as losses in motor efficiency or drive train losses, can also contribute to energy loss in a pump.
- **Inlet and outlet losses**: Inlet and outlet losses, such as turbulence at the entrance and exit of the pump, can result in energy loss as well.
- **Inefficient design**: Poor design of the pump, including improper sizing or selection of components, can lead to energy loss.
- **Cavitation**: Cavitation occurs when the pressure in the pump drops below the vapor pressure of the fluid, resulting in the formation of vapor bubbles that collapse and cause damage to the pump, as well as energy loss.

Details of some key energy losses

What is hydraulic loss in a pump?

Hydraulic loss is a large proportion of the total energy loss in a pump. Hydraulic loss is a type of energy loss that occurs in a pump when fluid moves through it, and is caused by friction within the fluid or in the pipe walls.

Volumetric loss

Volumetric efficiency is the ratio of the actual amount of fluid flowing through a pump to its theoretical maximum. It's calculated by dividing the quantity of liquid discharged per second from the pump by the quantity of liquid passing through the impeller.

References

Research Direct
Wikipedia

CHAPTER 41

Heat Transfer

Heat transfer and thermodynamics are two related fields of study. While thermodynamics deals with equilibrium states by defining the relationship between energy and work, heat transfer explains the physical mechanism to understand the transport of energy. Heat transfer provides a comprehensive illustration of thermodynamic processes. From the behaviour of heat in engines and refrigerators to the thermal properties of materials, the principles of thermodynamics play a critical role in various engineering and scientific disciplines. In this context, the study of heat from the perspective of thermodynamics allows us to explore the fundamental laws and principles that govern the behaviour of heat and its interactions with matter. By delving into the mechanisms of heat transfer, work, and energy conversion, we can gain a deeper insight into the underlying principles that drive natural processes and technological innovations.

ROLE OF REYNOLDS NUMBER AND PRANDTL NUMBER IN HEAT TRANSFER

Some other dimensionless numbers used in heat transfer are mentioned below.

Grashof Number (Gr) Biot Number (Bi) Sherwood Number (Sh) Stanton Number (St) Fourier Number (Fo) Eckert Number (Ec) Peclet Number (Pe) Prandtl-Grashof Number Stanton-Prandtl Number

This note is focused on Re and Pr numbers

The Reynolds number and Prandtl number are both dimensionless numbers that play imortant roles in heat transfer in a fluid medium.

The Reynolds number (Re) is defined as the ratio of inertial forces to viscous forces in a fluid flow. It is given by the equation:

$$Re = \rho VL/\mu$$

where:

ρ is the density of the fluid

V is the velocity of the fluid

L is a characteristic length scale of the flow (such as the diameter of a pipe)

μ is the dynamic viscosity of the fluid

The Reynolds number helps determine whether the flow is laminar or turbulent. For heat transfer, the Reynolds number influences the rate at which heat is transferred within the fluid. Turbulent flows have greater mixing and thus higher rates of heat transfer compared to laminar flows.

The Prandtl number (Pr) is defined as the ratio of momentum diffusivity to thermal diffusivity in a fluid. It is given by the equation:

$$Pr = \mu cp/\kappa$$

where:

cp is the specific heat of the fluid

κ is the thermal conductivity of the fluid

The Prandtl number helps characterize the relative importance of conduction and convection in heat transfer. A low Prandtl number indicates that thermal diffusivity dominates over momentum diffusivity, leading to rapid mixing and high rates of heat transfer. In contrast, a high Prandtl number indicates that momentum diffusivity is more significant, resulting in slower and less efficient heat transfer.

Therefore the Reynolds number influences the flow regime and mixing characteristics, while the Prandtl number indicates the relative dominance of conduction and convection in heat transfer.

Heat transfer coefficient

The Dittus-Boelter equation is an empirical correlation used to predict the convective heat transfer coefficient in forced convection flows.

The equation relates the Nusselt number (Nu), Reynolds number (Re), and Prandtl number (Pr) in the form:

$$Nu = c(Re)^\wedge m(Pr)^\wedge n$$

where c, m, and n are constants that depend on the flow regime and geometry of the system. For turbulent flow in smooth tubes, the Dittus-Boelter equation is typically written as:

$$Nu = 0.023(\text{Re})^\wedge 0.8(\text{Pr})^\wedge 0.4$$

Nu is Nusselt number

Nusselt number

The Nusselt number (Nu) is a dimensionless number used in fluid dynamics to describe the ratio of convective to conductive heat transfer across a boundary.

It is defined as the ratio of convective heat transfer to conductive heat transfer, and is given by the formula:

$$\text{Nu} = \text{hL/k}$$

where:

Nu = Nusselt number

h = convective heat transfer coefficient

L = characteristic length scale

k = thermal conductivity of the fluid

A higher Nusselt number indicates a more efficient heat transfer process.

Details

Prandtl number

Why Prandtl number (Pr) is important?

- Relates viscosity and thermal conductivity
- Controls boundary layer thickness
- Indicates heat transfer efficiency

The Prandtl number (Pr) is a dimensionless number that represents ratio of momentum diffusivity to thermal diffusivity in a fluid.

It's a fundamental parameter in fluid mechanics and heat transfer that indicates the relative importance of heat transfer and fluid flow

The Pr is calculated using fluid's properties:

Formula: The Pr is calculated by dividing the kinematic viscosity by the thermal diffusivity

$$Pr = nu/\text{alpa} = cp \times mu\ /k$$

Variables

nu is the kinematic viscosity, alpa is thermal diffusivity, mu is dynamic viscosity and k is thermal conductivity.

The relative importance of momentum and thermal diffusivity is characterized by the Prandtl number.

Significance

High Pr number

A high Prandtl number indicates that thermal diffusion happens slower than momentum diffusion. This results in a thinner thermal boundary layer than the velocity boundary layer.

Low Pr number

A low Prandtl number indicates that thermal diffusion happens faster than momentum diffusion. This results in a thicker thermal boundary layer than the velocity boundary layer.

Thermal diffusivity

Determines how quickly a material heats up or cools down. Materials with a high thermal diffusivity heat up or cool down faster. Fast heating or cooling refers to the rate at which an object's temperature changes:

Thermal diffusivity is directly proportional to thermal conductivity. Thermal diffusivity is a measure of how quickly a material can transfer heat and spread temperature, and is calculated by dividing thermal conductivity by the product of density and specific heat capacity.

The formula for thermal diffusivity

$$\alpha = k/(\rho cp)$$

k is thermal conductivity, alpha is thermal diffusivity, ρ is density and cp is specific heat.

Thermal diffusivity is an important factor in determining which material to use for a particular application, such as in heat

Momentum diffusivity

Determines how well components in a mixture interact and combine. A higher momentum diffusivity means more efficient mixing. Momentum diffusivity is also known as kinematic viscosity, momentum diffusivity is the ratio of dynamic viscosity to fluid density. It's represented by Greek letter nu (ν) and has units of m2s-1 Kinematic viscosity = Dynamic viscosity/density

The momentum diffusivity depends on specific heat and viscosity.

$$m^2 s^{-1}$$

Kinematic viscosity is a measurement of how fast a fluid can move when a force is applied to it. It's used to prevent corrosion and lubricate objects

Here are some examples of Prandtl numbers for different materials:

- **Gases:** Gases such as air and helium have a Prandtl number of around 0.7.
- **Liquids:** Liquids such as water and oil have a Prandtl number of around 10.
- **Metals and other solids:** Metals and other solids typically have a very low Prandtl number.

Reynolds number

The Re number is a dimensionless number that measures the ratio of inertial force to viscous force in a fluid. The Re number is a parameter that determines if a fluid flow will be turbulent or steady. It is used to study the viscosity of a fluid in a quantitative way.

Formula

Re = D U rho/ mu, D is the pipe diameter U is fluid velocity rho is fluid density and mu is fluid viscosity.

The Re number is used to predict the type of flow in a pipe. The flow is considered laminar if the Re number is less than 2000, turbulent if the Re number is greater than 4000, and in a transition region if the Re. number is between 2000 and 4000.

Nusselt number combines Re and Pr numbers by Dittus Boelter's equation to predict the heat transfer coefficient:

$Nu = hL/k = 0.023$ Re^0.8 Pr^0.4

Explanation

Re = Inertial force / Momentum force

Inertial forces are only significant in turbulent flow, while viscous forces are the most important in laminar flow.

Inertial force

Inertial force is the force due to the momentum of the fluid. This is usually expressed in the momentum equation by the term $\rho(du/dt)$ or $(\rho v)\ v$. So, the denser the fluid is, and the higher its velocity, the more momentum (inertia) it has.

Momentum forces are inertial forces. Inertia is a property of matter by which it continues in its existing state of rest or uniform motion in a straight line unless that state is changed by an external force. The force that keeps fluid moving against viscous [viscosity] forces is the inertial force. The inertial forces are characterized by the product of the density rho times the velocity V times the gradient of the velocity dV/dx.

Viscous force

It is a measure of a fluid's resistance to flow. Viscous forces are shear forces. The viscous forces are the forces due to the friction between the layers of any real fluid. Since the fluid particles are very closely packed so necessarily there is friction between layers of fluid. When a body moves in a fluid, it can experience two types of forces. Forces normal to the surface are called pressure forces and forces tangential to the surface are called shear forces. Pressure forces change linear momentum and shear forces are described by 'viscosity effects' and the forces become viscous forces that oppose the flow.

Inertial force (in the direction of flow) and viscous force (in the reverse direction of flow) act in opposite direction.

The viscous force counteracts or counterbalances the inertial force. In the case of fluid flow, this is represented by Newton's law, $\tau x = \mu dvdy$. This is only dependent on the viscosity and gradient of velocity.

Turbulent flow in pipes

When a flow is turbulent, it comprises eddying motions of various sizes, and a significant portion of the mechanical energy in the flow is used to generate these eddies, which eventually lose their energy as heat.

As a result, the drag of a turbulent flow is greater than the drag of a laminar flow.

THE ROLE OF SPECIFIC HEAT [SENSIBLE HEAT TRANSFER] TO REGULATE OUR BODY TEMPERATURE

$Cp\ dT$ = Total energy

Cp is specific heat and dT is temperature gradient.

Even at extreme hot and cold surrounding temperature the high specific heat of our body keeps our body temperature fairly constant. Staying hydrated helps no depletion of our body's heat storage capacity.

Remember, the amount of sensible heat that can be transferred from one material to another is limited by the specific heat capacity of the receiving material.

Therefore, specific heat at the receiving end has a critical role in sensible heat transfer.

The same applies to human body.

The specific heat of the human body plays a role in how our body responds to changes in temperature. This plays a key role in how we feel and experience sensations of hot and cold.

Specific heat refers to the amount of heat required to raise the temperature of a unit mass of a substance by one degree Celsius.

The specific heat of the human body helps regulate our body temperature and how we perceive hot or cold sensations. When our body is exposed to changes in temperature, the specific heat of our body influences how quickly or slowly our body temperature changes.

Our body's high specific heat, which is due to the presence of a large amount of water in our bodies, allows our body to resist quick changes in temperature. This means that our body does not easily heat up or cool down, which helps to regulate our internal temperature and maintain homeostasis.

The high specific heat of our body helps to keep our body temperature relatively stable in response to external temperature changes, which influences how we perceive sensations of hot and cold.

In summary, specific heat plays a significant role in how our body perceives and responds to temperature changes. The higher the specific heat of a substance, the more energy is required to change its temperature. In the case of our bodies, the high specific heat due to the presence of water helps to regulate our internal temperature and prevent rapid changes in response to external temperature fluctuations. This plays a key role in how we feel and experience sensations of hot and cold.

SPECIFIC HEAT *vs.* THERMAL CONDUCTIVITY

Thermal conductivity and heat capacity are two important properties related to heat transfer in materials, but they describe different aspects of how materials respond to heat.

Thermal Conductivity

- **Definition**: Thermal conductivity (k) is a measure of a material's ability to conduct heat. It quantifies how easily heat can flow through a material.
- **Units**: The SI unit is watts per meter-kelvin (W/m·K).
- **Dependence**: It depends on the material's composition, structure, and temperature. Metals typically have high thermal conductivity, while insulators like wood or rubber have low thermal conductivity.
- **Application**: High thermal conductivity materials are used in heat exchangers and cooking utensils, while low thermal conductivity materials are used for insulation.

Heat Capacity

- **Definition**: Heat capacity (C) is the amount of heat energy required to change the temperature of a given quantity of a substance by one degree Celsius (or one Kelvin). It reflects how much heat a material can store.
- **Units**: The SI unit is joules per kelvin (J/K). Specific heat capacity, which is heat capacity per unit mass, is expressed in joules per kilogram-kelvin (J/kg·K).
- **Dependence**: It depends on the material's mass, temperature, and phase (solid, liquid, or gas). Different substances have different heat capacities; for example, water has a high heat capacity, which is why it can absorb a lot of heat without a significant temperature change.
- **Application**: Heat capacity is important in thermal management, climate science, and cooking, as it affects how substances heat up or cool down.

Summary

- **Thermal Conductivity**: Measures how quickly heat moves through a material.
- **Heat Capacity**: Measures how much heat a material can store and how much its temperature will change in response to heat added or removed.

Details

Heat capacity, also known as specific heat capacity, is the amount of heat energy required to raise the temperature of a unit mass of a substance by one degree Celsius. It is a measure of how much heat a material can store or absorb before its temperature changes. The heat capacity of a material is determined by the interactions between the molecules in the material. In a material with a high heat capacity, the molecules are able to store more internal energy before their temperature increases, while in a material with a low heat capacity, the molecules quickly increase in temperature with the addition of heat energy. Just the opposite happens with high specific heat. At the molecular level, heat capacity and thermal conductivity are intrinsic properties of a substance that govern its ability to store and transfer heat energy.

Heat Storage in Molecules

The equipartition theory in statistical mechanics helps us understand how molecules store energy through different degrees of freedom, such as translational, rotational, and vibrational motions. Each degree of freedom contributes 0.5 kT of energy to the total energy of the system, where k is the Boltzmann constant and T is the temperature in Kelvin. This energy is related to the specific heat capacity of the substance and influences its ability to store heat.

The degree of freedom of a molecule determines the ways in which it can store and transfer heat energy. Molecules with more degrees of freedom have more ways to store energy, impacting their heat capacity. For example, a molecule with multiple translational and rotational degrees of freedom will store more energy than a molecule with fewer degrees of freedom.

In summary, understanding the molecular motions and degrees of freedom within a substance is essential for comprehending how heat energy is stored and transferred within the material, affecting both its heat capacity and thermal conductivity. The equipartition theory provides a theoretical framework for explaining these concepts and their relationship to specific heat.

Thermal conductivity, on the other hand, is a measure of how well a material can transfer heat energy through conduction. It describes the ability of a material to conduct heat and is determined by the molecular structure of the material. In materials with high thermal conductivity, such as metals, the molecules are closely packed together and can easily transfer heat energy through vibrations and collisions. In materials with low thermal conductivity, such as insulators, the molecules are more spread out and have difficulty transferring heat energy.

At Molecular Level

At the molecular level, heat capacity and thermal conductivity are related to the interactions between the molecules in a material. Heat capacity is related to the energy stored within the molecules, while thermal conductivity is related to the ability of the molecules to transfer this energy to neighboring molecules. In materials with strong intermolecular forces, such as metals, the heat capacity is high and the thermal conductivity is also high because the molecules can store and transfer heat energy efficiently. In contrast, materials with weak intermolecular forces, such as gases, have low heat capacity and low thermal conductivity because the molecules are not as efficient at storing and transferring heat energy.

In summary, heat capacity measures the amount of heat energy a material can store, while thermal conductivity measures the ability of a material to transfer heat energy through conduction. These properties are determined by the molecular structure and interactions in a material, with strong intermolecular forces leading to high heat capacity and thermal conductivity, and weak intermolecular forces leading to low heat capacity and thermal conductivity.

HEAT AND HEAT TRANSFER

These are two different activities in a heat exchanger.

Heat is thermal energy that is transferred from a hotter object to a cooler object. The amount of heat transferred is determined by the mass of the material (m), the specific heat capacity of the material (c), and the temperature change (ΔT). Heat Transfer Rate, on the other hand, is the amount of heat transferred per unit time. It is typically measured in watts (W) or kilowatts (kW) and is influenced by factors such as the temperature difference between the two objects, the thermal conductivity of the materials, and the surface area of contact.

Heat transfer is an essential phenomenon in various processes, where thermal energy is transferred from one medium to another. One of the critical parameters in understanding heat transfer is the heat transfer rate, which quantifies the rate at which heat is exchanged between the different mediums. In a heat exchanger, which is a common device used for heat transfer between two fluids at different temperatures, several resistances need to be overcome for efficient heat transfer.

On the conduction side of the heat exchanger, the thermal conductivity of the material and the boundary layers on either side of the heat transfer surface play a crucial role in determining the rate of heat exchange. Moving to the convection side, the mode of heat transfer and critical heat flux [CHF] can significantly affect the heat transfer coefficient as a mode of boiling switches to film boiling from nucleate boiling. In film boiling the vapor generation near the heating wall can significantly reduce the heat transfer coefficient.

While the temperature difference (ΔT) acts as the driving force for heat transfer, it can also act as a resistance, especially when ΔT between wall temperature and the surface temperature of the fluid reaches such high that there is vapor generation on the heat transfer wall. The vapor begins to act as heat transfer resistance. Heat transfer coefficient a measure for the resistance in the path of heat transfer. The heat transfer coefficient (h) is calculated as the ratio of heat transferred (Q) to the temperature difference (ΔT),

$$h = Q/A \times \Delta T$$

where (in SI units):

Heat transfer rate (W)

h is the heat transfer coefficient (W/m²K)

The surface area is A where the heat transfer takes place (m²).

ΔT is the difference in temperature between the solid surface and surrounding fluid area (K).

Nucleate Boiling

It enables efficient heat transfer by utilizing the latent heat of vaporization as the bubbles carry away heat, providing enhanced cooling compared to other heat transfer modes. During nucleate boiling, small vapor bubbles or "nuclei" form on the surface due to local temperature variations. These bubbles grow and detach from the surface, carrying heat away from it. As the heat flux applied to the surface increases, the number and size of vapor bubbles also increase. As the bubbles expand and reach the liquid's surface, convective or evaporative processes cause them to release the heat they have absorbed into the surrounding air. The entire rate of heat transfer between the heated surface and the liquid is accelerated by the processes of bubble formation, growth, and departure.

Film Boiling

The entire liquid interface boils on the heating wall, and a continuous vapor film develops on the surface, indicating film boiling. Film boiling is characterized by a stable vapor film formation on the heated surface, providing poor heat transfer.

In film boiling a layer of vapor bubbles forms on the heated surface close to CHF, preventing additional heat transfer. The local wall temperature rises quickly when a heated surface approaches near CHF conditions because the vapour bubbles' presence decreases the area of contact between the heated surface and the coolant. Consequently, the wall temperature rises and the heat transfer coefficient falls.

The creation of an insulating vapor film on the heated surface is the cause of this decrease in heat transfer.

Critical heat Flux [CHF]

Critical heat flux (CHF) is the heat flux at which boiling ceases to be an effective form of transferring heat from a solid surface to a liquid. The use of boiling as a means of heat removal is limited by a condition called critical heat flux (CHF).

The most serious problem that can occur around CHF is that the temperature of the heated surface may increase dramatically due to significant reduction in heat transfer.

Because the heat transfer mechanism shifts from single-phase convective heat transfer to boiling heat transfer at CHF conditions, the near critical heat flux (CHF) temperature of a heated surface increases significantly due to a significant reduction in heat transfer.

In Summary, These are Key Points

1. The formation of a low-density vapor film in film boiling reduces the heat transfer rate by creating a thermal barrier between the heat transfer surface and the bulk fluid. This reduction in heat transfer is due to the lower thermal conductivity of the vapor compared to the liquid. As a result, the heat transfer capacity of the fluid is reduced, leading to a decrease in overall heat transfer rate.
2. As the heat transfer rate decreases in film boiling, more heat is absorbed by the solid wall, leading to a larger temperature difference between the wall temperature and the fluid surface temperature. This increase in temperature difference exacerbates the thermal resistance at the interface between the wall and the fluid, further hindering heat transfer efficiency.

The elevated wall temperature can also lead to safety concerns such as overheating or thermal damage to the heat transfer surface if not properly managed.

WHAT DRIVES HEATING AND HEAT TRANSFER?

In thermodynamics, the equation for heating is often represented by the formula $Q = m\ Cp\ \Delta T$, where Q is the amount of heat energy transferred, m is the mass of the substance being heated, c is the specific heat capacity of the substance, and ΔT is the change in temperature.

On the other hand, heat transfer is often described by the formula $Q = k\ A\ \Delta T/\text{l}$, where Q is the amount of heat transferred, k is the thermal conductivity of the material, A is the cross-sectional area through which heat is transferred, ΔT is the temperature difference across the material, and l is the thickness of the material.

So while it is correct to say that heating is driven by factors such as mass, specific heat capacity, and temperature change ($m\ Cp\ \Delta T$), and heat transfer is driven by factors such as thermal conductivity, surface area, temperature difference, and material thickness ($k\ \Delta x/\Delta t$), it's important to note that these formulas can vary depending on the specific circumstances and properties of the materials involved.

Newton's Law of Cooling: *dT/dt* of Temperature

How to find the cooling dT/dt of a substance with lapse of time is often asked by many. Newton's law cooling gives the answer.

Newton's law of cooling states that the rate of heat loss of a body is directly proportional to the difference in temperature between the body and its surroundings. This can be described using the formula:

$$dT/dt = -k(A)(T - Ts)$$

Where:

- dT/dt is the rate of change of temperature with respect to time
- k is the cooling constant
- A is the surface area of the object
- T is the temperature of the object
- Ts is the temperature of the surroundings.

The law can be applied to both liquids and solids.

For Liquids

In the context of heat transfer in liquids, the cooling constant "k" is often replaced by the heat transfer coefficient "h" in the formula for Newton's law of cooling. The heat transfer coefficient "h" is a measure of the amount of heat transferred per unit area per unit temperature difference between the object and the surrounding fluid. So, the formula for heat transfer in liquids using Newton's law of cooling would be:

$$dT/dt = -h(A)(T - Ts)$$

Where:

- dT/dt is the rate of change of temperature with respect to time
- h is the heat transfer coefficient
- A is the surface area of the object
- T is the temperature of the object
- Ts is the temperature of the surroundings

Exponential Decay of Temperature with Time

Newton's law of cooling suggests that the rate of change of temperature of an object is directly proportional to the difference between the temperature of the object and the temperature of its surroundings. This relationship can be expressed in the form of an ordinary differential equation, giving an exponential decay of temperature with time.

The formula for Newton's law of cooling can be expressed as:

$$dT/dt = -k \times (T - Ts)$$

Where:

- dT/dt is the rate of change of temperature with respect to time
- k is the cooling constant (or heat transfer coefficient "h" in the context of liquids)
- T is the temperature of the object
- Ts is the temperature of the surroundings

This equation can be solved to find the temperature of the object at any given time, resulting in an exponential decay in temperature over time. The general solution to this first-order ordinary differential equation is given by:

$$T(t) = Ts + (T_0 - Ts) \times \exp(-kt)$$

Where:

- $T(t)$ is the temperature of the object at time t
- Ts is the temperature of the surroundings
- T_0 is the initial temperature of the object
- k is the cooling constant

This formula demonstrates the exponential decay of temperature with time, as the temperature difference ($T - Ts$) decreases over time, causing the rate of temperature change to decrease exponentially.

CONDUCTION OR CONVECTION: WHICH IS MORE EFFICIENT?

The answer is convection.

In conduction, heat is transferred through direct contact between materials, causing the vibration and movement of molecules within the material. This transfer of heat is more uniform as it occurs in a straight line from the warmer object to the cooler object. However, in convection, it is a counter-current mixing of a hot stream rising and a cold stream descending during heat transfer.

What Makes the Difference

There are two points to consider.

Point 1: The counter-current mixing of hot and cold streams of fluids leads to uneven temperature distribution and the formation of convection currents. These currents can help distribute heat more effectively but also result in non-uniform temperatures within the fluid that make the heat transfer more efficient than conduction.

Point 2: In the case of convection, the rising hot stream and descending cold stream have always different heat capacities. It is unlikely that the enthalpy of these two streams can ever be equal. Therefore, the cold stream draws heat continuously from the hot stream. This is another distinction between conduction and convection that makes convection a more efficient process than conduction.

ROLE OF BOUNDARY LAYER IN HEAT TRANSFER

Step by Step

How Flow Develops in a Pipe?

The flow of a liquid in a pipe develops from a developing flow to a fully developed flow as the boundary layer expands to fill the entire pipe:

The Process of Flow Developing

The flow is developing in the region where the boundary layer is expanding, and the pressure difference is larger. The length of the pipe where the boundary layers intersect is called the entrance length.

Fully Developed Flow

The flow has a fully developed velocity profile, and the flow characteristics no longer change with increased distance along the pipe.

The flow of a liquid in a pipe is affected by many factors, including:

- **Friction**: Friction from the pipe wall causes the fluid to move slower near the wall.
- **Viscosity** and density: The viscosity and density of the fluid are affected by changes in temperature.
- **Pipe properties**: The length, inner diameter, and internal roughness of the pipe affect the flow.

What is Boundary Layer?

A boundary layer is a thin layer of fluid near a surface where the fluid's flow is affected by the wall's friction.

Role of Boundary Layer in Heat Transfer

The thin boundary layer of fluid near a surface where the fluid's flow is affected by the wall's friction.

The fluid velocity changes from zero at the wall to a freestream velocity further away from the wall.

What affect heat transfer?

Thickness

A thicker boundary layer can negatively impact heat transfer performance.

Temperature gradient

A thinner boundary layer usually has a larger temperature gradient, which means a higher rate of heat transfer.

Turbulence

When a boundary layer is fully turbulent, heat transfer is dominated by turbulent eddies. However, molecular conduction still occurs near the wall because the eddies are inhibited by the wall.

Laminar flow

In laminar flow, the thermal boundary layer thickness increases linearly with distance from the surface.

Turbulent flow

In turbulent flow, the thermal boundary layer is thicker than in laminar flow due to increased mixing and turbulence.

References

Wikipedia

Research Gate

SHELL AND TUBE HEAT EXCHANGER: A SHORT REVIEW

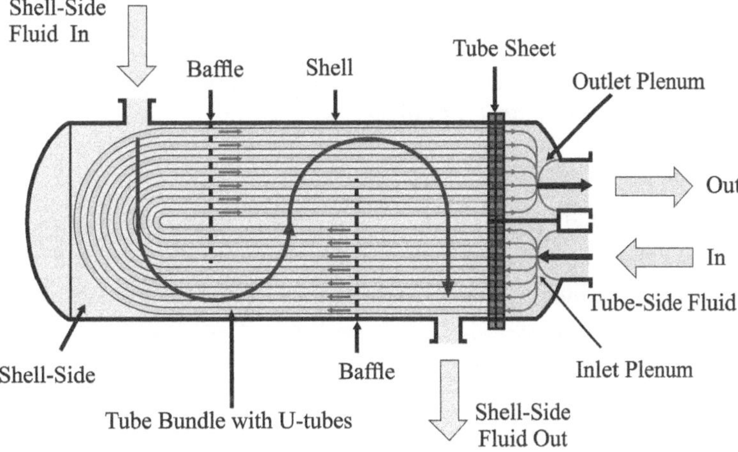

Figure 1: U-tube Heat Exchanger

A shell and tube heat exchanger is a device in which tubes contained within a cylindrical shell are used to exchange heat through thermal contact between two working fluids. The difference in fluid temperatures inside the shell and tube propels the process of temperature exchange. The set of tubes is called a tube bundle, and may be composed of several types of tubes: plain, longitudinally finned, etc.

The shell and tube design can take many different forms. Usually, tubes' ends are joined to plenums, also known as water boxes, via perforations in tube sheets. U-tubes are made up of tubes that can be bent or straight.

Important Features and the Implications

Tube Diameter

The heat exchanger is compact and cost-effective when it uses a small tube diameter. Nevertheless, the heat exchanger is more likely to foul up more quickly, and because of its small size, mechanical fouling cleaning is challenging. To prevail over the fouling and cleaning problems, larger tube diameters can be used. Thus, to determine the tube diameter, the available space, cost and fouling nature of the fluids must be considered.

Thickness of Tube

Typically, the tubes' wall thickness is determined to ensure that:

- there's adequate space for corrosion.
- there is resistance to that flow-induced vibration.
- axial strength.
- availability of spare parts.
- hoop strength (to sustain the pressure inside the tube).
- buckling strength (to tolerate excessive shell pressure).

Tube Length

Heat exchangers with longer tubes are more cost-effective than those with smaller shell diameters. Consequently, the goal is usually to extend the heat exchanger's physical length as long as possible without exceeding production capacity. This has several limitations, though, such as the installation site's space requirements and the requirement to make sure that tubes are available in lengths twice as long as needed (to allow for withdrawal and replacement). Long, thin tubes are also more challenging to remove and replace.

Tube Pitch

It is practical to make sure that the tube pitch, or the center-to-center distance between adjacent tubes, is at least 1.25 times the outside diameter of the tubes when designing the tubes. A more costly heat exchanger results from a larger overall shell diameter caused by a larger tube pitch.

Tube Corrugation

This kind of tube, which is primarily used for the inner tubes, improves performance by increasing the fluids' turbulence, which has a significant impact on heat transfer.

Tube Layout

The arrangement of tubes inside the shell is referred to as the "tube layout." Triangle (30°), square (90°), rotated square (45°), and square (60°) are the four primary forms of tube layouts. The use of triangle patterns causes the fluid to flow around the piping in a more turbulent manner, which increases heat transfer. In areas with high fouling and more frequent cleaning, square patterns are used.

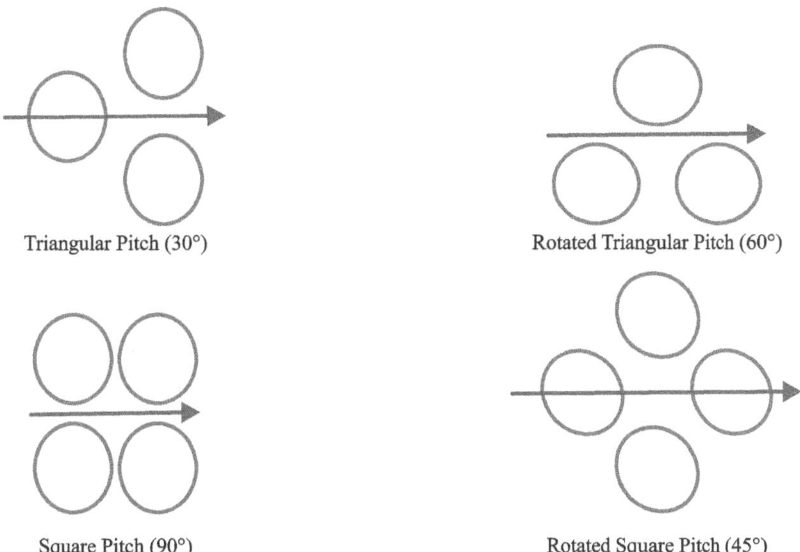

Figure 2: Various Types of Tube Pitch

Baffle Design

When designing shell and tube heat exchangers, baffle spacing is a major thermodynamic concern. Baffles must be spaced with consideration for the conversion of pressure drop and heat transfer. For hermoeconomic optimisation, it is suggested that the baffles be spaced no more than 180 degrees apart. The most common type of baffle is the segmental baffle. The semicircular segmental baffles are oriented at 180 degrees to the adjacent baffles forcing the fluid to flow upward and downward between the tube bundle.

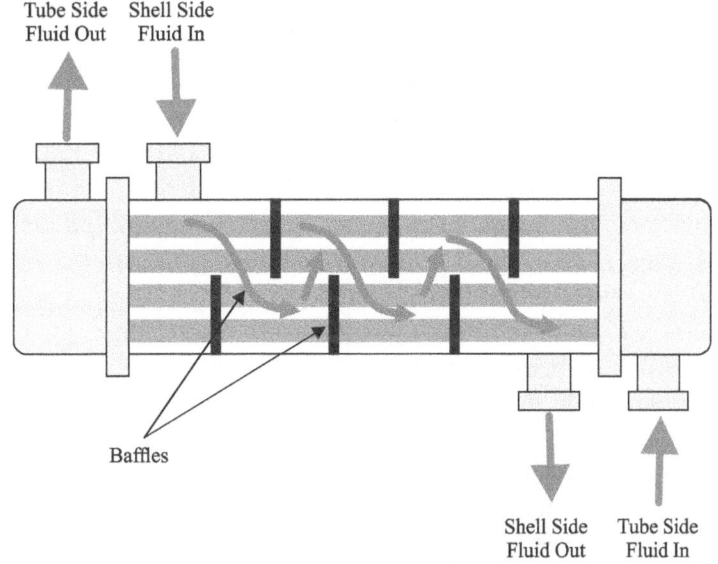

Figure 3: Baffles in Heat Exchanger

It is recommended that the baffles be positioned no closer than 20% of the inner diameter of the shell for thermoeconomic optimisation. Because of flow redirection, baffles spaced too closely result in a higher pressure drop. There may therefore be colder spots in the corners between baffles if the baffles are placed too widely apart. Making sure the baffles are spaced closely enough to prevent the tubes from sagging is also crucial. The disc and doughnut baffle, which is made up of two concentric baffles, is the other major kind of baffle. The inner baffle is shaped like a disc, while the outer, wider baffle resembles a doughnut. This kind of baffle compels the fluid to move through and around the disk's edges.

Fins and tubes Design

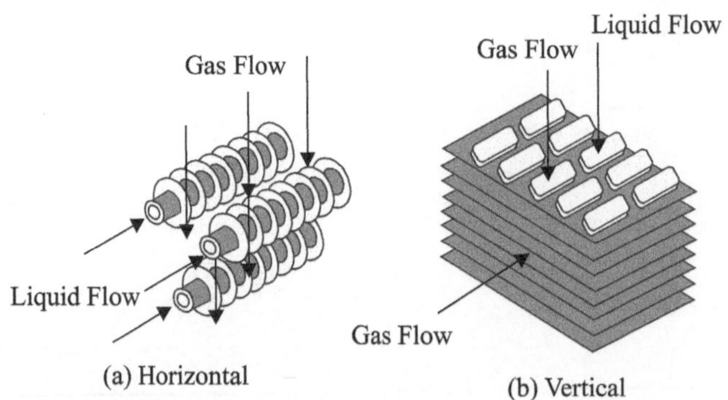

Figure 4: Components of Finned Tube Heat Exchanger

The difference in heat transfer between air and cold fluid in an application involving shell-and-tube technology (such as an intercooler or charge air cooler for combustion engines) may require an increase in the heat transfer area on the air side. Fins can be added to the tubes for this purpose in order to create a tubes & fins configuration and increase the heat transfer area on the air side. Fixed tube liquid-cooled heat exchangers with forged brass integral end hubs, copper tubes, brass baffles, and brass shells are particularly well-suited for harsh and maritime applications.

PINCH TEMPERATURE AND TEMPERATURE CROSS

These two are important phenomena in shell and tube heat exchangers.

Pinch temperature is a concept used in process design and heat exchanger design to optimize the performance of a heat exchanger. It refers to the temperature difference between the hottest stream entering the heat exchanger and the coldest stream leaving the heat exchanger. The pinch temperature is an important parameter for determining the minimum approach temperature in a heat exchanger system, which is the smallest temperature difference between the hot and cold streams at any point in the heat exchanger.

Temperature cross, on the other hand, occurs when the temperature of the colder fluid exceeds the temperature of the hotter fluid at some point in the heat exchanger.

This phenomenon can occur in a heat exchanger when the flow rates and temperatures of the hot and cold streams are not optimized, leading to decreased efficiency and potential energy loss.

The pinch temperature and temperature cross are important in determining the efficiency of a heat exchanger.

In a well-designed heat exchanger system, the goal is to minimize the pinch temperature and avoid temperature cross, as these factors can greatly impact the heat transfer efficiency.

When the pinch temperature is minimized, it increases the heat transfer efficiency of the heat exchanger. This is because a smaller temperature difference between the hot and cold streams allows for more effective heat transfer and a more compact heat exchanger design. Minimizing the pinch temperature also reduces the temperature cross, which further enhances the heat transfer efficiency.

On the other hand, when a temperature cross occurs, it decreases the heat transfer efficiency of the heat exchanger. This is because the colder fluid may be exiting the heat exchanger at a higher temperature than the hotter fluid entering the heat exchanger, resulting in reduced heat transfer effectiveness and potential energy wastage.

Mitigation of Temperature Cross by Extra Surface Area

Fundamental Reasons for Temperature Cross

When there is a temperature cross, it typically indicates that the heat exchanger's design doesn't have sufficient surface area or isn't configured optimally to efficiently transfer heat between the hot and cold fluids. At the point of temperature cross, it can be visualized that the heat exchange surface area is practically zero, leading to the temperature cross phenomenon.

In summary, insufficient heat transfer surface area can lead to a temperature cross under certain conditions, where the temperatures of the hot and cold fluids intersect within the heat exchanger, potentially causing reduced heat transfer efficiency and performance. When the extra surface area is added to a heat exchanger, there is an increase in the heat transfer due to the additional contact area between the hot and cold fluids that allows for more efficient heat transfer.

HOW A SMALL PINCH TEMPERATURE IMPROVES HEAT TRANSFER

A smaller pinch temperature difference can result in a larger heat transfer area but can also lead to higher capital costs. When designing heat exchangers, it's important to balance the thermodynamic performance with the economics.

Thermodynamics

A small pinch temperature can reduce the heat exchanger's irreversibilities and entropy loss. In a heat exchanger, the area around the pinch point contributes the most to entropy production and the required heat transfer coefficient. By overlapping the hot and cold composite curves and separating them by the minimum temperature difference, the maximum process heat recovery can be achieved. This indicates minimum hot and cold utility requirements for the process.

What is Pinch Technology?

Pinch technology is a process engineering technique that aims to reduce the amount of energy used in chemical processes. It's based on the idea that heat can be transferred from higher to lower temperatures, which can reduce the need for fuel and steam.

The Origin of the Word "Pinch"

The term "pinch" in pinch technology comes from the pinch principle, which states that heat can be transferred from hotter to cooler temperatures to recover energy. The pinch method is a two-step process that calculates the maximum energy recovery in a system to minimize the amount of external energy needed.

Improved Heat Transfer

A small pinch temperature can improve heat transfer by minimizing the amount of heat that needs to be transferred across the pinch:

Pinch Point

The pinch point is the minimum temperature difference at which heat can be transferred between two streams.

Pinch Analysis

Pinch analysis is a technique that identifies the pinch point and helps to optimize heat transfer.

Heat Transfer

Heat is always transferred from the higher-temperature object to the lower-temperature object.

Reduces Energy Consumption

Pinch analysis can help reduce the amount of energy required from outside, which can lower operating costs.

How to Design Pinch Temperature?

To design pinch temperature, you can use the pinch method, which involves two steps:

Step by Step

Calculate maximum energy recovery

Calculate the maximum energy recovery in the system to minimize the external energy needed.

Find the Pinch

Use a graphical approach to find the pinch, which is the point where the hot and cold composite curves are closest in temperature on a temperature-enthalpy graph.

Here are some other things to consider when designing pinch temperature.

Minimum allowable temperature approach

The minimum allowable temperature approach, or ΔT_{min}, ensures that equipment sizes are reasonable.

A trade-off between equipment and operating costs

The selection of ΔT_{min} is based on the trade-off between the equipment and operating costs. Finally, pinch technology can help maximize heat recovery, which can lead to energy savings, reduced operational costs, and lower greenhouse gas emissions

Finally, pinch technology can help maximize heat recovery, which can lead to energy savings, reduced operational costs, and lower greenhouse gas emissions

APPROACH TEMPERATURE AND TEMPERATURE CROSS

Briefly, what is temperature cross?

Temperature cross is a loss of efficiency.

Heat transfer occurs in a heat exchanger as a result of three unequal thermal resistances. Cold and hot side convective heat transfer and the thermal resistance provided by the metal wall separating hot and cold fluid streams provide these resistances.

These three unequal resistances prevent the hot-cold liquid temperature crossover and generate delta t between the hot and cold fluids, which acts as a driving force for heat transfer.

When these three resistances are equal, there is a temperature crossover between cold and hot fluid. This temperature crossover is a temperature cross.

Temperature cross occurs when the temperatures of hot and cold stream fluids intersect.

Approach Temperature

The approach temperature of a heat exchanger is the smallest difference between the temperatures of the hot and cold fluids.

For example, if a hot fluid at 105°C is used to heat a cold fluid from 80°C to 100°C, the approach temperature is 5 degc.

A lower approach temperature means a higher heating area. For example, a plate heat exchanger can have an approach temperature as low as 1°C, while a shell and tube heat exchanger requires an approach of 5°C or more.

Approach temperature affects the initial cost and size of the heat exchanger. The initial cost and size of the heat exchanger increases as the two temperatures get closer together.

Plate heat exchangers:

The approach temperature can be as low as 1°C.

Reasonable Approach Temperatures

Reasonable approach temperatures are usually 15°F for water cooled heat exchangers, 20°F for steam heated heat exchangers, and 25°F for air cooled heat exchangers.

Calculation

The approach temperature is calculated by subtracting the outlet temperature of the cold fluid from the outlet temperature of the hot.

Use

Approach temperature is used to monitor condition of a heat exchanger and determine when it needs cleaning. A decrease in approach temperature over time indicates that heat exchanger is becoming fouled or scaled. Cleaning the heat exchanger improves performance and reduces operating costs

Temperature Cross

Temperature cross is a phenomenon that can occur in heat exchangers, where hot outlet temperature is lower than cold outlet temperature. It can happen in multiple shell and tube heat exchangers in the chemical process industries.

To handle temperature cross, multiple heat exchangers may need to be connected in series. The temperature cross must be within an allowable limit.

Factors that may affect the design of a heat exchanger include: Pressure, Temperature, Flow rate, Fouling, and material limitations.

The temperature difference between hot side outlet temperature and cold side outlet temperature is called the "temperature approach". A good rule of thumb is to design a single shell and tube heat exchanger with a minimum temperature approach of 10F.

HEAT EXCHANGER TEMPERATURE PROFILE

In a heat exchanger, the metal wall acts as a barrier between the hot and cold fluids, and it plays a crucial role in transferring heat from one fluid to the other. The temperature at the outer wall of the exchanger is influenced by the hot fluid temperature and the resistance offered by the boundary layer.

The outer wall of the heat exchanger absorbs heat from the hot fluid, and the inner wall delivers the heat to the cooler fluid. Due to the thermal conductivity of the metal wall, there is a temperature difference between the inner and outer walls.

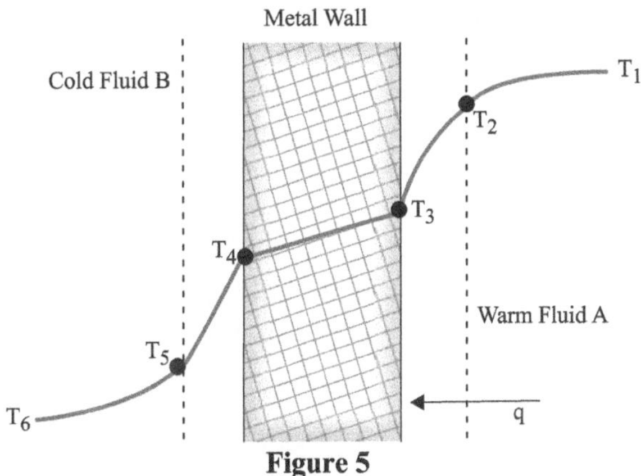

Figure 5

Details are explained below.

Please refer to the image above.

- **Hot Fluid Path:** The temperature profile of warm fluid is represented by T_1, T_2 and T_3. The path T_1 to T_2 is convection mode of heat transfer. Path T_2 to T_3 represents conduction mode of heat transfer through the boundary layer. T_3 is the outer wall temperature. It may be noted that the outer wall temperature $T_3 <$ is less than hot fluid bulk temperature T_1.
- **Path of Heat in the Plate:** This is represented by T_3 to T_4. The drop in temperature is caused by the thermal conductivity of the metal that acts as a resistance to heat transfer. The mode of heat transfer in the plate is conduction.
- **Cold Fluid Path:** The temperature profile of cold [cooler] fluid is represented by T_4, T_5, and T_6. The temperature reduces from T_4 to T_5 in the inner boundary layer by conduction thermal resistance. T_5 to T_6 represents the convection mode of heat transfer with $T_6 < T_5$.

If you study the temperature profile in the image you will observe a sharp temperature, drop where the mode of heat transfer is conduction than convection. This means a much sharper resistance to heat transfer in the conduction path of heat transfer than convection due to the counter-current mixing of hot and cold fluids in convection which makes convection a more efficient mode of heat transfer compared to conduction.

KINETICS [RATE] AND THERMODYNAMICS OF HEAT EXCHANGERS

Kinetics focuses on the rate at which heat is transferred between two bodies, considering factors such as the temperature gradient, material properties, and surface area for heat transfer. Thermodynamics, on the other hand, deals with the overall process of heat exchange, including optimizing the design and operation of heat exchangers to maximize efficiency and effectiveness. By understanding both the kinetics and thermodynamics of heat transfer processes, engineers can design and operate heat exchangers to achieve the desired level of performance and energy efficiency.

Generalized Facts

The kinetics work predominantly when there are resistances to heat flow. Resistance determines the rate of heat transfer in these situations, $Q = U A\, dT$. In processes where resistances do not play a critical role, $Q = m\, Cp\, dT$, thermodynamics may become more important in determining the behavior of the system. Examples of such processes include cooling towers and refrigerators, where thermodynamic principles govern the efficiency and effectiveness of heat transfer. In a heat exchanger, there is an approach temperature that arises from the unequal resistances in a heat exchanger. The interplay of the resistances with the heat transfer process to get the minimum pinch temperature and no temperature cross is a very critical factor. The resistances in the path of heat transfer are kinetics-controlled factors.

Basic Facts

Simple Cold and Hot Fluid Mixing

When hot and cold miscible fluids are mixed, heat transfer occurs from the hot fluid to the cold fluid. During the process, the temperature difference between hot and cold fluids drives the heat transfer until both fluids reach the same temperature. The rate of heat transfer is dependent in addition to the temperature difference between the fluids, on the specific heat capacity of the fluids, and the mass or volume of the fluids being mixed. In such a process the specific heat is the only resistance. In such cases, $Q = m\, Cp\, dT$.

Heat Transfer in a Heat Exchanger

The same process as above becomes different when hot and cold fluid are separated by a metal plate. The problem is multiple resistances in the path of heat flow with nonlinear thermal properties.

The temperature gradient in forced convection

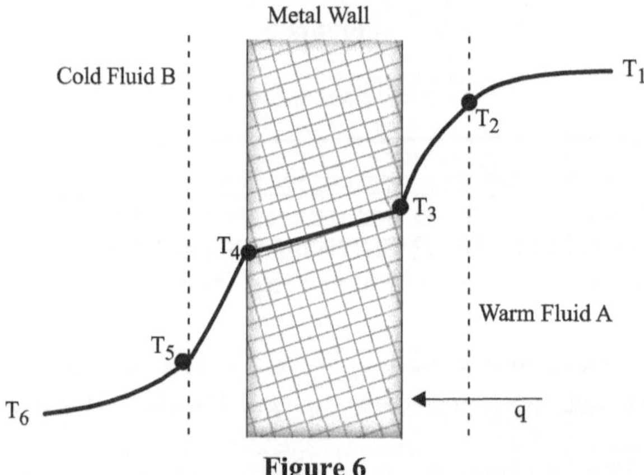

Figure 6

Simply count the resistances along the heat flow path. Each resistance provides its barrier to heat flow. The specific heats of the fluids on either side, cold and hot, differ. The thermal conductivities of the boundary layers on either side of the metal plate where conduction is the mode of heat transfer differ. Furthermore, the metal has its thermal resistance. Besides all these resistances there are fouling on both in and outside of the tube surfaces that heat has to overcome. As a result, heat transfer in a heat exchanger is a rather complex process.

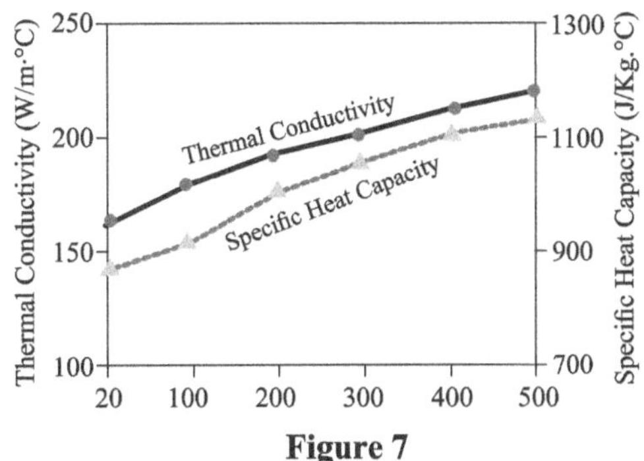

Figure 7

Nonlinear response of thermal conductivity and specific heat with temperature further compounds heat transfer. In such cases, $Q = U A dT$. U represents the inverse of the sum of all resistances.

To summarize, in heat transfer, the two main forms of heat transfer resistance that need to be overcome are convection and conduction, with a metal wall serving as a barrier between them. Heat, Q moves from hot to cold. For a fluid flow, when there is no external resistance to heat flow, only *Cp* [specific heat] offers the internal resistance to heat flow. *Cp* takes away Q/mass- degc and supplies it for molecular motions, the equation is $Q = m\ Cp\ dt$.

When there are external resistances like the metal wall between the hot and cold fluid in a heat exchanger with boundary layers, the equation gets modified to, $Q = U A dt$, "*U*" is called the overall heat transfer coefficient. In the equation $Q = U A dt$, the overall heat transfer coefficient U takes into account various modes of heat transfer, such as conduction, convection, and radiation.

Thermodynamics and Kinetics in Heat Exchangers

Both kinetics and thermodynamics play a role in controlling a heat exchanger. Thermodynamics determines the overall energy transfer and heat exchange in the system, while kinetics governs the rate at which these processes occur. The design and operation of a heat exchanger must consider both thermodynamic principles, such as heat transfer efficiency and temperature gradients, as well as kinetic factors, like flow rates and reaction rates, to optimize performance.

Role of thermodynamics

Determining how well heat exchange devices perform in terms of thermodynamics and heat transfer is crucial. The devices that allow thermal energy to move between two or more fluids at different temperatures are called heat exchangers. There are numerous uses for heat exchangers. In a typical heat exchanger, losses come in four basic categories.

- losses brought on by heat transfer across a limited temperature differential,
- fluid friction-related losses,
- losses resulting from heat exchanger material and manufacturing,
- heat losses as a result of surrounding heat exchange. Since the heat exchanger surface is insulated to prevent this kind of heat exchange, the fourth heat exchanger losses are typically negligible.

Heat exchanger thermodynamic analysis is used to assess additional losses.

Irreversibilities in heat transfer reduce the efficiency of energy systems, and entropy generation is a gauge of how much irreversibilities are present during that process. The generation of entropy will increase with the degree of irreversibilities. As a result, one can quantify the irreversibilities connected to a process using entropy generation. Therefore, when designing a system to improve performance, we must minimise this inevitable entropy generation during the process.

Role of Kinetics [rate] in Heat Transfer

In a heat exchanger with multiple resistances $Q = U A dT$.

The overall heat transfer coefficient, U takes into account all resistances contributing to the heat transfer rate in the path of heat transfer and U is inversely proportional to *dT*.

The resistances in a heat exchanger, including those across solid walls and fluid boundary layers, are expressed in units of watts per meter per Kelvin, reflecting the relationship between temperature difference, heat transfer rate, and material/flow properties in determining the efficiency of heat transfer processes.

In this concept, the overall heat transfer resistance in a heat exchanger is typically expressed in terms of thermal conductivity (W/m*K) and thickness (m) for solid walls and convective heat transfer coefficients ($W/m^2 \times K$) for fluid streams.

The thermal resistance for conduction through a solid wall can be calculated using the equation:

$R\_wall = L\_wall/(k\_wall\ A)$ where:

- *R*_wall is the thermal resistance of the solid wall (in K/W),
- *L*_wall is the thickness of the wall (in meters),
- *k*_wall is the thermal conductivity of the wall material (in W/mK), and
- *A* is the surface area of the wall (in square meters).

Similarly, the thermal resistance for convection across a fluid boundary layer can be calculated as:

$R\_fluid = 1/(h*A)$ where:

*R*_fluid is the thermal resistance of the fluid boundary layer (in K/W),

h is the convective heat transfer coefficient (in $W/m^2 \times K$), and

A is the surface area for convective heat transfer (in square meters).

The overall thermal resistance in a heat exchanger is obtained by summing up the individual resistances for conduction through solid walls and convection across fluid boundary layers. This concept helps in analyzing and optimizing heat exchanger performance by identifying and minimizing the dominant resistances to heat transfer.

HEAT EXCHANGER: TWO-PHASE FLOW HEAT TRANSFER

What is a Phase of a State of Matter?

A phase is a cluster of molecules. There are forces of attraction and repulsion that exist between molecules of all substances. These intermolecular forces allow molecules to pack together in the solid and liquid states. All phase changes involve either an increase or decrease of intermolecular forces. The presence of a two-phase flow enhances heat transfer in a heat exchanger by increasing the effective surface area, utilizing latent heat, promoting convective heat transfer, and improving mixing, all of which contribute to a higher overall heat transfer coefficient.

One disadvantage of two-phase flow in a heat exchanger is the erosion of the metal surface. A two-phase flow can cause erosion of metal in a heat exchanger. When a heat exchanger operates with a two-phase flow, such as a mixture of liquid and vapor, the velocity and impact of the fluid on the metal surfaces can lead to erosion. This erosion can be particularly pronounced in areas where the flow changes direction or encounters obstacles, leading to localized damage to the metal surfaces. Proper design and material selection, as well as regular maintenance and inspection, are important to mitigate the effects of erosion in heat exchangers operating with two-phase flow.

A phase is a region of material that is chemically uniform, physically distinct, and (typically) mechanically separable and is referred to in the physical sciences as a phase. The ice cubes are one phase, the water is a second phase, and the humid air is a third phase over the ice and water in a system that consists of ice and water in a glass jar. Another distinct phase is the jar's glass. More specifically, a phase is a space-time region (a thermodynamic system) in which a material's physical properties are nearly constant. Physical characteristics include things like chemical composition, density, magnetization, and index of refraction. Although the terms "phase" and "state of matter" are sometimes used interchangeably, a state of matter can have multiple immiscible phases (for example, when water and oil separate into different phases while still in the liquid state). Additionally, it can occasionally be used to allude to the equilibrium states depicted on a phase diagram.

Why and How Does Phase Change Occur?

The four main states of matter are solid, liquid, gas, and plasma. Phase changes occur when matter transforms from one state to another. When a significant quantity of energy is gained or lost, this process takes place. In addition, temperature and pressure affect phase change. Heat is either absorbed or released when matter changes phases. There are three ways in which this process, known as heat transfer, can take place: conduction, convection, and radiation.

Phase changes happen when an external force, like heating or cooling, alters the intermolecular energy between the molecules, causing the bonds between them to break. The declustered molecules migrate to a stable phase where they reorganize into new molecules. Water turns into ice when it cools and vapors when it heats up.

Heat Transfer in a Heat Exchanger with Phase Change

There is a phase change on the warm side, the cold side, or both in a two-phase heat exchange process. The following describes what happens when a gas or liquid changes phases.

A liquid's temperature will rise until it reaches its boiling point if heat is added to it. It is not possible to raise the temperature further. Rather, it makes the liquid more gaseous, forming a two-phase mixture of gas and liquid. As the liquid boils, the gas produced forms bubbles. It won't get any warmer until all of the liquid has evaporated. A gas is referred to as superheated when its temperature rises above its boiling point.

Further Explanation

The Following are Sequences

The heat transfer at a boundary takes place due to a temperature difference between bulk and wall temperatures.

- $Q/A = h\,[T\text{bulk} - T\text{wall}]$, h is the heat transfer coefficient, T bulk is bulk temperature, and T wall is wall temperature

- At low *dt*, few bubbles form, and heat transfer is by natural convection. As *dt* increases, more bubbles form, increasing convection (flow) in the liquid phase, and increasing *h* [heat transfer coefficient]
- As *dt* increases further, a film appears and grows, increasing resistance to heat transfer.

In a two-phase heat transfer the phase change causes, one of the phases, usually a liquid, to evaporate or vaporize. This creates a two-phase flow where both liquid and vapor phases coexist, and this flow is important for efficient heat transfer in the heat exchanger. The vapor phase carries the heat away from the hot surface, while the liquid phase helps in maintaining contact with the surface for further heat transfer. The presence of a two-phase flow enhances the overall heat transfer coefficient and improves the efficiency of the heat exchanger. Therefore, it is essential to consider and optimize two-phase flow in heat exchanger design and operation.

The Presence of Two Phases Improves the Heat Transfer Coefficient

The presence of a two-phase flow in a heat exchanger improves the heat transfer coefficient in several ways:

1. *Increased Surface Area*

The formation of vapor bubbles increases the effective surface area available for heat transfer. As the bubbles move through the liquid, they create turbulence and increase the contact area between the liquid and the heated surface, enhancing heat transfer.

2. *Latent Heat*

As the vapor condenses back into liquid, it releases the latent heat it had absorbed during vaporization. This additional heat is transferred to the cooler surface, aiding in heat transfer. This latent heat transfer significantly increases the overall heat transfer coefficient.

3. *Convective Heat Transfer*

The movement of the two-phase flow, particularly the vapor phase, can lead to convective heat transfer effects, further enhancing the overall heat transfer coefficient.

4. *Enhanced Mixing*

The presence of vapor bubbles in the liquid promotes mixing and circulation, ensuring that fresh liquid is continuously brought into contact with the heated surface, improving heat transfer.

To summarize, the presence of a two-phase flow enhances heat transfer in a heat exchanger by increasing the effective surface area, utilizing latent heat, promoting convective heat transfer, and improving mixing, all of which contribute to a higher overall heat transfer coefficient.

HEAT EXCHANGER IRREVERSIBILITY

Determining how well heat exchange devices perform in terms of thermodynamics and heat transfer is crucial. The devices that allow thermal energy to move between two or more fluids at different temperatures are called heat exchangers. There are numerous uses for heat exchangers. In a typical heat exchanger, losses come in four basic categories.

- losses brought on by heat transfer across a limited temperature differential,
- fluid friction-related losses,
- losses resulting from heat exchanger material and manufacturing,
- heat losses as a result of surrounding heat exchange. Since the heat exchanger surface is insulated to prevent this kind of heat exchange, the fourth heat exchanger losses are typically negligible.

Heat exchanger thermodynamic analysis is used to assess additional losses.

The first law of thermodynamics addresses changes in the energy stored in the system as well as the quantitative conservation of energy in various forms transferred between the system and its surroundings. It views interactions between heat and work as equivalent forms of energy in motion. It discusses energy's quantity and states that energy cannot be produced or destroyed. On the other hand, energy quality is covered by the second law. More precisely, it is focused on lost work opportunities, entropy generation, and energy degradation during a process. To determine the quality difference between mechanical and thermal energy, the second law of thermodynamics is necessary and has proven to be an incredibly effective tool in the optimisation of complex thermodynamic systems.

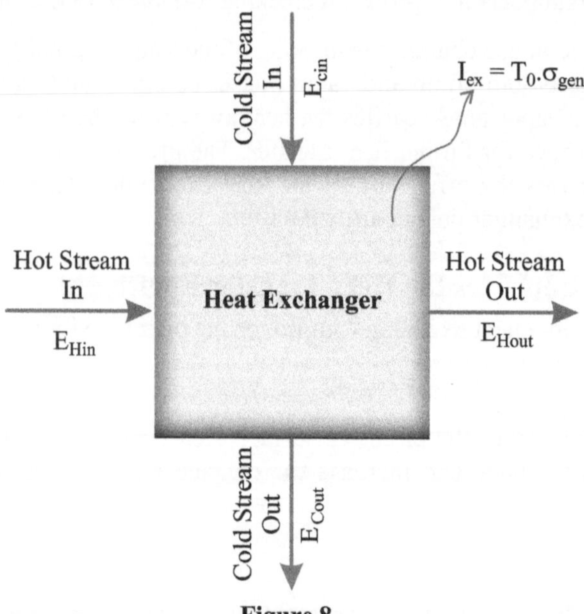

Figure 8

The first law of thermodynamics addresses changes in the energy stored in the system as well as the quantitative conservation of energy in various forms transferred between the system and its surroundings. It views interactions between heat and work as equivalent forms of energy in motion. It discusses energy's quantity and states that energy cannot be produced or destroyed. On the other hand, energy quality is covered by the second law. More precisely, it is focused on lost work opportunities, entropy generation, and energy degradation during a process.

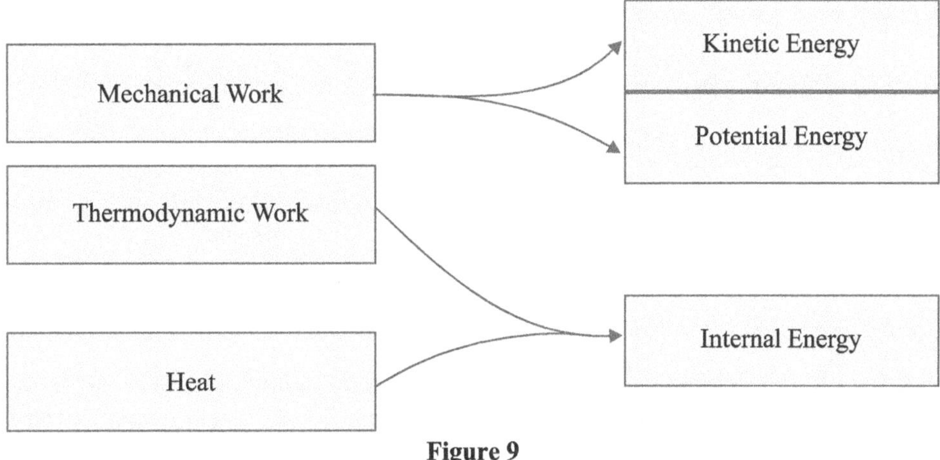

Figure 9

When optimising intricate thermodynamic systems, the second law of thermodynamics has proven to be an extremely effective tool. It is necessary to determine the disparity in quality between mechanical and thermal energy.

A process that can be reversed without leaving any trace in its surroundings is called reversible. Irreversible processes are those that cannot be reversed. Irreversibilities are the conditions that make a process irreversible. These include mixing of fluids, friction, unrestricted expansion, heat transfer across a finite temperature difference, etc. Process irreversibility is the inability to reverse actions that take place during a process. When a heat transfer dQ crosses the system boundary where the local temperature is T, an entity $\int dQ/T$ is transferred along with it. This definition of entropy as a system property is stated as follows. As a result, this abstract system property suggests that entropy must accompany heat transfer.

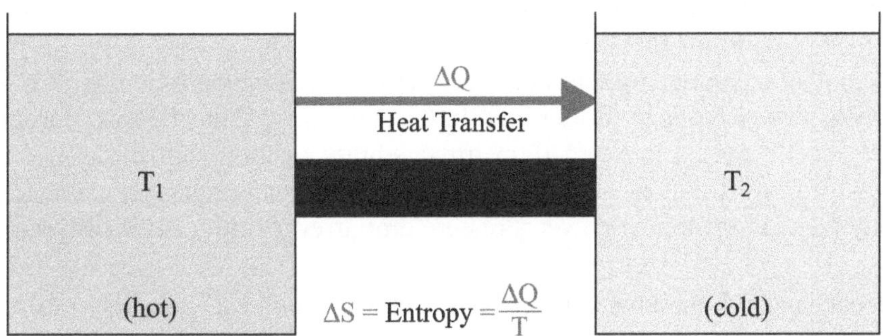

There exists a useful thermodynamic variable called entropy (S). A natural process that starts in one equilibrium state and ends in another will go in the direction that causes the entropy of the system plus the environment to increase for an irreversible process and to remain constant for a reversible process.

$S_f = S_i$ (reversible) $S_f > S_i$ (irreversible)

Figure 10

As a consequence, a reversible adiabatic process can be identified by zero entropy change. If a process is not reversible (as with any heat transfer across a finite temperature difference), the situation is radically different. Entropy change ΔS is either equal (reversible process) or larger (irreversible process) than the entropy transfer ($\int dQ/T$) that accompanies heat transfer dQ, the difference being attributed to entropy generation S gen.

The quantitative indicator of the quality of energy transfer is the quantity of entropy generation. When entropy generation is zero, it indicates that energy transfer and/or energy conversion (a reversible process) is of the highest quality; when it is greater than zero, it indicates a lower quality. The entropy generation of all real processes is larger than zero. Energy-wise, irreversibility can be written as the product of entropy generation and T_0, the reference temperature. Irreversibility = $T_0 \times S$ gen.

Heat exchanger performance should be assessed using both first- and second-law.

The change in internal energy of a system is equal to the heat added to the system minus the work done by the system.

$$\Delta U = Q - W$$

Change in internal energy | Heat added to the system | Work done by the system

Figure 11

First law of Thermodynamics

Heat exchangers are generally inefficient from an energy conservation point of view because they have been designed in the past on the basis of low cost that dictates a minimum-size unit. To achieve the small-sized heat exchanger, the temperature difference between the fluid streams is maximized. Nonetheless, the loss during heat transfer in a heat exchanger will increase with the size of the temperature differential. Thermodynamic irreversibility is influenced by temperature and temperature difference distributions inside the heat exchanger. The difference in fluid temperatures between the two heat exchanger streams is related to the generation of entropy. Hence, the difference between mean temperatures of two fluids directly influences the entropy measure of the irreversibility manifested within the heat exchanger.

$$\frac{\delta Q}{T} \leq dS$$

Therefore, by lowering the temperature differentials between the fluids, the heat exchanger irreversibility for a given heat transfer rate can be decreased, increasing the effectiveness of the exchanger. Compared to a heat exchanger with a larger temperature difference between the fluids (for the same heat transfer rate), one with a smaller temperature difference produces less irreversibility in a given system.

Capacity mismatch is another technique used to improve performance between streams. It is difficult to determine this from the first law analysis, which leads to further loss. In the same way, fluid streams' pumping pressure is raised to attain improved performance-increased pressure drop irreversibility results from this. The fluid encounters fluid friction and other pressures drop contributors as it travels through the heat exchanger passage. This causes a fluid pressure drop that is directly related to pumping power. Pressure drop irreversibility is the irreversibility brought on by a fluid pressure drop.

This is the reason heat exchanger thermodynamic analysis is crucial, and heat exchanger evaluation should also take into account the first law.

Because of this, heat exchanger thermodynamic analysis is crucial, and heat exchanger performance should be assessed using both first- and second-law based standards. The only method available for calculating the losses resulting from process irreversibility is second law analysis.

Production and Manufacturing Irreversibilities

These are the irreversibilities related to the use of materials in a system. It is the irreversibilities resulting from material production, component construction, or component assembly. These irreversibilities may also be included when a component is recycled at the end of its useful life. It can be computed using cumulative exergy analysis and the cumulative exergy of a component production process. The total exergy comprises the total exergy of electricity used, the equipment utilised in the production process, and the exergy of raw materials. The energy consumption is minimal in the production of heat exchangers because machining operations account for the majority of the process.

Consequently, the primary contributor to the overall value is the cumulative exergy of the raw materials. For instance, the total amount of energy destroyed during the extraction of raw materials from ore, primary processes like rolling, casting, and extrusion, secondary processes like welding and bending, and final manufacturing steps like assembling make up the cumulative energy destruction associated with the production of material and the manufacturing of tubes. Therefore, in addition to the irreversibilities related to heat transfer and pressure drop, the second law of thermodynamic analysis of heat exchangers also takes into account the irreversibilities related to material production and manufacturing. This goes a step further in expanding the research that leads to life cycle analysis.

Irreversibility, the amount of entropy created, and the amount of work that is lost during the process are all directly correlated. The goal of second law analysis is to reduce these losses as much as possible by limiting the amount of entropy produced. A practical method for designing systems is to start with minimum entropy production. Stated differently, one can quantify the quality of thermal energy transformation using entropy generation. As a result, it is important to take into account both the available energy or energy destruction process and the rate of irreversible entropy generation when analysing and designing heat exchangers. Irreversibilities reduce the efficiency of energy systems, and entropy generation is a gauge of how much irreversibilities are present during that process. The generation of entropy will increase with the degree of irreversibilities. As a result, one can quantify the irreversibilities connected to a process using entropy generation. Therefore, when designing a system to improve performance, we must minimise this inevitable entropy generation during the process.

CONDENSER: CONSTANT TEMPERATURE *VS.* CONSTANT PRESSURE

Constant Temperature Condenser

Let us use the refrigeration process as an example.

- In a constant temperature condenser, the refrigerant undergoes a phase change from vapor to liquid at a constant temperature. This means that the temperature of the refrigerant remains constant throughout the condensation process.
- Since the temperature remains constant, there is no change in the internal energy (U) of the refrigerant during the condensation process. The enthalpy (H) of the refrigerant also remains constant because there is no work being done on the system since at constant *T*, *PV* remains constant if we assume it is an ideal gas. Heat is only being transferred at a constant temperature.
- The entropy (S) of the refrigerant decreases as it changes from a high entropy vapor to a low entropy liquid. This process is non-equilibrium as the system moves away from equilibrium during the phase change at constant temperature.

Constant Pressure Condenser: In a constant pressure condenser, the refrigerant also undergoes a phase change from vapor to liquid, but this process occurs at a constant pressure. This means that the pressure of the refrigerant remains constant throughout the condensation process.

- As the refrigerant condenses at constant pressure, the vapor compresses and generates heat. This heat is absorbed by the cooling medium (such as water or air) at a constant pressure, allowing for more efficient heat transfer.
- The enthalpy (H) of the refrigerant decreases as heat is removed and work is done on the system during the compression process in a constant pressure condenser. This allows for more heat extraction compared to a constant temperature condenser where enthalpy remains constant.
- The entropy (S) of the refrigerant also decreases as it changes from a high entropy vapor to a low entropy liquid. This process goes away from equilibrium as the system undergoes a phase change at constant pressure.

In summary, a constant temperature condenser maintains a constant temperature during the phase change process, resulting in no change in internal energy or enthalpy but a decrease in entropy.

On the other hand, a constant pressure condenser maintains a constant pressure during the phase change process, allowing for more efficient heat extraction or enthalpy extraction due to work being done on the system, resulting in a decrease in both enthalpy and entropy.

Therefore, it may be observed that there is no change in enthalpy in a constant-temperature condenser while one can set the condenser pressure like in the Rankine cycle and run the condenser at constant pressure with corresponding saturated temperature and extract more enthalpy.

Carnot cycle and Rankine cycle: Carnot and Rankine had different missions when they conceived their heat engines. Carnot wanted to conceive an ideal machine that could be the benchmark for efficiency. Rankine conceived a practical machine that could run in real life.

Carnot chose constant temperature condenser. Carnot conceived a machine that could run while it was in equilibrium with its surroundings. From the first law of thermodynamics, $U_2 - U_1 = Q - W$, U is internal energy, Q is heat and W is work. Carnot conceived a hypothetical slow cyclic thermodynamic process that cannot transfer internal energy outside the system, $dU = 0$, $Q = W$ and the system remains always in equilibrium with outside surroundings in a constant temperature condenser.

Rankine's approach was more practical.

Rankine Opted for Constant Pressure Condenser

Rankine converted Carnot's idea into a practical working heat engine. He realized a real-life machine cannot be reversible. He also realized that one major constraint of the Carnot engine is its constant temperature condenser. Constant temperature heat rejection limits the cold reservoir temperature and thus limits the capability of an engine to produce the amount of work. Rankine conceived a constant pressure condenser.

A constant pressure condenser does maximum heat rejection in the condenser at the set condenser pressure [vacuum in Rankine cycle] it receives energy from outside in proportion to the work it does to keep the pressure constant.

HEAT EXCHANGER EFFECTIVENESS: ROLE OF TRANSFER UNITS [NTU]

The number of transfer units (NTU) method is used to calculate the rate of heat transfer in heat exchangers when there isn't enough information to calculate the log mean temperature difference (LMTD). It's also used to determine the expected heat exchanger effectiveness from the known geometry. The NTU is a dimensionless parameter that summarizes the dimensional parameters of heat transfer coefficients, transfer area, fluid flow rate, and heat capacity. A higher NTU value usually implies a more efficient heat transfer system. The effectiveness-NTU method is preferred when the heat transfer surface area of the heat exchanger is not known, but the outlet temperatures are. In summary, the number of transfer units (NTU) method is used to calculate the rate of heat transfer in heat exchangers (especially parallel flow, counter current, and cross-flow exchangers) when there is insufficient information to calculate the log mean temperature difference (LMTD).

Details

Heat transfer in a heat exchanger occurs through the process of conduction, convection, and/or radiation. Heat is transferred from one fluid to another through a separating surface, such as a metal wall, in a heat exchanger. The resistance to heat transfer is represented by the overall heat transfer coefficient U.

Role of dT

The driving force for heat transfer in a heat exchanger is the temperature difference (dT) between the two fluids. The temperature difference, or *dT*, is what allows heat to flow from a higher-temperature fluid to a lower-temperature fluid, and it is a critical factor in determining the rate of heat transfer.

The heat transfer rate, Q, can be expressed as $Q = U A\, dT$, where U is the overall heat transfer coefficient, A is the surface area for heat transfer, and dT is the temperature difference between the two fluids.

The effectiveness of heat transfer depends on how much heat (Q) can be transferred through the heat exchanger with a minimum approach temperature, which is the smallest temperature difference between the two fluids. The design of the heat exchanger must consider the pinch temperature, which is the temperature at which the hottest stream entering the heat exchanger matches the coldest stream exiting the heat exchanger. Operating near the pinch temperature ensures efficient heat transfer.

However, there can be extreme situations such as temperature cross, where the temperature difference between the two fluids changes direction along the length of the heat exchanger. In such cases, careful consideration must be given to the design and operation to ensure efficient heat transfer.

Temperature cross

In a given heat exchanger design, if too much heat (Q) is pushed through without a sufficient surface area (A), the temperature difference (dT) diminishes to a point where it becomes zero, reducing the heat transfer rate. This highlights the importance of proper design considerations to ensure optimal heat transfer.

Role of transfer units [NTU]

The effectiveness of a heat exchanger is closely related to the number of transfer units (NTU). The NTU is a dimensionless measure of the size of the heat exchanger and the rate of heat transfer. It is related to the heat exchanger's size, flow configuration, and overall heat transfer coefficient.

The effectiveness (ε) of a heat exchanger can be calculated using the NTU, along with the heat capacity rate of the cold fluid (*Cp* cold) and the heat capacity rate of the hot fluid (*Cp* hot). The formula for the effectiveness of a heat exchanger, in terms of NTU, is:

$$\varepsilon = 1 - \exp(-NTU \times (1 + C\_r))$$

Here, *C_r* is the capacity ratio, which is defined as the ratio of the heat capacity rate of the cold fluid to the heat capacity rate of the hot fluid. The effectiveness of a heat exchanger provides a measure of how well the heat exchanger is performing in transferring heat from the hot fluid to the cold fluid.

Target Effectiveness of a Heat Exchanger

The target effectiveness (ε) for a heat exchanger depends on the specific requirements of the heat transfer application and the desired level of thermal efficiency. In general, a higher effectiveness indicates better heat transfer performance and is often a desirable objective in heat exchanger design.

The choice of target effectiveness is influenced by various factors, such as the temperature approach (minimum temperature difference) between the hot and cold fluids, the desired heat transfer rate, and the overall energy efficiency goals. In many cases, the target effectiveness is determined based on the thermodynamic and operational requirements of the heat transfer process.

For parallel-flow and counter-flow heat exchangers, the maximum possible effectiveness is limited by the temperature approach, which is the minimum temperature difference between the hot and cold fluids at the end of the heat exchanger. The target effectiveness is commonly set to achieve the highest possible thermal performance while considering practical constraints and economic considerations.

The relationship between effectiveness and NTU is crucial in the design and analysis of heat exchangers. By calculating or determining the NTU, engineers, and designers can assess the performance and efficiency of the heat exchanger, allowing them to optimize its design to achieve the desired heat transfer characteristics.

ORIGIN OF LMTD

The logarithmic mean temperature difference (LMTD) is used in thermal engineering to calculate the temperature driv ing force for heat transfer.

The origin of LMTD lies with the non-linear thermal properties of fluids in a heat exchanger with temperature.

Detail

Why LMTD?

The non-linear motion of molecules with temperature does contribute to the non-linear changes in thermal properties such as thermal expansion, thermal conductivity, and specific heat. This non-linearity can significantly affect thermal systems, making the temperature difference (ΔT) non-linear

Here comes the use of logarithmic mean temperature difference (LMTD) to calculate heat transfer in heat exchangers and other systems.

The concept of logarithmic mean temperature difference (LMTD) is derived from the non-linear changes in temperature that occur during heat transfer processes. LMTD is used to calculate the heat transfer rate in heat exchangers, where the temperature of the fluid streams may not change linearly. By using the LMTD method, the non-linear temperature profiles of the hot and cold fluids in the heat exchanger can be accounted for, allowing for an accurate estimation of the rate of heat transfer. This concept is essential in the design and analysis of heat exchangers and other thermal systems, where non-linear temperature changes are prevalent.

Linear and nonlinear functions

Fundamentally ' what is linear and what do nonlinear functions look like?

What is exponential change?

Exponential change occurs when the rate of change of something is proportionate to the subject itself. In other words, in a straight-line equation y = mx +c, when the exponent of x is not one. Then the line becomes exponential. The change becomes quicker and quicker as the subject itself increases or decreases. This is what happens in a heat exchanger as temperature increases.

What is linear change?

Any function that graphs as a straight line is considered linear. Any function that is not linear is referred to as non-linear. The slope-intercept form is the general form of a linear function that manifests: y = mx + c. Because there is only one x variable and the power of x is always one, this equation produces a straight line. The graph is linear in nature as long as x has an exponent of one. Different values will result in nonlinear curves. The correlation between variables in linear data is constant, while nonlinear data changes from point to point.

Why log scale?

A logarithmic scale is used to display non-linear data because it can help make large amounts of data more manageable and easier to understand. Logarithmic scales can display data with large value differences on the same chart. For example, a logarithmic scale can easily display values from 10 to 100,000 on the same chart.

ENTROPY GENERATION IN HEAT EXCHANGERS

In one line, the fundamental point is that heat transfer in a heat exchanger is an irreversible process. The cold fluid in a heat exchanger acts as the surroundings, and entropy loss to the cold fluid cannot be reversed

Most irreversible processes occur in heat exchangers due to the heat transfer in finite heat transfer area and pressure drop in frictional pipes.

Entropy is a measure of the disorder or randomness in a system. In a heat exchanger, entropy is generated as a result of heat transfer between two fluids at different temperatures. This process is governed by the second law of thermodynamics, which states that in any energy conversion process, the total entropy of a closed system will always increase.

Entropy is a thermodynamic property that measures the dispersal of energy in a system. It's a measure of how much energy has been dispersed in a process, and it's related to the Second Law of Thermodynamics. Entropy is expressed in units of joules per kelvin (J/K)

Entropy Generation

A measure of the magnitudes of the irreversibilities present during a process. It's not a property of the system, but depends on the path of the process. The more irreversible a process is, the larger the entropy generation.

Heat Exchanger Irreversibilities

The fundamental point is that heat transfer in a heat exchanger is an irreversible process. The cold fluid in a heat exchanger acts as the surroundings, and entropy loss to the cold fluid cannot be reversed.

The difference in fluid temperatures between the two heat exchanger streams is related to the generation of entropy. Hence, the difference between mean temperatures of two fluids directly influences the entropy measure of the irreversibility manifested within the heat exchanger.

Compared to a heat exchanger with a larger temperature difference between the fluids (for the same heat transfer rate), one with a smaller temperature difference produces less irreversibility in a given system.

Entropy generation depends on the temperature difference between the cold and hot work fluids when a heater operates in a steady state. During a transient process, additional entropy generation occurs due to the variation on heat storage in the work fluids and heat transfer metal slabs.

When hot fluid enters a heat exchanger, it transfers heat to the cold fluid. This heat transfer results in an increase in the entropy of both fluids. As heat is transferred, the hot fluid loses some of its energy and becomes cooler, while the cold fluid gains energy and becomes warmer. This increase in temperature leads to an increase in the molecular movement and disorder within the fluids, contributing to the overall entropy generation in the system.

The rate of entropy generation in a heat exchanger can be quantified using the entropy balance equation, which relates the change in entropy of the system to the heat transfer and the temperature difference between the hot and cold fluids.

The entropy generation in the heat exchanger is a measure of the irreversibility in the process, as some of the energy is lost as waste heat due to the temperature difference between the two fluids.

In practical applications, minimizing entropy generation in heat exchangers is important for optimizing energy efficiency and reducing operating costs. This can be achieved through the use of efficient heat exchanger designs, appropriate temperature control, and insulation to minimize heat losses. Additionally, using fluids with similar temperature ranges and maximizing heat transfer surface area can also help reduce entropy generation in heat exchangers.

In conclusion, entropy generation in heat exchangers is a natural consequence of heat transfer processes that occur in accordance with the second law of thermodynamics. Understanding and controlling entropy generation is essential for efficient operation and design of heat exchangers in various industrial and commercial applications.

Role of Pinch Temperature

The pinch temperature plays a significant role in the generation of entropy in heat exchangers. Entropy generation is a measure of the irreversibilities or losses that occur during heat transfer processes and is closely related to the efficiency of the system. The pinch temperature affects entropy generation in heat exchangers

Higher temperature differences: Operating a heat exchanger with a large temperature difference between the hot and cold streams can lead to increased entropy generation. This is because a larger temperature difference results in a higher driving force for heat transfer, leading to increased heat transfer rates but also higher irreversible losses. By minimizing the temperature difference around the pinch point, known as the pinch temperature, the entropy generation can be reduced, improving the overall efficiency of the heat exchanger.

CHAPTER 42

Evaporation

In thermodynamics, evaporation is considered as an important unit process that involves the conversion of a liquid into vapor through the input of heat energy. Evaporation occurs due to the increase in kinetic energy of the liquid molecules, leading to the breaking of intermolecular forces and transition into the gaseous state. The process of evaporation is driven by the principles of thermodynamics, specifically the second law which states that heat will flow from a higher temperature region to a lower temperature region. As heat is supplied to the liquid, its temperature increases, and at a certain point, the energy is enough to overcome the forces holding the molecules together, resulting in evaporation. Overall, evaporation as a unit process in thermodynamics highlights the relationship between heat transfer and phase change, demonstrating the fundamental principles of energy transformation within a system.

EVAPORATION AND BOILING: KINETIC *VS.* POTENTIAL ENERGY

Evaporation and boiling are two related but different processes that involve the conversion of a liquid to its gaseous form. Both processes involve the molecules of a liquid gaining enough energy to overcome the intermolecular forces that hold them together in the liquid phase.

One important difference between evaporation and boiling is evaporation can occur at any temperature, as long as there is enough kinetic energy for molecules to escape from the liquid surface. Boiling, on the other hand, occurs at the saturation temperature of the liquid, where the vapor pressure of the liquid is equal to the atmospheric pressure.

Boiling involves the conversion of potential energy (from heat) into kinetic energy (increased molecular movement), while evaporation is solely driven by the kinetic energy of the molecules in the liquid

Who Provides Energy for Evaporation?

The surroundings provide energy for evaporation.

How?

Molecules are always in motion with different amounts of kinetic energy. Refer to Maxwell – Boltzmann distribution of kinetic energy.

It says the following:

> Maxwell-Boltzmann. energy distribution shows that while the average kinetic energy (or speed) of molecules in a gas remains constant at a given temperature, individual molecules within the gas can have different speeds. This variation in speeds is due to the random nature of molecular collisions and the transfer of momentum during these collisions. The distribution describes the probability of finding a molecule in the gas with a particular speed, with the majority of molecules having speeds around the average value.

Detail

Evaporation occurs at the surface of a liquid and is driven by the kinetic energy of some of the molecules in the liquid. These molecules gain enough energy to break free from the liquid and enter the gas phase. As these molecules leave the liquid, they take away energy in the form of heat, causing the remaining liquid to cool. Evaporation can occur at any temperature below the boiling point of a liquid.

Boiling, on the other hand, occurs throughout the bulk of the liquid and requires the supply of external heat. When a liquid reaches its saturation temperature, it is said to be saturated with energy. Additional heat supplied to the liquid causes the molecules to gain even more energy, breaking the intermolecular bonds and causing the liquid to expand rapidly into a vapor.

This Turbulence in the Liquid is What We Refer to as Boiling

In summary, evaporation is a surface phenomenon driven by the kinetic energy of molecules, while boiling is a bulk phenomenon that requires the supply of external heat. Both processes involve the conversion of a liquid to a gas, but they differ in the way they occur and the conditions under which they take place.

EVAPORATION: HEAT AND MASS BALANCE

The note covers [1] General discussion of evaporators [2] Key controls [3] Heat and mass balance concept [4] Performance of evaporators [capacity] and steam economy [5] Boiling point elevation and Duhring graph and [6] Problem solution.

General: Evaporation is a unit operation that separates a liquid from solids by means of heat transfer via vaporization or boiling. The purpose of evaporation is to concentrate a solution of a non-volatile solute (i.e., solids) and a solvent (i.e., liquid), which is typically water. Evaporating a portion of the solvent concentrates the solute into a more viscous liquid product.

Evaporation Process – Short Description

Single Effect Evaporator

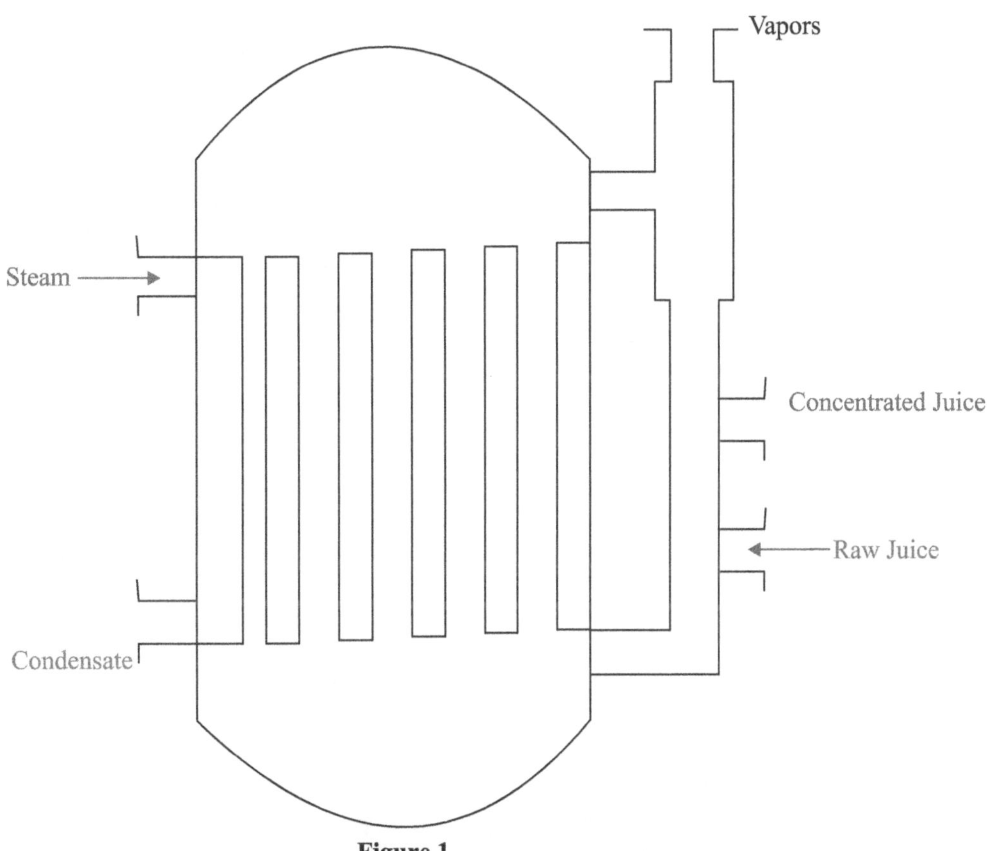

Figure 1

All evaporators are comprised of two sections: a heating section, calandria (called a steam chest), and a vapor/liquid separation section. These sections can be located within a single vessel (body), or the heating section may be external to the vessel that houses the vapor/liquid separation section.

The solution containing the desired product is fed into the evaporator and passes across a heat source. The applied heat converts the water in the solution into vapor. The vapor is removed from the rest of the solution and is condensed while the now-concentrated solution is either fed into a second evaporator or is removed.

Double Effect Evaporator

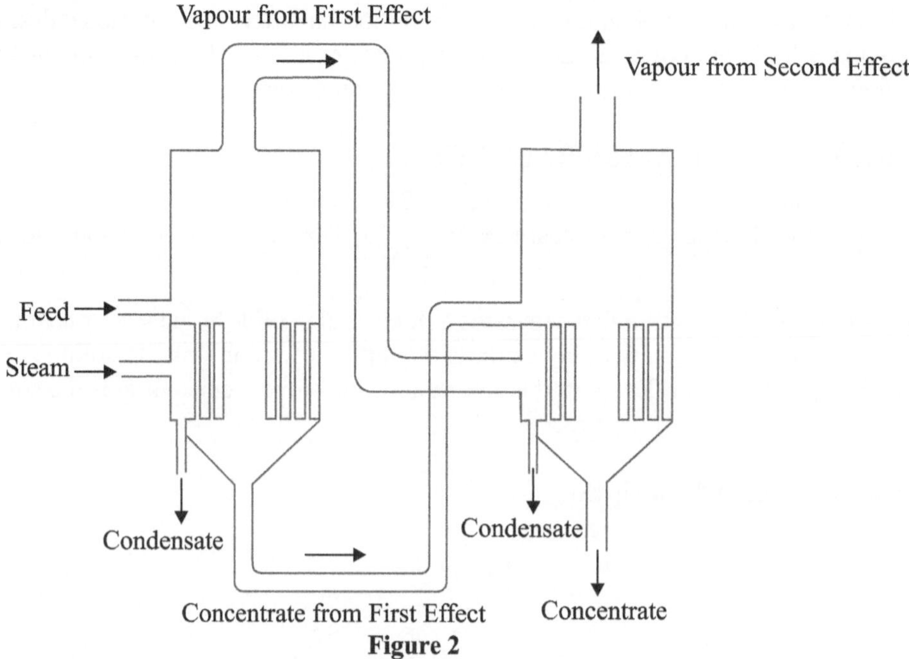

Figure 2

Multiple Effect Evaporators

Evaporators are classified by the number of effects. In a single-effect evaporator [above left side image], steam provides energy for vaporization, and the vapor product is condensed and removed from the system. In a double-effect evaporator [above center image], the vapor production of the first effect is used to provide energy for a second vaporization unit. The image on the right side is a triple effect evaporator. The vapor production of the second effect is used to provide energy for the vaporization of water in the third effect.

Multiple-effect evaporators can remove much larger amounts of solvent than is possible in a single effect.

Triple Effect Evaporator

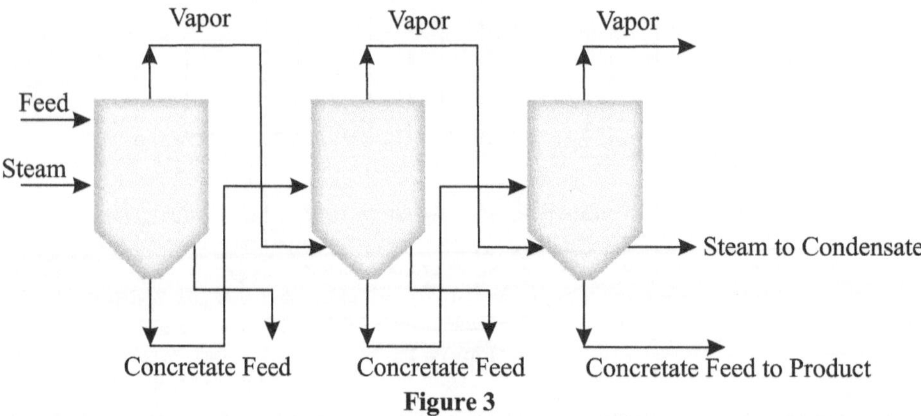

Figure 3

In a multiple-effect arrangement, the latent heat of the vapor product of an effect is used to heat the following effect. Effects are thus numbered beginning with the one heated by steam. It will have the highest pressure. Vapor from Effect I will be used to heat Effect II, which consequently will operate at lower pressure. This continues through the train: pressure drops through the sequence so that the hot vapor will travel from one effect to the next. Normally, all effects in an evaporator will be physically the same in terms of size, construction, and heat transfer area. Unless thermal losses are significant, they will all have the same capacity as well.

Evaporators are typically operated under a vacuum to reduce the temperature of boiling. Steam ejectors or mechanical vacuum pumps are often used to create a vacuum.

Key Controls

Once through evaporator vs. circulation evaporator

In once-through evaporation, the feed liquor passes through the calandria tubes only once and releases vapor, and leaves the unit as thick liquor. All evaporation is accomplished in a single pass. This is how multiple effect evaporators work where the total concentration is spread over several effects Once through evaporation is especially useful for heat-sensitive materials.

In circulation evaporation, a pool of liquid is held within the equipment. Incoming feed mixes with the liquid from the pool and the mixture passes through the tubes. Unevaporated liquid discharged from tubes returns to the pool so that only a part of total evaporation occurs in one pass. Circulation evaporators do not suit heat-sensitive materials. Circulation through tubes can operate by natural circulation because of density difference or by forced circulation by the pump.

Effect of Feed Temperature

The volume of liquid in an evaporator is always very large and in its final concentration and therefore at the boiling temperature of the final solution. The rate at which feed is added to such an evaporator should be such that the average temperature of the liquid boiling in the evaporator is not appreciably affected by such additions.

Choice of Steam Pressure

The general choice with some exceptions is about 1 atmosphere. The choice is low-pressure steam for large latent heat.

Pressure in Vapor Space

Evaporators don't necessarily have to work in a vacuum. The reason for the vacuum is it is most economical to apply vacuum to feed steam at a modest pressure into an evaporator which gives an economical Δt.

Heat and Mass balance on a single effect evaporator

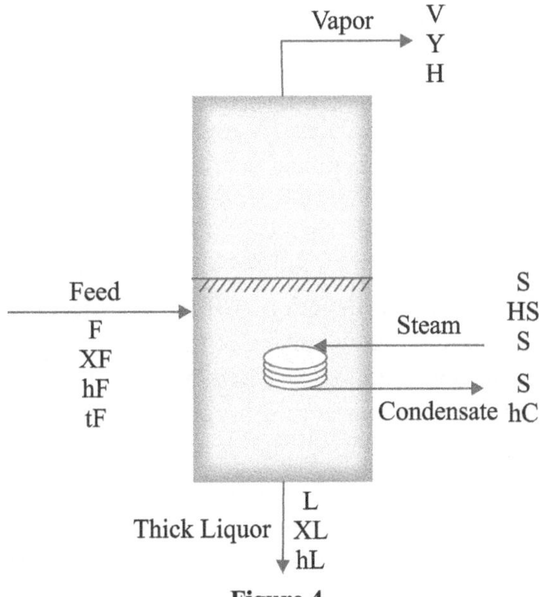

Figure 4

The capacity of an evaporator like any heat transfer process is given by

$q = U A \Delta tm$ [q is the rate of heat transfer, U is overall heat transfer coefficient, A is heat transfer area and Δtm is temperature difference].

Performance of an Evaporator

The above diagram is a simple diagram for an evaporator. The heating surface is represented by a coil. F lb/hr of feed enters the evaporator with a solid content XF. Symbol × stands for weight fraction which is percentage divided by 100. Let the enthalpy of feed is Hf/lb. L is thick liquor taken out from evaporator with XL weight fraction of solute and hL enthalpy in Btu/lb. V lb is vapor with a solute concentration of y and enthalpy H Btu/lb. In most evaporators, the vapor is water, and therefore $Y = 0$.

Mass balance

$$F = L + V$$

For solute alone,

$$F \times XF = L \times XL + V \times Y \quad \dots (1)$$

Let S lb of steam is supplied with enthalpy HS Btu/lb. Let S lb condensate leaves the heating surface with enthalpy hC Btu/lb.

One assumption is usually made in evaporators is that there is very little cooling of condensate. This is never more than a few degrees in practice and the sensible heat recovered from cooling the condensate is so small compared to latent heat of the steam supplied to the heating surface that this sensible heat is usually neglected. This amounts to assuming that the condensate will leave at the condensing temperature of the steam.

Heat balance

Heat in = Heat out

[Heat in feed] + [Heat in steam] = [Heat in thick liquor] + [Heat in vapor] + [Heat in condensate] + [Heat loss by radiation]

Neglecting heat loss by radiation,

$$F\times hF + S\times HS = L\times hL + V\times H + S\times hC \quad \dots (2)$$

Performance of Evaporators [capacity] and Steam Economy

The performance of a steam-heated evaporator is measured in terms of its capacity and economy. Capacity is defined as the number of kilograms of water vaporized per hour. Economy (or steam economy) is the number of a kilogram of water vaporized from all the effects per kilogram of steam used. For a single effect evaporator, the steam economy is about 0.8 (<1). The capacity is about n-times that of a single effect evaporator and the economy is about $0.8n$ for an n-effect evaporator. However, pumps, interconnecting pipes, and valves are required for the transfer of liquid from one effect to another effect which increases both equipment and operating costs.

Boiling point Elevation (BPE)

Most evaporators produce concentrated liquor having a boiling point considerably higher than that of pure solvent (or water). This phenomenon is called boiling point elevation (BPE). BPE occurs as the vapor pressure of a solution (usually aqueous solution) is less than that of the pure solvent at the same temperature. The boiling point of a solution is a colligative property. It depends on the concentration of solute in the solution for a pair of solute and solvent. BPE of the concentrated liquor reduces the effective temperature driving force compared to the boiling of the pure solvent. Equilibrium vapor generated from a solution exhibiting boiling point elevation is superheated with respect to vapor generated during boiling of the pure solvent. The vapor is generated at the solution boiling point, which is higher than the pure component boiling point. The vapor, however, is solute-free, so it won't condense until the extra heat corresponding to the elevation is removed, thus it is superheated. Therefore, the BPE of the concentrated solution must be known for evaporator design.

Determination of BPE

For strong solutions, the BPE data is estimated from an empirical rule known as the Dühring rule. This states that the boiling point of a given solution is a linear function of the boiling point of pure water at the same pressure. Thus, if the boiling point of the solution is plotted against the corresponding boiling point of pure water at the same pressure, a straight line is generated. Different lines are obtained if such plots made for solutions of different concentrations. The main advantage is that Dühring lines can be drawn if boiling points of a solution and water at two different pressures are known. This line can be used to predict the boiling point of a solution at any pressure. A Dühring plot for the NaOH-water system is given below

Duhring graph

Duhring graph is on the left side. RHS graph stands for the Enthalpy concentration chart for caustic soda.

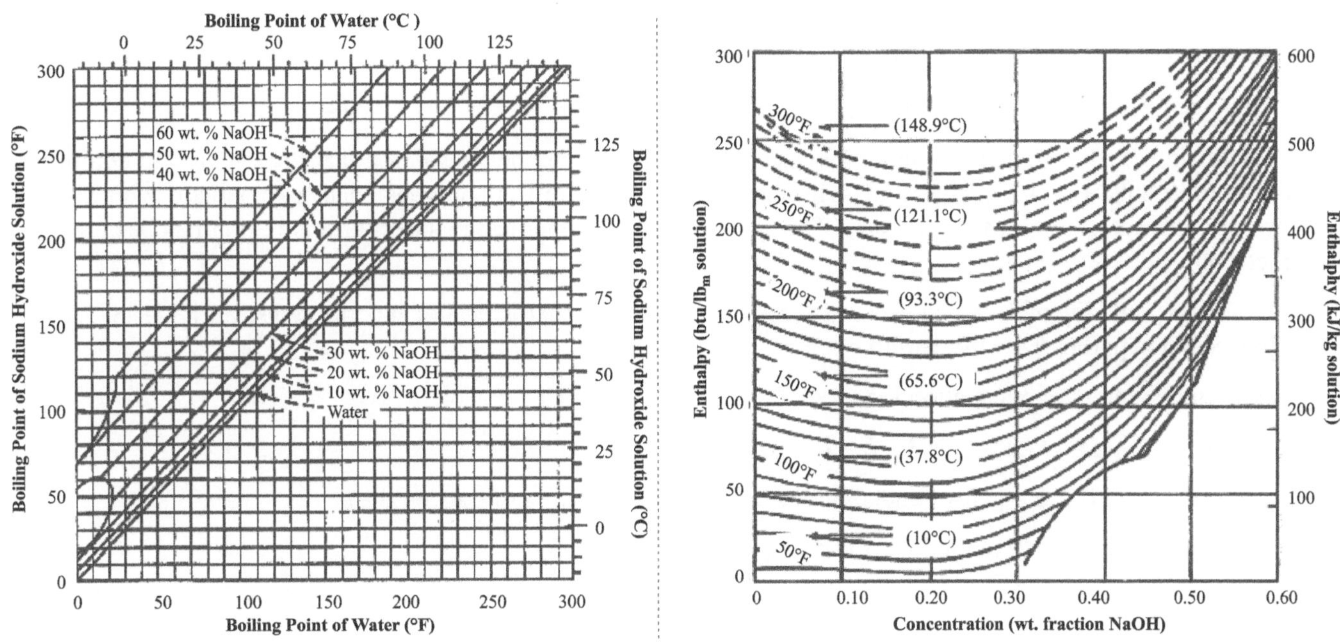

Figure 5: Dühring Lines for Aqueous Solutions of Sodium Hydroxide

Figure 6: Enthalpy-Concentration Chart for the System NaOH-water [Reference state liquid water at 0°C (273 K) or 32°F .] [From W.L. McCabe, Trans. A.I.Ch.E., 31, 129 (1935). With permission]

Solution of Problem

Problem statement: A given evaporator is to be fed with 10,000 lb/hr of a 20% solution of sodium hydroxide at 100 deg F. This is to be concentrated to 50% solution. The evaporator is supplied with saturated steam at 5 psig and operates with vapor space at a vacuum of 26 in. There is negligible entrainment. Condensate may be assumed to leave at steam temperature and losses by radiation are neglected. What is the steam consumption? If overall heat transfer coefficient $U = 400$, what is the heating surface required?

Given [For symbols refer to heat and mass balance diagram above]

$F = 10{,}000$ lb/hr, $XF = 0.20$, $tF = 100$ degF, $XL = 0.50$, $y = 0$.

From steam table, $HS = 1156$ Btu/lb, $hC = 196$ Btu/lb, $tS = 228$ deg F

Vapor space at a vacuum of 26 in.

The boiling point of water at 4-in abs pressure [26 -in vacuum] = 125.4 deg F

Enthalpy of saturated steam at 125 deg $F = 1116$ Btu/lb [from steam table]

The boiling point of concentrated solution = 198 deg F at the boiling point of water at 125.4 deg F from Duhring graph below on the left side. RHS graph stands for the Enthalpy concentration chart for caustic soda.

From Enthalpy concentration chart for caustic soda, on RHS,

$hF = 56.5$ Btu/lb, $hL = 222$ Btu/lb.

Material Balance

$F \times XF = L \times XL + V \times Y$

$[10,000] \times [0.2] = 0.5 \times L$

$L = 4000$ lb/hr, therefore $V = 6000$ l b/hr

Enthalpy Balance

Steam Consumption

From equation [2]

$F \times hF + S \times HS = V \times H + L \times hL + S \times hC$

$$\text{Or, } F \times hF + S \times [HS - hC] = V \times H + L \times hL \quad \ldots (3)$$

In calculating H, the enthalpy of vapor leaving the solution it must be considered that the vapor is in equilibrium with the boiling solution at a pressure of 4 in absolute and therefore is superheated in comparison with vapor in equilibrium with water at the same pressure. The specific heat of superheated steam at this region has been taken as 0.46, therefore

H = 1116 [enthalpy of saturated vapor at 125.4 degF] + 0.46 [specific heat of superheated steam] × [198–125.4] [degree of superheat] = 1149 Btu/lb.

Substituting value of H in equation [3]

$F \times hF + S \times [HS - hC] = V \times H + L \times hL$

[10,000] × [56.5] + S × [1156 (enthalpy of steam) – 196 (enthalpy of condensate)]

= [6000 (vapor) × 1149 (enthalpy of vapor) + [4000 (liqor) × 222] [image 4]

$S = 7520$ lb/hr

Steam Economy

Water evaporated:

Initial water = 10000 × 0.80 = 8000 lb/hr [2000 lbs solid and 8000 lb water, 20% solution]

Final water = 4000 × 0.5 = 2000 lb solid and 2000 lb/hr water, 50% solution

Water evaporated = 8000 – 2000 = 6000 lb/hr

Steam economy = 6000/7520 = 0.8

Heating Surface

$Q = U \times A \times \Delta t$

[7520] × – [1156–196] = [400] (U = 400) × A × [228 (steam temperature) – 198] [image 3, Duhring lines for sodium hydroxide]

$A = 600$ sq ft.

SIZING OF VACUUM EQUIPMENT

There are two factors that decide broadly the size of a vacuum system [1] Vacuum requirement in the process and [2] compression ratio of the ejector. A key performance measure is the compression ratio: the ratio of the discharge pressure to the suction pressure (note that the pressure of the motive steam is not included). A single ejector stage can achieve compression ratios up to 8:1, although values in the 3:1 to 5:1 range are more typical. The discharge pressure is set by the condenser pressure—minimum pressure is the condensing pressure of steam at the vapor outlet temperature.

Imagine a typical vacuum system like in the image on the left side. The process vacuum requirement is 748 mm or 12 mm pressure. Suppose the compression ratio of the ejector is 4:1. Then there is a requirement of three ejectors. You can see how the ejectors are connected in series with a barometric condenser between two ejectors. The barometric condenser is employed as a means of removing air and other vapors from vacuum equipment. The principal feature of the barometric condenser is it is a direct gas-water condenser. It is located at a barometric height of 34 feet above the ground. One atmosphere 760 mm is equivalent to the pressure caused by the weight of a column of freshwater of

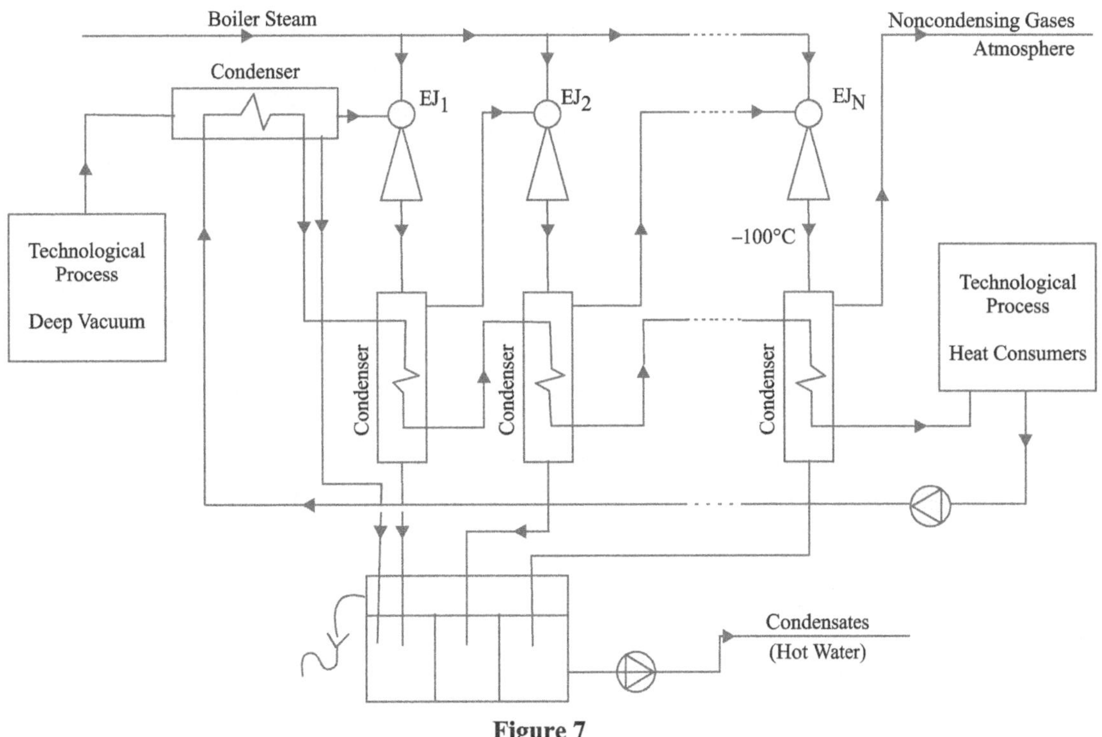

Figure 7

approximately 33.8 ft. Even at 0 mm pressure, a barometric condenser can discharge water to atmospheric pressure freely. The important point is that the discharge pressure of the last stage ejector should be greater than 760 mm. The last stage ejector finally discharges the non-condensable into the atmosphere. In order to travel from the first ejector to 3rd ejector to discharge non-condensable to the atmosphere, you require vacuum seals between the process vessel and 1st, 1st and 2nd, and 2nd and 3rd ejectors. Barometric condenser acts as a vacuum seal, and so you have three barometric condensers each has a leg discharging water into a common hot well.

First ejector

Multiplies pressure at outlet = 12 × 4 = 48 mm.

Second ejector

Multiplies pressure at outlet = 48 × 4 = 192.

Third ejector

Multiplies pressure at outlet

$$192 \times 4 = 768 \text{ mm}$$

At the final 3rd ejector at 768 mm pressure steam is discharged to the atmosphere.

You can notice as you go from process vessel onwards, then progressively the pressure in the barometric condenser increases or the vacuum creation by ejectors diminishes. Ejector needs to do less work as you go from left to right of the above image. Correspondingly mass flow of steam/mass flow of water in barometric condenser reduces. Each ejector is designed to cope with inlet-outlet pressure.

STEAM EJECTOR: A TWO-PHASE VACUUM DEVICE

An ejector is a two-phase flow device. In most typical ejectors, steam is used as the motive fluid, while non-condensable gases are drawn from the vessel as the entrained fluid.

A working fluid usually steam, enters an ejector through a nozzle into a converging and diverging tube that initially has a smaller cross-sectional area and then increases in size. Due to Bernoulli's principle, the fluid leaving the nozzle moves at a high velocity, which causes it to have low pressure and create a vacuum. In the outer converging tube, the steam and non-condensable gases drawn from the vessel gases mix together. The diverging tube imparts enough velocity for it to be ejected at the throat. The tube then typically expands to reduce the velocity. The reduced velocity provides the pressure to smoothly increase to the external pressure.

Is steam Expansion Through the Nozzle Adiabatic/Isentropic?

In a best-case scenario, it is assumed that the expansion of steam from a convergent nozzle is adiabatic and isentropic.

Adiabatic expansion means that no heat or energy is transferred to or from the surrounding environment during the expansion process. This calls for a system that is completely heat-insulated. Isentropic expansion denotes a change in entropy that is equal to zero during the expansion. This suggests that the process of expansion is reversible and does not result in any energy losses as a result of friction, turbulence, etc.

However, in practice, energy is always lost during the expansion process as a result of friction, turbulence, and other undesirable circumstances. As a result, the expansion process might not always be adiabatic or isentropic in real life.

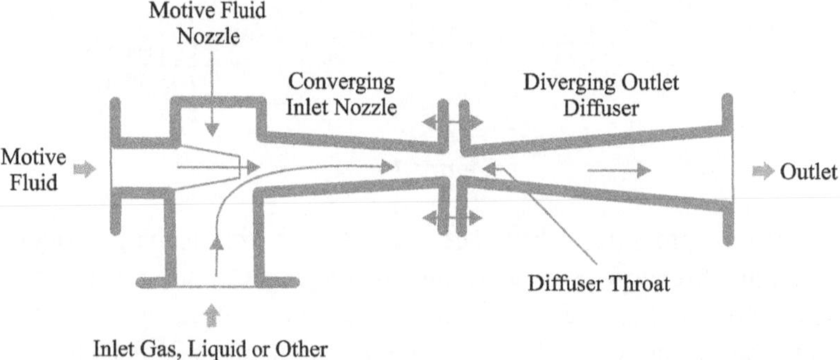

Figure 8

Details

The Purpose of the Nozzle

There are two things.

In order to transform high-pressure steam energy into kinetic energy in the form of high-velocity steam, a nozzle is used in a steam ejector. A low-pressure area is then created in the mixing section by the high-velocity steam, which draws gases or vapours from the process into the ejector.

The nozzle also serves to maintain a constant vacuum level and regulate the rate of steam flow into the ejector. The operating conditions of the ejector can be controlled by adjusting the nozzle, enabling process optimisation and effective operation.

Nozzle Selectively Draws Non-condensable Gases

In order to avoid entraining water vapour and only be effective for sucking non-condensable gases, ejectors are made to maintain a pressure above the vapour pressure of water.

The Purpose of the Convergent Tube

It's a mixing tube. Entrainment of steam-noncondensable gases takes place in the convergent part of the ejector, away from the nozzle. Steam enters through the nozzle and noncondensable gases are sucked directly into the convergent tube [image]. They mix together in the convergent tube

Post-entrainment Steam-noncondensable behaviour

The velocity of the combined mass reduces.

There is more than one reason.

The combined mass flow rate of the non-condensable gases and steam rises as they combine. As a result, the gas mixture's total momentum rises, necessitating a larger cross-sectional area of the convergent section to keep the mass flow velocity constant. Since the cross-section remains constant the velocity reduces.

The steam moves slower than non-condensable gases

Another element that affects the velocity of the fluids in the convergent tube is density. Because steam is much more dense than non-condensable gases, it expends more energy (and moves at a slower speed) to move through a given space. Non-condensable gases, on the other hand, have a lower density and can move more readily at greater speeds.

How do two different velocities steam and noncondensable gases help to mix?

There are two things

As a result of this velocity difference, turbulence is produced, causing steam and non-condensable gases to mix more thoroughly. The chance for the high-velocity steam to entrain the low-velocity non-condensable gases increases with the thoroughness of the mixing, increasing the ejector's overall suction capacity.

Additionally, the convergent tube makes sure that the steam flow is constant and axial, which improves the mixing of the two fluids.

Steam does not condense in the convergent tube

To avoid any condensation of the steam, the steam ejector system is designed with a pressure that is always higher than atmospheric pressure. This is due to the fact that steam can only exist in a gaseous state when its saturation pressure—which is correlated with temperature and can be found in steam tables—is at or above that level.

The steam would start to condense (into water) due to the drop in temperature if the pressure in the ejector system dropped below the saturation pressure. In the ejector system, condensation can lead to a variety of problems, including decreased efficiency and potential component damage.

In order to avoid this, the steam is typically supplied to the ejector system at high pressure and velocity. The pressure is then maintained at this high level throughout the ejector system to keep the steam in a gaseous state.

What Happens at the Throat

The stream flowing through the convergent section of the ejector reaches its maximum velocity and a corresponding decrease in pressure at the throat of the ejector. Due to the conservation of energy, which states that fluid velocity increases as the cross-sectional area between nozzle walls decreases, this happens.

The fluid enters a zone of intense mixing and turbulence as it moves through the throat. This mixing effect results from the fluid's sudden expansion as it moves from the converging section to the diffusing section's larger diameter.

The Purpose of the Divergent Tube

Divergent Tube carries out two essential tasks.

The fluid's velocity is decreased and its kinetic energy is transformed into pressure energy in the divergent section. As a result, the fluid's pressure rises, entrained fluid accelerates, and overall efficiency is increased.

Pressure restoration in order to recover pressure, an ejector's divergent section is crucial. The fluid's velocity decreases and pressure rise as it expands in the divergent section. Due to friction and turbulence losses, some of the pressure that was lost in the convergent section is recovered in part by this increased pressure.

CHAPTER 43

Mass Transfer

Mass transfer and thermodynamics play an important role in unit processes like distillation and leaching/extraction. In distillation, mass transfer is essential for separating components based on their boiling points, while thermodynamic principles determine the efficiency of separation. Leaching and extraction processes rely on mass transfer to transfer solutes between phases, with thermodynamics governing the equilibrium and selectivity of the process. The design and optimization of distillation columns or leaching/extraction equipment involve balancing mass transfer rates with thermodynamic considerations to achieve desired separation efficiencies. Understanding the interplay between mass transfer and thermodynamics is essential for improving the performance and efficiency of unit processes like distillation and leaching/extraction.

MASS TRANSFER: THE MINIMUM ONE SHOULD KNOW

What is Mass Transfer?

Mass transfer is the movement of mass from one place to another, often involving the diffusion of a substance from an area of higher concentration to an area of lower concentration. This process is important in many industrial and natural processes, such as distillation, absorption, and adsorption.

Explanation

Mass transfer is a result of random molecular motion or circulating currents in a fluid. It can occur in a single phase or across phase boundaries.

Types

The two main types of mass transfer are diffusion and convection.

- **Diffusion:** The microscopic result of random molecular motion.
- **Convection**: The transport of material between a boundary surface and a moving fluid.

Importance

Mass transfer is important in many industrial processes, such as distillation, evaporation, drying, and membrane filtration.

One should know about mass transfer because it is essential in many fields of engineering, such as chemical engineering, environmental engineering, and materials.

Explanation

The how of mass transfer involves understanding the mechanisms by which mass is transferred, such as diffusion, convection, and mass transfer coefficients. It also involves knowing the various factors that can affect mass transfer, such as temperature, pressure, and concentration gradients.

Equations that are commonly used in mass transfer include Fick's law of diffusion, which relates the rate of diffusion to the concentration gradient, and the mass transfer coefficient, which relates the rate of mass transfer to the driving force.

Mass Transfer Coefficient

Explanation

Mass transfer coefficient is a measure of the rate at which a substance is transferred from one phase to another (e.g. from a gas to a liquid, from a liquid to a solid). It is expressed in units of m/s or cm/s.

It is a proportionality constant that describes how quickly a substance moves from one phase to another. It's a diffusion rate constant that relates the mass transfer rate, mass transfer area, and concentration change.

The mass transfer coefficient is important because it helps us understand how variables like temperature, concentration gradient, and fluid velocity affect the transfer process. It's also used to design and manufacture separation process equipment, and to estimate what will happen in real-world situations.

The significance of mass transfer coefficient lies in its ability to quantify the efficiency of mass transfer processes. By knowing the mass transfer coefficient, one can better understand and optimize processes such as distillation, extraction, absorption, and adsorption.

The mass transfer coefficient can be estimated using theoretical equations, correlations, and analogies. The most applicable model depends on the materials and the system being studied

Mass transfer applications

Common examples of mass transfer are evaporation, distillation, extraction, leaching, etc.

MASS TRANSFER: THREE FUNDAMENTAL LAWS

The three fundamental mass transfer laws are following:

(1) Fick's law of diffusion, (2) Fourier's law of conduction, and (3) Newton's law of viscosity.

What is common in them?

The common in these laws in a sense is that they all describe the rate of transfer of a specific property (e.g. mass, heat, viscosity) through a material or medium.

Ficks law

Fick's law of diffusion relates the rate of diffusion of a species (mass transfer) to the concentration gradient in a medium. It states that the rate of diffusion is directly proportional to the concentration gradient and the diffusion coefficient.

Fourier's law

Fourier's law of conduction describes the rate of heat transfer through a material. It states that the rate of heat transfer is directly proportional to the temperature gradient and the thermal conductivity of the material.

Newton's law of viscosity

Newton's law of viscosity describes the relationship between the shear stress and the velocity gradient in a fluid. It states that the shear stress is directly proportional to the velocity gradient and the viscosity of the fluid.

Equations and Explanation

Fick's Law of Diffusion

Fick's first law of diffusion describes the rate of diffusion of a substance through a medium. The equation is given by:

$$J = -D \times dC/dx$$

Where:

J is the mass flux (amount of substance passing through a unit area per unit time),

D is the diffusion coefficient,

C is the concentration of the substance,

x is the distance.

The negative sign indicates that the mass transfer is from higher concentration to lower concentration. Fick's law of diffusion essentially states that the rate of diffusion is directly proportional to the concentration gradient.

Fourier's Law of Conduction

Fourier's law of heat conduction describes the rate of heat transfer through a material. The equation is given by:

$$q = -k\, dT/dx$$

Where:

q is the heat flux (amount of heat passing through a unit area per unit time),

k is the thermal conductivity of the material,

T is the temperature,

The negative sign indicates that heat flows from higher temperature to lower temperature. Fourier's law of conduction states that the rate of heat transfer is directly proportional to the

Newton's Law of Viscosity

Newton's law of viscosity describes the flow behavior of a fluid under shear stress. The equation is given by:

$$tau = mu \times du/dy$$

Where:

tau is the shear stress,

mu is the dynamic viscosity,

du/dy is the velocity gradient.

Newton's law of viscosity states that the shear stress is directly proportional to the velocity gradient, with the dynamic viscosity of the fluid being the proportionality constant.

MASS AND HEAT TRANSFER COUPLING

Heat and mass transfer are coupled.

Mass and heat transfer coupling refers to the simultaneous transfer of mass and thermal energy between two bodies or substances. When a system undergoes a process involving both mass transfer (such as diffusion or convection) and heat transfer (such as conduction or convection), the two phenomena are often interconnected and can influence each other.

For example, in situations where a substance is being heated or cooled while also undergoing a phase change (such as evaporation or condensation), the mass transfer rate and heat transfer rate are closely related. The amount of heat required for a phase change to occur is directly related to the mass transfer rate of the substance.

Similarly, in processes involving chemical reactions, the transfer of mass and heat are usually coupled together as the chemical reactions release or absorb heat. This can result in changes in the temperature and concentration profiles within the system.

Fick's and Fourier's laws are fundamentals to mass and heat coupling

Both Fick's law and Fourier's law involve concept of gradients.

Fick's law states that rate of mass transfer through a substance is directly proportional to concentration gradient, i.e., the difference in concentration of the substance between two points. This law describes how molecules or particles move from regions of high concentration to regions of low concentration.

On the other hand, Fourier's law states that the rate of heat transfer through a substance is directly proportional to the temperature gradient, i.e., the difference in temperature between two points. This law describes how heat flows from regions of high temperature to regions of low temperature.

Therefore, both laws rely on the concept of gradients to explain the movement of mass and heat in a system.

Example

A typical most simple example is a distillation column

In distillation columns, simultaneous mass and heat transfer occur at each tray.

Vapor and liquid are at equilibrium at each tray. Hot vapor flowing in an upward direction loses heat to liquid on each tray and because of this heat transfer there will be a mass transfer of key components from liquid to vapor and the vice versa heavy component will condense out from vapor and get into liquid due to cooling effect.

Fluid systems, in contrast to solid systems, allows for diffusive transport of molecules which often are coupled with temperature gradients.

Gases transfer heat only by the transport of the molecular species since it is the collision of mostly non-interacting gas molecules that transfers heat.

The energy of a gas is nearly all in the form of kinetic energy that each molecule possesses from its momentum, $K = p^2/(2m)$.

For gases, there is therefore a fundamental coupling of mass and heat transfer. For this reason, mass transfer coefficients and heat transfer coefficients tend to be of a similar order in gas systems.

HENRY'S, RAOULT'S, AND DALTON'S LAWS: THREE CLOSE LAW CALCULATIONS

To describe the chemical behaviour of solutions in the presence of vapour pressure, both Henry's and Raoult's laws are used. The total pressure exerted by a gas mixture equals the sum of the partial pressures exerted by each individual gas in the mixture, according to Dalton's partial pressure law. In some ways, Dalton's and Raoult's laws are very similar. Raoult's law is concerned with solids or liquids' vapour pressure, whereas Dalton's law is concerned with the partial pressure of non-reacting gases. When it comes to applying these three laws, they frequently lead to misunderstanding.

Henry's law

According to Henry's law, the amount of dissolved gas in a liquid is directly proportional to its partial pressure above the liquid. Henry's law constant is the proportionality factor.

$$C = KH \text{ Pgas}$$

Where,

C is the solubility of a gas at a fixed temperature in a particular solvent (in units of M or mL gas/L).

KH is Henry's law constant (often in units of atm-m^3/mol).

Pgas is the partial pressure of the gas (often in units of atm).

Henry's Law Constant [K]

Henry's law constant (KH) (also known as the air-water partition coefficient) is the ratio of a compound's partial pressure in air to its concentration in water at a given temperature. Values for Henry's law constants are expressed in atmospheres for air to moles per cubic metre for water (atm-m^3/mol) or in a dimensionless unit described as $KH' = KH/(RT)$, where KH' is the dimensionless unit. K is Henry's law constant (atm-m^3/mol), R is the ideal gas constant (8.20575 10^{-5} atm-m^3/mol-K), and T is the water temperature (Kelvin). Compounds with Henry's law constant greater than 10^{-3} atm-m^3/mol and a molecular weight less than 200 grammes per mole are considered volatile.

A compound with Henry's law constant less than about 5×10^{-5} atm–m^3/mol is considered soluble and tends to remain in the water.

Henry's law constant of solubility of gases in water at 298.15 K

Table 1

Species	K_H^{cc}	$1/(1+K_H^{cc})$	$1/(1+1/K_H^{cc})$
He	1.514×10^{-5}	0.99999	1×10^{-5}
O_2	5.32×10^{-4}	0.9995	5×10^{-4}
N_2	2.66×10^{-5}	0.99997	3×10^{-5}
O_3	4.91×10^{-5}	0.99995	5×10^{-5}
NO	7.77×10^{-5}	0.99992	8×10^{-5}
NO_2	1.68×10^{-3}	0.998	2×10^{-3}
N_2O	1.02×10^{-3}	0.999	1×10^{-3}
NO_3	7.37×10^{-2}	0.93	7×10^{-2}
N_2O_5	8.59×10^{-2}	0.92	0.08
H_2	3.19×10^{-5}	0.99997	3×10^{-5}
OH	1.19	0.457	0.543
HO_2	233	4×10^{-3}	0.996
H_2O_2	2905	3×10^{-4}	0.9997
HNO_2	2.01	0.33	0.67
HNO_3	8593	1×10^{-4}	0.9999

Example

Determination of solubility of oxygen in water using Henry's Law at 298K

Henry's law states that the solubility of a gas in a liquid is directly proportional to the partial pressure of that gas above the liquid. The mathematical expression for Henry's law is:

$$C = k \times P$$

where:

C = concentration of the dissolved gas in the liquid (mol/L)

k = Henry's law constant (mol/L-atm)

P = partial pressure of the gas above the liquid (atm)

For oxygen (O_2) in water at 25°C, the Henry's law constant is approximately 1.3×10^{-3} mol/L–atm.

Let's assume the partial pressure of O_2 above the liquid is 0.2 atm. We can calculate the solubility of O_2 in water using Henry's law:

$$C = (1.3 \times 10^{-3}) \times 0.2$$

$$C = 2.6 \times 10^{-4} \text{ mol/L}$$

Therefore, the solubility of O_2 in water at 25°C and a partial pressure of 0.2 atm is 2.6×10^{-4} mol/L.

Limitations of Henry's Law

Henry's law only works if the molecules are at equilibrium.

Henry's law does not work for gases at high pressures.

Henry's law does not work if there is a chemical reaction between the solute and solvent (e.g., HCl (g) reacts with water by a dissociation reaction to generate H_3O^+ and Cl^- ions).

Raoult's Law

Where two volatile liquids A and B are mixed with each other to form a solution, the vapor phase consists of both components of the solution. Once the components in the solution have reached equilibrium, the total vapor pressure of the solution can be determined by combining Raoult's law with Dalton's law of partial pressures to give.

[Raoult's and Dalton's laws combine here]

$$P = P{\times}a\, Xa + P{\times}b\, Xb$$

P is the vapor pressure of the solution. $P{\times}a$ and $P{\times}b$ are the partial pressures of components, a and b. $X\,a$ and B are the mole fractions of a and b.

In other words, the vapor pressure of the solution is the mole-weighted mean of the individual vapor pressures:

If a non-volatile solute B (it has zero vapor pressure, so does not evaporate) is dissolved into solvent A to form an ideal solution, the vapor pressure of the solution will be lower than that of the solvent. In an ideal solution of a non-volatile solute, the decrease in vapor pressure is directly proportional to the mole fraction of the solute:

$$P = P \times a\, Xa$$

Fundamentals

Raoult's law is a phenomenological relation that assumes ideal behavior based on the simple microscopic assumption that intermolecular forces between unlike molecules are equal to those between similar molecules, and their molar volumes are the same: ideal solution conditions. This is analogous to the ideal gas law, a limiting law that holds true when the interactive forces between molecules approach zero, such as when the concentration approaches zero. Raoult's law, on the other hand, holds true if the physical properties of the components are identical. The more similar the components are, the more their behavior resembles

Raoult's law, for example, is essentially exact if the two components differ only in isotopic content.

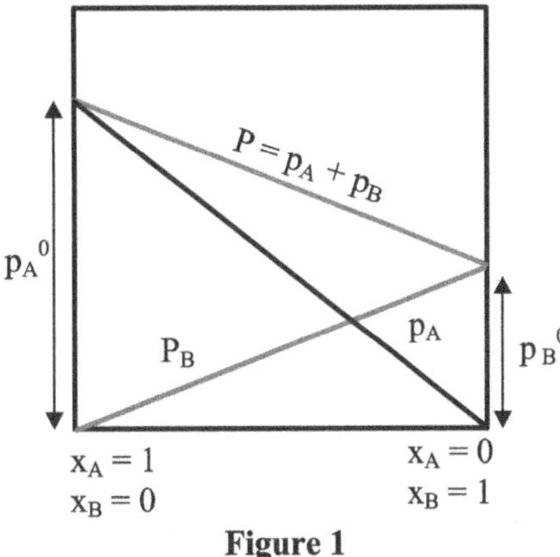

Figure 1

Vapor pressure of a binary solution that obeys Raoult's law. The black line shows the total vapor pressure as a function of the mole fraction of component B, and the two green lines are the partial pressures of the two components.

When measured vapour pressures are compared to predicted values from Raoult's law, the true relative strength of intermolecular forces is revealed. If the vapor pressure is lower than predicted (a negative deviation), indicates that the forces between unlike molecules are stronger. Positive deviations have the opposite effect.

Calculation of Vapor Pressure Using Raoult's Law

Let's assume a hypothetical binary system consisting of two ideal liquids A and B. According to Raoult's law, the vapor pressure of a component in a mixture is directly proportional to its mole fraction in the liquid phase. Mathematically, Raoult's law can be expressed as follows:

$$P = P\_0A \times x\_A + P\_0B \times x\_B$$

where P is the total vapor pressure of the mixture, P_0A and P_0B are the vapor pressures of pure components A and B, respectively, and x_A and x_B are the mole fractions of components A and B in the liquid phase.

Let's assume that the mole fraction of component A is 0.6 and the mole fraction of component B is 0.4. The vapor pressures of pure components A and B are 100 kPa and 50 kPa, respectively. Using Raoult's law, we can calculate the total vapor pressure of the mixture as follows:

$$P = (100 \text{ kPa} \times 0.6) + (50 \text{ kPa} \times 0.4)$$

$$P = 60 \text{ kPa} + 20 \text{ kPa}$$

$$P = 80 \text{ kPa}$$

Therefore, the total vapor pressure of the mixture is 80 kPa.

Dalton's Law

Dalton's law, the statement that the total pressure of a mixture of gases is equal to the sum of the partial pressures of the individual component gases. The partial pressure is the pressure that each gas would exert if it alone occupied the volume of the mixture at the same temperature. Raoult's law is concerned with the vapor pressure of solids or liquids, whereas Dalton's law is concerned with the partial pressure of non-reacting gases. Dalton's law is important for understanding how gases move down their pressure differences by diffusion. Henry's law helps explain how the solubility of a gas relates to its diffusion. Each gas in a mixture of gases exerts its own pressure as if no other gases were present.

RAOULT'S AND DALTON'S LAW

These are two important basic concepts in distillation. According to Raoult's law, in an ideal solution of two (example) volatile solvents, the partial pressure of each component equals its mole fraction in the liquid multiplied by its vapor pressure.

Raoult's law

According to Raoult's law, the following will apply.

If A is one such solvent in a solution with another volatile solvent B, the partial vapor pressure of A = Mole fraction of A in A's liquid × Vapor pressure. Let's say it is Pa.

Similarly, let's say the partial vapor pressure of B is Pb

By using Raoults law one can find out the partial pressure of each component in an ideal solution of volatile solvents.

Dalton's Law

According to Dalton's law, the total pressure is the sum of partial pressures.

Combination of Raoult's and Dalton's law

If you apply the two laws in the above case then you get the total pressure, let's call it $P = Pa + Pb$.

Let's Take an Example

" Let's take a non-ideal solution of water and ethanol at a 1:1 mole ratio. The vapor pressure of water is 760 mm and the vapor pressure of ethanol is 45 mm Hg. The question is: predict total pressure and how it will deviate from Raoult's law in terms of positive or negative deviation"

Had this been the ideal solution following would have applied.

In an ideal solution, the total vapor pressure of the solution would be the sum of the partial pressures of each component, as calculated using Raoult's Law.

Using Raoult's law, the partial pressure of water in the solution would be 0.5 × 760 mm Hg = 380 mm Hg, and the partial pressure of ethanol would be 0.5 × 45 mm Hg = 22.5 mm Hg. Therefore, the total vapor pressure of the solution according to Raoult's Law would be 380 mm Hg + 22.5 mm Hg = 402.5 mm Hg.

However, it's a non-ideal solution as water and ethanol have structural dissimilarities. Their polarities are different with water being more polar than ethanol.

Here, the total vapor pressure may deviate from Raoult's Law due to interactions between the molecules of water. Remember in a solution if you reduce moles of water, you reduce H bonds and increase vapor pressure.

P = Mole fraction × Pa as explained above does not apply. Pa does not remain vapor pressure of pure solvent anymore. As explained, the following happens:

As a result, in this non-ideal solution of water and ethanol at a 1:1 mole ratio, the interactions between molecules in the mixture are weaker than what would be expected based on Raoult's law for ideal solutions.

This suggests that the total pressure in the system would deviate positively from Raoult's law.

How and Why?

Due to the lesser polarity of the water and ethanol mixture, there will be fewer *H* bonds in the solution. This will increase the vapor pressure because there are fewer *H* bonds to break.

What is a Positive Deviation from Raoult's Law?

Positive deviation from Raoult's law occurs when the observed vapor pressure of a solution is higher than what is expected based on Raoult's law. This happens when the *A–B* molecular interaction forces are weaker than the average of the attraction in the pure component of the mixture.

HENRY'S LAW AND RAOULT'S LAW

Both these laws apply to distillation. At small solute concentrations, Henry's law plays a role. Raoult's law becomes dominant at higher solute concentrations.

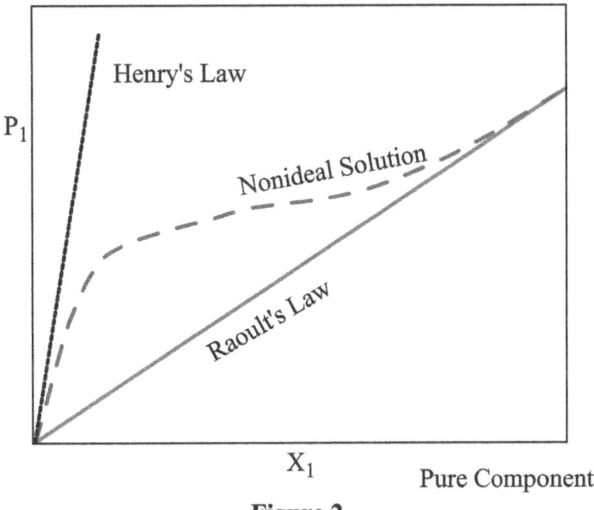

Figure 2

Both Henry's law and Raoult's law are related to the behavior of mixtures of substances. Henry's law describes the behavior of dilute solutions, where the solute is present in small amounts compared to the solvent.

It states that the solubility of a gas in a liquid is directly proportional to the partial pressure of the gas above the liquid.

Raoult's law, on the other hand, describes the behavior of ideal solutions, where the interactions between the molecules of the solute and solvent are similar. It states that the vapor pressure of a component in a mixture is directly proportional to its mole fraction in the mixture.

While they describe different behaviors, both Henry's law and Raoult's law can be seen as limiting cases of ideal behavior. In a real system, deviations from these laws may occur due to factors such as non-ideal interactions between molecules.

Therefore, although Henry's law and Raoult's law describe different aspects of the behavior of mixtures, they can be seen as related in that they both represent idealized behaviors that may not always be observed in real systems.

Henry's law and Raoult's law are often referred to as limiting laws because they represent idealized behaviors that are only observed under specific conditions. In the case of Henry's law, it is a limiting law because it applies specifically to dilute solutions where the solute is present in small amounts compared to the solvent. At higher concentrations, the interactions between the solute and solvent molecules may become significant and lead to deviations from the linear relationship described by Henry's law. Similarly, Raoult's law is a limiting law because it applies only to ideal solutions where the interactions between the solute and solvent molecules are similar.

In real systems, deviations from Raoult's law can occur due to non-ideal interactions between molecules. Therefore, Henry's law and Raoult's law are considered limiting laws because they represent idealized cases that may not always be observed in real-world systems.

Despite their limitations, these laws provide important insights into the behavior of mixtures.

RELATIVE VOLATILITY: EFFECT OF TEMPERATURE AND PRESSURE

Summary

Relative volatility decreases with temperature and pressure. When you reduce the pressure the change in the boiling point of the lower boiling point component is large compared to the high boiling point component so the difference in the boiling point of the low boiling point component and high boiling point component increases, so relative volatility increases.

When you increase the pressure then the lower boiling point component's boiling point increase significantly but the higher boiling point component's boiling point increase by a negligible amount so the difference in the boiling point of the higher boiling component and lower boiling component decreases so relative volatility decreases.

When relative volatility decreases, it requires more stages for the same purity of the product. As a result, the height of the distillation column will rise.

We choose a packed bed column for lesser pressure and a tray column for higher pressure. So, if we increase the pressure to a particular point, we must adjust the internals of the column. If we raise the pressure, we must also increase the reflux ratio in order to achieve the desired purity.

General

What is Relative Volatility?

Relative volatility is a measurement that indicates how easily a liquid mixture can be separated into its components through distillation. It's used in the design of all types of distillation processes, as well as other separation and absorption processes.

Definition

Relative volatility is the ratio of the vapor pressures of the components in a liquid mixture at a given temperature.

Use

Relative volatility is used in the design of distillation processes to separate more volatile components from less volatile components. It's also used in other separation and absorption processes.

Significance

A relative volatility greater than 1 means that one component is more volatile than the other, making separation.

Calculation

Relative volatility is calculated by multiplying the pure component vapor pressures by their activity coefficients.

How relative volatility is expressed?

Relative volatility is expressed in a few different ways:

As a Ratio

Relative volatility is the ratio of the vapor pressure of one liquid component to another in a mixture. The symbol for relative volatility is α.

.As a Mole Fraction

For a binary mixture, relative volatility can be expressed in terms of the mole fraction of the more volatile component in the liquid and vapor phases.

As a Percentage

The Relative Volatility Index (RVI) is a normalized percentage between 0 and 100 that measures the standard deviation of price changes over a period of time. The RVI is calculated by dividing the standard deviation of "up" periods by the standard deviation of "down" periods, and then multiplying by 100.

What Does High or Low Relative Volatility Mean?

In distillation, high relative volatility means that it is easier to separate components of a liquid mixture by distillation.

Explanation

Relative volatility is the ratio of the vapor pressure of one component to another at a given temperature. A high relative volatility indicates that one component is more volatile than other, making separation easier.

What is the Minimum Relative Volatility for Distillation

The minimum relative volatility for distillation is 1.05.

When relative volatility is low, a solvent-based approach may be used instead of ordinary distillation. Distillation columns that perform difficult separations are called superfractionators.

How does Solvent Based Distillation Work?

In extractive distillation, a non-volatile solvent with a high boiling point is added to a distillation column. The solvent increases the relative volatility of mixture by increasing the activities of the components.

How Distillation Process is Adjusted to Suit Relative Volatility

A higher relative volatility means fewer stages are needed in a distillation column to separate the components. For low relative volatility extractive distillation is used that increases relative volatility between components by adding an entrainer.

Details

Relative Volatility

The relative volatility of mixtures of two or more components is equal to their relative vapour pressures for ideal mixtures. In a binary system. The difference between the vapour pressures of the more volatile components of a liquid mixture and the vapour pressures of the less volatile components of the mixture is measured as relative volatility. This metric is commonly used in the construction of large-scale industrial distillation plants.

It basically indicates how easy or difficult it is to separate the more volatile components from the less volatile components in a combination using distillation. To put it another way, the higher the relative volatility of a liquid mixture, the easier it is to distil the mixture components apart. Relative volatility is usually indicated by the letter α (alpha).

$$\alpha_{ij} = P_i/P_j$$

α_{ij} = relative volatility of component i and j

P_i = vapor pressure of pure component i

P_j = vapor pressure of pure component j

The relative volatility is temperature-dependent, although for similar components (two alkanes, for example) it is fairly constant over the temperature range of a distillation column. Strictly, the relative volatility is the ratio of K factors for the components, where for each component, i, $K_i = y_i/X_i$. For ideal components, $K_1/K_2 = P_1/P_2$.

Although relative volatility varies with temperature, it is fairly consistent for similar components (two alkanes, for example) over the temperature range of a distillation column. The relative volatility is defined as the ratio of K factors for the components, where $K_i = y_i/x_i$ for each component. $K_1/K_2 = P_1/P_2$ for perfect components.

K factors in Distillation: Explanation

K values, also known as equilibrium ratios or distribution coefficients, are ratios of one phase's mole fraction to another phase's mole fraction and are temperature and pressure-dependent (and composition as well in non-ideal systems). It is the ratio of the vapour phase to the liquid phase in vapour-liquid systems.

The Relative Volatility of Binary Liquid Mixtures

For a liquid mixture of two components (called a binary mixture) at a given temperature and pressure, the relative volatility is defined as.

$$\alpha = [y_i/x_i/[y_j/x_j] = K_i/K_j$$

α = the relative volatility of the more volatile component i to the less volatile component j

y_i = the vapour-liquid equilibrium concentration of component y in the vapor phase

x_i = the vapor-liquid equilibrium concentration of component x in the liquid

y_j = the vapor-liquid equilibrium concentration of component j in the vapor phase

x_j = the vapor-liquid equilibrium concentration of component j in the liquid phase

y/x = Commonly called the K value or vapor-liquid distribution ratio of a component α.

When their liquid concentrations are equal, more volatile components have higher vapour pressures than less volatile components. Thus, a K value = y/x for a more volatile component is larger than a K value for a less volatile component. That means that $\alpha \geq 1$ since the larger K value of the more volatile component is in the numerator and the smaller K of the less volatile component is in the denominator.

The quantity α is unitless. When the volatilities of both key components are equal, it follows that $\alpha = 1$ and separation of the two by distillation would be impossible under the given conditions. As the value of α increases above 1, separation by distillation becomes progressively easier.

When distilling a binary liquid mixture, full separation of the two components is rare. Typically, the more volatile component dominates the upper fraction of the distillation column with a tiny amount of the less volatile component, while the less volatile component dominates the bottom fraction with a small amount of the more volatile component.

Effect of Pressure and Temperature on Relative Volatility

The relative volatility of the components in a liquid mixture change dramatically as pressure rises. Because an increase in pressure necessitates an increase in temperature, a rise in temperature has an impact on relative volatility.

The image is based on the vapour-liquid equilibrium of a hypothetical binary liquid combination and shows how increasing pressure or temperature reduces the mixture's relative volatility. For a hypothetical binary mixture of liquids (for example, acetone and water or butane and pentane), the image displays four sets of vapour-liquid equilibrium, each set at a different temperature and pressure.

Because pressure does not have a linear relation with the boiling point, relative volatility decreases when pressure is increased. When you increase the pressure then the lower boiling point component's boiling point increase significantly but the higher boiling point component's boiling point

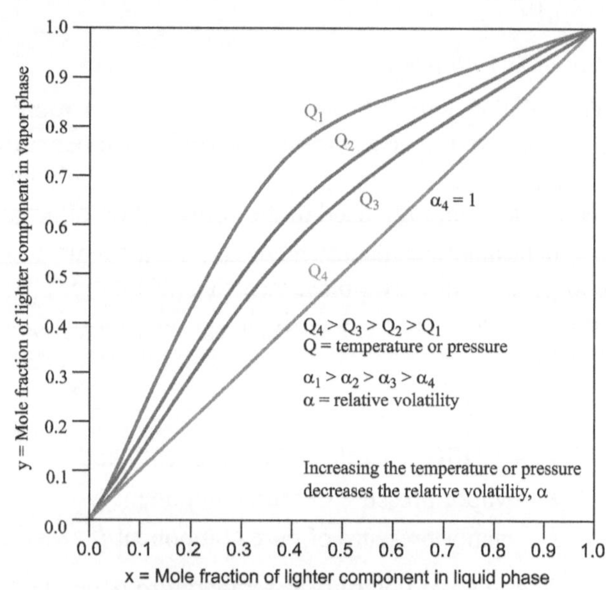

Figure 3: Effect of Temperature or Pressure on Relative Volatility

increase in a negligible amount so the difference in the boiling point of the higher boiling component and lower boiling component decrease so relative volatility decrease. Since the relative volatility of a liquid mixture varies with temperature and pressure, each of the four sets is also at a different relative volatility. Condition 1 is at a temperature and pressure where the relative volatility of the hypothetical binary solution is high enough to make the distillation quite easy. As the temperature and/or pressure (Q in the diagram) increases from condition 1 to condition 4, the relative volatility (α in the diagram) decreases to a value of 1, and separation of the components by distillation is no longer possible since at that point there is no difference in the vapor pressures of the binary components. As can be seen in the diagram at $\alpha = 1$, the mole fraction of the lighter (more volatile) component in the liquid phase is exactly the same as in the vapor phase which means that no separation is possible by distillation.

VLE DIAGRAM AND RELATIVE VOLATILITY [TUTORIAL]

What is a VLE?

The boiling point properties of the components in the mixtures being separated are used to design distillation columns. As a result, the sizes, particularly the heights, of distillation columns are determined by the mixtures' vapor liquid equilibrium (VLE) data.

How does Vapor-Liquid Equilibrium Occur?

The term phase equilibria refer to when two or more phases coexist (in equilibrium). When two phases coexist, they are in a state of saturation. The chemical potential can predict phase stability because the most stable form of the substance has the lowest chemical potential at the given temperature and pressure. The chemical potential of a substance introduced by Gibbs is simply the Gibbs free energy or chemical energy per mole of that substance.

The vapor-liquid equilibrium (VLE) describes the distribution of a chemical species between the vapor and liquid phases. A component distributes itself between vapor and liquid phases in the ratio of mole fractions y/x. This is called the K value of the component. The ratio of K of the more volatile phase to the less volatile phase is relative volatility.

The pressure in a distillation column is generally constant. The temperature varies from bottom to top. At equilibrium, both phases have near equal Gibbs free energy at a minimum while the components of each phase can have a different composition.

Consider a tray in a distillation column with a liquid phase that has its own vapor pressure derived from the sum of two unequal partial pressures of two components [binary system]. The tray also has a vapor phase made up of two vapor components with unequal partial pressures. The total pressure is the sum total. As a result, there are four variables. Because such an unstable system cannot exist, they all work towards the same goal of minimizing chemical potential and achieving equilibrium.

The most convenient thing that happens is the liquid phase gives up more volatiles and thus becomes more concentrated with high boilers on each plate, whereas the vapor phase gives up less volatiles and thus becomes more concentrated with low boilers on each plate. This continues until the two phases have separated. Finally, both phases have reduced their chemical potential and achieved equilibrium.

To summarize, distillation takes advantage of differences in concentrations of components in the liquid and vapor phases on each tray.

Draw a vapor-liquid equilibrium curve with the following Benzene-Toluene vapor pressure data.

Table 1

Toluene		*n-Heptane*	
T (°C)	*V.P. (torr)*	*T (°C)*	*V.P. (torr)*
18.4	20	9.5	20
31.8	40	22.3	40
40.3	60	30.6	60
51.9	100	41.8	100
69.5	200	58.7	200
89.5	400	78.0	400
110.6	760	98.4	760
136	1520	124	1520

Step by Step

Write Raoult's Law for a binary mixture of A and B;

$$pA = Pvp, A\ xA;\ pB = Pvp, B\ xB = Pvp, B\ (1 - xA)$$

pA and pB are the partial pressures of A and B. Pvp, A and Pvp, B are the vapor pressure of A and B

The total pressure, $PT = pA + pB$

Replacing for the partial pressures and re-arrange, we have:

$$P\text{ tot} = P\ vpA\ Xa + [PvpB - PvpB\ Xa]$$

$$P\text{ tot} = P\ vpA\ Xa + PvpB - PvpB\ Xa$$

$$P\text{ tot} = Xa\ [P\ vp\ A - PvpB] + PvpB$$

Re-arrange, we have the expression for xA.

Fundamental equations

$$Xa = [P\text{ tot} - pvpB]/[pvpA - pvpB]$$

Since $Pvp, A\ Xa = yA\ P$ tot; we have the expression for yA: [combined Daltons + Raoult's law]

$$Ya = [PvpA/Ptot]\ Xa$$

The VLE data can be calculated at each temperature by substituting for total pressure (101.32 kPa) and the appropriate vapor pressures. For example, at 85.0 degc,

Let A is benzene and B is toluene

$$Xa = [101.32 - 46]/116.9 - 46] = 0.7803$$

$$Ya = [116.9/101.32]\ 0.7803 = 0.9002$$

$$K1 = 0.9002/0.7803 = 1.15\ [y/x]$$

We have the following final results:

Table 3

Temperature (°C)	*Mole Fraction Benzene in Liquid*	*Mole Fraction Benzene in Vapour*
80.1	1.000	1.000
85.0	0.780	0.900
90.0	0.581	0.777
95.0	0.411	0.632
100.0	0.258	0.456
105.0	0.130	0.261
110.6	0.000	0.000

Similarly for Toluene

$$Xb = [P\ tot - PvpA]/[PvpB - PvpA]$$

$$Yb = [PvpB/\text{Ptot}]\ Yb$$

$$Yb = [101.32 - 116.9]/[46 - 116.9] = 15.58/70.9 = 0.2197$$

$$Yb = [46/101.32]\ 0.2197 = 0.0997$$

$$K_2 = 0.0997/0.2197 = 0.45\ [y/x]$$

Relative volatility = 1.15/0.45 = 2.55

Definition of Relative Volatility

Alpha = the relative volatility of the more volatile component to the less volatile component .

Alpha = K_1/K_2 = [y/x] in benzene/[y/x] in toluene.

VLE diagram

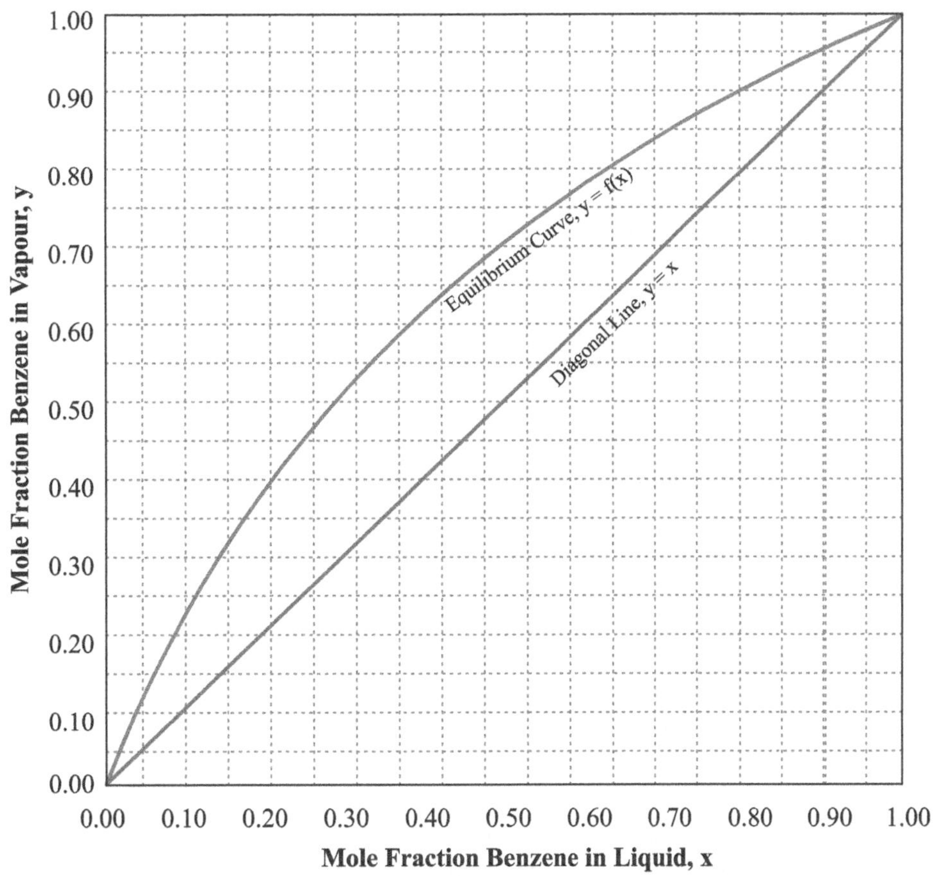

Figure 4: Equilibrium Curve for Benzene-Toluene System at 1 atm

MCCABE-THIELE METHOD OF DISTILLATION: AN OVERVIEW

General Reading

Heat and Mass Transfer in a Distillation Column

To quickly comprehend this process, let's focus on one stage or tray at a time. The vapor moving from the bottom tray to the top tray is obviously at a higher temperature than the liquid on the top tray. Therefore, when the top tray liquid and bottom vapor make contact, heat will be transferred from the vapor to the liquid. As a result, the temperature of the liquid will increase, and the volatile components in the liquid will become vapor and rise as a result. However, as a result of the heat exchange, the vapor's temperature will drop and become less intense, and less volatile components in the vapor will condense into the liquid. Thus, the vapor and liquid will simultaneously transfer heat and mass, which results in the enrichment of the vapor in more volatile components and the enrichment of the liquid in less volatile components. This was only for one tray. All of the trays go through the same process. Keep in mind that until the equilibrium composition is reached, the aforementioned process will continue and enrichment will take place. Increasing the temperature after the equilibrium composition of the components has been archived won't alter the composition unless and until there is a reflux provision.

Two important aspects of the distillation column [1] Composition of feed and the location of the feed plate.

Effect of composition of feed on the feed plate location

It is not necessary for the feed will be always liquid.

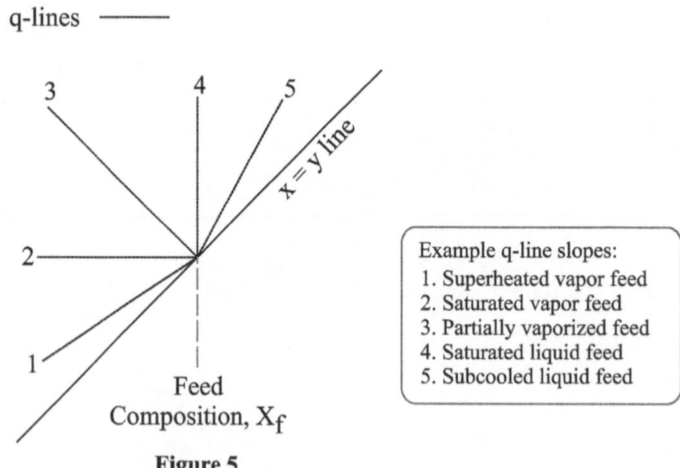

Figure 5

As the image shows it can just be a saturated liquid [4] vertical line an extension of the feed line. It can be partially vaporized feed [3]. It can be saturated vapor [2] It can be superheated vapor [1] It can also be subcooled liquid feed [5]. It may be noticed in each case the slope of the feed line is different and its intersection with the rectification operating line is different. Thus, each composition will have a different location in the feed tray. For example, a saturated liquid will have a feed plate somewhere at the center of the column. A partially vaporized feed will have a feed tray more towards the bottom of the column meaning more rectification trays and fewer stripping trays. If the feed is subcooled liquid there will be more stripping plates and fewer rectification plates.

Location of Feed Plate

This is another important aspect of a distillation column.

As discussed, the location of the feed tray is an important aspect and this requires a rigorous tray-to-tray calculation to identify a tray where the feed composition is the same as the tray composition.

In short, the definition is an ideal feed location is feeding to a section of the distillation column where column internal liquid traffic composition is similar to feed stream composition. It means, it's the location of a plate or tray in the distillation column where entering feed can make minimum composition gradient between feed stream and distillation internal fluids and thus don't break plate to plate equilibrium. The optimum location of the feed plate would be the one represented by the triangle that has one corner on the rectifying line and the other corner on the stripping line.

What is McCabe's plot

McCabe-Thiele plot is an easy and condensed graphical tool to understand distillation. It is a technique for estimating the number of theoretical trays needed for the distillation of a binary mixture.

McCabe – Thiele method assumptions

[1] Molal heats of vaporization of the components are roughly the same.
[2] Heat effects (heats of solution, heat losses to and from column, etc.) are negligible.
[3] For every mole of vapor condensed, 1 mole of liquid is vaporized.
[4] It is not used when the relative volatility is less than 1.3 or greater than 5.
[5] It is not used when more than 25 theoretical stages/plates are required, and.
[6] It is not used when the operating reflux ratio is less than 1.1 times the minimum reflux ratio.
[7] It applies to ideal binary mixtures of liquids.

[8] 100% tray efficiency. Trays calculated using the McCabe-Thiele method are called "theoretical trays". In practice, fluids do not reach equilibrium on each tray, and so the "tray efficiency" is used to determine the number of actual trays required to make a given separation.

The above-cited assumptions lead to a concept of constant molal vapor flow and constant molal liquid flow in any section of the column (i.e., the molal flow rates of vapor and liquid are constant in any section of the column). In other words, for one mole of liquid vaporized, one mole of vapor is condensed in any section (rectifying or stripping).

What is a theoretical plate?

A theoretical plate in a distillation column is a plate in which two phases, such as the liquid and vapor phases of a substance, establish an equilibrium with each other. In other words, the average composition of vapor rising up and the average composition of liquid leaving the plate are in equilibrium. Such equilibrium stages are also referred to as an equilibrium stage, ideal stage, or theoretical tray.

An actual, physical plate can never be a 100% efficient equilibrium stage, the number of actual plates is more than the required theoretical plates.

$$Na = Nt/E$$

Na is a number of actual plates, Nt is a number of theoretical plates E is the plate or tray efficiency. Typical values for tray efficiency range from 0.5 to 0.7 and depends on a number of factors, such as the type of trays being used, and internal liquid and vapor flow conditions. Sometimes, additional trays are added (up to 10%) to accommodate the possibility that the column may be under-designed.

Construction of McCabe – Thiele Diagram Step by Step

Step 1 is to draw equal-sized vertical and horizontal axes of a graph. The horizontal axis will be for the mole fraction (denoted by x) of the lower-boiling feed component in the liquid phase. The vertical axis will be for the mole fraction (denoted by y) of the lower-boiling feed component in the vapor phase.

Step2 is to draw a straight line from the origin of the graph to the point where x and y both equal 1.0, which is the $x = y$ line. This 45-degree line is used simply as a graphical aid for drawing the remaining lines.

Step3 is to draw the equilibrium line using the VLE [vapor liquid equilibrium] data points of the lower boiling component, representing the equilibrium vapor phase compositions for each value of liquid phase composition.

Table 4: Equilibrium Data for Benzene-Toluene System at 101.3 kPa

T (°C)	*Vapor Pressure (kPa)*		*Mole Fraction of Benzene at 101.325 kPa*	
	Benzene	*Toluene*	X_A	V_A
80.1	101.32		1.000	1.000
85	116.9	46.0	0.780	0.900
90	135.5	54.0	0.581	0.777
95	155.7	63.3	0.411	0.632
100	179.2	74.3	0.258	0.456
105	204.2	86.0	0.130	0.261
110.6	240	101.32	0	0

What is the Equilibrium Line?

It is a curved line that describes the compositions of the liquid and vapor in equilibrium at some fixed pressure. For example, the Benzene-toluene VLE plot shows a binary mixture that has a uniform vapor-liquid equilibrium and it is relatively easy to separate.

The table above is showing equilibrium data for benzene-toluene at 101.3 kPa.

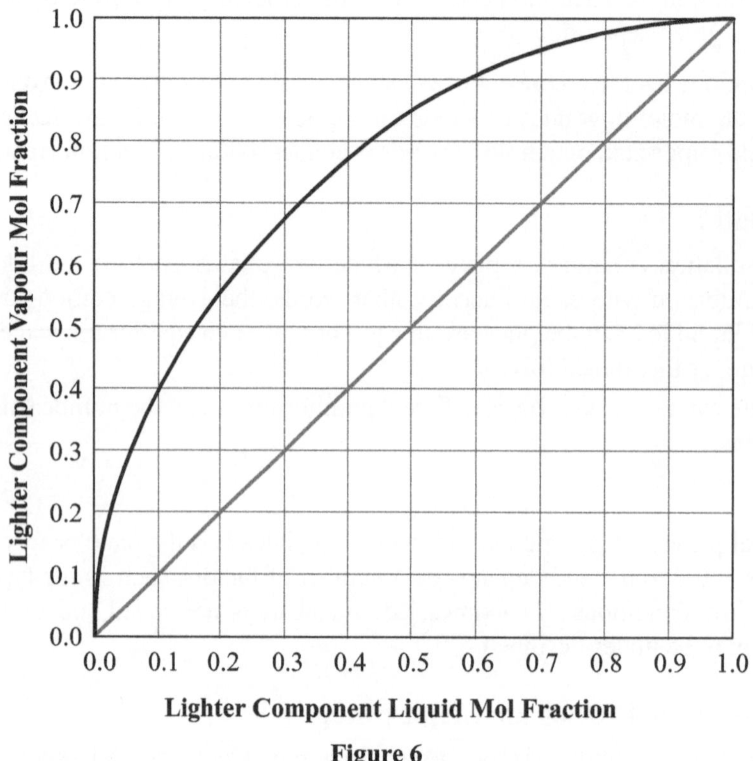

Figure 6

Step 4, Choose Product Compositions.

Mark the composition of the feed, top, and bottom products on the x-axis. Then plot a vertical line from each of these points to the 45° line as shown below.

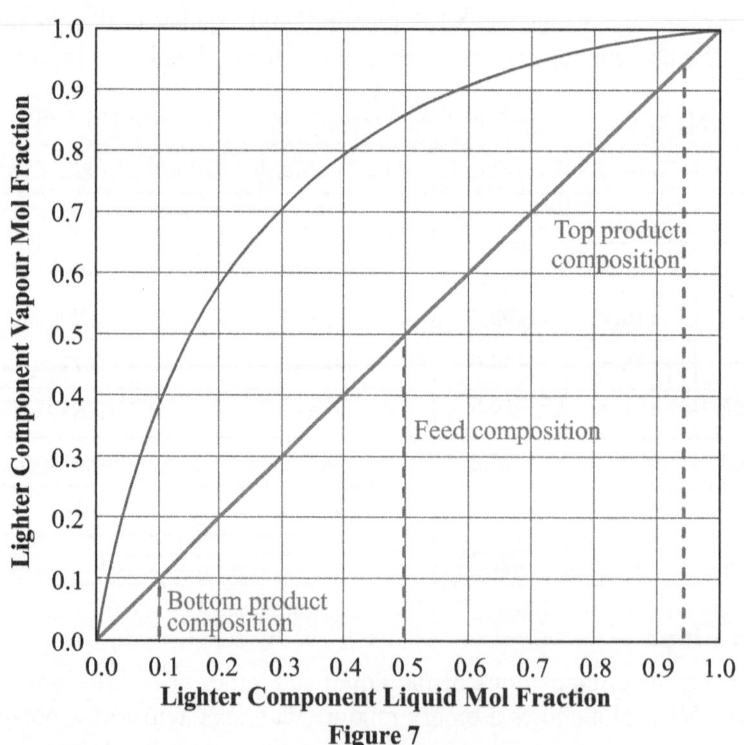

Figure 7

Step 5, extend *XB* [bottom] *XF* [feed], *XT*[Top] to diagonal line in the image.

Step 6, draw feed line. Where the feed is saturated liquid, the line is vertical. Where the feed is a saturated vapor, the line is horizontal.

Depending on the state of the feed, the slope of the feed line on $x = y$ line called q line will be different.

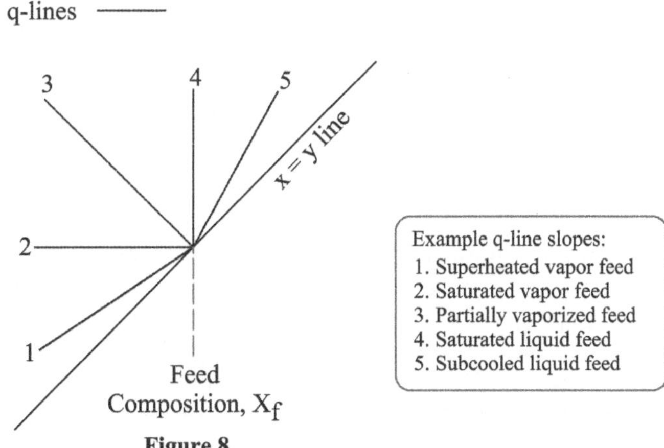

Figure 8

Where the feed is saturated liquid, the line is vertical. Where the feed is a saturated vapor, the line is horizontal. This has been discussed on page 1.

For feeds lying between a saturated vapor and a saturated liquid we can use a simple formula to determine the slope of the feed condition line:

$$m = q/q-1$$

Where q is the mole fraction of liquid in the feed.

Once the slope of the line feed condition line is calculated it can be drawn on the McCabe-Thiele plot from the point where the feed line meets the 45° line until it intersects the VLE curve. For a saturated liquid-like binary mixture of benzene-toluene q is a. straight line.

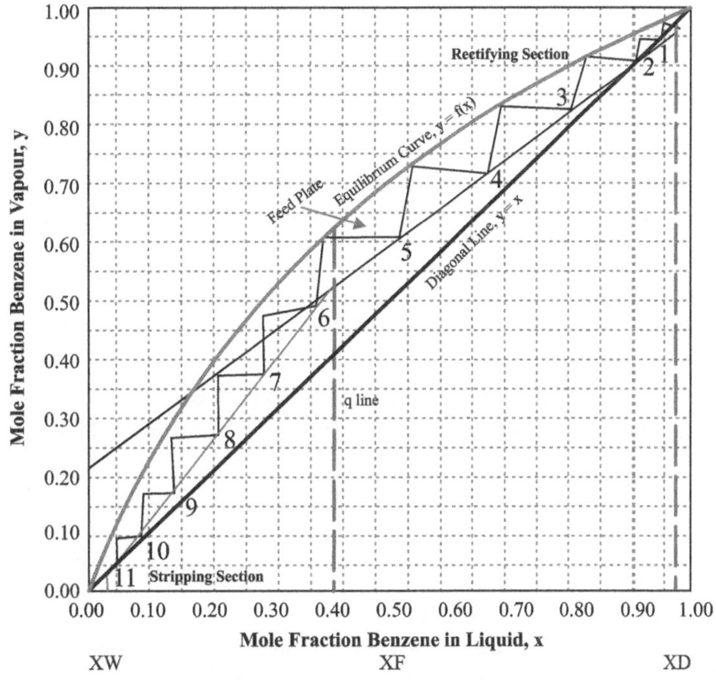

Take blank Take blank *x-y* diagram from internet to fill your data

Figure 9: Equilibrium Curve for Benzene-Toluene System at 1 atm

Step 7 is to plot the operating line or rectifying part of the column. The intercept of the operating line on the y-axis is given $XD/[R+1]$. Draw an upper operating line from the intersection of XD with $x = y$ diagonal line such that its intercept is $XD/[R+1]$ on Y-axis. This is a rectification operating line.

What is Rectifying Section of the Distillation Column?

The rectification section of a distillation column is the portion of the column above the feed plate where volatiles are stripped from vapor.

Step 8, the next step is joining the intersection of the upper operating line with q line with XB [bottom]. This line is an operating line for the stripping section.

What is the Stripping Section of the Distillation Column?

The stripping section of a distillation column is the portion of the column below the feed plate where volatiles are stripped from the liquid.

The feed plate triangle is located with its base both in the rectification and stripping sections.

Step 9, the final step is to draw triangular steps between two operating lines and the equilibrium diagram. By counting the triangular steps, you can see between 11 to 12 perfect plates are required for the desired separation in the above image with feed on the 6th plate from the top of the column. If the reboiler can be considered as a perfect plate, between 10–11 perfect plates are required in the column.

DISTILLATION COLUMN: HOW TO FIX TEMPERATURE AND PRESSURE?

General Reading

Temperature Profile

A distillation column's basic temperature distribution is warmer at the bottom and colder at the top. For a simple two-component distillation the temperature at the bottom is just lower than the boiling point of the heavier component. The temperature at the top of the column is just above the boiling point of the lighter component.

At the bottom of the column, we would like the heavy component to remain as a liquid and the lighter component to stay as a gas. So, we set the temperature at the bottom to match this requirement. This temperature is set by adding heat via a reboiler. Typically, the heat added to the bottom of the column is easy to control, via steam or hot oil flow rates.

At the top of the column, the situation is reversed. We would like the light component to remain a gas while the heavier component is condensed to a liquid and falls back down the column. The top temperature is set just above the boiling point of the lighter component. The temperature control situation is different here from the bottom of the column because we usually want the top product to be a liquid when we send it for storage. So, we condense all of the gas coming out of the top of the column to liquid. This liquid stream is split with some returning to the column and some going to storage. The top temperature is often controlled by changing the reflux rate, i.e., the flow rate of liquid sent back to the top of the column. A higher reflux rate means cooler liquid falls down the column against the rising warmer gas, and the top temperature is lower.

Overall heat is added at the bottom of the column and heat is extracted at the top of the column. Inside the column, the temperature balance is created between the hot gas rising up the column and the cooler liquid falling down the column.

Pressure Profile

There is typically a pressure gradient across the column with the pressure being higher at the bottom of the column than at the top. This pressure gradient occurs as the liquid coming down the column impedes the flow of vapor up the column and imposes a pressure loss on the flow. In steady-state distillations, the pressure in the column is held constant, and the temperature is varied to control the composition of the product streams.

Detail

Pressure

Under normal, usual conditions it is the vapor pressure of the liquid on the top tray that fixes the pressure at that location before the generated vapors exit to enter the overhead condenser. This is the basic parameter that fixes the column pressure. The pressure at other sites within the column depends on the ability of the vapors and liquids to distribute themselves up and down the column with minimum pressure drops.

Therefore, the composition – or purity – of the liquid on the top tray rather defines what pressure the column is expected – or designed – to operate at. The external reflux ratio (L/D) has a bearing on fixing that composition – as the various *L/V*'s (internal reflux ratios) that are generated down the column have on the various trays' compositions.

What is *L/D*?

It is the external reflux ratio. It is defined as the ratio of the liquid returned to the column divided by the liquid removed as product, i.e., $R = L/D$.

External reflux vas internal reflux

As the external reflux cools the top of the tower, vapors made of heavier fractions condense and liquid made of heavier faction flows down the tower it's referred to as internal reflux.

What is L/V Internal Reflux Ratio?

It is the liquid/vapor flow ratio inside the upper section of the column.

After the top pressure is specified, the bottom pressure will be dictated by pressure drop along the column which depends on the relevant selected technology and the load of vapor and liquid inside the column. The corresponding bubble point of bottom pressure will specify the bottom temperature.

The purpose of reflux is to provide down-flowing liquid throughout the rectification section to contact with the up-flowing vapor in order to achieve stage-by-stage equilibrium heat and mass transfer and, hence, purification of the top product. When sub-cooled reflux is introduced to the top tray, it must be heated up to its bubble point before the lighter components will vaporize.

Where does the Heat Come From?

The only place it can come from is from condensing vapor that is approaching the top tray from below. When this vapor condenses, it adds to the total liquid flowing from tray 1 down the column. In other words, sub-cooled reflux introduces a greater volume (or mass or molar) flow of reflux than is delivered to the column by the external reflux flow controller.

If the degree of sub-cooling was constant, then this wouldn't be such a big source of disturbance; however, this is usually not the case. The amount of sub-cooling will vary with the temperature of the cooling medium (ambient air, cooling water, another process stream, etc.).

Temperature: The Column Gets Cooler as You Go Up

There are two control points while fixing top and bottom boundary temperatures in a distillation column. At the bottom it is reboiler and at the top it is reflux. Within the column, the temperature gets set by relative volatility or more precisely by the partial pressure of HC [feed] according to Rault's law. The volatility of components in the column is different. If P is the total pressure in the column, this is equal to the sum of $P_1 + P_2 + P_3 + \ldots$ partial pressures at a different height. While total pressure is constant in the column, the partial pressures of HC components are different along with the height. Partial pressure = Mole fraction × Vapor pressure of pure component at a height [Rault's law. Since the volatility of hydrocarbon becomes more as you go up, the vapor pressure becomes more, and consequently, saturation temperature gets lower. Therefore, the column gets cooler as we go up. In a steady state, the partial pressures do not change much.

Sub-cooled reflux will eventually trigger internal reflux and can substantially improve the quality of the top cut. The downside is it may destabilize the column in exceptional cases if it adds resistance to rising vapor in a given column diameter.

COMMON ISSUES IN CRUDE DISTILLATION AND TROUBLESHOOTING: TYPICAL CASES

Column performance issues are frequently mistakenly attributed to issues with column internals. However, there are additional reasons.

The crude oil distillation column is a critical piece of processing equipment that must be run under optimal and stable conditions in order to produce on-spec products. Operating personnel must have sufficient knowledge and experience in the crude oil distillation process to troubleshoot successfully. Process engineers must be analytical in order to diagnose the problem and implement appropriate actions to restore the crude distillation process to normal.

Many Factors Influence Distillation

The most common factors are [1] feed conditions, such as feed state and feed composition, and [2] reflux conditions. [3] Internal liquid and fluid flow conditions [4] condition of trays (packings), etc. This post will concentrate on three internal flow-related issues: [1] entrainment, [2] flooding, and [3] weeping.

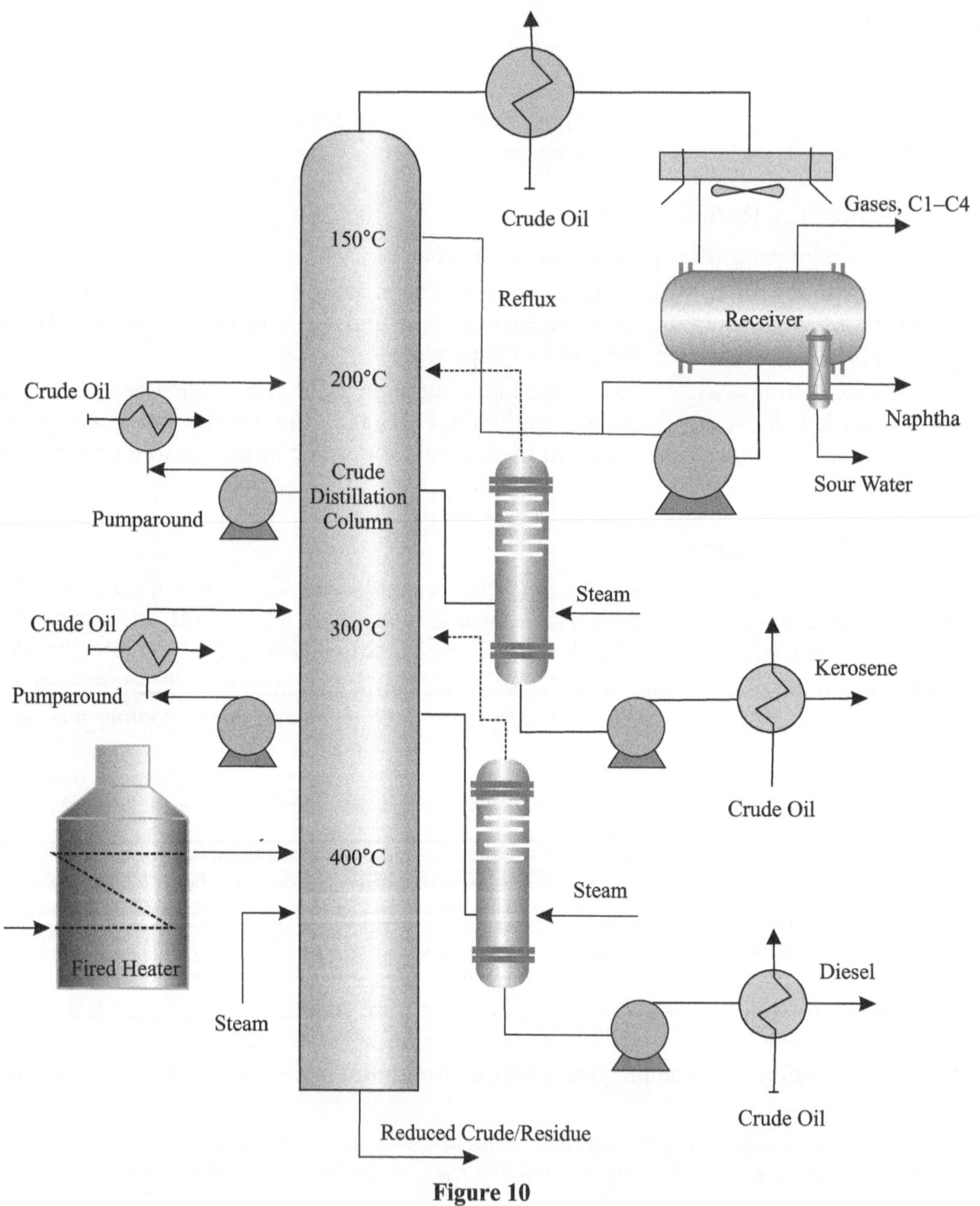

Figure 10

Some of the key issues are as follows:

Flooding of the Crude Distillation Column

Flooding of crude oil distillation towers occurs when vapor loads exceed the tower's maximum capacity and levels at tower trays are high. The interaction between vapor and liquid is so strong as a result of flooding that stable counter-current operation is no longer possible. Pressure drop and liquid holdup skyrocket, while distillation column efficiency plummets. Flooding reduces the quality of petroleum products.

Flooding symptoms include 1. a very high-pressure drop that prevents vapors from rising, 2. unstable levels and indicators, and 3. a decrease in temperature or a low-temperature gradient across the flooded section of the tower, 4. a change in product specifications or poor product quality, 5. One section (for example, diesel) is flooded, while the downstream section is dry.

Troubleshooting

As the above-mentioned conditions confirm the flooding, the column becomes less capable of handling higher vapor rates and high feed rates. Control actions to troubleshoot the flooding problem include lowering the liquid rates to the tower, reflux rates, stripping steam to the tower, and lowering the crude heater outlet temperature based on the unit feed rate. Furthermore, if flooding is observed in a specific section, the product draw-out rate can be increased.

High Level in a Distillation Column

The levels of the tower bottom and draw-off trays in the distillation column are kept constant by adjusting the product rate. Distillation towers are outfitted with level indicators that help control room operators adjust the conditions as needed. If the level is too high, the liquid will rise to the trays section of the tower and behave like flooding in the column, exhibiting symptoms of low-temperature gradient, high-pressure drops, and poor fractionation efficiency.

Because water quickly vaporizes, causing a pressure surge and level fluctuations, the stripping steam should be dry before use. Reduce the feed rate while increasing the product draw-out rate from the tower.

Troubleshooting

Verify the levels at the level glasses where the problem is observed physically. Reduce the crude heater outlet temperature or increase the product draw-out rate to control the high-level problem.

Dry Trays

The dry trays scenario is the inverse of flooding in that the trays in flooding are filled with liquid, whereas dry trays are empty of liquid. This situation can occur when the product withdrawal rate is greater than the amount of internal reflux and the feed rate in the column is lower. In this case, the trays below the draw-off point run dry, and no fractionation occurs. At the same time, there is insufficient liquid to keep the levels in the stripper draw-off trays and side-cut strippers constant. A low-pressure drop and low-temperature gradient will exist in the tower or section of the tower with dry trays.

Troubleshooting

To control the dry tray issue, either reduce the product withdrawal rate or increase the internal reflux to the column. Furthermore, by increasing the crude heater outlet temperature or feed rate to the tower, the internal reflux can be increased. Furthermore, the rate of the next lightest product draw-off rate can be reduced to increase the reflux rate. The dry trays caused by superheated vapors increase the tower's reflux rate while also lowering the feed inlet temperature.

Trapped Water

The trapped water reduces fractionation and limits oil and vapor contact, resulting in poor product quality. The presence of water is indicated by a low temperature at the top trays or tower top. Furthermore, water evaporation may result in pressure increases in the tower. Trapped water can be gradually removed by raising the top temperature or decreasing the naphtha reflux flow. Water in the tower may also be caused by a poor crude oil desalting process, as explained further below.

Additionally, the stripping steam to the column is condensed at tower overhead condensers and removed from the overhead receiver as sour water.

If there is too much water in the reflux, it will upset the crude column. It will build up on the upper trays, flooding the column.

Troubleshooting

The solution is to reduce the reflux rate, raise the tower's top temperature, and allow the water to evaporate. Tower pressure rise can be controlled by increasing condensation at the tower overhead system and increasing compressor capacity. Crude tower, inlet temperature, and stripping steam can all be reduced to reduce the vapor load on the tower, resulting in normalized pressure.

Water carryover from desalter: Source of corrosion

Water carryover from a poor desalting process will result in an immediate temperature drop at the tower's inlet and bottom. The low temperature of the crude feed could be the cause of water carryover from the desalter. Maintaining the Desalter inlet temperature between 125 and 130°C improves desalting efficiency. Salts in water cause corrosion in the column overhead system. Ca and Mg salts will migrate to the top of the tower and convert to HCl. The pressure rise of the pre-flash vessel is the first sign of water carryover.

Troubleshooting

The efficient operation of crude Desalter is the solution to this problem. Restore the Desalter's water level to normal. Increase the flow of corrosion inhibitor and neutralizer at the tower's top. Water carryover also causes energy loss and poor tower efficiency, which has a negative impact on product specifications.

References

Google.

Troubleshooting of Crude Oil Distillation Column: NASIR HUSSAIN.

MASS AND HEAT TRANSFER IN A DISTILLATION COLUMN

Key Points in Distillation

Energy flows from high [reboiler] to low [condenser]

Entropy increases as liquid converts to vapor on each tray

Gibbs free energy remains minimum at each tray

Vapor-liquid remains at thermodynamic equilibrium on each tray due to minimum free energy and maximum entropy.

Total vapor pressure = Partial pressure A + Partial pressure B [A is more volatile]

The vapor pressure of A reduces and the mole fraction increases with increasing height

Relative volatility increases with height due to lower temperature

While every tray is in thermodynamic Vapor = Liquid equilibrium, the composition is different

Specific heat of A decreases with height

The density of A increases with height

At condenser temperature, A becomes a saturated vapor and changes phase to liquid. No further cooling. Specific heat becomes infinity.

Consider a case of a continuous fractionating column operating at 14,7 psia is to be designed to separate 30,000 lb/hr of a solution of benzene and toluene, containing 0.400 mass fraction benzene into an overhead product containing 0.97 mass fraction benzene and a bottom product containing 0.98 mass fraction toluene. A reflux ratio of 3,5 lb of reflux per lb of product is to be used. The feed will be liquid at its boiling point and the reflux will be returned to the column at 100 deg F.

Determine the quantity of top and bottom products and calculate the condenser duty and rate of heat input to the reboiler.

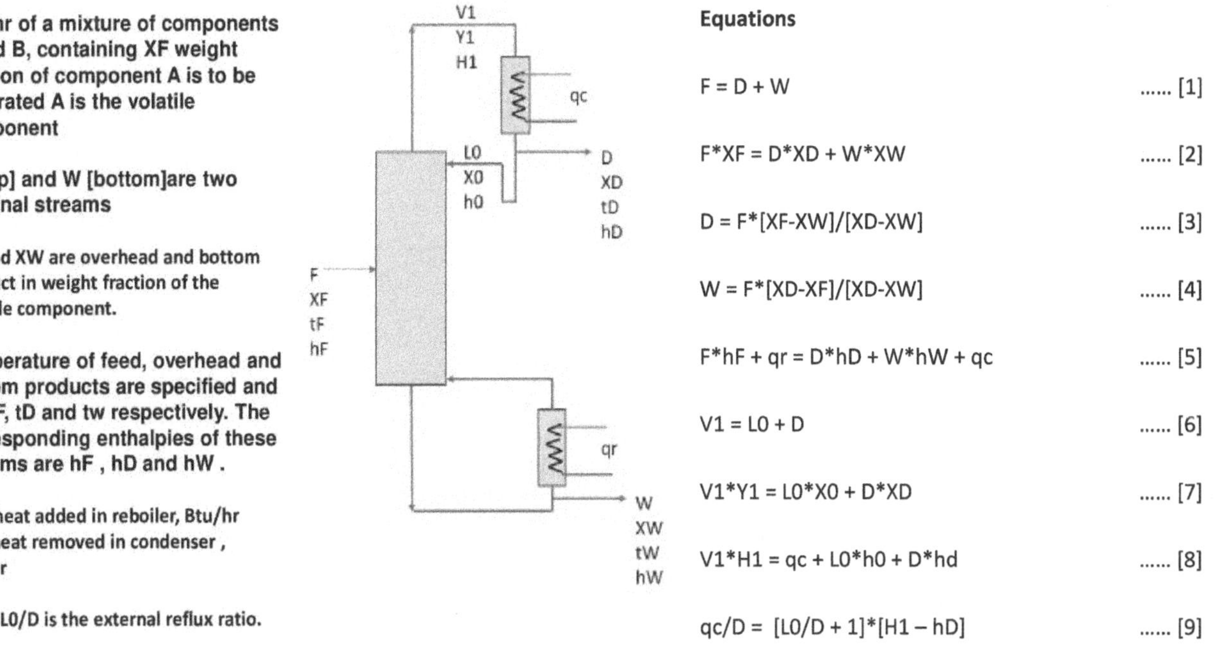

Figure 11: Mass and Heat Transfer in Distillation Column

Mass Balance

System = Benzene – Toluene [Benzene is more volatile]

Feed rate, $F = 30{,}000$ lb/hr

Benzene mass fraction in feed = 0.400

$XF = 0.400$

Mass fraction of benzene at bottom = [1–0.98] = 0.02

$XW = 0.02$

Mass fraction of benzene in overhead product = 0.97

$XD = 0.97$

Apply equation [3], $D = F\times[XF-XW]/[XD-XW]$

$D = [30{,}000[\times[0.400-0.02]]/[0.97-0.02]$

Top product = 12,000 lb/hr

$W = F-D = [30000-12000]$ [W is bottom product]

Bottom product = 18000 lb/hr

Energy Balance

Enthalpy concentration diagram for Benzene – Toluene at 1 atmosphere. Reference state: liquid benzene at 32 deg F and toluene liquid at 32 deg *F*.

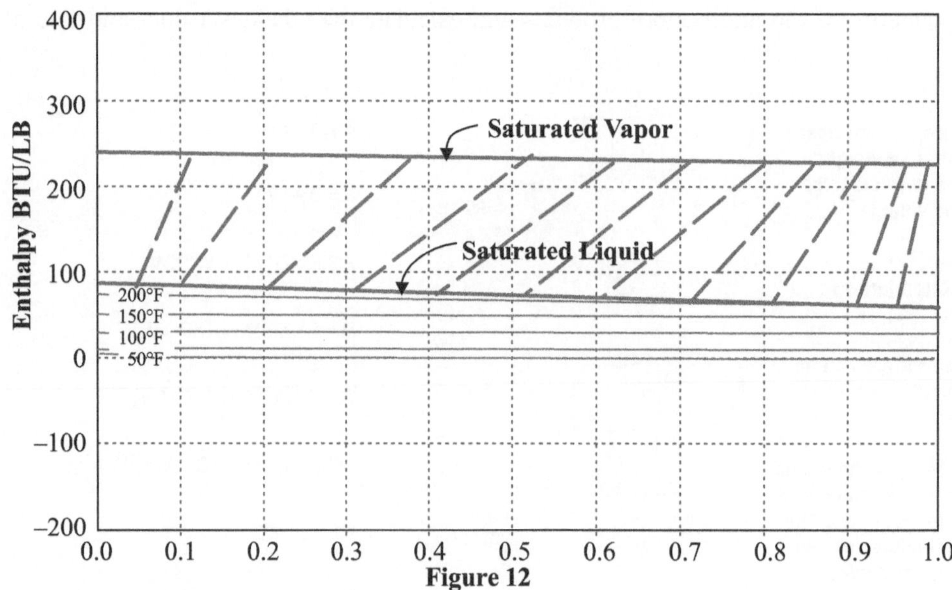

Figure 12

X axis = Weight fraction benzene Y axis = Enthalpy Btu/lb.

The heat removed by the condenser

Column operating pressure = 14.7 psi

Please refer to the benzene – toluene enthalpy concentration diagram.

Enthalpy of feed as saturated liquid, hF [at benzene mass fraction 0.400] = 73.5 Btu/lb

Enthalpy of vapor as saturated vapor, H_1[intersection diagonal line at 0,400 mass fraction between saturated liquid and saturated vapor] = 232 Btu/lb

$h_0 = hD$ [both stand for enthalpy of overhead product, D]

Reflux temperature = 100 deg F

Enthalpy of reflux at 100 deg F at benzene mass fraction 0.97 = 28.7 Btu/lb

The liquid leaving the reboiler is saturated liquid.

The mass fraction of benzene at the bottom product is 0.02 [explained above]

At mass fraction 0.02, saturated liquid benzene enthalpy, hW = 86.6 Btu/lb

From equation [9], $qc/D = [L_0/D + 1]\times[H_1 - hD]$,

L0/D = reflux ratio = 3.5 [given]

H_1 = vapor enthalpy = 232 Btu/lb from enthalpy diagram

hD = liquid distillate enthalpy = reflux enthalpy at 100 deg F, this is from the enthalpy diagram above is 28.7 Btu/lb, explained above.

qc/D = [3.5+1]×[232.0–28.7] = 915, D = 12,000 lb/hr [calculated above], therefore, heat removed by the condenser, qc = [12000]×[915] = 11,000,000 Btu/hr

Heat Added by Reboiler

From equation [5]

$$F\times hF + qr = D\times hD + W\times hW + qc, \text{ or}$$

$$qr = D\times hD + W\times hW + qc - F\times hF$$

We have calculated, D = 12000 lb/hr, hD = 28.7 Btu/lb, W = 18000 lb/hr, hW = 86.6 Btu/lb, qc = 11,000,000 Btu/hr, F = 30,000 lb/hr and hF = 73.5 Btu/lb

qr = [12000 28.7] + [18000 86.6] + 11,000,000 – [30,000×73.5]

Heat added by reboiler, qr = 10,698,000 Btu/hr.

Summary

Top product = 12,000 lb/hr.

Bottom product = 18,000 lb/hr.

Heat removed by condenser = 11,000,000 Btu/hr.

Heat added by reboiler = 10,700,000 Btu/hr.

AZEOTROPE: AN OUTPUT OF RAOULT'S LAW DEVIATION

Contents

What is liquid?
What is solution?
Raoult's law-key to distillation
Ideal solution
Non-ideal solution
Azeotrope

Summary

A succinct summary is that when intermolecular interactions - vander walls forces of attraction – prohibit two distinct molecules with different polarity from escaping into the vapor phase, both molecules acquire a constant boiling point, which is referred to as azeotrope. Azeotropes occur when fraction of the liquids cannot be altered by distillation.

Azeotrope forms only when a solution deviates from Raoult's law. An azeotrope is a constant boiling point mixture. It is a mixture of two or more liquids whose proportions cannot be modified through simple distillation. This occurs because when an azeotrope is boiled, the vapour contains the same chemical proportions as the un-boiled combination. A non-ideal solution does not follow Raoult's law. Two substances will form non-ideal solution when intermolecular forces within each substance and between two substances are not the same.

The boiling point of each azeotrope is unique. An azeotrope's boiling point is either lower than the boiling point temperatures of any of its constituents (positive azeotrope) or higher than the boiling point temperatures of any of its constituents (negative azeotrope) (negative azeotrope). When the deviation is great enough to cause a maximum or minimum in the vapor pressure versus composition function, Azeotrope formation is a mathematical consequence.

What Exactly is Liquid?

Solid, liquid, and gas are the three main states of matter. All matter, according to the kinetic theory of matter, is made up of particles with a specific amount of energy that permits them to travel at varying speeds depending on temperature (energy). Between the particles, there exist gaps. When particles are close together, attractive forces between them play a role. The molecules in a liquid are closer together than in a solid.

Liquid – Liquid Solution

The Key to Distillation is Raoult's Law

A solution is a mixture of two substances that is homogenous. When one liquid dissolves in another, the solution's saturated vapour pressure is determined by the components' saturated.

vapour pressures as well as the solution's composition. A mole fraction of each component can be used to indicate the composition of a mixture of liquids. The mole fraction of A in a mixture of A and B is defined as the number of moles of A divided by the total number of moles. The mole fraction $XA = NA/(NA+NB)$ is represented by the symbol X. NA and NB are A and B moles, respectively. Raoult's law says that the saturated vapor pressure of each component in the mixture is equal to the product of the mole fraction of that component and the saturated vapor pressure of that component when pure.

This can be expressed as $PA = XA(\text{l}) \times PA$ (pure), PA is the saturated vapor pressure of A, PA (pure) is the saturated vapor pressure of pure A, XA(l) is the mole fraction of A in solution. Similarly, for B, $PB = XB(\text{l}) \times \text{PB}$ (pure).

What Causes some Liquids to Follow Raoult's Law?

Raoult's law is obeyed by a mixture of similar compounds. They are said to form ideal solutions. The substances A and B would form an ideal solution if intermolecular forces A.... A, A.... B and B......B are all equal. In other words, A sticks to B with roughly equal energy as A does with A and B does with B. Since the interaction energies between A and B are almost equal there is neither a volume change nor an enthalpy [heat] change on mixing.

The vapor above the mixture of liquids does not have the same composition as the liquid. If XA (v) and XB (v) are the mole fraction of A and B in the vapor phase, then

$XA(v)/XB(v) = PA/PB = XA(\text{l})\ PA\ (\text{pure})/XB(\text{l})\ PB\ (\text{pure})$

PA and PB are partial pressures of A and B generated by the solution

Total vapor pressure = $PA + PB$ [sum of partial pressures of A and B]

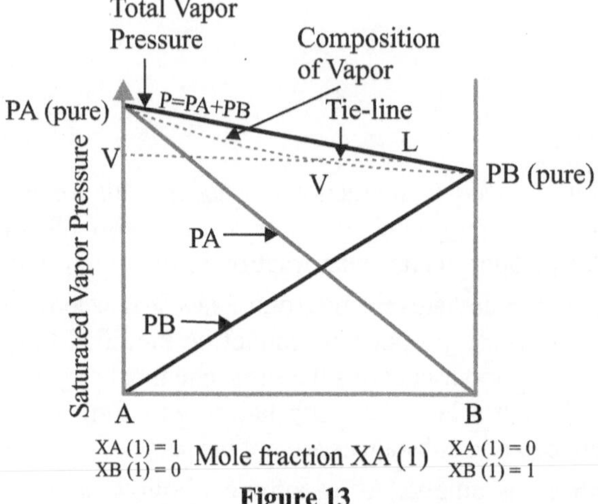

Figure 13

The image is showing the vapor pressure curve for an ideal mixture consisting of liquid A and B. Vapor pressure of this binary solution obeys Raoult's law. The black line shows the total vapor pressure as a function of the mole fraction of component B, and the two green lines are the partial pressures of A and B. When the total vapor pressure is v, the composition of the liquid is represented by point L and the composition of vapor by point V. The line V....L is called tie – line. The vapor is richer than the liquid in the more volatile component, A. The difference in composition between the liquid and vapor is an estimate in the figure.

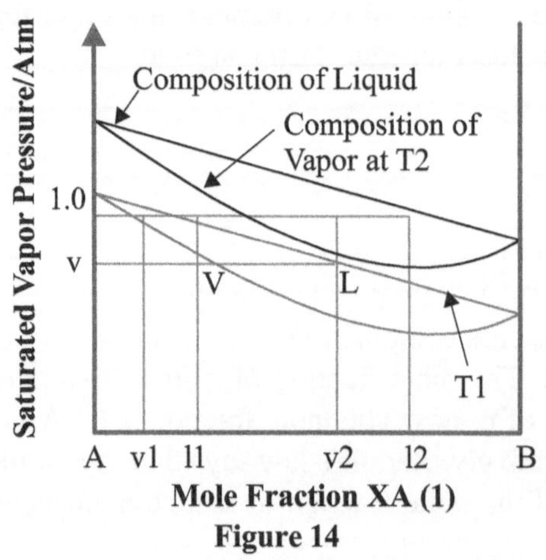

Figure 14

When the vapor pressure of a liquid reaches atmospheric pressure, the liquid boils. T_1 is the boiling temperature of a liquid with liquid composition l_1 in equilibrium vapor composition v_1. T_2 is the boiling temperature of a liquid that has liquid composition l_2, in equilibrium with a vapor composition v_2, figure, below.

Summary

To summarize, an ideal solution is one that obeys Rault's law. Raoult's law states that the vapor pressure of a solution is dependent on the mole fraction of a solute added to the solution. Raoult's law is obeyed by a mixture of similar compounds. Two substances will form an ideal solution when intermolecular forces within each substance and between two substances are the same. Although there are no ideal gases, ideal solutions may exist If the molecules are almost identical chemically, e.g., 1-butanol and 2-butanol, then the solution will be almost ideal. Since the interaction energies between A and B are almost equal, it follows that there is a very small overall energy (enthalpy) change when the substances are mixed. The more dissimilar the nature of A and B, the more strongly the solution is expected to deviate from ideality. The highly diluted solution also behaves as ideal solutions due to less interactions between solute molecules and solute-solvent molecules.

Ideal Liquid

It assumes that a mixture of two liquids will obey Raoult's law when they are chemically similar with the mean strength of the interactions being the same between all the molecules of the solution. The highly diluted solution also behaves as ideal solutions due to fewer interactions between solute molecules and solute-solvent molecules. Benzene C6H6 and toluene C7H8 form near-ideal solution and obey Raoult's law.

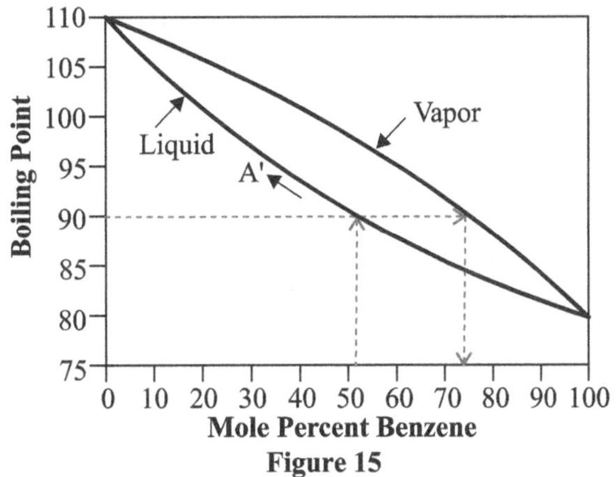

Figure 15

Typical ideal solutions

Benzene and Toluene

n-hexane and *n*-heptane

Bromoethane and Chloroethane

CCl_4 and $SiCl_4$

Chlorobenzene and Bromobenzene

Nonideal Solutions

What pushes a liquid molecule towards Nonideal behavior? At the molecular level what makes 'Azeotrope'.

Deviation from Raoult's Law

Solutions that have vapor pressure greater than that predicted by Raoult's law are said to show positive deviation from the law. Those with a vapor pressure lower than the calculated value are said to show negative deviation, see the figures below.

Typical of a pair of liquids showing a slight deviation is hexane and ethanol. The molecules are very different. Molecules of ethanol can more easily escape from a mixture of ethanol and hexane than from pure ethanol because, in a mixture of ethanol with hexane, hexane interferes with hydrogen bonding between ethanol molecules. This is against when ethanol is a single liquid where hydrogen bonding holds ethanol molecules together. A slight negative deviation from Raoult's law is shown by a mixture of trichloromethane $CHCl_3$ and ethoxyethane, $C_2H_5OC_2H_5$. Here both molecules are practically non-polar. The intermolecular forces – Vander walls forces of attraction between these molecules tend to prevent the molecules from escaping into the vapor phase. This is an important point and needs to be understood to have a good understanding of how Azeotropes form.

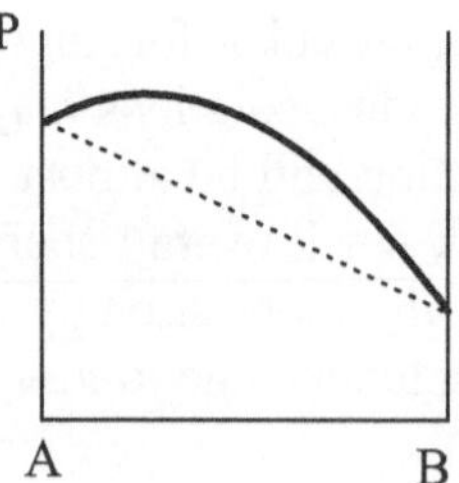

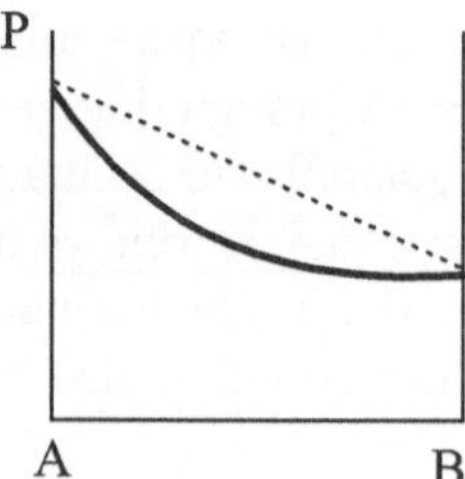

Positive deviation: ΔH_{mix} is positive

A & B interact weakly with one another in the liquid (compared to one another)

Negative deviation: ΔH_{mix} is negative

A & B interact strongly with one another liquid (compared to one another)

Figure 16

Azeotrope

As explained, a constant boiling point mixture is known as an azeotrope. An azeotrope is a mixture of two or more liquids whose proportions cannot be adjusted or changed by simple distillation.

Only when a mixture deviates from Raoult's law can azeotropes arise.

Minimum boiling azeotropes and maximum boiling azeotropes are the two types of azeotropes. The boiling point of each azeotrope is unique. An azeotrope's boiling point is either lower than the boiling point temperatures of any of its constituents (positive azeotrope) or higher than the boiling point temperatures of any of its constituents (negative azeotrope) (negative azeotrope). *When the deviation is great enough to cause a maximum or minimum in the vapor pressure versus composition function, Azeotrope formation is a mathematical consequence.*

Minimum boiling azeotrope

At a given composition, **a solution with a bigger positive departure from Raoult's law forms a minimal boiling azeotrope.** For example, fractional distillation of an ethanol-water mixture forms minimum boiling azeotrope in a solution containing approximately 95 percent by volume of ethanol. The liquid and vapour cannot separate once this composition has been attained.

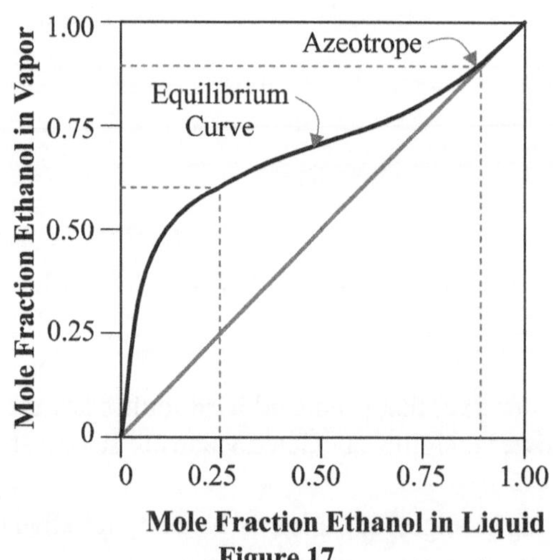

Figure 17

Maximum boiling azeotrope

At a given composition, a solution with a substantial negative deviation from Raoult's law generates a maximum boiling azeotrope. This type of azeotrope includes things like nitric acid and water. Figure below shows an azeotrope with a bulk composition of 68 percent nitric acid and 32 percent water, with a boiling point of 393.5 K (120.4 °C).

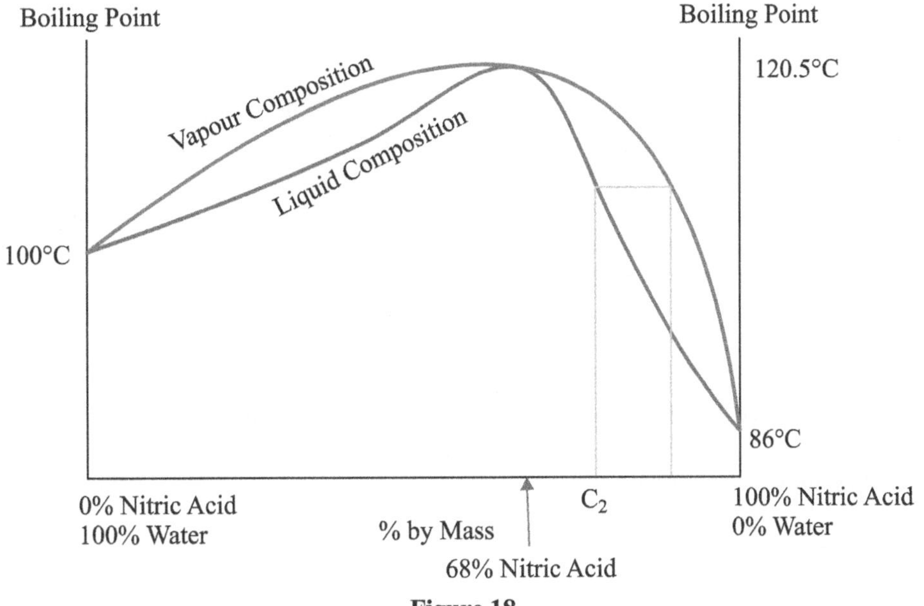

Figure 18

CHAPTER 44

Compressor

A compressor is a device that increases the pressure of a gas by reducing its volume in a confined space. In the field of thermodynamics, compressors are used in many applications to increase the efficiency of power cycles and refrigeration systems. Compressors are essential components in air conditioning systems, refrigerators, and gas turbines. The compression process in a compressor is typically adiabatic, meaning there is no heat transfer involved.

WHAT IS A COMPRESSOR? HOW DOES IT WORK?

Definition

A compressor is a device that increases the pressure of a gas by adding energy to it and extracts work from the gas. It is a polytropic process. This process is described by the equation PV^n = Gamma, where P is the pressure, V is the volume, n is a constant exponent, and Gamma represents the work done on the gas. This definition unique telling everything about what a compressor is in five lines.

How Does Internal Energy Increase?

When a gas is compressed against its intermolecular forces, several changes occur that affect the energy of its molecules. **Firstly, the positional energy** of the molecules increases. This is because as the gas is compressed, the distance between the molecules decreases, causing an increase in the potential energy resulting from the proton-proton repulsion. This increase in positional energy contributes to the overall increase in internal energy. **Secondly, the kinetic energy** of the molecules also increases. Compression leads to a higher density of molecules, resulting in a higher frequency of collisions between them. These collisions lead to an increase in the kinetic energy of the molecules, as they bounce off each other with greater force and speed. Overall, the increase in positional energy due to proton repulsion and the increase in kinetic energy from more collisions both contribute to an increase in the internal energy of the gas. This increase in internal energy is a result of the compression process.

Compressors are approximated as an adiabatic process, $dQ = 0$.

Explanation

The term "adiabatic" is often used to describe the compression process in compressors, including both reciprocating and centrifugal compressors. In the context of compressors, the term "adiabatic" refers to the ideal assumption that no heat is exchanged with the surroundings during the compression process. It implies that the compression process is rapid, occurring so quickly that there is no time for heat transfer to take place. While this assumption does not hold true in real-world situations where there is always some heat transfer and energy loss, considering the compression process as adiabatic is a common simplification used in compressor analysis and design. It allows for easier calculation of compressor performance and efficiency. It is essential to note that in real-world compressors, there will be heat transfer and some potential for energy losses due to factors such as mechanical inefficiencies, heat exchange with the surroundings, and incomplete sealing. However, the adiabatic assumption provides a theoretical ideal for understanding and analyzing compressor performance.

How Does a Real-World Compressor Work?

in real-world compressors, the work done is typically described by the polytropic equation PV^n = const, where n is the polytropic exponent. The polytropic exponent, also known as the index of compression or the index of polytropic efficiency, is related to the ratio of specific heat capacities (Cp/Cv), often denoted as γ (gamma). The ratio of specific heat capacities, γ, is a thermodynamic property of the working fluid (such as air) and is a measure of the fluid's ability to store heat. It is defined as the ratio of the heat capacity at constant pressure (Cp) to the heat capacity at constant volume (Cv). Hence, while compressors are commonly analyzed based on the idealized adiabatic process, the actual compression process is considered polytropic, accounting for factors such as heat transfer and mechanical inefficiencies.

Role of *Cv* in a Compressor

The specific heat capacity at constant volume (Cv) plays a significant role in the compression process. *Cv* represents the amount of heat required to raise the temperature of a substance by one unit when the volume remains constant. *Cv* is internal energy. In compression processes, the work done by the compressor results in an increase in the temperature and pressure of the fluid. This increase in temperature leads to an increase in the internal energy of the fluid. The specific heat capacity at constant volume (Cv) determines how much the internal energy of the fluid will increase for a given increase in temperature. A higher Cv value indicates that the fluid can store more heat energy per unit mass, resulting in a higher increase in its internal energy. Therefore, in compression processes, a higher *Cv* value implies that the fluid can store more heat energy, leading to a greater increase in its internal energy. This relationship between *Cv* and internal energy is essential to understanding the thermodynamics of compressors.

What Does it Mean?

The specific heat ratio (γ), which is the ratio of specific heat capacities (Cp/Cv), plays an important role in determining the efficiency of a compressor. In compressible flow [gases], the specific heat ratio represents the relationship between the heat capacities at constant pressure (Cp) and constant volume (Cv) for a gas. For monoatomic gas with three degrees of freedom, $Cp = 5/2R$ and $Cv = 3/2R$. $Cp/Cv = 5/3 = 1.665$. For diatomic gas with degrees of freedom of five, $Cp = 7/2R$ and $Cv = 5/2R = 7/5 = 1.4$.

Helium, a noble monoatomic gas with Cp/Cv = 1.665 heats up quickly if compressed adiabatically and ends up at a higher pressure and needs more energy to compress than air.

Therefore, it can be justified that, all other factors being equal (such as compression ratio and compressor design), a compressor working with air would potentially have higher efficiency compared to a compressor working with noble gases. The internal energy of helium, $U = 3/2\ KB \times T$ and air, $U = 5/2\ KB \times T$, here U is internal energy, KB is Boltzmann constant and T is temperature. The specific heat ratio, y is the minimum pressure ratio at a constriction for choked flow to occur.

Isentropy

What is it?

A compressor is approximated as an isentropic process, meaning adiabatic and internally reversible. Any compression squeezes the volume and reduces entropy by reducing the number of microstates.

In the context of a purely theoretical scenario with no friction and other sources of irreversibilities meaning for an internally reversible process, the compressibility and the resulting decrease in microstates can indeed affect the isentropic efficiency of a gas. Under the same conditions, if we compare air and helium, air typically has a higher compressibility compared to helium. This means that air can be compressed more easily, resulting in a larger decrease in the number of microstates in an adiabatic and internally reversible compression process.

Since isentropic efficiency is related to the ratio of actual work to ideal work, a gas with a larger decrease in microstates (such as air in this case) can potentially have higher isentropic efficiency compared to a gas with fewer compressible microstates (such as helium). However, it is essential to note that there might be other factors at play that can also impact the isentropic efficiency, such as specific heat capacities. So, in summary, under the same conditions and without considering other sources of irreversibilities, air's higher compressibility can lead to a larger decrease in microstates and potentially higher isentropic efficiency compared to helium. This explains what is 'isentropic'.

RECIPROCATING COMPRESSOR VS. CENTRIFUGAL COMPRESSOR

The fundamental point is both reciprocating and centrifugal compressors add internal energy but their working mechanisms are different.

Reciprocating compressors and centrifugal compressors are two common types of compressors used in various industries for increasing the pressure of a gas. These compressors operate on different principles of thermodynamics and have distinct advantages and disadvantages.

Overall, the choice between reciprocating and centrifugal compressors depends on factors such as the desired pressure and flow rate, maintenance requirements, noise and vibration levels, and energy efficiency. Both types of compressors have their own advantages and disadvantages, and the selection of the appropriate compressor depends on the specific requirements of the application.

In terms of thermodynamics, both reciprocating and centrifugal compressors follow the principles of the first law of thermodynamics, which states that energy cannot be created or destroyed, only transferred or converted. The compression process in both types of compressors involves the input of energy in the form of work to increase the pressure of the gas. The efficiency of a compressor can be calculated using thermodynamic principles, such as the isentropic efficiency, which measures the actual work input compared to the ideal work input for a given compression process.

Details

Reciprocating compressors use a piston-cylinder arrangement to compress the air. The piston moves back and forth inside the cylinder, decreasing its volume and thus increasing the pressure and temperature of the air. This process adds energy to the air by increasing its internal energy when the piston does positive work [+ W] on the gas.

Reciprocating compressors are typically used for low-flow, high-pressure applications, such as in industrial settings or for small-scale home appliances.

Centrifugal Compressors are Somewhat Similar to Centrifugal Pumps

The centrifugal compressors are inherently high-volume machines. Centrifugal compressors differ from reciprocating compressors cause the pressure developed is a function of the density of air or gas, the speed of the impeller and of the restriction of the flow. The gas is accelerated in passing through the rapidly rotating impeller, and the velocity is converted into pressure in the volute casing surrounding the impeller.

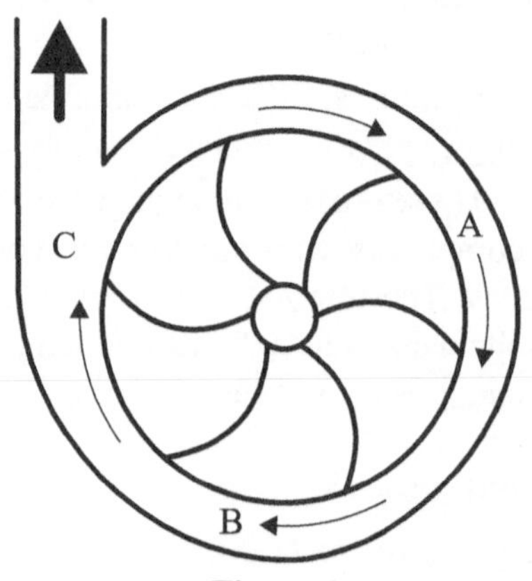

Figure 1

The centrifugal force utilized by the centrifugal pump is the same force utilized by the centrifugal compressors. The air particles enter the eye of the impeller designated *D* in image below. As the impeller rotates, air is thrown against the casing of the compressor. The air becomes compressed as more and more air are thrown out to the casing by the impeller blades. The air is pushed along the path designated *A*, *B*, and *C* in the image. The pressure of the air is increased as it is pushed along this path.

The centrifugal compressors utilize a high-speed rotating impeller to accelerate the air. As the air enters the compressor, it is directed to the center of the impeller where it gains kinetic energy due to the high rotational speed. The kinetic energy is then converted into pressure energy in the diffuser. The diffuser and the increasing diameter of the volute casing cause the air to slow down and increase in pressure that's why we call centrifugal compressors constant pressure variable flow. This is how centrifugal compressors use Bernoulli's equation.

Some other benefits of centrifugal compressors

- Centrifugal compressors are energy efficient and use less energy for large flows.
- Centrifugal compressors are oil-free, which means they don't contaminate equipment lines or the finished product with trace oil.
- Centrifugal compressors have fewer rubbing parts and longer-lasting bearings, which means they require less maintenance.
- Centrifugal compressors produce higher discharge pressures than axial compressors.

RECIPROCATING COMPRESSION OF AIR – THERMODYNAMICS ON *PV* AND *TS* DIAGRAM

An air compressor is a pneumatic device that converts power into potential energy stored in pressurized air. Positive-displacement compressors draw a volume of gas into the compressor cylinder's compression chamber where it is trapped, compressed, and pushed. The reference point for a compressor is its pressure-volume relationship on the [P-V] diagram.

Credits

- Greg Phillipi, Director Process compressor Marketing and Sale
- Ariel Corporation
- Wikipedia

  Direct Science.com

Contents

Capacity Control

Suction unloader, Bypass, Variable volume control pockets.

Efficiency

Volumetric efficiency, Isentropic efficiency on *TS* diagram
Volumetric efficiency, clearance, and compression ratio relationship

Adiabatic compression, how?

Concept of isentropy
Power to drive a reciprocating compressor

  The valve loss power on PV diagram

  Definitions of terms used in the post

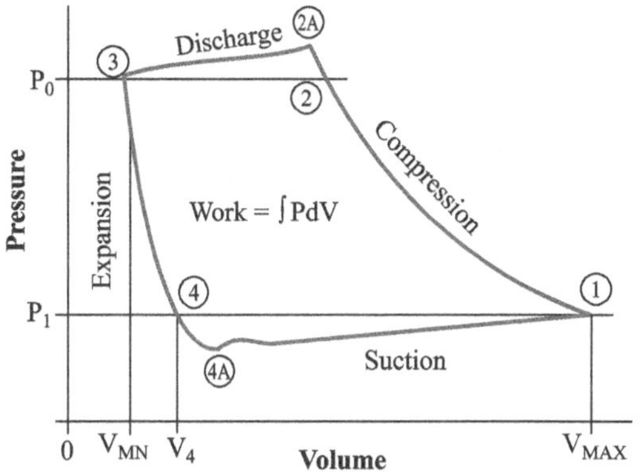

Compression ratio: It is defined as the ratio of the volume of air before compression to the volume of air after compression.

Compressor capacity: It is the quantity of air actually delivered by a compressor in m3/minute or m3/sec.

Swept Volume: The volume displaced or swept by the piston when it moves between top dead center and bottom dead center.

Figure 2

Clearance volume: It is the difference between the total volume and the swept volume, basically the gap that remains between the piston head and the cylinder head when at the top dead center.

Variable volume control pockets: A VVCP is a device on the outer end of a reciprocating cylinder that can add clearance to a cylinder.

Valve loss power: This valve loss power represents all of the inefficiency in the *P-V* diagram.

Volumetric efficiency: Volumetric efficiency (VE) is the percentage of stroke that fills with gas at suction pressure and suction temperature.

Overall efficiency: It is the ratio of isothermal power (*P* iso) to the shaft (Brake) Power (P shaft).

Quick Re-visit to Background

Ps, suction pressure, represents the pressure of the gas at the inlet to the compressor cylinder. *PD*, discharge pressure, represents the pressure of the gas at the outlet from the compressor cylinder. V_1 represents the maximum volume of gas trapped in the compression chamber and V3 the minimum. The difference between V_1 and V_3 is known as piston displacement, or how much volume is displaced in one stroke length of the piston.

***Piston Displacement* = $V_1 - V_3$**

P-V diagram is made up of four basic parts, 1–2, 2–2A–3, 3–4, and 4–4A–1. At position 1 on the *P-V* diagram, the volume of gas trapped in the compression chamber is maximum [Vmax] at suction pressure and temperature.

At position 1 all of the compressor valves are closed and the piston is at rest. All of the compressor valves are simple spring-loaded check valves and are not actuated by any outside means. As the crankshaft rotates the piston will move a*nd* the volume inside the compression chamber decreases (moving from Position 1 to 2 in the *P-V* diagram). From position 1, as the volume decreases, the gas pressure increases (the gas is trapped in the compression chamber with the volume decreasing). When the pressure inside the compression chamber becomes slightly higher than discharge pressure (PD, at position 2), the discharge compressor valve will open. This completes the first part of the diagram and is called the compression event.

At position 2, the discharge valve opens and, as the piston continues to move, gas at discharge pressure and the temperature is pushed out of the compression chamber through the discharge compressor valve and into the discharge gas passage of the cylinder. At Position 3, the piston comes to rest, the discharge compressor valve closes and this part ends. Part 2-2A-3 is called the discharge event. Moving from Position 1 to 3 the piston has moved through one stroke length and one-half revolution of the crankshaft. Position 3 represents the minimum volume of gas trapped inside the compression chamber (V3). Note this is not zero volume.

At position 3 the piston will reverse and travel in the opposite direction. As it moves the volume inside the compression chamber increases and the trapped gas (V3) will increase in volume and decrease in pressure—the gas will expand. Part 3–4 is called the expansion event.

At position 4, the pressure inside the compression chamber will be slightly less than suction pressure (PS) causing the suction compressor valve to open. With the suction valve open, the compression chamber is open to the suction gas passage and as the piston continues to move, the volume continues to increase and the compression chamber fills with gas at suction pressure and temperature. The piston returns to position 1, comes to rest, and the process repeats. Part 4–4A–1 is called the suction event. This completes the basic reciprocating compression process. One cycle around the P-V diagram represents one revolution of the crankshaft and two-stroke lengths of the piston, one from Position 1 to 3 and another from Position 3 to 1. Four events make up the P-V diagram—compression, discharge, expansion, and suction

Capacity

It is the volume of gas compressed by this *P-V* diagram is the difference in volume between Positions 1 and 4. Capacity $= V_1 – V_4$,

How Capacity is Controlled?

There are two main reasons why compressor capacity regulated. The most common reason is to adjust the suction flow to match the process demand. The second reason is to save energy. As a rule, capacity control is determined by the compressor discharge pressure. There are several methods. These methods are bypass recycling, unloader valves, clearance pockets, stepless valves, and variable-speed drives. Every method has some advantages and disadvantages as well. Bypass control is preferred over other control methods because of its smoothness, simplicity, and low initial costs.

We will discuss the following three methods.

Suction Unloader Valves

Unloader valves keep the suction valve open and let the gas return to the suction. In other words, suction valve unloaders are mechanisms that are held open or bypass suction valves of cylinders. This provides complete unloading of one or both ends of the cylinder. Unloaders are usually pneumatically operated by instrument air and equipped with positioner indication.

Bypass control

This control method uses an external bypass around the compressor to recycle gas from the compressor discharge to the inlet, or to the atmosphere in the case of an air compressor. It consists of piping that connects the discharge piping with the suction piping through a control valve—a recycle valve. When the valve is opened discharge, gas is allowed to flow back to suction and not proceed on down the discharge pipe thus reducing the amount of gas the proceeds down the pipe. The bypass should tie-in upstream of a suction knockout drum so that any condensate resulting from the expansion cannot enter the compressor. The problem is while the capacity is changed (lowered when the recycle valve is opened) the power is not, making recycling, therefore, this is an inefficient form of capacity control.

Variable Volume Control Pockets

A VVCP is a device installed on the outer end of a reciprocating cylinder that can add clearance to a cylinder by turning the pocket out. It is a device that uses only the differential operating pressures of the cylinder and allows easy adjustment of cylinder clearance while the unit is running. No external source of power is needed for the movement of this variable pocket piston.

How does control pockets work?

From point 1 to 2, the suction valve opens because the cylinder's internal pressure is less than external suction pressure; this draws gas into the cylinder. From point 2 to 3, gas compression begins when the suction valve closes. From point 3 to 4, compressed gas is discharged from the cylinder; when internal pressure is greater than external discharge pressure, the discharge valve opens. From point 4 to 1, the discharge valve closes and the suction valve remains closed until it reaches point 1. This is because of the trapped gas in the fixed clearance of the cylinder.

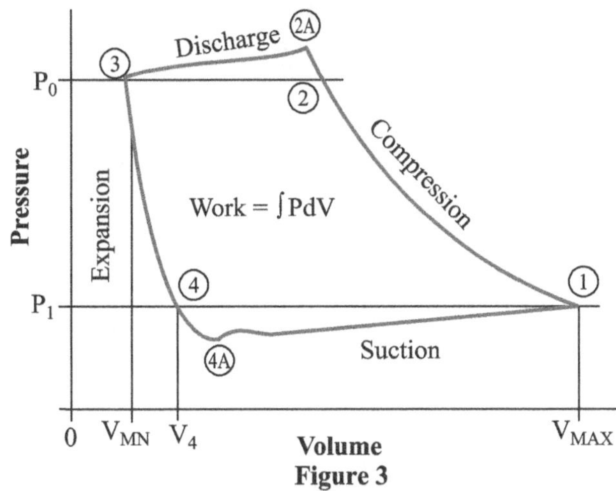

Figure 3

By opening the VVCP we add clearance to the end of the cylinder, effectively increasing the re-expansion period because the larger volume of gas trapped in the cylinder needs to move further during the piston stroke to re-expand out to suction pressure. On the compression cycle, the piston also needs to travel further along the stroke to reach discharge pressure as there is more volume to pressurize., Reducing the *VE*, reduces horsepower consumption (the area inside of the curve) and capacity moved.

Efficiencies

Volumetric efficiency (VE) is the percentage of stroke that fills with gas at suction pressure and suction temperature. Part 2-2A-3 is called the discharge event. Part 4-4A-1 is called the suction event. VVCP increases both discharge and suction events. That makes total stroke distance increase. The ratio of discharge stroke distance with the valve open (or suction stroke distance with the valve open) over total stroke distance is the discharge (or suction) volumetric efficiency. Volumetric efficiency (VE) is the percentage of stroke that fills with gas at suction pressure and suction temperature. Reduction of volumetric efficiency means a reduction in capacity.

Definition of Volumetric Efficiency

The volumetric efficiency represents the efficiency of a compressor cylinder to compress the gas. It may be defined as the ratio of the volume of gas actually delivered to the piston displacement, corrected to suction temperature and pressure.

Equation

$EV = 1 - (r^{1/k} - 1)\, CV$, EV = volumetric efficiency, fraction, r = cylinder compression ratio, CV = clearance, fraction.

As the compression ratio increases the capacity decreases as less gas is drawn into the compression chamber and compressed. As clearance fraction increases capacity decreases.

Theoretical volumetric efficiency versus clearance volume, compression ratio (pressure ratio *P* discharge/*P* suction), and *K* value. $K = Cp/Cv$.

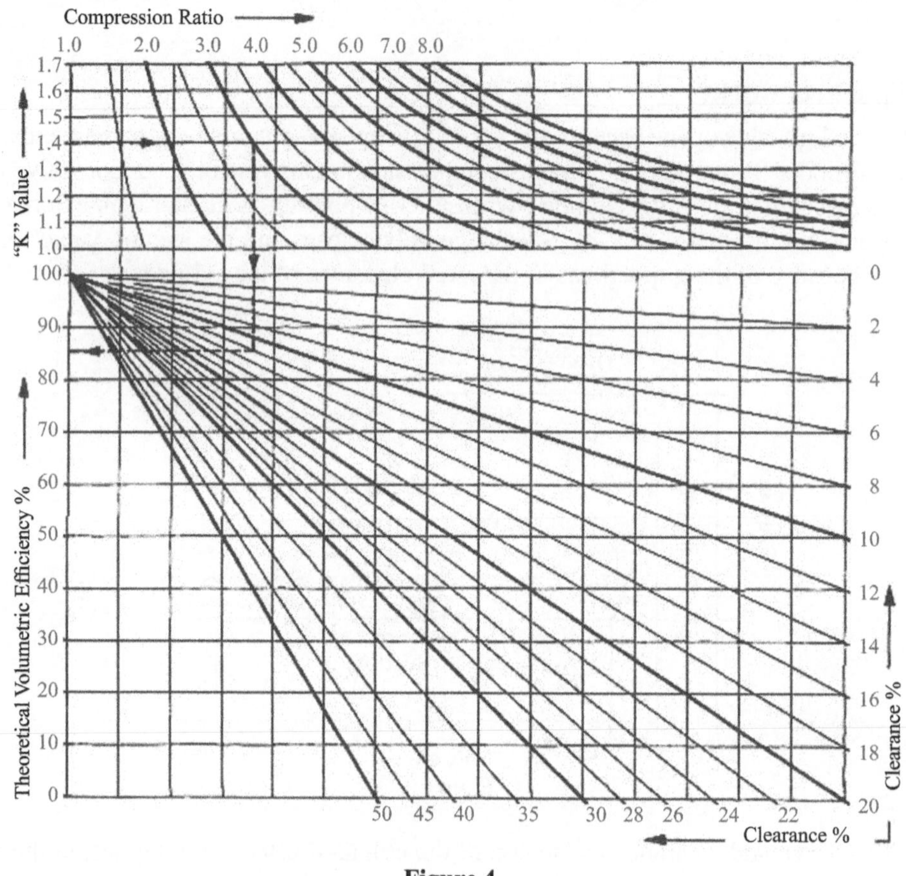

Figure 4

Example: *K*- 1.4; compression ratio- 3; clearance volume- 12%; theoretical volumetric efficiency- 85.5%. (Credit: Sulzer-Burckhardt, Winterthur and Basel, Switzerland.)

Efficiency Loses

Factors that decrease the volumetric efficiency, such as pressure losses across the valves increasing the pressure ratio, preheating of the gas in the suction chamber resulting in a decrease of its specific weight, deviation from ideal gas law (compressibility factors at both suction and discharge conditions), and internal gas leakages.

Power to drive a reciprocating compressor.

The power required to drive a reciprocating compressor can be divided into three pieces, adiabatic, valve loss. The power required to compress a volume of gas is represented by the area enclosed by the P-V diagram, work = ∫PdV. The compression and expansion events are expressed thermodynamically as adiabatic processes, meaning it is assumed that no heat is transferred to or from the gas during these events dQ = 0. An adiabatic thermodynamic process is an isentropic (constant entropy) process. The area of the P-V diagram in the figure below bounded by 1-2-3-4-1 is the adiabatic power.

Isentropic Process

An isentropic process is a thermodynamic process, in which the entropy of the fluid or gas remains constant. It means the isentropic process is a special case of an adiabatic process in which there is no transfer of heat or matter. It can be expressed with the ideal gas law as PV^k = Constant, $k = Cp/Cv$.

Concept of Isentropy and Isentropic Efficiency

An isentropic system must be adiabatic + reversible. Looking at the process if the system's interaction with the surrounding like a completely insulated system or a very fast energy transfer process when the energy for a process change comes from its internal energy because the system does not have any other option to draw energy, you can approximate the system to be adiabatic, $dQ = 0$. A compressor is an example as explained. But you cannot approximate how far the system is reversible. Therefore, you calculate the actual efficiency *W* act and you also calculate ideal isentropic efficiency *W* isentropic on assumption that your compression is isentropic with no change in entropy. You express the isentropic efficiency as % *W* actual/*W* isentropic. *W* stands for work done.

Though it is assumed compression efficiency is isentropic, in reality, the compression will always absorb more power, so isentropic efficiency is the ratio of isentropic to actual work of compression. Leakage losses are incurred from high to low pressure. There will also be heat transfer to and from external sources to the compressed fluid.

Isentropic Efficiency on *TS* Diagram

The Isentropic efficiency is a comparison of the actual power output with the Isentropic case. Typical Isentropic efficiencies range from 70–90%.

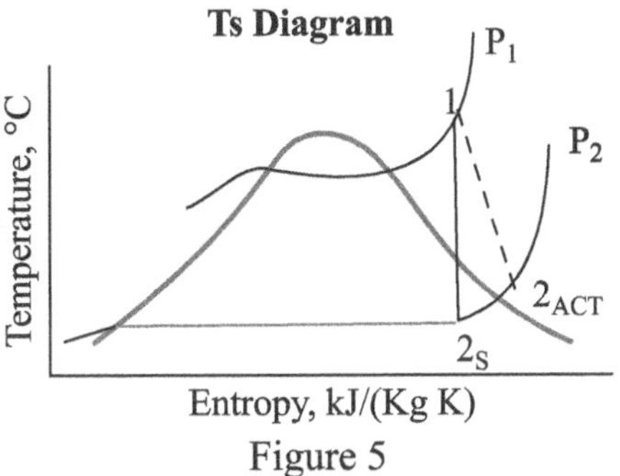

Figure 5

Isentropic Efficiency = *W* actual/*W* isentropic = [h_1[Enthalpy at state 1] – h_2 act [Actual enthalpy at state2] [/[h_1– h_2s [Isentropic enthalpy at state2]. *W* stands for work done.

How Compression and Expansion are Adiabatic?

How valid is the assumption that the compression and expansion events are adiabatic? For a compressor with a rotating speed of 300 rpm (a slow rotating speed), one *P-V* cycle takes only 0.2 seconds to complete. Assuming each of the four events of the P-V cycle take equal time, that's 0.05 seconds (or 50 milliseconds) per event. That's not much time for any significant amount of heat to transfer, therefore lending credibility to the adiabatic assumption. Yes, the gas does get hot as it's compressed but not from the heat being transferred to the gas. That heat is the heat of compression. Inefficiency in the P-V diagram is the pressure drop incurred in moving the gas from the inlet flange of the cylinder into the compression chamber and in moving the gas from the compression chamber to the outlet flange. Overcoming this pressure drop requires energy. This energy is represented by the areas 1-4-4A-1 (suction valve loss power) and 2-2A-3-2 (discharge valve loss power) in the figure below.

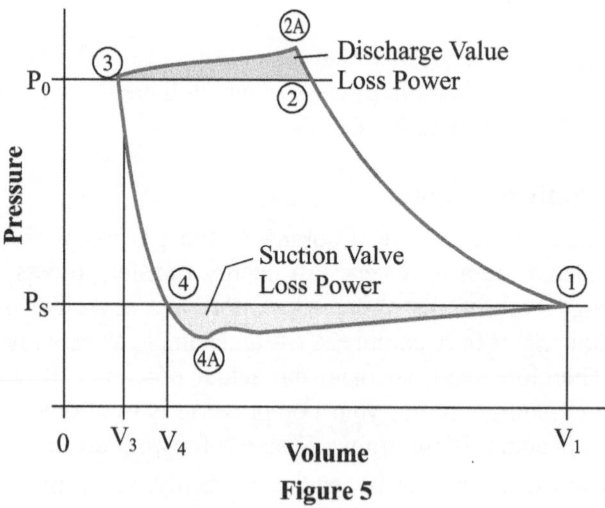

Figure 5

References

Greg Phillipi, Director Process compressor Marketing and SaleAriel Corporation

Wikipedia

Direct Science.com

SURGE: COMPRESSOR *VS.* CENTRIFUGAL PUMP

A surge is a sudden large increase in something that has previously been steady.

Surge is a phenomenon that occurs in a fluid-transporting pipe as a result of a change in flow velocity. When the flow drops, the energy of the fluid moving forward in the pipe causes a sudden rise in pressure upstream of the valve due to compression. Because the fluid is still moving, negative pressure will develop immediately downstream of the valve, causing a temporary separation of the liquid column followed by a reverse flow back towards the valve if the pressure gradient is high enough.

Compressor

Surge is defined as a condition in which the flow through the entire compressor is reversed intermittently, i.e., gas flows from the discharge nozzle to the suction nozzle. A surge is a disruption of the airflow through the compressor and this is caused by flow separation.

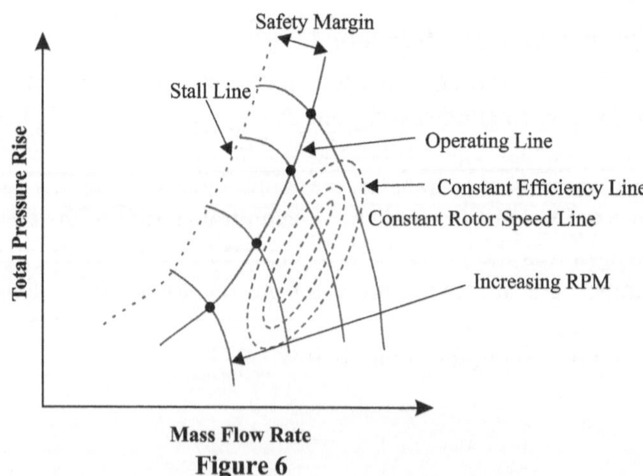

Figure 6

Only up to a certain pressure ratio will a compressor pump air in a stable manner. Beyond this point, the flow will degrade and become unstable. On a compressor map, this is referred to as the surge line. The entire engine is designed to keep the compressor operating a small distance below the surge pressure ratio on a compressor map, which is known as the operating line, see the image below. On a compressor map, the distance between the two lines is known as the surge margin. Various events can occur during engine operation to reduce the surge pressure ratio or increase the operating pressure ratio. When the two coincide, there is no longer any surge margin, and a compressor stage or the entire compressor can surge.

Concept of stall

There is a concept known as stall. Stall is a compressor aerodynamic phenomenon. A stall that results in the complete disruption of the airflow through the compressor is referred to as a compressor surge. The flow disruption is caused by flow separation. If the flow through a constant-speed compressor is reduced, the losses in all aerodynamic components increase.

Why does This Happen?

When dirt, saltwater mist, or other air contaminants enter the inlet air, the efficiency of the air compressor can be drastically reduced in a very short period of time. Contaminants build up in the air compressor, reducing its compression efficiency. As a result, the mass flow will be reduced, the compressor discharge pressure will be reduced, horsepower will be reduced, and engine temperatures will be higher than normal.

Components of Aerodynamics

Aerodynamics is the movement of objects through the air. The inlet, inlet guide-vane (IGV), impeller, diffuser, return channel, volute, and side-stream (or side entry/exit) are the key aerodynamic components in the centrifugal compressor.

The flow in one of the aerodynamic components, usually the diffuser, but sometimes the impeller inlet, will eventually separate. It should be noted that stall usually occurs in the first stage of a compressor.

Stall can be a precursor to surge, but it doesn't have to be. Despite the fact that stall occurs in one of the stages, multistage compressors, in particular, may still show increased head at reduced flow. Increased vibrations of the compressor rotor are frequently used to detect stall.

Flow separation in a vaneless diffuser means that all or a portion of the flow does not exit the diffuser at the discharge end, but instead forms areas where the flow stagnates or reverses direction back to the diffuser's inlet (i.e., the impeller exit).

Stalls in the impeller inlet or a vaned diffuser occur because the direction of the incoming flow (relative to the rotating impeller) changes with the flow rate through the compressor. When compared to a vaneless diffuser, vanes in the diffuser typically reduce the operating range of a stage. As a result, a decrease in flow will result in a greater mismatch between the direction of the incoming flow for which the impeller was designed and the actual direction of the incoming flow. This mismatch eventually becomes so significant that the flow through the impeller breaks down.

How does Fluid Separate?

If the fluid is slowing down in the streamwise direction, then the pressure is increasing in that direction. That is called an adverse pressure gradient with regard to the boundary layer because it tends to cause the boundary layer to separate. The sluggish fluid deep in the boundary layer near the wall is moving slowly and has little momentum, so the increasing pressure slows it down and even pushes it back upstream. Then it collides with the fluid coming downstream towards it and that deflects the flow upwards away from the wall. That outward deflection could be interpreted as a boundary layer with the thickness that is increasing rapidly. It's really more like a shear layer at that point, but grossly, ignoring the details, the region near the wall that is not moving at the freestream velocity just got a whole lot thicker.

The phenomenon of boundary layer separation can occur at a number of locations on a compressor. In such cases, the boundary layer is incapable of negotiating a local adverse pressure gradient, which is then relieved as it breaks away from the surface and induces a strong interaction with the inviscid stream.

Compressor Stalls are Classified into Two Types

Rotating Stall

A rotating stall is a local disruption of airflow within the compressor that continues to provide compressed air but at a lower efficiency. Rotating stall occurs when a small proportion of airfoils stall, disrupting local airflow but not destabilizing the compressor.

A compressor airfoil has pressure and suction sides that extend from the root to the tip and between the leading and trailing edges.

Stall cells are pockets of relatively stagnant air created by stalled airfoils that rotate around the circumference of the compressor rather than moving in the flow direction.

The stall cells rotate with the rotor blades, but at 50 to 70% of their speed, affecting subsequent airfoils around the rotor as each encounters the stall cell.

Axi-symmetric Surge

This is known as a compressor surge, or a pressure surge. It is a complete breakdown in compression that causes a reversal of flow and the violent expulsion of previously compressed air through the engine intake due to the compressor's inability to continue working against the already-compressed air behind it. The compressor is either subjected to conditions that exceed its pressure rise capabilities or is so heavily loaded that it is unable to absorb a momentary disturbance, resulting in a rotational stall that can spread to include the entire compressor in less than a second.

Surge Pressures

Surge pressures, also known as pressure surges, occur in a fluid-transporting pipe as a result of a change in flow velocity, such as when a valve is closed or opened too quickly. When a valve is quickly closed, the energy of the fluid moving forward in the pipe causes a sudden rise in pressure upstream of the valve due to compression.

Because the fluid is still moving, negative pressure will develop immediately downstream of the valve, causing a temporary separation of the liquid column followed by a reverse flow back towards the valve if the pressure gradient is high enough. This could result in the valve and pipe being destroyed.

When the maximum number of pumps at a pumping station stop operating due to a power outage, the highest surge pressures frequently occur. Because the change in pressure and velocity is not limited to the point of disturbance, it continues up- and downstream with the velocity of a pressure wave propagation.

The boundary conditions govern the extent to which waves are reflected at non-steady flow locations (e. g. pipe branches, valves, changes in cross-sections, tanks), resulting in a phase or amplitude reversal.

COMPRESSOR CURVE

Reading compressor curves can be essential to make informed decisions when selecting or operating a compressor.

1. **Flow Rate:** The x-axis on the compressor curve represents the flow rate of the compressor, typically in cubic feet per minute (cfm) or cubic meters per hour (m3/h). Determine the flow rate you require for your application.
2. **Pressure Ratio:** The y-axis on the compressor curve represents the pressure ratio, which is the outlet pressure divided by the inlet pressure. Identify the pressure ratio that corresponds to your system's needs.
3. **Efficiency:** Compressor curves often include efficiency lines that show the efficiency of the compressor at different flow rates and pressure ratios. Look for the efficiency line that corresponds to your desired operating conditions.
4. **Surge and Choke Regions:** Compressor curves may also indicate surge and choke regions where the compressor may experience instability or performance limitations. Avoid operating the compressor within these regions.
5. **Operating Point:** The intersection of the desired flow rate, pressure ratio, and efficiency lines on the compressor curve indicates the operating point of the compressor. This point represents the compressor's performance under specific operating conditions.
6. The horizontal axis of a centrifugal compressor performance curve usually represents the inlet gas volume flow rate, and the vertical axis represents the pressure ratio or head of the compressor.

7. **Operating point:** The operating point of a centrifugal compressor is where the head required by the process intersects with the head produced by the compressor
8. **Surge:** Surge is a dangerous phenomenon that occurs when a centrifugal compressor can't produce the amount of head needed by the process system. Surge can damage a compressor and make it inoperable.
9. **Stonewall:** Stonewall is when the physical limitations of the compressor prevent it from allowing more air through the stages. Operating at or beyond stonewall can cause cavitation-like action, which can damage the compressor.
10. **Most energy-efficient point:** For many centrifugal compressors, the most energy-efficient point is just before stonewall.

THERMODYNAMICS OF AIR COMPRESSION

Thermodynamics is the study of thermal energy, and it's fundamental to understanding how air compressors work.

Kinetic Energy

In a compressor, the kinetic energy of gas changes as it passes through the compressor, increasing and decreasing in velocity: The reason is the kinetic energy converts to pressure energy at the diffuser and it changes with pressure ratio.

- **Increase in velocity**: As the gas's velocity increases, so does its kinetic energy.
- **Decrease in velocity**: As the gas's velocity decreases, its kinetic energy is converted into increased pressure.

The gas velocity increases and decreases as the gas flows through the compressor, and these changes affect the compressor's performance:

Other Effects

Gas Velocity Increases

When the gas velocity increases, the flow rate through the compressor increases. However, the work input to the compressor also increases, which reduces the compressor's efficiency.

Gas velocity decreases

When the gas velocity decreases, the gas is forced to decelerate in a diffuser, which transforms the velocity into static pressure.

Gas Velocity Reaches Sonic Velocity

When the gas flow velocity reaches sonic velocity, the compressor reaches its upper flow rate limit.

Potential Energy

As the air gets compressed it can store more energy so the potential energy increases.

Internal Energy Change

Adiabatic Compression

In adiabatic compression, the internal energy of the air increases because work is done on the air. The increase in internal energy is equal to the external work done on the air. This increase in internal energy causes the temperature and pressure of the air to increase.

Adiabatic Expansion

In adiabatic expansion, the internal energy of the system decreases, causing the temperature and pressure to decrease.

Entropy Changes

- **Isothermal compression:** The entropy change is always negative.
- **Isentropic compression:** The entropy of the system remains constant.

What is Isothermal Compression?

Isothermal compression is a thermodynamic process that involves compressing a gas while keeping its temperature constant.

How it Works

In this process, heat is added or removed to a system at same rate as it's produced. This process is equivalent to an infinite number of intercoolers.

Applications

Isothermal compression is used in centrifugal compressors, gas turbines, and the process industry.

What is Isentropic Compression?

Isentropic compression is an idealized process that is used as a model for real processes:

Definition

Isentropic compression is an adiabatic process which is internaly reversible with no entropy generation either internally or externally.

Free Energy Change

The compression adds internal energy and the volume reduction reduces entropy in a compressor. Therefore, this makes a positive free energy change in the compression process.

CHAPTER 45

Acentricity of Molecules

In summary, acentricity is a measure of the non-spherical shape of a molecule and it can affect the strength of intermolecular forces such as van der Waals forces or hydrogen bonding. These intermolecular forces play a significant role in determining the thermodynamic properties of a substance, such as its boiling point, melting point, and vapor pressure.

Therefore, acentricity can impact various thermodynamic state variables.

The acentricity of a molecule is a measure of how asymmetrical or non-spherical the molecule is. It is a dimensionless quantity that is often used in thermodynamic calculations to describe the shape and interactions of molecules. Acentricity values range from –1 to 1, with values closer to 0 indicating a more spherical or symmetric molecule, and values further from 0 indicating a more asymmetrical or non-spherical molecule.

WHY ARE SOME MOLECULES ACENTRIC?

Some molecules are acentric due to their asymmetric shape or distribution of electron density. Various factors, such as the presence of different types of atoms or functional groups in the molecule, different bond angles or bond lengths, or the overall geometry of the molecule can cause this asymmetry.

For example, molecules with multiple substituents or functional groups not symmetrically arranged around the central atom are likely to be acentric. Additionally, molecules with lone pairs of electrons that distort the electron cloud around the central atom can also be acentric.

The table below shows some typical values of the acentric factor.

You may find in the table that Xenon which is an inert gas has an acentric factor w = 0 [almost]. Being an inert molecule it has the maximum sphericicity. While water has the maximum value of w in the table that shows it is the most polar and distorted molecule. Hydrogen has negative w because its compressibility Z > 1 dominated by repulsive forces.

Table 1

Substance	p_c, *MPa*	T_c, *K*	Z_c	ω
CO_2	7.38	304.1	0.274	0.239
H_2O	22.1	647.3	0.235	0.344
Ethane	4.88	305.4	0.285	0.099
Ethene	5.04	282.4	0.280	0.089
Propane	4.25	369.8	0.281	0.153
Xenon	5.84	289.7	0.287	0.008
O_2	5.04	154.6	0.288	0.025
H_2	1.3	33.2	0.303	-0.216
N_2	3.4	126.3	0.29	0.039

WHAT DOES ACENTRIC FACTOR *W* SIGNIFY?

The acentric factor (ω) can vary between –1 and +1, representing different molecular shapes and behaviors:

1. **$\omega < 0$:** This indicates that the molecule is more compact or spherical in shape compared to a reference spherical molecule. In other words, molecules with a negative acentric factor tend to be more symmetric and have a more ideal behavior.

2. **ω = 0:** A value of zero for the acentric factor usually indicates that the molecule has a relatively simple or spherical shape compared to the reference molecule. This value implies a moderate deviation from ideal behavior.
3. **ω > 0:** A positive acentric factor suggests that the molecule is more elongated or asymmetric in shape compared to the reference spherical molecule. Molecules with a positive acentric factor often exhibit more complex behavior, such as associating or exhibiting larger deviations from ideal gas behavior.

Examples

ω < 0: Methane (CH4) is an example of a molecule with a negative acentric factor. Methane is a simple, tetrahedral molecule with a relatively spherical shape compared to the reference molecule. As a result, its acentric factor is less than zero.

ω = 0: Xenon (Xe) is an example of a molecule that can have an acentric factor close to zero. Xenon is a noble gas known for its relatively simple spherical shape and low reactivity. While the acentric factor of xenon may not be exactly zero, it can be considered to have a value close to zero due to its symmetric and non-polar nature.

ω > 0: Acetone (CH_3COCH_3) is an example of a molecule with a positive acentric factor. Acetone has a more elongated, asymmetric shape compared to spherical reference molecules, leading to a positive acentric factor. This asymmetry can result in more complex behavior, such as deviations from ideal gas behavior or interactions with other molecules.

ROLE OF ACENTRIC FACTOR *W* IN THERMODYNAMICS

The acentric factor (ω) is a dimensionless number in thermodynamics. It describes the deviation of a pure substance from ideal gas behavior. It provides information about the intermolecular forces and the shape of the substance's phase diagram. The acentric factor is often used in equations of state, such as the Peng-Robinson equation, to predict the behavior of a substance over a range of temperatures and pressures.

Substances with higher acentric factors typically have stronger intermolecular forces, which can lead to deviations from ideal gas behavior at lower temperatures and higher pressures.

Examples

The acentric factor (ω) plays a significant role in determining a substance's compressibility and vapor pressure behavior. Additionally, the acentric factor plays a critical role in the reduced properties of gases.

Here's how it affects these properties:

COMPRESSIBILITY FACTOR *Z VS.* ACENTRIC FACTOR

Typically, a compressibility factor less than 1 ($Z < 1$) indicates that the interactions between gas molecules are attractive (more like a liquid), leading to a higher pressure than predicted by the ideal gas law. On the other hand, a compressibility factor greater than 1 ($Z > 1$) indicates that the interactions are repulsive and the gas molecules occupy more volume than expected.

So, in general, if the acentric factor is high ($\omega > 1$) and the compressibility factor is less than 1 ($Z < 1$), it means there are strong attractive forces between molecules leading to more liquid-like behavior and less repulsive forces.

Vapor Pressure

- The vapor pressure is the pressure at which a substance transitions from the liquid phase to the vapor phase at a given temperature. The acentric factor influences the strength of intermolecular forces, which in turn affects the tendency of a substance to escape into the vapor phase.
- Substances with higher ω values typically have higher vapor pressures compared to substances with lower ω values. This is because higher ω values indicate stronger intermolecular forces, making it easier for the substance to escape into the vapor phase.
- Additionally, substances with higher ω values tend to have a more pronounced boiling point curve, indicating a greater influence of intermolecular interactions on vapor pressure behavior.

The theory of the corresponding state applies to molecules only when acentric factor w is identical

The theory of corresponding states suggests that substances with the same reduced properties will exhibit similar behavior in terms of phase transitions, compressibility, and intermolecular forces, regardless of the specific molecular structure or chemical composition. This makes it easier to predict the behavior of substances under different conditions and to generalize their properties based on their acentric factor.

In the context of the theory of corresponding states, the concept of reduced properties plays an important role in comparing the behavior of different substances at their respective critical points.

Reduced properties are dimensionless quantities obtained by dividing the actual property of a substance by its value at the critical point. These reduced properties can then be plotted against each other to create dimensionless curves that describe the behavior of substances in a universal way.

When molecules have identical acentric factors (w values), their reduced properties will be the same, as the acentric factor is used to calculate the reduced temperature (Tr) and reduced pressure (Pr) values. This means that substances with the same acentric factor will have identical reduced temperature and pressure values, allowing for a direct comparison of their behavior at critical and near-critical conditions.

By using reduced properties, substances with identical acentric factors can be compared and analyzed in a universal manner, as the theory of corresponding states suggests.

CHAPTER 46

Chemical Reactions, Kinetics and Thermodynamics

Chemical reactions, kinetics and thermodynamics are inseparable. Chemical reaction refers to the transformation of reactants into products through the breaking and forming of chemical bonds.

Kinetics is the study of the rate at which chemical reactions occur, including factors that affect reaction rates such as temperature and concentration. Thermodynamics, on the other hand, deals with the overall energy changes that occur during a chemical reaction and whether the reaction is spontaneous or non-spontaneous.

The concept of Gibbs free energy allows us to calculate the maximum amount of useful work that can be obtained from a reaction. Thermodynamics also helps us understand the stability of chemical compounds and how different factors, such as temperature and pressure, affect the equilibrium of a reaction.

EQUILIBRIUM CONSTANT

What is Equilibrium Constant?

The equilibrium constant, denoted as K, is a measure of the extent to which a reaction will proceed to reach equilibrium. It is defined as the ratio of the product concentrations to the reactant concentrations, each raised to their respective stoichiometric coefficients in the balanced chemical equation.

In the context of thermodynamics, the equilibrium constant is related to the Gibbs free energy change (ΔG) of a reaction according to the equation:

$$\Delta G = -RT \ln(K)$$

where R is the gas constant, T is the temperature in Kelvin, and ln denotes the natural logarithm. This equation shows that the equilibrium constant is influenced by the temperature of the reaction and indicates whether the reaction will proceed in the forward or reverse direction to reach equilibrium.

The equilibrium constant is widely used in chemical kinetics and equilibrium calculations, as it provides important information about the state of a reaction system. It is commonly used in industries such as pharmaceuticals, petrochemicals, and environmental science to predict the outcome of chemical reactions and optimize reaction conditions.

Example

Calculation

Consider the reaction:

$$2A + 2B \rightarrow C + D$$

If the equilibrium constant for this reaction is $K = 0.5$, and the concentrations of A and B are each 0.1 M, while the concentrations of C and D are both 0 M, the equilibrium expression for the reaction would be:

$$K = [C][D]/[A]^2[B]^2$$

Substitute the given concentrations into the equation and solve for the concentrations of C and D:

$$0.5 = (x)(x)/(0.1)^2(0.1)^2$$

$$0.5 = x^2/0.0001$$

$$x^2 = 0.00005$$

$$x = 0.00707 \text{ M}$$

Therefore, at equilibrium, the concentrations of C and D would both be 0.00707 M.

Explanation (Very interesting to note).

What do We Learn?

If the concentrations of C and D at equilibrium are both 0 M, it means that the reaction has not proceeded significantly in the forward direction to produce these products. This suggests that the equilibrium lies strongly towards the reactants A and B, meaning that there are higher concentrations of A and B compared to C and D at equilibrium.

In this case, the equilibrium constant (K) for the reaction would be very small, indicating that the reaction favors the formation of the reactants over the products at equilibrium.

The value of K would be much less than 1, suggesting that the reaction does not proceed to a significant extent in the forward direction.

If the concentrations of C and D are both 0 M at equilibrium, it may be necessary to adjust the reaction conditions, such as temperature, pressure, or the concentrations of the reactants, in order to drive the reaction towards the production of C and D. This could involve manipulating the reaction conditions to favor the forward reaction and increase the concentrations of C and D at equilibrium.

RATE CONSTANT *VS.* EQUILIBRIUM CONSTANT

What is rate constant? How is rate constant different from equilibrium constant?

What is rate constant?

Rate constant, denoted as k, is a proportionality constant that relates the rate of a chemical reaction to the concentrations of reactants. It signifies how quickly a reaction will proceed at a specific temperature and concentration of reactants. A larger rate constant indicates a faster reaction, while a smaller rate constant indicates a slower reaction.

How different from equilibrium constant?

Rate constant is different from equilibrium constant (K_c or K_p) in that rate constant describes the speed at which a reaction occurs, while equilibrium constant describes the extent to which a reaction proceeds towards completion. Equilibrium constant is a reflection of the ratio of products to reactants at equilibrium, whereas rate constant is a reflection of the reaction rate at a particular point in time during the reaction.

Example

For example, in the reaction $A + B \rightarrow C$, the rate constant k may be 0.05 mol/L/s. This means that the reaction is proceeding at a rate of 0.05 mol of product C formed per liter per second at the given conditions of temperature and concentration.

VAN'T HOFF EQUATION AND VAN'T HOFF FACTOR

In one line, the Van't Hoff equation relates the effect of temperature on an equilibrium constant, while the Van't Hoff factor measures the extent of dissociation or ionization of a solute in a solution.

Van't Hoff Equation

What is it?

The Van't Hoff equation is an equation used to relate the change in equilibrium constant of a reaction with temperature. It is expressed as:

$$\ln(K_2/K_1) = -\Delta H/R\,(1/T_2 - 1/T_1)$$

Where:

- $\ln(K_2/K_1)$ is the natural logarithm of the ratio of equilibrium constants at different temperatures

- ΔH is the enthalpy change of the reaction
- R is the gas constant
- T_1 and T_2 are the temperatures at which the equilibrium constants are measured.

Significance

The Van't Hoff equation signifies the relationship between temperature and equilibrium constants, showing how changes in temperature affect the equilibrium position of a reaction. It is based on the principle that the equilibrium constant is temperature dependent.

Uses

The Van't Hoff equation is used in various fields such as chemical kinetics, thermodynamics, and biochemistry to study the temperature dependence of chemical reactions. It can be used to predict how changes in temperature will impact the equilibrium position of a reaction.

Example

Calculation

Consider the reaction

$$2A + B \rightarrow C$$

If the equilibrium constant K at 25°C is 10, and the enthalpy change ΔH is 50 kJ/mol, we can use the Van't Hoff equation to calculate the equilibrium constant at 35°C.

$$\ln(K_2/10) = -(50 \text{ kJ/mol})/(8.314 \text{ J/mol-K}) \times (1/308 \text{ K} - 1/298 \text{ K})$$

$$\ln(K_2/10) = -3000$$

$$K_2/10 = e^{(-3000)}$$

$$K_2 = 10 \times e^{(-3000)}$$

By solving the equation, we can find the equilibrium constant at 35°C using the Van't Hoff equation.

Van't Hoff factor

What is it?

The Van't Hoff factor, is a measure of the extent to which a solute dissociates or ionizes in a solution. It is denoted by the symbol "*i*" and is defined as the number of particles produced per molecule of a substance that dissolves in a solvent.

Uses

The Van't Hoff factor is used in colligative properties of solutions, such as boiling point elevation, freezing point depression, and osmotic pressure. It is particularly important for solutions containing substances that dissociate into ions when dissolved, such as ionic compounds.

Example

An example of the Van't Hoff factor can be seen in the dissociation of table salt (sodium chloride, NaCl) in water. When NaCl dissolves in water, it dissociates into sodium ions (Na^+) and chloride ions (Cl^-). Since NaCl dissociates into two ions per molecule, the Van't Hoff factor for NaCl in solution is 2.

In general, the Van't Hoff factor is used to predict the impact of solute particles on the colligative properties of a solution, based on the degree of dissociation or association of the solute in the solvent. It helps in understanding and calculating the effects of solutes on solution properties in various chemical and physical processes.

ROLE OF THERMODYNAMIC LAWS IN THE JOURNEY OF REACTANTS TO PRODUCTS

The journey of A and B to C and D

$$A + B = C + D$$

This note lists only a few of them.

Arrhenius Equation: The Arrhenius equation relates the rate constant of a reaction (k) to the activation energy (Ea), temperature (T), and pre-exponential factor (A). It helps in understanding the kinetics of the reaction and how the rate of the reaction changes with temperature.

Rate Law Equation: The rate law equation expresses the rate of a chemical reaction in terms of the concentration of reactants. It allows us to determine the order of the reaction with respect to each reactant and the overall rate expression.

Equilibrium Constant (K): The equilibrium constant (K) relates the concentrations of products and reactants at equilibrium. It helps in determining the extent of the reaction towards the formation of products or reactants and provides information on the spontaneity of the reaction.

Gibbs Free Energy Equation: The Gibbs free energy equation ($\Delta G = \Delta H - T\Delta S$) combines the enthalpy (ΔH) and entropy (ΔS) changes of a system to calculate the change in free energy during a reaction. It determines the feasibility and spontaneity of the reaction at a given temperature.

Van't Hoff Equation: The Van't Hoff equation relates the change in equilibrium constant (ΔK) to the temperature change (ΔT) and enthalpy change (ΔH) of a reaction. It helps in predicting the effect of temperature on equilibrium position and the direction of the reaction.

Kinetic Gas Equation: The kinetic gas equation (PV = nRT) relates the pressure (P), volume (V), amount of substance in moles (n), gas constant (R), and temperature (T) of a gas. It provides a quantitative relationship between these variables during a chemical reaction involving gases.

Barriers

Activation Energy (Ea) Barrier

- The activation energy barrier represents the minimum energy required for a reaction to occur. It is a key kinetic parameter that influences the rate of the reaction.
- A higher activation energy barrier implies a slower reaction rate, as a larger fraction of collisions between reactant molecules will not have sufficient energy to overcome this barrier. A high activation energy often requires catalysts.

Thermodynamic Barriers

- The thermodynamic barrier is related to the free energy change (ΔG) of the reaction, which determines the feasibility and direction of the reaction.
- If the reaction has a positive ΔG value, it faces a thermodynamic barrier in the form of an unfavorable energy change that hinders the conversion of reactants to products.

ACTIVITY, ACTIVITY COEFFICIENT AND CHEMICAL POTENTIAL

These concepts are often confused.

1. **Activity**: Activity is a measure of the "effective concentration" of a species in a solution. It takes into account not only the concentration of the species but also any interactions or deviations from ideal behavior in the solution.

It is defined as the ratio of the chemical potential of the species in solution to the standard state chemical potential.

$$\text{Activity} = \gamma \times [X]$$

Where:

γ is the activity coefficient

[X] is the concentration of the species

2. **Effective Activity**: Effective activity is a concept used in complex systems where multiple interactions are present. It takes into account the activity coefficients of all species involved in the system to determine the overall effective activity of a species.

$$\text{Effective Activity} = \prod(\gamma i) \times [X]$$

Where:

γi is the activity coefficient of each species i

[X] is the concentration of the species

3. **Activity Coefficient**: The activity coefficient is a measure of the deviation of a solution from ideal behavior. It takes into account the effects of interactions between different species in a solution. It is defined as the ratio of the activity of a species to its concentration.

$$\gamma = \text{Activity}/[X]$$

4. **Chemical Potential**: Chemical potential is a thermodynamic property that represents the potential energy of a species in a system. It is related to the Gibbs free energy and is a measure of the tendency of a species to react or move from one phase to another.

The chemical potential of a species in a solution is related to its activity by the equation:

$$\mu = \mu° + RT \times \ln(\gamma \times [X])$$

Where:

μ is the chemical potential of the species in the solution

$\mu°$ is the standard state chemical potential

R is the gas constant

T is the temperature

Example

Consider the equation for the ionization of a weak acid (HA) in water:

$$HA \rightleftharpoons H^+ + A^-$$

The equilibrium constant expression for this reaction can be written as:

$$Ka = [H^+][A^-]/[HA]$$

In this equation, $[H^+]$, $[A^-]$, and $[HA]$ represent the activities of the hydrogen ion, conjugate base, and weak acid, respectively.

Activity is used in the equilibrium constant for liquids when the conditions are non-ideal, such as in concentrated solutions. Activity is a measure of the effective concentration of a species under these conditions.

All these concepts are used to describe and quantify the behavior of species in solution, taking into account factors such as concentration, interactions, and deviations from ideal behavior.

Calculations involving these concepts are used in various fields of chemistry and chemical engineering to understand and predict the behavior of complex systems.

WHEN CONCENTRATION AND WHEN ACTIVITY USED IN EQUILIBRIUM CONSTANT?

When is concentration used and when is activity used in equilibrium constant calculation:

1. Concentration is typically used when describing the amount of a substance dissolved in a solution or present in a mixture. It is commonly used in chemical reactions and equilibrium calculations to quantify the amount of reactants and products. Concentration is expressed in units such as molarity (mol/L), percent by weight, or parts per million.
2. Activity is a concept used in thermodynamics and chemical equilibrium to account for the non-ideality of solutions. Activity is defined as the effective concentration of a species in a solution, taking into account the interactions

between particles in the solution. The activity coefficient is a dimensionless factor that corrects for deviations from ideality.

In the calculation of the equilibrium constant, the activity of a species is used instead of its concentration. The relationship between activity and concentration is described by the activity coefficient, which is multiplied by the concentration to obtain the activity of a species. This correction factor accounts for the deviations from ideal behavior and ensures that the equilibrium constant is calculated accurately for non-ideal solutions.

ACTIVITY COEFFICIENT, CHEMICAL POTENTIAL, AND FUGACITY

What is the Activity Coefficient?

Activity coefficients are dimensionless values that indicate the deviation of a solution's behavior from ideal behavior. In an ideal solution, the interactions between molecules are considered to be the same between all species, but in real solutions, interactions can vary leading to non-ideal behavior. The activity coefficient takes into account these non-ideal interactions and is used to correct the concentrations of the components in a solution to accurately predict the behavior of the solution. The activity coefficient can be defined as: $\gamma i = ai/ci$ where, γi = activity coefficient of component *i* ai = activity of component *i* ci = concentration of component *i*.

What is Chemical Potential?

Chemical potential is a thermodynamic potential that measures the potential energy of a substance to undergo a chemical reaction or physical change. It represents the change in free energy of a system when a unit amount of a component is added to the system. In simple terms, it can be thought of as the work required to transfer a substance from a reference state to a particular state.

The chemical potential equation

$$\mu = \mu^\circ + RT\ln(\gamma a)$$

where μ° is the standard chemical potential, R is gas constant, T is the temperature, and a is the activity of the specie.

What is Fugacity?

Fugacity is a thermodynamic concept that represents the tendency of a substance to escape or vaporize from a mixture. It is often used in context of gases and liquids to describe the effective pressure of a substance in a non-ideal mixture. Fugacity is related to chemical potential of a substance and is used to calculate equilibrium conditions and properties of mixtures.

Relationship between activity coefficient, chemical potential, and fugacity

The activity coefficient (γ), chemical potential (μ), and fugacity (f) are all related in the context of non-ideal solutions. The activity coefficient is a measure of how the effective concentration of a species in a mixture deviates from its ideal behavior. It accounts for interactions between molecules in the solution.

The chemical potential represents the free energy per particle in a system and is related to the activity of a species in a non-ideal solution. The chemical potential of a species in a non-ideal solution can be expressed as $\mu = \mu^\circ + RT\ln(\gamma a)$, where μ° is the standard chemical potential, R is the gas constant, T is the temperature, and a is the activity of the species.

Fugacity is a measure of the escaping tendency of a substance in a non-ideal mixture. It is related to the chemical potential through the equation $f = \gamma P$, where P is the total pressure. **In Summary**, the activity coefficient, chemical potential, and fugacity are all related in that they all describe the behavior of species in non-ideal solutions and are used to account for deviations from ideal behavior.

WHAT DOES NERNST EQUATION SIGNIFY?

In one line the Nernst equation signifies the thermodynamic work done by electrons in a cell. Nernst equation predicts the direction of an electrochemical reaction. So, in simple terms Nernst equation is a measure of thermodynamic work in a cell.

When the cell potential is positive, the free energy change dG is negative, the corrosion is spontaneous.

When the cell potential is negative, the free energy change dG is positive, the corrosion is not spontaneous.

Remember the above lines.

Nernst equation tells us the direction of an electrochemical reaction like corrosion.

Nernst equation $dG = -nFE$ signifies thermodynamic work done by electrons to flow from anode to cathode. The work done by electrons consume energy. When there is consumption of free energy the reaction moves forward to reach equilibrium. This makes an electrochemical reaction moving forward spontaneously to reach equilibrium.

In other when there is consumption of energy and dG is negative the reaction becomes spontaneous.

Details

In an electrochemical cell, the Nernst equation is used to determine the cell potential and the feasibility of the redox reaction. The cell potential is a measure of the driving force behind the reaction, and it represents the maximum amount of work that can be extracted from the cell.Electrons do work when they flow from anode to cathode.

When the cell potential is positive, the reaction will proceed in the forward direction (spontaneous reaction) and work can be done by the electrons flowing through the external circuit. In this case, the Gibbs free energy change (ΔG) for the reaction will be negative, indicating that the reaction is thermodynamically favorable.

On the other hand, if the cell potential is negative, the reaction will proceed in the reverse direction (non-spontaneous reaction) and work will need to be done on the system to drive the reaction in the desired direction. In this case, the Gibbs free energy change (ΔG) will be positive, indicating that energy needs to be supplied to the system for the reaction to occur.

$$dG = -n\,FE$$

G is free energy, n. Is the number of electrons transferred, F Is Faraday constant, which is the charge on a mole of electrons. and E is cell potential.

You may notice in the above equation that when the cell potential E is positive, dG is negative. Meaning a spontaneous process. It is just the opposite when the cell potential is negative.

Overall, the Nernst equation allows us to quantify the relationship between the cell potential, work done in the system, and the thermodynamic feasibility of the redox reaction. A positive cell potential implies a spontaneous reaction with negative ΔG, while a negative cell potential implies a non-spontaneous reaction with positive ΔG. Work is done and energy is consumed when the reaction moves towards equilibrium, which is reflected in the sign of ΔG.

CHEMICAL REACTION *VS.* DISSOLUTION: HOW AND WHY A SOLUTE DISSOLVES IN A SOLVENT?

The dissolution and chemical reaction are different processes.

Dissolution

A physical change in which a substance dissolves in a solvent to form a solution. The substance's composition does not change, and it can be returned to its original state by evaporating the solvent. For example, dissolving sugar in water is a physical process.

Chemical reaction

A chemical change in which two or more substances, called reactants, interact to form a new substance. The chemical composition of the substances changes, and energy is involved.

How and why a solute dissolves in a solvent?

There are two requirements (1) kinetics and (2) thermodynamics. Until these requirements are met a solute can't dissolve in a solvent.

In the case of a solid or liquid solute, the interactions between the solute particles and the solvent particles are so strong that the individual solute particles separate from each other and, surrounded by solvent molecules, enter the solution.

Kinetics

The kinetics of dissolution involves the lattice energy of solute and the hydration energy of the solvent.

Lattice Energy

The energy required to separate the ions in a solid lattice.

Explanation

It's a measure of the strength of ionic bonds that hold the compound together. Lattice energy is related to physical properties of ionic solids, such as their solubility, hardness, and volatility.

Hydration Energy

The energy released when ions from a solid become surrounded by water molecules.

The hydration energy and lattice energy of a solute determine whether it will dissolve in water and how well it will dissolve.

Explanation

It is the energy liberated when something is dissolve in water. If hydration energy is greater than lattice energy substance will soluble.

High Solubility

If the hydration energy is greater than the lattice energy, the solid dissolves well and the solution is exothermic

Low Solubility

If the lattice energy is greater than the hydration energy, the solid does not dissolve easily. increasing the temperature of a solution increases the solubility of many solid solutes in liquid water.

Explanation

The higher kinetic energy of the solute molecules at higher temperatures makes it easier for the solvent molecules to break apart the solute molecules. This is because the intermolecular attractions that hold the solute molecules together are less effective at higher temperatures.

Thermodynamics

A solute dissolves in a solvent when the resulting solution has a small free energy of dissolution (Gibbs energy) and large. entropy. The dissolution process increases entropy and minimises the free energy. $\Delta G = \Delta H - T\Delta S$.

As entropy change ΔS. increases, the free energy changev ΔG minimises and the solution moves towards equilibrium.

Entropy

Dissolving a solute in a solvent usually increases the entropy of system. This is because the solute molecules spread out through the solvent, which increases the randomness and disorder of the system.

Free Energy of Dissolution

A smaller free energy of dissolution means greater solubility.

The free energy of dissolution is dependent on temperature and is given by the relationship: $\Delta G = \Delta H - T\Delta S$.

As entropy increases with dissolution the free energy change ΔG minimises.

When a solute dissolves in a solvent, the molecules of the solvent arrange around the solute molecules. This process is called solvation. The chemical properties of the solvent and solute, such as hydrogen bonding, dipole moment, and polarizability, do this arrangement.

HOW A CHEMICAL REACTION, A + B = C TAKES PLACE?
HOW NEGATIVE FREE ENERGY CHANGE MAKES A REACTION SPONTANEOUS?

The note explains, [1] how chemical energy produces work and [2] what makes a negative free energy change spontaneous forward-moving reaction.

A chemical reaction is a rearrangement of atoms of reactants into a new format. The single key point is that every substance wants to stay in a state of minimum energy and maximum stability. *A* and *B* may have many options and many pathways to react and produce a product but the final call is taken by Gibbs free energy. The molecule with the lowest energy and maximum stability becomes the choice when two reactants create a new molecule.

Here, the concept of Free energy becomes important. While this is the requirement of thermodynamics, the kinetics of the reaction or the speed of the reaction is equally important.

Coming to Gibbs's free energy change requirement in a reaction, dG must be negative. Only when the free energy change is negative it pushes the reaction forward to reach $dG = 0$ or equilibrium thus making the reaction spontaneous. In other words, in a reaction with negative free energy change, $-dG$ always strives to move forward spontaneously to achieve equilibrium, $dG = 0$. This is the explanation in general of why negative free energy change makes a reaction a forward-moving spontaneous reaction. In making free energy change negative, $dG = dH - TdS$, the reaction must produce enough entropy S [disorder or new microstates] to make G negative.

Coming to the kinetics of a reaction, it is primarily controlled by the temperature or the collision frequency of molecules. Here, the concept of activation energy plays its role.

In other words, how easily the interatomic bonds are broken and make the atoms available to make new bonds by doing 'Work 'against the proton-proton repulsions between two atoms become important.

Incidentally, this is the 'Work' a chemical reaction does. Interestingly, it is not visible, unlike thermodynamic PV work.

Summary

Thermodynamics and kinetics both play roles in a chemical reaction. At the primary level, three factors influence a reaction from going from reactants to products: [1] temperature, [2] pressure, and [3] concentration. At the secondary level, these three factors generate three requirements: [1] activation energy, [2] enthalpy, and [3] entropy. The activation energy determines how fast or slow a reaction can proceed. This is the kinetics part of the reaction. Thermodynamics is the study of equilibrium and essentially free energy. The change in free energy in a chemical reaction is the result of three factors: [1] enthalpy, [2] entropy, and [3] temperature.

Who Controls a Reaction?

Who controls, thermodynamics or kinetics is decided by the composition of a reaction product mixture when competing pathways lead to different products. If in a reaction there are two products *X* and *Y*. If *X* has lower activation energy the kinetics would favor *X* since *X* would form faster than *Y*.

Yet *X* may not be a stable product unless the Gibbs free energy change [dG] is negative. So not necessarily even if kinetics meets the activation energy level the thermodynamics can still put a stop on the reaction if dG is not negative,

So, in a chemical reaction, there could be a kinetic product and the other could be a thermodynamic product. Thermodynamics can favor a reaction yet it could be such a slow reaction that it would not proceed.

Thus, thermodynamics and kinetics are inextricably linked.

With the above background let's look at two key controlling factors for a chemical reaction.

- Kinetics
- Gibbs free energy change

Activation energy

The activation energy (Ea) of a reaction is measured in joules per mole. The activation energy of a particular reaction determines the rate at [kinetics] at which it will proceed. The higher the activation energy, the slower the chemical reaction will be.

Even though a chemical reaction may already be thermodynamically feasible, activation energy is the minimal amount of energy required for the reaction to occur. The potential barrier that stands between the reactants and products in chemical kinetics is known as the activation energy. It establishes how the reaction rate varies with temperature. By offering a different pathway for the reaction, a catalyst can lower the activation energy of a reaction.

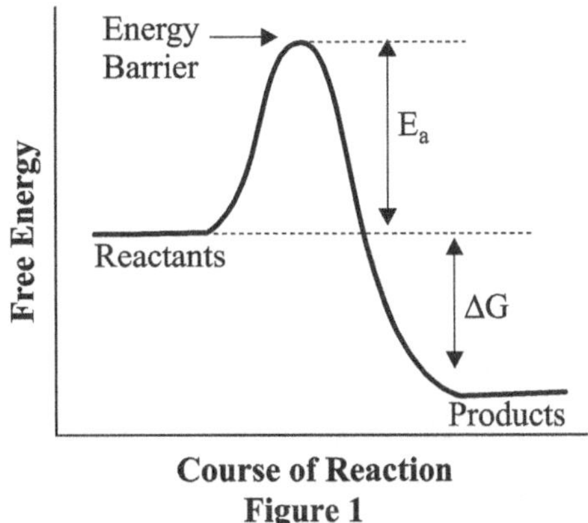

Course of Reaction
Figure 1

The activation energy *(Ea)* and the free energy difference (ΔG) between the reactants and products are two different quantities, as the diagram that goes with it illustrates.

Activation energy and reaction rate relation

Ea vs. Rate constant

$K = A\, e^{(-Ea/RT)}$, this is the Arrhenius equation. It shows the rate constant of reaction K is the exponential inverse of activation energy *Ea*, meaning that the rate of reaction decreases exponentially with an increase in activation energy.

Arrhenius equation explanation

$$K = A\, e^{(-Ea/RT)}$$

Meaning of Arrhenius equation

If we neglect the A factor for the time being. It may be observed in the above equation that what is "decaying" here is not the concentration of a reactant as a function of time, but the magnitude of the rate constant K as a function of the exponent $-Ea/RT$. K is the rate constant.

And what is the significance of this quantity? Recalling that RT is the average kinetic energy, it becomes apparent that the exponent is just the ratio of the activation energy Ea to the average kinetic energy. The larger this ratio, the smaller the rate (hence the negative sign). This means that high temperature and low activation energy favor larger rate constants, and thus speed up the reaction. Because these terms occur in an exponent, their effects on the rate are quite substantial.

Summary

What controls the kinetics of a chemical reaction?

It is the ratio of Activation energy [Ea]/Kinetic energy [*RT*] that drives the rate constant of reaction. The more the kinetic energy the less is the activation energy for a rate constant.

Thermodynamics

Gibbs free energy change, in short controls the thermodynamics or the feasibility of a chemical reaction.

What is free energy?

The free energy of a system is a measure of the energy that is available to do work. In the context of chemical reactions, the free energy change (dG) indicates whether a reaction will proceed spontaneously or non-spontaneously. The fundamental Gibbs equation is, $dG = dH - TdS$, G is free energy, H is enthalpy, S is entropy and T is temperature.

- **Positive ΔG**: A positive ΔG indicates that the products of the reaction have more free energy than the reactants. In this case, the reaction is non-spontaneous, and energy must be supplied to drive the reaction forward.
- **Negative ΔG:** A negative ΔG indicates that the products of the reaction have less free energy than the reactants.

This implies that the reaction is spontaneous and will proceed without the need for additional energy input. Therefore, the sign of the free energy change (ΔG) determines the spontaneity of a reaction. Negative ΔG indicates a spontaneous reaction, while positive ΔG indicates a non-spontaneous reaction.

Explanation

Only when the free energy change is negative it pushes the reaction forward to reach $dG = 0$ or equilibrium thus making the reaction forward going and spontaneous. In other words, a reaction with negative free energy change, $-dG$ always strives to move forward spontaneously to achieve equilibrium, $dG = 0$. This is the explanation in general why negative free energy change make a reaction a forward moving spontaneous reaction. In making free energy change negative, $dG = dH - TdS$, the reaction must produce enough entropy S [disorder or new microstates] to make G negative.

RELATIONSHIP BETWEEN RATE CONSTANT AND TEMPERATURE

The Arrhenius equation is a mathematical expression that describes the relationship between the rate constant (K) of a chemical reaction, the activation energy (Ea) required for the reaction to occur, and the temperature (T) at which the reaction takes place.

The equation is written as: $K = A\ e^{(-Ea/RT)}$

Where: $-K$ is the rate constant of the reaction

- A is the pre-exponential factor, which represents the frequency of collisions between reactant molecules
- e is the base of the natural logarithm (approximately equal to 2.71828)
- Ea is the activation energy, which is the minimum amount of energy required for the reaction to occur
- R is the gas constant (8.314 J/molK)
- T is the temperature in Kelvin Each term in the Arrhenius equation relates to the reaction rate as follows:
- The pre-exponential factor (A) accounts for the frequency of collisions between reactant molecules.

Higher values of A indicate a greater frequency of collisions, which can increase the reaction rate.

The activation energy (Ea) represents the minimum energy required for the reaction. Higher values of Ea result in a slower reaction rate, as more energy is needed for the reactants to overcome the energy barrier and form products.

The temperature (T) affects the reaction rate by influencing the kinetic energy of the reactant molecules.

As the temperature increases, the kinetic energy of the molecules increases, leading to more collisions and a higher reaction rate. This effect is captured in the exponential term $e^{(-Ea/RT)}$ in the Arrhenius equation.

Overall, the Arrhenius equation provides a quantitative relationship between a chemical reaction's rate constant, activation energy, and temperature.

RELATIONSHIP BETWEEN ACTIVATION ENERGY, EA, FREE ENERGY CHANGE, DG AND RATE CONSTANT *K*

Definitions

What is activation energy Ea?

Activation energy is the minimum amount of energy required to start a chemical reaction.

What is free energy dG?

Free energy, also called as Gibbs free energy, is the amount of energy in a system that can be used to do work.

What is rate constant k?

A rate constant is a proportionality constant that shows the relationship between the rate of a chemical reaction and the concentration of reactants.

All above are related

Case 1

Relationship between Ea and K

The rate constant is related to temperature, activation energy, and frequency factor. Arrhenius equation, which relates rate constant (k) to temperature, activation energy (Ea), and the gas constant (R), is given by: $k = A\ e^{(-Ea/RT)}$

Where:

- k is the rate constant
- A is the frequency factor (or pre-exponential factor), which represents the frequency of collisions between reactant molecules
- Ea is the activation energy
- R is the gas constant
- T is the temperature in Kelvin

What does it mean?

This equation combines kinetics and thermodynamics because it incorporates the energy barrier (activation energy) that reactant molecules must overcome for a reaction to occur. As the temperature increases, the rate constant increases exponentially due to the temperature dependence of the exponential term in the Arrhenius equation. This reflects the fact that, at higher temperatures, the reactant molecules have more kinetic energy and are more likely to overcome the activation energy barrier.

Case 2

Relationship between k and dG

The rate constant (k) is related to the free energy change (ΔG) through the Arrhenius equation:

$$k = A\ e^{(-\Delta G/RT)}$$

where:

k is the rate constant,
A is the pre-exponential factor,
e is the base of the natural logarithm,
ΔG is the free energy change,
R is the gas constant,
T is the temperature.

What does it mean?

This equation shows that the rate constant is inversely proportional to the free energy change. As the free energy change increases, the rate constant decreases and vice versa. In general, reactions with a lower free energy change will have a higher rate constant, indicating that they will occur more quickly.

Case 3

Relationship between Ea and dG

The activation energy (Ea) and the free energy change (ΔG) are related through the **Eyring equation,** which describes the rate of a chemical reaction

$$k = (kB\ T/h)\ e\text{^}(-\Delta G‡/RT)$$

where:

k is the rate constant,

kB is the Boltzmann constant,

T is the temperature

h is the Planck constant,

$\Delta G‡$ is the Gibbs free energy of activation,

R is the gas constant.

From the Eyring equation, we can see that the activation energy (Ea) is related to the Gibbs free energy of activation ($\Delta G‡$) as $Ea = \Delta G‡ + RT$

What does it mean? This relationship shows that the activation energy is directly related to the Gibbs free energy of activation.

Activation energy

The amount of energy required for a reaction to reach its transition state. It's the energy difference between the reactants and the activated complex, also known as the transition state.

Effect of a catalyst

A catalyst changes the activation energy of a reaction, but it doesn't change the overall enthalpy change of the reaction. This is because a catalyst changes the reaction rates without changing the reaction products.

CHEMICAL ENERGY: AN OVERVIEW

A chemical reaction is essentially a transfer of potential energy between reactants and products through electron relocation from a higher energy to a lower stable energy product. This is not present in gases. Gases have only kinetic energy, and the kinetic energy of reactants and products at a given temperature is the same. This is why, in order for N_2 to react with H_2 to form NH3, 400 degrees Celsius and 200 atmospheres of pressure are required. This means that you must provide a large amount of potential and kinetic energy externally to make this reaction occur. When you pressurize a gas, it simply implies supply of potential energy to the gas.

Chemical energy, in a nutshell, is the redistribution of potential energy between reactants and products caused by electron movement from a high to a lower energy level.

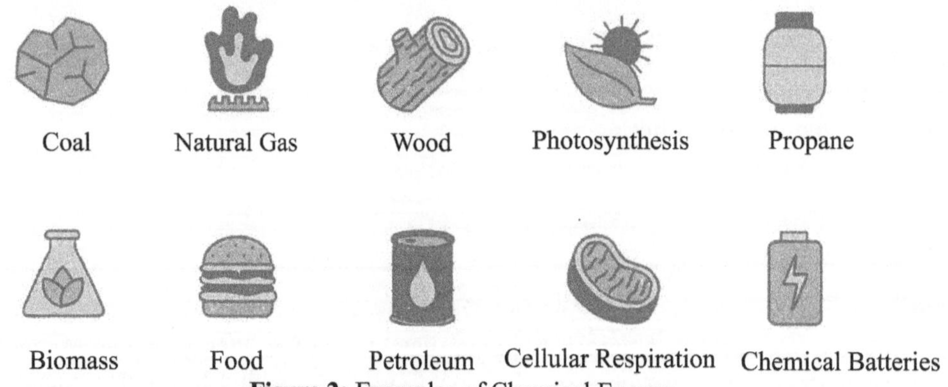

Figure 2: Examples of Chemical Energy

Definition of chemical energy: Collins dictionary

"Chemical energy means involving or resulting from a reaction between two or more substances or relating to the substances."

It simply means that chemical energy results from the interaction of two substances when they react to produce a new substance.

$A + B \rightarrow C$ [new substance]

The interesting thing is that every *A* and *B* do not combine to produce energy. The next obvious question is why chemical energy is so selective in selecting the right partners to combine with, as opposed to thermal energy, which has only one criterion, Δt. When an acid and a base are combined, they immediately react. However, no reaction occurs when caustic soda is mixed with water. The reactants either must have a natural affinity for each other to react spontaneously or require an external supply of energy. In the case of acid-base reaction, it is the strong affinity of + *H* [acid] and – *OH* [base] make them react instantly. *+H* is the electron acceptor and -OH is the electron donor. Water and caustic soda do not have an affinity to react. In the absence of natural affinity, you need to supply energy externally to make them combine.

How does a Chemical Reaction Take Place?

Except for gases, liquids, and solids, all of which have a lot of potential energy stored in their bonds. There are no intermolecular attractions or repulsions between gas particles. Because the particles have no potential energy, their total energy is simply equal to their kinetic energies, according to this assumption.

Therefore, to summarize, A chemical reaction is essentially a transfer of potential energy between reactants and products through electron relocation from a higher energy to a lower stable energy product. Reactions frequently consist of a series of individual sub-steps, known as elementary reactions and information on the precise course of action is stored as part of the reaction mechanism. Chemical equations are used to represent the starting materials, end products, and sometimes intermediate products and reaction conditions in chemical reactions.

Chemical reactions occur at a predictable rate at a given temperature and chemical concentration. Typically, reaction rates increase as the temperature rises because there is more thermal energy available to reach the activation energy required to break atom bonds.

Fundamentals

Every substance in nature is an energy carrier. They do not have the capacity to perform thermodynamic work to be called energy.

Activation Energy Graph

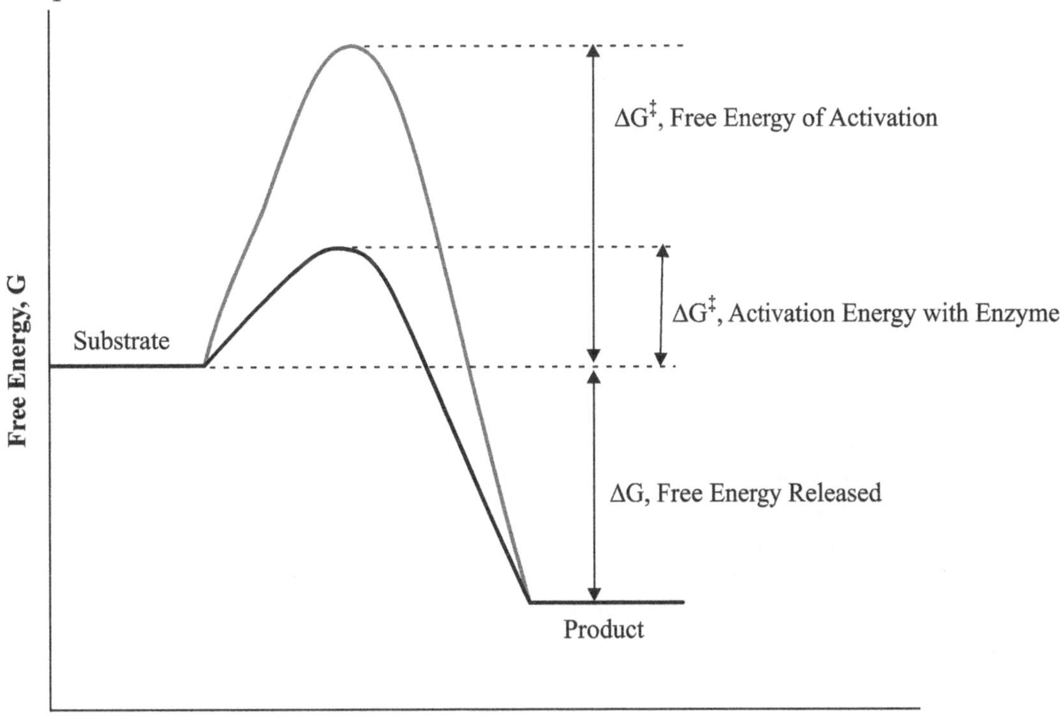

Figure 3

For any chemical reaction to occur, energy must be consumed, in other words, the products must be always at a lower energy level than the reactants in a spontaneous forward moving reaction. When the products are at a lower energy level than the reactants, products pull the energy from the reactants to achieve equilibrium. A negative free energy [G] changes between products and reactants, G reactant – G products < 0 dives a chemical reaction.

$$dG = dH - TdS$$

[H is enthalpy or total energy, T is temperature, and S is entropy]. To put it another way, the component TdS in the preceding equation must be > total energy change dH to keep dG negative and the reaction moving forward. As a result, any spontaneous reaction generates positive entropy.

Fundamentally, for any chemical reaction to occur, there is one requirement, and it is a critical requirement from the second law of thermodynamics.

Energy must flow from high to low. This implies that negative dG, or negative free energy, attracts fresh energy from the environment to propel a reaction forward. This new energy is heat, which replenishes the energy of a chemical reaction. This energy is commonly referred to as activation energy. The movement of electrons during the redistribution of potential energy between reactants and products is the thermodynamic work in a chemical reaction.

How does it Work on the Ground?

The energy released (or absorbed) as a result of a chemical reaction between two substances ("reactants") is equal to the difference in energy content between the products and the reactants. This energy change is referred to as the change in the internal energy of a chemical system. It is calculable from ΔfU [reactants], the internal energy of formation of the reactant molecules related to the bond energies of the molecules under consideration, and ΔfU [products] the internal energy of formation of the product molecules. The change in internal energy is equal to the heat change in the reaction at constant volume [STP] condition.

ΔfU [products] – ΔfU [reactants = Heat of reaction

However, at constant pressure, as in reactions in vessels open to the atmosphere, the measured heat is usually not equal to the internal energy change, because pressure-volume work also releases or absorbs energy. (The heat change at constant pressure is called the enthalpy change; in this case, the widely tabulated enthalpies of formation are used.)

THERMODYNAMICS AND KINETICS OF CHEMICAL REACTIONS

At fundamental Level

Thermodynamics and kinetics both play roles in a chemical reaction. At the primary level, three factors influence a reaction from going from reactants to products: [1] temperature, [2] pressure, and [3] concentration. At the secondary level, these three factors generate three requirements: [1] activation energy, [2] enthalpy, and [3] entropy. The activation energy determines how fast or slow a reaction can proceed. This is the kinetics part of the reaction. Thermodynamics is the study of equilibrium and essentially free energy. The change in free energy in a chemical reaction is the result of three factors: [1] enthalpy, [2] entropy, and [3] temperature.

Kinetics and thermodynamics: Chemical reactions are usually characterized by a chemical change, and they yield one or more products. Chemical reactions are described with chemical equations, which symbolically present the starting materials, end products, and sometimes intermediate products and reaction conditions. Chemical reactions happen at a characteristic reaction rate at a given temperature and chemical concentration. Typically, reaction rates increase with increasing temperature because there is more thermal energy available to reach the activation energy necessary for breaking bonds between atoms. Reactions may proceed in the forward or reverse direction until they go to completion or reach equilibrium. Reactions that proceed in the forward direction to approach equilibrium are often described as spontaneous, requiring no input of free energy to go forward. Non-spontaneous reactions require the input of free energy to go forward.

In summary, thermodynamics would control a reaction or kinetics is decided by the composition of a reaction product mixture when competing pathways lead to different products. If in a reaction there are two products X and Y. If X has lower activation energy the kinetics would favor X since X would form faster than y. Yet X may not be a stable product unless the Gibbs free energy change [dG] is negative. So not necessarily even if kinetics meets the activation energy level the thermodynamics can put a stop on the reaction if dG is not negative. So, in a chemical reaction there could be a kinetic product and the other could be a thermodynamic product. The opposite is also true. Thermodynamics can favor a reaction yet it could be such a slow reaction [kinetics] that it would not proceed. Thus, thermodynamics and kinetics are inextricably linked.

Example: A simple typical example is graphite and diamond are both forms of carbon, but graphite has a lower free energy. That is low irreversibility. Therefore, thermodynamics favor the conversion of diamond into graphite. Even though the reaction is thermodynamically favorable, it is slow. It's just plain too difficult to get the diamond to break all of its bonds and re-form them in the different, more stable graphite configuration. The kinetics does not support diamond going back to graphite.

CHEMICAL REACTORS ENGINEERING [CRE]

Hurdles in the journey of a reaction

$$A + B = C + D$$

Any chemical reaction has two aspects [1] 'How fast' and [2] 'How long'. The first one is dealt with by kinetics and the second by thermodynamics. Thermodynamics can tell you only that a reaction should go because the products are more stable (have lower free energy) than the reactants. Thermodynamics has nothing to do with time. Kinetics, on the other hand, can tell you how fast the reaction will go but doesn't tell you anything about the final state of things once it gets there. Thermodynamically favorable can be kinetically unfavorable:

- **Kinetics:** It is the speed at which reactions take place. The rate of a typical first-order reaction = (–) $d[A]/dt = Kx\,A$. Here K is the first-order rate constant having dimension 1/time, $[A](t)$ is the concentration at a time t, and [A]0 is the initial concentration. [Kept it simple]
- **Some key factors**: [1] Reactant concentration, increasing the concentration of one or more reactants will often increase the rate of reaction, [2] the physical state of the reactants and surface area, temperature, and [3] the presence of a catalyst.
- **Thermodynamics**: Thermodynamics simply says, substances combine chemically when their total energy contents change from a higher level to a lower level. In other words, there is the consumption of energy. In terms of Gibbs free energy $\Delta G < 0$ from the point of thermodynamics if there is a spontaneous reaction.
- **Some key factors**: Types of reactants, pressure, and temperature.

Reactor

A chemical reactor facilitates a chemical reaction to occur in the most efficient way to achieve the kinetics of the reaction within given thermodynamic conditions

Choice of reactors

The basic purpose in the design of a reactor is to produce a specified product at a given rate from the given reactants.

The first question arises if the process will be conducted as a batch or continuous process or it should part batch and part continuous—a hybrid type? The next question is more thermodynamics related, if the reaction would be conducted isothermally or adiabatically? The next question comes from the physical condition of reactants and their compositions to start with? Processing conditions, temperature, pressure, etc.

Two broad groups

There are two broad groups [1] homogeneous and [2] heterogeneous. Inhomogeneous reactors only one phase, usually a gas or a liquid, is present. If more than one reactant is involved, provision must of course be made for mixing them together to form a homogenous whole. In heterogeneous reactors, where there is more than one phase, it could be any of these systems, gas-liquid, gas-solid, liquid-solid, and liquid-liquid. In a heterogeneous reactor, the chemical reaction may itself be heterogeneous. Because of these complexities, heterogeneous reactors are more complex than homogeneous reactors with a greater variety of configurations and contacting patterns.

Batch vs. continuous reactors Image X

The above image shows diagrams of the three ideal reactor types, namely the batch stirred tank reactor (BSTR), the continuous plug-flow reactor (CPFR) or continuous tubular flow reactor, TFR in the middle, and the continuous stirred tank reactor (CSTR) at the bottom.

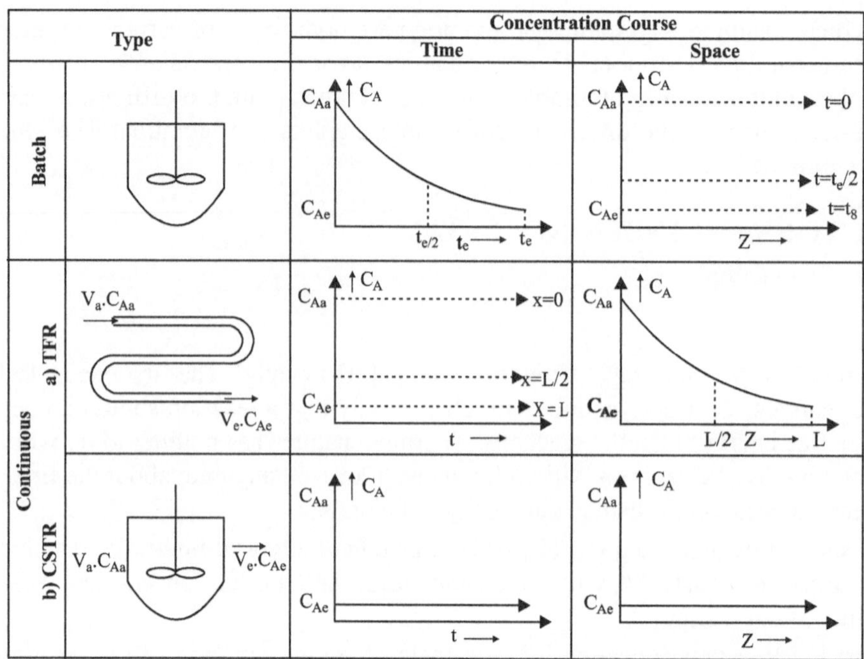

Figure 4

In the tubular-flow [plug flow] reactor at the middle [CPFR], here the aim is to pass the reactants along a tube so that there is as little intermingling as possible between the reactants entering the tube and the products leaving at the far end. In the continuous stirred-tank reactor (C.S.T.R.) at the bottom of the image, an agitator is mixing disperse the reactants thoroughly into the reaction mixture immediately they enter the tank. The product stream is drawn off continuously and, in the ideal state of perfect mixing, will have the same composition as the contents of the tank. Because as soon as the reactants enter the tank, they are mixed and a portion leaves, it is often advantageous to employ a number of stirred tanks connected in series in the product stream flowing out. The stirred-tank reactor [BSTR] at the top of the image is by its nature well suited to liquid-phase reactions. The tubular reactor, although sometimes used for liquid-phase reactions, is the natural choice for gas-phase reactions

BSTR [Batch stirred tank reactor at top]

A BSTR is a mechanically stirred vessel. It is first filled with substrate and catalyst to initiate the reaction, after which no material is removed until the reaction is stopped. BSTRs are well-mixed reactors, meaning that concentrations are the same regardless of location within the reactor. BSTRs are commonly used for biocatalytic reactions due to their simplicity and flexibility. Additionally, *pH* changes in a reaction can be neutralized through the addition of an acid or base to maintain the optimal *pH*.

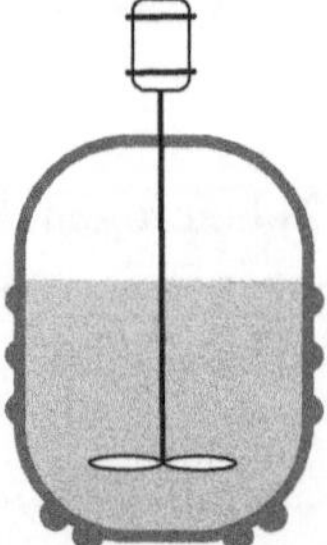

Figure 5

CSTR [Continuous stirred tank reactor at the bottom of image]

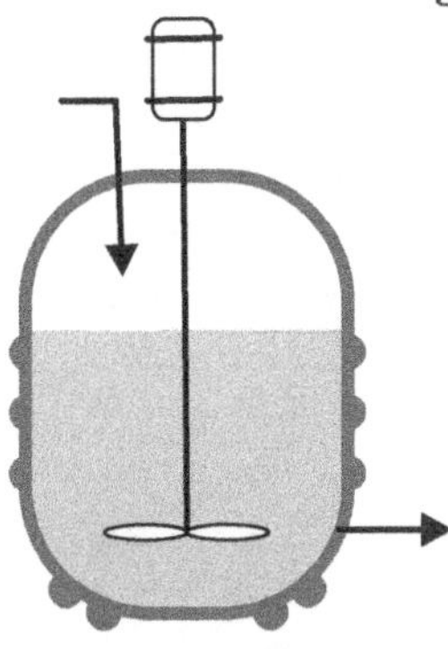

Figure 6

The design of a CSTR is similar to that of a BSTR, except that material is continuously added to, and removed from, the reactor such that the working volume remains constant. Like BSTRs, CSTRs are well-mixed and so the reactor contents. However, since there must be enough reactants in the reactor to achieve an adequate reaction rate, the products will always contain some unconverted reactants. Therefore, full reactants conversion is not possible. This trade-off between reaction rate and conversion is an important characteristic of CSTRs.

CPFR [Continuous plug flow reactor in the middle of the image]

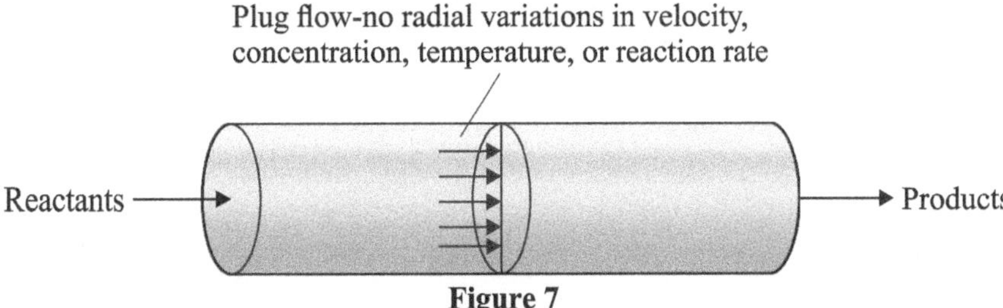

Figure 7

In a CPFR, reactants are pumped into a long tubular reactor where, unlike stirred tanks, material flowing through does not mix with any material flowing ahead of it, or behind it. This results in concentration gradients over the length of the reactor, identical to the concentration gradients over time in a BSTR. Therefore, if the reactor is sufficiently long, the reactants can be fully converted. For this reason, in the image at the middle,]Image *X*], the concentration profile is given with respect to length since the time material spends in a CPFR is simply a function of the reactor length and volumetric flowrate.

Two important considerations while choosing a reactor

[1] The residence time distribution (RTD) and
[2] The kinetics of a catalyzed reaction.

Advantage of a continuous process

BSTRs and CPFRs both have identical kinetic behavior and both afford good control over RTD. Consequently, it would appear that there is little motivation to invest in the shift from batch operation to continuous since both give the same result. The key advantage of a continuous process is, at production scales, where mixing and heat transfer in large batch reactors are less efficient, shifting to continuous operation could be beneficial. Additionally, the steady-state operation of a continuous reactor affords simpler control and greater consistency than a dynamic batch process at a large scale

Residence-Time Distribution (RTD) in a reactor

What is the residence time distribution for a chemical reaction in a reactor?

In simple words, the residence time describes the time spent by a molecule inside a reactor, i.e., the time at which a molecule enters the reactor subtracted from the time at which it leaves the reactor, $[t + \Delta t] - t = \Delta t$.

Why RTD is important?

The chemical reaction rates depend on the retention time of the reactants and their co-presence, which requires good mixing of the reactant streams in a reactor. Any short-circuiting of specific streams may reduce the retention time on average, and finally reduce the reaction extent, which results in incomplete reactions. Dead volume may be present in the reactor, which in general will waste the effective volume for reactions and the mainstream will have a shorter retention time. In any reactor, the distribution of residence times can significantly affect its performance.

Residence-Time Distribution (RTD)

RTD for PFR and BSTR

In an ideal plug-flow reactor, all the molecules of material leaving the reactor have been inside it for exactly the same amount of time. Similarly, in an ideal batch reactor, all the molecules of materials within the reactor have been inside it for an identical length of time. The time the molecules have spent in the reactor is the residence time of the molecules in the reactor. The idealized plug-flow and batch reactors are the only two classes of reactors in which all the molecules in the reactors have the same residence time. In all other reactor types, the various molecules the feed spend different times inside the reactor. There is no axial mixing in a plug-flow reactor, and this omission is reflected in the RTD.

RTD for CSTR

In the CSTR; the feed introduced at any given time becomes completely mixed with the material already in the reactor. In other words, some of the molecules entering the CSTR. leave it almost immediately because the material is being continuously withdrawn from the reactor; other molecules remain in the reactor almost forever because all the material is never removed from the reactor at one time. Many of the molecules, of course, leave the reactor after spending a period of time somewhere in the vicinity of the mean residence time.

How to find the RTD

The RTD is determined experimentally by injecting an inert chemical, molecule, or atom, called a tracer, into the reactor at some time t_0 and then measuring the tracer concentration, C, in the product stream as a function of time. In addition to being a nonreactive species that is easily detectable, the tracer should have physical properties similar to those of the reacting mixture and be completely soluble in the mixture. It also should not adsorb on the walls or other surfaces in the reactor. The latter requirements are needed so that the tracer's behavior will honestly reflect that of the material flowing through the reactor. Colored and radioactive materials along with inert gases are the most common types of tracers. The two most used methods of injection are pulse input and step input.

Pulse Input

Experiment In a pulse input: An amount of tracer N_0 is suddenly injected in one shot into the feed stream entering the reactor in as short a time as possible. The outlet concentration is then measured as a function of time. First, we choose an increment of time Δt sufficiently small that the concentration of the tracer, $C(t)$, exiting between time t and $t + \Delta t$ is essentially the same. The amount of tracer material, ΔN, leaving the reactor between time t and $t + \Delta t$ is then $\Delta N = C(t)$ $v\Delta t$.

Where v is the product volumetric flow rate. In other words, ΔN is the amount of material exiting the reactor that has spent an amount of time between t and $t+\Delta t$ in the reactor. If we now divide by the total amount of material that was injected into the reactor, N_0, we get $\Delta N/N_0 = vC(t)/No \ x \ \Delta t$ which represents the fraction of material that has a residence time in the reactor between time t and $t + \Delta t$. For pulse injection we define $E(t) = vC(t)/N_0$,

Therefore, $\Delta N/N0 = E(t)\ \Delta t$

The quantity $E(t)$ is called the residence time distribution function. It is the function that describes in a quantitative manner how much time different fluid elements have spent in the reactor. The quantity $E(t)dt$ is the fraction of fluid exiting the reactor that has spent between time t and $t + dt$ inside the reactor. The chemical reaction rates depend on the retention time of the reactants and their co-presence in the reactor.

Kinetics of Reactor Models

Mass balance [General]

The key equation governing processes on the reactor level is the mass balance. In order to inherently account for the proper stoichiometry, this is most typically written as a mole balance. The general mole balance for a species i is given as

$$Fi_0 - Fi + Gi = dNi/dt$$

where Fi_0 is the input molar flow rate, Fi is the output molar flow rate, Gi is the generation term, and the differential term is the accumulation. If the system variables are uniform throughout the system volume, then $Gi = ri\text{V}$.

where ri is the reaction rate of species i and V is the system volume. More generally speaking, if ri changes with position in the system volume, then $Gi = \int ri\ d\text{V}$

such that the mole balance can be written as $Fi0 - Fi + \int ri\ dV = dNi/dt$

Two other useful expressions that should be kept in mind are as follows. For a uniform concentration of i across the system volume $Ni = CiV$ Additionally, for a given flow rate $Fi = Ci\ \dot{v}$, where Ci is the concentration of species I and $\dot{v}$ is volumetric flow rate.

Reaction rate [Specific]

Batch reactor

A batch reactor is a constant volume reactor that has no input or output when the chemical reaction is occurring. The batch reactor is often a good reactor model for slow reactions. With this information, it is clear that the batch reactor has $Fi0 = Fi = 0$. As such, the mole balance is

$$dNi\ dt = \int ri/d\text{V},$$

If the reaction mixture is perfectly mixed (i.e. spatially uniform) so that ri is independent of position (a common assumption for the batch reactor), then we can state.

$$ri = 1/V\ dNi/dt = dCi/dt$$

where ri is the reaction rate.

Continuous stirred tank reactor

The continuous stirred tank reactor (CSTR) has an inlet and outlet flow of chemicals. CSTRs are operated at a steady-state (such that the accumulation term is zero) and are assumed to be perfectly mixed. As such, the mole balance for the CSTR can be written as $Fi0 - Fi + riV = 0$ Solving for the reaction rate

yields $ri = -\ Fi0 - Fi/\text{V}$

Plug-flow reactor

A Plug-flow reactor (PFR) is a tubular reactor operated at steady-state and has axial gradients but no radial gradients. These types of reactors are useful for fast reactions that could not be as easily observed in a batch environment. Since the concentration varies continuously down the reactor tube, so does the reaction rate (except for zeroth-order reactions). For a PFR, the design equation can be solved by differentiating the mole balance with respect to volume, but an easier way is to perform a mole balance on species i in a different segment of the reactor volume, ΔV. This differential balance can be written as

$$Fi\ (V) - Fi\ (V + \Delta V) + ri\Delta V = 0$$

since the system is in a steady-state. Solving for the rate, dividing by ΔV, and letting it approach zero (while applying the definition of the derivative)

yields, $ri = dFi/d$

CHAPTER 47

Electrochemical Processes

Thermodynamics plays an important role in electrochemical reactions as it determines the feasibility and direction of the reaction. In electrochemical systems, thermodynamics governs the flow of electrons and ions between electrodes, as well as the energy changes associated with these processes. Thermodynamic principles such as Gibbs free energy, enthalpy, and entropy are used to analyze the stability and spontaneity of electrochemical reactions. The Nernst equation, which relates the electromotive force (EMF) of a cell to the standard electrode potential, is based on thermodynamic principles. Additionally, thermodynamics helps in predicting the voltage performance of electrochemical cells and in designing efficient energy conversion devices such as batteries and fuel cells. By understanding the thermodynamics of electrochemical reactions, researchers and engineers can optimize the performance and efficiency of these systems.

WHAT IS ELECTROCHEMICAL REACTION?

The main advantages of the electrochemical process are based on the fact that this technique does not consume chemical reagents. However, it requires large capital investment, consumes a lot of electric energy and presents long-term running and maintenance costs

Electrochemical reactions are red-ox reactions.

An electrochemical reaction is a chemical reaction that involves the transfer of electrons between reactants. This transfer of electrons occurs through an electrical circuit, which allows for the generation of an electric current.

Electrochemical reactions happen due to the difference in the electrical potentials of the reactants involved. When two different materials with different electron affinities come into contact, electrons may transfer from one material to another in order to achieve a more stable electron configuration. This transfer of electrons results in the formation of electrical potential energy, which can be harnessed to do work or generate electricity.

Process

Electrochemical reactions involve the exchange of electrons between two substances, typically in an electrolyte solution. This process occurs at the interface of two conductive materials, such as metal electrodes immersed in an electrolyte solution.

The process of an electrochemical reaction involves two half-reactions: oxidation and reduction. During oxidation, one substance loses electrons and becomes positively charged, while during reduction, another substance gains electrons and becomes negatively charged. The transfer of electrons from the oxidized substance to the reduced substance creates an electric current.

In a basic electrochemical cell, oxidation occurs at the anode (negative electrode), where electrons are released into the external circuit, and reduction occurs at the cathode (positive electrode), where electrons are accepted from the external circuit. The movement of electrons through the external circuit generates an electrical potential difference, driving the electrochemical reaction.

The rate and direction of an electrochemical reaction are controlled by factors such as the concentration of reactants, temperature, surface area of electrodes, and the presence of catalysts. The electrical potential applied to the electrodes can also influence the outcome of the reaction. Overall, electrochemical reactions play a crucial role in various technologies, including batteries, fuel cells, and corrosion prevention systems.

Example

Let's take a typical electrochemical reaction involving the oxidation of zinc metal (Zn) to zinc ions (Zn^{2+}) and the reduction of hydrogen ions (H^+) to hydrogen gas (H_2) in an acidic medium:

$$Zn(s) + 2H^{+}(aq) \rightarrow Zn^{2+}(aq) + H_2(g)$$

In this equation:

Zn(s) represents solid zinc metal.

H^{+}(aq) represents hydrogen ions in the acidic electrolyte solution.

Zn^{2+}(aq) represents zinc ions in the solution.

$H_2(g)$ represents gaseous hydrogen produced during the reduction reaction.

This equation shows the transfer of electrons from zinc to hydrogen ions, resulting in the oxidation of zinc and reduction of hydrogen ions to form zinc ions and hydrogen gas, respectively.

Kinetics and Thermodynamics

The kinetics and thermodynamics of electrochemical reactions are controlled by several factors.

Electrode Material and Surface Area

The type of electrode material used and its surface area can significantly influence the kinetics of electrochemical reactions. Different materials have varying catalytic properties that can enhance or hinder the reaction rates.

Electrolyte Composition

The electrolyte composition (concentration and type of ions present) affects the conductivity of the electrolyte solution, which in turn influences the rate of electron transfer and overall reaction kinetics.

Temperature

Temperature plays a crucial role in both the kinetics and thermodynamics of electrochemical reactions. Higher temperatures generally increase reaction rates due to higher thermal energy and faster molecular motion.

Overpotential

Overpotential is the additional voltage required to drive an electrochemical reaction at a desired rate. It affects the kinetics of the reaction as it determines the activation energy required for the reaction to proceed.

Activation Energy

The activation energy is the energy barrier that reactants must overcome to form products, and it directly influences the rate of the reaction. Lower activation energies lead to faster reaction rates.

Concentration of Reactants

The concentration of reactants in the electrolyte solution can also impact the reaction kinetics. Higher concentrations typically lead to faster reaction rates due to increased collision frequency between reactant molecules.

Cell Potential and Gibbs Free Energy

The cell potential (voltage) of an electrochemical cell is related to the Gibbs free energy change of the reaction. Thermodynamic considerations, such as the Gibbs free energy change (ΔG) and the equilibrium constant (K), can determine the feasibility and spontaneity of the reaction.

Activation Overpotentials

Activation overpotentials represent the additional potential needed to initiate redox reactions and affect the kinetics of the reaction by influencing the rate of charge transfer at the electrode/electrolyte interface.

WHAT MAKES H_2 MAKING BY ELECTROLYSIS EXPENSIVE

This note highlights 'Why and How 'high energy consumption in water electrolysis is a challenge.

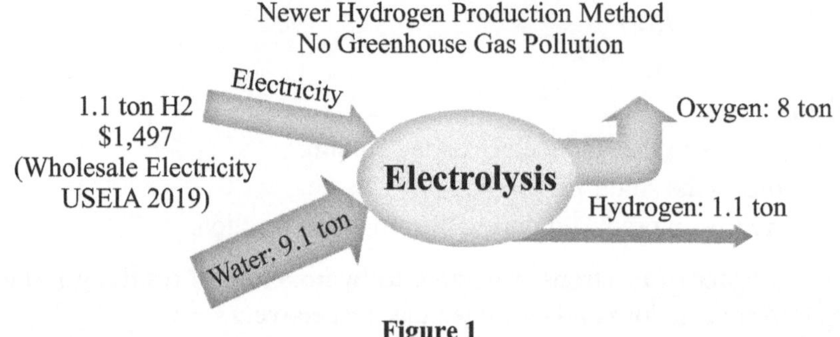

Figure 1

The main cost issues in producing hydrogen gas from water electrolysis include the high energy consumption required for the electrolysis process, the high cost of electricity needed to power the electrolyzers, and the high initial capital costs associated with purchasing and installing the necessary equipment. Additionally, the cost of maintaining and operating the electrolysis system, as well as the cost of transporting and storing the hydrogen gas, can also be significant factors contributing to the overall cost of producing hydrogen from water electrolysis.

Electrolysis is an energy-intensive process for producing hydrogen. It takes 50 megawatt hours (MWh) of electricity to produce one ton of hydrogen through electrolysis using PEM or alkaline electrolysis. If the water needs to be desalinated, an additional 9 MWh of energy is required, bringing the total energy demand to 59 MWh.

Efficiency

A theoretical electrolysis system that's 100% efficient would require 39.4 kWh of electricity to produce 1 kg of hydrogen. However, commercial devices are less efficient, typically requiring around 50 kWh per kg

The process of making H_2 by water electrolysis is made up of two half-cell reactions:

- **Hydrogen evolution reaction (HER)**: Water is reduced at the cathode to produce hydrogen gas.
- **Oxygen evolution reaction (OER)**: Water is oxidized at the anode to produce oxygen gas.

Both thermal and electrical energy are used in water electrolysis because each has its own advantages and applications.

Thermal Energy

Thermal energy is cheaper than electricity, and high-temperature electrolysis (HTE) is more efficient economically than traditional room-temperature electrolysis. HTE can be used when high-temperature heat is available as waste heat from other processes.

Electrical Energy

Electrolysis is technologically mature when applied at low temperatures, but it requires large quantities of electrical energy. Electricity is the major component in the cost of hydrogen production via electrolysis.

High-temperature electrolysis (HTE) is a method that uses both heat and electricity to generate hydrogen. HTE can convert up to 50% of the initial heat energy into chemical energy, which is double the efficiency of low-temperature electrolysis.

- **Thermal energy reduces the need for electricity**: At high temperatures, the voltage required for electrolysis is reduced, which means less electricity is needed. This can make electrolysis cheaper because electricity is more expensive than heat
- **Thermal energy improves reaction kinetics**: Heat improves the reaction kinetics for splitting water molecules.
- **Thermal energy can come from various sources**: Thermal energy can come from sources like nuclear reactors, solar concentration power plants, and geothermal energy.

Process

Water electrolysis is a process in which an electric current is passed through water to split it into its constituent elements, hydrogen and oxygen. This process involves the use of an electrolyzer, which is a device that contains two electrodes (anode and cathode) separated by an electrolyte solution.

When an electric current is applied to the electrolyzer, the water molecules at the cathode are reduced to form hydrogen gas (H_2) and hydroxide ions (OH^-). At the anode, water molecules are oxidized to form oxygen gas (O_2) and protons (H^+).

The overall reaction for water electrolysis can be represented as:

$$2H_2O(l) \rightarrow 2H_2(g) + O_2(g)$$

Hydrogen gas collected at the cathode can be used as a clean and efficient energy source for various applications, such as fuel cells and transportation. Oxygen gas produced at the anode is typically released into the atmosphere.

Water

About 9000 litters of water is required to produce one ton H_2 gas.

Water purification

The water used for electrolysis needs to be purified first. Purifying water typically requires 2 tons of impure water to produce 1 ton of purified water.

Desalination

If the water needs to be desalinated, it can add an additional 9 MWh of energy demand.

Kinetics and Thermodynamics

In water electrolysis for hydrogen gas production, kinetics and thermodynamics play significant roles in determining the efficiency and feasibility of the process.

Kinetics

- Kinetics refers to the rate at which a reaction occurs. In the context of water electrolysis, the kinetics of the electrode reactions at the anode and cathode determine how quickly hydrogen and oxygen gas are produced.
- The kinetics of the hydrogen evolution reaction (HER) at the cathode and the oxygen evolution reaction (OER) at the anode are critical in determining the efficiency of the electrolysis process.
- Factors such as the electrode material, surface area, concentration of electrolyte, temperature, and applied current density can influence the kinetics of the electrode reactions. Optimizing these parameters can improve the efficiency of hydrogen gas production.

Thermodynamics

- Thermodynamics refers to the study of energy and heat transfer in a system. In the context of water electrolysis, thermodynamics governs whether the overall reaction is spontaneous or not.
- The standard Gibbs free energy change ($\Delta G°$) of the water splitting reaction determines the feasibility of the electrolysis process. For water electrolysis to occur spontaneously, the $\Delta G°$ value must be negative.
- The cell voltage required for water electrolysis can be calculated using the Nernst equation, which considers the standard electrode potentials of the anode and cathode reactions. The cell voltage should be higher than the theoretical voltage required for the electrolysis reaction to overcome overpotential losses and ensure efficient hydrogen gas production.

Role of Nernst Equation and Gibbs Free Energy

The Nernst equation and Gibbs free energy play crucial roles in understanding and predicting the behavior of chemical reactions, including water electrolysis for hydrogen gas production.

Nernst Equation

- The Nernst equation relates the standard electrode potential of a half-reaction to the actual electrode potential under non-standard conditions. It allows for the calculation of the cell voltage (Ecell) of an electrochemical cell based on the concentrations of the reactants and products involved.

- In the context of water electrolysis, the Nernst equation helps determine the cell voltage required to drive the overall reaction forward. This voltage must be higher than the standard potential difference for the water splitting reaction to proceed.
- The Nernst equation is given by: Ecell = E°cell – $(RT/nF) \times ln(Q)$, where $E°$cell is the standard cell potential, R is the gas constant, T is the temperature, n is the number of moles of electrons transferred in the reaction, F is Faraday's constant, and Q is the reaction quotient.

Gibbs Free Energy ($\Delta G°$):

- Gibbs free energy is a measure of the thermodynamic feasibility of a chemical reaction. It determines whether a reaction is spontaneous ($\Delta G < 0$) or non-spontaneous ($\Delta G > 0$) under standard conditions.
- In the context of water electrolysis, the standard Gibbs free energy change (ΔG°) for the overall reaction ($2H_2O \rightarrow 2H_2 + O_2$) must be negative for the process to be thermodynamically favorable.
- The relationship between Gibbs free energy and the equilibrium constant (K) of a reaction is described by the equation: $\Delta G° = -RT \times \ln(K)$, where R is the gas constant, T is the temperature, and ln denotes the natural logarithm. This equation provides insights into the direction and extent of a chemical reaction at equilibrium.
- By understanding the Gibbs free energy change of the water splitting reaction and comparing it to the Nernst equation-derived cell voltage, researchers can assess the thermodynamic feasibility and efficiency of water electrolysis for hydrogen gas production.

WHY WATER ELECTROLYSIS IS THERMODYNAMICALLY UNFAVOURABLE

The electrolysis of water can be written as:

$$2H_2O\ (l) \rightarrow 2H_2\ (g) + O_2\ (g)$$

This equation shows that for every 2 moles of water that are electrolyzed, 1 mole of oxygen gas and 2 moles of hydrogen gas are produced. In order for this reaction to occur, 2 moles of electrons are required to split the water molecules into their constituent elements.

Overall, the electrolysis of water can be represented as:

$$2H_2O\ (l) \rightarrow 2H_2\ (g) + O_2\ (g) + 4e^-$$

This equation shows that 4 moles of electrons are required for the electrolysis of 2 moles of water.

The potential required for electrolysis of water is about 1.23V. Actual is about 1.49V for various inefficiencies.

Since each mole of water requires two moles of electrons, and given that the Faraday constant F represents the charge of a mole of electrons (96485 C/mol), it follows that the minimum voltage necessary for electrolysis is about 1.23

The negative cell potential of water electrolysis makes it thermodynamically unfavorable, which impacts the cost of water electrolysis.

How and Why?

Additional Voltage Required

The low concentration of ions and interfaces that electrons must cross require an extra voltage of about 0.6V at each electrode.

Minimum Voltage

The minimum cell voltage required to start water electrolysis is 1.229 V, which results in a loss of at least 21% efficiency.

What Is Cell Potential?

Cell Potential is the difference in electrical potential between an electrochemical cell's two electrodes.

The differential in electron affinities between the two electrodes and the electrolytes' reactivity are what cause the cell potential.

Electrons will move from the metal with lower electron affinity to the metal with higher electron affinity when two different metals are in contact, creating an electrical potential difference.

What is positive and negative cell potential?

The cell potential is a measure of the ability of the cell to do work and is related to the Gibbs free energy change (ΔG) of the electrochemical reaction.

Positive and negative cell potential

A positive cell potential indicates that the reaction is spontaneous and can generate electrical energy, while a negative cell potential indicates that the reaction is non-spontaneous and requires an external source of energy to occur.

Why water has negative cell potential?

Water has a negative cell potential because of the presence of solutes in the water:

What is solute potential of water?

When a solute is dissolved in water, the amount of free water molecules decreases, which reduces the kinetic energy of the molecules. This causes the solute potential to decrease and become negative.

Typical Energy Losses

The process of producing hydrogen is always associated with energy losses. Electrolyzers need higher voltage than the standard potential to separate water into hydrogen and oxygen. It also involves energy losses due to side reactions, overpotential, and the bubble effect.

Side Reactions

Bubble effect

Lower voltage efficiency

Side reactions

In many cells, competing side reactions occur, resulting in less-than-ideal faradaic efficiency. Electrolyzers need higher voltage than the standard potential of hydrogen oxidation to separate water into hydrogen and oxygen.

Typical Side Reactions

Preferential reduction of anions

For example, in the presence of stainless steel, the preferential reduction of NO_3^- anions can prevent hydrogen from evolving.

Stainless steel is used as an anode in alkaline water electrolysis. Stainless steel can catalyze the reduction of nitrate anions (NO_3^-) to other N-containing species, such as NO_2^- and NH_2OH. The reduction of nitrate is autocatalytic, meaning that its reduction product (nitrous acid) can promote its reduction reaction.

In water electrolysis, the side reactions of nitrate reduction include the formation of nitrite and ammonium:

- **Nitrite**: The hydrogen formed during water hydrolysis reduces nitrate to nitrite.
- **Ammonium**: The hydrogen formed during water hydrolysis reduces nitrite to ammonium.

Bubble Effect

Bubbles on the electrode surface and in the electrolyte cause high ohmic voltage drop and large reaction over potential. The bubble effect in electrolysis is the impact that bubbles have on the efficiency of the electrolysis process:

Bubble Growth

Bubbles grow as dissolved hydrogen diffuses into the gas phase. The growth of bubbles can be accelerated by high current densities, which deplete reactants rapidly.

Bubble Detachment

Bubbles eventually detach from the electrode and move away, which mixes species locally and decreases concentration gradients.

Bubble Blocking

Bubbles can block ion conduction pathways, which increases the electrolyte's effective resistance.

Bubble Slugs

Bubbles can coalesce and form slugs between the electrodes and gas separators. These slugs act as insulators, which impedes ion movement and increases resistance

Bubble Coverage

Bubbles can cover the electrode surface, which acts as an electrical insulator and disrupts current distribution.

Other effects of bubbles include: Changing the local electrolyte concentration and pH, causing excessive overpotential, causing unstable currents, Blocking active sites on electrode surfaces, and Physically degrading catalysts.

Lower Voltage Efficiency

Membraneless electrolysis cells have lower voltage efficiency at high operating current densities. This is due to the larger gap required between the two electrodes.

CHAPTER 48

Surface Tension

Surface tension is an intermolecular interaction related phenomenon in fluids. Thermodynamics plays an important role in predicting intermolecular interplay in fluids. Surface tension is a result of the cohesive forces between molecules at the surface of a liquid, a property that is governed by thermodynamics. The intermolecular forces that give rise to surface tension contribute to the overall energy balance of a system.

In oil-gas industries, surface tension plays a crucial role in processes such as oil extraction and gas separation. For example, in the case of oil extraction, surface tension helps to hold oil droplets together, facilitating their removal from the reservoir. In gas separation, surface tension aids in the separation of different components based on their differential affinities for the liquid phase, allowing for the efficient purification of gases.

SURFACE TENSION'S FUNDAMENTALS

What is surface tension?

The most important one-line answer is that surface tension is a property of a liquid that holds its molecules on the surface from the threats of surface deformation.

Whenever you see a liquid flowing through a pipe, remember at the molecular level liquid spheres are moving with limited contact with the substrate [wetting] overcoming the pressure at the interface generated by the surface tension. Now visualize the consequence of such a flow on the unit operation you are going to use the liquid.

When you conceive a drop of liquid consisting of spheres held by molecular forces and protected by the surface tension by two opposing forces generating a net pressure at the interface and preventing surface deformation half of your problems in understanding " What are the consequences of surface tension of a liquid on a process " would be over. Whether it is heat transfer or mass transfer or fluid flow you would be able to visualize the effect of surface tension.

Brief introduction

The surface of a liquid behaves as though it is covered by a stretched membrane. The strength of the membrane depends on the liquid and its surface tension. It is, for example, much higher for pure water than for soapy water. Cohesive forces that act between the molecules of the liquid without chemical bonding give rise to the surface tension. [Molecules in liquids experience strong intermolecular attractive forces. When those forces act between like molecules, they are referred to as cohesive forces. When the attractive forces act between unlike molecules, they are said to be adhesive forces.]

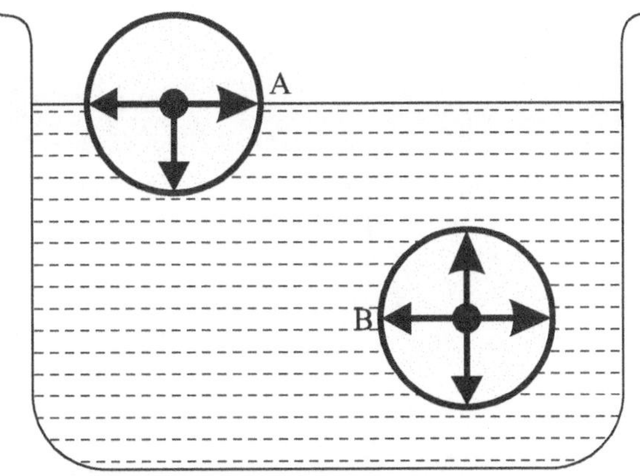

Figure 1: Phenomenon of Surface Tension

The molecules at the surface do not have other neighbors of the same kind on all sides. They experience a net attractive force toward the bulk of the liquid. This force results in a surface "film" which makes it more difficult to move an object through the surface than to move it when it is completely submerged.

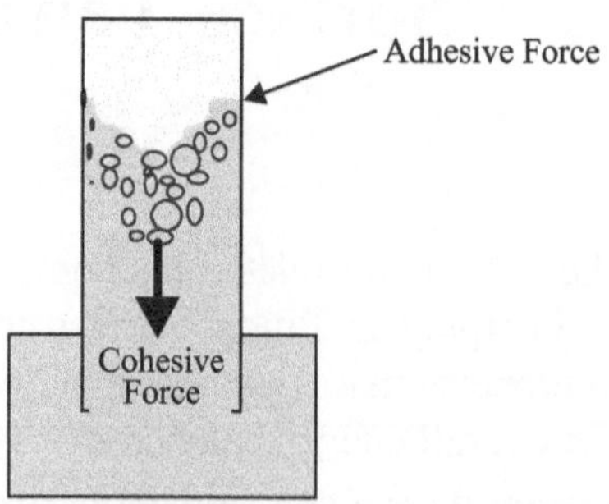

Figure 2

If there are no competing external forces present, a liquid system will minimize its surface area. Soap bubbles and raindrops are spherical because the sphere is the shape with the smallest surface-to-volume ratio.

The surface tension γ is defined as the force along a line of unit length. The force is parallel to the surface and perpendicular to the line. Surface tension is therefore measured in units of N/m in SI units.

Therefore, it is coming out clearly unless countered, a liquid surface would like to shrink into spheres by giving up its surface energy. The reason is the sphere is the shape with the smallest surface to volume ratio and the sphere is the most stable geometry of a liquid.

But this has serious consequences. Spheres have point contact with the substrate. As liquid shrinks into spheres, its wettability reduces depending on the contact angle of the substance and the substrate.

What is the contact angle?

A curved liquid surface can't wet a flat solid surface as much as a flat liquid surface can do.

The complexity of surface tension: Liquid – solid contact angle and wetting

Imagine a drop of water sitting on a solid surface. The adhesive forces between air and liquid working on the drop, pushing the drop to spread across the surface. while the cohesive forces within the liquid [liquid-solid] preventing the drop to spread over the surface by making its meniscus convex [image]. When the contact angle between liquid and solid interface is zero [extreme right], there is maximum contact between liquid and solid, and that gives the maximum wetting. When contact angle, $\theta = 90$ deg, there is moderate wetting, and when contact angle > 90, practically very small liquid-solid contact and no wetting. The substate behaves like a hydrophobic surface.

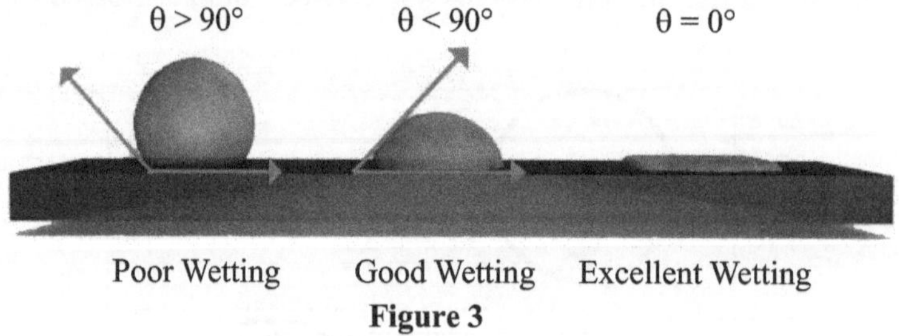

Figure 3

Cos θ = [YSG-YSL]/YLG

[Young's equation]

What Does it Mean?

As *YLG* [liq-gas interface surface tension] reduces, as the liquid spreads with the reduction of surface tension Cos θ which is Base/Hypotenuse, increases until the contact angle becomes zero. This gives maximum wetting of the substrate.

G stands for air. *YSG* = solid-gas interface surface tension, *YSL* = solid-liq interface surface tension and Θ is liquid-solid contact angle. Just opposite happens when liquid-solid surface tension increases. Liquid –solid contact area reduces.

Therefore, every surface tension has its own wettability character. Consequently, the surface tension influences every thermodynamic property of a liquid

Application of the Surface Tension on Fluid Flow

Take Water as an Example

Just imagine a pipeline carrying a flow of incompressible spheres with very little wetting because the spheres have point contact with the substrate with limited wetting. The flow is generating at the molecular level a pressure difference at the interface in dynes/cm^2 equivalent to the surface tension of the liquid. This comes from the definition of surface tension. For example, water has ST 72 dynes/cm this can be one case. The other case could be an oil with ST 30–35 dynes/cm.

The oil with a smaller surface gives more wetting on the substrate. Their behavior in terms of flow through a pipeline would completely be different

Water with high surface tension would generate more pressure caused by cohesive forces than oil at the interface.

Now, let us consider a gas

What does this mean to compressible gases?

A new equation has been derived which relates the surface tension (σ) to a liquid's isothermal compressibility (κ) and mass density (ρ).

The equation is $\sigma(\kappa/\rho)^{1/2}$ = constant in the normal liquid range.

If we take a cue from this equation, the surface tension varies with the square root of compressibility at constant density. This means as the surface tension increases the compressibility reduces.

This has implications for the mass flow rate of gases.

Mass Flow Rate = Area × Velocity × Density

The conservation of mass (continuity) tells us that the mass flow rate through a pipe is a constant and equal to the product of the density, velocity, and flow area. Considering the mass flow rate equation, it appears that for a given area, we can increase the mass flow rate to any desired point by increasing the velocity. However, in real fluids, compressibility effects limit the speed at which a flow can be forced through a given area.

Mass Flow Rate for an Ideal Compressible Gas

For a compressible, ideal gas, the mass flow rate is a unique function of the flow area, total pressure, the temperature of the flow, properties of the gas, and the Mach number.

The Problem is Here

If we have a pipe with changing area, the maximum mass flow rate through the system occurs when the Mach number is one at the smallest area. This limits the amount of mass flow rate that can be passed through the constriction called choking.

ROLE OF SURFACE TENSION IN FLUID FLOW, HEAT TRANSFER AND MASS TRANSFER WITH IMPORTANT EQUATIONS

What is Surface Tension?

The surface tension is a liquid's ability to withstand forces from outside that might cause it to separate is a result of its surface tension. Simply put, it is the cohesive force that forms a thin layer on a liquid's surface and allows it to keep its shape and form. Surface tension is an important subject in chemical engineering many unit processes because it generates pressure called Laplace pressure at the interface between air-liquid systems and modifies flow, heat trans fer and mass transfer. **Droplet formation of saturated steam has profound contribution to large heat transfer coefficient of saturated steam.**

What does the surface tension do?

A few examples

- **Wetting**: Surface tension affects how a liquid wets a solid, which is important for cleaning, printing, and coating.
- **Droplet size**: Surface tension determines the size of droplets in sprays and emulsions.
- **Floating**: Surface tension allows objects denser than water, like insects and razor blades, to float on water.
- **Condensation**: Surface tension is responsible for water droplets condensing on surfaces as rain or fog.
- **Heat transfer**: Surface tension affects the heat transfer performance of thermal systems, including heat pipes, thermosyphons, and boiling.
- **Cleaning**: Soaps and detergents decrease the surface tension of water, allowing them to soak into dirty areas of clothes.

Some Specific Cases

Role of surface tension in fluid flow and heat transfer

- **Surface tension affects** the behavior of fluids at interfaces: Surface tension causes liquids to form droplets or beads when placed on a surface as it minimizes the surface area of the liquid. This behavior affects the flow of fluids over surfaces and through tubes.
- **Capillary action**: Surface tension plays a crucial role in capillary action, a phenomenon where liquids flow against gravity in narrow tubes. Surface tension pulls the liquid up the tube and creates a concave meniscus at the surface of the liquid.
- **Heat transfer**: Surface tension affects heat transfer by influencing the formation of thin liquid films and liquid bridges. These structures can enhance or impede heat transfer depending on the specific conditions.
- **Boiling and condensation**: Surface tension affects the formation and behavior of bubbles during boiling and condensation processes. It influences the nucleation of bubbles and the spreading of condensate droplets on surfaces.

Role of Surface Tension in Mass Transfer

- **Enhancing separation efficiency**: Surface tension can create a barrier between two phases, promoting the selective transfer of components based on their solubility and interactions with the interface. This can improve the efficiency of separation processes, such as liquid-liquid extraction or gas absorption.
- **Facilitating transport across interfaces**: Surface tension can aid in the transportation of mass across interfaces by reducing the resistance to mass transfer. This is particularly important in processes like distillation, where the movement of components between liquid and vapor phases is essential for separation.
- **Controlling droplet formation** and coalescence: Surface tension influences the formation of droplets in dispersed phase systems, such as emulsions or foams. By regulating the size and stability of droplets, surface tension can impact the mass transfer rates within these systems. **Droplet formation of saturated steam has profound contribution to large heat transfer coefficient of saturated steam.**
- **Modifying interfacial properties**: Surface tension can alter the properties of interfaces, such as the surface area and curvature, which can influence the rate of mass transfer. Changes in surface tension can also affect the wetting behavior of surfaces, leading to modifications in the contact angle and spreading of liquids.
- **Enabling capillary action**: Surface tension is responsible for capillary action, the phenomenon where liquids move against gravity in narrow spaces. This effect can enhance mass transfer in porous materials or microfluidic devices, where surface tension-driven flow can help transport substances within confined spaces.

Important Equations of Surface Tension

The Young–Laplace Equation

The Young–Laplace equation is an algebraic equation that describes the pressure difference across the interface between two static fluids:

Delta $P = y\ (1/r_1 + 1/r_2)$

Here,

gamma is the surface tension

r_1 and r_2 are the two principal radii of curvature of two static fluids at the interface.

The sign convention usually dictates positive curvature when convex and negative when concave.

When $r_1 = r_2$, $\Delta P = y = 0$

ΔP is Called Laplace Pressure

Young's Equation

Young's equation is a formula that describes the relationship between the surface tensions of a solid and a liquid when they come into contact:

$$\cos \theta = [\gamma\ SV - \gamma\ SL]/\gamma\ LV$$

Where:

$\gamma\ SL$: is the surface tension of the solid-liquid interface

$\gamma\ SV$: is the surface tension of the solid-gas interface

$\gamma\ LV$: is the surface tension of the liquid-gas interface

θ: is the contact angle when the three phases reach equilibrium

ROLE OF SURFACE TENSION IN OIL AND GAS INDUSTRIES

How does Surface Tension Help?

Liquids' surface tension gives them the ability to withstand forces from outside that would otherwise try to separate them, such as stretching, bending, or compressing. The following are a few ways surface tension benefits these sectors:

- **Prevents two-phase flow**: Surface tension helps to prevent multiphase flow.
- **Controlling oil spills**: Surface tension helps to keep oil spills in rivers, lakes, and oceans from spreading too far. By reducing surface tension, chemical dispersants, and surfactants also make oil break up into smaller droplets that are simpler to spread.
- **Increased oil recovery**: During oil extraction procedures, surface tension makes it easier for oil to move through reservoir rock pores.
- **Drilling fluids**: Drill cuttings are kept suspended and prevented from settling and obstructing the drill pipe by the surface tension of drilling mud. The surface tension of the drilling mud can be reduced by adding surface-active chemicals.

How surface tension can pose some challenges?

While surface tension has many advantages for the oil and gas industries, there are some drawbacks as well. Surface tension can present difficulties in these industries in a number of ways, including:

- **Emulsions:** Oil and gas products may develop emulsions due to surface tension, which can make it challenging to separate them. Emulsions can cause problems with production like corrosion, increased viscosity, and decreased effectiveness.
- **Fluid flow**: Surface tension can obstruct the movement of fluid in pipes and other machinery.
- **Capillary forces**: Surface tension can produce capillary forces in porous media, such as oil-bearing rock formations, which keep water and oil in the pores. These forces may make it challenging to extract oil and result in ineffective production.

CHAPTER 49

Phase Change

Phase changes in thermodynamics refer to the transitions between different states of matter, such as solid, liquid, and gas. During a phase change, there is a transfer of energy between the system and surroundings, which can result in changes in temperature and pressure. The process of a phase change is characterized by specific heat capacities and latent heats, which represent the energy required to change the phase of a substance. Thermodynamics describes phase changes using concepts such as enthalpy and entropy, and provides insights into the equilibrium conditions at which phase transitions occur.

WHEN DOES PHASE CHANGE OCCUR?

Phase changes occur when energy is either added or removed from a substance, causing it to transition to a different state of matter. When a substance is at equilibrium, the particles are in a stable arrangement based on the temperature and pressure of the surroundings. Any deviation from this equilibrium state can lead to a phase change.

Adding energy to a substance can disrupt the intermolecular forces holding the particles together, causing the substance to transition to a higher energy state. This often results in an increase in entropy, as the particles become more disordered and move more freely. For example, when water is heated, the energy added breaks the hydrogen bonds between water molecules, causing the liquid water to turn into vapor (steam).

On the other hand, removing energy from a substance can cause the particles to slow down and come closer together, leading to a decrease in entropy and a transition to a more ordered state. For example, when water is cooled, the energy is removed from the system, allowing the particles to come closer together and form a solid (ice) through the reformation of hydrogen bonds.

In summary, phase changes occur when a substance's energy balance is disrupted, either by adding energy to break intermolecular forces or by removing energy to strengthen these forces. These changes lead to alterations in the arrangement and movement of particles, resulting in transitions between different states of matter.

WHY DOES WATER EVAPORATE AT AMBIENT TEMPERATURE?

The one-line answer is that the free energy change of water at 298.15 K and 1 atm is 8.6 kJ/mol.

This means water has a positive free energy. Water in your glass at ambient temperature is not in a state of thermodynamic equilibrium of minimum free energy and maximum entropy change.

Gibbs free energy is expressed by equation

$dG = dH - TdS$, here G is free energy, H Is enthalpy, T is Kelvin and S is entropy.

In the case of liquid water $dG = 8.6$ KJ/mol at 298.15 k and 1 atm pressure.

Water while trying to reach a stable thermodynamic equilibrium state always try to minimize it's free energy to go towards $dG = 0$ by increasing entropy, please refer back to Gibbs equation.

How?

Evaporation of water and phase change increase entropy because they increase the randomness of the system's particles.

Explanation

Entropy is a measure of disorder in a system. When water evaporates, it changes from a liquid to a gas, which increases the randomness of the water molecules and the entropy of the system.

This balancing act of water of reaching a stable thermodynamic state by generating entropy therefore, leads water to evaporate even when it is in subcooled state at 25 degc or 298.15 K.

So what evaporates from your glass of water at ambient temperature is in fact the irreversible entropy losses to surroundings that water rejects in its attempt to reach stability.

WHAT IS SUBLIMATION AND DEPOSITION? HOW THEY OCCUR?

In one line the sublimation is type of phase transition that involves phase change of solids to directly vapor with no intermediate liquid phase. This happens with solids those have weak intermolecular forces holding the molecules and have vapor pressure greater than atmospheric pressure at a given temperature.

Solid CO_2 sublimes. Solid CO_2 molecular crystal, are held together by weak intermolecular interactions weaker than ionic or covalent bonds.

Deposition if just the opposite of sublimation.

Details

Sublimation is a phase transition where a substance turns from a solid directly into a gas, without passing through a liquid state. It's a type of process that occurs when a substance's vapor pressure is higher than its atmospheric pressure.

The point at which sublimation occurs rapidly is called critical sublimation point.

The reverse process of sublimation is deposition in which a substance passes directly from a gas to a solid phase, without passing through the liquid state.

How and Why?

Sublimation is caused by absorption of heat which provides enough energy for some molecules to overcome the attractive forces of their neighbors and escape into the vapor phase. Since the process requires additional energy, sublimation is an endothermic change. The enthalpy of sublimation (also called heat of sublimation) $dH = dH$ fusion $+dH$ vaporization.

In sublimation, the crystal lattice of the solid has to break in addition there should be enough internal energy for a solid to transition to a gas.

The solids that sublime have vapor pressure > atmosphere pressure.

This makes the solids go directly from solid to gas without becoming liquid. The solids vapor pressure more than atmosphere pressure does not allow it to become liquid.

Explanation

Sublimation occurs when a solid's vapor pressure > atmospheric pressure at a specific temperature. This allows the solid to transition directly into a gas, bypassing the liquid phase.

Example: Dry ice (Solid CO_2)

Dry ice: Solid carbon dioxide sublimes at –78.5°C under normal atmospheric pressure.

Dry ice is solid carbon dioxide (CO_2) is an example that changes directly from a solid to a gas through sublimation. This happens when the temperature is below –56.4 °C and the pressure is below 5.13 atm. Therefore, at 1 atm pressure the dry ice is a gas.

Sublimation *vs.* Boiling

Sublimation is similar to boiling, but involves a solid's vapor pressure instead of a liquid's.

Most solids don't have a significant vapor pressure at normal temperatures, so the ability to sublime is uncommon.

Deposition

Deposition if just the opposite of sublimation when the vapor deposits into solids with no intermediate liquid phase.

How?

When temperature is lowered enough and the kinetic energy of molecules falls below that the intermolecular forces of the corresponding liquid phase and the vapor goes straight to solid phase.

ENTROPY GENERATION IN TWO-PHASE VAPORIZATION AND CONDENSATION

Temperature-Enthalpy [T-S] diagram

The diagram below is a temperature entropy diagram of water. The x-axis is entropy and the y-axis is temperature. The pink outline of the dome on the left represents saturated water while the red outline of the come on the right side represents the saturated water vapor line. At the topmost point of the dome where saturated water line insects the saturated vapor line is the critical point. On the left of the dome, there is water and on the right of the dome, there is superheated steam. Above the critical point, there is the supercritical region. Blue lines are pressure lines in Mpa [1 Mpa = 10 bar]. q represents the dryness fraction of steam + water inside the two-phase dome where they coexist together. On the extreme left on the saturated water pink line, the dryness fraction = 0 which shows it is all water. On the right $q = 1$ on the saturated vapor line meaning it is dry steam with no water. Inside the dome, both water and steam coexist as two phases. In a phase transition when both phases coexist in equilibrium, the difference in Gibbs free energy is equal to zero, $\Delta G = 0$. By the definition of entropy, the heat transferred to or from a system equals the area under the *T-s* curve of the process.

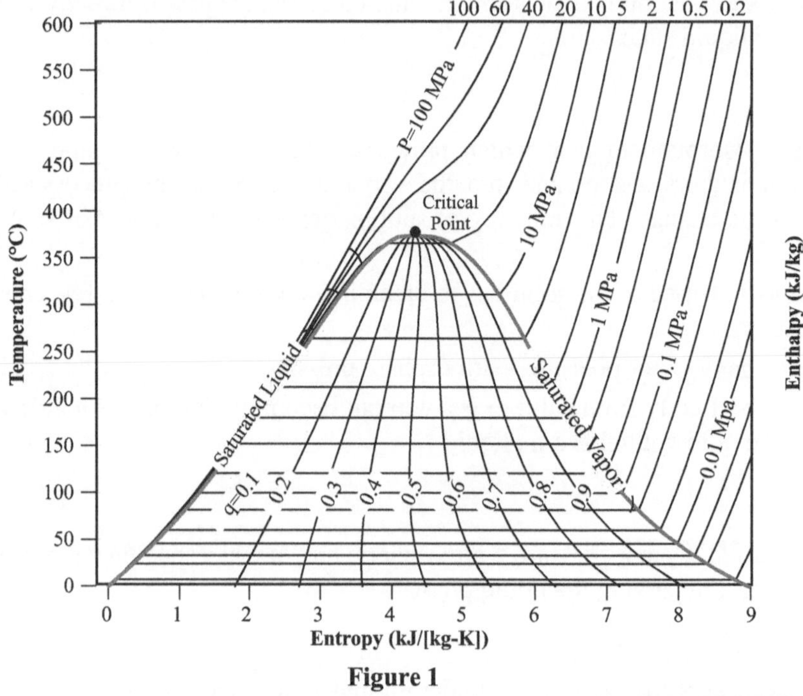

Figure 1

Entropy of Vaporization

Let us follow the horizontal line corresponding to 100 degc and 0.1 Mpa which is 1 bar from left to right on the *TS* diagram. As we proceed to right on this isothermal and isobaric line from saturated water to saturated vapor contour of the dome, the vaporization simultaneously proceeds with dryness fraction '*q*' increasing from zero to 1. This is the line where the two phases coexist in equilibrium. As we proceed to right the supply of energy breaks intermolecular bonds. As more and more water converts to vapor on this line the dryness fraction '*q*' increases with simultaneous generation of more and more vapor. This vapor generation is the source of entropy in the process of vaporization. It may be observed, as we move from the saturated water line on the dome towards the saturated vapor line on the right the entropy increases. The total entropy increase is about 6.5 kj/kg-k if we consider the intersection of isothermal/isobaric line corresponding to 100 degc, 0.1 Mpa with saturated water and saturated vapor curves. This is an output of huge expansion [> 1600 times volume] and disorder created by steam. Entropy generation in the vaporization is always positive, since the degree of disorder increases in the transition from a liquid in a relatively small volume to a vapor or gas occupying a much larger space.

Entropy of Condensation

When water vapor condenses to liquid water, the same latent heat is released however the entropy of the water is decreased, so something must have its entropy increased to compensate. This can be observed in the *TS* diagram. As we proceed from saturated vapor to saturated water outline on the dome on 100 degc/0.1 Mpa line entropy decreases by the same amount of 6.5 kj/kg-k what was the entropy creation in vaporization. Primarily the decrease in entropy takes place due to the conversion of more disordered vapor to more ordered liquid water molecules.

Summary

To summarize, while in both cases of vaporization and condensation, from saturated water to saturated vapor during vaporization and from saturated vapor to saturated water the enthalpy remains the same, in the case of vaporization positive entropy creation adds about 6.5 kj/kg-k more heat requirement for conversion of water to saturated steam. In the case of condensation since the entropy creation is negative, the net heat released as steam converts to saturated water = [Enthalpy of vaporization – Entropy gain]

Vaporization [saturated water to saturated steam]

Enthalpy = 2256.7 kj/kg

Positive entropy = 6.5 kj/kg-k

Condensation

Enthalpy = 2256.7 kj/kg

Negative entropy = 6.5 kj/kg-k

THERMAL AND THERMODYNAMIC EQUILIBRIUM IN PHASE CHANGE

With the beginning of phase change at a constant temperature, T and constant pressure, P. The two phases reach a thermal equilibrium when there is no more heat transfer between the two phases. Over time the two phases in contact with each other reach a state of minimum free energy, G a stage defined by thermodynamic equilibrium.

At phase change, the combined mass of the two phases remains constant but the volume increases due to the expansion of liquid into vapor.

As a consequence, while P and T reach constant values, the molar volume and molar entropy increase at the interface.

The two phases reach thermodynamic equilibrium when in Gibbs equation, $dG = dH - TdS$ (H is enthalpy, G is free energy, T is temperature and S is entropy) TdS becomes equal to dH. This makes, $dG = 0$. At dG zero or near zero the two phases reach thermodynamic equilibrium.

In conclusion, a phase change occurs when at the phase boundary both phases have identical P and T. This makes the phase change an isothermal process. When P and T get fixed in the isothermal condition the supply of heat breaks intermolecular bonds and generates vapor. This follows an expansion of gas and generation of entropy which begins consumption of the total energy called enthalpy at constant P and T and the gain in enthalpy dH is completely negated by the entropy. This is the stage when the two phases reach thermodynamic equilibrium. In other words when $TdS = dH$, the Gibbs free reaches zero. At $dG = 0$, the system reaches thermodynamic equilibrium.

WHAT MARKS THE BEGINNING OF PHASE CHANGE? WHY DOES PHASE CHANGE AT CONSTANT TEMPERATURE?

Phase change begins at the set point when the two phases establish thermodynamic equilibrium. The thermodynamic equilibrium involves the transition of a substance from one state to another. The thermal equilibrium at thermodynamic equilibrium between two phases is the compelling factor in fixing the phase change temperature constant.

In a single-component system, the degree of freedom (F) is determined by the number of components (C), the number of phases present (P), and a constant value of 2. This equation, $F = C - P + 2$, allows for the prediction that F is fixed at one. Consequently, the intensive independent variable, typically temperature, dictates the phase change by setting the temperature and pressure at the phase change point. Pressure, as a temperature dependent variable at constant volume, automatically adjusts accordingly. At saturation temperature a phase runs out its heat capacity. At saturation temperature the phase does not have any more sensible heat. Therefore, any heat addition at saturation temperature and pressure is used in breaking intermolecular bonds and the expansion of the vapor phase with consequent increase in entropy. This entropy generation pushes the vapor and the liquid phase into a thermodynamic equilibrium by minimising the Gibbs free energy, $dG = dH - TdS$ [G is free energy, H is enthalpy, T is temperature, and S is entropy.

The set point temperature of phase change is the saturation temperature of the substance.

Details

Phase change happens at a constant temperature due to the saturation of the existing phase with its total energy, leading to a halt in temperature increase. This halt occurs because at saturation point the liquid phase has no more heat capacity to absorb heat to raise its temperature further. This saturation point is crucial for the phase transition to occur.

At saturation point, the thermal equilibrium between the two phases does not allow any heat transfer between the two phases. It is the saturation temperature of the liquid phase that decides the phase change temperature.

During a phase change, the substance undergoes significant internal rearrangement as molecules shift from one phase to another. This rearrangement requires a precise amount of energy to break existing intermolecular bonds and form new ones. This energy, known as the latent heat of fusion or vaporization, sustains the phase change process and is essential for the transition to occur.

Moreover, during the phase change from liquid to vapor, the expanding vapor requires additional energy, contributing to the overall energy budget of the phase change process. As the intermolecular bonds break and the vapor expands, entropy is generated as the vapor phase expands, further driving the transition from one phase to another.

At phase change, with the onset of thermodynamic equilibrium, the free energy change reaches a minimum.

Thermodynamic equilibrium ensures that both phases involved in the phase change remain in balance, with minimal free energy and constant temperature. This equilibrium allows for the efficient exchange of energy between the two phases, facilitating the transition while maintaining thermal stability.

Three Important Points

1. The beginning of thermodynamic equilibrium is the set point for phase change.
2. Phase change always occurs at thermodynamic equilibrium at minimum free energy
3. It is the thermal equilibrium between two phases that makes the phase change temperature constant.

PHASE CHANGE: ROLE OF TEMPERATURE AND PRESSURE

In the context of more likely phase change on the hot side of a heat exchanger, the discussion is restricted to a single component system to explain the role of temperature and pressure in the phase change. The simplest example of a single component is water-vapor phase change.

This is a pressure-enthalpy [*P-H]* diagram of water. The Y-axis is pressure. X-axis is enthalpy. There are three distinct regions in the diagram. The left side of the dome is the water region. The dome represents the two-phase region when water transforms into vapor. The right side is the steam region. The left-side outer line of the dome represents the saturated water line. The right-side outer line of the dome represents the saturated steam line. The meeting point of these two lines shown by a small red dot on the top of the dome is the critical point.

Phase change occurs between the saturation point of water on the left outer line and the saturation point of steam on the right outer line of the dome. In the case of water saturated water exists at 100 degc/1 bar pressure with a dryness fraction, $X = 0$, and saturated steam forms at 100 degc/1 bar pressure with a dryness fraction, $X = 1$. The steam dryness fraction is used to quantify the amount of water within the steam. If steam contains no water, it has a dryness fraction of $X = 1$.

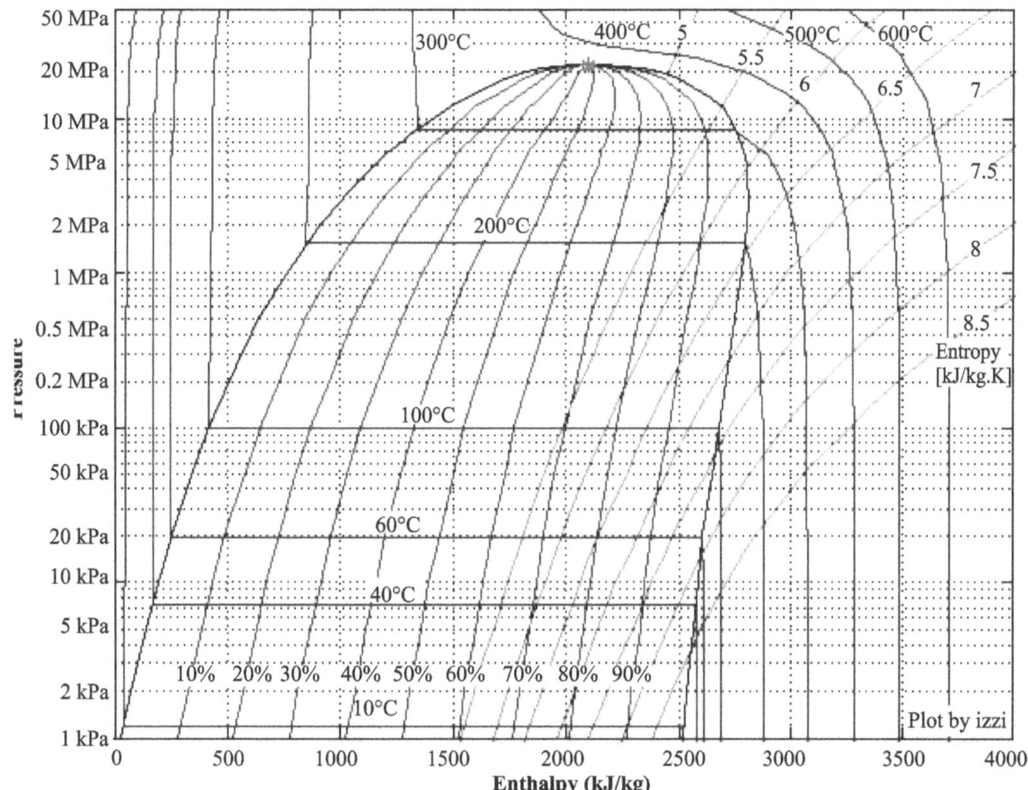

Figure 2: P-h Diagram for Water

Why and when a phase change occurs?

The phase change occurs by external intervention when the phase is saturated with energy and has no space to accommodate any more energy. Depending on the type of external intervention cooling or heating a saturated phase converts into a new phase. When saturated water is further heated, for example, it converts to vapor, and when saturated steam is cooled it converts into a new water phase.

Why does phase change occur at constant temperature?

Phase change occurs at a constant temperature because water [liquid] beyond saturation point has no more heat capacity left to increase its temperature. The heat applied beyond the saturation point breaks intermolecular bonds, generates vapor, expands vapor [PdV work], and does entropy generation as the vapor expands. This supplied heat remains latent and expressed as enthalpy of vaporization or latent heat. Entropy is accounted for separately. These entire phenomena happen inside the dome until two phases of water vapor combine into a single dry steam called saturated steam on the right-side outer line of the dome.

Why does phase change occur at constant pressure?

A phase change happens at a constant saturation pressure fixed by the temperature. For example, the phase change of water to vapor happens at 100 degc/1 bar pressure.

DEGREES OF FREEDOM FOR PHASES AND THE GIBBS PHASE RULE: WATER AS AN EXAMPLE

A phase diagram begins at –273.15 degc and ends at the critical point.

It is an interesting concept to analyze the phase behaviors of phases in a system. It tells us at every step how PVT interacts with phases on the equilibrium line of a phase diagram. Gibbs phase rule provides a theoretical foundation in thermodynamics for characterizing a system's chemical state and predicting the equilibrium relationships of the phases present as a function of physical conditions such as pressure and temperature. Gibbs' phase rule also enables the creation of phase diagrams to represent and interpret phase equilibria in heterogeneous systems. Gibbs's phase rule tells you why in all one component system the phase can be described by pressure and temperature. This explains why all one-component phase diagrams are essentially a P-T diagram.

The Gibbs phase rule states that the number of degrees of freedom is given by the equation F = C-P+2, where C represents the number of chemical components and P represents the number of phases if the equilibrium in a heterogeneous system is not affected by gravity or electrical and magnetic forces.

What are degrees of freedom [thermodynamics]

Degrees of freedom [DOF] are independent variables required to specify the thermodynamic state of a system containing components and phases.

The phase rule is a general principle in thermodynamics that governs "*pVT*" systems, whose thermodynamic states are completely described by the variables pressure (p), volume (V), and temperature (T) in thermodynamic equilibrium.

Gibbs Equation

$$F = C - P + 2$$

if F is the number of degrees of freedom, C is the number of components, and P is the number of phases.

A phase is a type of matter that has a uniform chemical composition and physical state. The most common phases are solid, liquid, and gas. Two immiscible liquids (or liquid mixtures with different compositions) separated by a distinct boundary, as well as two immiscible solids, are counted as two distinct phases. The number of components (C) is the number of chemically independent constituents of the system, i.e., the bare minimum of independent species required to define the composition of all system phases.

In this context, the number of degrees of freedom (F) refers to the number of intensive variables that are independent of one another.

Application of Phase Rule in Water

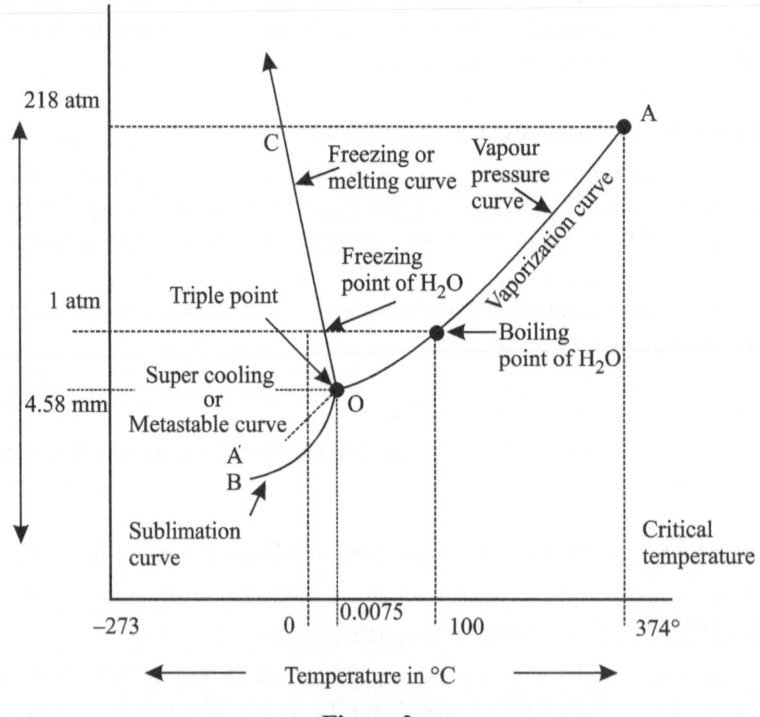

Figure 3

Explanation

Phases in Water

Fundamentals

The water system consists of three phases.

Ice, liquid water, and water vapor.

Since H_2O is the only chemical compound involved, there is a single component.

From phase rule, when $C = 1$

From Gibbs equation, $F = C - P + 2$, or, $1 - P + 2 = 3 - P$

This means that the degrees of freedom F depends on the number of phases

The three different cases are possible

P = 1; F = 2 (bivariant system)

P = 2; F = 1 (univariant system)

P = 3; F = 0 (invariant system)

You may observe that for any one component system, when $C = 1$, the maximum number of degrees of freedom is two. Therefore, you would find that all one component system phase diagram is essentially *P-T* diagram.

Curves on the Phase Diagram

The most convenient variables are pressure and temperature.

Curves

OC Ice & Water, F = 1 – 2 + 2 = 1, (univariant)

It is a fusion curve. The curve *OC* terminates at *C*, the critical pressure. The two phases solid-ice & liquid-water coexist in equilibrium. The curve indicates the melting point of Ice decreases with pressure increases. One atm line meets the curve at 0 degc.

OA Water & Vapor, F = 1 – 2 + 2 = 1, (univariant)

Curve *OA* is the vaporization curve. The curve OA terminates at *A*. It is the critical point is 218 atm. & temperature is 374 degc. It represents the vapor pressure of the liquid at different temperatures. Two phases of water & water vapors coexist in equilibrium along the curve $P = 2$, $C = 1$. Vapor pressure is 1 atm. The corresponding temperature in degree centigrade is the Boiling point of water 100 degc

OB Ice and Vapor (univariant)

It is a sublimation curve. The curve *OB* terminates at *B*, the absolute zero –273 degc temperature. It shows the vapor pressure of solid ice at different temperatures. The two phase's solid-ice & water-vapor coexist in equilibrium.

Triple point

All the three curves *OA*, *OB* & *OC* meet at pt *O* called a triple point, where all three phases solid, liquid & vapor are simultaneously in equilibrium. $F = C - P + 2 = 1 - 3 + \ = 0$

Critical point

This is interesting. At the critical point there are three phases [1] subcritical liquid phase [2] subcritical vapor phase and [3] coexistence of liquid-vapor phase as one single phase. $F = C - \mathrm{P} + 2 = 1 - 3 + 2 = 0$. Therefore, the degree of freedom at critical point is zero. This tells us that there is no independent variable at the critical point for the two phases to exist together. Therefore, the critical point is a unique point on the phase diagram.

WHAT MAKES WATER TO CHANGE INTO VAPOR?

The answer is entropy change.

How?

At 100 degc the saturated water has an enthalpy of 419 kJ/kg. Its entropy is 419/373 = 1.12 KJ/kg/K. The sensible temperature increase stops when water gets saturated in liquid phase and its heat capacity does not allow any more temperature increase.

As heating of water continues the entropy wants to increase. To accommodate increased entropy the water expands into vapor by generating more volume or microstates and hence more disorder.

The vapor generation and vapor expansion require 2240 KJ/kg heat which we call latent heat. The enthalpy of saturated vapor at 100 degc = 2240 + 419 = dH = 2659 KJ/kg.

The entropy changes increase to 2659/373 = 7.12 KJ/kg. This is the maximum entropy increase possible in a [water-vapor] system. This makes the free energy change $dG = 0$

$$dG = dH - TdS = 2659 - 7.12 \times 373 = 2659 - 2659 = 0.$$

The free energy dG change becomes zero. There is no energy to drive the process further. The saturated water stops absorbing any heat. Both liquid and vapor phases reach an equilibrium.

ADIABATIC PROCESS: CAN IT CAUSE PHASE CHANGE?

The answer is no.

The phase change does not happen arbitrarily. It's a state when the two phases go into thermodynamic equilibrium with minimum Gibbs free energy and maximum entropy.

At phase change of a single component system by Gibbs phase rule, $F = C - P + 2$, there is one component, $C = 1$, and there are two phases, $P = 2$, therefore, the degree of freedom, $F = 1$, meaning for a single component system a phase change is dictated by one intensive independent variable. It could either be pressure or temperature. When one of them is fixed the other dependent variable gets automatically fixed. For every saturation temperature on a phase (PT) diagram, there is a fixed saturation pressure and vice-versa, simply because one intensive independent variable controls a phase change.

How Does Phase Change Occur?

A phase change occurs when the substance in its current phase is saturated with energy. This means the existing phase has no more heat capacity to absorb any more heat and increase its sensible heat and temperature. At this point, the temperature rise stops despite supplying heat. The phase goes into a new phase, say from liquid to vapor by breaking its intermolecular bonds.

At the point of the phase change the phase has maximum enthalpy, for example for water to steam phase change the liquid water called saturated water has a maximum enthalpy 419 kJ/kg. It has maximum entropy and minimum free energy for two phases to go into thermodynamic equilibrium. This equilibrium is maintained during the phase change process when the dryness fraction, $X = 0$ at the start point of phase change reaches $X = 1$ at the end point and saturated water at 100 degc/1 bar becomes saturated steam at 100 degc/1.

During the entire phase change process temperature and pressure remain constant.

Now let us look at an adiabatic process.

In an adiabatic process, since $Q = 0$, $[U = Q - W]$, $U = W$ (U internal energy and W work) keep compensating each other. In adiabatic expansion, U supplies energy to W. In adiabatic compression, W supplies energy to U. in either case U changes and therefore, the temperature is not constant internally in an adiabatic process. Therefore, an adiabatic process does not meet the requirements of thermodynamic equilibrium for phase change.

PHASE CHANGE AT CRITICAL POINT: SECOND-ORDER PHASE CHANGE

The difference between first-order and second-order phase transitions is that there are large fluctuations before a second-order phase change, which act as a 'warning' that unusual behavior is about to occur. For example, the enthalpy of vaporization gradually diminishes as you approach a critical point. That indicates the gradual diminishing of van der Waals attractive intermolecular forces. At the critical point, the enthalpy of vaporization is zero. However, first-order phase changes occur abruptly and do not have any prior fluctuations. It's an abrupt switch of sensible heat to latent heat at the saturation temperature when the liquid phase runs out of heat capacity.

A phase change can occur in two ways [1] when two existing phases in a system find out a point of pressure and temperature where they have maximum molecular symmetry and [2] when two existing phases in a system find out a point of temperature and pressure where both phases have the minimum free energy.

The first one is called the second-order phase change. The second one is called the first-order phase change.

What exactly is molecular symmetry?

Energy levels differ between the two phases. Both phases of a phase molecule have the same energy level at maximum symmetry.

Details

First-order phase change

Any transition involving latent heat in which the system absorbs the heat but the temperature remains constant is referred to be first order.

- The system does not fully shift to the new phase in a seamless manner (regions of both phases exist simultaneously)
- The state variables of the system have a discontinuity (such as temperature or pressure)

Melting ice and boiling water are two common instances of first-order phase transitions (not all ice melts instantaneously nor does all water turn to steam immediately).

Second-order phase change

Second-order phase transitions occur when a new state of reduced symmetry develops

- These transitions occur continuously.
- The entire system transitions to the new phase continuously.
- This order of transitions encompasses those where thermodynamic quantities diverge (such as susceptibility, heat capacity, etc.)

Typical examples of second-order phase transition are super-fluids. The critical point is a second-order transition. In any system containing liquid and gaseous phases, there exists a special combination of pressure and temperature, known as the critical point, at which the transition between liquid and gas becomes a second-order transition.

Distinction between 1st order and 2nd order phase change

The 1st order phase change, e.g., a water-vapor phase change **occurs on an equilibrium line between two phases** with one degree of freedom [phase diagram]. That is, by varying either temperature or pressure independently while holding the other constant, there are infinite possibilities for a phase change

The second order phase, e.g. the phase change at a critical point occurs at one particular temperature and pressure. **It's a point.**

Phase change at critical point: Second-order phase change

A critical point is a property of a pure substance. Every pure substance has a critical composition at a fixed temperature of pressure for example water has a critical point at 374 degc and 217.7-atmosphere pressure. The critical point is a point where the two phases have minimum molecular structural symmetry.

Let us elaborate on the second-order phase change.

Let's start with the basics. When the energy difference between two phases is the smallest, a phase change occurs. They achieve equilibrium. Entropy or disorder is important whether the phase change is first or second order.

There are two points to be made here. The first is that a phase is more disordered at high temperatures than at low temperatures due to molecular motions.

Due to spontaneous symmetry breaking, the high-temperature phase typically contains fewer symmetries or more entropy than the low-temperature phase. This minimizes the free energy for phase change.

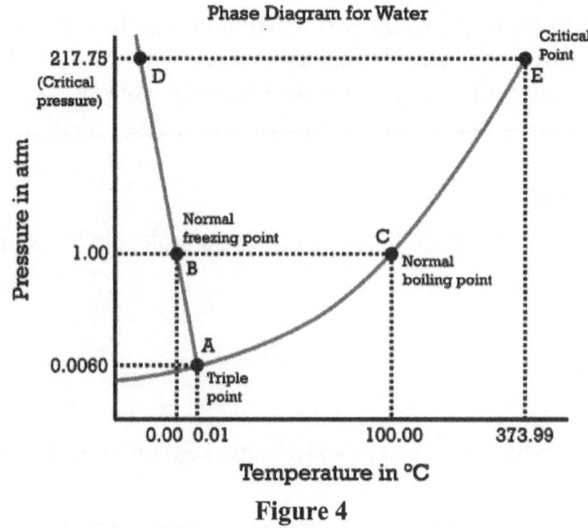

Figure 4

This happens abruptly when two phases interact and reach critical temperature and pressure. Critical point phase change is an example of 2nd order phase change. Phase change at a critical point does not involve the role of latent heat. At a critical point, all phases exist at minimum free energy with degrees of freedom $F = 0$. [$F = C - P + 2$, at the critical point there are 3 phases [1] Sub-critical vapor, [2] Sub-critical liquid, and [3] critical fluid, $F = C - P + 2 = 1 - 3 + 2 = 0$]. Critical point is a point where the sub-critical liquid and sub-critical vapor intersects at the boundary of the water liquid-vapor equilibrium phase diagram with emergence of critical fluid.

Therefore, the critical point does not have an option or freedom to reverse its phase. A superfluid is a unique collection of particles that has no kinetic energy and no viscosity. The superfluid component has zero viscosity and zero entropy. Since all the atoms in the superfluid are in the same energy level the total entropy of the system is zero.

To summarize, the distinction between first and second-order phase transitions is that in a first-order phase transition, the thermodynamic system's entropy, volume, and energy change abruptly, but in a second-order phase transition, the system's entropy, volume, and energy change gradually.

TWO-PHASE FLOW AND SLUG FLOW

Difference between Phase Change and Multiphase

Phase Change

It is physical. process. Phase change is an energy-driven process. Change can take place in either direction. Water has so little energy at 0 degrees Celsius that it converts to ice, a lower energy phase. Water has so much energy at 100 degrees Celsius that it turns to vapour, a higher energy phase. Water's kinetic energy is so low at 0 degrees Celsius that molecules coalesce into solid ice, which acts as a centre for potential energy via H-bonds. Similarly, the kinetic energy is so high at 100 degrees Celsius that molecules break H-bonds and move to a stable high energy centre for kinetic energy, which is vapour.

In summary, a phase change occurs in order for a phase to maintain maximum stability in a given situation. At the phase boundary, both phases stay at equilibrium. For, a single component system as water with two phases, $C = 1$, $P = 2$, the degree of freedom $F = C - P + 2 = 1 - 2 + 2 = 1$. Only one independent variable, temperature or pressure, decides the location of the phase boundary in the phase diagram.

Multiphase

A multiphase flow is defined as one in which more than one phase (i.e., gas, solid and liquid) occurs.

Different regimes of two-phase flows

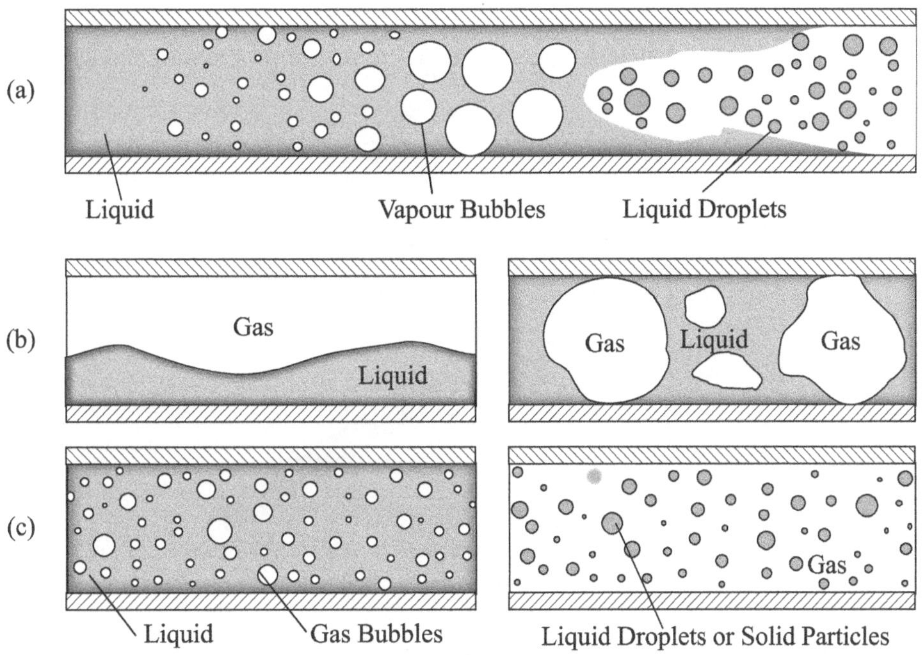

Figure 5

(a) transient two-phase flow with the transition from liquid to vapour flow, (b) separated or segregated two-phase flow with large regions of gas, (c) dispersed two-phase flow with gas bubbles (or liquid droplets) dispersed in a liquid or solid particles or droplets dispersed in gas.

Almost all oil and gas wells, as well as the associated pipeline systems that collect produced fluids and transport them to treatment facilities, exhibit multiphase flow. Because these fluids have significantly different physical properties, multiphase flow is much more complicated than single-phase flow. The following major issues are associated with the main difficulties encountered in describing multiphase flow: (1) Concurrently flowing liquid and gas in a pipe can have different geometrical arrangements, known as flow patterns; (2) the flowing mixture is compressible, so its density changes with changes in pressure and temperature; and (3) in addition to fictional losses, a new type of energy loss is occurring. Slippage loss occurs as a result of the gas phase slipping through the liquid due to the large specific gravity difference between phases.

Most multiphase correlations assume a homogeneous flowing mixture and employ the general energy equation written for this hypothetical fluid.

Significance of prediction of flow patterns: Flow in any channel requires design, development, and optimization. It is vital to predict the flow phases, like flow patterns primarily based on the interfaces formed between the phases. This know-how enables the prediction of the pressure drop and heat transfer characteristics based totally on the flow rate, channel size and operating conditions. From the pressure drop data, the acceptable float characteristics can be decided to decrease the incidence of corrosion, erosion, or scale formation, all of which can lead to excessive friction.

Slug flow

A Slug flow is a two-phase flow, more particularly a liquid-vapor flow. Slug flow occurs due to two different densities of the phases.

In a multiphase flow in a pipe, the lighter fluid stays as bubbles dispersed within, and pushing along, the heavier fluid. The word slug normally refers to the heavier, slower-moving fluid, but sometimes to the bubbles of lighter fluid.

There are also small bubbles within the liquid, but many of these have coalesced to form large bubbles until they span much of the pipe. In gas-liquid mixtures, slug flow is similar to plug flow, but the bubbles are generally larger and move faster. As flow rates increase, slug flow becomes churn flow.

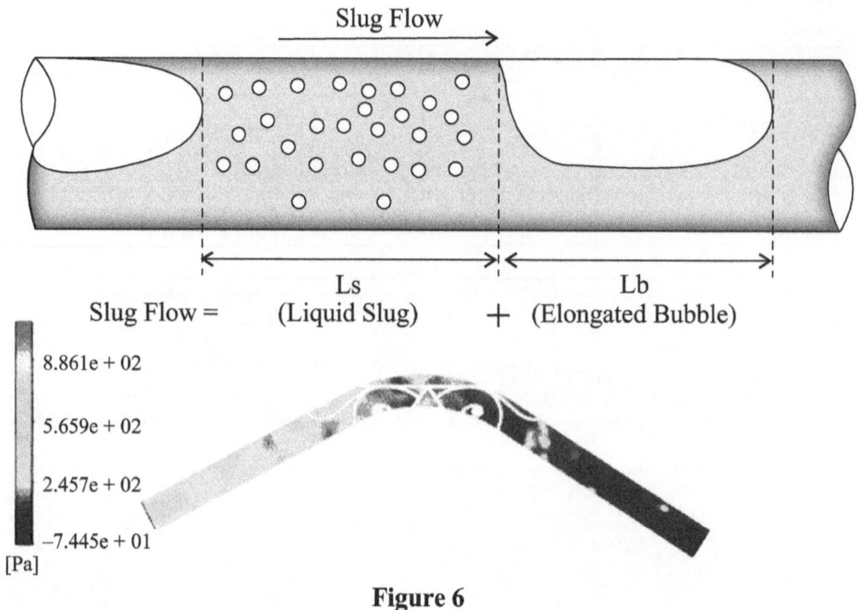

Figure 6

To summarize, the intermittent gas-liquid flow could be divided into two distinct systems: plug and slug flow. Each regime can provide its particular shape and hydrodynamics properties. The main characteristic of the slug flow is the intermittent passage of air-laden liquid eruptions followed by elongated air bubbles called cells. Interactions with neighbouring cells form or break liquid slugs

Different dimensions of the slug flow

Every fluid wants to reach the shape of a sphere because a sphere has the minimum energy. A sphere is the most stable shape for any fluid.

The problem is created by the surface tension at the interphase of two fluids. Surface tension is nothing but a stretch of the surface by the intermolecular forces at the interphase of two fluids. In the bulk, a liquid molecule experiences 360-degree pull from all sides.

Slug flow

Interestingly you can see bubbles of gases floating in the liquid causing a slug flow.

The obvious question arises how the light bubbles survive the hydrostatic pressure of the liquid inside the liquid. The answer is a gas has no option for any different shape. It will break and again will reform the sphere.

The answer is the surface tension

Surface tension

Gases have no surface tension because molecules are far apart to cause any wall tension.

This is making air or gas fold itself into its minimum energy shape which is a sphere.

Water cannot fold itself because of its large intermolecular forces pulling in different directions at 360 degree

Viscosity

There is another dimension. That is the effect of viscosity on gas bubbles. The shearing of the surface of the bubble by non-Newtonian liquids elongates the bubbles like a bullet that pushes a slug.

Consequences of slug flow

Oil and gas transportation is an important activity for the oil industry worldwide. Pipelines are primarily used to transport oil and gas from wells or production areas to refinery facilities. These pipelines are subjected to severe environmental and operational stresses that can result in severe dynamic reactions that lead to pipe failure and fatigue.

Slug flow is one of the greatest threats to the pipeline industry

This flow system can cause rapid fluctuations in mass distribution and pressure disturbances along the pipe, which can lead to violent vibrations. These vibrations create transient stresses that lead to fatigue failure, localized buckling, and even excessive bending and failure. These fluctuating stresses in pipelines can significantly disrupt normal operation, accelerate pipeline corrosion, and cause significant damage to pipe walls. In fact, it has been reported that many accidents have occurred due to vibration caused by internal two-phase flow.

Effect of temperature on slug flow

The temperature would primarily influence the viscosity and solubility of the gas.

The following events are anticipated:

1. **Viscosity:** The viscosity of the liquid phase decreases as temperature rises. As a result, the liquid flows more easily and forms smaller slugs as the temperature rises.
2. **Gas fraction:** As the temperature rises, so does the gas fraction in slug flow. This is because the solubility of gas in liquids decreases with increasing temperature. The gas bubbles congregate and form larger pockets as the temperature rises, resulting in larger slugs.
3. **Slug length and frequency:** As temperature rises, the length of liquid slugs shortens, while slug formation frequency increases. This is due to the lower viscosity of the liquid phase, which allows it to flow. more easily and form smaller slugs.
4. **Flow instability:** At extremely high temperatures, the liquid evaporates rapidly, causing the liquid phase to decrease and the gas phase to increase. This destabilises the flow, which can lead to flow reversal or other system instabilities.

WHAT DRIVES TRANSITION OF LIQUID TO SOLID

In all phase changes either liquid to vapor or liquid to solid entropy change plays an important role.

The change in entropy (ΔS) at constant temperature and pressure is given by the equation ΔS = qrev/T, where qrev is the heat absorbed or released during a reversible process and T is the temperature.

Liquid to Vapor

Briefly, in the case of liquid to vapor when the liquid reaches its heat capacity and maximum internal energy and consequently the maximum entropy the liquid converts to vapor to accommodate the increased entropy.

Liquid to Solid

When a liquid transitions to a solid, the decrease in entropy (ΔS) plays a significant role in the overall process.

The reduction in entropy, or an increase in orderliness, is essential for the transition to a solid state, which requires a decrease in the randomness of the molecular arrangement.

This decrease in entropy is driven by the release of heat energy (enthalpy) during the transition.

Explanation

Freezing is a phase transition that occurs when a liquid's temperature drops below its freezing point.

When liquid freezes, it releases heat of fusion to the surrounding area. This process is spontaneous because the entropy of the surroundings increases, while the entropy of the water decreases.

Overall, there is an increase of entropy in a freezing process due to exothermic nature of the freezing process. This makes $dG = dH - TdS$ negative and the process spontaneous.

So, in summary, it is the change in entropy that is important in driving the transition of phase be it liquid to vapor or liquid to solid.

CHAPTER 50

Ideal and Real Gases

Ideal gas thermodynamics assumes that gas molecules have no volume and do not interact with each other, simplifying calculations for pressure, volume, and temperature relationships. Real gas thermodynamics accounts for the volume occupied by gas molecules and the interactions between them, leading to deviations from ideal gas behavior at high pressures and low temperatures. Ideal gas equations such as the ideal gas law ($PV = nRT$) are useful approximations in many practical applications, while real gas equations like the van der Waals equation provide more accurate predictions for real gas behavior. The compressibility factor Z is used to quantify deviations of a gas from ideal behavior, with $Z = 1$ indicating ideal gas behavior. Understanding the differences between ideal and real gas thermodynamics is essential for predicting the behavior of gases in various conditions and engineering applications.

IDEAL GASES AND BEYOND

The volume, pressure, temperature, and moles of a gas in a closed system can be related by the following equation, which is referred to as the **ideal gas law**:

$PV = n\,RT$

$PV = nRT$R is referred to as the **Universal Gas Constant**, and it has the following values for different units of P, V, n, and T:

$R = 0.0821\ \text{L} \times \text{atm/mol} \times K = 8.31\ \text{J/mol} \times K = 8.31\ \text{Pa} \times \text{m}^3\text{/mol} \times K$

Background Information on Ideal Gas Law Theory

The classical Kinetic Theory of Gases, which forms the basis of the ideal gas law, is predicated on several rather strong assumptions about the nature of gases. These presumptions are:

1. That there is no interaction between gas molecules.
2. That all molecular collisions with the container's walls or with each other are entirely elastic, maintaining the molecules' mean kinetic energy constant.
3. The size of gas molecules and their separation from one another is minuscule.

Important Points and Facts About Ideal Gases

Ideal gases have several characteristics that make working with them easy and beneficial:

1. An ideal gas can also be an ideal mixture of ideal gases. As a result, the ideal gas law remains valid whether applied to the mixture as a whole or any one of its constituent parts.
2. The equation, where is the mole fraction of component A in the mixture, relates the partial pressure of a component to the total pressure in an ideal gas mixture.
3. The ideal gas equation is independent of the gas's characteristics, assuming it is valid.
   Consequently, you only need to know three of the unknowns (P, V, T, and n) and you can solve for the fourth one.
4. An ideal gas's enthalpy and entropy are only dependent on temperature; they are not dependent on pressure or volume.
5. Many gases are near ideal at high temperatures and low pressures. As a result, it can serve as a useful benchmark state to which the actual real gas can be compared.

Thermodynamics will cover a plethora of additional beneficial characteristics of ideal gases.

Ideal Gas Deviations under High Pressure

Let's say you have a water bottle filled with water vapour. What happens if you reduce the volume of that vapour to apply pressure? The gas molecules inside the bottle get closer to one another as pressure is increased.

However, dispersion forces and, if relevant, polar forces between molecules of a substance increase with increasing molecular proximity. The above-mentioned kinetic theory of gases eventually loses validity as the dispersion forces grow.

As a result, the ideal gas law is either no longer valid or is only a rough approximation at best.

A vapour may even condense into a liquid at extremely high pressures and thus can invalidate the use of the law.

Ideal Gas Deviations at Low Temperatures

According to the definition of temperature, a substance's average kinetic energy of its molecules decreases as its temperature drops.

Because they have less momentum, molecules with lower kinetic energy are more susceptible to the influence of gravity and dispersion forces from other molecules.

Eventually, the temperature might become low enough that the forces from other molecules cause significant deviations in the path of a molecule, and in this case, the ideal gas law becomes less valid. As with pressure, a very low temperature can cause a gas to condense.

Rule of Thumb for the Application of Ideal Gas Law

The ideal gas law is a reasonable assumption at high temperatures and low pressures, but it shouldn't be applied outside of those parameters, according to the two points mentioned above.

The Concept of Equations of State

There are other ways to express relationships between system properties besides the ideal gas law.

Although thermodynamics offers some guidelines about how many variables to correlate and what kinds of correlations we should look for, there are countless ways in which you could suggest to correlate system variables.

Correlations between state variables, in particular, are particularly helpful since they indicate changes in the variables' characteristics independent of the method of change.

An equation of state is any equation that connects variables of the state. Since pressure, temperature, volume, and number of moles (P, T, V, and n) are all measurable whereas many other possible variables are not directly measurable [such as enthalpy]

Compressibility

$$Z = P \times V/n\,RT$$

For an ideal gas, since

$$PV = n\,R\,T, Z = 1$$

Therefore, any deviation of the compressibility from 1 is a nonideality. Therefore, any deviation of the compressibility from 1 is a nonideality.

The compressibility of a liquid is very small, due to a small volume per mole of substance compared to that of a gas.

Ideal Gas Law Extension

One application of compressibility is that it permits a straightforward generalisation of the ideal gas law.

Unfortunately, Z varies with temperature and pressure and is not constant for any material. However, there is a technique called the generalized compressibility method with which you can estimate the value of Z for any substance, given certain data. Once it is known, this extension can be used to calculate an unknown system property.

Alternatives to the Ideal Gas Law

1: *van der Waals Equation*

One of the oldest equations that was developed to take non-ideality into account is called the van der Waals equation which has two substance-dependent parameters, a and b. One takes into account the interactions between particles, and the other the fact that particles have volume, sometimes substantial.

The van der Waals equation is as follows:

$$[P + n^2 a/v^2]\,[v - nb] = n\,R\,T$$

a and b are constants

The van der Waals equation is significantly more accurate than the ideal gas law and can be used to crudely predict when a gas will condense. However, it is not sufficiently accurate for many industrial purposes, and therefore other methods have been sought since then.

Alternatives to the ideal gas law 2: Virial Equation [Brief introduction]

The Virial equation is a formula with a potentially infinite number of parameters that vary according to the characteristics of the constituent materials. It is significant because, unlike the other equations of state, which have been derived semi-empirically, it can be demonstrated to be a legitimate extension of the ideal gas law using statistical theories.

Alternatives to the ideal gas law 3: Peng-Robinson equation [Brief introduction]

The Peng-Robinson equation, one of the more recent equations of state, is most helpful in characterising nonpolar molecules like nitrogen or hydrocarbons. Similar to the van der Waals equation, the Peng-Robinson equation has two parameters, but one of them is not constant and is dependent on both the system's temperature and the characteristics of the substance inside.

COMPARISON OF IDEAL AND REAL GAS

The behavior in the study of gases, both ideal and real gases are important considerations. Ideal gases are conceptual, theoretical entities that adhere strictly to the ideal gas law, $PV = nRT$, where P represents pressure, V is volume, n is the number of moles, R is the gas constant, and T is temperature. In this ideal scenario, gases do not perform thermodynamic work, and they obey the ideal gas law under all conditions.

***PV* = Constant** for ideal gases. Ideal gases are incapable of doing thermodynamic work. This does not apply to real gases for real gases.

$P5V^n$ **= Constant** with n can vary from 0 to infinity (explained below).

Real gases deviate from ideal behavior due to factors such as molecular size, intermolecular forces, and non-ideal conditions. For real gases, the relationship between pressure and volume is more accurately described **by the van der Waals equation, where**

$(P + an^2/V^2)\,(V - nb) = nRT$, and a and b are empirical constants.

Furthermore, real gases also follow the polytropic process equation, where

PV^n **= constant**, emphasizing that the relationship between pressure and volume is not strictly linear as in the ideal gas case.

The polytropic index (n) can vary from 0 to infinity, representing different thermodynamic processes and the work performed by real gases. It's important to recognize these distinctions when considering the behavior of gases in various practical and theoretical applications.

IDEAL GAS EQUATION *PV = N RT*

Summary

$PV = n\ RT$ signifies work done by an ideal gas. The three important equations that have origin in the ideal gas equation like the van der Waals equation, Redlich-Kwong equation, and Peng-Robinson equation are equations of state that can be applied to calculate the thermodynamic work done by an ideal gas under different conditions. These equations are modifications of the ideal gas equation, incorporating additional terms to account for the deviations from ideal behavior observed at high pressures and low temperatures. By using these equations, you can accurately calculate the work done by gases in real-world scenarios where ideal gas assumptions do not hold true.

Details

The equation $PV = nRT$, also known as the ideal gas law, describes the relationship between the pressure (P), volume (V), amount of substance (n), and temperature (T) of an ideal gas. It is primarily used to understand the behavior of gases under various conditions, assuming they follow ideal gas behavior.

Significance

The ideal gas equation signifies that an ideal gas does not have the capacity to do thermodynamic work. $PV =$ Constant.

Explanation

In an ideal gas at a constant temperature T and constant pressure P, the product PV should remain constant (assuming one mole of gas). In such a scenario, the gas does not perform any work on the surroundings, as there is no change in volume. Similarly, since there is no change in volume or any other variables, the entropy change and the free energy change would be zero.

Some of the other important equations of state have their origin in the ideal gas equation.

There are several equations of state that have their origin in the ideal gas law, $PV = nRT$ (where P is pressure, V is volume, n is the number of moles, R is the ideal gas constant, and T is temperature). Some commonly used equations of state include:

- **van der Waals equation of state**: $(P + a(n/V)^2)\ (V - nb) = nRT$ – This equation accounts for intermolecular forces and non-zero particle volume. The constants 'a' and 'b' are specific to each gas.
- **Redlich-Kwong equation of state**: $P = (RT)/(V - b) - (a/\sqrt{(T)}V\ (V + b))$ – This equation incorporates corrections for gas compressibility and intermolecular forces. The constants 'a' and 'b' are specific to each gas.
- **Peng-Robinson equation of state**: $P = (RT)/(V - b) - (a\alpha/\sqrt{(T)}\ ((V(V + b) + b(V - b)))) - (RTa/(2\sqrt{2})(V(V + b) + b(V - b)))$ – This equation is a modified form of the van der Waals equation and is commonly used for hydrocarbons and mixtures. The constants 'a' and 'b' are specific to each gas, and 'α' represents a temperature-dependent parameter.

How to Select an Equation of State

The selection of an equation of state depends on the particular fluid or mixture being analyzed, as well as the conditions under which the analysis is being conducted. Here are some general considerations for selecting an equation of state.

- **van der Waals equation of state**: This equation is most suitable for simple gases that have relatively low pressures and temperatures. It provides a reasonably accurate description of the behavior of nonpolar gases, but may not be as accurate for polar gases or those at high pressures and/or temperatures. It can be applied in cases where intermolecular forces are relatively weak.
- **Redlich-Kwong equation of state**: This equation takes into account the size of the molecules in addition to intermolecular forces and is suitable for gases and mixtures under moderate to high pressures and temperatures. It provides improved accuracy compared to the van der Waals equation, particularly for polar gases.
- **Peng-Robinson equation of state:** This equation is an improvement over the Redlich-Kwong equation and provides a more accurate description of real gas behavior, especially for polar and nonpolar compounds at high pressures and/or temperatures. It can be used for pure substances as well as complex mixtures.

AVOGADRO'S MATHEMATICS: HOW AND WHY?

Avogadro is known for his two earth shaking contributions

Avogadro's Hypothesis

At STP one mole of an ideal gas occupies 22.41 liter volume.

Avogadro's hypothesis is an experimental gas law relating the volume of a gas to the amount of substance of gas present. The law is a specific case of the ideal gas law. Avogadro's law states that "equal volumes of all gases, at the same temperature and pressure, have the same number of molecules. For a given mass of an ideal gas, the volume and amount (moles) of the gas are directly proportional if the temperature and pressure are constant.

Mathematics

The law can be written as:

$$V \propto n$$

$$V/n = k$$

Where V is the volume of the gas; n is the amount of substance of the gas (measured in moles); k is a constant for a given temperature and pressure.

This law describes how, under the same condition of temperature and pressure, equal volumes of all gases contain the same number of molecules. For comparing the same substance under two different sets of conditions, the law can be usefully expressed as follows:

$$V_1/n_1 = V_2/n_2$$

The equation shows that, as the number of moles of gas increases, the volume of the gas also increases in proportion. Similarly, if the number of moles of gas is decreased, then the volume also decreases. Thus, the number of molecules or atoms in a specific volume of an ideal gas is independent of their size or the molar mass of the gas.

Derivation from the Ideal Gas Law

The derivation of Avogadro's law follows directly from the ideal gas law.

$$PV = n\,RT$$

where R is the gas constant, T is the Kelvin temperature, and P is the pressure (in pascals).

Solving for V/n, we get, $V/n = RT/P$

RT/P = Constant, for a fixed pressure and a fixed temperature.

$V/n = RT/P$ = [8.314 J/mole- k × 273.15 k]/101.321 kPa = 22.41 liters/mole

For 101.325 kPa and 273.15 K, the molar volume of an ideal gas is 22.41 liters

Avogadro's Number

Avogadro's number tells us the number of particles in 1 mole (or mol) of a substance. These particles could be electrons or molecules or atoms. The value of Avogadro's number is approximately 6.022140857×10^{23} mol−1.

Avogadro's number $NA = 6.02 \times 10^{23}$ mol−1

Accurate determinations of Avogadro's number became possible for the first time when American physicist Robert Millikan measured the charge on an electron. The charge on a mole of electrons had been known for some time and is the constant called the Faraday. The best estimate of the value of a Faraday, according to the National Institute of Standards and Technology (NIST), is 96,485.3383 coulombs per mole of electrons. The best estimate of the charge on an electron based on modern experiments is $1.60217653 \times 10^{-19}$ coulombs per electron. If you divide the charge on a mole of electrons by the charge on a single electron you obtain a value of Avogadro's number of 6.02×10^{23} particles per mole.

Avogadro number $= 96{,}485.3383/1.60217662 \times 10^{-19}$
$= 6.022140857 \times 10^{23}\ mol^{-1}$.

WHY AND HOW GASES DEVIATE FROM IDEAL GAS BEHAVIOUR NEAR CRITICAL POINT?

Explanation

Near the critical point the assumptions of the ideal gas law are no longer valid: (1), Gas molecules have negligible volume and (2) Intermolecular interactions are negligible.

As a gas approaches its liquefaction point, the deviation from ideal behavior increases. This is because the volume will not be zero, even if the pressure increases, because the molecules occupy some space that cannot be compressed.

What Do We Know?

Ideal gas behavior.

Gases generally behave like ideal gases at high temperatures and low pressures.

Deviations at High Pressures

At high pressures, the finite particle volumes lower the actual volume available to the gas particles, which results in a higher pressure than the ideal gas value.

Compression and Cooling

Compression brings gas molecules closer together, and reduced temperature slows their random movement. This combination of compression and cooling leads to intermolecular interactions, which cause the molecules to move closer together and change state.

Application of van der Waals Equation at Critical Point

The van der Waals equation applies to the critical point of a gas by defining the critical point as the point where the first and second derivatives of pressure with respect to volume are zero: $(\partial p/\partial v)\ T = 0$ and $(\partial^2 p/\partial V^2)\ T = 0$.

The critical point is the boundary between two different behaviors of the equation:

- Above the critical point: The curves are monotonic.
- Below the critical point: The curves exhibit maxima and minima.

The van der Waals equation of state is: $[P + a \times (n^2/V^2)]\ [(V\text{–}nb)] = RT$.

The critical constants are related to the van der Waals constants as follows:

$T_c : 8a/27\ Rb$

$V_c : 3b$

$P_c : a/27b2$

In summary, gases deviate from ideal gas behavior near the critical point due to the increased intermolecular forces that are present at high pressures and temperatures.

As a gas is compressed and heated towards its critical point, the molecules are forced closer together and begin to interact more strongly with each other. This causes deviations from the ideal gas law, leading to non-ideal behavior.

Near the critical point, the gas molecules are no longer able to move freely and independently of each other, as they do in ideal gases. Instead, they begin to exhibit more complex behavior, such as clustering and forming intermolecular bonds. This results in the gas behaving more like a liquid than a typical gas.

Additionally, at the critical point, the distinction between the liquid and gas phases becomes blurred, as the properties of both phases merge together. This can lead to significant deviations from the ideal gas law, as the gas no longer conforms to simple ideal gas behavior.

Overall, gases deviate from gas laws near the critical point due to the increased intermolecular forces and complex interactions that occur at high pressure and temperature conditions.

IDEAL GAS EQUATIONS IN ONE PLACE

Isothermal, isochoric, [constant volume] isobaric, [constant pressure] and polytropic are the basic processes with ideal gas. table below contains a description of each process as well as the relevant equation.

Table 1

Process	*Relationship between ρ, V, T*	*Work (W)*	*Internal-energy change (U_2–U_1)*	*Heat (Q)*
Constant pressure	ρ = constant $\frac{V_2}{T_1} = \frac{V_2}{T_2}$	$\rho(V_2 - V_1)$	$mc_v(T_2 - T_1)$	$mc_p(T_2 - T_1)$
Constant volume	V = constant $\frac{\rho_1}{T_1} = \frac{\rho_2}{T_2}$	0	$mc_v(T_2 - T_1)$	$Q = U_2 - U_1$ $Q = mc_v(T_2 - T_1)$
Isothermal	$T = c$ $\rho_1 V_1 = \rho_2 V_2$	$\rho_1 V_1 \ln\left(\frac{V_2}{V_1}\right)$	0	$Q = W$ $Q = \rho_1 V_1 \ln\left(\frac{V_2}{V_1}\right)$
Polytropic	ρV^n = constant $\frac{T_2}{T_1} = \left(\frac{V_1}{V_2}\right)^{n-1}$ $\frac{\rho_2}{\rho_1} = \left(\frac{T_2}{T_1}\right)^{\frac{n}{n-1}}$	$\frac{\rho_1 V_1 - \rho_2 V_2}{n-1}$	$mc_v(T_2 - T_1)$	$Q = W + U_2 - U_1$
Adiabatic	As for polytropic with $n = \gamma$	As for polytropic with $n = \gamma$	$mc_v(T_2 - T_1)$	0

Ideal Gas Theory

An ideal gas is one in which both the volume of molecules and the forces between the molecules are so small that they have no effect on the gas's behaviour.

The ideal gas theory is very important for process analysis. The volume of an ideal gas, the pressure it exerts, and its temperature are the three parameters that define it.

The definition of ideal gas is a special state of matter that can be delimited by a system boundary.

Key points

- **All particles are assumed** to have rest mass ($m > 0$; the particles are not photons).
- **The particle count is low** in comparison to the system volume.
- **Particle volume is negligible** in comparison to system volume.
- **Collisions between particles are much less likely than collisions with the system boundary.**

The practical benefit of treating real gases as ideal (under certain conditions) is that it results in a simple equation of state with only one constant. The ideal gas equation of state can be written as follows:

$$PV = m\,RT \text{ or } P\,v = RT \qquad \dots (1)$$

where P is the pressure in Pa, V is the gas volume in m^3, m is mass of gas in kg, T is gas temperature in K, R is known as the gas constant and is given in J/kg K, v is mass specific volume in m^3/kg, v is molar specific volume in m^3/k mol, and $\mathcal{R}$ is the universal gas constant of 8.134 J/mol K.

Take note that the gas constant is unique to each gas and is determined by the universal gas constant and the gas's molecular mass (M).

$$R = \mathcal{R}/M \quad \dots (2)$$

The thermal equation of state" of the ideal gas expresses the relationship between pressure, specific volume, and temperature.

The ideal gas equation can be expressed in terms of internal energy, specific volume, and temperature. The equation of state in this case is known as the **Caloric equation of state**. The internal energy of an ideal gas, in particular, is temperature dependent. The following is the caloric equation of state for a monoatomic ideal gas:

$$u = 1.5\ RT \quad \dots (3)$$

[this equation can also be written as $du/dT = 1.5\ R$ or $Cv = 1.5\ R$]

u denotes molar specific internal energy.

Because $h = u + P\ v$, the enthalpy of a monoatomic ideal gas is given by

$$h = 2.5\ RT \quad \dots (4)$$

[this equation can also be written as $dh/dT = 2.5\ R$] OR $Cp = 2.5\ R$]

Combining the two caloric equations of state from equations (3) and (4), the known Robert Meyer equation for ideal gas can be derived as follows:

$$Cp - Cv = R \text{ or}$$

$$Cp = Cv + R \quad \dots (5)$$

It should be noted that the internal energy of an ideal gas is only a function of temperature. As a result, the ideal gas's specific heat is Cv = 1.5 R and Cp = 2.5 R. The adiabatic exponent is the ratio of specific heat at constant pressure and constant volume.

$$Y = Cp/Cv \quad \dots (6)$$

The adiabatic exponent for ideal gas is 1.4 for diatomic gas and it is 1.67 for monatomic gases. If one of P, v, or T is constant, there are some special cases. At a constant temperature, the volume of a given quantity of ideal gas varies inversely with pressure (this is known as Boyle's law), describing a volume change as follows:

$$P_1\ V_1 = P_2\ V_2 \quad \dots (7)$$

where the subscripts denote the initial and final states.

Analysts use eq. (7) in a variety of situations, including selecting an air compressor, calculating the consumption of compressed air in reciprocating air cylinders, and determining the time required to store air. Charles' law applies if the process is under constant pressure or volume:

$$v_1/T_1 = v_2/T_2 \text{ and } p_1/T_1 = p_2/T_2 \quad \dots (8)$$

If the number of moles of an ideal gas in an enclosed volume does not change, the combined ideal equation of state is as follows:

$$P_1\ v_1/T_1 = p_2\ v_2/T_2 \quad \dots (9)$$

CHAPTER 51

Piping

Thermodynamics plays an important role in determining the flow rate and pressure drop in piping systems, helping engineers design efficient and safe pipelines for transporting fluids. By applying thermodynamic principles, engineers can optimize piping layouts to reduce energy losses and minimize heat transfer during fluid transport. Thermodynamics also guides the selection of materials for piping to withstand temperature and pressure changes, ensuring the integrity and longevity of the system. Overall, thermodynamics provides valuable insights into the behavior of fluids in piping networks, leading to improved performance and cost-effective solutions.

EROSION IS ENGINEERS' NIGHTMARE: API RP 14E EROSIONAL VELOCITY EQUATION

In heat transfer, in particular the velocity has a key role. The irony is that, in addition to adding pressure drop, this velocity is a potential source of erosion. This balancing act that an engineer performs, whether consciously or unconsciously, is a nightmare.

Pipe erosion is the process by which the internal surface of a pipe deteriorates as a result of the abrasive action of multiphase flow. Corrosion occurs after erosion of a surface breaks the corrosion-protection layer on the surface. The combined corrosion - erosion damage to a surface is greater than the sum of the two effects. Erosion, like corrosion, cannot be zero in a pipeline. It can only be reduced. Corrosion would normally follow erosion, but the opposite is not true.

During the extraction of hydrocarbons from underground reservoirs, corrosion of carbon steel piping and equipment is a major issue. When electrochemical corrosion is involved, the situation becomes even more complicated. To maintain production rates, operators continuously dig deeper into the reservoirs or use proppants and reservoir fracturing techniques. As a result, deeper aquifers are encountered, water cuts increase, more multiphase streams are produced, and more solids and corrosive species are introduced into production, transportation, and processing systems, resulting in increased erosion-corrosion problems.

Conflict

High fluid velocities are often desirable in multiphase systems to reduce the chances of slug flow or particle settling. High velocities, on the other hand, increase the energy with which liquid droplets impact on pipe and vessel walls, which, if strong enough, can cause erosion or accelerate corrosion.

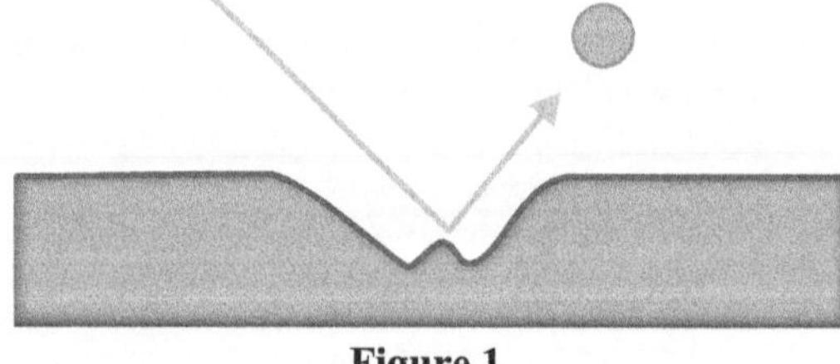

Figure 1

How do Particles Settle at Low Velocity

Explanation

Setting Velocity

A drag force, *Fd*, a buoyant force, *Fb*, and a gravitational force, *Fg* all act on a particle in a fluid flow. The particle settles when the gravitational force, *Fg* exceeds the drag force, *Fd*, and buoyant force, *Fb*.

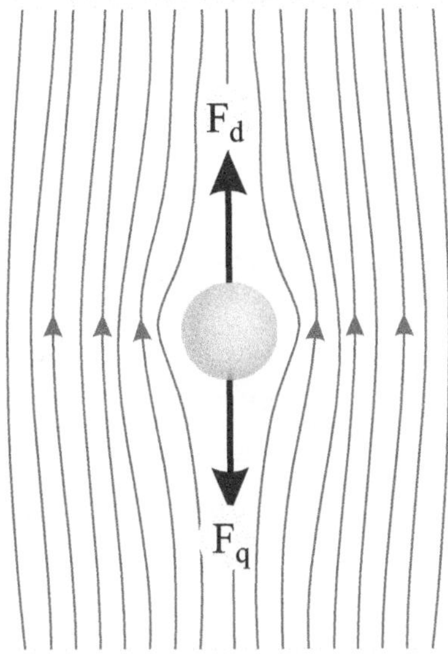

Figure 2

The process by which particulates settle to the bottom of a liquid and form a sediment is known as settling. Particles that experience a force, whether due to gravity or centrifugal motion, tend to move uniformly in the direction of that force. This means that the particles will tend to settle to the bottom of the vessel, forming sludge or slurry at the vessel's base.

Any particle is subject to two main forces. The primary force is an applied force, such as gravity, and a drag force caused by the particle's motion through the fluid. The applied force is usually unaffected by the particle's velocity, whereas the drag force is.

A particle at rest will exhibit no drag force, causing the particle to accelerate due to the applied force. When a particle accelerates, the drag force acts in the opposite direction of the particle's motion, slowing it down further; in the absence of other forces, drag directly opposes the applied force. As the particle's velocity increases, the drag force and the applied force will approximate equivalence, resulting in no further change in the particle's velocity. This velocity is referred to as the particle's terminal velocity, settling velocity, or fall velocity.

Synergistic effects of erosion-corrosion rate > [erosion + corrosion rates]

The terms erosion and erosion-corrosion are frequently used interchangeably. To be clear, erosion is defined as pure mechanical removal of the base metal, usually caused by impingement by solid particles, though impingement by liquid droplets can cause the same type of damage. Corrosion is a type of (electro) chemical metal loss in which iron dissolves in an aqueous solution, a process that can be accelerated by turbulent flow. Erosion-corrosion is a chemo-mechanical mode of attack that involves both erosion and corrosion. Due to synergistic effects, the resulting erosion-corrosion rate may be greater than the sum of erosion and corrosion rates.

Explanation

Steel corrosion is an electrochemical process in which metal atoms are first oxidised into cations before leaving the metal interface. Erosion, on the other hand, is a mechanical wear process that physically separates metal atoms from the steel substrate. The metal loss caused by the coupled effect of erosion and corrosion on a finely polished steel surface can sometimes be significantly greater than the sum of the metal losses caused by individual pure erosion and pure corrosion.

Corrosion effects on erosion at low particle velocity, with particles settling on the pipe surface and causing under deposit corrosion. While high particle velocity erosion affects corrosion by impingement of particles on the pipe surface, breaking the protective corrosion barrier and exposing the bare metal to oxygen attack. In general, erosion affects corrosion performance because turbulence accelerates mass transfer and increases flow shear force, which can remove corrosion product films or corrosion inhibitor from the steel surface.

What is under deposit corrosion by deposited solids?

Under deposit corrosion is a differential 'aeration corrosion'.

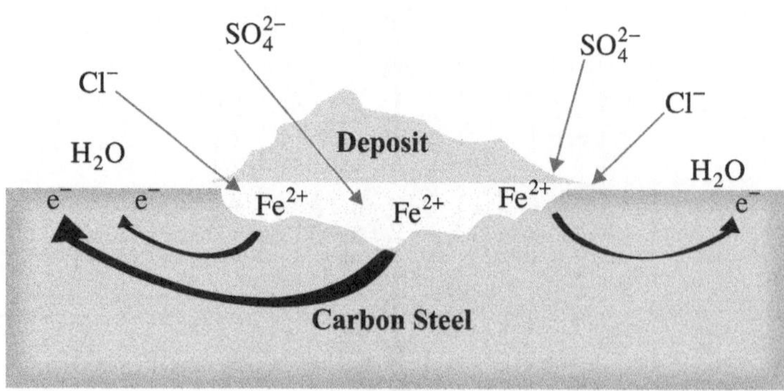

Figure 3

Explanation

The part of the metal surface where the deposits of particles are sitting is cut off from the contact of air or oxygen. While most of the metal surface is exposed to air or oxygen.

This uneven level of oxygen between two parts of the metal creates anodic and cathodic sites in the corrosion cell. This sets up a differential aeration cell is also known as an oxygen concentration cell. The part of the metal exposed to higher oxygen concentration becomes cathodic and the part exposed to lower oxygen concentration where the deposits are sitting becomes anodic. The corrosion begins.

The following reactions occur at the respective electrodes:

At the anode (lower oxygen level): Metal $\rightarrow M^+ + n\,e^-$

At the cathode (higher oxygen level): $H_2O + \frac{1}{2}O_2 + 2e^-$

The API RP 14E

The *API RP* 14*E* specifies minimum requirements and guidelines for the design and installation of new piping systems on offshore production platforms.

What does AP1 RP 14E Say about Single Phase Flow?

Liquid flow lines in a single-phase Flow velocity and pressure drop are the primary factors in sizing single-phase liquid lines. It is recommended that the pressure always be greater than the vapour pressure of the liquid at the given temperature to avoid cavitation, which could lead to erosion. On the other hand, it is recommended that the velocity not be less than 3 ft/s in order to minimise sand and other solids deposition, which may result in under-deposit corrosion attack. There is no mention of any other limiting criteria for determining flow velocity that are related to erosion or erosion-corrosion. Gas flow lines in a single phase. The primary basis for sizing single-phase gas lines is pressure drop.

There is only a passing mention of a velocity limitation related to "stripping a corrosion inhibitor film from the pipe wall," which clearly points to erosion-corrosion. However, no specific guidance on how to determine this limitation is provided.

What does AP1 RP 14E Say about Gas/Liquid Two-Phase Lines?

Erosion Velocity

The velocity at which erosion may start to occur can be calculated using the average density and velocity of the flow and the following empirical equation:

$$Ve = c/\sqrt{\rho m}$$

Where Ve is fluid erosional velocity in ft/s, c is empirical constant in $\sqrt{(lb/(ft \cdot s^2)}$ and ρm is gas/liquid mixture density at flowing pressure and temperature in lb/ft^3.

$$\rho m = [12409SlP + 2.7RSgP]/[198.7P + RTZ]$$

Where P is operating pressure in psia, Sl is liquid specific gravity at standard conditions (water = 1), Sg is gas specific gravity at standard conditions (air = 1), R is gas/liquid ratio at standard conditions, T is operating temperature in Rankine scale (R) and Z is gas compressibility factor. Once the erosional velocity (Ve) is determined, the minimum cross-sectional area required to avoid erosion can be calculated using the following equation

$$A = [(9.35) + (RTZ/21.25P)]/Ve$$

Where A is minimum pipe cross-sectional flow area required in in2/1000 barrels liquid per day.

TRANSIENT VIBRATIONS OF PIPING AND WAVE GENERATION

Vibration can be defined as the simple oscillation or back-and-forth movement of machinery and parts. Transient pressure surge is one of the key factors for liquid pipeline vibrations.

Transient Surge Pressure

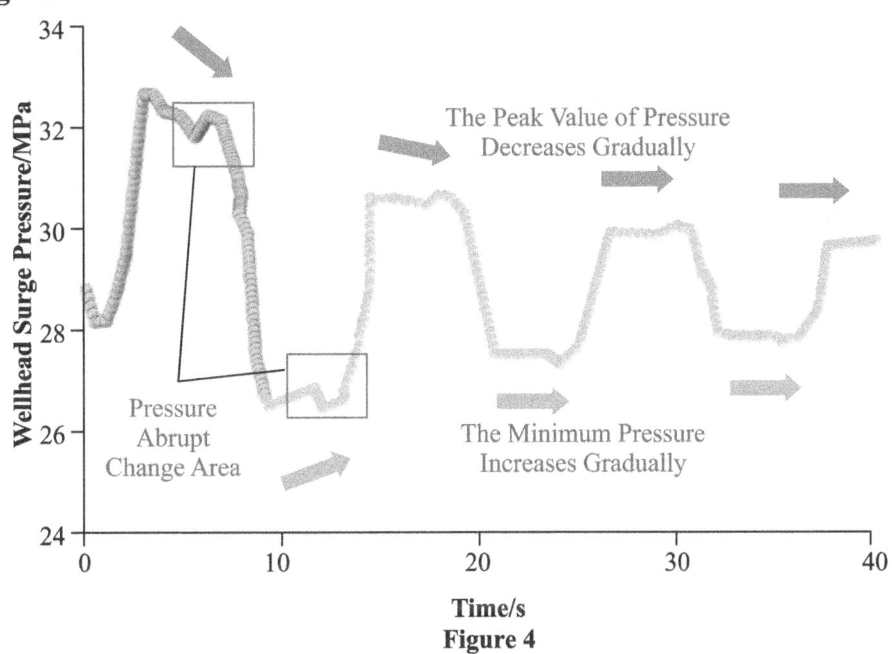

Figure 4

Most Typical Reasons for Vibration in Machines

Many factors, either by themselves or in combination, can cause vibration. Transient pressure surge in liquid piping is one of them.

The following are some of the key reasons.

Imbalance

Unbalanced weight rotating around the machine's axis produces centrifugal force, which in turn causes vibrations in rotating components that have a "heavy spot". The consequences of imbalance get worse as machine speed rises. Both excessive machine vibration and a significant reduction in bearing life can result from imbalance.

Alignment Errors

When machine shafts are misaligned, vibration may occur. When a motor and pump's axes are not parallel, for instance, there is an angular misalignment.

Wear

When parts like drive belts, gears, ball or roller bearings, or other components get worn down, vibrations may result. Vibration can also be caused by a heavily worn-out or chipped gear tooth or a failing drive belt.

Looseness

If a component is vibrating and has loose bearings or is loosely mounted, a vibration that would not normally be noticeable could become destructive.

Water Hammer and Transient Pressure Surge

Water hammer, also known as transient surge, is a phenomenon where large, abrupt pressure surges caused by transient momentum changes in the liquid piping system can result in ruptured pipes, cracked fittings, and fluid release.

Pipelines can sustain significant damage from the effects of water hammer on the piping system caused by transient events like pump startup, shutdown, emergency shutdown (ESD), relief valve activation, valve swings, and check valve closures.

Liquid Pipeline Systems

Different fluids (densities, viscosities, bulk moduli), pressures, and flow rates are frequently handled by large pipeline systems. As an illustration, certain stations have several pumps connected in series, and the system may switch between 0, 1, 2, or 3 pumps depending on what the operations demand.

Numerous transient operating conditions, including starting, stopping, and changing fluids, can excite the small-bore connections.

Why do Waves Form? Entropy's Role

When an exit or intake valve is opened, a pressure wave is produced. The wave travels through the pipe's fluid at the sound speed. A wave with the opposite sign will be reflected from the pipe's end when this wave reaches a change in cross-sectional area, such as the pipe's end. The length of the pipe can be calculated based on how long it takes for this wave to return to the valve.

This is how the water hammer takes place.

Summary

When the linear flow rate of fluid in a pipe changes rapidly, such as when pumps start or stop, valves open or close with quick-acting actuation devices, or trapped air moves within the system, a surge in pressure known as a "water hammer" or "fluid hammer" occurs within the piping system. The risk of shock increases with the length of the pipe line and the fluid's velocity. Pressure spikes can cause the piping system to physically move and put stress on the joints and materials in the system.

What do waves do?

Waves are a mechanism for energy dissipation by generating compression and expansion by the creation of entropy. As waves propagate through a medium, compression, and expansion occur due to changes in pressure and density. This process involves the conversion of energy from the wave into heat, which can be related to the concept of entropy. This compression and expansion of water on the ocean surface generate an upward and downwards, crest, and trough movement of water which we call a wave.

Shockwave

Let us take the case of a water hammer. In the case of a water hammer, the excess energy or pressure surge in the fluid wants to dissipate to reach a state of equilibrium. Just like in ocean waves, this energy is dissipated through various mechanisms, including entropy. The compression and expansion of the fluid play an important role in dissipating the excess energy. When the flow is suddenly stopped or slowed down, the fluid experiences compression, similar to the compression regions in waves. This compression leads to an increase in the pressure and density of the fluid.

Entropy's Role

As the excess energy is dissipated, the fluid gradually expands, resulting in expansion regions. The process of compression and expansion in the water hammer helps disperse and distribute the excess energy throughout the fluid, reducing the pressure surge. This dissipation of energy through compression and expansion is ultimately governed by entropy, which drives the system towards a state of higher disorder and equilibrium.

SURGE IN PIPELINES

Special Thanks to Jason Modisette for his Excellent Paper and Explanation of "The Joukowsky Equation."

A surge is a sudden large increase in something that has previously been steady.

Surge is a phenomenon that occurs in a fluid-transporting pipe as a result of a change in flow velocity. When the flow drops, the energy of the fluid moving forward in the pipe causes a sudden rise in pressure upstream of the valve due to compression. Because the fluid is still moving, negative pressure will develop immediately downstream of the valve, causing a temporary separation of the liquid column followed by a reverse flow back towards the valve if the pressure gradient is high enough.

Transient flow is usually referred to as a surge or water hammer.

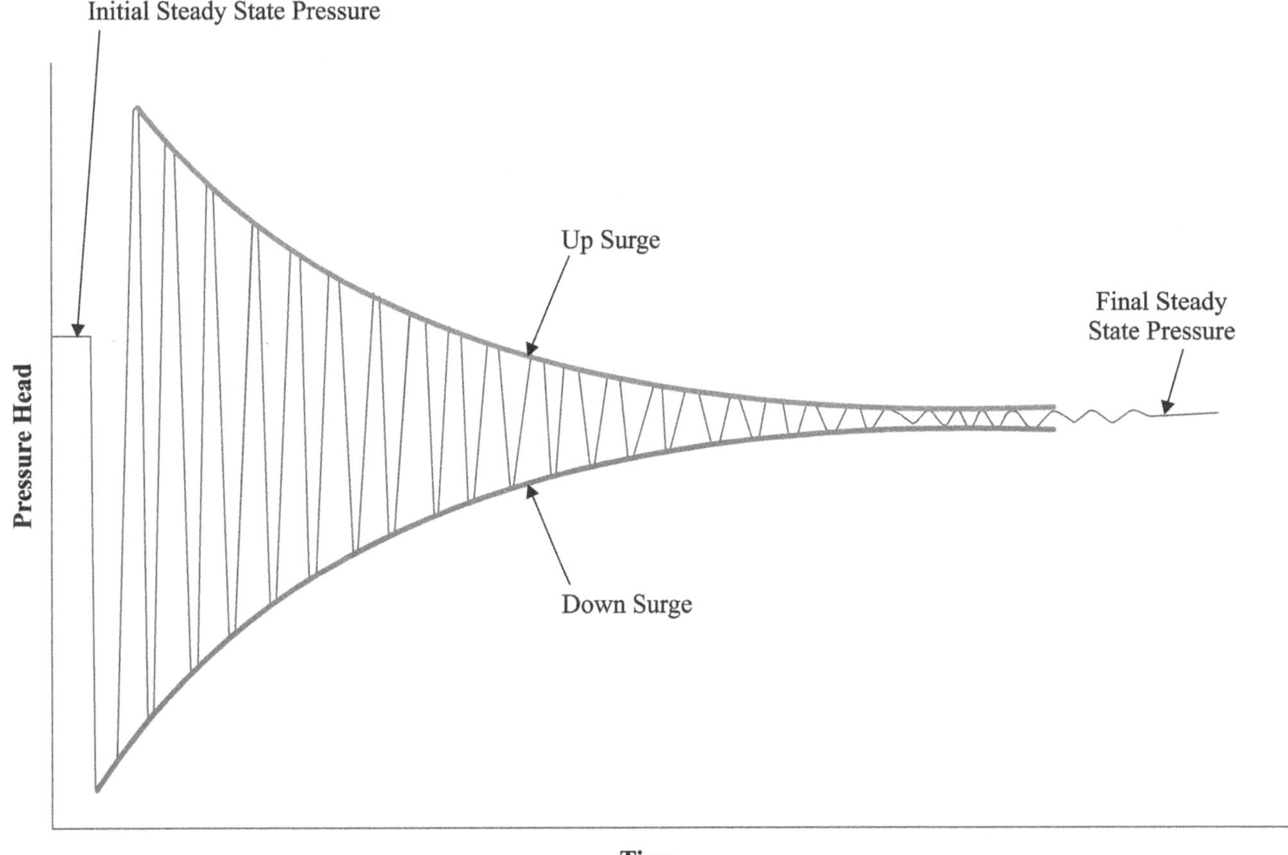

Figure 5

Transient flow is a flow where the flow velocity and pressure are changing with time. When changes occur to a fluid system such as the starting or stopping of a pump, closing or opening a valve, or changes in tank levels, then transient flow conditions exist: otherwise, the system is steady state system.. Often transient flow conditions persist as oscillating pressure and velocity waves for some time after the initial event that caused it.

Surge and water hammer, and the terms are commonly interchanged. But strictly speaking, surge is where the mass oscillation of the fluid is the dominating force and the compressibility of fluid is not significant, for example, two connecting reservoirs oscillating up and down.

Water hammer on the other hand is where the compressibility of the fluid has the dominant effect such as the sudden closure of a valve.

Summary

Transient flow can result in significant transient pressures that may exceed the design limit of the pipes and fittings. Or it may be important to maintain a steady flow for example in the process industry.

On the other hand, in a water hammer, the fluid's compressibility dominates the action, leading to unexpected valve closures. Significant transient pressures caused by transient flow may be greater than the pipes and fittings' design limits. Alternatively, it might be crucial to keep things moving smoothly, as in the case of the process industry.

Details

Surge is a transient flow event that happens in internal flow systems, like the flow in piping systems or compressor flow, in the context of fluid dynamics. It describes a flow transient event where a sudden drop in inlet pressure or sudden rise in outlet pressure causes a fluid stream to be forced to move in the opposite direction due to an unfavorable pressure gradient.

For instance, if the rotor of a centrifugal compressor suddenly stops rotating, surge—a reversal in flow direction—may occur. In this instance, the pressure that has accumulated at the compressor's outlet cannot be supplied by the compressor. As a result of the adverse pressure gradient (which is reversed in the direction of the flow stream), the flow within the gas passages will begin to move in the opposite direction.

A sudden blockage in the compressor's exit can also cause a surge. In this instance, pressure builds up at the exit until it reaches a high enough level to cause a surge and reverse the flow direction in the gas passages if the compressor keeps running at the same speed (assuming there is enough power to keep it running).

Although it may have an impact on forecasting the start of the surge event or its depth, the flow's compressibility is not a determining factor in whether a surge will occur.

Although it could have an impact on when or how the surge event is predicted to begin, the flow's compressibility is not a criterion for predicting whether or not a surge will occur.

Surge Pressure

Joukowsky Equation

By using the Joukowsky Equation, we can manually compute surges. $\Delta P = -\rho\, c\, \Delta V$ is the Joukowsky Equation. Where ΔP is the pressure surge magnitude resulting from the velocity change, and this is the value we are attempting to compute:

The velocity change that causes the surge is represented by ΔV, where p is the fluid's density and c is the wave speed.

These quantities must be in a standard set of units, such as SI: pressure in Pascals, density in kg/m^3, wave speed, and velocity change in m/s, in order for there to be no unit conversion factors.

Using the Joukowsky equation, you can determine the velocity change that results from an abrupt pressure change or the pressure surge that goes along with a specific velocity change.

Solved Example

How much pressure surge can I anticipate, for instance, if I suddenly shut off a valve in a gasoline pipeline that is flowing at a speed of 2 m/sec [two metres per second]?

Apply Joukowsky Equation as explained above.

According to the references, the density of petrol is approximately 800 kg/m^3, and the sound speed is approximately 1380 m/s. When the valve closes, the gasoline abruptly stops, meaning that ΔV is –2 m/s. The pressure change is then ΔP

= –1380 m/s × 800 kg/m^3 × –2 m/s = 22,08,000 kg/m/sec^2 = 22,08,000/101325 = 21.9 bar on the downstream side. This means, that in the event that the valve truly closes instantly, a negative surge and a 21.8 bar pressure surge will both travel at 1380 m/s down the pipe and back up the pipe, respectively.

PRESSURE-DROP IN PIPELINE BENDS

When frictional forces caused by flow resistance act on a fluid as it flows through a tube, a pressure drop occurs.

What is Special About Bends

Single Phase Flow

The presence of a radial pressure gradient created by the centrifugal force acting on the fluid is the primary feature of flow through a bend. As a result, the fluid in the center of the pipe moves to the outer side and then returns along the wall to the inner side. This results in a double spiral flow field, as depicted schematically in Figure below.

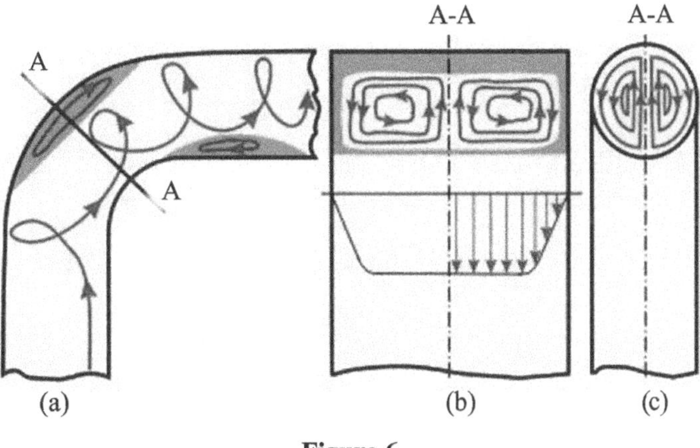

Figure 6

If the bend curvature is strong enough, the adverse pressure gradient near the outer wall in the bend and near the inner wall just after the bend may lead to flow separation at these points, giving rise to a large increase in pressure losses. Even for fairly large-radius bends, the flow field in the bend will be severely distorted as shown in the figure below.

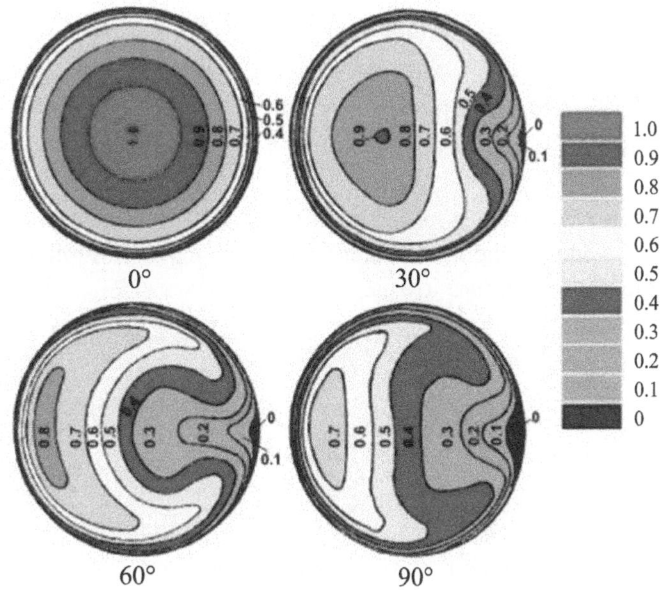

Figure 7

Pressure losses in a bend are caused by both friction and momentum exchanges caused by a change in the flow direction. Both of these variables are affected by the bend angle, curvature ratio, and Reynolds Number. The overall pressure drop can be expressed as the sum of two components: 1) that resulting from friction in a straight pipe of equivalent length which depends mainly on the Reynolds number (and the pipe roughness); and 2) that resulting from losses due to change of direction, normally expressed in terms of a bend-loss coefficient, which depends mainly on the curvature ratio and the bend angle. Thus, the pressure loss in a bend can be calculated as follows:

$$\Delta P = \frac{1}{2} f_s \rho u^2 \frac{\pi R_b}{D} \frac{\theta}{180°} + \frac{1}{2} k_b \rho u^2$$

where *fs* is the Moody friction factor in a straight pipe; ρ, the density; u, the mean flow velocity; *Rb* the bend radius; *D*, the tube diameter; θ, the bend angle; and *kb*, the bend loss coefficient obtained from Figure below.

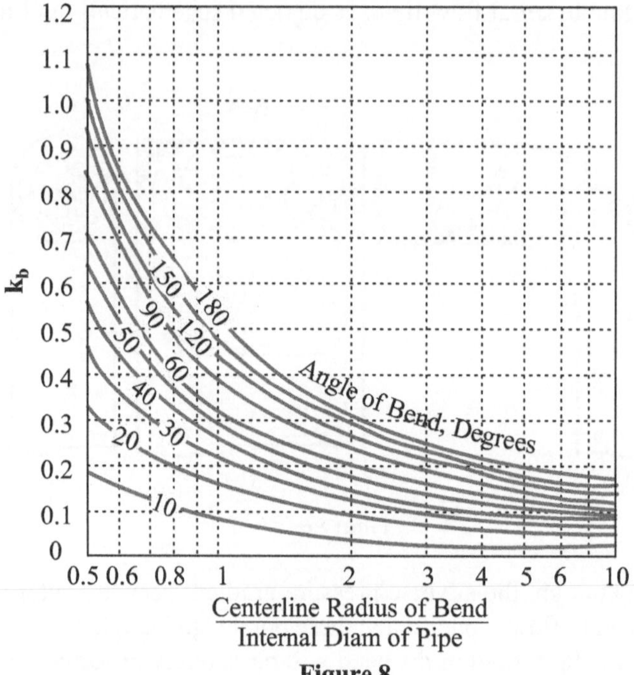

Figure 8

Pressure Drops in Bends for Two-Phase Flow

Two-phase flow in bends is rendered more complex by the centrifugal-force-induced stratification of the two phases. Under certain conditions, this manifests itself in the migration of bubbles in the bubbly flow towards the inner side of the bend, and, more paradoxically, in the migration of the heavier phase towards the inner side in separated flows (a process known as film inversion).

The pressure losses suffered in two-phase flow-through bends are influenced by a number of parameters, and no generalized method is available to calculate them accurately. The two-phase multiplier, which is empirically correlated, is usually multiplied by the single-phase pressure losses.

$$\frac{\Delta p_{TP}}{\Delta p_{LO}} = 1 + \left(\frac{\rho_L}{\rho_G} - 1\right) \times \left[\left\{1 + \frac{2.2}{k_{LO}\left(2 + \frac{R_b}{D}\right)}\right\} \times (1 - x) + x^2\right],$$

where ΔpTP is the pressure drop in two-phase flow, ΔpLO is that in a single-phase flow of the total mass flux and liquid properties, *kLO* is the bend loss coefficient for single-phase flow, and x is the quality.

LINE SIZING: ENERGY LOSSES BY 'VORTEX'

List of Contents

Why vortex is important for engineers?

What is a vortex in a pipeline?

What does vortex do? What is the mode of energy loss by vortex?

What is a boundary layer?

Vortex and boundary layer separation: Why and how?

What harm does vortex cause?

Vortex shedding

Vortex source: Pump

Why Vortex is Important for Engineers?

The three common unusual flow events in a pipeline are: [1] Vortex [2] Vibration [3] Water Hammering.

From the engineering design point, besides energy losses, the shedding of vortices produces alternating forces, which occur more frequently as the flow velocity increases. In fluid dynamics, vortex shedding is an oscillating flow that generates when the vortex collapses. This is a major cause of vibration in a pipe.

Rotational symmetry with tangential components characterizes the vortex flow. Flow velocity in vortex flow in pipes frequently has both tangential and axial components. Particles carried by the flow follow helical paths, and when compared to vortex-free flow, they must negotiate a longer path along the pipe wall for the same length of piping at the same flow velocity, resulting in a higher-pressure loss. The tangential component (vu) at radius (r) in a vortex flow usually follows the law applicable to potential flow,

$$Vu \times r = \text{Constant}$$

Fluid Flow is Irreversible

The most important thing to remember when it comes to line sizing is that a flow is irreversible. Any energy lost due to an abnormal flow event in a pipeline cannot be recovered. All unusual flow events cause irreversible energy losses.

Because fluids have viscosity, fluid flow is irreversible, and whenever velocity gradients appear in the flow, friction and energy dissipation due to viscous stresses occur. Only if there are no velocity gradients anywhere in the flow can it be reversible.

The three common unusual flow events in a pipeline are:

[1] Vortex [2] Vibration [3] Water Hammering

What is a Vortex in a Pipeline?

A vortex is a rapidly spinning, circular, or spiral flow of fluid around a central axis. The swirling motion draws everything in the fluid toward the centre. The fluid's speed and rate of rotation are greatest in the centre and gradually decrease as one moves away from it. A vortex is wider on the top, and it narrows gradually while moving towards the bottom. This is because of the pressure difference. The pressure is less on the top and is maximum at the bottom.

What does Vortex Do? What is the Mode of Energy Loss by Vortex?

Vortices in an intake can cause operational issues such as increased energy loss, decreased discharge coefficient, pulling floating particles, and increased noise and vibration. The discharge coefficient is defined as the ratio of actual discharge to ideal discharge. The swirling liquid maintains its angular momentum by continuously rotating around a central point while simultaneously losing a portion of energy due to the force of friction existing between the moving and the stagnant liquid.

Detail

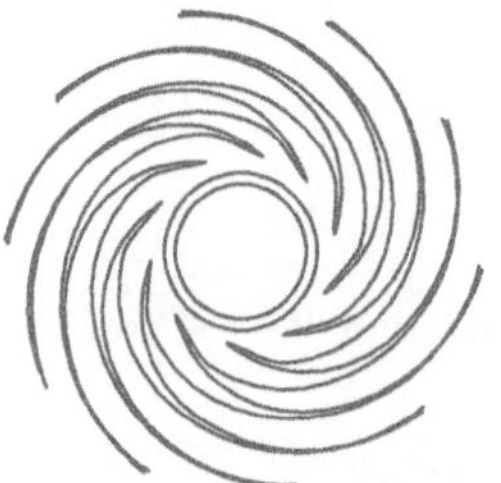

Figure 9: Vortex Shape

Shape of Vortex

The vorticity, or the local rotation rate of fluid particles, defines vortex structures. A vortex is a region in a fluid where the flow revolves around an axis line, which can be straight or curved. Vortices are defined by the distribution of velocity, vorticity (the curl of the flow velocity), and the concept of circulation. The fluid flow velocity in most vortices is greatest near the axis and decreases in inverse proportion to the distance from the axis. A vortex is a physics phenomenon that occurs when a gas or liquid moves in a circular motion. The matter swirls around a vortex line in the centre.

They are formed when the velocity of what surrounds the line differs. Turbulent flow is dominated by vortices. In the absence of external forces, viscous friction within the fluid tends to organize the flow into a collection of irrotational vortices, which may be superimposed on larger-scale flows and vortices. Vortices, once formed, can move, stretch, twist, and interact in complex ways. A vortex in motion carries angular and linear momentum, energy, and mass with it.

The Boundary Layer Generates a Vortex

Vortex has a relationship with turbulent flow and turbulent flow has a relationship with boundary layer separation.

What is a Boundary Layer?

Because of the viscosity, forces act between the fluid and the surface when it flows around a body. These intermolecular forces, among other things, are responsible for drag. However, as a result of the forces at work, the surface of the solid body attempts to bind the fluid to itself. This also causes the fluid to adhere to the surface (no-slip condition). Because intermolecular attraction and pressure forces act between the layers, the layer above this adherent fluid layer will not be able to simply tear itself away from it. As a result, any flow flowing around a body is enticed to follow the surface profile.

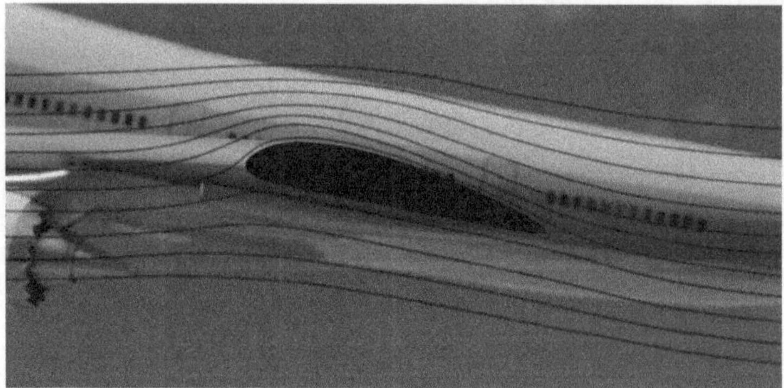

Figure 10

A flow can follow the contour of a body as long as the contour has smooth transitions and the internal cohesion of the fluid (viscosity) is large enough to resist inertial forces. Around the body, a typical hydrodynamic boundary layer forms, the thickness of which is heavily influenced by viscosity.

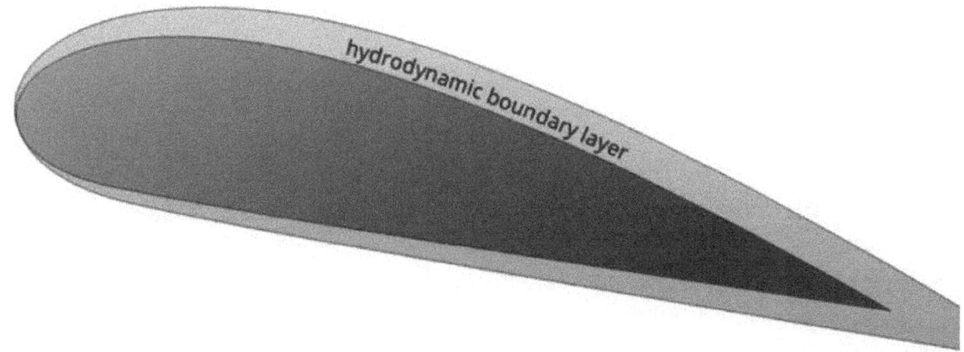

Figure 11

A flow can follow the contour of a body as long as the contour has smooth transitions and the internal cohesion of the fluid (viscosity) is large enough to resist inertial forces. Around the body, a typical hydrodynamic boundary layer forms, the thickness of which is heavily influenced by viscosity.

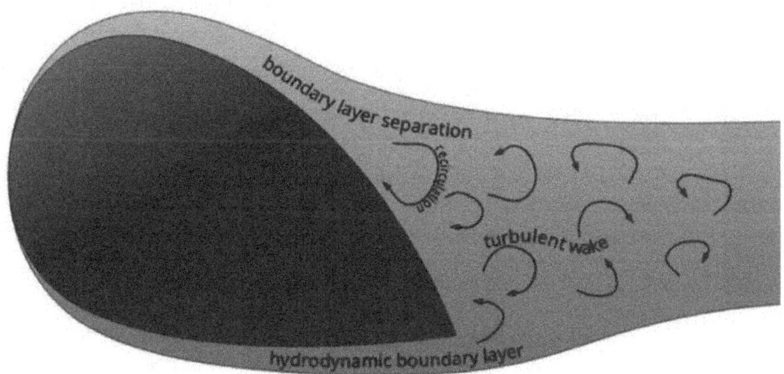

Figure 12

When a boundary layer separates (flow separation), the flow is no longer able to follow the profile of the body it is flowing around and separates turbulently from it.

When separation occurs, frictional effects are always important. It is worth noting that sharp corners almost always produce a separated flow.

Vortex and Boundary Layer Separation: Why and How?

A vortex can form as a result of the boundary layer separation phenomenon, which occurs when a fluid moves over a surface and experiences a rapid acceleration from the fluid velocity to zero due to the no-slip condition. This rapid negative acceleration forms a boundary layer, causing a local rotation of fluid at the wall (i.e., vorticity), which is measured as the wall shear rate. If the diameter or thickness of the vessel or fluid is less than the boundary layer thickness then the boundary layer will not separate and vortices will not form. However, when the boundary layer does grow beyond this critical boundary layer thickness then separation will occur which will generate vortices.

What Harm does Vortex Cause?

Energy Losses

The swirling liquid maintains its angular momentum by continuously rotating around a central point while simultaneously losing a portion of energy due to the force of friction existing between the moving and the stagnant liquid. Part of the mechanical energy of the fluid is dissipated as heat. Flow velocity in vortex flow in pipes frequently has both tangential and axial components. Particles carried by the flow follow helical paths, and when compared to vortex-free flow, they must negotiate a longer path along the pipe wall for the same length of piping at the same flow velocity, resulting in a higher-pressure loss. The tangential component (vu) at radius (r) in a vortex flow usually follows the law applicable to potential flow,

$$Vu \times r = \text{Constant}$$

Vibration and Noise

Vortex shedding: The shedding of vortices is important in engineering design because the alternate formation and shedding of vortices produces alternating forces, which occur more frequently as the flow velocity increases. Vortex shedding is a natural effect in which circular movements of fluid come off a solid surface, and often cause vibration. Vortex shedding causes vibration and noise in the pipeline. In simpler terms: Vortex shedding is when fluid flows over an object, creating an oscillating flow. In certain circumstances, this cross-flow can cause the object to vibrate.

How to minimize vortexing? Reduce fluid velocity by increasing pipe size or by changing process conditions. Increase system rigidity by increasing pipe wall thickness. Reduce the number of turbulent sources like elbows, reducers, etc. Tighten the clearances in-between pipes and supports.

One common source of a vortex is the pump.

Vortexing can occur if a pump suction is not placed at an acceptable distance from the liquid's surface. Vortexing, like cavitation, is caused by too much water being drawn into the suction line relative to the pit/sump depth, resulting in loss of efficiency and pump damage. A depression forms on the water's surface, which is commonly seen as a whirling vortex.

Vortexes form between the pump suction and the fluid surface of the intake tank. This allows air into the pump suction, which reduces pump capacity and efficiency. Entrainment also causes excessive shaft deflection. The vortex condition can damage the pump's mechanical seals, bearings, piping, couplings, and impeller.

CHAPTER 52

Oil and Gas

Thermodynamics plays an important role in the oil and gas industries, as it helps in understanding and optimizing the processes involved in exploration, production, refining, and transportation of oil. Some ways thermodynamics is important in these specific areas in the oil and gas industry: reservoir modelling, refining processes, transportation and storage and environmental impact.

METHANE HYDRATE THERMODYNAMICS

Hydrates of Natural Gas

Natural gas and water combine chemically under pressure at a temperature that is significantly higher than the freezing point of water to form natural gas hydrates, which are solid, crystalline compounds. For natural gas hydrates, the chemical formulae are as follows:

Methane $CH_4 \cdot 7H_2O$

Ethane $C_2H_6 \cdot 8H_2O$

Propane $C_3H_4 \cdot 18H_2O$

Carbon Dioxide $CO_2 \cdot 7H_2O$

When free water is present and the temperature drops below what is necessary for hydrate formation, hydrates typically form. This is typically a result of sudden expansion-related pressure drop in the surface line. Usually, when fluids pass through chokes, back pressure regulators, or orifices, this occurs.

Natural gas hydrates will undoubtedly form more readily in the system if there is a tiny "seed" crystal of hydrate or acid gas (H_2S or CO_2) present and the flow rate is high with agitation.

Hydrate formation is an unusual existence of hydrocarbon in which molecules of natural gas, typically methane, are trapped in ice molecules well above the freezing point of water at high pressures. Clathrate hydrates are non-stoichiometric mixtures of water and natural gas. The gas molecules are trapped in a polygonal crystalline structure made of water molecules. The water molecules arrange themselves in an orderly fashion around the gas molecules, thus entrapping them.

High-pressure and low-temperature conditions favor hydrate formation.

Issues of Gas Hydrate

Gas hydrate deposits are found wherever methane occurs in the presence of water under elevated pressures and at relatively low temperatures. Methane that forms hydrate can be both biogenic, created by biological activity in sediments, and thermogenic, created by geological processes deeper within the earth.

It has been reported that the maximum fast time of hydrate formation at $T = 1°C$ and $P = 10$ MPa is about 2.5 hours. It has been determined that the greater the methane concentration in the mixture, the greater must be the pressure in the system for the gas hydrate formation.

Hydrates have a strong tendency to agglomerate and adhere to the pipe wall thereby plugging the pipeline.

Thermodynamic conditions favoring hydrate formation are often found in pipelines. The sequence of events leading to hydrate formation in gas pipelines involves water vapor condensation, accumulation of water at lowered sections of the pipeline, nucleation, and growth of hydrate particles ultimately blocking the pipeline.

Natural gas pipelines always have some moisture content as the gas reaches the wellhead saturated with water. During the cold season, the ground temperatures are lower resulting in the temperature drop of the pipeline content. If the temperature of the gas falls below the saturation temperature corresponding to the dew point of water, condensation of water vapors will start. In straighter sections of the pipeline, the condensate starts to accumulate at the base of the pipe due to gravity. The accumulation may increase at the base of the sagging/uphill sections of the pipe. If enough water is present and the thermodynamic conditions are suitable, hydrate nucleation processes are initiated. Generally speaking, compression increases the dew point, and expansion lowers the dew point of the air. Please refer to the above dew point graph. At atmospheric pressure, the dew point is about 10 degc. At 3 bar pressure, the dew point is about 35 degc and at 7 bar pressure the air dew point can increase to about 50 degc.

Methane Hydrate Phase Diagram: Key Points to Note

When hydrocarbon contacts water, the two components separate into two phases in which the mutual component solubility is less than 1.0 mol% at ambient conditions. This splitting of phases affects almost all treatments of mixed water and hydrocarbon systems and is caused by the different molecular attractions within water and hydrocarbons. Hydrocarbon molecules have a weak, noncharged attraction for each other, while water attracts other water molecules through a strong, charged hydrogen bond. Because hydrogen bonds are significantly stronger than those between hydrocarbon molecules, hydrocarbon solubility in water (and that of water in hydrocarbons) is very small.

This phase diagram shows water depth (pressure) on the vertical y-axis and temperature on the horizontal x-axis. The dashed lines separate stability fields of water, water ice, gas, and gas hydrate. The line labeled "hydrate to gas. transition" is significant.

Conditions for the formation of methane hydrate occur below this line. Above this line, methane hydrate will not form. The red line traces a geotherm (the change of temperature with depth at a specific location). Note how, as depth increases, the geotherm crosses the hydrate-to-gas transition line.

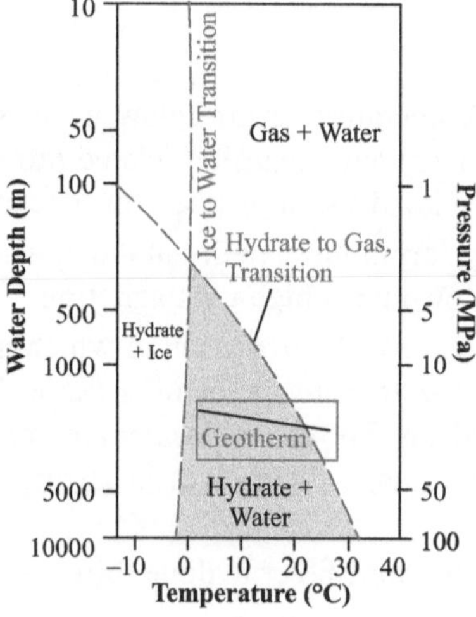

Figure 1: Gas Hydrate

Hydrate Formation, Prevention, and Mitigation

Hydrates have a strong tendency to agglomerate and adhere to the pipe wall thereby plugging the pipeline. Once formed, they can be decomposed by increasing the temperature and/or decreasing the pressure. Even under these conditions, the clathrate dissociation is a slow process.

Therefore, preventing hydrate formation appears to be the key to the problem. Hydrate prevention could typically be based on three levels:

[1] Avoid conditions that might cause the formation of hydrates. Depress temperature using glycol dehydration;

[2] Change operating conditions to avoid hydrate formation;

[3] Add chemicals to (a) shift the hydrate equilibrium conditions towards lower temperatures and higher pressures or (b) increase hydrate formation time by inhibitors.

NATURAL GAS DEHYDRATION BY SOLID DESICCANTS

Water vapor in a natural gas stream can cause line plugging because of the formation of hydrates, reduce line capacity because of the accumulation of free water in the line, and raise the possibility of pipeline damage because of the corrosive nature of water. Consequently, in order to avoid hydrate formation and condensed water corrosion, water vapor from natural gas must be eliminated.

The process of removing water vapor from a gas stream in order to reduce the temperature at which water will condense from the stream—referred to as the gas's "dew point" - is known as gas dehydration. One of the most important materials for desiccant use in industrial natural gas dehydration is thought to be molecular sieves.

Hydrate Challenges

A challenge to the natural gas industry's transition to high-pressure natural gas production, processing, and transportation is hydrate formation. Hydrocarbon molecules like ethane and methane combine with water to form hydrates, which are crystalline, solid substances.

Apart from one significant difference, they resemble regular ice in terms of appearance and properties. Above 32°F, or the freezing point of water, temperatures form hydrates.

Hydrate Structure

In a hydrate, the water molecules form a three-dimensional cage-like structure, with another molecule trapped within the cage. Examples of gases that form hydrates include methane, ethane, propane, butane, carbon dioxide, and hydrogen sulfide. These are all common components of natural gas. Only small molecules form hydrates typically, gases larger than butane does not form hydrates. The water molecules were referred to as the "host" molecules and the other compounds, which stabilize the crystal, are called the "guest" molecules, Figure above shows water methane hydrate structure.

How Hydrates Form?

The formation of a hydrate requires the following three (necessary) conditions:

- Lowering gas temperature.
- Elevation of gas pressure.
- natural gas composition {Water vapors "host" and hydrocarbon percentage "guest"}, a sufficient amount of water should not too much, not too little and free water is not required.

It is necessary to prevent the condensation of liquid water and hydrocarbons to ensure trouble free operation of a natural gas transmission system. Apart from the risk of hydrate formation, is the liquids can reduce the volumetric capacity of the system and interfere with the operation of pressure regulators and filters. Condensed liquids accumulated in pipelines, caused an increase in operating pressures and potential damage to equipment due to liquid carry over.

Dehydration Technologies

While there are a variety of methods used to dehydrate natural gas, the two most widely used methods nowadays are adsorption by solid desiccants and absorption by liquid desiccants.

Solid bed, or adsorption dehydration is the process of extracting water vapor from a gas stream using a solid desiccant. Regenerating solid desiccants are frequently utilized for multiple adsorption-desorption cycles in gas dehydration processes.

Types of Solid Desiccants

The following three fundamental materials are most frequently utilized because they satisfactorily possess these qualities:

(i) Activated alumina. (ii) Silica gel and silica-alumina gel. (iii) Molecular sieves.

Adsorption Bed

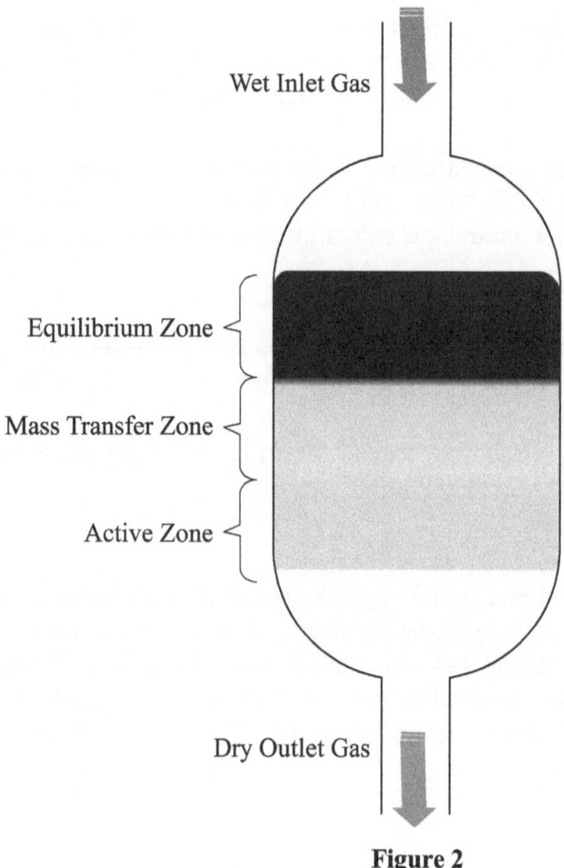

Figure 2

The fundamental actions of an adsorbent bed in a gas dehydration service are depicted in the figure above. Three distinct zones exist in the bed during regular drying (adsorbing) cycle operation.

Zone of Equilibrium

The desiccant is saturated with water in the equilibrium zone, having reached its equilibrium water capacity based on the conditions of the inlet gas and losing its ability to adsorb water.

Zone of Mass Transfer (MTZ)

Virtually all of the mass transfer takes place in the MTZ, a concentration gradient exists across the MTZ.

Active Zone

In the active zone the desiccant has its full capacity for water vapor removal and contains only that amount of residual water left from the regeneration cycle. When the leading edge of the MTZ reaches the end of the bed, breakthrough occurs.

PHASE BEHAVIOR OF THE WATER-HYDROCARBON SYSTEM

There are two phase diagrams below, [1] methane a typical hydrocarbon, and [2] phase diagram of a water-hydrocarbon system.

Methane phase properties: It's a straight forward phase diagram. Methane has a boiling point of −164 degc. Its triple point is –182 degc and critical temperature and pressure are –82 degc and 46 bar respectively. Other properties are self-explanatory.

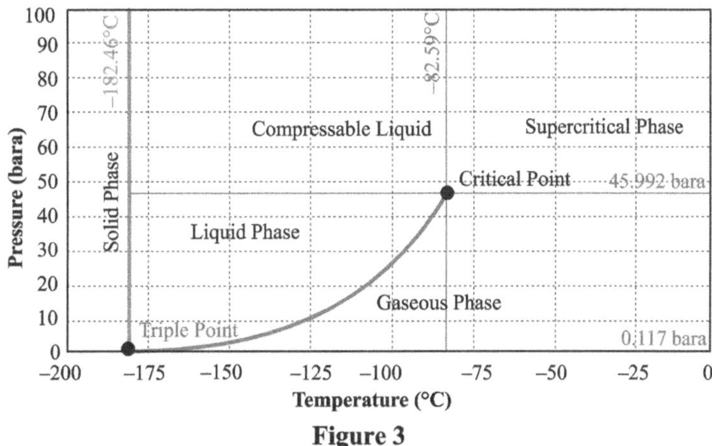

Figure 3

The phase behavior of water and hydrocarbon mixtures diagram differs significantly from the vapor/liquid equilibria of normal hydrocarbons diagram two ways: [1] Aqueous and hydrocarbon components usually separate, with very low mutual solubility and [2] Hydrates often form with water and hydrocarbons smaller than n-pentane.

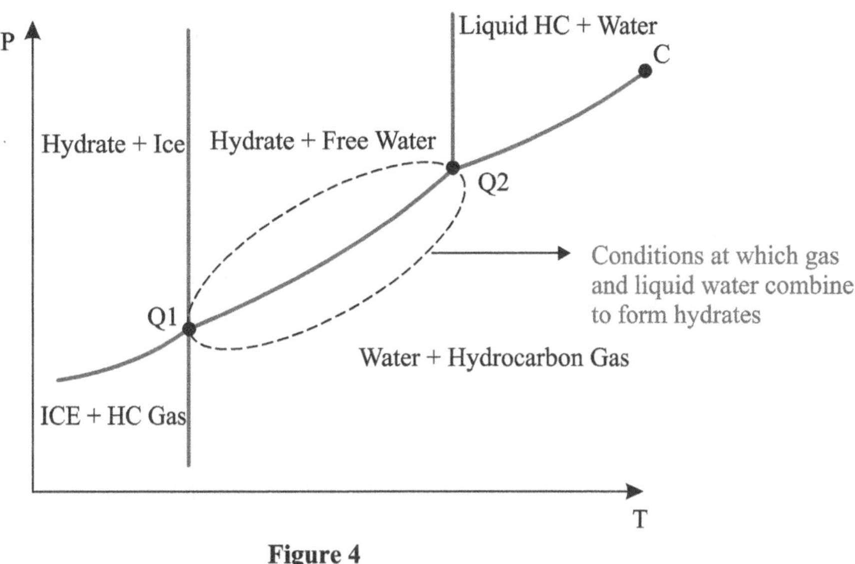

Figure 4

When hydrocarbon contacts water, the two components separate into two phases in which the mutual component solubility is less than 1.0 mol% at ambient conditions. This splitting of phases affects almost all treatments of mixed water and hydrocarbon systems and is caused by the different molecular attractions within water and hydrocarbons. Hydrocarbon molecules have a weak, noncharged attraction for each other, while water attracts other water molecules through a strong, charged hydrogen bond. Because hydrogen bonds are significantly stronger than those between hydrocarbon molecules, hydrocarbon solubility in water (and that of water in hydrocarbons) is very small. Water's very high heat of vaporization, which absorbs large amounts of heat and buffers many hydrocarbon reservoir temperatures. Another significant point is water has a very high normal boiling point relative to its molecular weight.

Phase Definitions and the Gibbs Phase Rule

Listing of Different Phases

HC = hydrocarbon, typically with a very low (<1%) water concentration. *I* = pure ice. *L* = liquid that is either water (*LW*) or hydrocarbon (*LHC*). *V* = vapor that is always a single-phase, never splitting. *W* = liquid water, usually of high (>99%) purity, except where indicated. *H* = hydrate or solid.

The order of phase listing is by decreasing water concentration. For example, the listing order $LW > H > V > LHC$ means that hydrates (H) contain less water than the liquid water phase (LW), but more water than vapor (V), which in turn contains more water than liquid hydrocarbon (LHC).

The Gibbs Phase Rule for Non-reacting Systems States

$F = C - P+2$, where: F = number of intensive variables (e.g., pressure, temperature, single-phase composition) required to define the system (known as the degrees of freedom), C = number of components, P = number of phases

For example, when excess gas (excess so that its composition does not change) contacts water to form hydrates, there are three phases ($P = 3$, namely $LW + H + V$) and two components ($C = 2$, namely water and gas of constant composition), so that $F = 1$; only one intensive variable (either pressure, temperature, or one phase composition) is needed to define the system. If this system is uniquely specified at a given pressure, for example, the identical temperature, and same individual phase compositions always will occur for the initial components at that pressure, when three phases are present. If gas also condenses ($P = 4$, with the addition of LHC), however, or the gas quantity is small enough to change composition (such that $C > 2$), then the $F = 1$ restriction no longer holds.

As a second example, consider again the case of a constant composition hydrocarbon vapor in equilibrium with water, so that $C = 2$. With two phases ($V + LW$), by the Gibbs phase rule the degrees of freedom are two ($F = 2$) so that for this example, both temperature and pressure are needed to determine the water composition in the hydrocarbon at conditions under which the two phases coexist. In contrast to the single variable required for three-phase systems in the above paragraph, with this case of two phases and two components, the saturated water concentration in the hydrocarbon (or water dewpoint) is determined by two variables. The water content in a hydrocarbon vapor is correlated with temperature and pressure in the Equilibrium of water and hydrocarbon systems without hydrates. Identical restrictions apply to the hydrocarbon content in the water phase.

The same restriction of $F = 2$ applies when a constant composition liquid hydrocarbon exists in equilibrium with liquid water. However, because both phases ($LW + LHC$) are dense, very large pressure changes are required to influence the water content of the hydrocarbon. In general, when condensed phases coexist, temperature and concentrations have a much greater influence than does pressure. With liquid hydrocarbon and water, the temperature controls the mutual solubility (i.e., the concentration of the water in the hydrocarbon liquid, as well as the concentration of the hydrocarbon in the liquid water). The mutual solubility of water and liquid hydrocarbons is correlated with temperature in the Equilibrium of water and hydrocarbon systems without hydrates.

Finally, you get the following phase behavior of HC and water

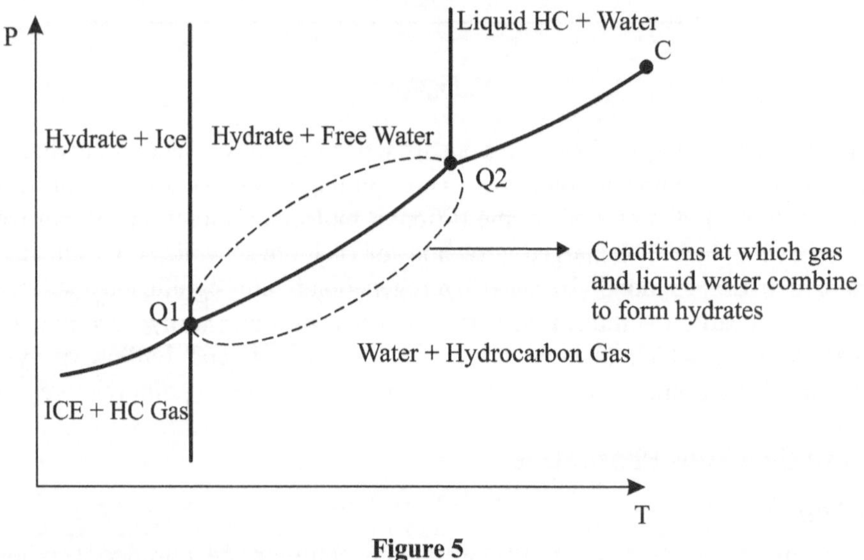

Figure 5

First of all, hydrate formation is clearly favored by low temperature and high pressure, so $C = 2$, $P = 2$ and $F = 2$ [temperature and pressure]. The three-phase critical point is the point C on the diagram that represents the condition where the liquid and gas hydrocarbon merge into a single hydrocarbon phase in equilibrium with liquid water, $C = 1$, $P = 1$ $F = 2$. $F = 2$ applies when a constant composition liquid hydrocarbon exists in equilibrium with liquid water. Point Q_2 is the upper quadruple point, where four phases (liquid water, liquid hydrocarbon, gaseous hydrocarbon, and solid hydrate) are found in equilibrium, $C = 4$ and $P = 4$, therefore $F = 2$. Point Q_1, the lower quadruple point, typically occurs at 32°F (ice freezing point) where four phases (ice, hydrate, liquid water, and hydrocarbon gas) are found in equilibrium. In this context, phases are not pure as they contain some amount of the other substances according to their mutual solubility.

For practical applications, the most important equilibrium line is the Q1Q2 segment. It represents the conditions for hydrate formation or dissociation, a critical piece of information for most industrial applications where hydrates are involved.

OIL AND GAS RESERVOIR'S PHYSICS AND CHEMISTRY: A REVIEW

Petroleum reservoir characterization simulates hydrocarbon behaviour under several circumstances. An oil and gas reservoir is essentially a naturally occurring storage area of hydrocarbons. Petroleum reservoirs may contain oil, natural gas, or both. The fluid type is determined very early in the life of a reservoir (often before sampling or initial production) because the fluid type is the critical factor in many of the decisions that must be made about producing the fluid from the reservoir. Reservoir fluids are composed of various hydrocarbon components with wide-ranging properties. Therefore, reservoir-fluid properties play a key role in the design and optimization of injection/production strategies and surface facilities for efficient reservoir management.

The most common classification of petroleum reservoirs is based on the type of hydrocarbons it stores and produces. Petroleum reservoirs are classified as follows:

- Dry gas
- Wet gas
- Gas condensate
- Light oil
- Black oil of intermediate composition
- Heavy oil
- Extra-heavy oil, bitumen

Origin of Oil and Gas

Millions of years ago, algae and plants lived in shallow seas. After dying and sinking to the seafloor, the organic material mixed with other sediments and was buried. Over millions of years under high pressure and high temperature, the remains of these organisms transformed into what we know today as fossil fuels. Coal, natural gas, and petroleum are all fossil fuels that formed under similar conditions.

Challenges of Oil and Gas Reservoirs

Geology

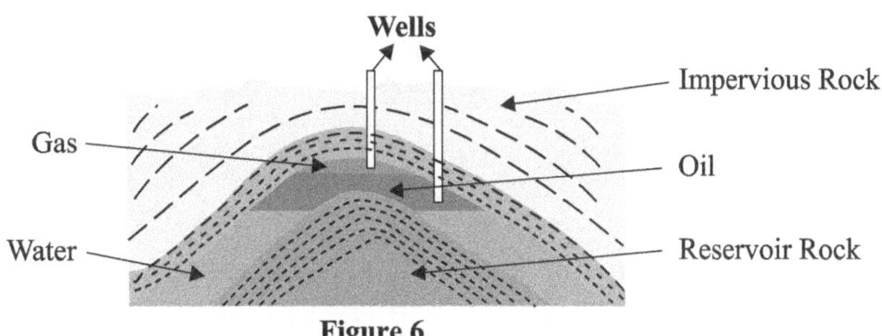

Figure 6

Petroleum reservoirs are normally found in sedimentary rocks. Several key ingredients must be present for a hydrocarbon reservoir to develop. First, a source for the hydrocarbon must be present. It is most commonly thought that oil forms from the remains of single-celled aquatic life. When the remains are baked as they become buried, oil and gas are formed. As with baking a cake, the mixture can be underdone or overcooked. In either case, the formation of oil will not be optimal. Second, there needs to be a conduit from the source rock to a reservoir rock. For hydrocarbons to be producible, they must be able to flow into wells. The flow rate must be large enough to make the wells economically viable. Two important characteristics control the economic viability of the reservoir: porosity and permeability.

Porosity is the ratio of the volume of void space to the total volume. Porosity is a factor that defines the capacity of the reservoir to store fluid. Permeability is a measure of the ability of fluids to flow.

Physics and Chemistry: PT Behaviour

Reservoir fluids can exist as a liquid (oil) phase, a gas phase, or both phases together. The conditions in which the two phases appear are illustrated on a pressure-temperature (PT) phase diagram, as shown below.

PT Phase Diagram Concept

It very similar to *TS* diagram of steam. Those who are familiar with *TS* diagram of steam can find it very east to understand the *P-T* diagram of oil-gas.

On the left side of the bell curve, there is one single phase that is unsaturated liquid. The outline of the bell curve on the left side is the bubble point line. The first bubble of gas forms on this line at a P and T in a liquid at equilibrium. The bubble point line is the saturated liquid line. The area bell curve is the region where both gas and oil coexist at a certain ratio as you can see in the figure below. As you go from left to right inside the bell curve more liquid converts to gas as temperature increases and pressure reduces. The extreme right-side outline of the bell curve stands for the dew point line at given P and T. The dew point line is the saturated gas line. The dew point is the temperature at which the first drop of liquid is formed again at equilibrium at a given pressure. All liquid gas converted to gas on this line. Outside the dew point line, there is only one phase that is gas. The point where the bubble point line intersects the dew point line is the critical point. Above the critical point, a liquid and gas phase cannot be distinguished, and this single phase is usually termed a fluid phase. If you look at the bell curve from the right side, the outline stands for where the first drop of liquid forms if you reduce the temperature at constant P. If you see the bell curve from the left side, the outline stands for the formation of the first bubble of gas at a given P when you increase the temperature.

PT Behaviour of a Gas Reservoir

At relatively high pressures and low temperatures, a liquid phase is formed. At relatively low pressures and high temperatures, a gas phase is formed. At intermediate temperatures and pressures, both liquid and gas phases may be present. The range of pressures and temperatures at which both phases are present defines a two-phase envelope on the PT diagram.

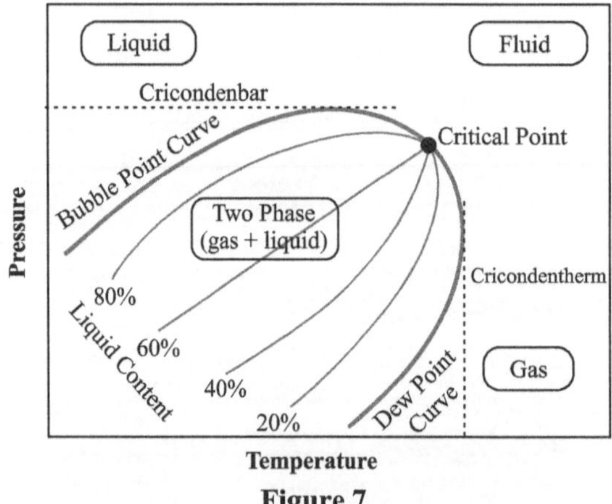

Figure 7

Pressure–Temperature Phase Diagram of a Multicomponent Mixture

The diagram is very similar to T-S diagram of steam except the terms are different.

The boundary of the two-phase envelope defines the bubble point and dew point curves of the reservoir fluid. The bubble point is the temperature at which the first bubble of gas is formed in a liquid at equilibrium at a given pressure. The dew point is the temperature at which the first drop of liquid is formed again at equilibrium at a given pressure. The dew point and bubble point curves meet at the critical point. The critical point is the temperature and pressure at which the composition of the two phases is the same. Above the critical point, a liquid and gas phase cannot be distinguished, and this single phase is usually termed a fluid phase.

The size and shape of the two-phase envelope depend on the composition of the reservoir fluid. However, reservoir fluids can be classified according to the position of the initial reservoir pressure and temperature relative to the phase envelope. To aid in the classification, one more term is defined, the cricondentherm. The cricondentherm is the line of constant temperature on the PT at which the temperature is the maximum two-phase temperature. The various reservoir fluid classifications are undersaturated oil, bubble point oil, volatile oil, retrograde condensate, and gas.

Two Possibilities with a Completely Different Character of an Oil Reservoir

[1] The reservoir fluid is initially above the bubble point and below the critical temperature.

[2] The reservoir fluid is initially at the bubble point

Undersaturated oil above the bubble point: In some cases the reservoir fluid is initially above the bubble point and below the critical temperature. A fluid above the bubble point is an undersaturated liquid, and this type of reservoir fluid is termed an undersaturated oil. The path of an undersaturated oil as the reservoir is produced is shown as line A1A2 on the image below.

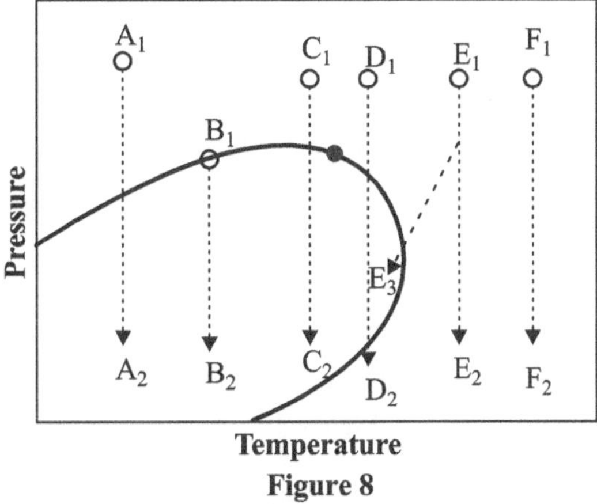

Figure 8

Location of reservoir fluid types and production paths on a *PT* diagram. (A_1A_2 = undersaturated oil reservoir; B_1B_2 = saturated oil or gas cap reservoir; C_1C_2 = volatile oil reservoir; D_1D_2 = retrograde condensate reservoir; E_1E_2 = wet gas reservoir, F_1F_2 = dry gas reservoir).

Note the lovely explanation below, how oil and gas form in the reservoir

Because the reservoir temperature is maintained by heat transfer from the surrounding rock, reservoirs are assumed to be isothermal. Hence, the reservoir pressure decreases at a constant temperature. As pressure decreases, the oil reaches the bubble point, and a gas phase begins to form in the reservoir. The gas is usually produced with the oil, but some gas may remain dispersed in the oil or rise to form a secondary gas cap above the oil.

Saturated (bubble point) oil: In some cases, the reservoir fluid is initially at the bubble point. This is an indication that the reservoir fluid is already in the two phase region and has separated into an oil layer (oil zone) and a gas layer (primary gas cap). The oil is saturated with gas and is therefore at its bubble point. The gas is saturated with liquid and is therefore at its dew point. The phase envelopes for the oil and gas intersect each other at the bubble point pressure and temperature, as shown below.

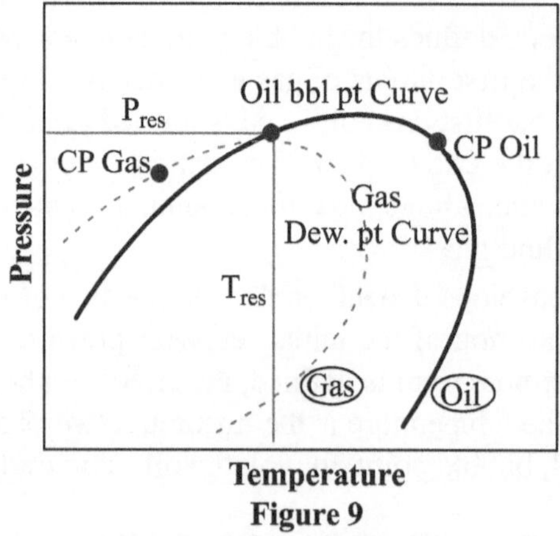

Figure 9

Phase Envelopes of a Bubble Point Oil and Its Gas Cap Gas

When a sample of fluid is withdrawn for testing, it is usually taken from either the oil zone or the gas zone and does not represent the total reservoir fluid composition. In most cases, the oil and gas zones are analyzed individually and it is not necessary to obtain a total fluid analysis. As a saturated oil is produced line B_1B_2 on the figure below gas will evolve immediately from the oil. This gas may be produced with the oil, remain dispersed in the oil zone, or rise through the reservoir to join the gas cap.

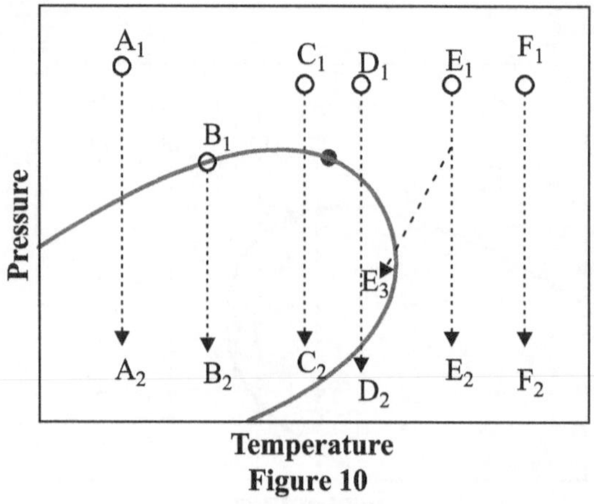

Figure 10

Volatile oil: If oil is near its critical point, its properties begin to resemble gas properties, and it is termed a volatile oil. Volatile oils, line C_1C_2 on the figure above evolve a substantial fraction of gas as pressure drops below the bubble point, and the composition and properties of the oil and evolved gas may change significantly. It is sometimes necessary to model the phase behaviour of a volatile oil with an equation-of-state-based composition model rather than the black oil model outlined later in this section.

An Important Concept

Cricondentherm

Cricondentherm can be defined as the maximum temperature at which liquids and vapour can coexist. At temperatures higher than cricondentherm only one phase occurs at any pressure. The liquid-vapour boundary terminates at a critical point with a critical temperature and critical pressure. At temperatures above the critical temperature, dewpoints are observed. At temperatures below the critical temperature, the mix will split into two phases at the bubble point pressure.

Explanation

Cricondentherm

When the reservoir temperature exceeds the cricondentherm in a gas well, only dry gas will exist in the reservoir throughout production.

If the reservoir temperature is below the cricondentherm, retrograde condensation occurs where oil condenses with decreasing rather than increasing pressure. In dewpoint reservoirs, the reservoir temperature lies between the critical temperature and the cricondentherm for the reservoir fluid. In bubble point reservoirs, the temperature is less than the critical temperature of the reservoir fluid.

Cricondentherm *vs.* Critical Point

The main difference between cricondentherm and critical point is that cricondentherm is the highest temperature at which two phases can coexist, while critical point is the temperature and pressure at which the liquid-vapor boundary ends. See the image below.

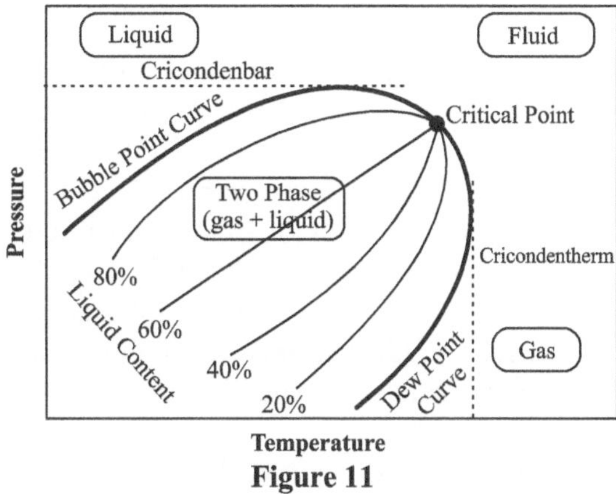

Figure 11

Cricondenbar

It is the maximum pressure above which no gas can be formed regardless of the temperature. The corresponding temperature is called cricondenbar temperature.

Retrograde Condensate

If the initial pressure and temperature lie in the fluid regime above the critical point but the temperature is below the cricondentherm, then the fluid is a retrograde condensate gas. When the pressure is reduced to the dew point, the fluid begins to condense a liquid phase line D_1D_2 on the figure. When a certain volume is reached, the amount of condensation will gradually increase upon further reduction in volume until the gases are liquefied. If the composition of the gases lies between their true and pseudo critical points the condensate formed will disappear on continued reduction of volume. This disappearance of condensation is called retrograde condensation.

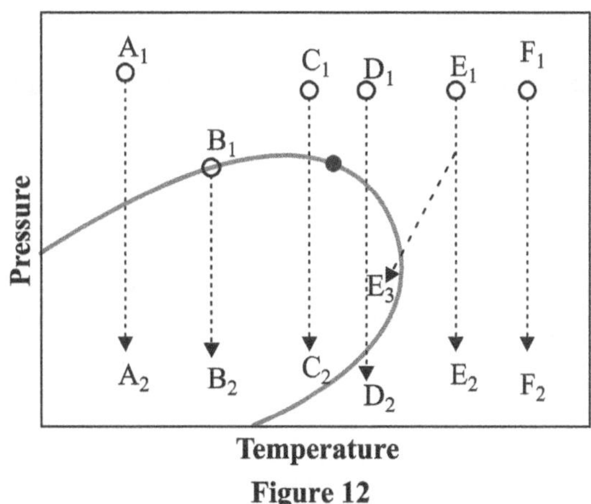

Figure 12

The liquids do not flow as readily as gas and usually inhibit the production and reduce fluid recovery from a reservoir. Compositional models are required to evaluate retrograde condensate reservoirs.

Gas: If the initial temperature is above the cricondentherm, the fluid is in the gas phase. As reservoir pressure decreases isothermally, the fluid remains in the single gas phase region, lines E_1E_2 and F_1F_2 on above. If the gas is near the two-phase envelope, liquids will condense from a reservoir gas in the wellbore or surface facilities as the temperature decreases, line E_1E_3. A reservoir gas in which condensation occurs on cooling is termed a wet gas. Condensation in the wellbore will always occur for gas cap production (unless there has been dry gas injection into the reservoir), because the gas is already at the dew point. If gas is far enough from the two-phase region, no condensation occurs, even on cooling, and the gas is termed a dry, line F_1F_2.

Summary

Reservoir-fluid properties play a key role in the design and optimization of injection/production strategies and surface facilities for efficient reservoir management Accurate and reliable phase behaviour, and volumetric data are essential elements for the proper management of petroleum reservoirs. Most reservoirs are produced by depletion, in which the reservoir pressure declines as fluids are recovered. The reservoir temperature stays practically constant in most recovery methods. The main variable that determines the behaviour of fluids, under reservoir conditions, during depletion is, therefore, the reservoir pressure. The main emphasis is on the volumetric data at the reservoir and surface temperatures that is known as pressure-volume temperature (PVT) data. The prime information from PVT tests is the ratio of phase volume at reservoir conditions to that at surface conditions and the solubility of a gas in the oil. The most commonly conducted PVT tests are empirical correlations used to estimate PVT properties from limited field data.

POROSITY AND PERMEABILITY OF OIL-GAS RESERVOIR ROCK – A REVIEW

According to the Society of Petroleum Engineers Glossary, a reservoir rock is a rock containing porosity, permeability, sufficient hydrocarbon accumulation, and a sealing mechanism to form a reservoir from which commercial flows of hydrocarbons can be produced. Porosity and permeability are the most significant physical properties of the reservoir rock. Porosity and permeability allow the migration and accumulation of petroleum under adequate trap conditions. Both of them are geometric properties that influence structural and compositional behavior (composition) reservoir rocks.

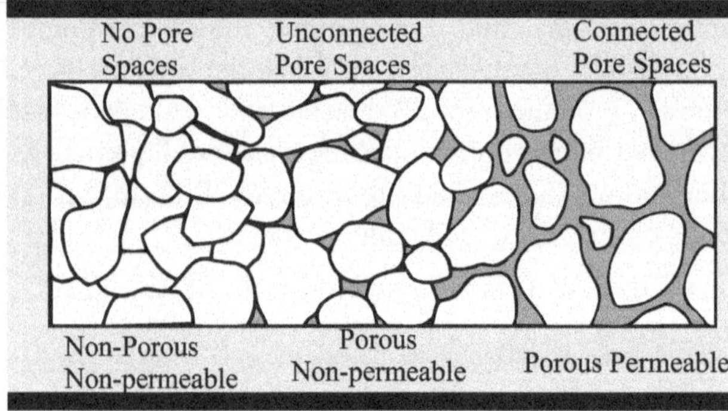

Figure 13

Porosity

Porosity is a measure in percentage of pore volume or size of holes per unit volume of rock. For example, well-sorted sand in a 300 ml container will hold about 100 ml of water in its pore space, or porosity of 33%. If petroleum is present, it can also occupy this pore space. During the burial of this sand, compaction reduces this porosity substantially to where only a small percentage of porosity is left.

Permeability

Permeability is a measure of the connectivity of pores in the subsurface. The sand in the container has narrow pore throats between the large pores that allow fluid to pass from one pore to another. Permeability is measured in millidarcy (md) or Darcy (1000 md) of these narrow throats. Commonly, permeability in the range of 100 to 500 md is a reasonable value for a petroleum reservoir rock. Fractures have infinite permeability.

Permeability Measurement

Darcy's fundamental law is given below if water was allowed to flow downward through the sand pack contained in an iron cylinder. Manometers located at the input and output ends measured fluid pressures, which were then related to flow rates to obtain the following fundamental Darcy's law:

$q = KA$ delta h/L

Where,

q = water flow rate, K = constant of proportionality that is characteristic of the sand pack
A = cross-sectional area of the sand pack
$\Delta h = h_1 - h_2$ = difference in height between the water levels in the manometers
L = length (cm)

The units in which permeability is typically expressed are the darcy (d) and millidarcy (md).

The concept is as follows: A permeability of 1 d allows the flow of 1 cm^3 per second of fluid with 1 cP (centipoise) viscosity through a cross-sectional area of 1 cm^2 when a pressure gradient of 1 atm/cm is applied.

Reservoir rocks that contain minor or lesser amounts of in-place hydrocarbons are the minor reservoir rocks. The major reservoir rock is the second name used in the name of the petroleum system. The major reservoir rock indicates the optimum migration path between the shell of the active source rock and the traps that include the major reservoir rock. The minor reservoir rock indicates the least effective migration path. Reservoir rocks are seldom if ever found to be homogeneous in physical properties or uniform in thickness. Variation in the geologic processes of erosion, deposition, lithification, etc. dictates that reservoir rocks be heterogeneous and nonuniform. Lithification is a process of porosity destruction through compaction and cementation.

Compressibility of Porous Rocks

Reservoir rocks are subjected to the internal stress exerted by fluids in the pores, and to external stress which is in part exerted by the overlying rocks. The depletion of fluids from the reservoir rocks results in a change in the internal (hydrostatic) stress in the formation, thus causing the rock to be subjected to an increased and variable overburden load, and the result is the compaction of the rock structure due to an increase in the effective stress. This compaction results in changes in the grain, pore, and bulk volume of the rock. The fractional change in the volume of solid rock constituent (grains) per unit change in pressure is defined as the rock matrix compressibility. The fractional change in the total or bulk volume of the formation per unit change in the reservoir pressure is called the rock bulk compressibility.

Reservoir Rock Properties

Storativity and Transmissibility

Reservoir rock characteristics are evaluated in terms of storativity and transmissibility to indicate the storage and flow potential of petroleum fluids, respectively. The two parameters combine various rock and fluid properties as in the following:

Storativity = Porosity × total compressibility × thickness

Transmissibility = [Permeability × thickness]/Fluid viscosity

Storativity indicates the amount of fluid that will be released from the porous medium when there is a unit drop in reservoir pressure. The unit of storativity is pounds per square inch (psi−1). The storativity of rock is directly proportional to effective porosity, net thickness, and total compressibility.

Larger values of the three properties lead to a greater storage of petroleum. Transmissibility is directly proportional to reservoir permeability and net thickness, and inversely proportional to fluid viscosity. The unit of transmissibility is mD-ft./cp. High values of rock permeability along with greater formation thickness and relatively low fluid viscosity lead to large volumetric flow in the porous medium that is ultimately produced through the wells. On the contrary, low reservoir permeability and viscous oil reduce the transmissibility of the rock and are hence detrimental to production.

References

Abdus Satter, Ghulam M. Iqbal, in Reservoir Engineering, 2016.

Leslie B. Magoon, in Encyclopedia of Energy, 2004.

Djebbar Tiab, Erle C. Donaldson, in Petrophysics (Fourth Edition), 2016.

WATER HAMMER EFFECT IN OIL AND GAS SUBSEA PIPELINES

The multi-phase subsea system is extremely complicated to analyze because of the presence of three phases and the sub-sea terrain of the pipeline. Normally, a two-phase water hammer or surge scenario can be analyzed and predicted by the famous Zhukovsky (Joukowsky) equation however things get complicated when a multi-phase situation arises. Joukowsky is used to predict the worst-case surge pressure for initial screening purposes if Hazop identifies the potential of surge within the pipeline. Joukowsky equation $\Delta P = -\rho a\ \Delta V$, where ΔP = pressure surgeρ = fluid density, a = wave speed, ΔV = change in velocity. The surge pressure is ΔP, which will result if the transient event happens instantaneously. While there may be limitations in this equation, the equation tells many things. That comes from relating the density of the medium to the predicted surge pressure. The surge pressure, I would guess would depend on the gas-liquid ratio of the medium. When the fluid in the reservoir is above the bubble point pressure in an unsaturated state, there is no gas, I would guess the fluid in the reservoir has the maximum incompressibility and least cushioning effect to absorb hammer waves. I would guess this unsaturated liquid has the maximum potential to do water hammer damage.

There are two things that are different in the sub-Sea. The deeper water imparts as much as 5,000 psi external hydrostatic pressure at 11000 feet depth upon the equipment, while the increased velocity of the control fluid results in a water hammer. Together they work to create situations that result in reverse water hammer. For every positive pressure spike, there is a corresponding negative pressure spike. The pressure on the downstream side of a suddenly closed valve could reach "0 psi absolute" or "negative (–) 5,000 psig such a huge pressure imbalance can be most destructive to controls equipment and materials.

In offshore, the subsea pipelines are the major modes of oil and gas transportation. Mixed oil, gas, and water pumped out from underwater (or offshore production platform) to other offshore facilities (processing platforms. Operating condition changes within the subsea oil-gas mixture pipeline, such as sudden downstream closure, start-stop of the pumps, valves, etc, cause a sudden rise of internal pressure leading to the water hammer effect.

What is a water hammer?

Water hammer is a commonly observed phenomenon in a fluid flow. The water hammer phenomenon is a special oscillation phenomenon caused by the flow's kinetic energy transforming into potential energy, which causes the pressure inside the pipeline to rise and forms the pressure wave which is transmitted through fluid contained in a piping system. The most basic explanation, in simple words, is that when a fluid in motion is suddenly forced to stop moving, the momentum of the fluid abruptly stopping creates a pressure wave that travels through the media within the pipe system, subjecting everything in that closed system to significant forces.

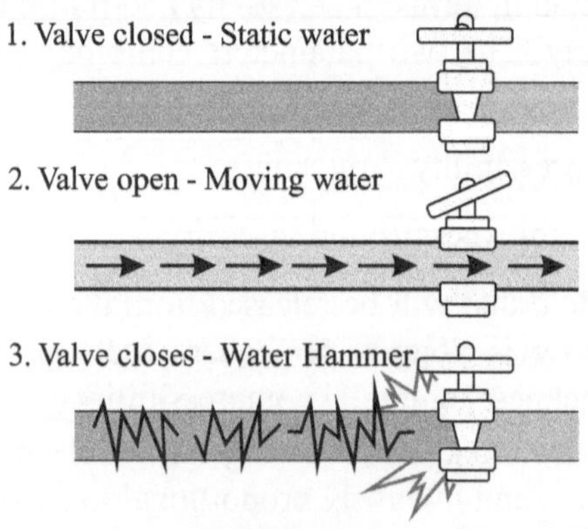

Figure 14

Normally the pressure wave is dampened or dissipated in a very short time, but the pressure spikes can do enormous damage during that brief period.

The presence of a water hammer can be easily detected by the noise it makes. Noise is not the final effect of a water hammer but just an indication of it. Water hammer has multiple adverse effects on steam systems. Water hammer can damage equipment like flow meters that are installed on the fluid network. Instances of rupture and disruption of piping on account of water hammer are also quite common. In a few cases, water hammer has resulted in catastrophic hazards.

Steam Lines

In steam lines, a water hammer is very common. As soon as steam leaves the boiler, it starts losing heat. As a result, steam starts condensing inside the pipework. The rate of condensate formation is high particularly during the start-ups when the system is cold. As a result of the condensation, droplets of water are formed. These droplets of condensate get built up along the length of steam pipework forming a slug. When this slug encounters an obstacle such as a bend, it will be brought to a halt abruptly. All the kinetic energy of the condensate slug will get converted into pressure energy which has to be absorbed by the pipework. This gives rise to the phenomenon of water hammer.

Formation of Water Hammer

After condensate is formed, the flow inside the pipe has two components, steam, and condensate.

The flow velocity of steam is much higher than that of the condensate. During such dual-phase flow, the heavy condensate that flows at the bottom of the pipe is pulled by high-speed steam. This results in the formation of a water slug which is much denser than steam traveling with the velocity of steam. When this slug is stopped by any abruption like a bend or equipment, the kinetic energy of the slug will be suddenly converted into pressure energy which will create a shock wave in the entire pipework. The pipework will keep on vibrating until this energy is dissipated in the structure.

The Impact of Water Hammer

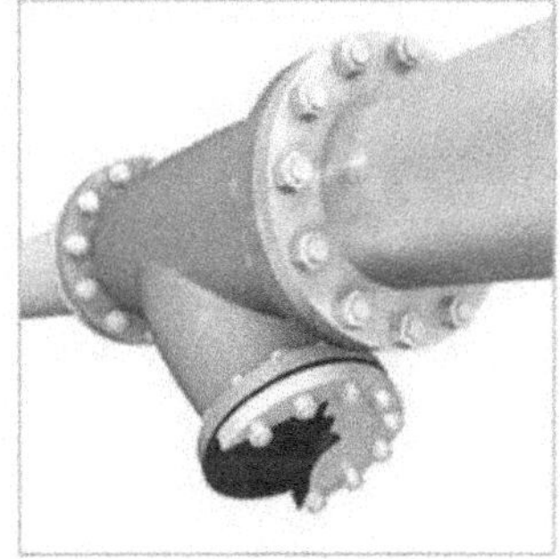

Figure 15

One might wonder why water hammers are thought to be a serious problem. The destructive nature of the water hammer can be realized through the following illustration:

The recommended velocity of saturated steam in pipe network = 20–35 m/s

The recommended velocity of water in pipe network = 2–3m/s

In the case of a water hammer, condensate is dragged by steam, and hence, the water slug travels with a velocity equal to that of steam which is around ten times more than the ideal water velocity. As a result, the total pressure impact exerted by the water hammer is very high.

Some representative incidents caused by water hammer are listed below:

Pressure rise:

[1] Pipe rupture.

[2] Damaged pipe fixtures.

[3] Damage to pumps, foundations, pipe internals, and valves.

Pressure fall:

[1] Buckling of plastic and thin-walled steel pipes.
[2] The disintegration of the cement lining of pipes.
[3] Dirty water or air being drawn into pipelines through flanged or socket connections, gland packing, or leaks.
[4] Water column separation is followed by high increases in pressure when the separate liquid columns recombine (macro-cavitation).
[5] Impact of Water Hammer on Inertia.

The sudden closure of a valve in a pipeline causes the mass inertia of the liquid column to exert a force on the valve's shut-off element. This causes the pressure on the upstream side of the valve to increase; on the downstream side of the valve, the pressure decreases.

Example

Let us consider a DN 200 pipe, L = 900 m, v = 3 m/s, the volume of water in the pipeline is calculated as mass = 0.2^2* δ/4 × 900 × 1000 = 28274 kg, this is more or less the same as the weight of a truck; v = 3 m/s corresponds to 11 km/h runs into a wall.

Subsea Production Systems

Example

Let us consider a DN 200 pipe, L = 900 m, v = 3 m/s, the volume of water in the pipeline is calculated as mass = $0.2^2 \times \pi/4 \times 900 \times 1000 = 0.04 \times 3.1428 \times 900 \times 1000/4 = 28252$ kg, this is more or less the same as the weight of a truck; v = 3 m/s corresponds to 11 km/h runs into a wall.

Subsea production systems consist of wells located on the seafloor, the petroleum is extracted at the seafloor, and then transported to an already existing production platform. This allows one strategically placed production platform to service many wells over a reasonably large area. Subsea systems are typically used at depths > 7,000 ft.

Operating condition changes within the subsea oil-gas mixture pipeline, such as sudden downstream closure, start-stop of the pumps, valves, and other unpredictable movements, will cause sudden flow stop in the pipe, resulting in a transient change of pressure and cause a sudden rise of internal pressure. The sudden closure of a valve in a pipeline causes the mass inertia of the liquid column to exert a force on the valve's shut-off element. A sharp, sometimes very high force and pressure load are created when a valve is closed too rapidly in a flowing stream. The major force occurs behind the valve. Most common in production well when sudden closure (slam closure) of a subsurface safety valve can create a load of over 50,000 lbs force in the tubing tensile/compressive loading The great water hammer pressure, especially for low gas-liquid ratio pipeline condition, may impact downstream equipment operation and even cause serious damages. The result of the sudden flow stop within the pipe, in turn, causes the change of flow state of the medium and sudden fluid momentum conversion into pressure.

The sudden stopping of the fluid flow will also create a new type of water hammer unnoticeable on surface controls, but present in subsea. A suddenly closed valve will create a high-pressure spike on the upstream side of the valve, and there is a reversed pressure occurrence downstream of the valve. In subsea conditions, the ambient conditions are extreme with hydrostatic pressures reaching 5,000 psi. Therefore, the pressure on the downstream side of a suddenly closed valve could reach "0 psi absolute" or "negative (–) 5,000 psi Gauge". Water hammer and reverse water hammer are the results of a fluid column momentum suddenly changing due to an obstruction and resulting in a pressure rise on the upstream side of the obstruction and a pressure loss immediately downstream of the obstruction. The deeper water imparting as much as 5,000 psi external hydrostatic pressure upon the equipment, while the increased velocity of the control fluid results in a water hammer, together with a reverse water hammer for every positive pressure spike. This can be most destructive to control equipment and materials.

For the gas-liquid mixture pipeline, the gas has obvious compressibility. This buffers the instantaneous increase of pressure, and thus the propagation velocity of a pressure wave in gas-liquid two-phase flow weakens greatly. In a certain gas holdup range, the propagation velocity of the pressure wave is far lower than that of pure liquid or pure gas. However, it has been found that, when the gas-liquid ratio is relatively low, the gas phase flow rate of the pipeline is almost zero under the operating pressure condition, and the gas is mostly dissolved in the liquid. In such a case, the sudden closure of the upper and lower valves will cause a significant surge of water hammer pressure.

To Summarize

Oil/gas water hammer is a special case. The deeper water imparts as much as 5,000 psi external hydrostatic pressure at 11000 feet depth upon the equipment. For every positive pressure spike, there is a corresponding negative pressure spike. The pressure on the downstream side of a suddenly closed valve could reach "0 psi absolute" or "negative (–) 5,000 psig such a huge pressure imbalance can be most destructive to control equipment and materials.

Steam Lines

If there is condensate in a steam line which is unavoidable, the water hammer cannot be completely eliminated. The recommended velocity of saturated steam in the pipe network = 20–35 m/s

The recommended velocity of water in the pipe network = 2–3 m/s

In the case of a water hammer, condensate is dragged by steam, and hence, the water slug travels with a velocity equal to that of steam which is around ten times more than the ideal water velocity. As a result, the total pressure exerted by the water hammer is very high.

EMULSION: BIG CHALLENGE IN OIL/GAS INDUSTRIES

It is a significant challenge in the oil/gas industry. The most challenging aspect in the petroleum industry is the high produced water accompanying crude oil extraction.

In simple and short, emulsions are dispersions of two or more nearly immiscible fluids, e.g., oil and water. One of these is dispersed as droplets in the other. Emulsions are widely used in industry and everyday life. Droplet-size distribution and the dispersed phase content influence the properties of an emulsion.

There are several factors that can contribute to formation of emulsions, such as presence of surfactants or surface active organic compounds either present in well or some chemicals added to enhance oil recovery is one major factor.

Addressing emulsification in oil-gas-water separation processes requires use of chemical demulsifiers, mechanical separation techniques, and specialized equipment to break down the emulsions and facilitate phase separation.

What is an Emulsion?

An emulsion is a mixture of two or more immiscible liquids, usually oil and water, where one liquid is dispersed as droplets throughout the other liquid. The resulting mixture is homogeneous and appears as a single phase, making it Oil and Gas difficult to separate the two liquids.

Emulsions fall under the broader category of colloids, which are two-phase systems of matter. Although the terms colloid and emulsion are sometimes used synonymously, emulsion is the preferred term when the dispersed and continuous phases are both liquids. One liquid (the dispersed phase) is dispersed in another (the continuous phase) in an emulsion.

Examples of emulsions include hair cream, homogenized milk, etc. In addition to a liquid-liquid emulsion, there are also emulsions possible with solid-gas forming smoke dust, liquidsolid forming cheese, butter, jellies, etc., and liquid gas phases forming fog, mist, cloud, etc.

How does an Emulsion form?

The quality of an emulsion depends on the droplet size of the dispersed phase. Generally, there are two ways in which an emulsion can form: (1) when two immiscible liquids with near density are vigorously mixed, and (2) by emulsifiers acting as a bridge, joining two immiscible liquids.

What is an Emulsifier?

An emulsifier is made up of an oil-loving hydrophobic tail and a water-loving hydrophilic head. Generally, these are surfactants or surface-active compounds. The oil phase is targeted by the hydrophobic tail and the aqueous phase by the hydrophilic head. The emulsifier positions itself at the oil/water or air/water interface, serves as a bridge, and stabilizes the emulsion by lowering surface tension.

How Emulsion Affects Viscosity? The viscosity of the continuous phase is increased by a solution of stabilizers in the continuous phase of oil-in-water (o/w) emulsions. The mobility of the droplets and their collision frequency decreases. Thus, stabilizers can improve the stability of emulsions by hydrodynamic effects.

Demulsification and Phase Inversion

Detail

Demulsification refers to the process of breaking down emulsions, which are mixtures of two immiscible liquids, such as oil and water, stabilized by an emulsifying agent. Phase inversion is a technique commonly used in demulsification processes.

In the demulsification process, phase inversion plays a crucial role in separating the oil and water phases of the emulsion. It involves changing the continuous phase (the phase in higher proportion) to the dispersed phase (the phase in lower proportion) and vice versa.

The role of phase inversion in demulsification can be described as follows:

1. **Breaking the emulsion**: Phase inversion helps break down the emulsion by changing the properties of the continuous and dispersed phases. This change destabilizes the emulsion, allowing for the separation of the oil and water phases.
2. **Enhancing coalescence**: Phase inversion promotes the coalescence (joining together) of dispersed droplets within the continuous phase. As the continuous phase is inverted, the dispersed droplets come closer to each other, facilitating their coalescence and subsequent separation.
3. **Oil/water separation**: Phase inversion facilitates the separation of oil and water by causing one phase to become the dispersed phase and the other to become the continuous phase. This change allows for the distinct separation of the two phases, enabling efficient recovery and purification of the desired components.
4. **Emulsion destabilization**: Phase inversion can destabilize the emulsion by altering the emulsifying agent's behavior. Emulsifying agents are typically surfactants, and their performance can be affected by changing the continuous and dispersed phases. Phase inversion can cause a change in the emulsifying agent's interfacial activity, reducing its effectiveness in stabilizing the emulsion.

Overall, phase inversion plays a vital role in demulsification by influencing the emulsion's stability, promoting phase separation, and facilitating the efficient recovery of desired components.

Thermodynamics

The thermodynamics of emulsions involves the study of the energy changes that occur during their formation, stability, and disruption. The key thermodynamic principles that govern emulsion behavior include interfacial tension, Gibbs free energy, and entropy.

Interfacial tension: Emulsions consist of two immiscible phases, such as oil and water, separated by an interface. The interfacial tension between the two phases influences the stability of the emulsion. Higher interfacial tension favors phase separation and destabilizes the emulsion, while lower interfacial tension promotes stability.

Gibbs free energy: Emulsion formation involves a change in the system's Gibbs free energy, which determines whether the emulsion is favorable or not. In an emulsion, the Gibbs free energy decreases due to reduced interfacial tension and increased mixing of oil and water. This decrease in energy contributes to the stability of the emulsion.

Entropy: Emulsions are thermodynamically driven by entropy, which is a measure of the system's disorder. When two immiscible phases are mixed to form an emulsion, the disorder, and entropy of the system increase due to the dispersion of droplets within the continuous phase. This increase in entropy enhances the stability of the emulsion.

Typically, emulsions tend to form spontaneously when there is enough energy input to disperse one phase into another. However, emulsion stability depends on various factors such as the composition, concentration of emulsifying agents, temperature, pressure, and shear forces. Understanding the thermodynamics of emulsions helps in controlling and manipulating their stability and behavior for various industrial applications.

Challenges of Emulsion in Oil/Gas Industries

Emulsions can indeed pose a significant challenge in the oil and gas industry. Emulsions can form during the production, transportation, and processing of oil and gas, particularly when water is present. The presence of emulsions can lead to difficulties in separating oil and water, increased viscosity of the fluid, reduced flow rates, equipment fouling, and decreased overall efficiency of the operations. Therefore, controlling and managing emulsions is a crucial aspect of the oil and gas industry

Detail

High produced water that results from crude oil extraction is the most difficult aspect of the petroleum industry. In the process of extracting crude oil, produced water is made up of a variety of inorganic, organic, and other compounds. However, a significant portion of produced water is an emulsion, which creates a significant issue with the handling and transportation of crude oil. Researchers have been looking into demulsification techniques for several years due to limitations in produced water treatment methods. Researchers also pointed out that a variety of factors, such as asphaltenes, resins, solid particles, water and oil content, PH, and others, affect the stability of emulsions and the interfacial film.

However, one method for enhanced oil recovery that has attracted attention is a chemical method by using surface active agents (surfactants).

ROLE OF SEPARATORS IN OIL-GAS INDUSTRIES: A SHORT NOTE

What?

Separators play a crucial role in the oil and gas industry by separating the produced fluids (oil, water, and gas) that come up from the well. A separator is a device that uses gravity to separate these fluids based on their different densities.

Why?

The main function of a separator is to separate the oil, water, and gas phases from each other. This allows for the efficient processing of the different components and ensures that the oil can be properly refined and sold.

How?

The separator works by allowing the fluids to enter a chamber where the differences in density cause them to separate naturally. The oil, being the lightest component, rises to the top, the water, being heavier, settles at the bottom, and the gas, being the lightest, is released at the top of the separator.

The key factors in the operation of a separator include the flow rates of the different fluids, the pressure at which they are produced, and the design of the separator itself.

Types

There are several general types of separators.

- **Horizontal separators**: These separators are the most common type and consist of a horizontal vessel with a series of internal baffles. The produced fluids enter the vessel and are separated based on their densities.
- **Vertical separators**: Vertical separators are similar to horizontal separators but are oriented vertically. They are often used when space is limited.
- **Three-phase separators**: Three-phase separators are designed to separate oil, water, and gas simultaneously. They have specialized internal components to facilitate the separation of all three phases.

- **Two-phase separators**: Two-phase separators are used to separate oil and gas or oil and water. They are simpler in design compared to three-phase separators and are often used in applications where only two phases need to be separated.

These are just a few general types of separators used in the oil and gas industry.

Issues

There are several common issues

- **Carryover: Carryover occurs** when one phase (e.g., water or gas) is not properly separated and carries over into the separated product.
- **Emulsions can form** in the separator due to factors such as high water cuts, high temperatures, and the presence of emulsifying agents
- **Foaming:** Foaming can occur in the separator when gas bubbles are entrained in the liquid phase.
- **Corrosion**: Corrosion can occur in the separator due to the presence of corrosive agents
- **Slugging:** Slugging refers to the intermittent flow of liquid and gas through the separator.
- **Pressure drops**: Pressure drops in the separator can result from factors such as fouling, blockages, or improper sizing.
- **Scaling**: Scaling occurs when solid deposits form on the internal surfaces of the separator. Scaling can reduce the separation efficiency.

CHALLENGES IN GAS SWEETENING IN GAS WELL OPERATIONS

In general, sweet gas has minimal or no hydrogen sulfide and carbon dioxide, while sour gas has significant amounts of these corrosive gases. The separation of sour and sweet gases is typically done through a process called gas sweetening or gas desulfurization.

Origin and Classification of Gases

The classification of gas as sweet or sour depends on hydrogen sulfide concentration and carbon dioxide.

Sweet gas: Sweet gas refers to natural gas that primarily consists of methane and has a low concentration of impurities, especially hydrogen sulfide (H_2S) and carbon dioxide (CO_2). Sweet gas formations are typically formed through the following process:

- **Organic matter accumulation:** Over millions of years, organic matter such as plants and microscopic organisms accumulates in sedimentary basins.
- **Heat and pressure:** As more sediment layers deposit above the organic matter, it gets buried deeper in the Earth's crust, subjecting it to high temperatures and pressures.
- **Thermal decomposition:** Due to the intense heat and pressure, the organic matter undergoes thermal decomposition, breaking down into simpler hydrocarbons, including methane (CH_4), which is the primary component of sweet gas.

Sour gas: Sour gas contains significant amounts of hydrogen sulfide (H_2S) and may also have elevated levels of carbon dioxide (CO_2) in addition to methane. Sour gas formations are formed through a slightly different process:

- **Presence of sulfur-rich compounds:** In some areas, the organic matter that accumulates and undergoes thermal decomposition contains sulfur-rich compounds, such as sulfates or organic sulfur.
- **Microbial activity:** As the organic matter decomposes, certain types of microorganisms can interact with sulfur compounds and produce hydrogen sulfide gas (H_2S) through biogenic processes.
- **Geological activities:** Apart from microbial activity, geological processes like hydrothermal activity or the interaction of groundwater with sulfur-rich rocks can also contribute to the presence of H_2S in sour gas reservoirs.

Separation of Sweet and Sour Gases

This process involves removing the corrosive and toxic hydrogen sulfide (H_2S) gas from the sour gas stream to produce sweet gas.

Gas sweetening can be accomplished using different methods, including:

Chemical Absorption: The most commonly used method is the use of chemical solvents to absorb the H_2S gas. The sour gas is passed through a tower or vessel where it comes into contact with the solvent. The H_2S molecules are selectively absorbed by the solvent, while the remaining gases, mainly methane and other hydrocarbons, pass through as sweet gas.

The most commonly used solvent in the chemical absorption process for gas sweetening is amine. Amines are organic compounds with nitrogen atoms that have the ability to react with acidic gases such as hydrogen sulfide (H_2S) and carbon dioxide (CO_2). Some commonly used amines include:

Monoethanolamine (MEA)

Diethanolamine (DEA)

Methyldiethanolamine (MDEA)

Diisopropanolamine (DIPA)

Diglycolamine (DGA)

These amines selectively react with H_2S, forming an amine sulfide complex, effectively removing H_2S from the sour gas stream. The choice of amine depends on factors such as the concentration of H_2S and CO_2 in the gas, desired sweet gas specifications, operating conditions, and economic considerations. Different amines have varying capacities, efficiencies, and operating parameters, so the selection of amine is based on the specific requirements of the gas-sweetening process.

Physical Solvent Extraction: In this method, specialized solvents with a high affinity for H_2S, such as N-methyl-2-pyrrolidone (NMP), are used to remove H_2S from the gas stream. The selective solvent is brought in contact with the sour gas, and it selectively extracts the H_2S. The gas that emerges is sweet gas.

Membrane Separation: Another method involves using semi-permeable membranes that allow the smaller molecules, such as methane and other hydrocarbons, to pass through while blocking the larger H_2S molecules. This separation process is based on the difference in molecular size. The gas mixture is passed through a series of membrane modules, and the final output is sweet gas.

Amine Treating: Amine solutions, such as monoethanolamine (MEA), can be used to remove H_2S from sour gas through a process called amine treating. The sour gas is brought into contact with the amine solution in an absorber tower. The H_2S is chemically absorbed by the amine, leaving the sweet gas to be separated.

After the separation process, the H2S-rich solution or concentrate is further processed to recover or neutralize the H2S, while the sweet gas is sent for further processing, distribution, or utilization.

Corrosion in Gas Sweetening Process

Corrosion is indeed a significant concern in the gas-sweetening process. Corrosion can occur due to several factors in the gas sweetening unit, including the presence of acids (such as H_2S and CO_2), high-pressure conditions, temperature variations, and the type of materials used in construction.

Fundamentals

Why Sour Gases are Highly Corrosive?

Both H_2S and CO_2 are weak acids yet they can do significant damage to metals. Interestingly, their mechanism of damage is different.

Two Fundamental Points

Carbon dioxide [CO_2]

CO_2 dissolved in water forms weak carbonic acid (H_2CO_3) which does not dissociate completely and can remain in the system, leading to corrosion.

$$H_2O + CO_2 \Leftrightarrow H_2CO_3$$

Hydrogen Sulfide [H_2S]

In the presence of water hydrogen sulfide (H_2S), it can react to form sulfuric acid (H_2SO_4) and hydrogen gas (H_2). H_2SO_4 is corrosive and can cause corrosion, while the release of H_2 gas can lead to embrittlement of certain materials.

$$H_2S + 4H_2O \rightarrow H_2SO_4 + 4H_2$$

Key Factors

Sour gases are highly corrosive due to several factors:

Acidic nature: Sour gases have acidic properties, which means they release hydrogen ions (H^+) when dissolved in water and lower the pH, increasing its corrosive nature.

Oxidation reactions: Sour gas H_2S can undergo oxidation reactions, with the oxidation number of sulfur going from –2 in H_2S to +6 in H_2SO_4.

This strong acid can attack metals, leading to corrosion. While H2 does the hydrogen embrittlement

$$H_2S + 4H_2O \rightarrow H_2SO_4 + 4H_2$$

Formation of aggressive ions: When sour gases dissolve in water, they can form aggressive ions such as sulfide (S^{2-}) and carbonate $(CO_3)^{2-}$. These ions can react with metal surfaces, accelerating corrosion processes.

Hydrogen embrittlement: Sour gases, particularly H_2S, can cause hydrogen embrittlement in metals. Hydrogen atoms can penetrate the metal lattice and weaken its structure, leading to increased susceptibility to corrosion and fracture.

Overall, the combination of acidic properties, oxidation reactions, formation of aggressive ions, and hydrogen embrittlement make sour gases highly corrosive to metals and other materials.

The corrosive nature of sour gases, particularly H_2S, can lead to corrosion of equipment and pipelines. This corrosion can result in damage, leaks, and reduced operational efficiency of the gas sweetening unit. Corrosion control measures are therefore crucial to maintain the integrity and safety of the system.

To mitigate corrosion in gas sweetening units, several strategies are employed, including:

Selection of corrosion-resistant materials: The equipment and piping used in the gas sweetening unit are often made of corrosion-resistant alloys or coated with materials that provide protection against corrosion.

Injection of corrosion inhibitors: Corrosion inhibitors, such as chemicals that form a protective film on metal surfaces, can be added to the gas stream to reduce corrosion rates.

Monitoring and control: Regular monitoring of corrosion rates and conditions in the equipment is essential to detect any potential issues. Control measures can then be implemented to mitigate corrosion, such as adjusting operating parameters or modifying the chemical treatment.

Maintenance and inspections: Scheduled maintenance, inspections, and cleaning of equipment are necessary to prevent corrosion accumulation and identify any signs of corrosion or damage.

CHAPTER 53

Corrosion

Corrosion is a natural process that occurs when metals react with their environment, such as water or oxygen, resulting in the degradation of the metal's surface. Thermodynamics plays a key role in corrosion processes by determining the energy changes associated with the oxidation of metals and the formation of corrosion products. The thermodynamic driving force for corrosion can be calculated using principles such as Gibbs free energy to predict the direction and rate of corrosion reactions.

A substance's potential energy is the energy that has the 'potential' to do work. A metal does work while oxidizing when it transfers charge [electrons] or ionizes. Ionization is a thermodynamic work. Corrosion occurs when metals ionize and lose metal ions. Electrode potential is a measure of the potential energy of metal as well as its ability to lose electrons and corrode. Platinum does not corrode due to its low potential energy or ability to do work. Low-energy platinum oxides are unstable and will break down and revert to metal. The same applies to gold. Gibbs free energy change, ΔG is used, as said above to predict corrosion.

WHAT IS CORROSION THERMODYNAMICS?

The driving force for corrosion is the potential difference between the anode and cathode. The potential difference is nothing but the difference in the potential energy between anode and cathode. Thermodynamics monitors the energy transfer in a corrosion reaction. Thermodynamics can help us understand the energy changes that occur during the corrosion reaction. In general, it can predict when corrosion is likely. Corrosion rates cannot be predicted using thermodynamics. Kinetics governs the rate at which the reaction occurs. Gibb's free energy, which is given by the following equation, gives us a tool for predicting whether a corrosion reaction is thermodynamically possible, $\Delta G = -nFE$. Three metals have been chosen for discussion in the is note. These are, iron which has potential for corrosion, copper which is a near noble metal, and platinum an inert noble metal to predict their corrosion behavior on the basis of thermodynamics principles.

It is well known that a metal that is easily oxidizable or more anodic, such as iron, corrodes more readily than a metal that is less oxidizable, such as platinum. The corrosion resistanceis related to the electrochemical potential of a metal. Metals with lower electrode potential values are more reactive and more susceptible to corrosion. **Corrosion potential:** The corrosion potential is the potential difference between a material and a reference electrode. A positive change in corrosion potential indicates that the surface is protected and corrosion is slowing down. A negative change indicates an increase in the corrosion rate

Negative Gibbs free energy change ΔG of a corrosion reaction signifies spontaneous corrosion. Zero change in free energy is an indicator of equilibrium between anode and cathode reactions. Positive free energy signifies no corrosion.

All metal corrosions involve the transfer of electrons, so all corrosions are electrochemical in nature. One anode, one cathode, an electrolyte, and a conductor to carry electrons from the anode to the cathode are required for corrosion to occur. At the anode electrons are lost from the metal, indicating an oxidation reaction, and at the cathode, electrons are gained, indicating a reduction reaction.

There are Many Questions

The first question is why, except for noble metals, we say corrosion is unavoidable. **The answer is simple.** For a corrosion cell to form on the metal surface, a very small ppb level of impurities on the metal surface is enough to generate a potential difference for current to flow from the anode to the cathode, causing the metal at the anode to ionize and be lost from the metal surface. Impurities could be present in the water or on the metal surface itself.

Table 1: Corrosion Predictability

Metal	*Gibbs Free Energy ΔG KJ*
Iron	(–)156 [Spontaneous corrosion]
Copper	(–) 12 [More noble metal than iron]
Platinum	(+) 154 [Noble metal. Corrosion resistant]

This is one of the reasons why you don't want any *TDS* or why you need very pure water in a power plant. This is why, in the semiconductor-making process, a large amount of ultra-pure water is required for washing the substrate.

This also explains why corrosion occurs.

The second question is why metal corrosion rates differ from one another. What is the **reason** that **iron is more corrosive than copper**? **Why** do **noble metals not corrode**, for example. To answer this question, you must first understand corrosion thermodynamics.

In this note, **we will** look at **the thermodynamics of corrosion** and answer two perplexing questions, such as [1] why **iron** is **more corrosive** in water **than copper**, [2] why **platinum** is **insensitive** to electrochemical **corrosion.**

Thermodynamics of Corrosion: Fundamentals

In electrical and electrochemical processes, electrical work is defined as the product of charges moved (***Q***) times the potential (***E***) through which it is moved. If this work is done in an electrochemical cell in which the potential difference between anode and cathode is ***E***, and the charge is that of one mole of reaction in which 'n' moles of electrons are transferred, then the electrical work (**-*W***) done by the cell must be ***nE***. In this relationship, the charge on the mole is converted to coulombs. The charge on one mole of electrons is 96,500 coulombs. This is also called 1 Faraday. No current flows at equilibrium between anode and cathode. At equilibrium, there is no corrosion. Gibb's free energy, which is given by the following equation, gives us a tool for predicting whether a corrosion reaction is thermodynamically possible: $\Delta G = -n\, FE$ [ΔG is Gibbs free energy]

What does it mean?

Corrosion occurs only when current flows due to the potential difference between anode and cathode. The energy change occurring in a reaction is expressed in the equation:

$$W = \Delta G = -nFE$$

Under standard conditions, the standard free energy of the cell reaction ΔG is directly related to the standard potential difference across the cell, E:

$$\Delta G = -nFE$$

What is the Standard Condition?

For solid or liquid compounds or elements, standard conditions are the pure compound or element; for gases, they are 100 kPa pressure; and for solutes, they are the ideal 1 molar (mol/L) concentration.

Electrode potentials can be combined algebraically as an equation.

$$E\text{ cell} = E\text{ cathode} + E\text{ anode}$$

Free energy Change

When the free energy change ΔG is negative there is corrosion and, in the case, when ΔG is zero, there is equilibrium between anode and cathode and there is no corrosion.

Entropy and Enthalpy Change

Other thermodynamic quantities can be derived from electrochemical measurements. For example, the entropy change (ΔS) in a cell reaction is given by the temperature dependence of (ΔG).

Corrosion Gibbs free energy and entropy

$$\Delta S = \left(\frac{\partial \Delta G}{\partial T}\right)_P$$

$$\Delta S = nF\left(\frac{\partial E}{\partial T}\right)_P$$

$$\Delta H = \Delta G + T\Delta S = nF\left[T\left(\frac{\partial E}{\partial T}\right)_P - E\right]$$

where ΔH is the enthalpy change and T the absolute temperature (K).

Entropy

If you know ΔH of a chemical reaction that is [product heat of formation – reactant heat of formation], if could find out ΔG for corrosion reaction as explained above, you can easily calculate ΔS at temperature T by using the following equation,

$$\Delta S = [\Delta H - \Delta G]/T$$

Predictability of Electrochemical Corrosion

- Iron
- Copper
- Platinum

Predictability of Iron Corrosion

Example 1: Steel is placed in aerated seawater with a neutral pH. Is corrosion of the steel possible, why? (Assume valence of 2)

Solution: Step by step

[1] Write the oxidation-reduction reaction [take help from the internet]. You will notice that there are two reactions, oxidation at the anode and reduction at the cathode. When you combine the two reactions [add] you get the overall reaction.

At anode $Fe \rightarrow Fe^{+2} + 2e^{-}$ [oxidation]

At cathode $O_2 + 2H_2O + 4e^{-} \rightarrow 4OH^{-}$ [reduction]

Overall reaction:

$2Fe + 2H_2O + O_2 \rightarrow 2Fe\,(OH)2$

[2] Calculate E which is the cell potential. E is the sum of oxidation potential and reduction potential.

$E = E$ oxidation + E reduction

[3] Go to the standard electrode potential chart on the internet and find out E oxidation and E reduction.

In this case,

E oxidation = 0.41v [Standard oxidation potential of iron electrode is + 0.41 v] [at anode]

E reduction = 0.40v [Standard reduction potential of O_2 is + 0.40v] [at cathode]

E = 0.41+ 0.4 = 0.81

[4] Find out the value of 'n'. 'n' is the number of electrons transferred from the anode to the cathode. Refer to the reactions above. In this case, there were two electrons transferred.

Insert the values of n, E and F in equation, $\Delta G = -n\,FE$ = –(2 mole) × 96,500 (J/v-mole^{-}) × 0.0.81(v) = –156330 J = –156.33 KJ. ΔG = –156.3 KJ

Negative Free Energy Signifies Spontaneous Corrosion of Iron

Predictability of Copper Corrosion

Example 2: Copper is placed in aerated seawater with a neutral *pH*. Is corrosion of the steel possible, why? (Assume valence of 2).

Calculation: Follow steps 1 to 4 written above.

Solution

At anode $Cu \rightarrow Cu^{+2} + 2e^{-}$ [oxidation] [at anode]

At cathode $O_2 + 2H_2O + 4e^{-} \rightarrow 4OH^{-}$ [reduction] [at cathode]

Overall reaction:

$$2Cu + 2H_2O + O_2 \rightarrow 2Cu\,(OH)2$$

E oxidation = –0.34v [Standard oxidation potential of copper electrode is –0.34v]
E reduction = 0.40v [Standard reduction potential of O_2 is + 0.40v]

$$E = E \text{ oxidation} + E \text{ reduction} = -0.34 + 0.4 = 0.06$$

$$\Delta G = -n\,FE = -(2 \text{ mole}) \times 96{,}500\ (\text{J/v-mole}^{-}) \times 0.06(v) = -11580 \text{ J} = -11.580 \text{ KJ}$$

$$\Delta G = -11.6 \text{ KJ}$$

The negative free energy change in the case of copper is much smaller than iron. This justifies why. copper is much more noble than iron.

Discussion/Clarifications: Standard Reduction Potentials

It is the tendency for a species to be reduced at standard conditions. Standard oxidation potential: It is the tendency for a species to be oxidized at standard conditions.

The more the positive oxidation potential the more its tendency to oxidize. The oxidation potential of iron is + 0.41v, the oxidation potential of copper is –0.34v, and the oxidation potential of platinum is – 1.2. In simple words, the potential of an element is its potential energy. The more the potential energy the more it is likely to oxidize or corrode. Iron's oxidation potential at 0.4v compared to –1.2v of platinum. Iron corrodes because it has high potential energy. Platinum is insensitive to corrosion because it has a very small potential energy.

Fe vs. *Cu*

Fe is more electropositive than copper. *Cu* is a more noble metal than *Fe*. You would notice that the oxidation potential of *Fe* is more than copper.

What does it mean?

Fe is more oxidizable than copper. With its oxidation potential much higher than copper, 0.4v *vs.* – 0.34v, iron has more potential energy to corrode. The same one could see with its Gibbs free energy change ΔG much more negative than copper, –156 *KJ vs.* –12 *Kj*.

Predictability of Platinum Corrosion

Calculation: Follow steps 1 to 4 written above.

Solution: At anode $Pt \rightarrow Pt^{+2} + 2e^{-}$ [oxidation][at anode]

At cathode $O_2 + 2H_2O + 4e^{-} \rightarrow 4OH^{-}$ [reduction][at cathode]

Overall reaction:

$$2Pt + 2H_2O + O_2 \rightarrow 2Pt\,(OH)2$$

E oxidation = –1.2v [Standard oxidation potential of platinum electrode is – 1, 2 v]
E reduction = 0.40v [Standard reduction potential of O_2 is + 0.40v]

$$E = E \text{ oxidation} + E \text{ reduction} = -1.2 + 0.4 = -0.8$$

$$\Delta G = -nFE = -(2 \text{ mole}) \times 96{,}500 \text{ (J/v-mole}^{-}) \times -0.8(v) = 154{,}400\text{J} = 154.4 \text{ KJ}$$

You can see a huge positive Gibbs-free energy. This explains why platinum is a corrosion-resistant noble metal.

ROLE OF GIBBS FREE ENERGY AND ENTROPY IN CORROSION

Like all chemical processes, the kinetics involved in corrosion obey the Arrhenius relationship:

$$k = k_0 \exp(-\Delta G/RT)$$

where k is the rate of reaction, k_0 is a fundamental rate constant, ΔG is the activation energy. R and T have their usual meanings of the ideal gas constant (8.3145 J K−1 mol 1) and temperature (in Kelvin) respectively.

Background Reading

What is Activation Energy and What is Free Energy?

All molecules possess a certain minimum amount of energy. The energy can be in the form of kinetic energy or potential energy. When molecules collide, the kinetic energy of the molecules can be used to stretch, bend, and ultimately break bonds, leading to chemical reactions. If molecules move too slowly with little kinetic energy or collide with improper orientation, they do not react and simply bounce off each other. However, if the molecules are moving fast enough with a proper collision orientation, such that the kinetic energy upon collision is greater than the minimum energy barrier, then a reaction occurs. The minimum energy requirement that must be met for a chemical reaction to occur is called the activation energy, Ea.

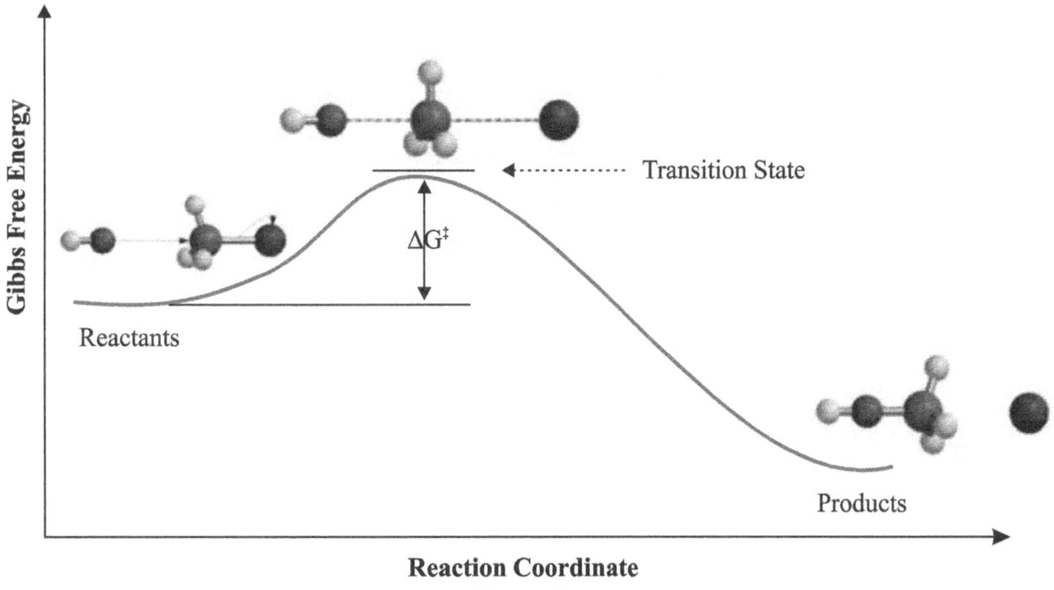

Figure 1

The reaction pathway is similar to what is shown in the above figure. To get to the other end of the road, an object must roll with enough speed to completely roll over the hill of a certain height. The faster the object moves, the more kinetic energy it has. If the object moves too slowly, it does not have enough kinetic energy necessary to overcome the barrier; as a result, it eventually rolls back down, the reaction reverses. In the same way, there is a minimum amount of energy needed in order for molecules to break existing bonds during a chemical reaction. If the kinetic energy of the molecules upon collision is greater than this minimum energy, then bond breaking and forming occur, forming a new product (provided that the molecules collide with the proper orientation).

The activation energy (Ea), labelled $\Delta G‡$ in the above figure is the energy difference between the reactants and the activated complex, also known as the transition state. In a chemical reaction, the transition state is defined as the highest-energy state of the system. If the molecules in the reactants collide with enough kinetic energy and this energy is higher than the transition state energy, then the reaction occurs and products form. In other words, the higher the activation energy, the harder it is for a reaction to occur and vice versa.

Let Us Go Back to Corrosion

The chemical nature of corrosion suggests that it is driven by a change in Gibbs free energy, ΔG but the electrical nature of corrosion leads to the conclusion that a voltage drives the reaction. Since both quantities can be considered as the driving force, they must be equivalent and, indeed they are related through the expression $\Delta G = -Z F E$, where Z is the stoichiometric number of electrons in the reaction, F is Faraday's constant, 96485 C mol−1 and E is the voltage driving the reaction.

Note the minus sign, used to correct for the conventions that a chemical reaction only proceeds if ΔG is negative but an electrical reaction only occurs if E is positive.

Since the absolute driving force of an applied voltage depends on what reaction is occurring, potentials are usually defined as the difference between applied voltage and the equilibrium potential of the reaction. The difference between applied potential and equilibrium potential is defined as the overpotential, η.

$$\eta = E - Ee$$

It is worth noting that the "equilibrium potential" is not necessarily the standard electrode potential of the reaction, as this has the added requirement that all reagents are in standard states. The equilibrium referred to is, in fact, the "equilibrium electrode potential", Ee, which is specific to every electrode individually.

If an electrode is at its equilibrium potential, both forwards and backward reactions occur at the same rate, so no net reaction will occur. Net reactions only occur when the potential is moved away from equilibrium.

Under equilibrium conditions, the energy landscape is symmetrical when free energy is plotted against the distance from a metallic surface.

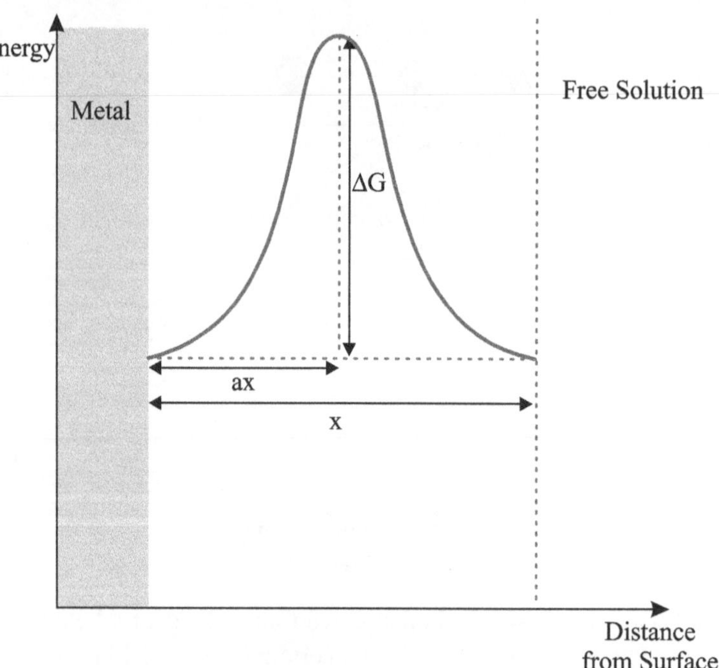

Figure 2

Corrosion and Entropy

According to a common misconception, "The entropy law describes the tendency for all objects to rust, break, fall apart, wear out, and otherwise move to a less ordered state."

One error arises in the very first instance. Does rusting lead to increased entropy?

The rust reaction is

$$4Fe + 3O_2 \rightarrow 2\ Fe_2O_3.$$

According to the available data, the entropy (at room temperature 298.15 K and at pressure 105 Pa) of one mole of Fe is 27.280 J/K, of one mole of O_2 is 205.147 J/K, and of one mole of Fe_2O_3 is 87.404 J/K. The entropy of the reactants is 724.561 J/K, the entropy of the products is 174.808 J/K, so the reaction results in an entropy decrease of 549.753 J/K.

It is easy to understand why this should be so: gases typically have much greater entropies than solids. And of course, this doesn't mean that during rusting, the entropy of the universe decreases: although the entropy of the iron plus oxygen decreases, the entropy of the surroundings increases even more. Overall, the Entropy of the universe increases.

What does it Mean?

What Happens to Free Energy When the Entropy Increases?

Free energy change dG becomes more negative, $dG = H - TS$, and Corrosion becomes spontaneous.

HOW DOES OXYGEN PASSIVATE METALS?

Contents

We often wonder "how and why metal passivation works." There are two obvious questions [1] How does an oxide layer form on the surface of a metal? [2] Why can't all metals be passivated by oxygen? The post has been written with reference to iron and chromium to explain why oxide does not protect iron while the oxide layer on chromium makes the metal impervious to oxygen and corrosion. The two key advantages of chromium compared to iron are [1] it is less electropositive than iron. It has less affinity for oxidation to form oxide. [2] The most important advantage of chromium is it has only three oxides while iron has seventeen known iron oxides and oxyhydroxides of different volumes. The creation of so many different types and different molecular size oxides of different volumes generate much more stress on the metal surface than chromium leading to flaking out of the metal. This post will try to answer the questions. I hope you will find this note useful.

What is Passivation?

Passivation is a metal finishing process that prevents corrosion. It is the process of forming an exterior layer on the surface of a metal part or component. Because of this exterior layer, the underlying metal will not be directly exposed to the surrounding atmosphere. Rather, the underlying metal will be sealed, making it more corrosion-resistant.

Metals are easily oxidized by oxygen. While a nonporous/no crack/fully developed oxide film can protect the metal from corrosion, oxygen diffusion through an undeveloped oxide film can destroy it.

Metal corrodes as a result of oxidation. When a metal, particularly iron or iron alloys, is exposed to oxygen, a chemical reaction occurs. Both the air and the water contain oxygen. When metal is exposed to air or water, a chemical reaction occurs that alters its physical properties. The iron will essentially oxidize, resulting in corrosion. By shielding metal parts and components with an external layer, passivation can protect them from corrosion.

Fundamentals

Oxygen has a strong affinity for metals.

The question is 'why'?

Oxygen

Oxygen is strongly electronegative with a strong desire/affinity for electrons.

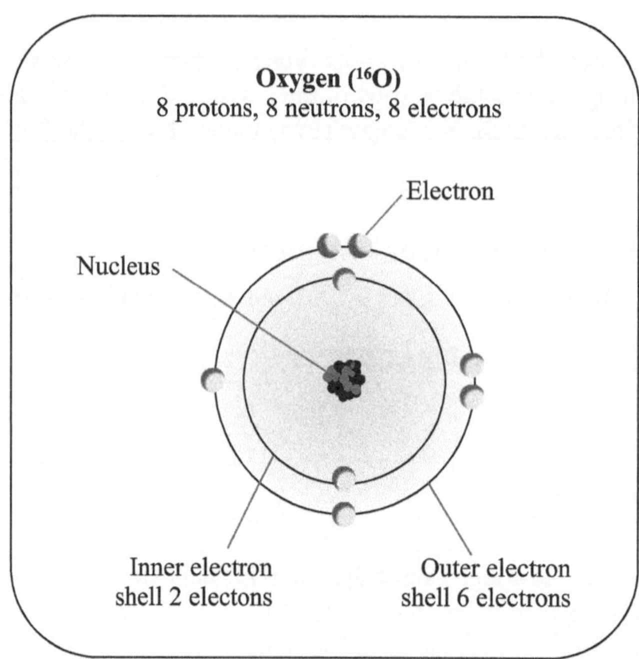

Figure 3

Oxygen is a very reactive element. It's a small atom. Its atomic number is eight. It has eight protons and eight neutrons in the nucleus and there are eight electrons distributed over two shells outside the nucleus. It is short of electrons to achieve a stable atomic structure and therefore it always looks for electrons to achieve a stable atomic structure and this is supported by its tiny size [see periodic table].

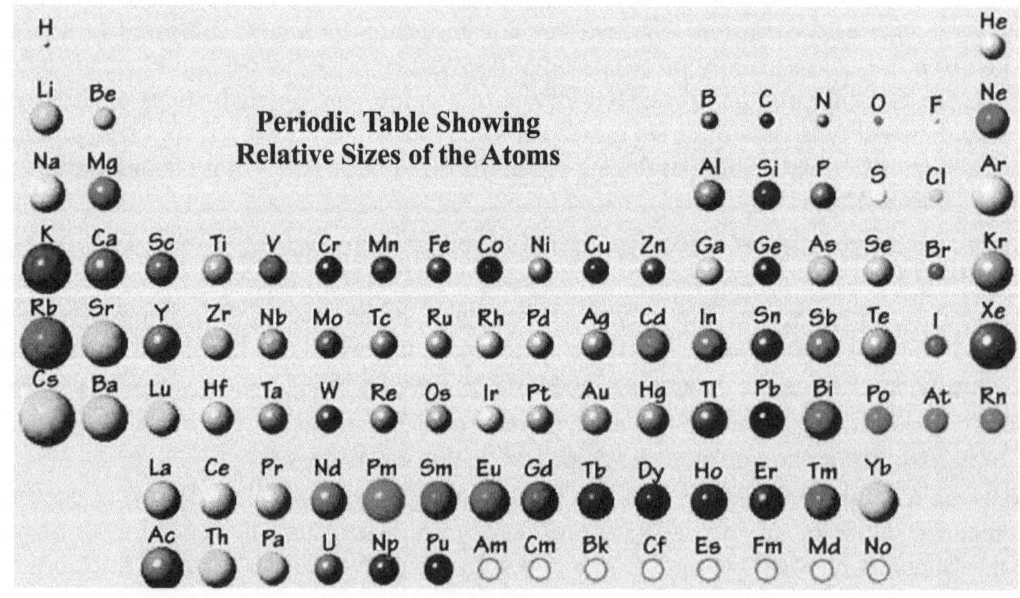

Figure 4

The oxygen atom, because of its small size with its outer shell electrons being very close to positive protons in the nucleus, has the strong attraction of protons to acquire electrons in the outer shell. This makes oxygen a very strong electronegative element next to fluorine and also makes it a very reactive element. It oxidizes almost every metal barring noble metals. It is a very strong oxidizing agent. By definition, an oxidizing agent is a substance that makes others lose electrons and get itself reduced by acquiring those electrons.

Metals

Metals are electropositive with aversion/dislike for electrons.

Opposite to oxygen, metals are much bigger in size with several layers of electrons outside the nucleus each repelling the other because of the same charge. This makes outer electrons of metals located far away from positive protons in the nucleus have practically no attraction for protons. Metals achieve their stable atomic structure by losing electrons, just opposite to oxygen.

Iron and Chromium Atom

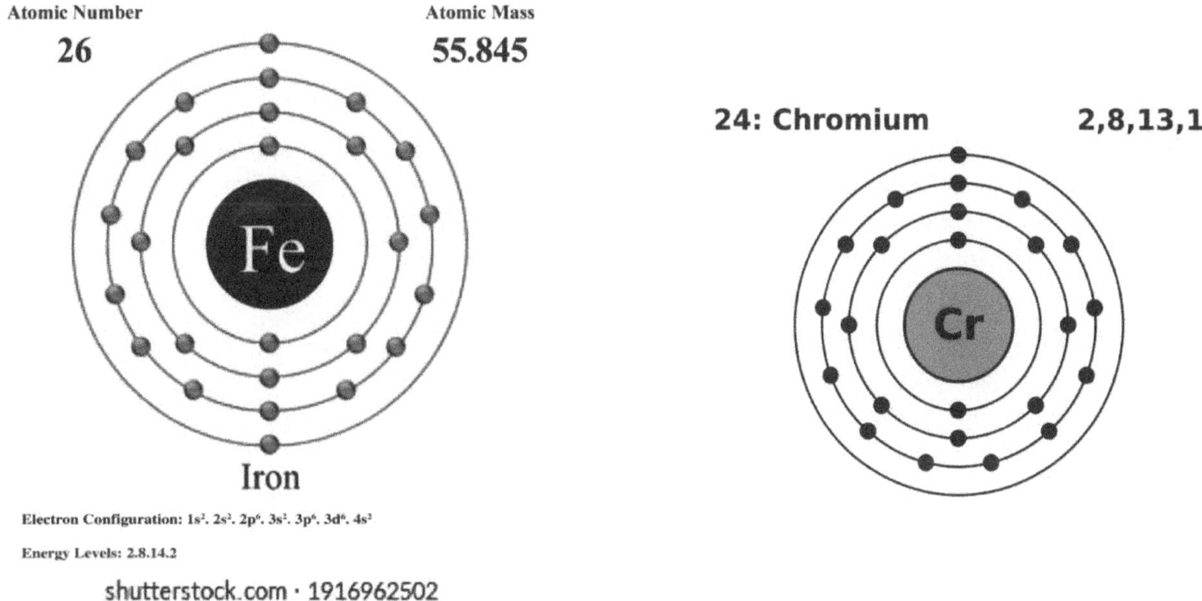

Figure 5

Key Properties

Iron-Atomic number 26, Electronegativity 1.8, hardness – 4.0 in mho scale.

Chromium-Atomic number 24, Electronegativity 1.6, hardness, 8.5 in mho scale.

What does it mean?

Iron has a greater tendency to lose electrons. Iron has a greater attraction for oxygen to form oxide. Let us take iron as an example. It is a big atom [see periodic table] Its atomic number is twenty-six. It has twenty-six protons in the nucleus and thirty neutrons. Positive protons in the nucleus are shielded by four layers of electrons making the outer electrons practically have no attraction for positive protons in the nucleus. Therefore, iron like other metals prefer to lose outer electrons to achieve a stable atomic structure. This is what makes metals electropositive.

Despite oxygen's strong affinity for metals to form oxides, there are only a few metal oxides that form a stable protective oxide layer on metals and protect them from corrosion.

Iron and Chromium Oxide Molecules

LHS image: Iron oxide molecule RHS image: Chromium oxide molecule [not on the scale].

Iron Oxide

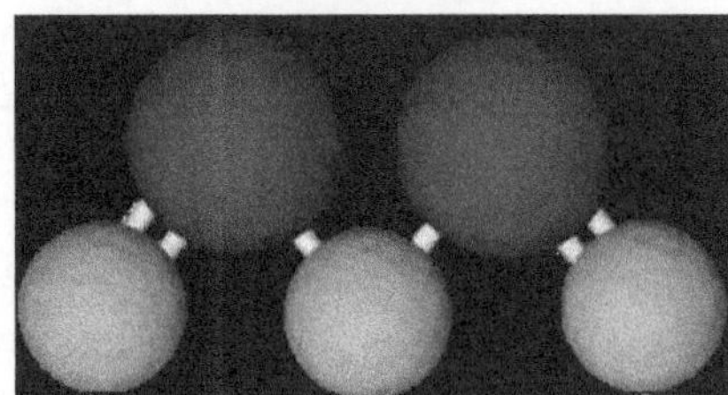

Chromium Oxide

Figure 6

Iron *vs.* Chromium

Summary

The chromium, as the images are showing has an atomic number of 24 compared to the atomic number of iron is 26. Therefore, iron has more affinity to attract oxygen to form oxides. Chromium is almost twice as harder as iron. Chromium oxide, CrO_3 is more compact and strongly bonded to oxygen. CrO_3 is more impermeable to oxygen than Fe_2O_3.

Mechanism of oxide passivation

Several factors can obstruct oxide growth on metal surfaces. The oxide-growth process occurs through either metal ion diffusion through the oxide or oxygen ion diffusion in the opposite direction. Several factors cause stress during the diffusion process. The higher the stress, the more fragile and porous the oxide deposit.

What does it Mean?

How and Why does Stress Develop on an Oxide Scale?

The generation of stress in an oxide scale can lead to scale cracking and spallation, [fragmenting the scale] affecting the protective oxide scale's maintenance. There are two kinds of stresses in oxide scales: growth stress, which occurs during the oxidation process, and thermal stress, which occurs as a result of the differential thermal expansion of the oxide scale and metal substrate.

Although the precise cause of the growth stress is complex and unknown, it is believed that the stress is strongly affected by the volumes of metal and oxide, the crystal structures of oxide and metal, and the growth mechanism of the oxide. When an oxide forms at the metal/oxide interface, the volume change due to the formation of the oxide can be expressed with the Pilling-Bedworth ratio (PBR).

Pilling-Bedworth Ratio (PBR)

PBR = Volume of metal oxide/Volume of metal in the oxide.

Since 1923, PBR has been used to predict the sign and magnitude of growth stress. It is generally accepted that when PBR >1, compressive stress develops in the oxide scale, whereas when PBR 1, tensile stress develops. The greater the difference in PBR from one, the greater the growth stress. However, it is also acknowledged that there is no direct relationship between PBR and the stress level on the scale. This means that the mechanisms of stress generation and release are complex, and the effects of PBR are not simple or well understood. Nonetheless, PBR has been frequently cited for explaining stress generation during oxidation, and it served as the foundation for the development of some recent models.

The Pilling-Bedworth ratio was established for use in the oxidation of metals. However, these are alloys that are widely used as high-temperature materials in practice.

Oxidation of Alloys

When an alloy is exposed to an oxidizing atmosphere at high temperatures, one or more of its elements will oxidize. *A* noble parent metal alloying with a base element and a base parent element alloying with a base element are the two types of alloys. Only *B* will oxidize in a binary alloy *A-B* composed of a noble element *A* and a base element *B*. If there are two base elements *A* and *B*, only *B* will oxidize if its concentration exceeds that of *A*.

Summary

If the PBR ratio is less than one, the oxide layer will be unprotective because the film that forms on the metal surface is porous and/or cracked.

Conversely, the metals with a ratio higher than 1 tend to be protective because they form an effective barrier that prevents the gas from further oxidizing the metal

General guidelines for PBR

RPB < 1: the oxide coating layer is too thin, likely broken, and provides no protective effect (for example magnesium)

RPB > 2: the oxide coating chips off and provides no protective effect (for example iron)

1 < RPB < 2: the oxide coating is passivating and provides a protective effect against further surface oxidation (for example aluminum, titanium, and chromium-containing steels).

Typical *P-B* Ratio

Provides strong protective oxide film:

Aluminium = 1.28

Chromium = 2.07

Iron (iii) oxide = 2.14

Advantages of Chromium

Chromium is a hard metal with a hardness twice of iron. Despite its high PBR which is due to essentially spreading oxide deposit film on the surface of the metal because it forms a very hard oxide deposit on the alloy surface. Chromium doesn't oxidize nearly as easily as steel. Chromium is passivated by oxygen, forming a thin protective oxide surface layer. This layer is a structure only a few atoms thick and very dense, preventing the diffusion of oxygen into the underlying material. The other advantages of chromium compared to iron are [1] it is less electropositive than iron. It has less affinity for oxidation to form oxide. The most important advantage of chromium is [2] it has only three oxides while iron has seventeen known iron oxides and oxyhydroxides of different volumes. The creation of so many different types and different molecular size oxides of different volumes generate much more stress on the metal surface than chromium leading to flaking out of the metal.

Key stresses that weaken oxide growth are as follows:

The geometry of Surfaces: The geometry of surfaces causes stress because the diffusion process can be modified, based on whether the surface is flat, concave, or convex. Thus, the geometry of a surface with the type of diffusing ion and its molar volume will generate stresses that can be tensile or compressive.

Thermal Stresses: These stresses are important as frequent heating and cooling happen in many industrial components. Thus, when a metal component is cooled from a rather high temperature, it can generate stresses that are generally compressive in nature.

When the oxide is unable to hold the stress, it needs to release these accumulated stresses. When the stresses are released, the protective oxide can form a microcrack or eventually lead to oxide failing.

WHY DISSOLVED OXYGEN IS CORROSIVE? A REVIEW

I doubt that any engineer, regardless of discipline, is unfamiliar with dissolved oxygen corrosion. I hope this note is useful to you. This should answer the majority of your questions about dissolved oxygen corrosion.

How many of us have ever wondered why oxygen is so corrosive to most metals? You'll be surprised to learn that all metal corrosion in your plants occurs between *pH* 4 and 9 is mostly due to oxygen.

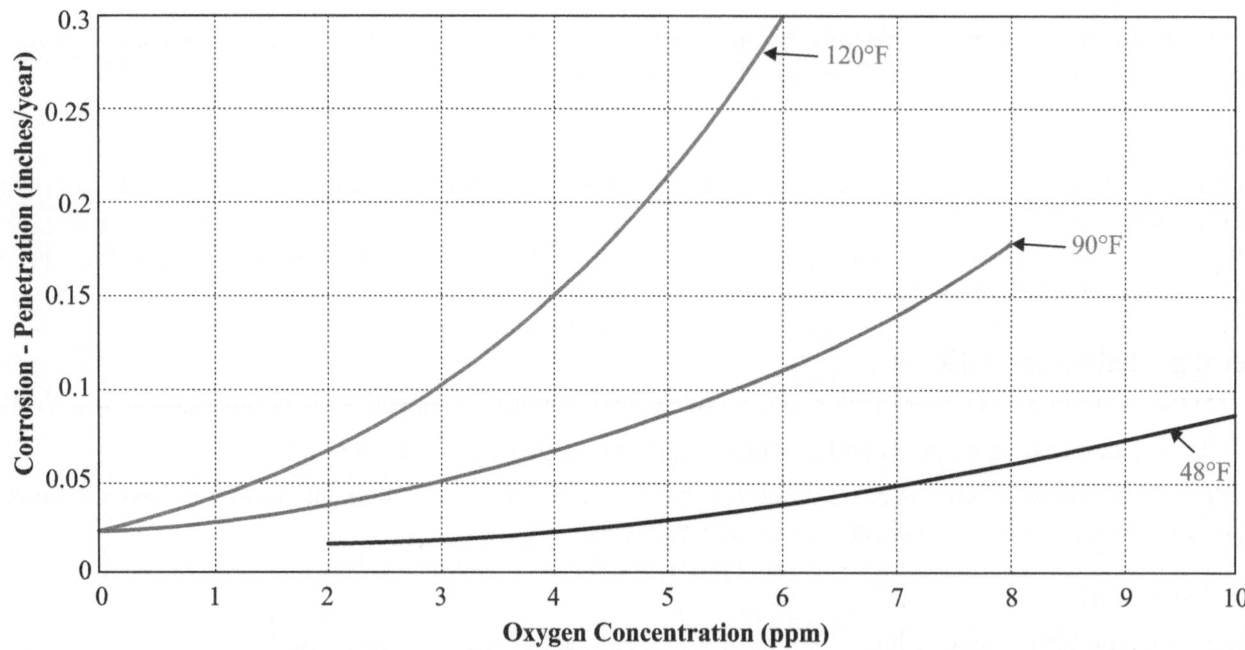

Figure 7: Low Carbon Steel Oxygen Concentration and Corrosion

Steel corrosion is strongly influenced by the amount of oxygen in the water as well as the temperature of the water. The diagram shows the effects of oxygen and temperature on low-carbon steel pipes with water flow.

What is Dissolved Oxygen?

The volume of oxygen dissolved in water is referred to as dissolved oxygen (DO). *DO* denotes the amount of free, noncompound oxygen (O_2) dissolved in water or other liquids. Water (H_2O) contains bonded oxygen, which does not contribute to dissolved oxygen levels.

Source of Dissolved Oxygen

Oxygen enters the water through photosynthesis in aquatic biology and oxygen transfer across the air-water interface. The amount of oxygen that water can hold depends on its temperature, salinity, and pressure. With decreasing temperature, gas solubility increases (colder water holds more oxygen). Gas solubility increases as salinity decreases (freshwater holds more oxygen than does saltwater). Both the partial pressure and degree of saturation of oxygen will change with altitude. Finally, as pressure decreases, so does gas solubility. As a result of the decrease in relative pressure, the amount of oxygen absorbed in water decreases as altitude increases.

Why does O_2 Attack Metal?

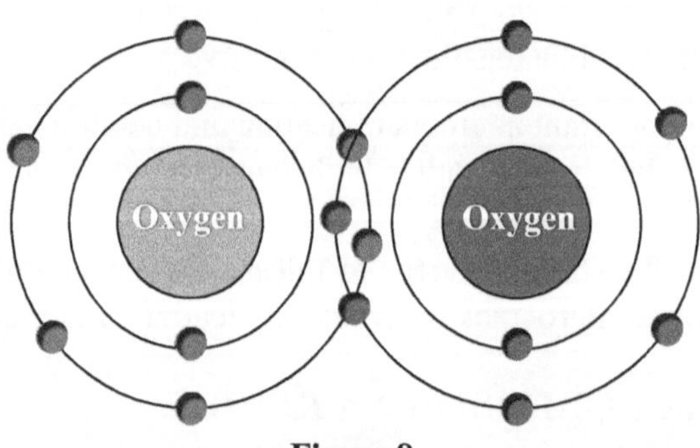

Figure 8

To understand the fundamentals, we must first understand the chemistry of the O_2 molecule. O_2 is a two-atom molecule. The atomic number of oxygen is eight in the periodic table. It consists of eight protons and eight electrons. Each O atom's valency shell contains six electrons. A stable structure is two electrons away. Each O shares two electrons from these four electrons, forming a double bond between two O atoms. This results in four free non-bonded electrons per oxygen atom or eight free non-bonded electrons per O_2 molecule. These electrons are a burden on the oxygen molecule the free non-bonded electrons of O_2 make oxygen very reactive. Because these electrons are so close to the protons, they cannot go away.

The O_2 molecule's only option is to find a partner who can share its electrons with oxygen and provide stability. Metals are an excellent oxygen target. Metals are much larger in size, with electrons in the outer shell are far away from the protons. This makes the metals electro-positive and susceptible to electron loss. This explains why oxygen has a such great affinity for metals.

Oxygen Corrosion Reaction

Corrosion reactions of metallic materials are of electrochemical nature, that is, the reaction can be divided into an oxidation reaction (anodic reaction) and a reduction reaction (cathodic reaction). In the case of corrosion, metal oxidation reaction (eqn [I]) is coupled with the reduction of species in the environment/electrolyte—the two most typical reactions taking place in aqueous environments are the reduction of protons (eqn [II]) or reduction of oxygen dissolved in water (eqn [III]) depending on the *pH*.

[I] $Me \rightarrow = Me^{n} + ne^{-}$

[II] $2H^{+} + 2e^{-} \rightarrow H_2$

[III] $O_2 + 2H_2O + 4e^{-} \rightarrow 4OH^{-}$

How does O_2 attack metal?

The most significant contaminant in terms of corrosion is dissolved oxygen (DO) from ambient air. During electrochemical corrosion, oxygen reacts with and removes the hydrogen ion film from the cathode surface, allowing the corrosion attack to continue. Corrosion rates in a closed vessel increase with temperature, emphasising the importance of removing *DO* from hot water systems and boilers. Steel and most other metals' corrosion rates can be expressed in terms of the amount of *DO* present in a *pH* range of about 5 to 9 (e.g., μm/y per mL of *DO* per litre of water).

What Influences Oxygen Attack

Effect of pH

At about *pH* 4.5, acid corrosion is initiated, overwhelming oxygen control. At about *pH* 9.5 and above, the deposition of insoluble ferric hydroxide [$Fe\ (OH)3$] tends to stifle the corrosion attack. In the range of *pH* 4 to *pH* 10, the corrosion rate of iron is relatively independent of the pH of the environment. In this *pH* range, the corrosion rate is governed largely by the rate at which oxygen reacts with absorbed atomic hydrogen, thereby depolarizing the surface and allowing the reduction reaction to continue. For *pH* values below 4.0, ferrous oxide (FeO) is soluble. Thus, the oxide dissolves as it is formed rather than depositing on the metal surface to form a film. In the absence of the protective oxide film, the metal surface is in direct contact with the acid solution, and the corrosion reaction proceeds at a greater rate than it does at higher *pH* values.

It is also observed that hydrogen is produced in acid solutions below a pH of 4, indicating that the corrosion rate no longer depends entirely on depolarization by oxygen, but on a combination of the two factors (hydrogen evolution and depolarization). For *pH* values above about *pH* 10, the corrosion rate is observed to fall as *pH* is increased. This is believed to be due to an increase in the rate of the reaction of oxygen with $Fe\ (OH)_2$ (hydrated FeO) in the oxide layer to form the more protective Fe_2O_3 (note that this effect is not observed in deaerated water at high temperatures).

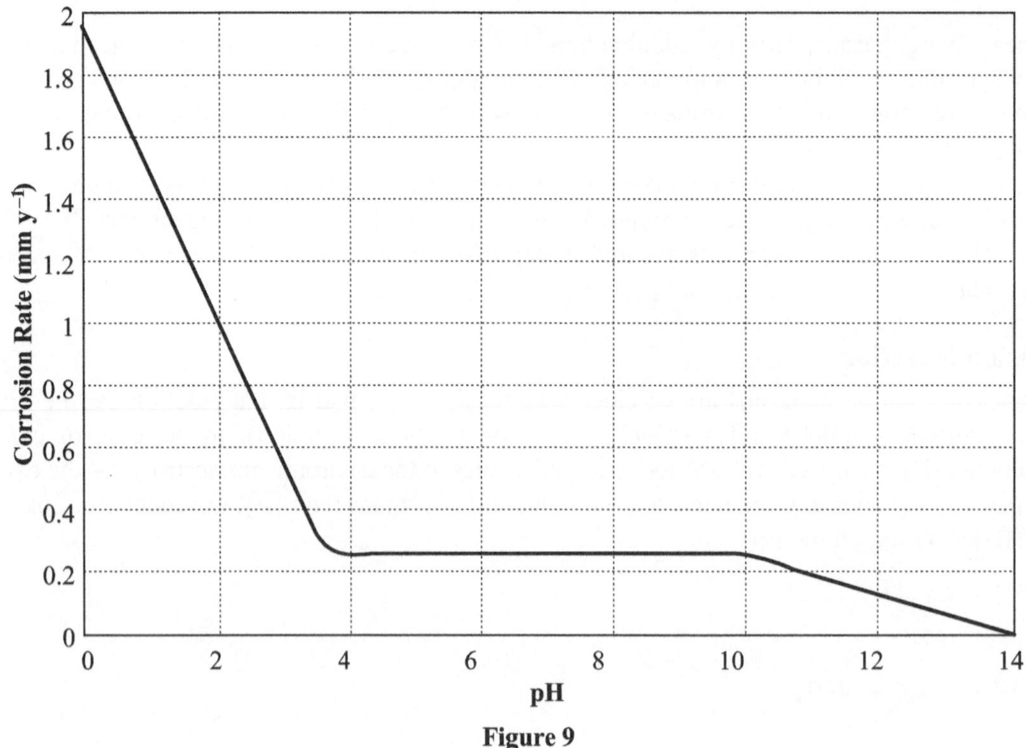

Figure 9

Effect of Temperature

A small amount of oxygen, up to about ten molecules of oxygen per million of water, is actually dissolved in water.

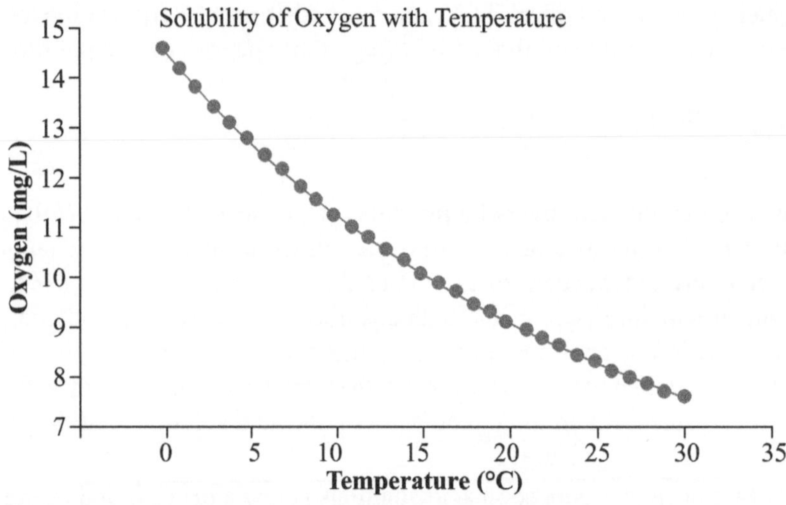

Figure 10: Temperature vs. Dissolved Oxygen

The maximum solubility of oxygen in water at 1 atm pressure (standard air pressure at sea level) ranges from about 15 mg/L at 0 degc to 8 mg/L at 30 degc-that is, ice-cold water can hold twice as much dissolved oxygen as warm water. Dissolved oxygen comes from the atmosphere and from photosynthesis by aquatic plants, and is depleted through chemical oxidation and respiration by aquatic animals and microorganisms, especially during the decomposition of plant biomass and other organic material.

Surface water, near the water-atmosphere interface and with sufficient light for photosynthesis, is generally saturated or even supersaturated with oxygen. Deeper water receives oxygen through mixing by wind, currents, and inflows. Mixing and aeration also occur at waterfalls and rapids. Dissolved oxygen can be reduced to very low levels during the winter months when water is trapped under ice.

FOULING: SOLID-LIQUID TWO PHASE FLOW

Fouling is the deposition of solids from a two-phase solid-liquid flow on solid surfaces. It is a big concern in industries because of its potential to cause big damage to the metal surfaces. The fouling materials can consist of either living organisms (biofouling) or a non-living substance (inorganic or organic). Fouling can be a potential source for under deposit corrosion of metal.

Fouling Deposition of Solids has No Solution

While scaling in a pipe can be controlled by inhibitors the fouling is unstoppable with suspended solids in a fluid flowing through a pipe. If the drag force < gravitational force, suspended particles will deposit. Since drag force increases or decreases with the velocity of the particle or the fluid the only control point is ensuring the recommended flow velocity in the pipe. Consequently controlling under-deposit corrosion, microbiological or inorganic is a big challenging job.

Cooling Tower Water is a Typical Case for Big Potential for Fouling

Generally, in solid-liquid two-phase flow like cooling tower water where the solids are not very high or it is not a slurry, we do not take into consideration of the unmeasurable gravitational forces of these solid particles while arriving at velocity. This, I would guess is a problem. We are more concerned with the flow velocity of two-phase liquid-vapor flow because of flow accelerated corrosion.

Fluid-Particle Interaction Force

The two dominant forces acting on a moving particle suspended in a flowing are [1] the buoyancy and [2] drag forces. Another force that is acting on the moving particle is the gravity force. The gravity force is usually not affected by the particle's velocity, whereas the drag force is a function of the particle velocity.

Buoyancy

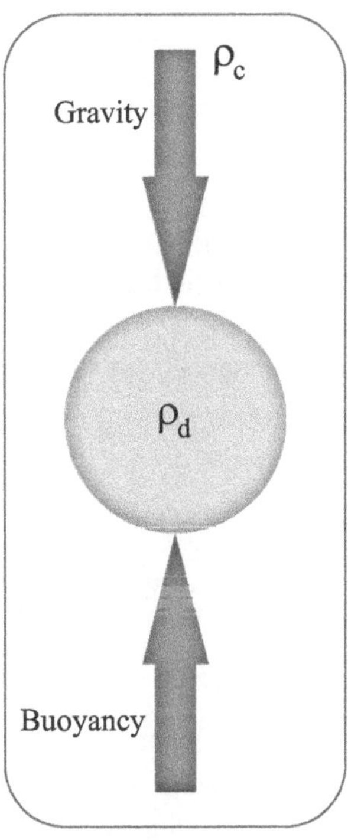

Figure 11

Buoyancy is the force that causes objects to float. It is the force exerted on an object that is partly or wholly immersed in a fluid. Buoyancy is caused by the differences in pressure acting on opposite sides of an object immersed in a static fluid. It is also known as the buoyant force.

Drag

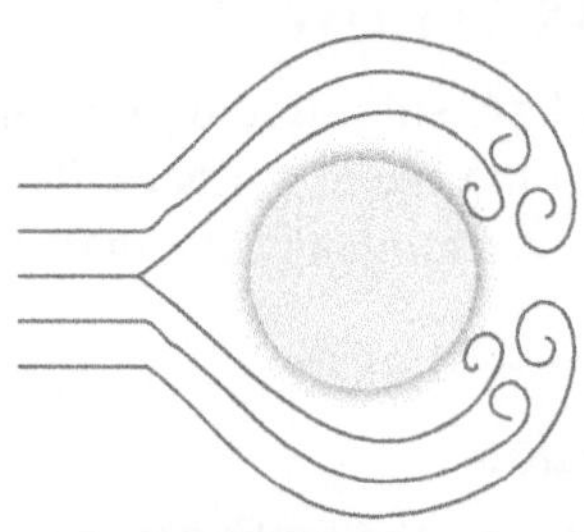

Figure 12

Drag is a force that tends to slow the movement of an object through a liquid or gas. As a moving object pushes the liquid or gas out of its way, the fluid pushes back on the object. This drag force is always opposite to the object's motion, and unlike friction between solid surfaces, the drag force increases as the object moves faster.

Settling Velocity

For a particle, at rest, no drag force will be exhibited, which causes the particle to accelerate due to the applied force. When the particle accelerates, the drag force acts in the direction opposite to the particle's motion, retarding further acceleration, in the absence of other forces drag directly opposes the gravity force. As the particle increases in velocity eventually the drag force and the applied force will approximately equate, causing no further change in the particle's velocity. This velocity is known as the terminal velocity, settling velocity or fall velocity of the particle.

Fouling in the water pipe and under deposit corrosion

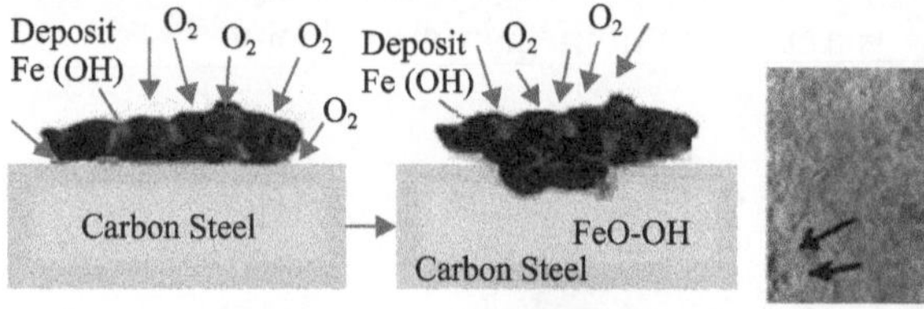

Figure 13

The journey of a suspended solid particle in a two-phase flow in a pipe

Imagine a suspended particle in water flowing in a pipeline. Two forces acting on it [1] upward frictional force *Fd* [2] a downward gravitational force, *g*, When $Fd > g$, the particle floats when $Fd < g$, the particle sinks on the pipe surface. This is the fundamental of fouling.

In turbulent flow, swirling of a fluid and reverse current generating eddies, keep suspended particles floating. Fouling begins typically in low-velocity areas in the pipeline network and then spreads.

Fouling can be best explained by Stoke's law.

Stoke's Law

$Fd = 6\eta\pi Rv$, *Fd* is the drag force, η is dynamic liquid viscosity, *R* is particle radius, *v* is the relative velocity of liquid and particle.

$Fd = 6\eta\pi Rv$, depends on [1] visc of liquid, [2] size of particle [3] relative liquid - solid velocity.

In a given system [1] and [2] are constant, *Fd*, then depends on *v*, the relative liquid-solid velocity.

An important point – it is the relative velocity between liquid and solid decides when a suspended particle will sink.

As *v* reduces, *Fd* reduces, and when $Fd < g$, particle deposits as foulant. Process continues.

BOILER ACID CORROSION

There can be both hydrochloric acid and sulfuric acid-induced corrosion in boilers.

Acid corrosion in boilers can be caused by magnesium chloride-containing cooling water that enters the condensate system.

This can occur, for example, via a power plant's seawater-cooled condenser. If such acid-forming salts occur in high local concentrations, hydrolysis produces small amounts of hydrochloric acid.

$$MgCl_2 + 2H_2O \rightarrow Mg(OH)_2 + 2HCl$$

Other chloride-containing contaminants can decompose and produce hydrochloric acid under certain conditions.

Even if the majority of the boiler water reacts alkaline, the acid does not need to be neutralized right away.

This is due to porous deposits, crevices, or poor boiler water circulation, which can prevent the water from mixing thoroughly. This allows for sharp decreases in pH at specific locations. The hydrochloric acid then attacks the steel in the following ways:

$$Fe + 2HCl \rightarrow FeCl_2 + H_2$$

A Word of Caution

"Magnesium chlorides sublime. They have sufficient vapor pressure to sublime. Therefore, the reaction to generate HCl can occur in the steam phase as well. Therefore, the solution lies in extremely strict control of steam purity, especially the conductivity."

A similar type of acid corrosion can occur in fire-tube boilers when sodium bisulfite is used for the chemical bonding of oxygen. When this is used in place of thermal deaeration, the boiler water will typically have an extremely high sulfate content. Under such circumstances, acid is most likely produced through hydrolysis.

Other sources of acid contamination in boiler water include the entrainment of regeneration acid in the makeup water or the presence of acid or acid-forming products in the condensation.

Finally, acid corrosion can occur when chemical cleaning treatments for boilers are not performed or inspected correctly.

GALVANIC CORROSION AND GALVANIC SERIES

There are five general types of corrosion: galvanic, stress cracking, general, localized, and caustic agent corrosion. Galvanic corrosion is extraordinarily common and occurs when two metals with different electrochemical charges are linked via a conductive path.

What is Galvanic Corrosion?

Galvanic corrosion occurs when two dissimilar metals are in contact with a solution. Both metals must be exposed to the solution. The driving force for galvanic corrosion is the electric potential difference between the two metals. Electrode potential is the voltage that the metal requires to oxidize or reduce.

Definition

Electrode potential: It is potential a metal requires to oxidize or reduce at standard condition, 1 M solution at 1 atm pressure and 298 deg K.

Galvanic corrosion depends on the charge on the metal or on the electronic configuration of the metal. Na^+ ion with a single positive charge has a different potential than Ca^{++} ion. Potential difference increases as the distance between the metals in the galvanic series increases.

Galvanic Corrosion is a Bimetallic Corrosion

The corrosion rate of the more active (anodic) metal increases and the corrosion rate of the more noble (cathodic) metal decreases. It is an electrochemical oxidation-reduction (redox) process. The metal having a lower value of electrode potential will oxidize (anodic reaction) and the metal having a higher value of electrode potential (nobler) will provide cathodic reaction (reduction) on its surface and the metal will be protected from corrosion. The metal which has lower electrode potential will corrode.

Table 2

Reduction Reaction	*E°/V*
$Au^+ + e^- \rightarrow Au$	+1.69
$Au^{3+} + 3e^- \rightarrow Au$	+1.40
$2Hg^{2+} + 2e^- \rightarrow Hg_2^{2+}$	+0.92
$Hg^{2+} + 2e^- \rightarrow Hg$	+0.86
$Ag^+ + e^- \rightarrow Ag$	+0.80
$Hg_2^{2+} + 2e^- \rightarrow 2Hg$	+0.79
$Fe^{3+} + e^- \rightarrow Fe^{2+}$	+0.77
$Cu^{2+} + e^- \rightarrow Cu^+$	+0.16
$Cr^{3+} + 3e^- \rightarrow Cr$	–0.74
$Na^+ + e^- \rightarrow Na$	–2.71

How does one know in a bimetal contact which metal will corrode?

You would require a standard reduction potential chart as shown above. The standard cell potential is the potential difference between the cathode and anode. Electrode potential is the potential a metal requires to oxidize or reduce at standard condition, 1 M solution at 1 atm pressure and 298 deg K. Therefore, the standard electrode potential is commonly written as standard reduction potential. The chart shows the potential required to reduce some of the elements compared to hydrogen. The electrode potential of hydrogen is taken as zero. The chart is showing all reduction reactions since there is a gain of electrons. All E0 [electrode potential] values on the right side of the chart area are relative to the hydrogen potential and not absolute values. The negative sign of the zinc E° value shows that it releases electrons more readily than hydrogen does. The positive sign of the copper E° value shows that it releases electrons less readily than hydrogen. Loss of electron is oxidation and gain of the electron is reduction. Therefore, zinc is more oxidizable than copper. The algebraic difference between the electrode potentials of the two metals determines the driving force of galvanic corrosion. Higher the electrode potential of an element more stable the element is – does not lose electron easily therefore, higher the electrode potential of an element nobler the metal is. Another way to express, higher the electrode potential of an element more cathodic the metal is and not corroded. Another way to express is higher the electrode potential of an element more difficult to oxidize it. Silver is the noblest metal in this chart and magnesium is the least. When two metals are in contact, the less noble metal will experience corrosion while the more noble metal will be protected. As you go up the chart elements are more oxidizing with fluorine at the top. As you go do down the chart, elements are more reducing with lithium at the bottom.

What is the Galvanic Series?

The galvanic series which is derived from electrode potential determines the nobility of metals and semi-metals [nonmetals]. When two metals are submerged in an electrolyte, while also electrically connected by some external conductor, the less noble (base) will experience galvanic corrosion.

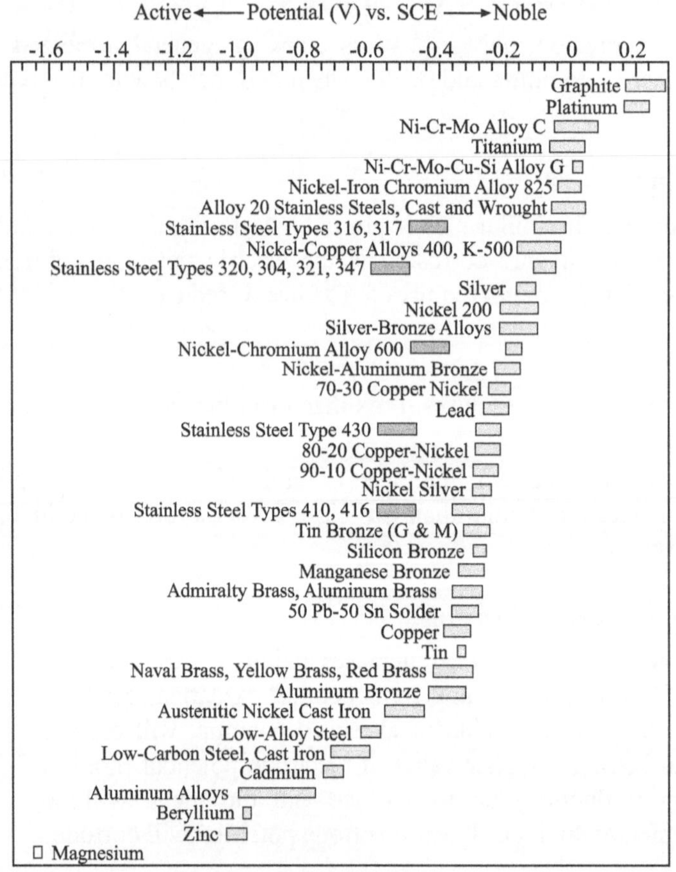

Figure 14

The table on the left side is the Galvanic series. You can see Graphite which is essentially carbon and a nonmetal at the top. Next comes platinum, then Ni-Cr-Mo alloy, and below this is titanium. SS 316 is much below in the Galvanic series. In any bimetal contact between SS 316 and Ti, since Ti is nobler than SS 316, SS 316 would get corroded.

Zinc is at the bottom. Zinc is the least noble metal. The less noble metal is the one with a lower (that is, more negative) electrode potential than the nobler one, and will function as the anode and will corrode.

Example: Iron – Zinc Bimetal Contact

The Standard electrode potentials of the metals: $E_0zn = -0.763$ V, and for iron $E_0fe = -0.44$ V.

Zn is more negative with more negative potential than *Fe*. Therefore, *Zn* will oxidize and act as an anode, and *Fe* will act as a cathode and will be protected.

Anodic reaction: $Zn = Zn_2^+ + 2e^-$ (corrosion of zinc). Zinc is anode.

Cathode Reaction

The electrons are given up by the anode flow to the cathode (iron) where oxygen is reduced to:

OH^- ions, $\frac{1}{2}O_2 + H_2O + 2e^- = 2OH^-$

Zinc combines with OH^- ions and precipitates zinc to zinc hydroxide on cathodic *Fe* surface. Iron is protected.

$Zn^{++} + 2OH^- \rightarrow Zn\,(OH)2$

$Zn\,(OH)2$ becomes a barrier layer on the cathode.

CORROSION UNDER INSULATION

Corrosion under insulation [CUI] is a serious and costly problem for many industries. It is a common issue in the refinery, pharmaceutical, maritime/marine, offshore, and petrochemical industries. It can cause equipment failure, leaks, downtime, and even explosions. The National Association of Corrosion Engineers (NACE) conducted a survey in 2008 and discovered that more than half of all facilities surveyed faced significant corrosion challenges.

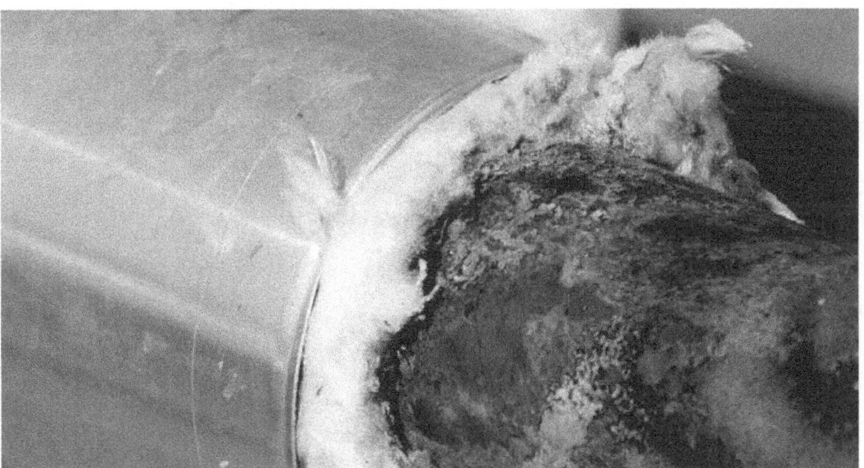

Figure 15

What Exactly is Corrosion Under the Insulation?

CUI is a Special Type of Corrosion

In general, as the amount of dissolved oxygen decreases with increasing temperature, so does the corrosion rate. When it is covered by insulation, however, the moisture closes the system and retains the dissolved oxygen. In some cases, chlorides and acids may be present in the insulation, which greatly promotes corrosion of the underlying surface when wet. In the presence of water, chlorides (or other halides) are transported to the hot surface.

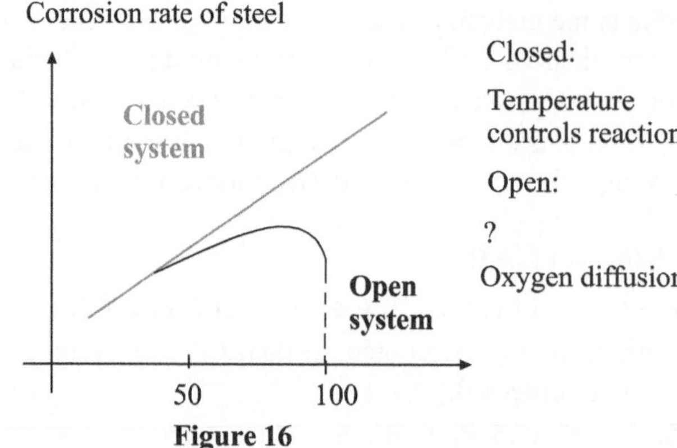

Figure 16

Generally, corrosion is a type of oxidation that happens when a material loses electrons to oxygen molecules in the air or water around it. This material is usually a metal, such as carbon steel or stainless steel, in the context of industrial manufacturing.

Corrosion under insulation (CUI) is oxidation on the surface of a metal pipe or vessel that has been insulated. The insulation material traps moisture beneath it, exposing the metal to oxygen in the water. This continuous contact aggravates the problem by increasing the rate of corrosion.

Why Does Corrosion Occur?

Moisture

Moisture exists naturally in the air as water vapor. The vapor undergoes a phase change and condenses into a liquid at the proper temperatures and pressures. This liquid can become trapped beneath the insulation material on the pipe's surface. Condensation on the metal surface can start and accelerate the corrosion process.

Oxidation

When atoms on the surface of a material (usually metal) lose electrons to oxygen in the surrounding air or water, this is referred to as oxidation. The oxygen molecules absorb the electrons, resulting in the formation of an "oxide layer" with the metal. This thin layer may offer some natural protection against future reactions, but the damage has already been done. Oxidation weakens materials, making them more prone to breaks and fractures when subjected to additional stress.

Mechanism

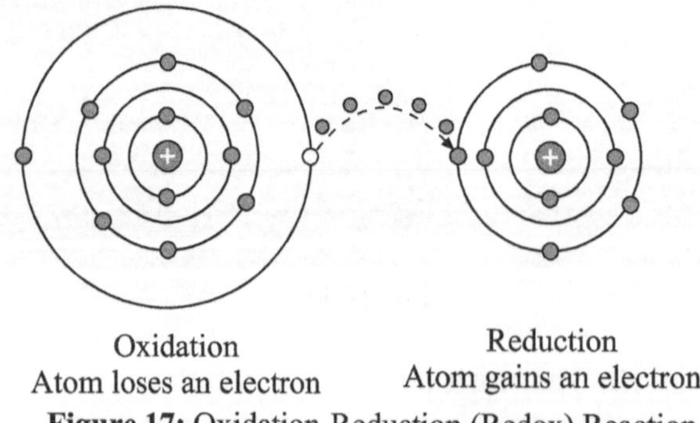

Figure 17: Oxidation-Reduction (Redox) Reaction

A corrosion reaction just needs moisture and oxygen on a metal surface.

A corrosion cell forms on the metal surface by a potential difference between two points. You just need is 100 mV potential difference between two points on an iron surface to set up a corrosion cell to create an anode and cathode. Generally, this potential difference arises from any defects on a metal surface or due to the presence of impurities. At the anode, the metal oxidizes by losing electrons and at the cathode, the oxygen in presence of water reduces by capturing those electrons. Electrons flow from the anode to the cathode. The metal surface acts as a conductor.

Process of Corrosion

Reaction at Anode

At the anode, the metal loses electrons and oxidized.

$$Fe = Fe^{++} + \text{electron}$$

Electron uses a metal plate as a conductor to travel to the cathode.

Reaction at Cathode

$$O_2 + e^- + H_2O \rightarrow \text{Oh}-$$ [this is not a balanced equation]

At the cathode, the oxygen in presence of water reduces to OH [hydroxide] ions. Water aids in the reduction of oxygen.

This is the Role of Water

Finally, *Fe* ions produced at the anode and *OH* ins produced at the cathode combine and form various iron oxides and hydroxides which are collectively called rust.

Stainless Steel CUI

The SS Mechanism CUI:

Step by step

Soluble chlorides leach from insulation material into the moisture

The chlorides are concentrated as a result of water evaporation.

The chloride concentration reaches a critical level, resulting in stress corrosion cracking (SCC) (The critical chloride concentration is temperature and alloy-type dependent.)

In addition, the insulation material may contribute to CUI in the following ways:

Creates a crevice for water retention

May absorb water

May leach contaminants (e.g. chlorides) that cause ESCC.

The presence of leachable chlorides causes problems with stainless steel. This is especially common when the insulation contains moisture and the operating temperature of the piping or equipment is between 60 and 175 degc. Furthermore, metal temperatures ranging from 50°C to 175°C are susceptible to external stress corrosion cracking.

Thermodynamics

$$2Fe + O_2 = 2FeO$$

It is an exothermic reaction. Therefore, it generates positive entropies. Since it is an exothermic reaction, the enthalpy, H of the reaction is negative.

$$dG = dH - TdS$$

The Gibbs free energy, *G* is negative. Therefore, the corrosion is a spontaneous forward reaction once it starts.

Corrosion Prevention

Maintenance is critical to lowering corrosion costs and avoiding long-term damage. Thermally insulated systems must be given special consideration in environments with high humidity or spill risk.

Preventing Water Ingress

Moisture trapped beneath the insulation material is a major contributor to CUI. However, preventing moisture ingress can be challenging. Wrapping, clamping, taping, or otherwise attaching conventional insulation materials to piping systems is required. Moisture can easily enter small spaces. Sealing every section of pipe can be difficult, especially in complex piping systems.

Usual Practices

Jackets with Insulation

Insulated piping systems may include a protective jacket wrapped around the insulation material. This jacket adds an extra layer to help keep moisture out. PVC jackets may suffice in less demanding conditions. Aluminum or stainless steel jacketing provides greater protection in more demanding applications.

Additional Material Layering

In some cases, additional materials are incorporated into the insulated piping system. Wrapping aluminum foil around the pipe and the insulation material, for example, can help prevent pitting and stress corrosion cracking

FLOW ASSISTED CORROSION: A THREAT TO POWER PLANTS

Flow-Accelerated Corrosion, also called erosion-corrosion, is a wear mechanism that results in wall thinning in piping systems and process components. This is a well-known source of problems in fossil fuel and nuclear power plants. known as FAC. This kind of corrosion can occur when carbon steel piping and its components become degraded through the presence of steam water or flowing water that has low-dissolved oxygen. The protective oxide layer consists mostly of magnetite and where the flowing medium is water. Its wear rate is greater than the usual corrosion rate since the dissolution of the magnetite by the flowing water can accelerate the thinning significantly.

There is a great surprise, unlike usual corrosion in FAC increasing the oxygen concentration in water reduces the FAC rate.

Flow-assisted corrosion or flow-accelerated corrosion (FAC), happens when the protective oxide layer on the metal surface is dissolved by fast-flowing water. The bare metal surface oxidizes to Fe++ [iron] ion at the anode. Anode gets created on the pipe surface due to the potential difference between the bare and unaffected metal surfaces. Fe^{++} is reduced to iron oxide at the cathode [unaffected metal surface] to recreate the oxide layer. This process goes on. The thinning of the metal surface continues. A typical anodic reaction is an anodic oxidation: Iron is oxidized to Fe++, ferrous ion at the anode, $Fe = Fe^{2+} + 2e^{-}$, at cathode the reduction reaction occurs when Fe^{+2} reduces to hydroxide and iron oxides [rust]. [Details I guess not necessary for this post]

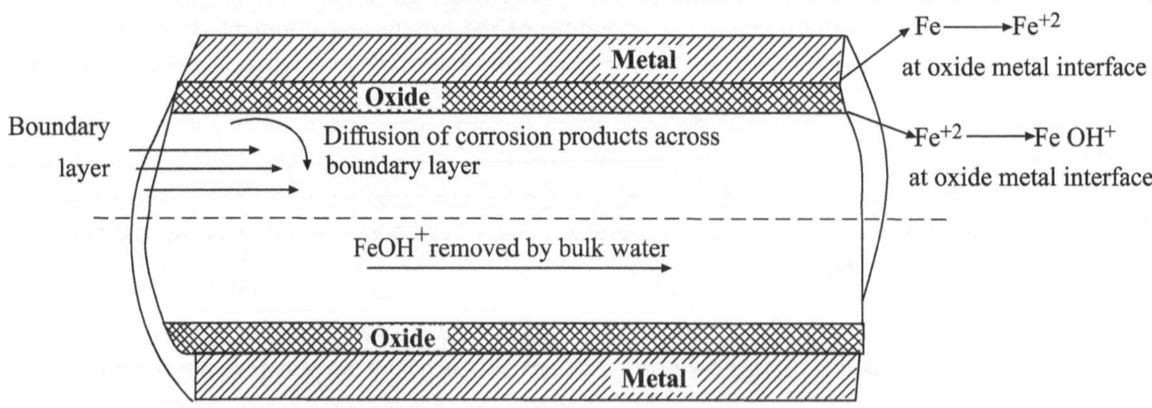

Figure 18: Visual Explanation of How FAC Sets in a Pipeline

FAC rate depends on flow velocity

It affects carbon steel in particular carrying deoxygenated ultra-pure water or wet steam. FAC does not happen in SS lines. Even a small amount of dissolved oxygen in water can prevent FAC. FAC happens when water is reducing in nature. Increasing pH can significantly reduce the FAC rate.

FAC is not erosion: Flow accelerated corrosion (FAC) is distinct from erosion and is primarily an electrochemical corrosion process aided by chemical dissolution and mass transfer. FAC has to be distinguished from erosion-corrosion because the fundamental mechanisms for the two corrosion modes are different. FAC does not involve impingement of particles, bubbles, or cavitation which cause the mechanical (often crater-like) wear on the surface. In contrast to mechanical erosion, FAC involves the dissolution of normally poorly soluble oxide by combined electrochemical, water chemistry, and mass-transfer phenomena.

Various dimensions of FAC

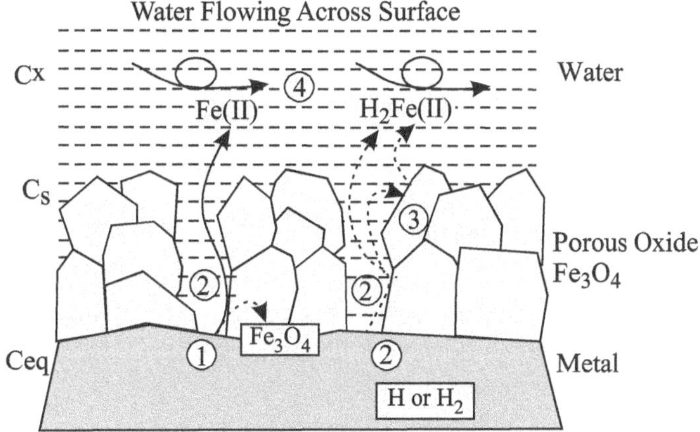

Figure 19: Grain of Carbon Steel

Mechanism

Two steps [1] High flow velocity uprooting protective metal oxide from pipe surface and [2] Iron corrosion begins at metal oxide rupture bare metal surface by red-ox reaction explained above the surface to recreate oxide. This pipe wall thinning goes on till the metal fails. FAC degradation mechanism results in thinning of large areas of piping and fittings that can lead to sudden and sometimes catastrophic failures, as well as a huge economic loss.

Different Pathways

If your steam distribution system generates two-phase flow below 130 degc and pH < 9, you are very much within the domain of FAC

1. **Flow conditions:** Both single- and two-phase flow conditions, with water or water-steam mixture as the flowing liquid, at a temperature of > 95 degC, and flow velocity is greater than zero – Typical for fossil power plant.
2. **Chemistry condition:** the flowing liquid should be such that a potential difference exists between the liquid and the carbon steel pipe wall. This difference will be responsible for the dissolution of the protective oxide layer in the flowing stream. A high protective oxide layer [called magnetite] solubility and subsequent rapid removal of the protective oxide from the pipe surface, is facilitated by either demineralized and neutral water or slightly alkalinized water under reducing conditions.
3. **Material:** the pipe material must be carbon steel or low-alloy steel. The general practice recommends that for a well-designed system, FAC will be effectively inhibited if steel components are made to contain at least 0.1% chromium.
4. **Flow Geometry:** FAC has been observed to occur downstream of flow-restricting or redirecting geometries like an orifice, sudden contraction, expansion, elbows, reducers, etc.

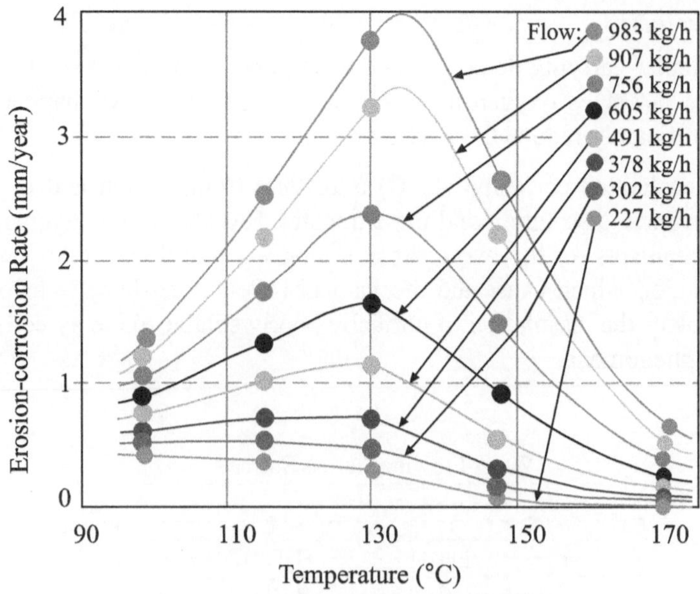

Figure 20: Effect of Temperature on FAC

Effect of Temperature: General Understanding

Effect of temperature on the flow accelerated corrosion rate of carbon steel in deoxygenated ammonia AVT power plant water at *pH*9.04. FAC peaks at 130 degc. The maximum FAC rate is in a temperature range of 140–160°C for single-phase flows and at about 180°C for two-phase flows. For temperatures below 50°C and above 150°C, the rate is an order of magnitude less than those at the peak temperatures.

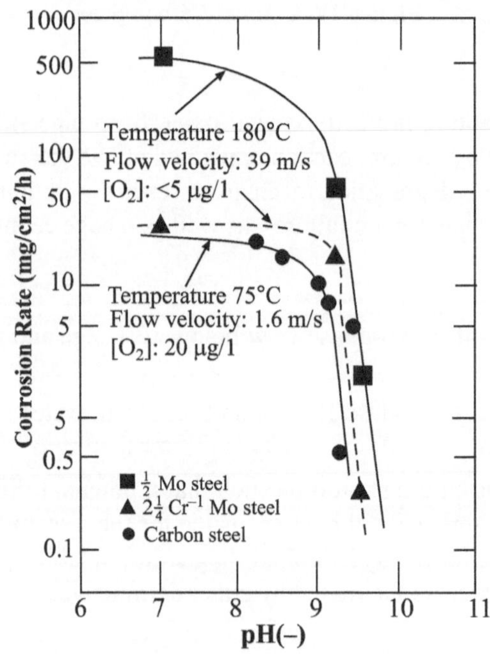

Figure 21: Effect of pH and Temperature on FAC

Effect of pH: General Understanding

Increasing the pH reduces the FAC rate. As seen from the figure, increasing the pH from 8.0 to 9.0 reduces the FAC rate significantly.

Effect of oxygen concentration in water: General understanding

Increasing the oxygen concentration in water reduces the FAC rate. This relationship is particularly strong for the low values of oxygen. For values greater than 150 ppb, the FAC rate is independent of the oxygen concentration.

SWEET CORROSION: SUBSEA CORROSION OF CARBON STEEL BY CO_2

CO_2 is made up of one carbon atom and two oxygen atoms. It is a corrosive compound that can be found in natural gas, crude oil, condensate, and produced water. It is one of the most common corrosion environments in the oil field industry. CO_2 corrosion is accelerated in the presence of oxygen and organic acids, both of which can dissolve iron carbonate scale and prevent further scaling.

How does CO_2 Become Corrosive?

Carbon dioxide is a weak acidic gas that, when dissolved in water, becomes corrosive. However, before CO_2 becomes acidic, it must hydrate to carbonic acid H_2CO_3, which is a relatively slow process. When carbonic acid comes into contact with steel, it lowers the *pH* of the water and causes corrosion.

Where is CO_2 Corrosion Most Common?

CO_2 corrosion is most common in flowing wells, gas condensate wells, water condensate areas, tanks filled with CO_2, saturated produced water, and pipelines, which corrode at a slower rate due to lower temperatures and pressures. CO_2 corrosion is accelerated in the presence of oxygen and organic acids, both of which can dissolve iron carbonate scale and prevent further scaling.

Types of CO_2 Corrosion

The maximum dissolved CO_2 concentration in water is 800 ppm. Uniform corrosion, pitting corrosion, wormhole attack, galvanic ringworm corrosion, heat-affected corrosion, mesa attack, raindrop corrosion, erosion-corrosion, and corrosion fatigue are the most common types of corrosion when CO_2 is present. The presence of carbon dioxide usually indicates that there is no H_2 embrittlement.

CO_2 Corrosion Rate

CO_2 corrosion rates outperform those of carbonic acid alone. Corrosion rates in a CO_2 system can be extremely high (thousands of mils per year), but they can be effectively inhibited. In the CO_2 system, velocity effects are extremely important; turbulence is frequently a critical factor in pushing a sweet system into a corrosive regime. This is because it either prevents the formation of or removes a protective iron carbonate (siderite) scale.

Corrosion products

Iron carbonate (siderite, $FeCO_3$), iron oxide, and magnetite are examples of CO_2 corrosion products. Colors of corrosion products range from green to tan to brown to black. Under certain conditions, this can be beneficial. Scale can be dissolved. Elevated temperatures, increased pH as occurs in bicarbonate-bearing waters, and a lack of turbulence favor the formation of a protective scale, allowing the scale film to remain in place. Turbulence is frequently a deciding factor in the formation or retention of a protective iron carbonate film. Iron carbonate is not electrically conductive. As a result, galvanic corrosion is impossible. Corrosion occurs where the protective iron carbonate film is absent and is fairly uniform across the exposed metal. When carbonic acid is formed, crevice and pitting corrosion occur. CO_2 (Carbon Dioxide) Carbon dioxide can also cause embrittlement, resulting in stress corrosion cracking.

Predictions for Corrosion

When liquid water is present, CO_2 corrosion of carbon steel used in oil production and transportation is influenced by a large number of parameters, some of which are listed below:

- Temperature;
- CO_2 partial pressure;
- Flow (flow regime and velocity);
- pH;
- Concentration of dissolved corrosion product ($FeCO_3$);
- Concentration of acetic acid;
- Water wetting;
- Metal microstructure (welds).

The precise impact of these parameters is still unknown, and some of them are inextricably linked. A minor change in one of them can have a significant impact on the corrosion rate.

Various prediction models have been developed and are used by various businesses. The de Waard *et al.* model (Shell), CORMED (Elf Aquitaine), LIPUCOR (Total), and a new electrochemically based model developed at IFE are among them. The corrosion models frequently overestimate the corrosion rate of carbon steel due to the complexity of the various corrosion-controlling mechanisms involved and a built-in conservatism.

Prediction of Corrosion Using a Shell Model

In the oil and gas industry, the Shell model for CO_2 corrosion is most commonly used. The de Waard equation serves as the foundation for the model. The model employs correction factors to quantify the impact of environmental parameters and corrosion product scale formation under different conditions. However, the model's first version was published in 1975, and it has been revised several times to make it less conservative by incorporating new knowledge and information.

The Corrosion Rate in Pipelines

CO_2 corrosion rates in carbon steel pipelines can be calculated using industry-accepted equations that ideally combine contributions from flow-independent kinetics of the corrosion reaction at the metal surface with contributions from flow-dependent mass transfer of dissolved CO_2.

The corrosion rate determined by the original formula and its correction factors is independent of liquid velocity. To account for the effect of flow, a new model was proposed that incorporates the effects of mass transport and fluid velocity via a resistance model:

$$Vcr = 1/[I/Vr + 1/Vm]$$

where Vcr is the corrosion rate in mm/year, Vr is the flow-independent contribution, denoted as the reaction rate, and Vm is the flow-dependent contribution, denoted as the mass transfer rate.

In multiphase turbulent pipeline flow, Vm depends on the velocity and the thickness of the liquid film, whereas Vr depends on the temperature, CO^+ pressure, and pH. For example, for pipeline steel containing 0.18% C and 0.08% Cr, the equations for Vr and Vm for liquid flow in a pipeline are:

$$\text{Log } Vr = 4.93 - 1119/[\text{temperature} + 273] + 0.58 \times \log(\text{p } CO_2)$$

where the temperature is the pipeline fluid temperature in °C, and the partial pressure pCO_2 of CO_2 in the bar.

AMINE CORROSION

H_2S [aq] $\rightarrow H^+ + HS^-$

It is an interesting subject. Essentially, there are two important steps. [1] HSS [high stable salts chelate on the metal surface and weaken the protective barrier of sulfide and carbonate layers. [2] The proton from de- protonated $H_2S = H^+ + HS^-$ does the rest.

Neither, amine is corrosive nor HSS is corrosive.

There are couple of important points

- Amine is not corrosive.
- During the operation of amine units, the solvent may degrade due to reactions with contaminants such as oxygen. As a result, corrosive species such as oxalates, formates, and acetates, which are also known as Heat Stable Salts (HSS), are formed.
- Several studies have shown that these degradation products are not corrosive and make excellent chelates. They can easily form complexes with iron. As a result, they reduce the stability of sulphide or carbonate protective layers formed on steel surfaces.
- Corrosion rate = f (acid gas concentration, temperature, velocity, HSS anion concentration).
- The key to understanding the acid attack is understanding the acid. First, H_2S enters the aqueous phase and is surrounded by water molecules. The H_2S then undergoes a two-step reaction sequence, losing an H^+ to form H^+ and HS^- followed by losing the other H^+ from the HS^- to form H^+ and $S =$.
- It is the proton that does the corrosion.

Brief description of amine plant

Traditional users of amine systems include the oil and gas and LNG industries. Another important amine user is Carbon Capture and Storage (CCS) technology, which can be associated with power generation or the production of chemicals and renewable fuel sources like hydrogen and biomethane. These applications require the removal of CO_2, H_2S, or both, from their gas streams. Purification may be required due to emission limits for environmental compliance or to protect downstream processing units. LNG transfers also have CO_2 and H_2S content specifications to prevent pipeline corrosion and to ensure a minimum heating value of the product.

Amine plants are made up of two basic units [see figure]. [1] the absorber, which removes unwanted constituents, and (2) the regenerator, which reactivates the solvent solution. In the absorber, gas flows up the column and amine cascades down, resulting in close contact. Sweet gas escapes from the top, while rich amine escapes from the bottom. Heat and steam stripping drive off the hydrogen sulphide and carbon dioxide in the regenerator. The absorber receives the lean amine solution.

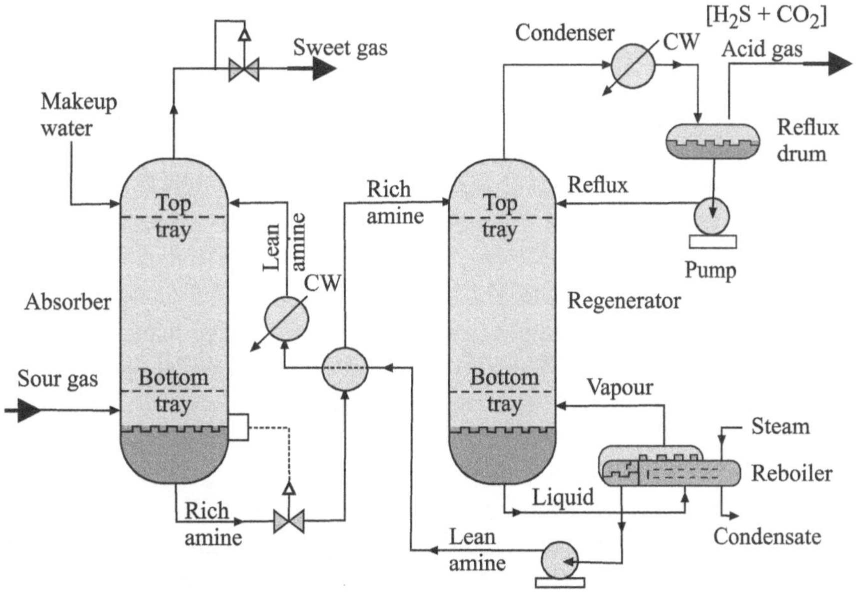

Figure 22

As a result, the amine operates in a nearly closed loop, continuously recycling and reusing itself. Top-up amine and make-up water compensate for losses at the regenerator's top.

The top of the absorber is filled with recycled amine. It is known as lean amine because it contains no CO_2 or H_2S. Simultaneously, process gas is rising from the tower's inlet at the bottom. In the absorber, packing or trays ensure maximum contact between the gas and the amine for CO_2 and H_2S absorption. The amine leaving the bottom of the tower as explained above is rich amine because it contains CO_2 and H_2S.

This rich amine is passed through a regenerator, where contaminants are removed. Clean process gas exits the absorber overheads for further processing or transfer, whereas acid gas (rich in CO_2 and H_2S) exits the regenerator overheads and is either combusted on-site or transported off-site to underground storage.

Complexities in Amine Plant

The complexity of amine systems makes them prone to corrosion in a variety of ways.

Basics

Generally, amines are not corrosive; however, during the operation of amine units, the solvent may degrade due to reactions with contaminants such as oxygen. As a result, corrosive species such as oxalates, formates, and acetates, which are also known as Heat Stable Salts (HSS), are formed. Corrosion problems may also be affected by other factors such as alkanolamine type and concentration, solvent temperature, or CO_2 loading.

The basic reactions between CO_2 and amines are given below.

Carbon dioxide reacts with amines by the following reactions where R designates an organic group. These reactions can lead to the formation of carbamates (R.1), bicarbonates (R.2) and carbonates (R.3):

$$2R2NH + CO_2 \Leftrightarrow R2NH_2^+ + R2NCOO^- \quad [R1]$$

$$R3N + CO_2 + H_2O \Leftrightarrow R3NH^+ + HCO_3^- \quad [R2]$$

$$H_2O + HCO_3^- \Leftrightarrow H_3O^+ + CO_3^- \quad [R3]$$

This reaction occurs only with primary and secondary amines, which possess a labile hydrogen. Bicarbonate formation (Reaction R.2) is possible with the three types of amines. Carbonate formation (Reaction R.3) is also possible but negligible in amine solutions because of its too high pKa of 10.33. Usually, a decrease of the pressure or an increase of temperature displaces the equilibrium from the right to the left. This principle is applied in the regenerator for stripping the CO_2 from the solvent.

Type of Amine

The choice of amine is primarily determined by the application. Primary amines (such as MEA) are generally acknowledged to be more corrosive than secondary amines (such as DEA). Tertiary amines (such as MDEA) are thought to have the lowest corrosivity. However, it is widely acknowledged that not all amines are intrinsically corrosive. Their corrosivity can only be increased to a high level when they react with CO_2 or form degradation by-products.

Acid Gas Loading

Different complexes are formed when alkanolamines react with acid gases. The main by-products of CO_2 are carbamate and bicarbonate, as well as protonated amine (Reactions R.1 and R.2). It is widely acknowledged that increasing CO_2 loading increases corrosion rate. However, the corrosion mechanisms have yet to be identified.

Heat Stable Salts (HSS) and Degradation Products

Alkanolamine reactions with CO_2 and O_2 produce degradation products. Several studies have shown that these degradation products are not corrosive and make excellent chelates. They can easily form complexes with iron. As a result, they reduce the stability of sulphide or carbonate protective layers formed on steel surfaces.

Acid degradation products primarily result from amine-oxygen (O_2) reactions or thermal degradation. The primary degradation products are carboxylic acid salts such as oxalic, formic, acetic, or glycolic acids. Because these acids are stronger than carbonic acid, they are not regenerated at the reboiler temperature. They are commonly referred to as Heat Stable Salts (or HSS).

Summary

Simply put, the corrosion rate in an amine system is determined by four factors. Corrosive acid concentration, temperature, fluid velocity, and heat stable salt (HSS) anion concentration are the variables. In the form of a mathematical expression:

Corrosion rate = f (acid gas concentration, temperature, velocity, HSS anion concentration).

There will also be some reliance on the type of acid gas and HSS anion.

The fundamental building blocks are the oxidation of iron by some agent and the interaction between corrosion products and the corrosive environment. The corrosive agent in the typical refinery H_2S amine system is a form of an acid gas in solution, either H_2S in solution or possibly CO_2 in solution. The "form" referred to here is the acid gas's molecular form. The acid gases mentioned are dibasic, which means they can give up two protons (hydrogen ions) in solution and form two different ions, as well as exist dissolved, hydrated, and unionised. H_2S exists in three forms: hydogen sulphide (H_2S), bisulphide (HS^-), and sulphide ($S =$).

Key

The key to understanding the acid attack is understanding the acid. First, H_2S enters the aqueous phase and is surrounded by water molecules. The H2S then undergoes a two-step reaction sequence, losing an H+ to form H^+ and HS^- followed by losing the other H+ from the HS- to form H^+ and $S =$. Only an extremely small portion of the original H_2S undergoes the second step. Eventually, an equilibrium distribution of the three forms is established.

CORROSION INHIBITORS: AN OVERVIEW

Corrosion is a broad topic. This discussion will be limited to metal corrosion caused by the electrochemical breakdown of metals. Metals are naturally prone to oxidation and corrosion. Corrosion inhibitors are chemicals that prevent metal corrosion. The inhibition of anodic or cathodic corrosion reactions can be due to the reduction of the active surface area of metal and/or to a change in the activation energy of the oxidation or reduction process in corrosion.

Why do Metals have a Tendency to Oxidize?

Metals are electropositive with little affinity for electrons. Metals want to lose electrons and form positively charged ions.

But why It is So?

There can be two ways we can explain it. [1] atoms [transition metal like Fe] have a big number of protons and big size with their outer valance electrons far away from the nucleus with relatively fewer attractions between protons in the nucleus and the far away located valance shell electrons. So, metals try to lose electrons. The second reason [2] is that transition metals have many electrons similar in energy, meaning that it is sometimes easy for many of those electrons to be lost. [This is also the reason why metals have a high electrical conductivity (i.e., their electrons can easily move from atom to atom]. So, to summarize, with little opportunity metal tends to lose electrons. The loss of electrons is oxidation.

Corrosion Inhibitor Definition

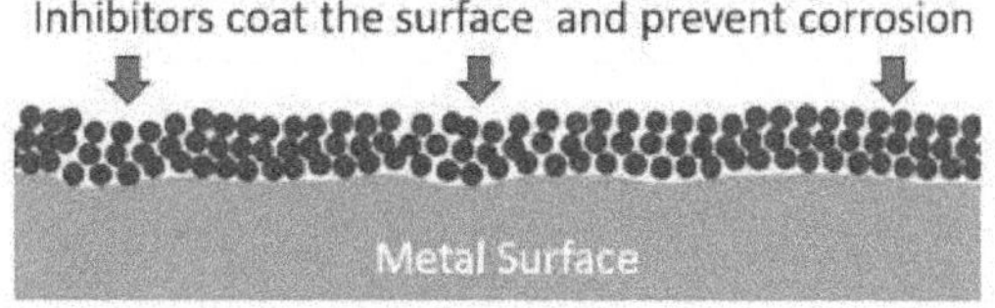

Figure 23

A corrosion inhibitor is defined as a "chemical substance that, when present in the corrosion system at a suitable concentration, reduces the corrosion rate while not significantly changing the concentration of any corrosive agent." In general, it is effective at low concentrations. This includes any chemical that reduces corrosion rate through significant pH variation, as well as oxygen and hydrogen sulphide scavengers that cause aggressive species to be removed from the solution.

Classification of Inhibitors

Corrosion inhibitors are classified as cathodic, anodic, or mixed, depending on whether their influence is primarily in slowing the cathodic or anodic reaction of the corrosion process, or both. As a result, they cause a shift in the corrosion potential of the inhibited metal toward either the cathodic or anodic directions, or they substantially leave the metal corrosion potential more or less unchanged. The inhibition of anodic or cathodic corrosion reactions can be caused by a decrease in a metal's active surface area and/or a change in the activation energy of the oxidation or reduction process in corrosion. The combination of cathodic and anodic corrosion inhibitors frequently results in improved protection and allows for lower inhibitor concentrations.

What Does it Mean? What is Corrosion Potential?

The electrode potential of a metal is an indication of the tendency of the metal to dissolve and corrode in a certain electrolyte. The magnitude of the electrode potential of a metal is a measure of its relative tendency to lose or gain electrons, i.e., it is a measure of the relative tendency to undergo oxidation (loss of electrons) or reduction (gain of electrons). Eventually, this provides energy for corrosion]

Reference is also made to the "nobility" of the metal. The nobler the metal, the higher the potential is, the less the tendency it has to dissolve in an electrolyte.

What is the Nobility of Metal?

By simple definition, "noble metal is one which resists oxidation and corrosion like gold, silver, and copper.

The electrode potentials of different metals can be specified in relation to one another in the galvanic series for different electrolytes. The galvanic series of various metals in seawater is shown below.

Galvanic Series

Considering the steel-copper example, it will be noted from the table above that copper has a higher potential (is nobler) than plain carbon steel. The steel will be the anode and corrode, whereas the copper will be the cathode and not corrode.

Table 3

Metal	*Electrode Potential, volt*
Gold	+0.42
Silver	+0.19
Stainless Steel (AISI 304), passive state	+0.09
Copper	+0.02
Tin	–0.26
Stainless Steel (AISI 304), active state	–0.29
Lead	–0.31
Steel	–0.46
Cadmium	–0.49
Aluminium	–0.51
Galvanized steel	–0.81
Zinc	–0.86
Magnesium	–1.36

What is Cathodic *vs.* Anodic Protection?

The underlying science is very interesting.

Cathodic protection converts all anodic areas on a metal surface to cathodes so that corrosion ceases. Anodic protection, on the contrary, makes the entire metal surface anodic. The surface becomes so anodic that the metal completely passivates.

Obviously, then, this technique is limited to metals that can form protective passive films.

How do Corrosion Inhibitors Work?

The inhibition of anodic or cathodic corrosion reactions can be due to the reduction of the active surface area of metal and/or to a change in the activation energy of the oxidation or reduction process in corrosion. The combination of cathodic and anodic corrosion inhibitors often determines improved protection and allows for reducing the inhibitor concentration.

What is the Activation Energy for Corrosion?

Activation is the changing of a passive surface of a metal to a chemically active state. Activation polarization is usually a controlling factor during corrosion in media containing high concentrations of active species or strong acids. The increased corrosion rate results from increased activation energy for chemical and electrochemical reactions, increased diffusion rates in the electrolyte and increased transport through the electrolyte or environment and across films that may be formed on the metal surface.

Types of Corrosion Inhibitors

Whichever the reaction is hindered by the corrosion inhibitor, it will interact at the metal/solution interface by forming a film, which may be of three different types: (1) passivating film; (2) precipitation film, and (3) adsorption film.

Passivating Film

Passivating oxide films are 30–200 Å thick and afford excellent corrosion protection. Among them, classical oxidizing inhibitors can be found, such as chromates and nitrites, now largely abandoned because of toxicity concerns. This class includes also nonoxidizing inhibitors, such as tungstates and molybdates, affording their passivating effect only in aerated solutions.

Precipitation Film

Precipitation film-type inhibitors are chemicals forming insoluble protective films by reaction with soluble species in the environments (e.g., phosphonates and polyphosphates, forming protective films with calcium ions in solution) or with the protected metal ions (e.g., copper-benzotriazole (BTA) salt film).

Adsorption Protective Film

Adsorption protective films are mainly organic substances. They often have the molecular structure of a surfactant, with a hydrophilic group capable of bonding with the metal surface and a hydrophobic part of the molecule protruding toward the solution bulk. Adsorbed inhibitor molecules limit oxygen diffusion and water access to the metal surface, so reducing the corrosion rate.

Vapor Phase Inhibitors/Volatile Corrosion Inhibitors

In short, Vapor phase corrosion inhibitors are volatile compounds introduced in a closed system for corrosion protection. These inhibitors possess a high vapor pressure at normal temperatures and protect the metal by forming a bond and barrier layer on the metal surface. In general, a weak volatile acid or base that easily hydrolyzes provides the most effective inhibition. Volatile alkaline compounds such as octadecylamine and morphine are introduced in boilers with steam to inhibit corrosion in condenser tubes by neutralizing the acidic carbon dioxide.

CHAPTER 54

Steam

WATER VAPOR

Steam is the gaseous state of water, where the water molecules have enough energy to move quickly and take up more space than liquid or solid water. Steam can be created through evaporation or boiling. Saturated or superheated steam is invisible, but wet steam is a visible mist of water droplets.

In thermodynamics, steam exists in a gaseous phase and a liquid phase at the same time, and the transition between the two phases is governed by the temperature and pressure of the system. This two-phase equilibrium is a fundamental concept in thermodynamics and is essential for understanding the behaviour of steam in various engineering applications such as power generation, refrigeration, and industrial processes. The study of steam and its thermodynamic properties is important for the design and operation of steam-based systems, and it plays a significant role in many fields of engineering and science.

STEAM PROPERTIES: CONSTRUCTION OF STEAM

Summary

Table 1: Construction of Steam

Summary	*Steam Table*	*Calculated Values*
uf - Internal energy of saturated water[KJ/g]	419	419
ug - Internal energy of saturated steam[KJ/kg]	2506	2506
hf - Specific enthalpy of Saturated water [KJ/kg]	419	419
hfg - Latent of evaporation [KJ/kg]	2256	2257
hg - Specific enthalpy of Saturated steam [KJ/kg]	2676	2675
sf - Entropy of water [KJ/kg/k	1.3	1.3
sfg - Entropy water + steam in the saturated region[KJkg/k]	6.05	6.05
sg - Entropy of steam[KJ/kg/k]	7.35	7.35

What Exactly are Steam Tables?

Steam tables are thermodynamic data containing the properties of water or steam. Engineers frequently use this information. It is typically used to obtain the following properties: the saturated steam temperature for saturated steam pressure and saturated temperature for saturated steam pressure. It is also commonly used for a specific enthalpy and volume.

Tables of properties range from saturated water to steam. This is due to the fact that water is a medium that can be either liquid or gas at saturation.

Subscript "*f*" refers to saturated water in the steam property table, while subscript "*g*" refers to saturated steam in the steam property table. Steam Tables can provide us with a variety of properties, such as steam temperature, saturated liquid heat, latent heat of steam, total heat of steam, and specific volume.

Nomenclatures steam tables

P - Pressure of the steam/Water

T - Saturation points of steam/water

vg - Specific volume of saturated steam
vf - Specific volume of saturated water
uf - Internal energy of saturated water
ug - Internal energy of saturated steam
hf - Specific enthalpy of saturated water
hfg - Latent of evaporation
hg - Specific enthalpy of saturated steam
sf - Entropy of water
sfg - Entropy water + steam in the saturated region
sg - Entropy of steam

Saturated Steam Table

The temperature and pressure of saturated steam are mutually dependent in saturated steam tables. The other is determined when one of them is given. Other related values, such as specific enthalpy (*h*) and specific volume, are usually included in these tables.

Because the saturated steam temperature and pressure are directly related, we can use two different formats. [1] Pressure [2] Temperature.

We must also consider pressure units when referring to any steam tables.

There are two types of pressure measurement systems: absolute pressure and sealed gauge pressure. The most basic distinction between the two is that absolute pressure has absolute zero as its zero point, whereas gauge pressure does not. Gauge pressure measurement is not precise due to varying atmospheric pressure, whereas absolute pressure is always definite.

Superheated Steam Table

If we heat saturated steam after it has reached its boiling points at constant pressure, it will be converted into superheated steam.

"Degree of Superheat" refers to the additional heat and degrees of temperature applied to steam above the saturation point.

Construction of Steam Table from 1st Principle

Below there is a typical steam table with two rows [1] 100 degc and [2] 180 degc. **We will construct this steam table column by column with an explanation.**

Basis: 1 kg water at, 0.1014 mPa at 0 degc, latent heat of saturated steam, 2675.6 kJ/kg, at 0.1014 mPa and 100 degc.

Two rows taken from a saturated steam table to explain the effect of temperature and pressure

Table 2: Saturation Properties for Steam-Temperature

Temp	*Pressure*	*Volume (m^3/kg)*		*Energy (kJ/kg)*		*Enthalpy (kJ/kg)*			*Entropy (kJ/k g.K)*		
°C	*MPa*	*vf*	*vg*	*uf*	*ug*	*hf*	*hfg*	*hg*	*sf*	*sfg*	*sg*
100	0.1014	0.00104	1.6718	419.06	2506	419.2	2256.4	2675.6	1.30	6.046	7.3541
180	1.0028	0.00113	0.1938	761.92	2582.8	763.1	2014.1	2777.2	2.13	4.444	6.584

Explanation

Column 1: Temperature, 100 degc.

Column 2: Saturated pressure corresponding temperature, 0.1014 mPa.

Column 3: Specific volume of water, vf in m^3/kg

Calculation

Density of water at 100 degc = 958.05 kg/m^3

Specific volume of water, vf = 1/958.05 = 0.001043 m^3/kg

Column 4: Specific volume of gas at 100 degc, vg

Calculation

Density of steam at 100 degc = 0.6 kg m^{-3}

Specific volume of steam, Vg = 1/0.6 = 1.670 m^3/kg

vfg is the saturated water-vapor specific volume = the difference between the specific volume of the saturated vapor and saturated liquid

Saturated specific volume of vapor-water, $vfg = (vg - vf)$ = [1.670–0.001043] = 1.6689 m^3/kg

Column 5: Internal energy of saturated water, uf at 100 degc.

What is the internal energy of saturated water?

The internal energy of saturated water at 100 degc is its stored energy which is mass × specific heat × delta t. The specific heat of water at 0 degc, = 4.191 kJ/kg/k, ug = 1 × 4.19 × 100 = 419.1 kJ/kg. This is the sensible heat that water contributes at 100 deg.

Therefore,

Uf = 419.1 kJ/k

Along with this sensible heat water also brings entropy $Sf = ln\ T_2/T_1$ [T_1 = 0 + 273.15 = 273.15 k] and [T_2 = 100 + 273.15, T_2/T_1 = 373.15/273.15 = 1.366]

$Sf = Cp \times ln$ 1.366 = 4.191 × 0.312 = 1.30 kJ/kg/k

TS Diagram of Saturated Steam

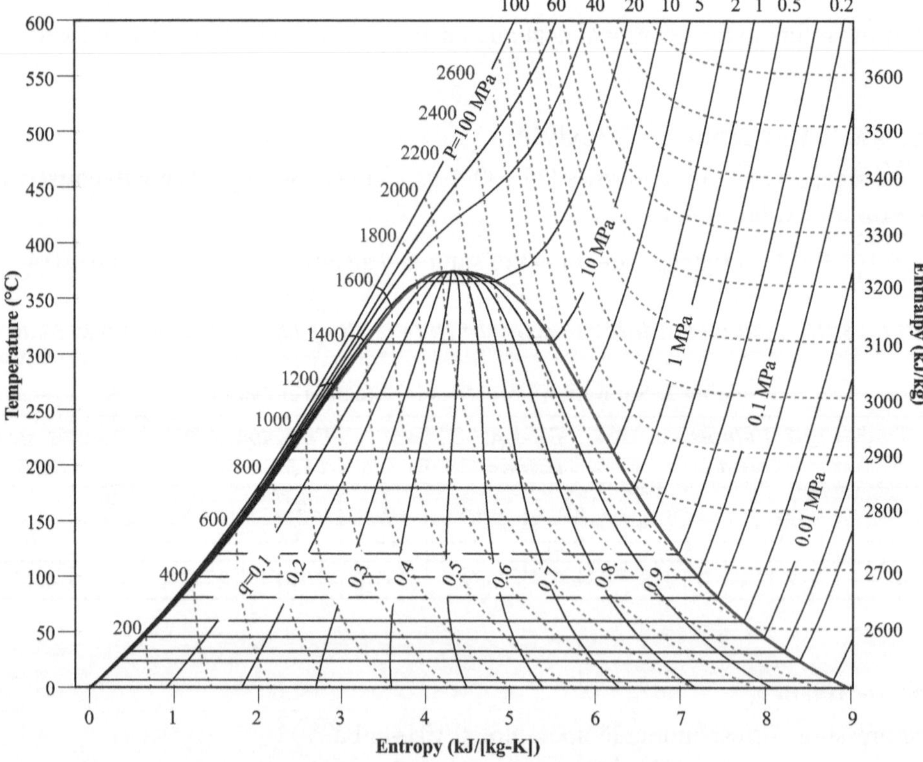

Figure 1: TS Diagram for Saturated Steam

Explanation: Combine Steam Table with *TS* Diagram

At this stage, the saturated water at 100 degc is on the saturated liquid outline of the above *TS* diagram. Next, the water enters the bell curve, in the saturated water-vapor region within the bell. Water breaks its intermolecular bonds at constant temperature and pressure as it travels from saturated liquid to saturated steam line on the right-side outer line of the bell curve. The dryness fraction of steam $X = 1$ on the saturated steam line of the bell curve. As water vaporizes, inside the bell curve, it generates entropy Q [liquid-vapor]/T or $vfg/T = sfg$. Water also generates heat of vaporization *vfg* inside the bell curve. The water finally leaves the bell curve with dryness fraction $X = 1$.

Its total energy at the right-side outer line consists of the sum of *uf* and $vfg = vg$, and entropy which is a sum of *sf* and *svg*. The left side outline of the bell curve is the saturated water line at 100 degc. The right side outline of the bell curve represents a saturated steam line. These two lines intersect at the critical point.

Column 6: *ug*, the internal energy of saturated steam on the saturated steam line of TS diagram.

Calculation

P = 0.1014 mPa, $vg = 1.670$ m^3/kg]

1 mPa = 1000000 Newton/m^2

0.1014 mPa = 1000000 × 0.1014

$PV = 1000000 \times 0.1014 \times 1.670 = 169338$ Newton-meter [PV = Work]

1 *kJ* = 1000 Newton-meters

Therefore, $PV = 169.338$ KJ,

$ug = hg - PV$

$hg = 2675.6$ kJ/kg

$ug = 2675.6–169.338 = 2506.3$ KJ/kg

Column 7: *hf*, enthalpy of water,

Calculation

Since $hf = uf + w$ and w [work] is zero for water, $hf = uf$, therefore,

$hf = 419.1$ kJ/kg

Column 8: *hfg*, the enthalpy of saturated water-steam vapor inside the bell curve.

Calculation

$hfg = hg\text{-}hf$ [difference of enthalpy of saturated steam on the saturated steam line of *TS* diagram and enthalpy of saturated water on the saturated water line of *TS* diagram.

$hfg = [2675.6–419.1] = 2256.5$ kJ/kg

Column 9: *hg*, the enthalpy of saturated steam on the steam saturation line of *TS* diagram. This is a property of water2675.6 KJ/kg at 100 degc and 0.1014 mPa pressure.

$hg = 2675.6$ kJ/kg

Column 10: *sf*, the entropy of water.

Calculation

Along with this sensible heat water also brings entropy $Sf = Cp \times ln\ T_2/T_1$ [$T_1 = 0 + 273.15 = 273.15$ k] and [$T_2 = 100 + 273.15$, $T_2/T_1 = 373.15/273.15 = 1.366$]

$Sf = Cp \times ln\ 1.366 = 4.191 \times 0.312 = 1.30$ kJ/kg/k

$Sf = 1.3$ kJ/kg/k

Column 11: *sfg*. Entropy of steam- water in equilibrium inside bell curve.

Calculation

$Sfg = vfg/T = 2256.4/373.15 = 6.0495$ kJ/kg/k

Column 12: *sg*, the entropy of saturated steam on the saturated steam line.

Calculation

It is a sum of *sf* + *sfg* [entropy of saturated water + entropy of water-steam in equilibrium]

sg = 1.3 + 6.0495 = 7.3495 Kj/kg/k

'HEAT' AND 'WORK' BY SATURATED *VS.* SUPERHEATED STEAM

Superheated steam is known for its tremendous energy for thermodynamic work while saturated steam is a source for heating. A simple calculation is presented below to highlight the 'work' and 'heat' by steam.

Summary

(Source of steam properties: Steam table)

Superheated steam [1 bar/150 degree-superheat]
Work capacity = 24.51 kJ
Heating capacity = 315 kJ/kg
Saturated steam 1 bar
Work capacity = 0
Heating capacity = 2240 kJ/kg

Calculation

Superheated steam

Superheated steam 1 kg/1.013 bar [0.1013 mPa]
Heating capability
150 - degree superheat
Specific heat = 2.1 KJ/kg/degc
= 2.1 × 150 = 315 KJ/kg

PdV work
150-degree superheat

The specific volume of saturated steam = 1.670 m^3/kg
The specific volume of superheated steam = 1.912 m^3/kg
Change in volume/kg = [1.912–1.670] = 0.242 m^3

PV work = 0.1013 × 0.242

1 mPa = 1000000 Newton/$meter^2$

0.1013 mPa = 0.1013 × 1000000

PdV work = 0.1013 × 1000000 × 0.242 Newton-meter work

1 KJ = 1000 Newton–meters

PdV = 0.1013 × 1000000 × 0.242/1000 = 24.51 KJ/kg

Saturated steam at 1 bar/100 degc
Heating capability
Latent heat = 2240 Kj/kg
Work capability
$PdV = 0$

SUPERHEATED STEAM

Superheated steam is known for its tremendous energy for thermodynamic work while saturated steam is a source for heating. A simple calculation is presented below to highlight the 'work' and 'heat' by steam.

When water is heated above its boiling point, superheated steam is created; when water is heated at its boiling point, saturated steam is produced. If the saturated steam produced in a boiler is exposed to a surface with a higher temperature, its temperature will increase above the evaporating temperature. The steam is then described as superheated by the number of temperature degrees through which it has been heated above saturation temperature. The superheated steam is produced by heating dry saturated steam at constant pressure. The temperature and pressure do not directly correlate with superheated steam. Consequently, superheated steam is able to exist at a variety of temperatures at a given pressure. This is the reason why superheating is done under constant pressure. The fundamental reason is that at very high temperatures in the superheated steam, the molecules scatter more than their collision leading to no rise in temperature and becoming independent of pressure.

Saturated steam transfers heat via latent heat. Superheated steam transfers heat via specific heat.

For superheated steam, the source of energy is its kinetic energy. The source of energy for saturated steam is its potential energy. Superheated steam has its applications in, for example, turbines where the steam is directed by nozzles onto a rotor. This causes the rotor to turn. The energy to make this happen can only have come from the steam

Strong point: The large specific volume of the superheated steam is its strongest point. The superheated steam's large specific volume allows it to be compressible and capable of thermodynamic PV work.

Weak point: The lack of latent heat and potential energy is the weakest point of the superheated steam. Superheated can only store a very small amount of heat as kinetic energy. Superheated steam has a very low specific heat of 2.1 kJ/kg/degc, which stores energy. This contrasts with the massive 2240 kJ/kg latent heat found in saturated steam. When 1 kg of saturated steam condenses, it can produce thermal energy equal to 2240 kJ/kg while, 1 kg superheated steam with 100 degrees of superheat can only transfer $2.1 \times 100 = 210$ kJ/kg. This is one of the reasons why superheated steam has very low heat transfer coefficient, compared to saturated steam. This makes the superheated steam a poor source of energy for heating purposes.

A point to note is that, unlike the specific heat capacity of water, the specific heat capacity for superheated steam varies considerably with pressure and temperature and cannot be taken as a constant.

The value of 2.1 kJ/kg/degc given above is therefore only the mean specific heat capacity over the specified temperature range for that pressure.

There is no direct relationship between temperature, pressure and the specific heat capacity of superheated steam. There is, however, a general trend towards an increase in specific heat capacity with increasing pressure at low degrees of superheat, but this is not always the case.

Superheated steam making: When the steam is still in contact with water, it cannot be given superheat because any additional heat will only cause the water to evaporate even more. It is necessary to run the saturated steam through one more heat exchanger. This could be an independent superheater unit or a second heat exchange stage in the boiler. The boiler's hot flue gas or a separate fire can serve as the main heating medium. In the superheater, the dry saturated steam is heated at constant pressure which increases the specific volume of the steam along with temperature. The temperature and pressure do not directly correlate with superheated steam. Consequently, superheated steam is able to exist at a variety of temperatures at a given pressure. This is the reason why superheating is done under constant pressure. The fundamental reason is that at very high temperatures in the superheated steam, the molecules scatter more than their collision that makes superheated steam independent of pressure.

Please see the steam table below.

Table 3: Superheated Steam

Absolute Pressure bar a	*Units*	*Temperature (°C)*					
		150	*200*	*250*	*300*	*400*	*500*
1.013	v_g (m^3/kg)	1.912	2.145	2.375	2.604	3.062	3.519
	u_g (kJ/kg)	2583	2659	2734	2811	2968	3131
	h_g (kJ/kg)	2777	2876	2975	3075	3278	3488
	s_g (kJ/kg K)	7.608	7.828	8.027	8.209	8.537	8.828

Pressure–temperature relationship: Unlike saturated steam, superheated steam does not have a direct relationship between pressure and temperature. This means superheated steam can exist at a wide range of temperatures. Typically, it's specified by pressure and a degree of superheat (= the degree above the saturated steam temperature).

Lowering the temperature: Compared to saturated steam, superheated steam also has a lower density. Lowering the temperature of saturated steam will revert it back to its liquid state including water droplets. This is not the case for superheated steam.

Superheated steam table: Similar to the saturated steam tables, the superheated steam tables show the characteristics of steam at different pressures. On the other hand, temperature and pressure do not directly correlate with superheated steam. Consequently, superheated steam might be able to exist at a variety of temperatures at a given pressure. Generally speaking, superheated steam tables provide absolute pressure, while saturated steam tables provide gauge pressure.

Refer to the above steam table at 1 bar. *v* is specific volume, u is internal energy, *h* is enthalpy and s is entropy of the superheated steam. You may notice that the temperature is directly related to a specific volume and the specific volume is directly related to the internal energy and enthalpy of the superheated steam. The entropy s is also related to the increase in specific volume as the degree of superheat increases.

SATURATED STEAM

The general belief is steam has a lot of kinetic energy [1/2 m V^2] The fact, since steam is about 1600 times lighter than water make, '*m*' very small to generate kinetic energy. According to reports, steam can hold 5–6 times more potential energy than water at equal mass.

One important point to note is that since at phase change, the two phases go into thermal equilibrium and thus remain at the same temperature, the internal energy increase if it is a liquid-to-vapor phase change, is primarily sourced from the potential energy.

Water converts to steam at constant temperature and pressure for two reasons [1] its predominant potential energy which is 5–6 times more than water at equal mass keeps the temperature of steam near constant and [2] the thermodynamic *PV* work at phase change.

These are the Sequence of Events

1. Water receives sensible heat and gets saturated at 1 bar/100 degc. At the saturated point, $Q = m\ Cp\ dt$, the three silos of kinetic energy, translational, vibrational, and rotational, all these three reservoirs of kinetic energy are full of energy. Water has reached its heat capacity. The temperature increases stop.
2. The burner, however, continues to supply the same amount of heat.
3. This marks the beginning of phase change. At this stage, the water is forced to change the phase into vapor and goes to the next higher energy level to accommodate the continuous supply of heat. The water has reached maximum entropy and minimum free energy. This stage also marks the beginning of water and vapor to go into thermodynamic equilibrium. At thermodynamic equilibrium, the two phases must be in thermal equilibrium, the two phases thus remain at the same temperature and pressure
4. Water turns into a lighter gaseous phase by expansion from liquid to vapor by about 1600 times at constant temperature and pressure by doing thermodynamic *PV* work. $H = U + W$, H is the enthalpy of steam, U is the

internal energy and *W* is the work at constant temperature and pressure. Since the two phases are in thermal equilibrium at phase change, the change in the internal energy of steam is primarily due to an increase in its potential energy. Steam can hold 5–6 times more potential energy than water according to reports at identical mass.

Thus the steam increases its enthalpy at constant temperature and pressure. The pressure is set by you and the temperature is set by the thermal equilibrium between the two phases.

5. The enthalpy of the steam thus has two components, internal energy which is predominantly potential energy, and work energy.
6. The process continues until the steam reaches saturation with the maximum energy it can hold at 1 bar/100 degc which is about 2675 kJ/kg

BENEFITS OF SUPERHEATED STEAM OVER SATURATED STEAM

Summary

The most important benefit of superheated steam is that you can extract 'work' from superheated steam over a wide range of temperatures until there is superheat in the steam. In saturated steam the ratio of pressure and temperature is constant. Saturated steam can perform thermodynamic work only at saturated temperature,

The high internal energy of superheated steam is its strength. This internal energy is mostly made up of kinetic energy or temperature. High entropy is its disadvantage. High latent heat is the advantage of saturated steam. The massive gain in the sensible heat of the water in equilibrium with steam is the source of the high latent heat of the water. This is the single largest factor for water's high latent heat. This is indeed unique

Because there is no space to store potential energy when all H-bonds have broken in superheated steam, it has very little potential energy. The specific heat of superheated steam is half that of water.

The degree of superheat is where superheated steam stores its internal energy. As superheated steam performs mechanical work, it transfers internal energy and cools without condensation as long as there is superheat. In other words, as long as superheated steam is above the saturation point, it can perform thermodynamic work. This is determined by the amount of superheat carried.

With saturated steam, this is not possible. Saturated steam condenses immediately when it expands and if the temperature falls below saturation temperature, limiting its ability to do mechanical work.

Saturated steam's strength is primarily due to the enormous potential energy of saturated water in the intermolecular bonds, which acts as a heat sink when saturated steam transfers heat, cools, and condenses into water. As a result, saturated heat is ideal for heating. While working in a superheated environment is beneficial.

Superheated Steam over Saturated Steam

Saturated steam: Saturated steam is the type of steam that occurs when the liquid and gaseous phases of water exist simultaneously at a given temperature and pressure. In simpler terms, the steam is in equilibrium with the heated water. During the saturated condition, the rate at which water is vaporized is equal to the rate at which it is condensed.

Superheated steam: Superheated steam is defined is steam at a temperature that is higher than its vaporization (boiling) point at the absolute pressure. It is steam that is formed at a temperature that exceeds that of saturated steam at the same pressure.

How saturated is steam made: When heat is applied to water at constant pressure, its temperature continues to rise until it reaches its boiling point at that pressure. This is the sensible heat added to water. As further heat is added, the water vaporizes at a constant temperature and converts to steam. The steam that exists at the same temperature as the water from which it is formed is known as saturated steam. The heat goes to break the H-bond in water as saturated water converts to saturated steam. This is called the enthalpy of vaporization.

The sum of the sensible heat and enthalpy of vaporization is the total energy of water at its boiling point at the given pressure is the latent heat of the water. This is the enthalpy of saturated steam, H at 100 degc/1 bar pressure.

Latent heat of water at 100 degc/1 bar pressure = Sensible heat + Enthalpy of vaporization.

How superheated steam is made: When dry steam is heated at constant pressure its temperature increases, saturated steam converts to superheated steam. The difference in temperature between saturated temperature and superheated steam temperature is called 'superheat'. Superheated steam stores its energy in the degree of superheat. One question is often asked why it is important to heat saturated steam at constant pressure to get superheated steam? The answer is to break the pressure/temperature ratio. As long as the pressure/temperature ratio in steam is constant it remains saturated steam.

Benefits of Superheated Steam over Saturated Steam

Table 4: Saturated Steam

Temperature (°C)	*Pressure MPa*	*Volume m^3/Kg*		*Energy KJ/Kg*		*Enthalpy KJ/Kg*			*Entropy KJ/Kg-K*	
		vf	*vg*	*uf*	*ug*	*hf*	*hfg*	*hg*	*sf*	*sfg*
100	0.1	0.001043	1.6718	419.06	2506	419.17	2675	2256.4	1.3075	7.3541
150	0.4761	0.00109	0.3924	631.66	2559	632.18	2745	2113.7	1.8418	6.8371

Table 5: Superheated Steam at 0.1 Mpa

Temperature (°C)	*vg* m^3/Kg	*ug* KJ/Kg	*hg* KJ/Kg	*sg* KJ/Kg-K
150	1.9367	2593	2777	7.6148

(This is case when saturated steam was superheated at 0.1 Mpa (same pressure as saturated steam) to 150°C)

Comparison of Saturated Steam at 100 degc and 150 degc

[Energy of saturated steam at 100 and 150 degc]

Analysis

- Internal energy of liquid water *uf* increased by 33.7%
- Internal energy of saturated steam *ug* increases by 2.1%
- Enthalpy of vaporization *hg* reduced by 6.3%
- Enthalpy of saturated steam *hfg* increased by 3.1%
- Entropy of saturated steam *sfg* reduced by 7.0%

Findings

- The massive gain in the sensible heat of water or potential energy in equilibrium with steam is the source of high latent heat of water.
- A reduction in the enthalpy of vaporization.
- A reduction of entropy of saturated steam.

Explanation

Liquid water has huge energy storage space due to *H* bonds Fewer *H* bonds at 150 degc reduced entropy and enthalpy of vaporization.

Analysis: Benefits of superheating [Please refer to the table above]

This is a case when saturated steam was superheated at 0.1 Mpa [same pressure as saturated steam] to 150 degc

- Volume of gas *vg* increased by 14.2%
- Internal energy *ug* increased by 3.1%
- Enthalpy *h* increased by 3.8%
- Entropy *sg* increased by 3.4%

Explanation

- Marginal increase in enthalpy by 3.8% from 2675 KJ/kg to 2777 KJ/kg = 102 KJ/kg.
- Energy lost over saturated steam by increased entropy at 150 degc = 2593 × 0.034 = 88. 2KJ/kg.
- Net benefit of enthalpy = 102–88.2 = 14.68 KJ/kg.
- Because of significant volume increase and correspondingly an increase of entropy there is a small benefit of enthalpy increase.
- The source of enthalpy increase is the increase in internal energy.

WHY STEAM HAS MORE ENERGY THAN WATER AT 100 DEGC/1 BAR?

The energy of saturated water (100 degc) = 418 kJ/kg (about).

Energy of saturated steam (100 degc) = 2240 kJ/kg (about).

Total energy of steam = 2658 kJ/kg (about).

The energy in saturated water is its sensible heat. It is the kinetic energy of atoms in motion in a water molecule at 100 degc. It is also the heat capacity of water at 100 degc.

The energy in the steam is primarily the potential latent energy which is 2240 kJ/kg. Potential energy is latent because it does not generate motion in a molecule as kinetic energy does and generates collisions which increase temperature.

Kinetic energy is temperature, $KE = 1/2KT$ (K is constant and T is temperature). Potential energy is not temperature.

At 100 degc, the temperature increase of water stops for not having any heat capacity to increase temperature. This also marks the end of kinetic energy in water. In other words, the water gets saturated with heat. There is no more kinetic energy left to raise the water temperature further. In other words, water as liquid has no more capacity to absorb heat.

At this stage, water transits into a new phase which is steam by breaking intermolecular bonds and doing PV work of expansion. Potential energy which is the stored energy gets released as water molecules come under stress when water expands to steam to meet the new energy demand of the vapor phase.

Remember, potential energy is specific to a location in an energy field. When water expands to steam the stress and the change of location of water molecules make water release its potential energy.

The potential energy converts to kinetic energy. Since it is the potential energy involved in the phase change there is no change in temperature at the phase change. Water silently converts into steam at 100 degc.

The above processes occur simultaneously and not sequentially as water is heated.

The line between the water and steam or the redistribution of heat by water between two phases is drawn by the saturation point of water. It is the saturation point of water at 100 degc that marks the end of kinetic energy and the beginning of potential energy taking over to convert water into steam.

The summary is water's kinetic energy is 418 kJ/kg and its potential energy is 2240 kJ/kg which generates steam. This makes the saturated steam carry more than five times the energy of water.

In a boiler, the fuel supplies the total energy of steam 2658 kJ/kg. Water distributes the energy between sensible (kinetic energy) and latent heat (potential energy). Both processes occur simultaneously.

WHY STEAM IS A MORE ENERGY-DENSE MATERIAL THAN AIR?

Steam is far superior to air on both counts, including internal energy and compressibility. Air's much smaller enthalpy is a big limitation while delivering mechanical work.

There are Two Reasons

[1] Small water [steam] molecules have smaller van der Walls forces than air molecules. The molecular weight of water is 18 g/mol *vs.* air [mean] 28 g/mol. This moves steam molecules more rapidly and consequently a lower density than air i.e., higher specific volume than air and

[2] Water molecules [mol weight = 18 against air mol weight 28 g/mol are much smaller. That makes superheated steam more compressible than air [emptier space/unit volume].

This implies that the specific volume of the steam is less than air during a compression process and higher than air in the expansion process.

Concept

Compressibility Z

Z in thermodynamics is the compression factor or the gas deviation factor, which describes the deviation of a real gas from ideal gas behavior. It is simply defined as the ratio of the molar volume of a gas to the molar volume of an ideal gas at the same temperature and pressure.

$$Z = V\text{ actual}/V\text{ ideal}$$

Steam is More Compressible

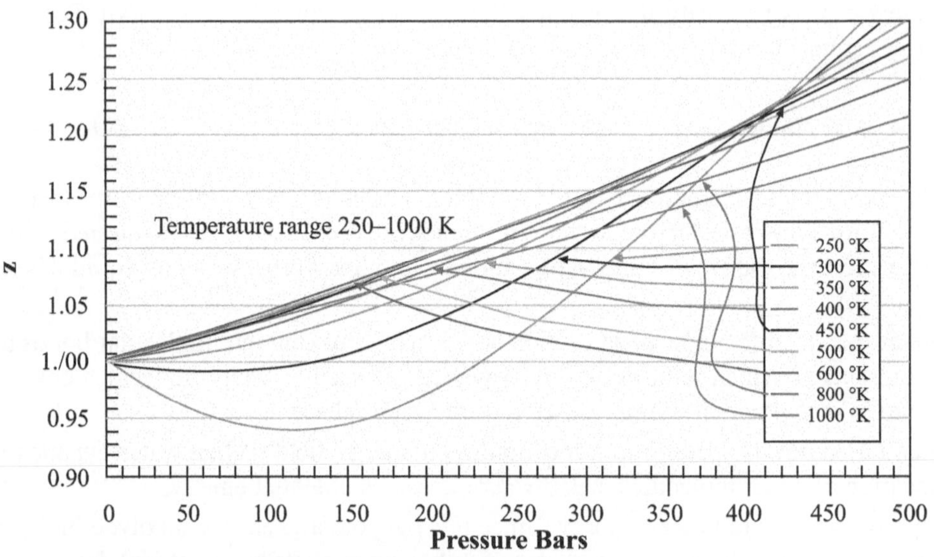

Figure 2: Compressibility Factor of Air

Superheated steam at 1500 kPa, has a specific volume of 2.75 m^3/k mol and a compressibility factor (*Z*) of **0.95.**

Thus, the compressibility of air $z > 1$ and compressibility of superheated steam < 1. In other words, under similar conditions, superheated steam is more compressible. Steam is the most compressible gas of all gases because water's stable form is liquid.

Steam always wants to go back to liquid. The specific volume of air at 100 bar/500 dec is 0.023073 m^3/kg while superheated steam specific volume is 0.032453 m^3/kg which is one and a half times more than air. This suggests, that under similar temperatures and pressure, superheated steam does more work than steam.

Steam has More Internal Energy

Specific heat [stored heat] of steam is twice of air. The internal energy of steam is 1.4 times of air. It is about 2.1 kJ/kg/K for superheated steam *vs.* 1.00 kJ/kg/k for air.

Arising from the above two points, steam has higher specific enthalpy than air.

Concept: Specific Enthalpy

The specific enthalpy (*H*) of a substance is its enthalpy per unit mass. It equals the total enthalpy (*H*) divided by the total mass (*m*).

Note that enthalpy is the thermodynamic quantity equivalent to the total heat content of a system. The specific enthalpy is equal to the specific internal energy of the system plus the product of pressure and specific volume.

$H = U + PV$, U = Internal energy and PV is the capacity to do work

As explained, the fact that the specific heat of superheated steam is more than twice of air under identical conditions, and steam has higher internal energy. A very conservative estimate is superheated steam has internal energy of more than 1.4 times air. The second point as explained steam is more compressible than air, $Z < 1$ at the same temperature and pressure gives the steam the ability to do more work. This explains why steam has more specific enthalpy or energy than air.

Enthalpy-Entropy Diagram for Air

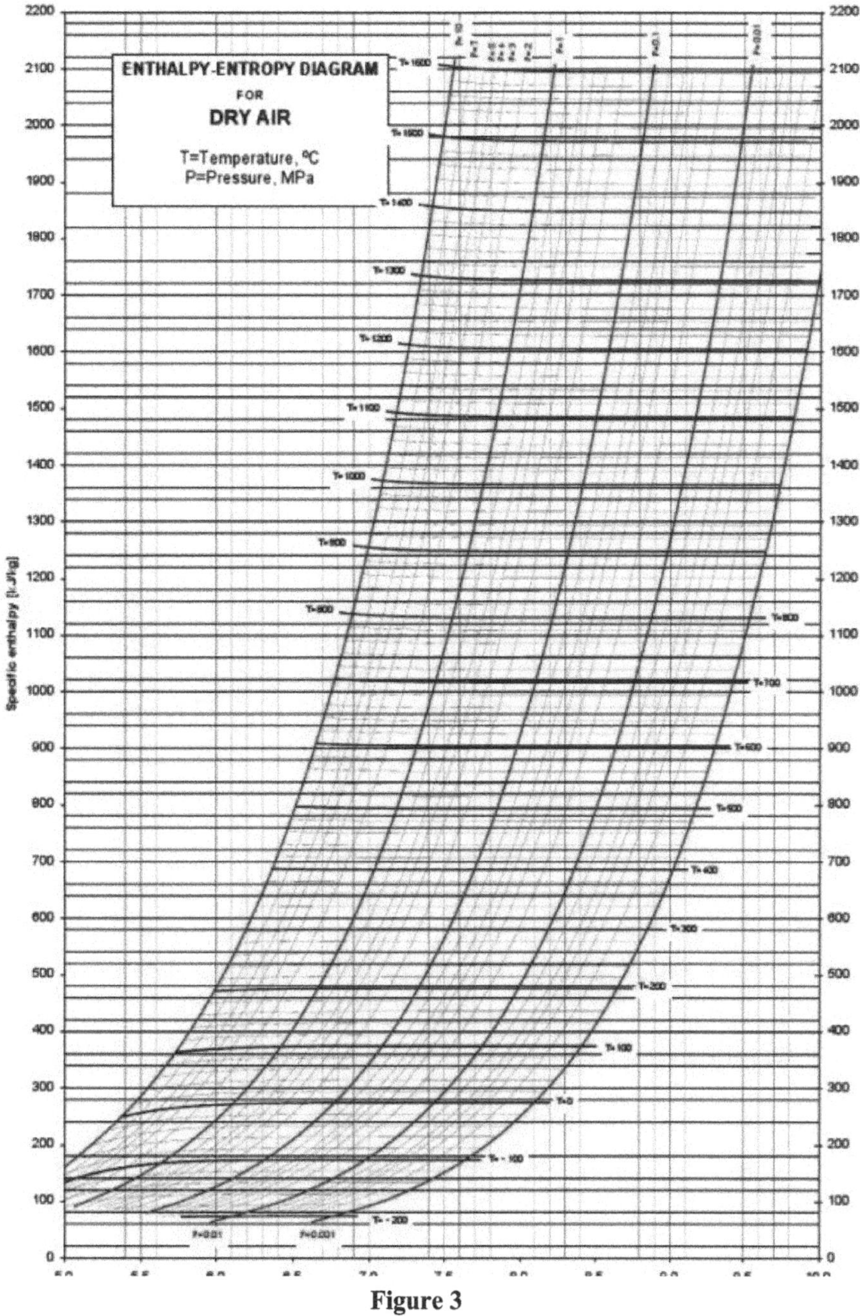

Figure 3

The X-axis is entropy in kj/kg-k. Red lines are pressure lines in Mpa. The extreme right-side red line stands for 0.001 Mpa and the extreme left-side red line stands for 10 Mpa. Black bold almost horizontal lines are temperature lines. Two important observations are emerging from this diagram [1] at a constant temperature as pressure increases, enthalpy remains practically constant while entropy reduces, and [2] at constant pressure as temperature increases, the enthalpy, and entropy both increase. The most important observation is specific enthalpy of air 100 bar/500 degc is only = 795.9 kj/kg.

Enthalpy-Entropy Diagram for Superheated Steam

At the initial condition (50 bar a/300°C): $h_g = 2.927$ kJ/kg

and $s_g = 6.212$ kJ/kg°C

For dry saturated steam 0.04 bar a: $s_f = 0.422$ kJ/kg°C

$s_{fg} = 8.051$ kJ/kg°C

and $s_g = 8.473$ kJ/kg°C

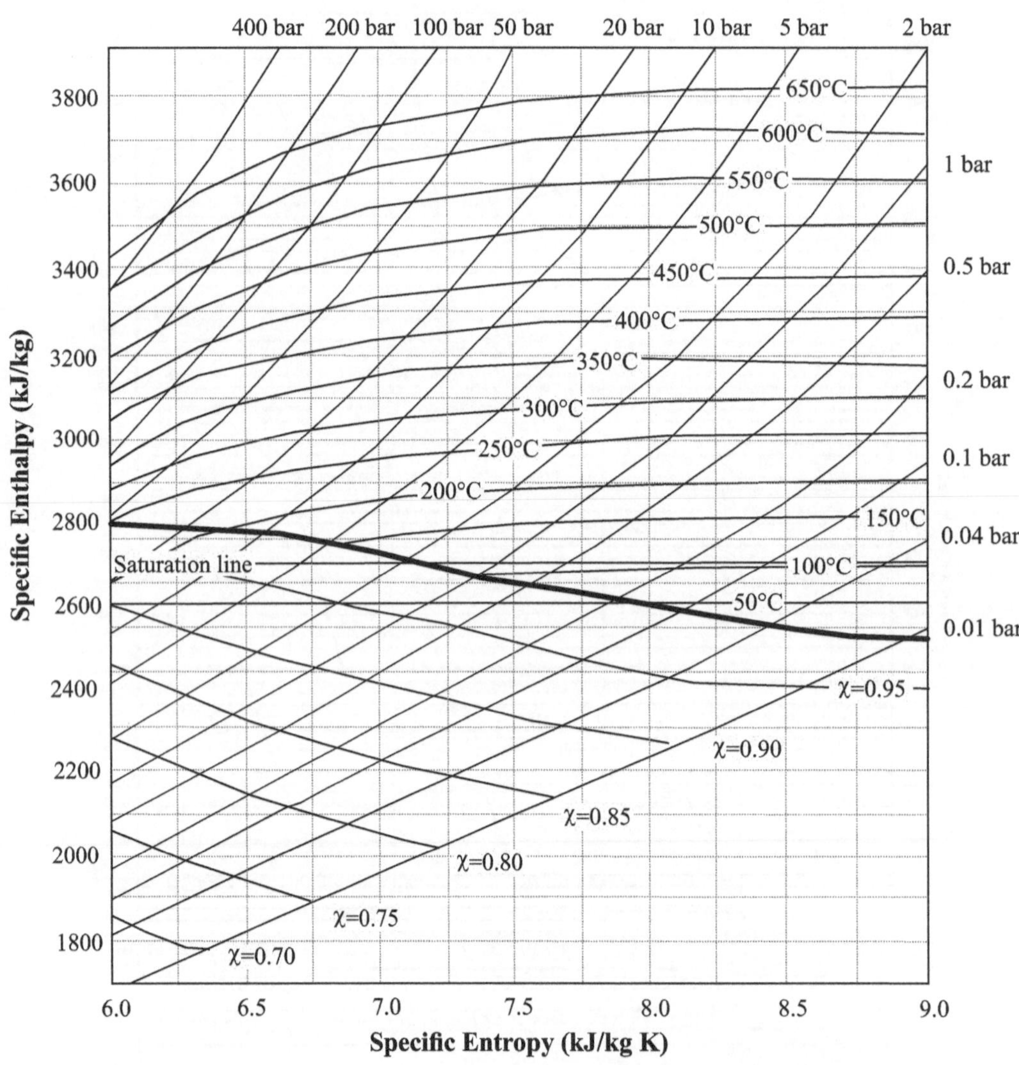

Figure 4: Enthalpy – Entropy or Mollier Chart for Steam – Example

Y-axis is enthalpy expressed as kj/kg. The X-axis is entropy expressed as kj/kg-k. There are two sets of curved lines in the image. These are well explained what they stand for. The curves rounded upwards are temperature lines expressed as degc. The curves rounded downwards are pressure lines. The image suggests the following: [1] at a constant temperature when pressure increases the enthalpy reduces and also entropy reduces [2] At constant pressure when the temperature is increased enthalpy and entropy both increase. The most important observation is the specific enthalpy of superheated steam at 100 bar/500 degc = 3373.81 kj/kg.

There is a huge difference in the enthalpy between superheated steam and air. Superheated steam has enthalpy at 100 bar/500 degc more than four times that of air. Air cannot achieve superheated steam's enthalpy even at 1700 degc and 100 bar.

So, air's enthalpy is a big limitation while delivering mechanical work.

WHY SUPERHEATED STEAM HAS A LOW HEAT TRANSFER COEFFICIENT?

The first point is superheated steam's higher enthalpy compared to saturated steam is primarily kinetic energy transfer in the absence of intermolecular interactions which are small. Superheated steam does primarily sensible heat transfer. Superheated steam has a small specific heat. In contrast, saturated steam has both kinetic and potential energy and the vehicle of heat transfer is its latent heat.

Explanation

Superheated steam possesses a lower heat transfer coefficient compared to saturated steam due to a variety of factors.

Firstly, superheated steam primarily carries kinetic energy, whereas saturated steam has both kinetic and potential energy. This means that the specific heat of superheated steam plays a key role in the heat transfer. The specific heat of superheated steam is small.

The lower specific heat of superheated steam is related to its lower density and lower mass per unit volume, which in turn leads to a reduced number of molecules per unit mass.

Furthermore, superheated steam's lower density means that there are fewer molecules per unit volume and therefore fewer opportunities for collisions and heat exchange.

As a result, the efficiency of heat transfer in superheated steam is decreased due to the limited number of molecules available for conducting energy from one location to another.

The Next Point is Important

Drop vs. Film Condensation

Due to the absence or low level of intermolecular attractions, superheated steam exhibits very low surface tension. As a consequence, it does not adhere to hot surfaces and forms a film that offers resistance to heat transfer. In contrast, saturated steam adheres to hot surfaces in the form of droplets with a lot of latent heat due to its higher surface tension and also a lot of bare metal surface, allowing for a greater area of direct contact and faster heat transfer.

In conclusion, the lower heat transfer coefficient of superheated steam can be attributed to its reduced density, low surface tension, etc. These factors collectively contribute to its reduced capacity for heat transfer in comparison to saturated steam.

SUPERCRITICAL *VS.* SUPERHEATED STEAM

Supercritical State

Supercritical steam is a supercritical fluid, which is a substance with properties of both a gas and a liquid. Supercritical steam is more efficient than superheated steam because it has a higher energy content per unit mass of water and can be used in more efficient steam-generating units

Super critical steam is not superheated steam because they are two different thermodynamic states of steam.

Steam is typically produced by boiling water in a boiler, which generates saturated steam at a specific temperature and pressure. When steam is at its saturation temperature and pressure, it is known as saturated steam. If more heat energy is added to saturated steam, it becomes superheated steam.

Superheated steam has higher energy content than saturated steam.

More specifically, superheated steam steam is created when a liquid evaporates and the remaining gas reaches a certain temperature. Superheated steam is created at pressure conditions below critical point.

Supercritical Steam

On the other hand, supercritical steam refers to steam that is at a critical pressure and temperature, where it exhibits properties of both a gas and a liquid. At this point, the distinction between liquid and vapor phases becomes less clear, and the properties of the steam become unique.

Super critical state is above critical point. For water it is 3206 psi and 706°F. At the critical point, the latent heat of steam is zero and its specific volume is the same whether it's liquid or gas.

In summary, superheated steam and supercritical steam are distinct thermodynamic states of steam, with superheated steam being steam that has been heated beyond its saturation temperature, and supercritical steam being steam that is at a critical pressure and temperature.

Thermodynamics of Supercritical *vs.* Superheated Steam

Supercritical steam has more energy than superheated steam because of elevated pressure.

Supercritical steam is denser than superheated steam. This tells a lot about their thermodynamics.

Preference of Supercritical Steam over Superheated Steam

The choice between supercritical steam and superheated steam depends on the intended application.

Supercritical Steam

Used for power generation and has several advantages, including:

- Lower fuel consumption.
- Supercritical steam can consume less fuel than other systems for a given output.
- Lower carbon emissions.
- Supercritical steam can produce lower carbon emissions than less efficient systems.

Superheated steam

- Higher temperature.
- Superheated steam has a higher temperature and lower density than other types of steam.
- Oxygen-free heating.

STEAM PURITY

What is Steam Purity?

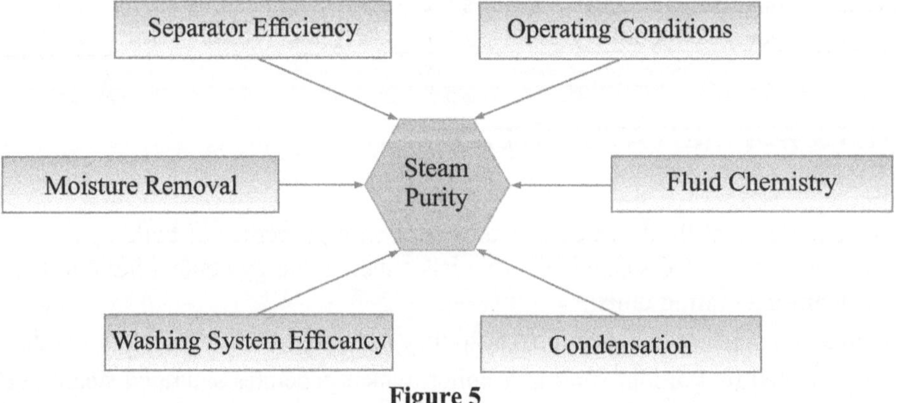

Figure 5

Steam purity represents the extent of non-water impurities present in steam. The usual impurities are dissolved or entrained particles in the steam. Steam purity is traditionally reported as solids contamination in mg/lit present in water.

Boiler Water Carryover is a Major Source of Contaminants?

Boiler water carryover causes deposition on the superheater and eventually leads to overheating failures. The primary superheater is the most common location for overheat failures. There can be many reasons for water carryover. Mechanical carryover could be a design issue of inefficient mechanical water-steam separation in the drum allowing higher moisture levels in steam. One unique reason is foaming [discussed in detail later] in the boiler water. Foaming is a chemistry-induced carryover that allows boiler water to bypass mechanical separating equipment in the drum. Superheated steam turbines are particularly prone to damage by carryover. Sticking of stop valves due to deposits can cause turbine overspeed and catastrophic damage. Solid particles in steam can erode turbine parts, while deposition on turbine blades can reduce efficiency and capacity. Losses of 5% in turbine efficiency and 20% in turbine capacity have been reported due to deposition.

How Steam Purity is Measured?

Measurement of sodium concentration in a condensed steam sample is often used to determine steam purity. Because sodium is abundant in boiler and is readily measurable on-line.

Steam purity Limits?

Acceptable steam purity for saturated steam for industrial boilers without superheater is 1 mg/lit total dissolved solids. When there is a superheater this restriction reduces to 33–67 microgram/lit of sodium in the steam considering a 3:1 ratio of TDS to sodium.

Source of Sodium in Steam?

The source of sodium is moisture in steam. The typical acceptable level of sodium requires an extremely low level of moisture in steam. Another big concern with moisture in steam is the moisture carries many non-volatile solids into the steam with corrosion and deposition potential.

Turbines Require Special Control of TDS?

Turbines are less tolerant of impurities and have far more potential to do a deposition of solids. Therefore, sodium for turbines is restricted to 10 microgram/lit.

Routes for Contaminants into Steam?

The key route of contaminants in boilers operation below 138 bar drum pressure [at 160 bar steam and water have same density] is different forms of mechanical carryover. The entrained water droplets in steam generate in saturated steam via foaming.

Silica Carryover is Unstoppable?

Silica carryover can happen even when there is no mechanical carryover. At operating, drum pressure > 41 bar silica has appreciable vaporous solubility. The only solution lies in limiting its concentration in boiler feed water.

Attemperator Spray Water Major Source of Contaminants?

Attemperator controls the maximum temperature of superheated steam by water injection. Often feed water is used for spray in attemperator because it is high-pressure water that is already available, however, it is not realized that the feedwater does not meet require purity standards and that puts the equipment at great risk. The standard requirement is attemperator spray water should beat specified steam purity limits.

The Highest Potential Contaminant is Silica?

Silica scaling preferentially takes place at the first opportunity on the first stage blades of geothermal turbines. In the case of saturated steam, this happens due to the evaporation of liquid films on turbine surfaces leading to supersaturation and then deposition. In the case of superheated heated steam, the deposition of solids takes place due to the pressure difference of superheated steam leading to a reduction of silica solubility. Once the silica gets supersaturated it deposits. Typically, the second stage turbine blades have more moisture therefore, generally silica deposition over second stage blades is near zero. Online silica analysis coupled with online sodium analysis is used to determine the quantity of dissolved silica in steam. Online steam purity analysis can assist operators in avoiding steam turbine damage.

Foaming a High Potential Route for Carryover?

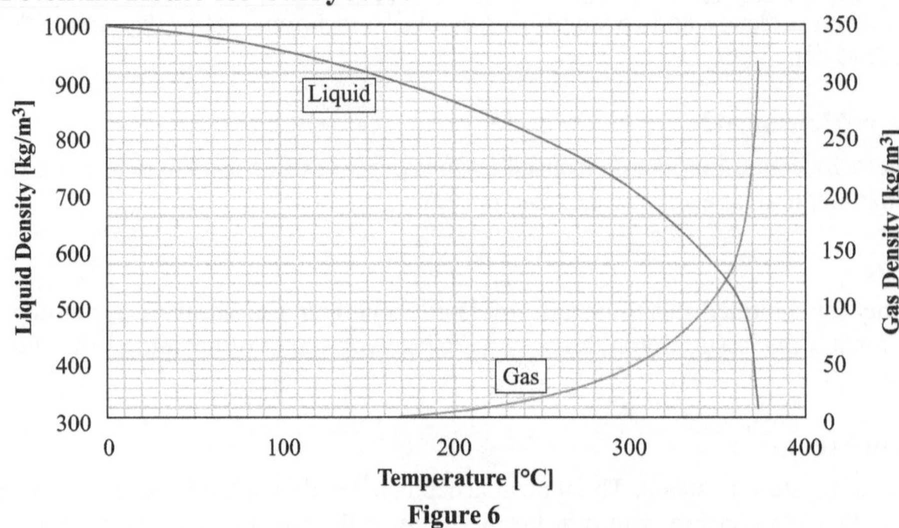

Figure 6

At about 350 degc and about 160 bar Steam and Water have the Same Density

Foaming happens more particularly in high-pressure boilers when bubbles have a density approaching that of steam, they are not readily removed by steam purifying equipment. Foaming has caused a variety of carryover problems and can cause erroneous water level readings that produce swings in feedwater flow. It has multidimensional effects. Foaming tendencies of boiler water are increased with increases in alkalinity and solids content. Boiler water solids have a dual effect on carryover. First, for a given boiler and static operating conditions, high solids in the boiler water result in high solids in each drop of boiler water carried over. Second, foaming potential increases with increasing boiler water solids. If boiler water solids are allowed to double (without any foaming), carryover will be doubled. If the higher solids cause foaming, the carryover may be increased even more.

FLASH STEAM THERMODYNAMICS: HOW FLASH DRUM WORKS?

Flash evaporation of a single component is an adiabatic and isentropic process.

Process

When a saturated liquid stream passes through a throttling valve or other throttling devices, it experiences partial evaporation, or flashing leading to the production of partial vapour. This is one of the most basic unit operations.

The term "flash drum" is frequently used to describe a pressure vessel that has a throttling valve or other device placed at the vessel's entrance, allowing flash evaporation to occur inside the vessel.

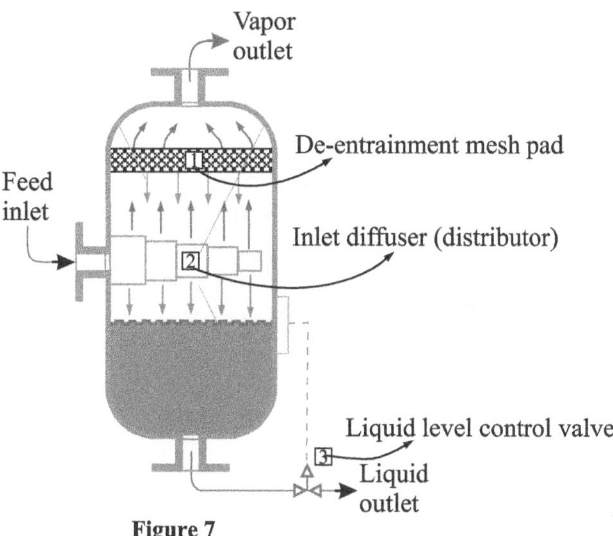

Figure 7

Part of the saturated liquid instantly "flashes" into vapor if it is a single-component liquid (hot water, for example).

At the lower pressure, the liquid's saturation temperature is reached by both the residual liquid and the vapor.

Explanation: How Flashing is Isenthalpic?

Let us take an example

Inlet to flash drum at 5 bar pressure

Let 1 kg saturated condensate enter the flash drum at 5 bar pressure

Enthalpy of condensate at 5 bar = 671 KJ/kg

Temperature of condensate = 159 degc

The outlet of the flash drum at 1 bar pressure

Enthalpy of water = 419 KJ/kg

Enthalpy of vapor = 2257 KJ/kg

Temperature = 100 degc

Let us do an enthalpy balance

What do We Know

[1] I kg hot condensate at a pressure of 5 bar g, enters the flash drum with enthalpy 671 kJ/kg of heat energy at its saturation temperature of 159 degc.

[2] Hot condensate flashes in the drum and cools to 100 degc at 1 bar and enthalpy 419 KJ/kg

[3] During the process of cooling the hot condensate releases latent heat = [671–419] = 252 KJ at 100 degc and generate vapor

[4] 252 KJ heat residual latent heat produces = 252/2257 = 0.11 kg vapor [heat of vaporization of water is at 1 bar 2257 KJ/kg]

Therefore 1 kg mixture at the outlet of the flash drum consists of

Vapor = 0.11 kg
Liquid water = 0.89 kg

Enthalpy of the liquid + vapor at the outlet of the flash drum

= 0.11 × 2675 + 0.89 × 419 [latent heat of steam at 1 bar is 2675 KJ/kg]
= 298 KJ + 393 KJ = 671 KJ/kg

Hence enthalpy in = Enthalpy out in the flash drum and the process is isenthalpic.

LP, MP, HP AND VHP STEAM

An important point is that pressure windows for *LP*, *MP*, and *HP* are the broad-use practices followed in the industry according to the process requirements.

The primary distinctions among low-pressure (LP), medium-pressure (MP), high-pressure (HP), and very high-pressure (VHP) steam are found in the pressure and temperature ranges. The pressure and temperature requirements of the different processes determine the particular applications for *LP, MP, LP*, and *VHP* steam. Chemical processes optimize the generation and distribution of these steam levels to meet process demands throughout the facility and maximize energy efficiency.

Typical pressure and temperature ranges are as follows:

- **LP steam:** 0–15 psig (pounds per square inch gauge) and 212–240°F.
- **MP steam:** 15–150 psig and 250–366°F.
- **HP steam:** 150–600 psig and 366–482°F.
- **VHP steam:** Over 600 psig and over 482°F.

The temperature of the steam increases with increasing pressure. This has an impact on the energy content and the possible uses of the steam.

The particular application and the need for temperature, energy density, and other variables will determine the steam pressure to use. In general, higher-pressure steam has more energy per unit volume, but handling the higher pressure also calls for more specialized equipment.

Because it has the lowest temperature and pressure, low-pressure industrial processes, water heating, and space heating are common uses for low-temperature LP steam. MP steam is utilized in more intensive industrial processes because of its higher energy density. With their extremely high energy content, HP and VHP steam are utilized in refineries, power plants, and other heavy industry applications that call for extremely high pressures and temperatures.

Typical Example: Petroleum Refineries

LP steam: Utilised in vacuum distillation units, crude oil distillation units, and certain fractionation towers for heating, stripping, and distillation operations that call for lower-temperature steam. Used in steam-driven machinery such as ejectors and steam turbines, as well as reboilers and steam heaters. It is also used in steam tracing.

MP Steam: Used in certain fractionation towers, hydrotreating units, and catalytic reforming units where higher temperature heating requirements are required than in LP steam. Used in equipment that requires intermediate steam pressure, such as reboilers, process heaters, and steam-driven machinery.

HP Steam: Used in the refinery's highest-temperature heating applications, including steam methane reformers, hydrogen production units, and fluid catalytic cracking (FCC) units. Used in steam turbines, process heaters, and other machinery that needs high-pressure steam.

EVIL OF WET STEAM

What is Wet Steam

Wet steam is the visible mist or aerosol of water droplets formed as water vapor condenses. Wet steam has two phases, water, and vapor. Wet steam cloud of vapor and water is visible while saturated or superheated steam is invisible.

How Wet Steam Forms

When a steam boiler heats up the water, bubbles breaking through the water surface will pull tiny water droplets in with the steam. Unless a superheater is used, this will cause the steam supply to become partially wet (wet steam) from the added liquid.

Steam Dryness Fraction

The steam dryness fraction is used to quantify the amount of water within the steam. If steam contains 10% water by mass, it's said to be 90% dry, or have a dryness fraction of 0.9.

Steam dryness is important because it has a direct effect on the total amount of transferable energy contained within the steam (usually just latent heat), which affects heating efficiency and quality.

Steam Dryness = 100% – [% Entrained Water] (by mass)

Journey of Water to Saturated Dry Steam along the Bell Curve

Explanation of steam quality and the thermodynamic properties of steam by a *T-S* diagram.

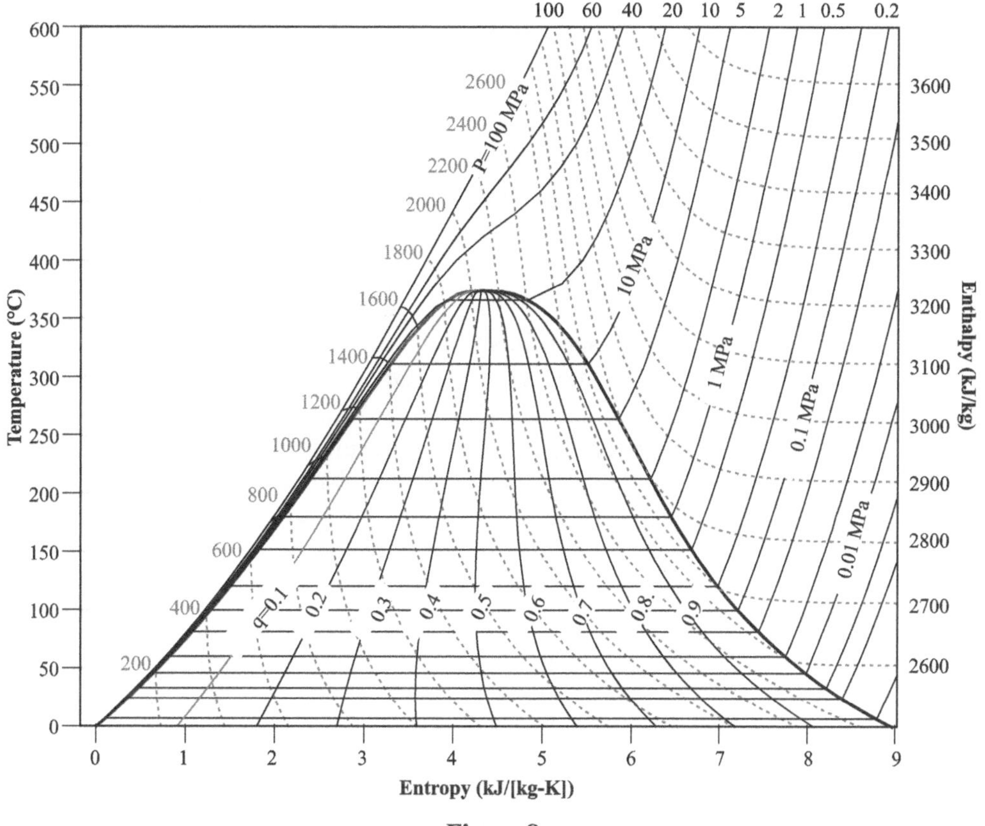

Figure 8

The image is temperature-steam entropy [TS] diagram with x-axis entropy and y-axis temperature. On the left side of the bell curve, there is single-phase water. *X* in red color is the dryness fraction for steam as steam proceeds to the right side in the bell. Inside the bell two phases, water and vapor coexist and on the right side, it is the one-phase superheated region for steam. The topmost point of the bell curve, the red sphere, is the critical point. Above critical point, there is the supercritical region

As water is heated at saturation temperature it begins conversion into vapor. This is the region within the bell curve.

The temperature gets fixed in the saturated region as heat goes to supply latent heat to water to convert to vapor. At $X = 1$ on the right contour of bell curve [dryness fraction = 100] Steam is 100% dry. The two-phase region ends. During the journey of two phases inside the bell curve entropy increases as more water converts to randomly moving vapor.

When steam is pressured outside the bell curve on right to generate enthalpy, the volume of steam reduces for the same mass of water and that limits the available space for steam particles to move around, the entropy reduces. The same happens till the critical point.

Definitions

Sub Saturated Water Region: This is on the left side of the bell curve.

At any pressure, water below its saturation temperature is said to be in a sub saturated state.

Saturated Water Line: This is the Bell Curve Line on the Left Side

The saturated water line shows water at its saturation temperature and depicts finite properties relating only to the saturated condition of the water.

Wet Steam Region: This is a Two-Phase Region Inside Bell

The wet steam region (also known as the two-phase region) depicts all values of steam in its wet condition. Its boundaries are the saturated water line and the dry saturated steam line. The addition of heat to steam at any point in the wet steam region causes the steam to dry, but always at the same temperature. The drier the steam, the closer it is to the dry saturated steam line.

Superheated Steam Region: This on the Right Side of the Bell Curve

The superheated steam region depicts steam at a temperature higher than its saturation temperature. Should saturated steam be heated at constant pressure, its temperature will rise, producing superheated steam.

Total Heat of Wet Steam

Steam tables contain values such as enthalpy (h), specific volume (v), entropy (s), etc. for saturated steam (100% dry) and for saturated water (0% dryness), but typically not for wet steam.

The wetter the steam, the lower the specific volume, enthalpy, and entropy will be because the dryness percentage is a factor of the 100% condition. Since steam dryness has a significant effect on all these values, to enable greater heating efficiency it is crucial to supply steam that is as close to being 100% dry as possible.

Saturated water

Dryness = 0%
Temperature = 100 degc
Total heat = 419 kj/kg

Wet steam

Dryness = X %
Temperature = 100 degc
Total heat < 2670 kj/kg

Dry steam

Dryness = 100 %
Temperature = 100 degc
Total heat = 2670 kj/kg

Disadvantages of Using Wet Steam

It is always advisable to use dry steam for heating applications. Dryness of steam is measured as expressed as dryness fraction. It is advised to use steam with maximum dryness fraction. Steam below the dryness fraction of 97–98% might lead to the following disadvantages.

Corrosion and Scaling

As steam evaporates from the boiler drum, it leaves back the impurities in the drum. When steam is wet, the water droplets have impurities dissolved in them. These impurities get carried into the steam network and process equipment. The impurities carried in damage the steam system by corroding it or by scale formation. The flow accelerated corrosion by high velocity two phase water-steam flow is one of the major issues.

Water Hammer

Wet steam can increase the chances of occurrence of water hammer. Water hammer is a phenomenon where a lump of condensate traveling at high speed hits the piping or some equipment fitted on the piping. The dryer the steam, the lesser are the chances of water hammer.

Reduced heat transfer

If a water film is formed over the heat transfer surface, it will reduce the rates of heat transfer as the thermal conductivity of water is much lower than that of steam. This will result in increased batch timings and uneven product heating.

Increased Load on Steam traps

Wet steam implies a large quantity of condensate formed. This increases the load on steam traps. Steam Traps are designed to discharge a certain amount of condensate. Wet steam might result in condensate quantities higher than trap capacity; this will result in undesired phenomena such as water hammers and waterlogging.

Damage to Equipment

Wet steam damages both process equipment and various equipment fixed on steam lines. Wet steam can damage equipment such as flow meters, control valves, etc. Hence, it is advisable to fix a moisture separator before flow meters, control valves, etc.

STEAM TRAP ISSUES

Steam traps are difficult to maintain. They are difficult to troubleshoot. They are numerous and often geographically inaccessible. Since the energy wasted by a trap is usually unseen, it is tempting to ignore maintenance. Unfortunately, neglecting faulty traps is terribly expensive.

The post has three parts [1] Steam trap – General [2] Failures of steam traps and consequences and [3] Why steam traps malfunction.

Steam trap General

A steam trap is a kind of automatic valve that discharges condensate generated during processes and in steam transport lines along with non-condensable gases with a negligible consumption or loss of live steam.

Types of Steam Trap: See the Images Below

There are three types of steam traps. Each works based on the thermodynamic properties of steam and condensate.

Mechanical Steam Traps

Thermodynamic Steam Traps

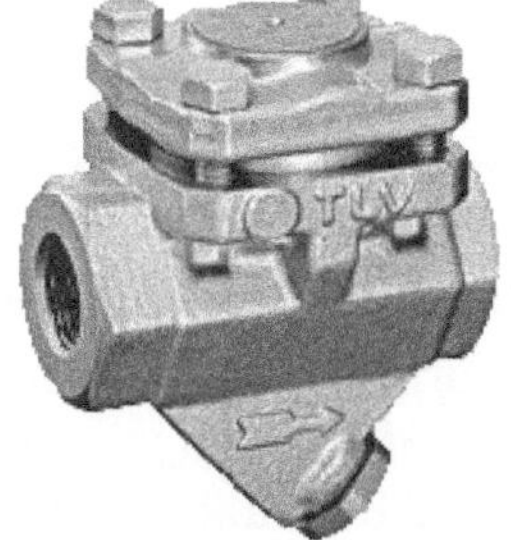

Thermostatic Steam Traps

Figure 9

[1] *Mechanical Steam Trap*

It traps condensate through the use of the mechanical properties of steam *vs.* condensate. Since the liquid is denser than the steam it will travel to the bottom of the system. Mechanical traps will have a bucket or float that rises and falls in relation to condensate level and this usually has a mechanical linkage attached that opens and closes the valve. Mechanical traps operate in direct relationship to condensate levels present in the body of the steam trap.

[2] ***Thermostatic Traps***

It traps and removes condensate through the temperature difference of steam *vs.* the liquid phase. The valve is driven through the expansion and contraction of an element that is exposed to the heat from steam or condensate. These traps take a temperature drop below the saturation curve to open and remove condensate. These traps can perform this by a filled element or bellows or bimetallic element.

[3] ***Thermodynamic Traps***

It works on the dynamic principles of steam *vs.* condensate and the use of Bernoulli's principle. When condensate released through an orifice the speed increases and a pressure drop occurs. This will flash steam to create higher pressure to close a valve (disc) or slow the discharge speed of the trap.

Consequences of Steam Trap Failures

Steam traps fail in two ways: Open and closed.

Steam Traps Fail-Open

When a steam trap fails open, it acts as an open valve, completely leaking the steam and condensate. When a steam trap fails closed, it acts like a closed valve not allowing the steam and condensate to pass through it. Whether a steam trap fails open or closed, it does equal damage to the steam system and hence, should be repaired or replaced.

Steam Traps Fail-Closed

If a steam trap fails-closed, the following are the different problems associated with the failure.

Water Hammer and Pressure Surges

The presence of water i.e., condensate in the steam system results in a water hammer. Condensate, if not removed by a steam trap accumulates at the bottom of the steam lines. In such situations, a bi-phase flow exists in the piping. During such bi-phase flows, condensate is dragged by the steam to travel along with the steam at speeds much higher than the maximum allowable speed for condensate. This results in slogs of condensate traveling at very high velocity in the steam lines. When this slug hits some bend or some equipment, a pressure wave is generated which is called a water hammer. Fail closed steam traps are the major reason for water hammer.

Water Logging in the Process

When a steam trap installed on process equipment fails closed, it stops removing condensate from the process. As a result, condensate accumulation takes place inside the process vessel. Waterlogging has multiple disadvantages with respect to the performance of the process. First of all, the product which is surrounded by water won't get heated. This results in non-uniform heating of the product. Waterlogging will also increase the time required to heat the product to the desired temperature.

Damage to Piping and Process Equipment

Condensate is pressurized fluid. If condensate remains in the steam system and there is a sudden drop of pressure, it will flash and will expand in its volume. Such sudden expansion can damage the piping. Water hammer, as well as moisture entrained in the piping, can damage various equipment fitted on the line.

Compromised Safety

As it can be seen from the points mentioned above, the presence of water in the system is a possible safety hazard. A failed steam trap compromises the safety of the steam system.

Steam Traps Fail-Open

When a steam trap fails open, the following are the problems that might be faced.

Increased Steam Consumption

When a steam trap fails open, it simply acts as an open valve, and hence, a lot of steam would leak through it. As a result, the overall steam consumption of the plant will go up, reducing the efficiency of the operation.

Increased Load on the Boiler

If a leaking trap is discharging to the drain, the entire steam and hence the heat energy that is being leaked will be wasted. This also increases the boiler load. Many times, it has been observed that the same size of boiler can be used to heat more processes if all the pipe leaks and leaking steam traps are repaired or replaced.

Why Steam Traps Malfunction?

When steam leaves the boiler house, it is mostly dry steam. As steam passes through the pipe-work, radiation losses take place and condensate is formed. During the heating applications, steam loses its heat to the products being heated and as a result, condensate is formed. Steam traps do the important task of removing this condensate as soon as it is formed.

Following are the key reasons for steam trap failures.

Feedwater Contamination

If feed water has dissolved gases like oxygen or carbon dioxide, corrosion of steam traps and other components will take place. Oxygen results in pitting whereas carbon dioxide dissolves in condensate to form carbonic acid which turns the condensate acidic. Improper feed water treatment also results in carryover which also contributes to steam trap failure.

Steam Trap Sizing

Steam traps should be sized correctly. Undersized steam traps tend to fail faster compared to correctly sized steam traps. It is a common mistake to size the steam traps based on condensate formed during the running mode. One must take into consideration the start-up condensate formation load while sizing the steam traps.

Incorrect Installation of Steam Traps

The operating life of steam traps gets significantly reduced if they are not installed correctly, keeping in mind the recommended installation guidelines. For instance, a strainer should be installed upstream of every steam trap. While some steam traps have inbuilt strainers, for some steam traps, a separate strainer needs to be installed before the steam trap. The absence of the strainer might result in the entrainment of rust or other debris inside the trap and the trap internals might get damaged.

Improper Maintenance Practices or Lack of Maintenance

Every steam trap needs the right kind of maintenance. It has been observed that at many plants, steam trap maintenance is either improperly carried out or is completely missing. In fact, every steam system should have its own maintenance calendar which should be strictly followed.

Contamination of Condensate

Leakage in the heat exchangers or use of multiple utilities results in condensate contamination. Contamination of condensate can reduce its pH and turn it acidic. Acidic condensate corrodes the internals of the steam traps and hence, is a reason why steam traps fail.

Selecting the Wrong Type of Steam Trap

Steam trap selection is probably the most crucial step which determines the performance as well as operating life of a steam trap. Many times, it is observed that the wrong kind of steam trap is installed on the wrong application. The result is lowered life of the steam trap.

DIFFERENCE BETWEEN CONDENSATION OF SATURATED STEAM AND SATURATED AIR

When saturated steam condenses, it occurs at a constant saturation temperature and pressure. This means that the steam will condense into liquid water without changing its temperature or pressure. On the other hand, condensation from saturated air can happen at the dew point when air is cooled below the wet-bulb temperature. It's a function of dry-bulb temperature and relative humidity as they exist at that point of time.

The dew point is the temperature at which air becomes saturated with water vapor and condensation occurs. This is because the air is cooled below its wet-bulb causing the water vapor to condense into liquid water.

The space between wet bulb temperature and dew point temperature is important because here intermolecular bonds between water molecules form for vapor to transition into liquid.

In the case of saturated air's water, condensation it can occur at various dry-bulb and relative humidity of air as explained above.

In the condensation of saturated steam, the boiler maintains constant temperature and pressure to supply latent heat, which is then converted to sensible heat during the heating process. This helps ensure a consistent energy transfer and temperature control in systems utilizing steam.

On the other hand, when saturated air condenses water through a cooling process, the vapor phase loses kinetic energy as it cools down, causing it to coalesce into a liquid phase.

The condensation process of saturated steam is reversible, meaning that the steam can transition back to its vapor phase under the same conditions. Condensation of steam is a classic example of thermodynamic reversibility. Latent heat of condensation and evaporation is same.

On the other hand, when saturated water vapor in the air condenses, it is an irreversible process involving a decrease in entropy. The reason is this condensation is a function of temperature and relative humidity which change continuously.

Condensation of water vapor in the air involves the transition from a disordered state (vapor phase) to an ordered state (liquid phase), resulting in a decrease in entropy. The condensation process leads to a decrease in the disorder of the water molecules in the air as they come together to form liquid water droplets. This decrease the entropy.

Therefore, while saturated steam condensation is an equilibrium process, saturated air condensation is a nonequilibrium irreversible process.

Almost all systems found in nature are not in thermodynamic equilibrium, for they are changing or can be triggered to change over time, and are continuously and discontinuously subject to flux of matter and energy to and from other systems.

Psychrometry is one such subject because RH is irreversible.

A fundamental and very important difference is the difficulty, in defining entropy at an instant of time in macroscopic terms for systems not in thermodynamic equilibrium.

CHAPTER 55

Psychrometry

In psychometrics, processes and conditions involving moist air are analysed using thermodynamics. In the field of heating, ventilation, air conditioning, and refrigeration (HVACR), a solid grasp of psychometrics is essential. In addition to evaluating and designing heating and cooling systems and guaranteeing building occupant comfort, psychometry is used in the construction of building materials (such as roofing and insulation) and in determining the stability and fire resistance of those materials. Heat exchangers, pumps, compressors, throttles, and turbines are the fundamental parts of HVACR systems. These are described, categorised, and thermodynamically examined.

WHAT IS PSYCHROMETRY?

Psychrometry is the science of studying the properties of moist air. Moist air is a mixture of dry air and water vapor. In atmospheric air, water vapor content varies from 0 to 3% by mass.

Properties of Air

Air-water equilibrium.

What is Air-water Equilibrium?

In the air-water system, we have a non-condensable gas, air, and a volatile liquid, water [small solubility of air in water is ignored] Due to the overlap in the energies in the molecules of liquid and water vapor, water molecules are continuously moving from one state to the other.

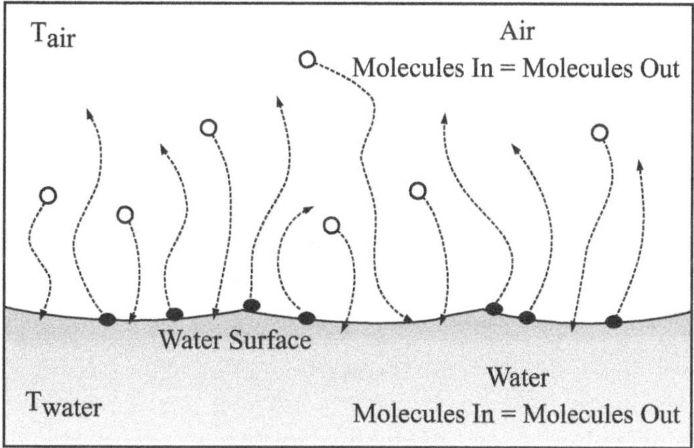

Figure 1: Equilibrium

The image shows, that when some molecules in the water near the surface [liquid] attain high enough kinetic energy to escape the water surface by overcoming liquid-phase intermolecular forces, they become molecules of water vapor. Only a fraction of liquid water can escape and become vapor. Most molecules do not have enough kinetic energy to escape, condense into liquid, and fall to join the liquid. When water is in equilibrium with moist air, there is an equal exchange of water molecules between liquid and vapor.

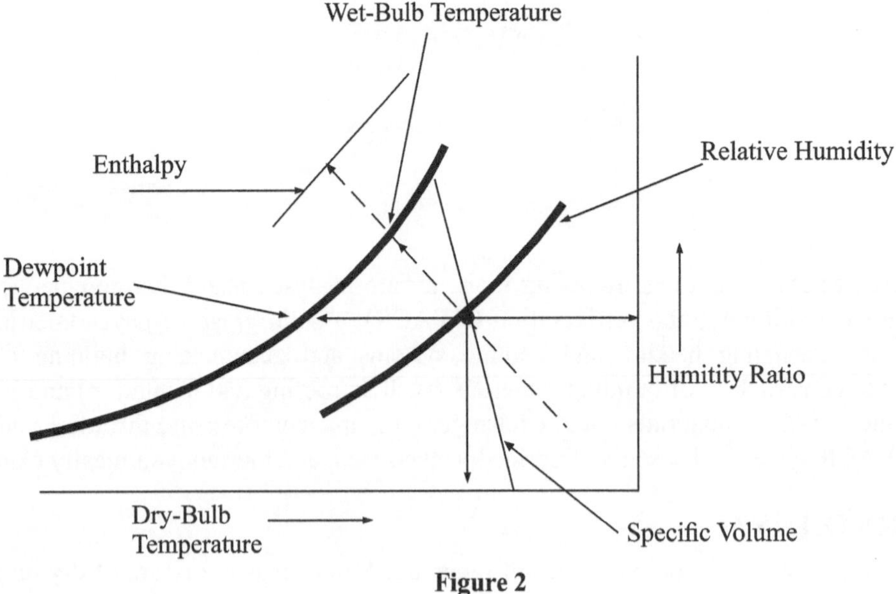

Figure 2

Psychrometric Chart

[One can also use Mollier's psychrometric chart-2nd image]

A psychrometric chart is a graphical representation of the psychrometric processes of air at constant pressure.

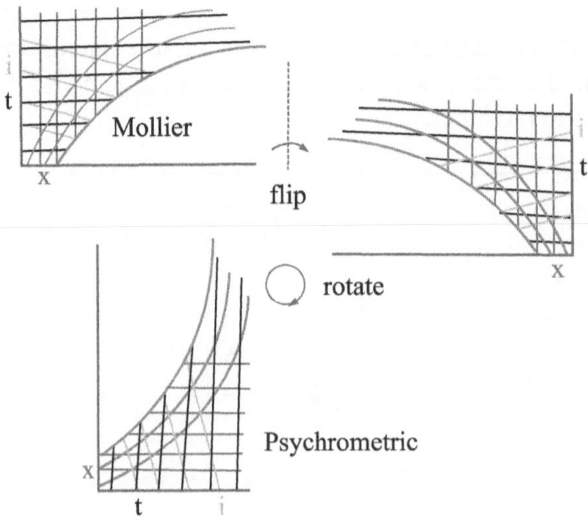

Figure 3

Psychrometric processes include physical and thermodynamic properties such as dry bulb temperature, wet bulb temperature, humidity, dew point, enthalpy, and air density, etc.

The various lines shown in the Figure are explained below.

1. **Dry-bulb [DBT] temperature lines**: DBT points are given on the x-axis. The vertical lines emerging from each DBT point on the x-axis are DBT lines. It is the ambient temperature of the air.
2. **Relative humidity [RH]:** Steady RH is represented by the bent lines running from the base left to the upper right of the graph. RH is the amount of water vapor present in air expressed as a percentage of the amount needed for saturation at the same temperature.

Question: Why These Lines are Curve Lines?

RH is a ratio of partial vapor pressure in air and the equilibrium vapor pressure of water at that temperature. These curve lines are vapor pressure lines of moisture in the air at the given DBT. Temperature and vapor pressure are non-linearly related (Clausius and Clapeyron relation). Therefore, all RH lines in the psychrometric chart are curve lines. WBT is the RH line at 100% saturation. DBT has nonlinear relation with RH or WBT.

3. **Humidity ratio:** The humidity ratio is given on the humidity ratio axis. It represents an air/moisture mixing ratio. The line parallel to the x-axis emerging from each point on the humidity ratio axis is the humidity ratio line [see the psychrometric chart]. It is moisture in the air per unit weight of dry air.
4. **Wet-bulb temperature [WBT] lines**: WBT line emerges from the intersection of DBT and RH and goes along the diagonal line. The point where this diagonal line intersects the saturation line is WBT. It is an indicator of moisture air at 100% saturation. WBT is the temperature of air if air gives off all its absorbed enthalpy [moisture] adiabatically and returns to its initial RH.
5. **Dew-point temperature [DPT] lines**: DPT lines originate from humidity ratio points on the humidity ratio axis and go parallel to the x-axis. The point where this line intersects the saturation line is DPT. The dew point is the temperature to which air must be cooled to become saturated with water vapor. When further cooled, the water vapor in the air will condense to form liquid water.
6. **Enthalpy line**: This is the extension of the WBT point along the diagonal. The point where the extension of WBT intersects the enthalpy line is the enthalpy in the air for a given DBT/RH/WBT. The enthalpy of moist air is the sum of the enthalpy of dry air and the enthalpy of water vapor which is latent heat stored in water vapor. The enthalpy line shows the heat content of dry air. There is a linear relation between DBT and enthalpy of dry air since the specific heat is constant. Heat or average kinetic energy is directly proportional to temperature, $KE = 5/2\ \text{Kb} \times T$ for diatomic air. *Kb* is the Boltzmann constant.

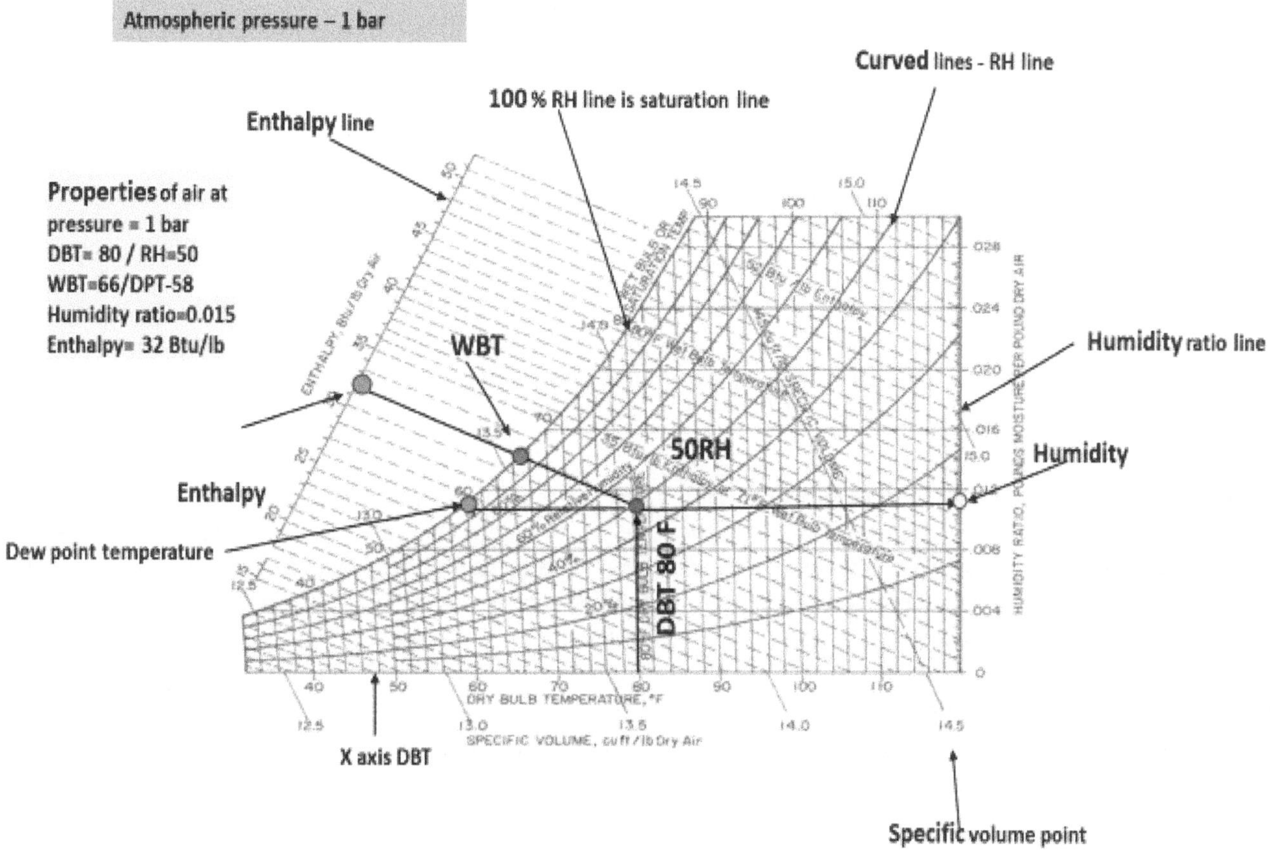

Figure 4

7. **Specific volume line**: On the psychrometric chart, lines of the constant specific volume are almost vertical lines with scale values written below the dry-bulb temperature scale and above the upper boundary's saturation temperature scale. Greater specific volume is an indicator of more moisture in the air. The specific volume of a substance is the ratio of the substance's volume to its mass. It is the reciprocal of density. Important point: The specific volume of dry air is about 17 ft^3/lb while for water vapor it is about 28 ft^3/lb. Dry air is denser than moist air. These numbers have important roles in air psychrometrics.
8. **Saturation line**: This is a 100% RH line when air is fully saturated with moisture. At saturation DBT, WBT and DPT are all same [see the psychrometric chart].

The intersection of any two variables can define all properties of air.

All data used are on an Imperial scale.

Calculation

Let us take a typical case, air DBT/RH, 80/50 at constant pressure, and calculate the properties of air from a psychrometric chart:

The Blue central dot is the intersection of 80 DBT and 50 RH. The diagonal extension of the blue dot to the red dot identifies WBT, 66F. The extension of the red dot along the same diagonal line to the green dot on the enthalpy line is enthalpy, 32 Btu/lb. Extension of the blue dot to the orange dot on the left side parallel to the x-axis intersecting 100 RH or saturation line is DPT, 58 F. Extension of blue dot to the right side parallel to x-axis intersecting humidity ratio line is humidity in-lb water/lb dry air, 0.015.

The Shift of Air-water Equilibrium

RH: Relative humidity is given as a percentage and tells you how close the air is to being saturated. If the relative humidity is 100%, the air is saturated. If the relative humidity is 50%, the air contains half the water vapor required for it to be saturated. If the amount of water vapor in the air increases, the humidity ratio increases.

RH is the ratio of the partial pressure of water vapor to the equilibrium vapor pressure of water at a given temperature. RH depends on the temperature and pressure of the system. As air temperature increases, air can hold more water molecules, and its relative humidity decreases. When temperatures drop, relative humidity increases. High relative humidity of the air occurs when the air temperature approaches the dew point value. Temperature therefore directly relates to the amount of moisture the atmosphere can hold.

What Happens when Two Key Psychrometric Variables, DBT/RH Change?

Constant DBT *vs.* variable RH: At DBT 80F and variable RH. Reduce RH. As RH reduces, there is less moisture and more air. The specific volume of water vapor is 28 ft^3/lb against 17 ft^3/lb for air at 1 bar pressure. Air is denser than water vapor. As moisture reduces and air increases, the volume of air reduces at constant pressure. This results in a drop in temperature and condensation of water. THIS IS DEW POINT. As a result, there is a drop in humidity ratio, DPT, WBT, and enthalpy of air.

Constant RH *vs.* variable DBT: At RH 50 and variable DBT: Increase the temperature. As temperature increases, air's capacity to hold more water triggers evaporation wherever it finds water. This results in an increase in humidity ratio, DPT, enthalpy, and WBT. DBT-WBT gap narrows. Air moves towards saturation. On a similar logic, if at 50 RH, DBT has reduced a point comes when there is a condensation of water because air has lost its moisture-holding capacity at a reduced temperature. Humidity ratio, DPT, enthalpy, and WBT reduce.

The General Relation between RH and Temperature

As temperature increases, air expands and it can hold more moisture, therefore an increase in temperature decreases the RH of air. Unsaturated air triggers evaporation. The same logic applies when the air cools. Air contracts and therefore has less capacity to hold water. Saturated air triggers condensation.

NON-EQUILIBRIUM THERMODYNAMICS: PSYCHROMETRY

The best application of nonequilibrium thermodynamics can be noticed in atmosphere.

Non-equilibrium thermodynamics is a branch of thermodynamics that studies physical systems that are not in thermodynamic equilibrium.

This includes systems that are changing or can be triggered to change over time.

Nonequilibrium thermodynamics arises from study of systems that are not in thermodynamic equilibrium, leading to non-uniform behavior of thermodynamic potentials. In systems experiencing nonequilibrium conditions, relationships between variables such as temperature, pressure, and entropy may not follow standard equilibrium thermodynamic relationships, leading to the need for additional theoretical frameworks to describe and understand their behavior.

A typical example of nonequilibrium thermodynamics is the psychrometry of wet air.

Wet air is always in a state of non-equilibrium. Temperature, vapor pressure, and the volume of air is continuously changing in ambient air and shifting internal energy, enthalpy, and entropy.

Non-equilibrium thermodynamics is a work in progress, not an established structure. A non-thermodynamics system is changing or can be triggered to change over time, and are continuously and discontinuously subject to flux of matter and energy to and from other systems.

Since composition of vapor is generally inhomogeneous in a moist atmosphere, vapor has a diffusive process. Energy transport is also associated with diffusion of vapor.

Diffusion is the net movement of anything (for example, atoms, ions, molecules) from a region of higher concentration to a region of lower concentration.

Diffusion in Psychrometry is a Nonequilibrium Thermodynamic Process

Diffusion refers to movement of water vapor molecules from areas of high concentration to areas of low concentration, driven by the concentration gradient.

This process is not at equilibrium because the system is constantly changing as water vapor molecules move and mix in the air.

Diffusion processes in psychrometry, involve changes in intermolecular interactions with time and are inherently unpredictable.

Entropy, Enthalpy, Free Energy of Wet Air is Changing Constantly

Explanation

Entropy, enthalpy, and free energy are related to relative humidity of air through Gibbs free energy equation, which is a fundamental concept in thermodynamics.

Entropy Enthalpy and Free energy are "Work in Progress" as wet air moves towards saturation and unpredictable.

Gibbs Free Energy Equation

$dG = dH - TdS$, G is free energy H is enthalpy T is temperature and S is entropy.

Relative Humidity

When the relative humidity of air reaches 100%, the entropy of the wet air reaches its maximum value. This is because the specific volume of air increases with its moisture content.

Equilibrium

When the relative humidity of air reaches 100%, the Gibbs free energy of the system reaches its minimum value, moving it into a state of equilibrium.

MOLLIER DIAGRAM AND PSYCHROMETRY

The psychrometric chart stems from the Mollier diagram.

The psychrometric chart is a graphical representation of the thermodynamic properties of moist air, including temperature, humidity ratio, relative humidity, enthalpy, and specific volume.

The Mollier diagram, also known as the h-s diagram or enthalpy-entropy diagram, is a graphical representation of the thermodynamic properties of a substance or mixture.

In the case of air-water mixtures, the Mollier diagram shows the enthalpy of moist air as a function of temperature and humidity ratio.

The psychrometric chart is a simplified version of the Mollier diagram that specifically focuses on the properties of moist air.

Psychrometry is the study of the thermodynamic properties of air-water mixtures. Psychrometric charts or Mollier diagrams are graphical representations of the thermodynamic properties of moist air at various conditions. These tools are essential for HVAC engineers, meteorologists, and other professionals who need to analyze and design systems that involve air-water mixtures.

The development of the psychrometric chart can be traced back to the work of Richard Mollier, a German engineer who in the early 20th century conducted extensive experiments on the properties of moist air. Mollier's research formed the basis for the development of the Mollier diagram, a graphical representation of the thermodynamic properties of moist air.

The Mollier diagram is a graphical representation of the enthalpy of moist air as a function of temperature and humidity ratio. It consists of a series of curves and lines that represent various properties of moist air, such as enthalpy, relative humidity, specific volume, and dew point temperature.

To derive a psychrometric chart or Mollier diagram, one must first conduct experiments to measure the properties of moist air at various conditions. These data are then used to plot the curves and lines that represent the properties of moist air on a graph.

The enthalpy of moist air is a key property that is commonly represented on a psychrometric chart. Enthalpy is a measure of the total energy content of a substance, and in the case of moist air, it includes both sensible heat (temperature) and latent heat (humidity). By plotting the enthalpy of moist air as a function of temperature and humidity ratio on a graph, one can easily analyze and any design systems that involve air-water mixtures.

In summary, the psychrometric chart stems from the principles of the Mollier diagram, which provides a more detailed and comprehensive analysis of the thermodynamic properties of moist air and other substances. The psychrometric chart is a practical and widely used tool that is based on the concepts of the Mollier diagram.

ROLE OF ENTHALPY, ENTROPY AND FREE ENERGY IN PSYCHROMETRY

Here comes the role of Mollier diagram.

A Moller diagram is a graphical representation that shows the thermodynamic properties of the mixture at different states or conditions.

In a Moller diagram, lines representing constant enthalpy, constant entropy, and constant free energy can be plotted to help analyze and interpret the behavior of the system. By studying the intersection points of these lines and the changes in state along a process path, one can gain insights into the energy exchanges, heat transfer, and phase changes that occur within the system.

By incorporating the concepts of enthalpy, entropy, and free energy within a Moller diagram framework, it becomes easier to make informed decisions and design choices that can maximize the performance and sustainability of air-water vapor mixtures in various applications.

Psychrometry

Psychrometry is an interplay of enthalpy, entropy, and Gibbs free energy.

Fundamentals

Enthalpy is the measure of the total energy of a system, including internal energy. In psychrometry, enthalpy is used to quantify the heat content of moist air, which is a combination of the dry bulb temperature and the moisture content of the air.

Entropy, on the other hand, is a measure of the disorder or randomness of a system. In psychrometry, entropy reflects the randomness of the air-water vapor mixture and is used to determine the energy required to change the state of the system, such as during evaporation or condensation.

Free energy, also known as Gibbs free energy, is a measure of the energy available to do work in a system. In psychrometry, free energy is used to determine the spontaneity of processes such as evaporation or condensation.

Role of Entropy

Entropy plays role in psychrometry in the phase change of gaseous water's condensation into liquid water. Phase change occurs when the system reaches maximum entropy and the free energy reaches minimum. The guiding equation is $dG = dH - TdS$, here G is free energy, H Is enthalpy T is temperature and S is entropy. When the change in entropy dS reaches maximum, the free energy change dG reaches minimum. When $dG = 0$, the two phases reach a thermodynamic equilibrium. Thermodynamic equilibrium is the ultimate stable state of a system which every system strives to achieve.

Explanation

In one line since entropy does not reach maximum and free energy never reaches minimum a cooling tower never reaches saturation in its prevailing conditions.

Dry bulb, wet bulb, and dew point temperature never become equal. The wet air in the cooling tower never gets saturated. We never see plumes inside a cooling tower.

Relative humidity refers to how close the air is to being saturated with moisture. When air is saturated, both its liquid and vapor phases are in thermodynamic equilibrium with identical state properties, maximum entropy, and minimum Gibbs free energy. This principle applies to wet air as well. Relative humidity (RH) measures how far the air composition is from minimum free energy and maximum entropy. For example, saturated air has the maximum entropy when the partial pressure of moisture equals the vapor pressure, resulting in 100% relative humidity. As relative humidity decreases, the air becomes less saturated and its entropy decreases.

Since air never reaches equilibrium the entropy never reaches maximum in a cooling tower.

ROLE OF HEAT CAPACITY OF AIR IN PSYCHROMETRY

This note gives an idea of how thermodynamics plays a crucial role in psychrometry.

The heat capacity of wet air plays a fundamental role in psychrometry, and it's sometimes overlooked when considering the properties of air and moisture mixtures. In psychrometry, understanding the heat capacity of wet air is crucial for properly analyzing and characterizing the thermodynamic properties of air and water vapor mixtures.

Dry Bulb Temperature

The dry bulb temperature is a measure of the heat content of ambient air at a given temperature and pressure. It is an important parameter in psychrometry and is often used as a reference point for determining other properties of air, such as relative humidity, dew point, and enthalpy.

Wet Bulb Temperature

At the wet bulb temperature, the heat capacity of air is at its maximum. This is because the wet bulb temperature represents the lowest temperature that can be achieved by evaporative cooling, and at this temperature, the air is at its maximum moisture content. During evaporative cooling, the air absorbs moisture as it evaporates water, and this process causes the air temperature to decrease. The heat capacity of the air at the wet bulb temperature is maximized because the air carries the highest amount of moisture, and as a result, it has the potential to release the most heat when it undergoes evaporative cooling.

Relative Humidity

Relative humidity is the ratio of the partial pressure of water vapor in the air to the vapor pressure of water at a given temperature. It measures the moisture content or the saturation level of the air. At any given moment, the heat capacity of the air is influenced by relative humidity because the presence of moisture affects the air's ability to absorb or release heat. Likewise, the heat capacity of the saturated air is also influenced by relative humidity as it determines the air's ability to absorb additional moisture.

Dew Point

The heat capacity of moist air can be calculated as a weighted average of the heat capacities of the individual components (i.e., water vapor and dry air). This is correct, as the overall heat capacity of moist air takes into account the contributions of both water vapor and dry air. At the dew point, the air becomes saturated with moisture, which means that the air's capacity to hold water vapor is at its maximum for that temperature. As a result, the heat capacity of the moist air at the dew point reflects the combined heat capacities of the water vapor and the dry air, which are now at their maximum for that temperature and pressure.

Enthalpy

The enthalpy H of wet air measures the total energy of the air, and it is expressed as the sum of the internal energy U and the product of pressure P and volume V. The internal energy U is a function of the heat capacity of wet air, as it represents the sum of the kinetic and potential energies of the molecules within the system.

DRY BULB AND WET BULB TEMPERATURE THERMODYNAMICS

Why does air not cool in a cooling tower? The answer is air receives potential energy in the form of latent heat and potential energy does not contribute to temperature. Latent heat is a form of potential energy which is held by the intermolecular bonds.

The post aims to delve deeply into these two fundamental temperatures in HVAC and seek an explanation from thermodynamics.

Definition

DBT

DBT is the temperature that is usually thought of as air temperature, and it is the true thermodynamic temperature. It is directly proportional to the mean kinetic energy of the air molecules.

WBT

Wet Bulb temperature can be measured by using a thermometer with the bulb wrapped in wet muslin. The adiabatic evaporation of water from the thermometer and the cooling effect is indicated by a "wet bulb temperature".

Thermodynamics

The dry bulb temperature is a measure of the actual temperature of the air, including the effects of air movement and humidity. It represents the kinetic energy of the air molecules, as it measures the speed at which they are moving.

The wet bulb temperature, on the other hand, is a measure of the lowest temperature that can be achieved by the evaporation of water into the air. This represents the potential energy of the air, as it measures the energy required to convert liquid water into water vapor by breaking intermolecular bonds.

DBT-WBT

The difference between the dry bulb and wet bulb temperatures, known as the wet bulb depression, reflects the potential for energy conversion within the air. When air expands, it can precipitate its moisture into water, releasing the stored potential energy as kinetic energy. This energy can then be used to do work, such as driving a wind turbine or powering a weather system.

In summary, the dry bulb temperature accounts for the kinetic energy of the air, the wet bulb temperature accounts for the potential energy of the air, and the difference between the two represents the conversion potential of energy within the air. The wet bulb temperature is a measure of the maximum cooling potential of saturated air through the process of evaporative cooling. When air comes into contact with a moist surface or water, the wet bulb temperature represents how low the air temperature can reach by evaporating water into it. This concept is important in areas such as HVAC and refrigeration, where understanding the wet bulb temperature is crucial for determining the efficiency of cooling systems.

In the process of evaporative cooling, potential energy in the form of heat is converted to kinetic energy as water changes from a liquid to a gas during the phase change. Then, when the water vapor condenses back into liquid form, the kinetic energy of air molecules is converted back to potential energy in the form of heat. This continuous conversion of energy helps in the cooling process.

BASIC PSYCHROMETRY: DBT, RH, AND WBT RELATIONSHIP

Humidity and Relative Humidity

Humidity

The amount of water vapour contained in one cubic metre of air under certain conditions is referred to as absolute humidity.

Relative Humidity

The ratio of the amount of water vapour in the air to the amount required for saturation at a specific temperature and pressure is known as relative humidity.

The definition of humidity is straightforward. The definition of RH often confuses.

RH in absolute terms is the moisture that is missing to saturate the air in a given temperature and pressure conditions.

In other words, RH signifies the extent of the unsaturation of the air. At 60 RH, the extent of unsaturation of air is less than 40 RH. It can quickly reach saturation over 40 RH air. That means 60 RH air needs a lesser amount of moisture or has less capacity to absorb moisture to reach saturation compared to 40 RH air.

RH and DBT Inverse Relationship

When the temperature of the air [DBT] increases its density reduces. This means the mass/volume reduces. In other words, there is more moisture [18 g/mol] than air [29 g/mol]. The RH, degree of unsaturation reduces. This explains the inverse relation between RH and temperature

Wet Bulb Temperature

The definition in books is confused often.

In simple words WBT is the minimum temperature air can achieve. The minimum temperature is the temperature, that air achieves when 100% saturated water gives up heat adiabatically at a given RH and DBT and cools.

DBT, RH and WBT relationship

DBT, RH and WBT are interrelated.

Explanation

Let us take a specific case.

'Why in winter WBT is low compared to summer?'

Step by Step

Because of the inverse relation, in winter at low temperatures, the RH is higher than in summer.

Explanation

In winter, the degree of unsaturation is higher [refer to the definition of RH] There is less moisture in the air. At saturation, the moist -air needs latent heat adiabatically which cools the air to WBT. When the saturation is 100% you get WBT which is the minimum temperature air can cool at a given RH and DBT.

The latent heat is the stored heat that depends on temperature. At lower temperatures, the latent is higher because of more intermolecular bonds. So in winter, 100% saturated air extracts more heat adiabatically from its internal energy to give up all moisture and cool more.

Therefore, WBT in winter is lower than in summer. In summer exactly the opposite happens. A lower WBT signifies larger cooling, typically in a cooling tower. This increases the ' range ' or the heat load to a cooling tower and increases the efficiency.

Efficiency = Range/[Range + Approach temperature] × 100

ASSUMPTIONS IN WET-BULB TEMPERATURE MEASUREMENT,

The Wet Bulb temperature is the air temperature of adiabatic saturation. This is the temperature indicated by a moistened thermometer bulb exposed to the air flow.

The big assumption here that air past the bulb is saturated and the bulb and wet sir are in thermal equilibrium. If it's not then wet bulb temperature is wrong. Also, it assumes zero approach temperature between the bulb and air..

More Specific

Evaporation is a part of the adiabatic saturation process, which occurs when air comes into contact with water without any external heat transfer.

Point 1

Each particle of water evaprates remove heat from air equal to the heat of vaporization of water which is about 2240 kj/kg. This is it what cools the air which is recorded by the wet-bulb. This is theoretical number. If it doesn't happen then wet bulb temperature is wrong.

Point 2

We assume.air in the surroundins is 100% saturated and the wet air is in thermal equilibrium with the bulb. If it doesn't happen then wet bulb temperature is wrong. Is this possible when the cold air in the immediate vicinity of the bulb is continuously being replaced by warmer air from the surroundings?

Point 3

The evaporation is first a mass transfer process then cooling.if 100% mass transfer doesn't happen then air is not saturated and wet bulb tempature is wrong.

Point 4

It assumes no approach temperature between the bulb and wet air. This implies perfect thermal equilibrium and infinite number of mass transfer stages between the bulb and wet air. Is it possible?

Question

Are these possible to achieve on ground?

WHAT IS RELATIVE HUMIDITY [RH]? ROLE OF ENTROPY

What is relative humidity [RH]?

The usual definition is RH = Partial pressure of moisture in air/Saturated vapor pressure × 100. This definition often confuses people. The definition has a hidden meaning.

RH of Air is a Thermodynamic Controlled Property of Air

Explanation

How do thermodynamics work in psychrometry? Role of entropy

Thermodynamics is the study of the relationships between heat, work, and energy. There are four fundamental laws of thermodynamics, which govern the behavior of energy and matter and are crucial to understanding the behavior of air as it reaches saturation and the relative humidity becomes 100%.

The first law of thermodynamics, also known as the law of conservation of energy, states that energy cannot be created or destroyed, only converted from one form to another. This law is important in understanding the energy exchange that occurs in the atmosphere as air reaches saturation.

The second law of thermodynamics states that the total entropy of an isolated system can never decrease over time. Entropy is a measure of disorder or randomness in a system. When the air becomes saturated with moisture, the entropy reaches its maximum value, and any further addition of moisture will result in the condensation of water vapor into liquid water, releasing latent heat and increasing the entropy of the system.

Additionally, the Gibbs free energy[G] equation, a fundamental concept in thermodynamics, describes the relationship between enthalpy changes, [H] entropy, [S] and the temperature, [T] of a system, $dG = dH - TdS$.

As the relative humidity of air reaches 100%, which happens at the dew point temperature, the entropy of the wet air reaches maximum because the specific volume of air increases with its moisture content. Correspondingly, the Gibbs free energy of the system reaches its minimum value, moving it into a state of equilibrium. Any further addition of moisture would result in the release of latent heat and the condensation of water vapor into liquid water, to maintain the equilibrium and minimize the free energy of the system.

Further explanation:

Role of Entropy

Relative humidity refers to how close the air is to being saturated with moisture. When air is saturated, both its liquid and vapor phases are in thermodynamic equilibrium as explained, with identical state properties, maximum entropy, and minimum Gibbs free energy. This principle applies to wet air as well. Relative humidity (RH) is a measure of how far the air composition is from minimum free energy and maximum entropy. For example, saturated air has the maximum entropy when the partial pressure of moisture equals the vapor pressure, resulting in 100% relative humidity. As relative humidity decreases, the air becomes less saturated and its entropy decreases.

SIGNIFICANCE OF ENTHALPY AND HUMIDITY RATIO IN PSYCHROMETRY

What Do These Two Lines Stand for?

You can see the amount of latent heat in the air by looking at humidity ratio lines. The total energy in the air is reported by the enthalpy line. The units of both are BTU/LB dry air. A psychrometric chart can be broadly divided into three sections. [1] The extreme right humidity ratio axis represents the air's latent heat. [2] The RH line in the centre of the grid represents the sensible heat exchange change. RH lines are vapor pressure lines.

$Q = m\ Cp\ dt$, caused by the temperature-dependent change in moisture vapor pressure.

[3] The enthalpy line on the extreme left represents the total energy transfer in a psychrometric process.

Why These Two Lines are Very Important Lines?

Simply glancing at these two lines will teach you about [1] air's latent heat, [2] sensible heat transfer in air, $Q = m\ Cp\ dt$, wet air specific heat occupies a very important part in sensible heat transfer, and [3] moisture evaporating or condensation from the air and [4] contribution of latent heat transfer in evaporation or condensation.

This note describes how engineers can reasonably analyze the heat movement in any psychrometric system by looking at the humidity ratio and enthalpy lines.

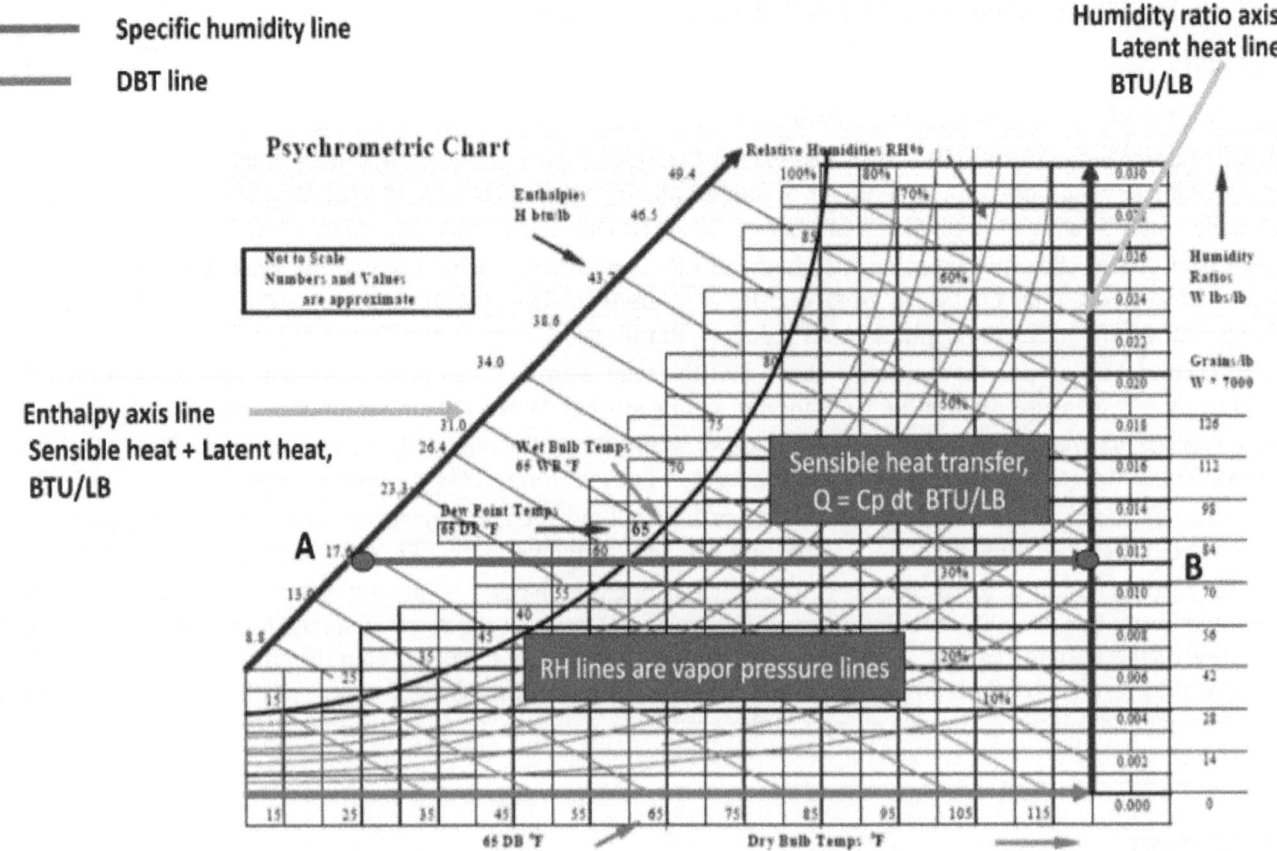

Figure 5: Specific Humidity and Dry Bulb Temperature

Humidity Ratio/Specific Humidity

The weight of water vapor in an air sample divided by the weight of dry air in the same sample yields the specific humidity, also known as the humidity ratio. The air's temperature as measured by a standard thermometer is known as the dry-bulb temperature.

"While the airstream's dry-bulb temperature stays constant, the thermometer bulb's temperature will drop if a wick that has been moistened with water is wrapped around it and process air is allowed to pass through it. This is because the water in the wick will evaporate. The air sample's wet-bulb temperature is the resultant indicated temperature. **As the humidity ratio rises, the difference between the dry-bulb and wet-bulb temperatures will get smaller** if more water vapor is added to the air sample without affecting the dry-bulb temperature.

There will come a time when no more water vapor can be added to the air without causing the dry-bulb temperature to change, at which point evaporation will stop.

After that, the air is said to be saturated. The temperatures of the wet and dry bulbs are the same. The dew-point temperature of this mixture of water vapor and the air is the temperature at which this happens.

Condensation will occur and the humidity ratio will drop when saturated air is cooled below this mixture's dewpoint temperature.

Two important points here [1] As the humidity ratio rises, the difference between the dry-bulb and wet-bulb temperatures will get smaller and **[2]** the humidity ratio will drop when saturated air is cooled below this mixture's dewpoint temperature.

Enthalpy Transfer in Psychrometric Chart

The air pressure of an air sample consists of the sum of the partial pressure exerted by the oxygen, nitrogen, and inert gas molecules, plus the vapor pressure of the water molecules contained in the air. The energy content of an air sample has two components. One component depends upon the kinetic energy (energy due to motion) of the dry air molecules. Changes in this energy component known as sensible heat $Q = m\ Cp\ dt$ are indicated by an increase or decrease in the dry-bulb temperature. A second component depends upon the energy required to produce the water vapor that has been evaporated into the air sample. Changes in this energy component known as latent heat are indicated by changes in the partial pressure of the water vapor in the sample. Changes in the water vapor pressure, the specific humidity, the humidity ratio, and the dew point of air and water vapor mixtures all depend only upon changes in the amount of water contained in the air sample. Changes in any one of these can be used as an indicator of a change in the latent heat of the air sample. Changes in the total energy possessed by an air sample are known as changes in the enthalpy of the air sample and consist of the sum of the sensible and latent heat changes.

Real-life Example: A Cooling Tower

Imagine a cooling tower working between the humidity ratio line at point B and the enthalpy line at point A. Humidity ratio lines are horizontal lines parallel to X-axis.

The two terminal points are:

Humidity ratio [extreme right] = 0.012 lb moisture/per pound dry air.

Enthalpy line [extreme left] = 17.6 BTU/LB.

Calculations

Latent heat from Humidity Ratio Line

1065 BTU/LB is a good average latent heat for the HAVAC temperature range.

[1] Latent heat of air = 0.012 × 1065 = 12.78 BTU/LB.

Latent Heat Cooling of Water

[2] Evaporation of water [since it is a cooling tower], dt = 12.78/1 = 12.78 F [Specific of water = 1 BTU/LB/F].

Sensible heat transfer from the enthalpy line

[3] Sensible heat transfer = [17.6– 12.78] = 4.82 BTU/LB.

% Cooling by Latent Heat

[4] Cooling by latent heat = 12.78/17.6 = 72.61%.

[5] Cooling by sensible heat transfer = 27.38%.

Thus, you can obtain a reasonable understanding of your system by simply looking at the humidity ratio and enthalpy.

THE DIFFERENCE BETWEEN DEW POINT AND WET-BULB TEMPERATURE

By definition they have different meaning.

Wet-bulb Temperature

The lowest temperature to which air can be cooled through evaporative cooling. It's influenced by both the air temperature and moisture content.

Dew Point Temperature

The temperature at which air becomes saturated with water vapor and condensation occurs. It's an indication of the amount of moisture in the air, with higher dew point temperatures indicating more moisture.

The wet bulb temperature is always between the dry bulb temperature and dew point temperature.

When the relative humidity of the air is 100%, the dew point temperature, wet bulb temperature, and dry bulb temperature are all equal.

The wet bulb temperature shows the relative amount of moisture in the air wrt saturated pressure and temperature condition i.e. relative humidity of the air at that particular pressure and temperature.

So, now if the air is 100% saturated, the wet bulb, dry bulb and dew point temperature are same.

Question

At 100% saturation when dry-bulb, wet-bulb and dew point temperatures are same why there is no condensation of water at wet-bulb temperature. Condensation begins only when you cool air below Wet-bulb temperature?

Answer

Explanation

The answer is the cooling you do is supposed to remove the heat of condensation which is exothermic.

Therefore, until you cool the air below Wet-bulb temperature to remove the heat of condensation to facilitate formation of H bonds there is no condensation.

What does it Mean?

It means while both Wet-bulb and dew point temperatures are same at 100% saturation they are in a different thermodynamic state.

Further Explanation

Thermodynamics

At WBT, Gibbs free energy change is zero. The system is in thermodynamic equilibrium.

At dew point Gibbs free energy is positive and not zero due to gain of entropy by cooling.

Dew point is a nonequilibrium process. WBT is an equilibrium process.

At WBT The dryness fraction $x = 1$

At dew point the dryness fraction $x = 0$

If you can visualise an enthalpy diagram of air-water system the above two points are two extreme points on the x- axis.

While the temperature remain constant the equilibrium shifts on the left as $x = 1$; approaches $x = 0$ on the dome of enthalpy diagram on x-axis.

This explains why Gibbs free change is positive in evaporation.

This is the difference between Wet-bulb and dew point temperature. Though they are same at 100% saturation they are at two different thermodynamic states of wet air.

WHY LARGER THE DIFFERENCE BETWEEN WET-BULB AND DRY-BULB TEMPERATURE THERE IS MORE POTENTIAL FOR EVAPORATION?

Let's analyse two situations.

1. Saturated air at 80 F with WBT 80 F
2. Dry air at 80 F with 50 RH

Let's look at enthalpy change in both cases

From psychrometric chart

Case 1

The enthalpy of saturated air at 80°F is 43.6 Btu/lbma.

Enthalpy is the heat energy in moist air, which is a combination of the air's temperature and moisture content. It's expressed in Btu per pound of dry air.

Case 2

The enthalpy of air at 80 F and 50 RH is about 32.5 Btu/lb

The difference between two enthalpy = (43.6–32.5) = 11.1 (about) Btu per pound

This delta enthalpy drives the cooling between DBT and WBT

How and Why? Here Comes Thermodynamics

In case 1 at saturation the Gibbs free energy is zero with much larger energy.

In case 2 the air is unsaturated away from saturation by 11.1 Btu/lb

The case 2 wants to reach equilibrium and to reach equilibrium it needs energy, 11.1 Btu per pound.

What it does is it makes the water to evaporate and to supply this enthalpy gap.

Therefore, the larger the difference between DBT and WBT the larger is the potential for evaporation.

HOW DOES DESERT COOLER WORK?

Unlike a cooling tower that cools water, a Desert cooler cools air. Desert cooler works best when the air is dry and hot. Cooling in a Desert cooler occurs in an isenthalpic/adiabatic manner.

There is a fine line between isenthalpic and isentropic processes. Both are adiabatic processes. Isenthalpic process cannot transfer work. Isenthalpic processes ard not internally reversible.

The difference between a cooling tower and a Desert cooler is that in a cooling tower, the water primarily provides its potential energy meaning enthalpy to air, and the water cools. In a desert cooler, the air's sensible heat is cooled by water by converting some of its kinetic energy into potential energy at constant pressure. The entire process is entropy-driven.

In evaporative cooling the water cools adiabatically by using its own internal energy. In a Desert cooler the wet air [the moisture carrying air from the cooler] cools adiabatically by taking heat from the hot space. In this entire process there is no consumption of thermodynamic work and therefore, the process is called 'Isenthalpic cooling'.

Since it is an isenthalpic process there is no heat and work transfer. The cooling occurs primarily by a change of entropy between in and out points by this equation, $dS = [U–W]/T$.

The Device

It's a device that cools the air. Desert coolers work well in hot summer in places where the humidity in the air is low. It consists of a water reservoir with a pump. The pump draws water from the reservoir and sprays water over the excelsior pads located on three sides of the cooler box to keep the pads wet. A centrifugal fan draws air from the atmosphere through wet pads and throws air into the room/building. The wet pads are about ½ inch thick and can provide very limited surface area for air to come to mass transfer equilibrium with water when air is drawn through these wet pads.

Working

Thermodynamics

The entire process is entropy-driven. The air cools primarily in the device by reducing its sensible heat by converting some of its kinetic energy into potential energy at constant enthalpy at constant pressure. There is a gain in entropy.

The wet air with added potential energy enters the hot surroundings with internal energy = U_1.

In the hot space air's added potential energy converts back to kinetic energy at constant enthalpy by taking heat from the hot space. Hot space cools. In hot space, there is a loss of entropy and this loss is made up by supplying energy from hot space that cools the hot space by reducing the internal energy.

The exit air leaves hot space with internal energy = U_2.

At constant pressure cooling in hot space occurs by using the equation from the first law of thermodynamics, $H = U + W$ [where H is enthalpy. U is internal energy and W is work]

At two points in and out $H_2 = H_2$. Since it is an isenthalpic process there is no heat and work transfer. The cooling occurs primarily by a change of entropy between in and out points by this equation, $dS = [U\text{-}W]/T$, U is internal energy, is W is work and T is temperature. $dS = ((U_1 – U_2) – (W_1 – W_2))/T.$, $[W_1 – W_2]$ is zero in isenthalpic process. Since there is no work. Therefore, $TdS = U_1 – U_2$. The change in internal energy, dU between the in and out points is a function of entropy change between these two points.

CHAPTER 56

Solubility, Scaling and Fouling

Solubility, scaling, and fouling are all related to thermodynamics in the context of chemical processes. Solubility refers to the ability of a substance to dissolve in a solvent. The thermodynamic aspect of solubility is governed by factors such as temperature, pressure, and the enthalpy of solution. The solubility of a substance can change based on these thermodynamic factors.

Scaling occurs when ions in a solution react to form insoluble compounds, which can deposit on surfaces and cause scaling. The thermodynamics of scaling involve understanding the equilibrium between dissolved ions and solid precipitates. Factors such as temperature, pressure, and the solubility product play a role in scaling behavior. Fouling is the accumulation of unwanted material on surfaces, which can hinder heat transfer and flow. Fouling can also be influenced by thermodynamic factors such as temperature, pressure, and the nature of the fouling material. Understanding the thermodynamics of fouling is important in designing systems to mitigate fouling effects.

SOLUBILITY: ROLE OF THERMODYNAMICS

There could be many questions about the solubility of a substance in water. A few common questions could be why common salt is so soluble in water. Why calcium carbonate has very little solubility in water? Why barium sulphate has practically no solubility in water.

Thermodynamics can answer all the questions.

A solute would dissolve in water until the solution is saturated with the solute. When the saturation reaches a thermodynamic equilibrium is established between the solute dissolved in the solution and the undissolved solutes in the solution. Both phases reach the maximum entropy and minimum free energy $dG = dH - TdS = 0$. This means both phases reach a stage when they do not have any useful energy to interact with each other.

Solubility

The solubility is the ability of a substance called solute to form a solution in a solvent.

The solubility of a substance fundamentally depends on the physical and chemical properties of the solute and solvent as well as on temperature, pressure, and the presence of other chemicals (including changes to the *pH*) of the solution. The extent of the solubility of a substance in a specific solvent is measured as the saturation concentration, where adding more solute does not increase the concentration of the solution and begins to precipitate the excess amount of solute.

The Molecular View of Solubility

The solubility of one substance in another is determined by the balance of intermolecular forces between the solvent and solute, and the entropy change that accompanies the solvation. Factors such as temperature and pressure will alter this balance, thus changing the solubility. Solubility may also strongly depend on the presence of other species dissolved in the solvent, for example, complex-forming anions in liquids. Solubility will also depend on the excess or deficiency of a common ion in the solution, a phenomenon known as the common ion effect, Ph, etc.

Solubility occurs under dynamic equilibrium, which means that solubility results from the simultaneous and opposing processes of dissolution and phase joining (e.g., precipitation of solids). The solubility equilibrium occurs when the two processes proceed at a constant rate.

Saturated Solution and the Onset of Equilibrium

This explains the obvious roles of temperature, pressure, and entropy in a process of solution. Under certain conditions, the equilibrium solubility can be exceeded to give a so-called supersaturated solution, which is metastable.

Role of Lattice Energy and Hydration Energy in a Solution

Two types of energy, lattice and hydration energy play fundamental roles.

Lattice Energy

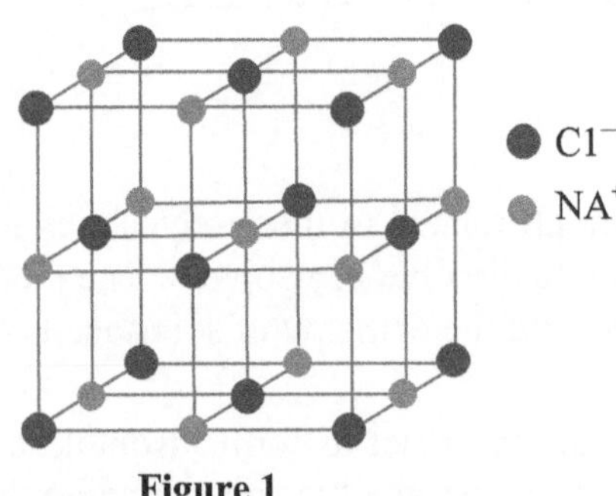

Figure 1

It is a measure of the cohesive forces that bind ions. Lattice Energy is a type of potential energy that may be defined in two ways. In one definition, lattice energy is the energy held by the Coulomb forces holding the ions.

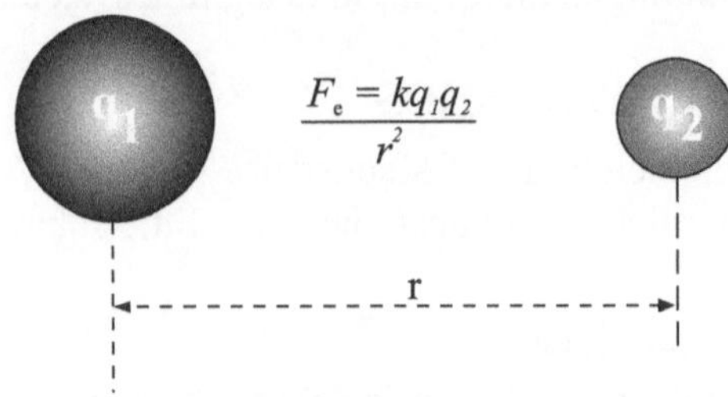

Figure 2

Any stable atom always contains the electrostatic force of attraction. An electrostatic force is also known as the Coulombic force. It is the force of attraction between two opposing charges, i.e., protons and electrons.

Coulomb force is directly proportional to the product of charge on the ions, q_1 and q_2 and inversely proportional to the square of the separated distance, r.

The lattice energy is always positive since this will always be an endothermic process to separate two ions.

Hydration Energy

If the attractive forces between the solvent and solute particles are greater than the attractive forces holding the solute particles together in the lattice, the solvent particles pull the solute particles apart and surround them. The surrounded solute particles then move away from the solid solute and out into the solution. Ions are surrounded by a solvent.

Next Step

This process is called solvation. Solvation is the process of reorganizing solvent and solute molecules into solvation complexes and involves bond formation, hydrogen bonding, and van der Waals forces. The solvation of a solute by water is called hydration.

When the hydration energy is equal or greater than the lattice the substance dissolves in water.

Thermodynamics

Why NaCl is highly soluble in water?

Gibbs equation: $\Delta G = \Delta H - T\Delta S$ [G is Gibbs free energy, H enthalpy, T is temperature and S is entropy]

For NaCl

Lattice energy = + 777 kj/mol
Hydration energy = –774 kj/mol

$T = 298$ k, $\Delta S = 40$ j/k-mol

ΔH = Lattice energy + Hydration energy = 777 + [–774] = 3 kj/mol

$\Delta G = [3 - 298 \times 0.04]$
$= -8.92$ kj/mol

Negative Gibbs free energy change makes the dissolution of *NaCl* in water spontaneous.

Why barium sulphate $BaSO_4$ is insoluble in water?

Lattice energy $BaSO_4$ = +2474 kj/mol
Hydration energy $BaSO_4$ = –2368 kj/mol

$T = 298$ k, $\Delta S = 102$ j/k-mol

ΔH = Lattice energy + Hydration energy = 2474 + [–2368] = +106 kj/mol

$\Delta G = [106 - 298 \times 0.102] = +75.6$ kj/mol (positive therefore $BaSO_4$ is insoluble!)

The point to note here is interesting

In barium sulphate, $BaSO_4$, there are two positive charges on each cation *Ba* ++ and anion SO_4^{-}, while in the case of common salt, *NaCl* there is a single charge on each sodium and chlorine Na^+ and Cl^-, this makes $BaSO_4$ much more tightly held lattice than *NaCl* which water cannot break.

Effect of temperature and pressure on the solubility: The solubility of most substances depends strongly on the temperature and, in the case of gases, on the pressure. The solubility of most solid or liquid solutes increases with increasing temperature. The components of a mixture can often be separated using fractional crystallization, which separates compounds according to their solubilities. The solubility of a gas decreases with increasing temperature. Henry's law describes the relationship between the pressure and the solubility of a gas.

SCALING AND FOULING OF HEAT EXCHANGERS: LANGELIER'S SATURATION INDEX AND STOKE'S LAW

Scaling and fouling are two unwanted events in every heat exchanger with practically no cure rather than prevention. Understanding scaling and fouling is crucial as they can have a significant impact on the performance and lifespan of heat exchangers. Regular monitoring, maintenance, and appropriate cleaning techniques can help minimise these issues and optimize the heat exchanger's efficiency and longevity.

Summary

Langelier's saturation index and Stoke's law are two fundamental concepts that can help explain the formation of scaling and fouling in water treatment processes or heat exchangers. Langelier's saturation index explains the instability of soluble calcium and magnesium salts in the water while Stoke's law explains sedimentation of insoluble suspended particles in water.

The deposit issues in heat exchangers can be grouped into three main categories:

- Scaling
- Fouling
- Under deposit corrosion

Scaling and fouling are two unwanted events in every heat exchanger with practically no cure rather than prevention. Understanding scaling and fouling is crucial as they can have a significant impact on the performance and lifespan of heat exchangers. Regular monitoring, maintenance, and appropriate cleaning techniques can help mitigate these issues and optimize the heat exchanger's efficiency and longevity.

How Scaling is Different from Fouling

Scaling occurs when dissolved solids, such as minerals and salts, precipitate and form a solid layer on the surfaces of heat exchangers. Fouling, on the other hand, refers to the deposition of suspended solids, such as particles or debris, onto the heat exchanger surfaces.

These terms may cause confusion. Most people may use the term "fouling" interchangeably to refer to any type of deposit or unwanted accumulation in a heat exchanger, whether it be scaling or other forms of fouling. It is important to note the distinction between scaling and fouling for those who work in the field or need to address specific issues related to heat exchanger performance. However, when communicating with the general public, it may be simpler to use a more general term like "deposits" or "unwanted accumulation" to avoid confusion.

Scaling *vs.* Fouling

Scaling: Scaling occurs when minerals and other impurities in the fluid being heated or cooled deposit onto the heat exchanger surfaces. These deposits gradually build up over time and form a thick layer, reducing heat transfer efficiency. Scaling can occur due to the presence of hard water, high mineral content in the fluid, or improper water treatment. The scaling layer acts as an insulator, causing an increase in the thermal resistance between the fluid and the heat exchanger surface, leading to reduced heat transfer rates and increased energy consumption. Ultimately, scaling can lead to a loss in heat exchanger efficiency, increased maintenance requirements, and even complete system failure if left untreated.

Fouling: Fouling refers to the accumulation of unwanted substances, such as dirt, sediment, organic matter, or biological growth, on the heat exchanger surfaces. Fouling can occur due to the nature of the fluid being processed, the presence of contaminants, or inadequate cleaning procedures. Similar to scaling, fouling reduces heat transfer efficiency by creating a layer of insulation on the heat exchanger surface. This layer increases the thermal resistance and reduces the heat transfer rates between the fluid and the heat exchanger, leading to decreased performance and increased energy consumption. Fouling also restricts the flow of fluid, reducing the overall flow rate and increasing pressure drop across the heat exchanger. This can further decrease the efficiency of the system and lead to increased maintenance and operational costs.

Scaling and fouling could be a potential source of under deposit corrosion

Two reasons

Corrosive deposits: Scaling in heat exchangers can also lead to under-deposit corrosion, which is another unwanted consequence. Scaling creates an uneven surface on the heat exchanger's inner walls, providing areas where corrosive agents can become trapped and accumulate. The trapped corrosive agents can initiate corrosion of the heat exchanger material, leading to metal loss, pitting, and potential leaks.

Differential oxygen concentration corrosion cells under the deposit

Discussed Later

Under-deposit corrosion is particularly challenging because the scale layer protects the corrosion underneath, making it difficult to detect and address. Over time, the corrosion can weaken the heat exchanger structure, compromise its integrity, and increase the risk of failure.

Detail

Scaling

Scaling is primarily associated with the precipitation and deposition of soluble salts, particularly calcium and magnesium salts, from water or other fluids. When water with high concentrations of dissolved calcium and magnesium compounds is heated or evaporated, these salts can form solid deposits on heat transfer surfaces, such as the inside of pipes or the outer surface of heat exchanger tubes. Over time, these deposits can reduce heat transfer efficiency, restrict flow, and impair equipment performance if not properly addressed.

Langelier's Saturation Index [LSI]

The Langelier Saturation Index (LSI) is a crucial concept in comprehending the instability of calcium and magnesium salts. The LSI is a calculated value used to predict the scaling potential of water and determine whether the water is under-saturated, saturated, or over-saturated with respect to calcium carbonate. It takes into account factors such as pH, temperature, alkalinity, and calcium and magnesium concentrations.

LSI is an empirically calculated number. It is used to predict the $CaCO_3$ stability of water, that is whether $CaCO_3$ *will precipitate, dissolve* or be in equilibrium with water. Langelier developed this index for predicting the *pHs* at which pH water is saturated in $CaCO_3$. LSI is expressed as, $LSI = pH - pHs$, where pH is the actual pH of water and pHs is the saturation pH of water with respect to $CaCO_3$. *pHs* is calculated as, $pHs = [9.3 + a + b] - [c + d]$, where a,b,c, and d are empirical constants based on total dissolved solids, [TDS, ppm], water temperature degc, calcium hardness [CH as $CaCO_3$ ppm] and total alkalinity [*TA* as *CaCO3* ppm] respectively.

If pH > pHs, the system water is supersaturated with respect to $CaCO_3$ and hence it has the urge to deposit [throw out] the excess $CaCO_3$ present in the water as scales. Hence, if LSI is positive, water is scale forming.

If pH < pHs, the system water is unsaturated with respect to $CaCO_3$ and hence there is an urge to dissolve the protective $CaCO_3$ layer already present on the pipelines and expose them to the corrosive attack. Hence if LSI is negative, water is corrosive.

If, pH = pHs, water is considered balanced.

The greater the deviation of actual pH from pHs, the more pronounced is the instability

By monitoring and understanding the LSI, individuals can recognize the saturation levels of calcium and magnesium salts and make necessary adjustments or treatments to prevent scaling problems in various industrial processes, cooling towers, boilers, and other applications that involve water heating and evaporation.

Fouling

Fouling is Caused by Suspended Solids

Imagine a suspended particle in water flowing in a pipeline. Two forces act on it [1] upward frictional force Fd [2] and a downward gravitational force, g, When Fd > g, the particle floats. when Fd < g, the particle sinks on the pipe surface. This is the fundamental of fouling.

In turbulent flow, swirling of fluid and reverse current generating eddies, keep SS particles floating. Fouling begins typically in low-velocity areas in the pipeline network and then spreads.

Fouling can be best explained by Stoke's law.

Stoke's law

$Fd = 6\eta\pi rv$, *Fd* Is the frictional force, η is dynamic liquid visc, R is particle radius, v is the relative velocity of liquid and particle.

$Fd = 6\eta\pi Rv$, depends on [1] visc of liquid,[2] the size of the particle [3] relative liquid-solid velocity.

In a given system [1] and [2] are constant, *F*d, then depends on v, the relative liquid–solid velocity.

An important point – it is the relative velocity between liquid and solid decides when a SS particle will sink.

As v reduces, Fd reduces, and when $Fd < g$, the particle deposits as foulant. Process continues.

Inter-molecular forces hold solid particles.

How and why suspended solids deposit inside a bend?

A liquid with suspended solids is basically a slurry. This is a two-phase flow. So conventional fluid dynamics has limitations. In this two-phase flow, there are two independent controls acting on the velocity of the slurry. Liquid velocity is controlled by the pressure while the velocity of solids is controlled by the viscosity drag between solids and liquid.

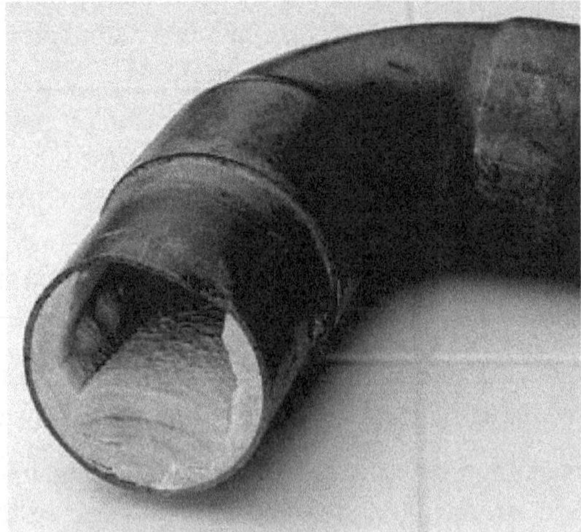

Figure 3

The bend is adding the additional frictional loss of the head to flow. Head = Pressure head + velocity head with static head being constant. Both pressure and velocity are falling in the bend. This is resulting in a drop in the relative velocity between liquid and solid particles. While there is a drop in liquid velocity, solid velocity remains unchanged because of no change in the viscosity of the liquid. Consequently, The drag force acting on the particle, $Fd = 6\eta\pi Rv$, is falling. When g exceeds Fd solid particles are deposited on the surface.

Summary- Fouling of solids is triggered when the gravitational force acting on solid particles exceeds the frictional drag acting on the particle.

Under Deposit Corrosion

Differential oxygen concentration cells can contribute to under-deposit corrosion in heat exchangers. When scaling deposits form on the heat exchanger's surface, they create small crevices or areas of trapped stagnant water. These trapped areas can result in different oxygen concentrations compared to the surrounding bulk water.

Differential oxygen concentration cells occur when there is a difference in oxygen concentration between two areas. This difference in concentration sets up an electrochemical cell, where the more oxygenated area acts as the cathode, the less oxygenated area acts as the anode, and the metal surface between them becomes the site of corrosion.

The stagnant water in the crevices of scale deposits can have reduced oxygen availability compared to the bulk of the water flowing through the heat exchanger. This creates areas of low oxygen concentration, which act as anodic sites for corrosion to occur.

CHAPTER 57

Hydrogen Gas

Hydrogen gas (H_2) is a key component in thermodynamic processes such as fuel cells and combustion engines, where it serves as a high-energy fuel source. The thermodynamics of hydrogen gas play a crucial role in determining the efficiency and performance of hydrogen-based energy systems. The use of hydrogen gas in fuel cells involves complex thermodynamic reactions that convert chemical energy into electrical energy with high efficiency. Understanding the thermodynamics of hydrogen gas is essential for developing sustainable and clean energy solutions that rely on hydrogen as a renewable fuel source. The thermodynamic properties of hydrogen gas, such as enthalpy and entropy, are critical factors in the design and operation of hydrogen-based technologies. Overall, hydrogen gas and thermodynamics are closely intertwined in advancing towards a more sustainable energy future.

HYDROGEN'S EXPANDING REACH

Although the use of hydrogen as a clean fuel is growing, it is still not up to par with what will be needed to contribute to achieving net zero emissions by 2050.

A quick summary

- Total hydrogen demand could grow as high as 105 Mt – compared with more than 200 Mt in the NZE scenario.
- Low-carbon hydrogen production could reach more than 17 Mt—one-eighth of the production level required in the NZE Scenario.
- Electrolysis capacity could rise to 90 GW—well below the nearly 850 GW in the NZE Scenario.
- Up to 6 million FCEVs could be deployed—40% of the level of deployment in the NZE Scenario (15 million FCEVs).

Demand and Supply

In 2020, there was a 90 Mt demand for hydrogen, almost all of which was produced from fossil fuels and used for industrial purposes, resulting in nearly 900 Mt of CO_2 emissions. However, encouraging progress is being made. Electrolysers, which are required to produce hydrogen from electricity, have doubled in capacity globally over the last five years, with a mid-2021 target of just over 300 MW. By 2030, there could be about 350 projects under development that could increase global capacity to 54 GW. Forty more projects with a combined capacity of over 35 GW are still in the planning stages. The amount of hydrogen produced by electrolysers worldwide could surpass 8 million tonnes by 2030 if all those projects come to fruition.

Even though this is noteworthy, it is still far less than the 80 Mt needed by that year in the IEA Roadmap for the Global Energy Sector's pathway to net zero CO_2 emissions by 2050.

Europe is at the forefront of the deployment of electrolyser capacity

With 40% of the world's installed capacity, Europe is at the forefront of the deployment of electrolyser capacity and is expected to maintain its position as the largest market in the near future due to the aggressive hydrogen policies of the United Kingdom and the European Union. According to plans, Australia may be able to catch up with Europe in a few years. Large-scale capacity deployment is also anticipated in Latin America and the Middle East, especially for export. Although it started slowly, the People's Republic of China ("China") is quickly announcing more projects, and the United States is raising the bar with its recently announced Hydrogen Earthshot.

Carbon Capture, Utilisation, and Storage (CCUS) Projects

Approximately sixteen carbon capture, utilisation, and storage (CCUS) projects are currently in operation, producing 0.7 million tonnes of hydrogen per year from fossil fuels. If completed, the 50 additional projects in development could push annual hydrogen production above 9 Mt by 2030. With over 80% of the world's capacity produced, Canada and the United States lead the world in the production of hydrogen from fossil fuels using CCUS, though the United Kingdom and the Netherlands are making a strong push to take the lead and contribute significantly to the projects that are currently in development.

Fuel Cell Electric Vehicles (FCEVs)'s Growth

There are a lot more uses for hydrogen than are currently common. Even though this still only makes up a small portion of the demand for hydrogen overall, there has been significant recent progress in spreading its use, especially in transportation. The cost of automotive fuel cells has decreased by 70% since 2008. With the help of Korea, the United States, China, and Japan, there were over 43 000 FCEVs on the road by mid- 2021, more than six times as many as there were in 2017.

Almost all FCEVs in 2017 were passenger vehicles. Currently, buses and trucks make up one-fifth of all vehicles, suggesting a shift to the long-distance market where hydrogen can more effectively compete with electric vehicles.

However, compared to the estimated 11 million electric vehicles currently on the road, the total number of FCEVs is still far lower. There are already a number of demonstration projects being developed for the use of hydrogen-based fuels in rail, shipping, and aviation; these projects should create new opportunities for the demand for hydrogen.

Decarbonisation Status

Although the majority of the technologies that can make a major contribution are still in their infancy, hydrogen is a crucial component of industry's decarbonisation. There are significant actions being taken. This year saw the start of the first pilot project in history to use low-carbon hydrogen to produce carbon-free steel in Sweden. At the end of 2021, a pilot project in Spain will begin using hydrogen derived from variable renewable sources to produce ammonia. Over the next two to three years, a number of projects with a capacity of tens of kilotonnes of hydrogen are anticipated to come online. Additionally, being developed are demonstration projects for the use of hydrogen in industrial processes like the production of glass, ceramics, and cement.

Investment Commitment

Governments must increase their goals and encourage the creation of demand.

At least USD 37 billion has been committed by nations that have adopted hydrogen strategies, and an additional USD 300 billion in private sector investment has been announced. However, in order to get the hydrogen industry on track for net zero emissions by 2050, USD 1200 billion must be invested in the supply and use of low-carbon hydrogen through 2030.

Cost Gap

The cost difference between hydrogen produced from unrestricted fossil fuels and low-carbon hydrogen is a major obstacle. For the most part of the world right now, the least expensive option is to produce hydrogen from fossil fuels. The levelized cost of producing hydrogen from natural gas ranges from USD 0.5 to USD 1.7 per kilogramme (kg), depending on local gas prices. When CO2 emissions from the production of hydrogen are reduced through the use of CCUS technologies, the levelized cost of production rises to approximately USD 1 to USD 2 per kg. Hydrogen production from renewable energy costs between $3 and $8 per kilogramme.

Scope for Cost Reduction

There is a great deal of room for cost reduction in production through increased deployment and technological innovation. The IEA's Net Zero Emissions by 2050 Scenario (NZE Scenario) illustrates this potential. In regions with excellent renewable resource availability, hydrogen from renewable sources can reach as low as USD 1.3 per kg by 2030; this is comparable to the cost of hydrogen from natural gas with CCUS. Longer-term costs of hydrogen from renewable electricity can be as low as USD 1 per kg (range USD 1.0–3.0 per kg) in the NZE Scenario, which means that even in some regions without CCUS, hydrogen from solar PV is cost-competitive with hydrogen from natural gas.

HYDROGEN'S TINY SIZE AN ADVANTAGE OR A DISADVANTAGE?

The answer is their unique atomic structures make H2 a gas and H_2O liquid. At ambient pressure (1 atm), hydrogen liquifies at a temperature of −253°C predominantly because of existence of repulsive forces in non-polar H_2 molecule. Therefore, at ambient temperature H_2 is a gas. The vapour pressure of water at room temperature is 23.8 mm Hg much smaller than atmospheric pressure of 760 mm, hence water is liquid at ambient temperature. The polar structure of H_2O and its large intermolecular interactions make water liquid with much higher vapor pressure than H_2 at ambient temperature.

Hydrogen

Hydrogen is a Unique Molecule

Hydrogen is a diatomic gas of two hydrogen atoms linked to each other. Hydrogen atom has one electron and one proton. It has no neutron. Its atomic number is 1 in the periodic table. It has the smallest size among all other elements in the periodic table. Its atomic structure and small size make it a unique molecule. This gives H2 some advantages and some disadvantages as well.

Advantages and Disadvantages of H_2 Gas

Advantages

Because of small mass and large velocity, H2 has huge kinetic energy stored in its atoms, $KE = ½\ mv^2$. This gives H_2 very high specific heat. The specific heats of hydrogen at constant pressure and constant volume are 14.307 kJ/kg K and 10.183 kJ/kg K, respectively. One big advantage of small molecular size of *H2* is it has the largest number of energy carrying molecules/gram than any other substances. This is why H_2 is an energy carrier and is an alternate to fossil fuels. When *H2* is burnt, it releases this energy with almost no GHG generation.

To summarize, Hydrogen gas is renewable and readily available. It is a clean fuel and flexible energy source to support Zero-Carbon energy strategies. It is more powerful and energy efficient than fossil fuels with almost zero emissions. Hydrogen production is possible from a variety of resources, including water, biomass, and natural gas. Hydrogen can be thought of as a potential renewable energy source if renewable energy sources are used to create it.

Challenges: Cost is One of the Big Issues

Since there is no free electron in H_2 molecule, H_2 is a non polar molecule. Because of small size and close proximity of H atoms in H_2 molecule, H_2 has predominantly repulsive [proton-proton repulsion] forces dominating in the molecule. This is the reason why the liquefaction of H_2 is so difficult and energy intensive. Because of small mass, H_2 has very high velocity, $KE = ½\ mv^2$ and that is one of the key disadvantages of H_2. At atmospheric pressure and 20°C, the volumetric mass density of gaseous hydrogen is 0.083 kg/m³. As a result, storing gaseous hydrogen normally requires a large amount of space. Because it is so inefficient, hydrogen is essentially never transported or stored in gaseous form at atmospheric pressure. This is one of the key issues of H_2's cost in its path to replace fossil fuels. Hydrogen gas storage requires 700 bar of pressure, and liquid hydrogen storage requires cryogenic temperatures, making hydrogen gas storage an expensive process.

An important point to remember is that Although Hydrogen has the highest gravimetric energy density of all known substances (120–142 MJ/kg), it falls short to natural fuel sources when it comes to volumetric energy density (9 MJ/L).

Due to its small size, hydrogen can easily enter metal structures where it can cause a number of problems, including hydrogen embrittlement and corrosion. Hydrogen can disrupt the internal structure of metals, which weakens the substance and increases its susceptibility to failure or cracking under stress. In high-stress settings like pipelines, pressure vessels, and some industrial machinery, this phenomenon is more prevalent. Therefore, one way to avoid hydrogen-induced damage is to use expensive metal alloys that are less reactive to hydrogen, but this comes at a higher cost.

GREEN HYDROGEN

What is Green Hydrogen?

Green hydrogen is hydrogen produced through electrolysis using renewable energy sources, such as solar or wind power. It is considered green because it does not produce any carbon emissions during its production process.

Green hydrogen is expensive primarily because of the high cost of producing it using renewable energy sources. This includes the cost of equipment and infrastructure needed for electrolysis, as well as the capital and operational costs associated with renewable energy sources like solar or wind power.

Process

The process of making green hydrogen involves electrolysis, where an electric current is passed through water (H_2O) to split it into hydrogen (H_2) and oxygen (O_2). The hydrogen gas is then collected and stored for use as a clean energy source.

Cost factors for green hydrogen production include the cost of renewable energy sources, the efficiency of the electrolysis process, the cost of equipment and infrastructure, and the scale of production.

The global status of green hydrogen production is still in its early stages, with several countries investing in research and development to scale up production. Countries like Australia, Germany, and Japan are leading the way in green hydrogen production.

Constraints

Major constraints to scaling up green hydrogen production include the high cost of production, the need for infrastructure to transport and store hydrogen, and the challenge of integrating hydrogen into existing energy systems. Additionally, the availability and cost of renewable energy sources can also impact the competitiveness of green hydrogen.

CHAPTER 58

Decarbonization

Carbon capture and storage technologies aimed at reducing emissions from fossil fuel plants present thermodynamic challenges in terms of energy consumption and the overall efficiency of the process. Overcoming these thermodynamic challenges of decarbonization requires innovative solutions that optimize energy conversion, storage, and distribution processes while ensuring sustainability and reliability in energy systems.

NET-ZERO CHALLENGES: A REVIEW

In simple words, net-zero means limiting global warming to 1.5°C or less by 2050. The two arms of Net-zero [1] as much as possible, switch to renewable energy sources in place of fossil fuels and [2] carbon capture technologies to remove any remaining CO_2 from the air and at the source. The goal of achieving net-zero emissions is unquestionably remarkable but it is also very challenging to accomplish. significant changes are needed in a number of industries, including energy, transportation, industry, and agriculture, to reach net zero globally. These changes entail a shift to low-carbon and renewable technologies, the adoption of carbon capture and storage technologies, as well as the implementation of energy efficiency measures. Significant financial commitments, policy adjustments, inventiveness, and global cooperation are necessary for the process.

What is Net Zero?

Net zero, in its simplest form, refers to reducing greenhouse gas emissions to as close to zero as possible, with any remaining emissions being reabsorbed from the atmosphere, for example by oceans and forests.

How Significant is Net Zero?

The evidence from science is unequivocal: the increase in global temperature must be kept to 1.5°C above pre-industrial levels in order to prevent the worst effects of climate change and preserve a planet that can sustain life. With rising emissions, the Earth's temperature has already increased by 1.1°C since the late 1800s. The Paris Agreement stipulates that emissions must be reduced by 45% by 2030 and to zero by 2050 in order to keep global warming to no more than 1.5°C.

Existing Patterns of Environmental Effects of CO_2 Emissions

Given the difficulties in collecting information on global warming, there should be some discrepancies in the trend of hard data. However, there is a clear trend that two important indicators, the earth's surface, and atmospheric temperatures, have been rising alarmingly over the past ten years, along with CO_2 emissions. Therefore, we should be more concerned with trends and not the absolute numbers.

Over the past decade, the concentration of carbon dioxide (CO_2) in the Earth's atmosphere has been steadily increasing. According to data from the Mauna Loa Observatory in Hawaii, which has been monitoring atmospheric CO_2 since the 1950s, the global average CO_2 concentration reached around 415 parts per million (ppm) in 2020. This represents a significant increase from around 390 ppm in 2010. Regarding global atmospheric temperature, it has also been rising over the past decade. According to the Intergovernmental Panel on Climate Change (IPCC), the average global surface temperature has increased by approximately 1.1 degrees Celsius (2 degrees Fahrenheit) since pre-industrial times (late 19th century).

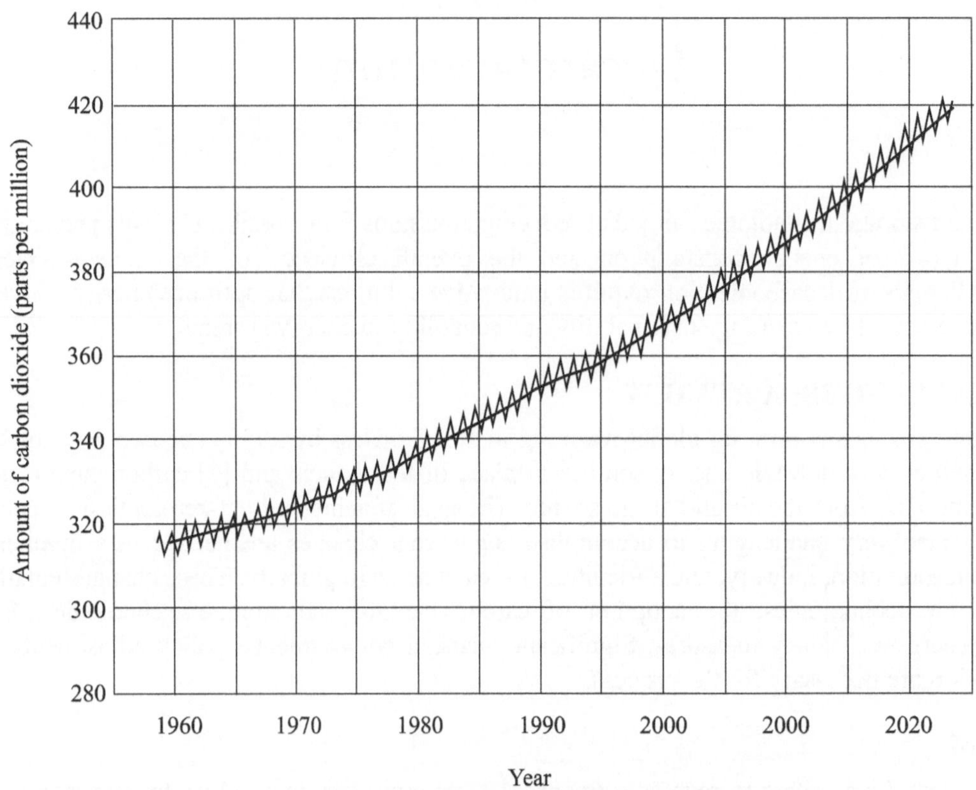

Figure 1: Atmospheric Carbon Dioxide

The level of CO_2 in the atmosphere has increased from about 390 ppm in 2010 to about 420 ppm in 2010. The current generation of CO_2 is about 20 billion tons/annum.

Global Temperature Change Pattern

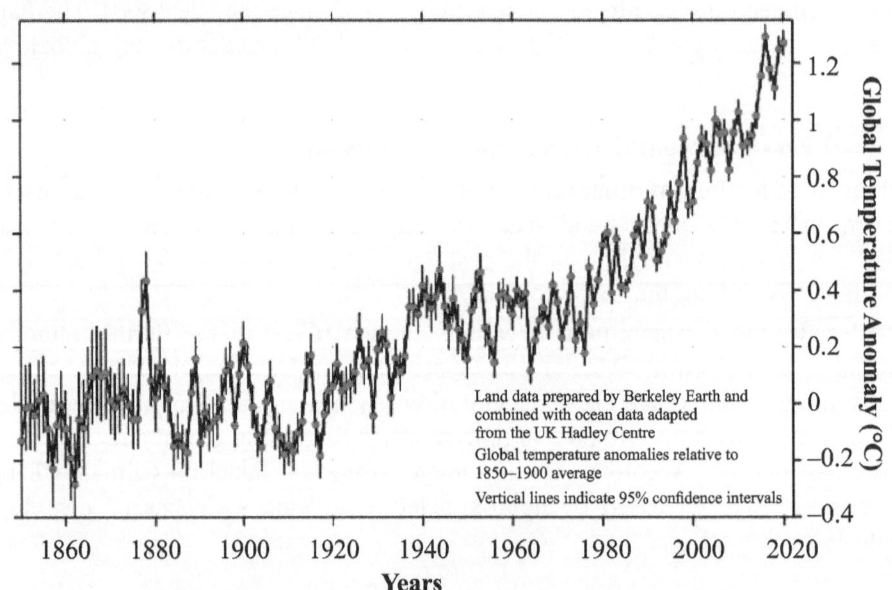

Figure 2: Global Average Temperature 1850–2020

Land and Ocean Average Temperature Increase Trend

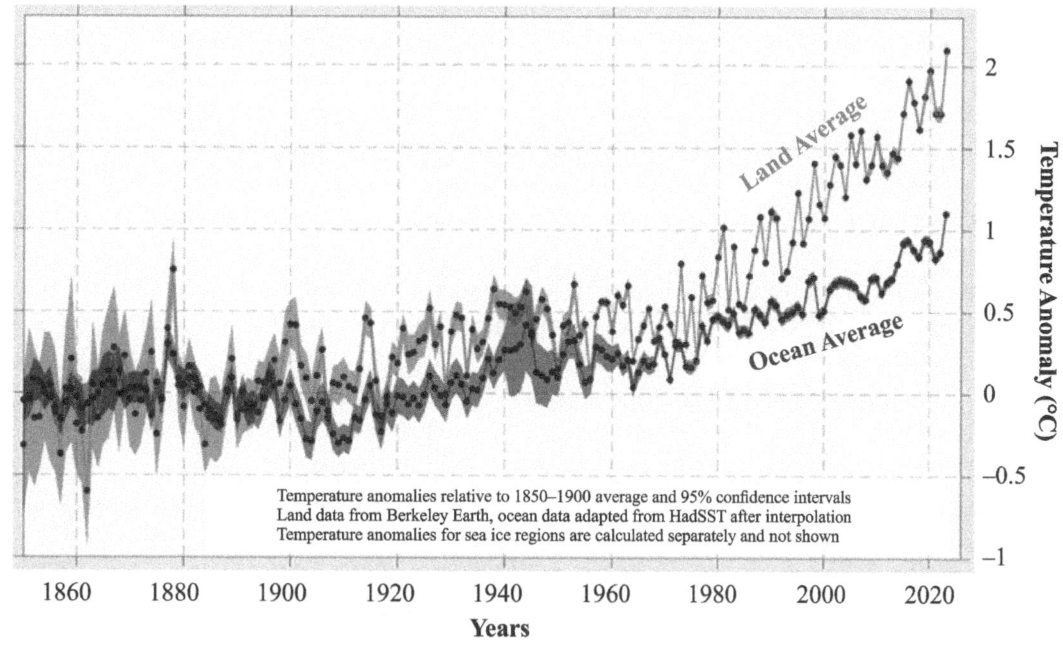

Figure 3: Land and Ocean Temperatures 1850–2023

Though it does not require explanation, the following are the key points.

The increase in global CO_2 concentration from 390 ppm to 415 ppm between 2010 and 2020 has several consequences for the Earth's climate system.

- **Climate change**: Rising levels of CO_2 contribute to global warming, leading to changes in temperature patterns and climate conditions. This can result in more frequent and intense heatwaves, extreme weather events (such as hurricanes and heavy rainfall), and shifts in precipitation patterns.
- **Sea-level rise**: Warmer temperatures caused by increased CO_2 levels lead to the melting of glaciers and ice sheets. This causes sea levels to rise, which poses a threat to coastal regions, increasing the risk of flooding, erosion, and salinization of freshwater resources.
- **Ocean acidification**: When CO_2 is absorbed by seawater, it undergoes chemical reactions that increase the acidity of the ocean. This can have damaging effects on marine ecosystems, particularly on calcifying organisms like coral reefs and shellfish, which rely on carbonate ions to build their structures.
- **Ecosystem impacts**: Climate change driven by elevated CO_2 levels can affect terrestrial and aquatic ecosystems. It can alter the timing of plant flowering, disrupt ecological interactions, impact species distribution and migration patterns, and increase the risk of extinction for certain species, leading to changes in biodiversity.
- **Human health** repercussions: Increased CO_2 concentrations can indirectly impact human health. Heatwaves and extreme weather events can cause heat-related illnesses, while changes in rainfall patterns can affect agriculture, water availability, and food security. Additionally, rising temperatures can worsen air quality and contribute to the spread of infectious diseases.

Technical

Achieving net-zero CO_2 emissions would typically require a multi-faceted approach that involves reducing CO_2 emissions at their source through measures like transitioning to renewable energy sources, improving energy efficiency, and adopting sustainable practices. This is the first and most crucial step in mitigating emissions.

However, even with significant reductions in emissions, it is likely that some residual CO_2 will still be emitted. In such cases, carbon removal technologies or negative emissions technologies can be employed to capture and store CO_2 from the atmosphere. These technologies, such as direct air capture or carbon capture and storage, aim to actively remove CO_2 from the air to compensate for the remaining emissions.

At the global level, several key technologies have the potential to significantly reduce CO_2 emissions. These include:

- **Renewable energy sources**: This includes solar, wind, hydroelectric, geothermal, and biomass energy. Transitioning from fossil fuels to renewable energy is crucial for reducing CO_2 emissions in the power generation sector.
- **Energy storage systems**: Implementing advanced energy storage technologies, such as large-scale batteries, pumped hydro storage, or thermal storage, enables better integration of intermittent renewable energy sources and helps stabilize the grid, thereby reducing reliance on fossil fuel-based backup power sources.
- **Nuclear energy**: Nuclear power is a low-carbon energy source capable of producing large amounts of electricity without direct CO_2 emissions. However, it also comes with challenges related to safety, waste management, and public acceptance.
- **Carbon capture and storage (CCS):** This technology captures CO_2 emissions from power plants and industrial facilities and stores it underground, preventing it from being released into the atmosphere. CCS can help reduce emissions from sectors that are difficult to decarbonize.
- **Electrification of transportation**: Increasing the uptake of electric vehicles (EVs) and supporting charging infrastructure is essential for reducing CO_2 emissions in the transportation sector. This requires transitioning away from fossil fuel-powered vehicles and adopting renewable energy sources for charging.
- **Improved energy efficiency**: Examining and enhancing energy efficiency in various sectors, including buildings, transportation, and industrial processes, can significantly reduce energy demand and, consequently, CO_2 emissions.
- **Sustainable agriculture and forestry practices**: Implementing sustainable land management practices, such as afforestation, reforestation, and improved agricultural techniques, can help sequester carbon in biomass and soils.

Present Status of Various Global Technologies for Renewable Energy

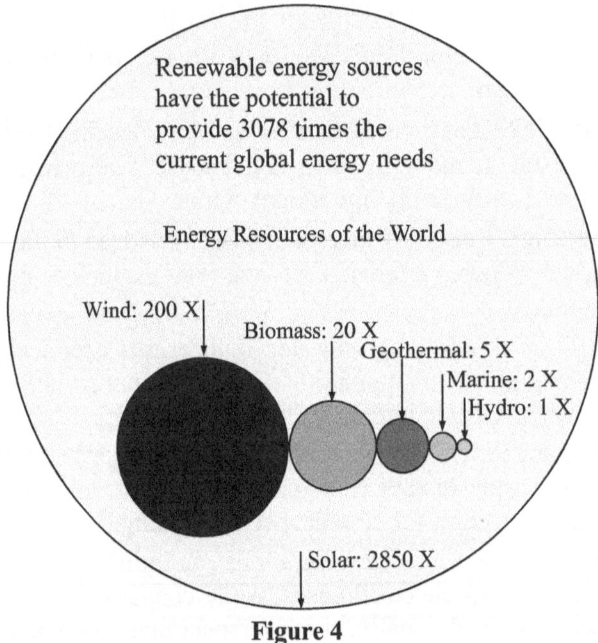

Figure 4

As of 2021, renewable energy production accounts for a growing portion of the total energy needed worldwide. The exact percentage varies by country and region, as well as the specific type of renewable energy.

In some countries, such as Iceland and Paraguay, the majority of their energy needs are already met by renewable sources. Many others have made significant progress toward increasing the share of renewables, particularly in the electricity sector. For example, in 2020, the European Union produced around 34% of its electricity from renewable sources.

It's important to note that the overall percentage of renewable energy in the global energy mix is still relatively small compared to fossil fuels. However, the transition to renewable energy is gaining momentum, and many countries have set ambitious targets to further increase their renewable energy production in the coming decades.

Carbon Capture and Storage

As of now, there are over 20 large-scale carbon capture and storage (CCS) facilities in operation globally. These plants are primarily located in North America, Europe, and Asia. Some examples include the Boundary Dam CCS facility in Canada, Petra Nova in the United States, Sleipner field in Norway, and Gorgon LNG project in Australia.

These operating CCS facilities utilize various technologies to capture carbon dioxide (CO_2) emissions from power plants, industrial facilities, or natural gas processing plants. The captured CO_2 is then compressed and transported via pipelines or ships to suitable storage locations such as deep geological formations or depleted oil and gas fields.

References

Google public domain articles.
Direct air capture [DAC] challenges.
Direct air capture [DAC] kinetics and thermodynamics.

DIRECT AIR CAPTURE: KINETICS AND THERMODYNAMICS

DAC is currently in the early stage of R and D. DAC costs 5–10 times the reforestation costs. Low reaction rate, exothermic adsorption, and entropy generation are three fundamental issues. One huge fundamental cost element in DAC is entropy. Just to pull 400 ppm CO_2 which is 0.04% CO_2, you draw 2500 times air that all contribute to entropy.

Fundamental Challenges

The entropy increases in direct air capture (DAC) because of the energy required to trap and release carbon dioxide. Entropy is a measure of disorder. Entropy measures a state of randomness, uncertainty, or disorder. DAC requires energy. The process of capturing carbon dioxide from the air requires a lot of energy, which increases entropy. This is known as the "entropy penalty". The second law of thermodynamics states that entropy increases over time in an isolated system. This means that if entropy is reduced in one place, it must increase somewhere else. Typically, DAC plants have several components that can contribute to entropy increases, including the air separation unit, calciner, etc.

Direct Air Capture: What Is It?

A technology called direct air capture extracts carbon dioxide from the atmosphere by means of chemical processes. When air passes over these substances, they react and trap CO_2 while letting the other elements of the air pass through. Leading systems of today use either liquid solvents or solid sorbents, which are made of regular chemicals already used in other applications. The carbon dioxide has been drawn from the atmosphere, it is usually released from the solvent or sorbent using heat. By doing this, the solvent or sorbent is replenished for another capture cycle.

Over the past few years, DAC has seen an increase in interest and investment, and more businesses are entering the market. This is because it is recognized that carbon removal will become more and more necessary to achieve national and international climate goals, as well as because DAC has advantages over other carbon removal strategies, such as few practical scaling constraints, minimal land area use, and siting flexibility.

DAC is Currently in the Prototype Phase

Climeworks, Carbon Engineering, and Global Thermostat are some of the businesses with the most cutting-edge technology today. Together, these businesses operate 18 plants, ranging in size from 1 tCO_2/yr to 4,000 tCO_2/yr (the largest plant currently in operation), capturing a combined total of just under 8,000 tCO_2/yr. The majority of that is sold for use in various products, with the other half being permanently sequestered, comparable to the annual emissions from 870 cars.

In the Southwest of the United States, a DAC plant with a capacity of one million tonnes annually is currently being built. It is anticipated to go into operation in late 2024. Other businesses that use sorbent, solvent, and other cutting-edge types of technology have been founded more recently.

On a scale of 1 to 9, DAC is currently rated as having a "technology readiness level" of 6, indicating that it is still in the large-scale and prototype stage and not yet prepared for full commercial deployment. However, it also means that there are lots of chances to enhance efficiency and cut costs by taking advantage of early technology iterations.

What Is the Price of Direct Air Capture?

Direct air capture, despite its advantages and adaptability, is more expensive per tonne of CO_2 removed than many mitigation strategies and non-manmade climate solutions because it requires more energy to separate carbon dioxide from ambient air. Depending on the technology chosen, the source of low-carbon energy, and the scale of their deployment, the cost range for DAC today ranges between $250 and $600; by contrast, the majority of reforestation costs less than $50/tonne.

Role of Kinetics and Thermodynamics of DAC

Kinetics of DAC

The rate at which CO_2 is drawn from outside air is referred to as the kinetics of direct CO_2 capture from air. In order to capture CO_2 from the air, it must first be absorbed by a liquid or solid substance before being released as purified CO_2. Several variables, including the material used for capture and the environment, have an impact on the rate of CO_2 absorption and release.

The concentration of CO_2 in the air is one of the most important factors affecting the kinetics of CO_2 capture from the air.

It takes more time to capture a sizable amount of CO_2 because the ambient air's CO_2 concentration is so low (around 400 ppm). Consequently, the process' kinetics are relatively slow compared to other methods of carbon capture, such as capturing CO_2 from industrial off-gases.

The kinetics of the process are also impacted by the effectiveness of the CO_2 capture materials and external factors like temperature. Depending on their chemical structure and composition, materials used for $CO2$ capture, like amine-based sorbents, can have varying rates of CO_2 absorption and release.

In general, the kinetics of direct CO_2 capture from air remains difficult, and significant research is being done to create better materials and systems that can increase the process efficiency and reduce its energy needs.

Thermodynamics of DAC

The thermodynamics of direct CO_2 capture from air involves assessing the process's viability and energy needs. Due to the low concentration of CO_2 in the air (400 parts per million) and low heat of absorption, capturing CO_2 from ambient air is a difficult thermodynamic process.

Materials with a high affinity for CO_2, such as amine-based sorbents or solid adsorbents, are typically used to absorb CO_2 from the air. These materials' thermodynamic performance is influenced by the system's working conditions, including temperature, pressure, and humidity.

Two important thermodynamic properties, the heat of adsorption and entropy of adsorption, can be used to assess the thermodynamic effectiveness of CO_2 capture from air.

The energy released or absorbed by the substance when it adsorbs CO_2 is referred to as the heat of adsorption. The degree of disorder in the system during the adsorption process is quantified by the entropy of adsorption.

Role of Entropy in DAC

When CO_2 is taken directly from air, the system's entropy typically decreases. The reason for this is that the adsorption process typically involves the binding of CO_2 molecules to a solid or liquid substrate and that the concentration of CO_2 in the air is low (only 400 parts per million). By aligning and attaching to particular binding sites on a surface, CO_2 molecules that have been adsorbed onto it increase the surface's order and reduce its entropy.

However, when CO_2 is captured directly from the air, the system's overall CO_2 concentration is so low and the volume of air is so high that the increase in entropy caused by the adsorbent's mixing with air typically outweighs the initial decrease in entropy caused by the capture process. [It is a very important point to note]

Therefore, during direct CO_2 capture from air, the overall entropy change is typically positive, or increases.

The feasibility and energy requirements of direct CO_2 capture from the air can be significantly impacted by changes in entropy, so thermodynamics is a key factor in the design and optimization of the process.

CO_2-AMINE: CO_2 STRIPPING REVIEW NOTE

Fundamental Points of CO_2 Absorption and Stripping

Absorption of CO_2

In general, there are three major issues in CO_2 mass transfer in amine [1] thermodynamics controlled reversible reaction [2] The reaction of CO_2 with water produces carbonic acid, which lowers pH and aids in the hydrolysis of carbamate, and [3] exothermic Amine-CO_2 reactions controlled by kinetics. Because CO_2 begins its reaction with H_2O after all of the amine has been consumed, its solubility is reduced by its low partial pressure.

Stripping of CO_2

$$Q = Q \text{ sensible heat} + Q \text{ vaporization heat} + Q \text{ absorption heat}$$

On the regeneration side, the cost of regeneration is a big issue. Depending on the operating system, the reported heat of regeneration in CO_2-Amine absorption ranges from 2.1 to 11.25 GJ/t. Capturing CO_2 raises the cost of electricity production by 70%, while the energy required for regeneration is estimated to be 15–30% of a power plant's output. Minimizing heat for regeneration has become a significant challenge, particularly for improving stripper structure, discovering new solvents, and improving operating conditions.

The reported heat of regeneration in CO_2-Amine absorption is in the range of 2.1–11.25 GJ/t, depending on the operating system. A simple stripper column is estimated to reduce the output of a coal-fired power plant by 25 to 30%.

The heat of regeneration

$$Q = Q \text{ sensible heat} + Q \text{ vaporization heat} + Q \text{ absorption heat}$$

Absorption/stripping with alkanolamine solvents is a cutting-edge technology for removing CO_2 from coal-fired power plants' flue gas. The lean solvent fed to the top of the absorber column countercurrently contacting the flue gas and chemically absorbing the CO_2. A typical coal-fired power plant emits 12% CO_2 in its flue gas, which the absorber captures 90% of, leaving 1.3%. The rich solvent is heated in a cross exchanger before being introduced into the stripper. Heat applied to the rich solvent reverses the reaction, releasing CO_2 and allowing the lean solvent to regenerate. The compressed gaseous CO_2 is sequestered, and the lean amine solvent is recycled to the absorber. Due to the energy requirements of the reboiler, pumps, and multi-stage compressor, the base case absorption/stripping flowsheet using Monoethanolamine (MEA) with a simple stripper column is estimated to reduce the output of a coal-fired power plant by 25 to 30%.

A Simple Stripper

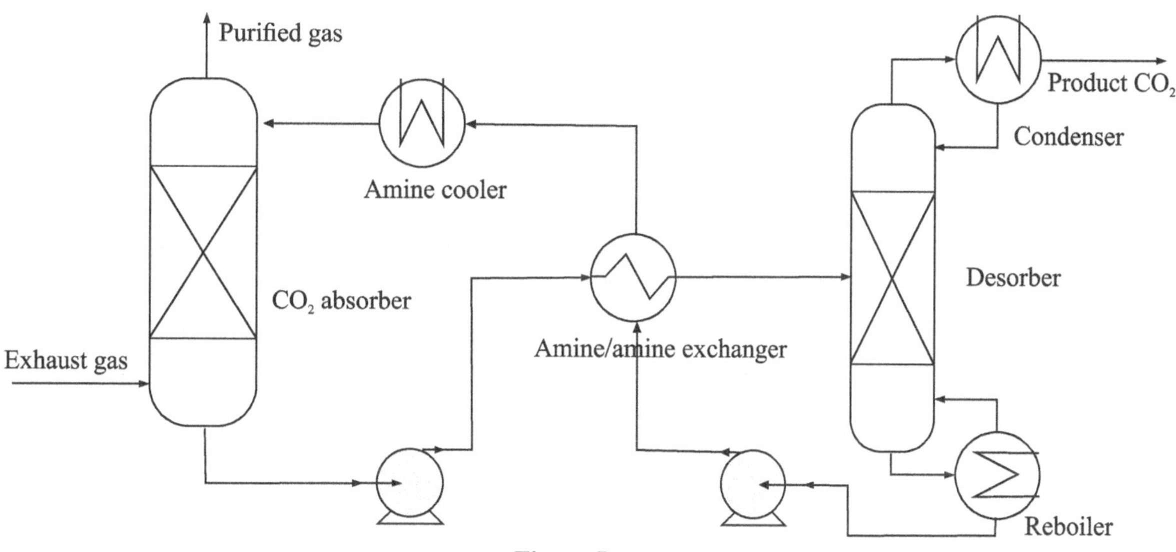

Figure 5

A variety of solvents have been used to capture CO_2. Amines are the most commonly used solvents in chemical absorption to capture CO_2. Because an amine has at least one OH and amine group in its chemical structure, the OH group can reduce the vapour pressure of the amine due to the formation of H bonds, and because an amine has alkaline properties, it can absorb acidic gases. Monoethanolamine (MEA) is the most widely used amine due to its high solution absorbability, high alkalinity, high reaction rate, regenerability, and low cost. However, some disadvantages have been identified, including high solvent regeneration energy, corrosion, and degradation. Capturing CO_2 raises the cost of electricity production by 70%, while the energy required for regeneration is estimated to be 15–30% of a power plant's output. As a result, effectively lowering the cost of electricity has become critical to the success or failure of carbon capture and sequestration (CCS).

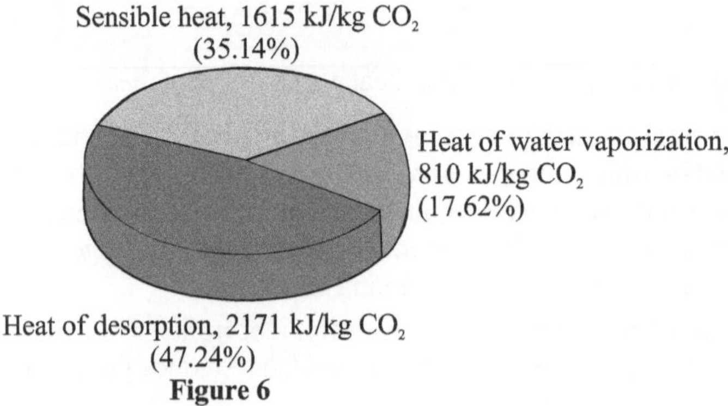

Figure 6

Vapor and Liquid Contact in the Stripper [More details]

Minimizing heat for regeneration has become a significant challenge, particularly for improving stripper structure, discovering new solvents, and improving operating conditions. Rich solvents are heated in the stripper to allow the release of CO_2 from scrubbed solutions during the stripping process. The stripping vapour is regenerated in the reboiler and rises through the column to the top of the stripper, containing water vapour, CO_2, and small amounts of solvents. The stripping vapour counter-current comes into contact with a rich-loading feed stream, which absorbs energy from the stripping steam and uses it to desorb CO_2. The remaining vapour condenses in the overhead condenser at the top of the column. The vapour and liquid contact system is depicted in the figure.

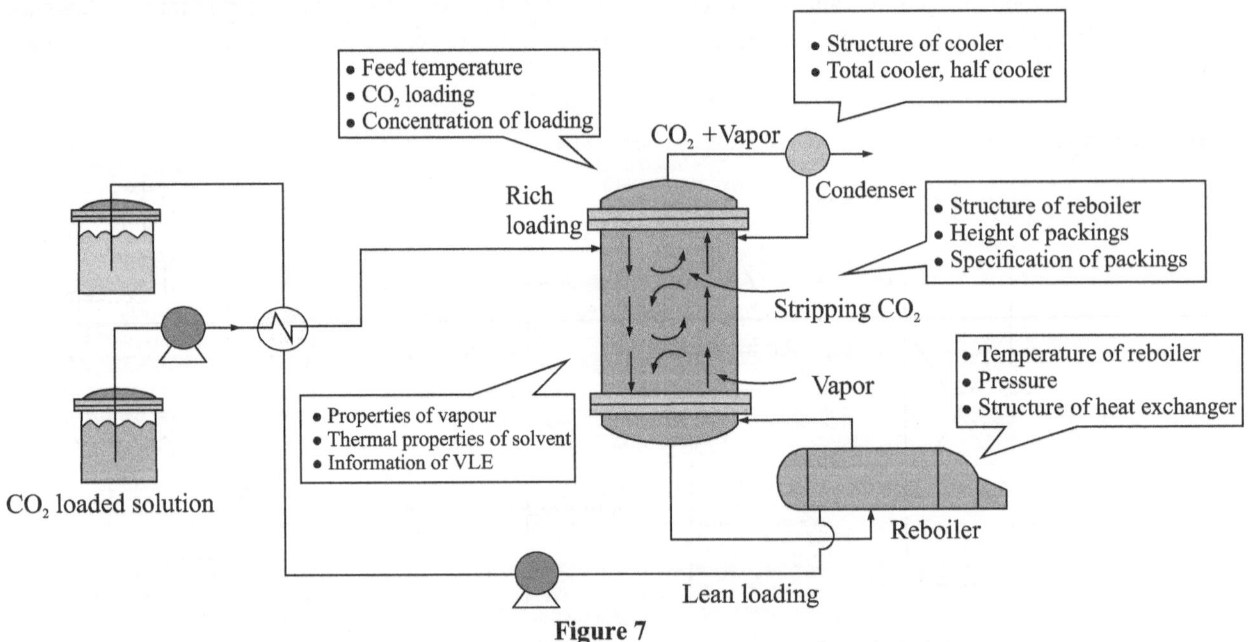

Figure 7

The heat of the solvent regeneration in the stripper of the CO_2 capture process can be described as follows: Q is equal to $Q = Q$ sen + Q vap + Q abs.

Where Qsen is the sensitive heat, Qvap is the evaporation heat, and Qabs is the absorption heat. When the relevant thermodynamic data is available, the three terms are generally evaluated separately.

The cost of regeneration is influenced by the solvent used, the structure of the stripping equipment, the operating temperature, and the cost of steam. Many studies have focused on more efficient solvents with lower heat of absorption in order to minimise the heat of regeneration. According to these studies, blended solvents have been extensively studied in the capture of CO_2 gas, while new solvents have also been actively tested.

As a result, many researchers have studied blended amines and effective solvents in order to improve regeneration energy. According to some studies, using a split-flow in the gas stripping column can save at least 20% of the energy. If thermodynamic data are available, an investigation of the heat mechanism of regeneration energy can also be explored in order to understand the contribution of individual heat duties. The reported levels for the heat of regeneration are in the range of 2.1–11.25 GJ/t, depending on the operating system. However, no empirical equations are available to predict the heat of regeneration under the given conditions.

Experimental Findings

A stripping system is depicted below. A packed column, a reboiler, a condenser, a heat exchanger, and a heating system were all part of it. The diameter of the column was 50 mm. It was stuffed with an 8 8 mm -ring. The packed column stood 800 mm tall, while the condenser stood 500 mm tall. Furthermore, a reboiler (12 L in volume) was heated with silicone oil via heating tubes. A pressure back valve was used as a suitable value at the top and bottom of the column to adjust the pressure in the column, as shown.

T1: Inlet Temperature of Rich Loading	T7: Reboiler Temperature	T12: Steam Temperature
T2: Temperature at outlet of heat exchanger	T8: Inlet temperature of rich loadin	T_{in}: Inlet temperature of cooling water
T4: Out temperature at the top of cooler	T9: Oil temperature at the inlet of reboiler (T_{in})	T_{out}: Outlet temperature of cooling
T5: Temperature of bed	T10: Oil temperature at the outlet of reboiler (T_{out})	P1: Pressure valve at the top of tower
T6: Liquid temperature at the inlet of heat exchanger	T11: Outlet temperature the reboiler	P2: Pressure valve at the bottom of tower

Figure 8: Temperature in Different Locations in Stripping Column

A continuous process was used to effectively investigate the effect of process variables on stripper performance. To begin, the temperature indicators, cooling water circulator, and oil-bath power supply were turned on and set to a predetermined temperature. The rich loading solution was then poured into the reboiler until it flooded. When the oil-bath temperature reached the preset level, the oil-bath pump power supply was activated, and the flow was adjusted using an oil-bath inlet valve. Third, the cooling water's inlet temperature was adjusted to the desired level.

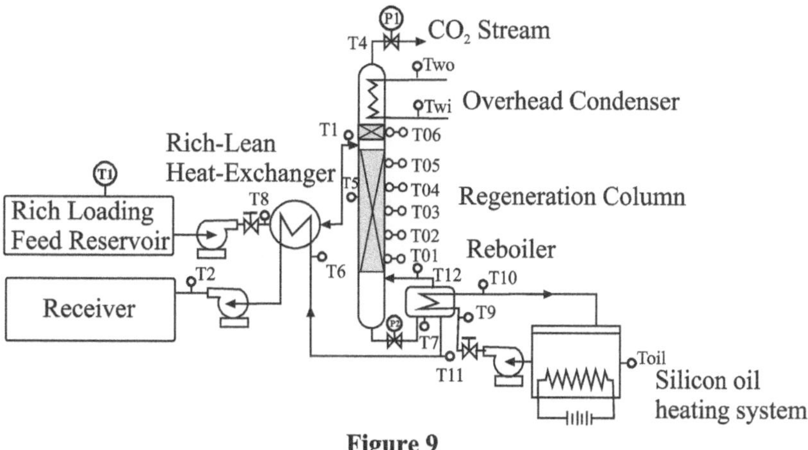

Figure 9

The reboiler was then adjusted to the desired experimental temperature. The experiment began when the temperatures of the cooling water and the reboiler vapour reached the predetermined levels. The rich loading in the storage tank passed through a heat exchanger and into a packed bed, where it came into contact with the vapour rising from the column's bottom. The heat was released to the rich loading as the input solvent after the lean loading was withdrawn at the bottom of the reboiler and passed through the heat exchanger. The lean loading was removed every 30 minutes

Steady-State in the Stripper

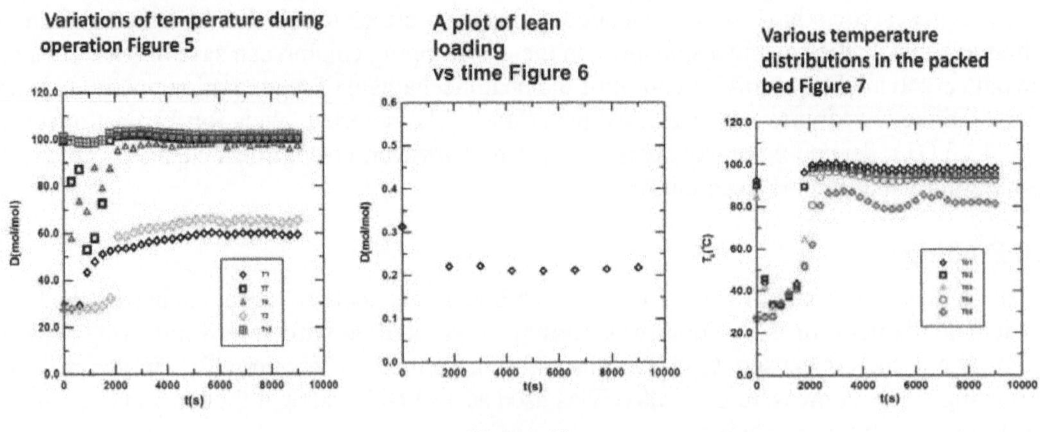

Figure 10

Image Credit: Science Credit

During the operation, the temperature at individual points was observed and recorded, as shown in Figure 4. As shown in Figure 5, several critical points were recorded, including T1, T2, T6, T7, and T12. When the operating time was greater than 80 minutes, the temperatures remained constant. Furthermore, the distribution of the lean loading was discovered, as shown in Figure 6, and it also remained constant after 80 minutes, which corresponded to Figure 5. In addition, as shown in Figure 7, the temperature distribution in the packed bed was recorded. Except for T05, which was near the location of the solvent input and was prone to perturbation during operation, the distributions approached steady-state operation after 80 minutes. The effect of the solvent input on other points (T01 to T04) was weak, however, because they were far from the solvent input. As a result, after 80 minutes, the system could be said to have transitioned to a steady state operation.

References

Wikipedia.

Science Direct.

Optimization in the Stripping Process of CO_2 Gas Using Mixed Amines.

References

- Thermodynamics: An Engineering Approach, by Yunus A. Cengel and Michael A. Boles.
- Introduction to Chemical Engineering Thermodynamics, by J.M. Smith, Hendrick C. Van Ness, and Michael Abbott.
- Thermodynamics: Concepts and Applications, by Stephen R. Turns.
- Fundamentals of Engineering Thermodynamics, by Michael J. Moran, Howard N. Shapiro, Daisie D. Boettner, and Margaret B. Bailey.
- Thermodynamics and Heat Power, by Kurt C. Rolle, Steven J. Skaves, and Edward F. Obert.
- Thermodynamics: Principles and Applications, by John D. Anderson.
- Thermodynamics: From Concepts to Applications, by Arthur Shavit and Chaim Gutfinger.
- Thermodynamics for Engineers, by D. Kumana and W.H. Collins.
- Engineering Thermodynamics, by P.K. Nag.
- Chemical, Biochemical, and Engineering Thermodynamics, by Sandler Stanley I.
- The International System of Units (SI) – Conversion Factors.
  https://www.govinfo.gov/content/pkg/GOVPUB-C13-f10c2ff9e7af2091314396a2d53213e4/pdf/GOVPUB-C13-f10c2ff9e7af2091314396a2d53213e4.pdf
- Energy Fundamentals: Ibrahim Dincer, Calin Zamfirescu, in Comprehensive Energy Systems, 2018.
- Ideal gas theory: Wikipedia.
- Oxtoby, Gillis and Campion. Principles of Modern Chemistry. 6th Edition. California: Thomson Brooks/Cole. 2008.
- Chang, Raymond. Physical Chemistry for the Biosciences. California: University Science Books. 2005.
- "The Nobel Prize in Physics 1910". Nobel Foundation. Retrieved 2008-10-09.
- Parsegian, V. Adrian (2005). van der Waals Forces: A Handbook for Biologists, Chemists.
- Engineers, and Physicists. (Cambridge University Press), p. 2. "The first clear evidence of forces between what were soon to be called molecules came from Johannes Diderik van der Waals's 1873 Ph.D. thesis formulation of the pressure p, volume V, and temperature T of dense gases."
- Johannes Diderik van der Waals - Biographical - Nobelprize.org
- van der Waals; J.D. (1873). Over de continuiteit van den gas- en vloeistoftoestand (On the Continuity of the Gaseous and Liquid States) (doctoral dissertation). Universiteit Leiden.
- Intermolecular bonds: Libre texts.
  http://www.ae.iitm.ac.in/~amitk/AS1300/Work.pdf
- Thermodynamics 5th edition, by Cengel Boles.
- Basic and applied thermodynamics by PK Nag.
- Thermodynamics e book.
- Direct air capture [DAC] challenges.
- Direct air capture [DAC] kinetics and thermodynamics.
- Optimization in the Stripping Proc Experimental and Thermodynamic Investigation on CO_2 absorption in Aqueous MEA Solutions.
- https://www.linkedin.com/pulse/co2-absorption-mea-thermodynamic-issues-short-review-note-mukherjee/ess of CO_2 Gas Using Mixed Amines.